T0366196

Continuum Mechanics and Thermodynamics of Matter

Aimed at advanced undergraduate and graduate students, this book provides a clear unified view of continuum mechanics that will be a welcome addition to the literature. Samuel Paolucci provides a well-grounded mathematical structure and also gives the reader a glimpse of how this material can be extended in a variety of directions, furnishing young researchers with the necessary tools to venture into brand new territory. Particular emphasis is given to the roles that thermodynamics and symmetries play in the development of constitutive equations for different materials.

Continuum Mechanics and Thermodynamics of Matter is ideal for a one-semester course in continuum mechanics, with 250 end-of-chapter exercises designed to test and develop the reader's understanding of the concepts covered. Six appendices enhance the material further, including a comprehensive discussion of the kinematics, dynamics, and balance laws applicable in Riemann spaces.

S. Paolucci is currently Professor of Aerospace and Mechanical Engineering and Director of C-SWARM at the University of Notre Dame.

Continuum Mechanics and Thermodynamics of Matter

S. PAOLUCCI
University of Notre Dame

CAMBRIDGE
UNIVERSITY PRESS

32 Avenue of the Americas, New York NY 10013

Cambridge University Press is part of the University of Cambridge.

It furthers the University's mission by disseminating knowledge in the pursuit of education, learning and research at the highest international levels of excellence.

www.cambridge.org
Information on this title: www.cambridge.org/9781107089952

First published 2016

A catalogue record for this publication is available from the British Library

Library of Congress Cataloguing in Publication data
Names: Paolucci, S., author.
Title: Continuum mechanics and thermodynamics of matter / S. Paolucci, University of Notre Dame.
Description: New York, NY : Cambridge University Press, 2016. | © 2016 | Includes bibliographical references and index.
Identifiers: LCCN 2015034141 | ISBN 9781107089952 (Hardback : alk. paper) | ISBN 1107089956 (Hardback : alk. paper)
Subjects: LCSH: Continuum mechanics. | Thermodynamics.
Classification: LCC QA808.2 .P36 2016 | DDC 531–dc23
LC record available at http://lccn.loc.gov/2015034141

ISBN 978-1-107-08995-2 Hardback

Contents

Preface

The goal of this text is to introduce students to the topic of continuum mechanics, with analysis of the kinematic and mechanical behavior of materials modeled under the continuum assumption. This includes the derivation of fundamental balance equations, based on the classical laws of physics, and the development of constitutive equations characterizing the behavior of idealized materials. Such background provides the starting point for the studies of thermoelasticity, fluid mechanics, and viscoelasticity that are provided in the text. Furthermore, the material covered also imparts students with sufficient background for studying more advanced topics in continuum mechanics, such as wave propagation, polar materials, mixture theory, shell theory, piezoelectricity, and electromagnetic and magnetohydrodynamic fluid mechanics.

A few years ago, I was involved in a project that required fundamental understanding of immiscible multiphase mixtures. I was not (and am still not) satisfied with the current formulations of continuum mechanics in this area, but this is a subject that will be taken up in future publications. Nevertheless, my extensive studies necessitated a deeper understanding of many aspects of single-phase continuum mechanics. Such studies provided me valuable insights and have enabled me to write the present book as an outgrowth of my efforts. At the same time, they have enabled me to become a better teacher of the subject. I hope that the results might be useful to other teachers and students as well. The book is intended for use by students in engineering, science, and applied mathematics. As pre-requisites, a student should have knowledge of multivariable calculus, linear algebra, and differential equations, which are standard in undergraduate programs of engineering and science.

I started writing this book in 2004 to fill a number of gaps that I felt limited my understanding and application of the beautiful theory of continuum mechanics – especially on the relation between continuum mechanics and thermodynamics. I became quite dissatisfied with existing textbooks. Some were delightful but superficial, others wonderful but ancient. Of course many excellent monographs existed, such as *The Classical Field Theories* by Truesdell and Toupin, *The Non-Linear Field Theories of Mechanics* by Truesdell and Noll, and *Mechanics of Continua* by Eringen. Unfortunately, such books were and have been out of print for quite a while and, in the case of the first two, they are challenging works that are not intended for use in a classroom. In the end, as I started to research the material, I fell in love with the subject. I sensed a unifying approach to teaching it that I wanted to develop and then to share. Since then, a number of good texts have appeared, but I feel that the need for the present book still exists. The present text is designed for a one-semester course in continuum mechanics. While covering the standard material, the book also provides a well-grounded mathematical structure and glimpses of how such material can be extended in a variety of directions. Thus, a major aim of the present text is not only providing a sound basis of continuum mechanics but, just as importantly, providing the tools for someone to venture into new territory. I hope that this aspect does not detract from the presentation and does not confuse the student.

Particularly in a subject such as continuum mechanics, many symbols and many fonts are used to refer to each specific quantity introduced. This is done to make the presentation clear. However, I have found this to be an absolutist approach that often burdens the reader to recall the meaning of way too many symbols. While I have retained the rigor, I have not tried to be an absolutist in this respect. Any symbol that is re-used, its meaning is made clear from the context. The text is divided into nine chapters, and each chapter includes exercise problems to test and extend the understanding of concepts presented.

Chapter 1 provides the essential understanding for the need of treating the behavior of common materials through the mathematical artifice of a continuum description. In addition, the different subject areas that make up continuum mechanics are introduced.

In Chapter 2, the essential mathematics for treating continuum problems is provided. Here, we define tensors and cover the algebra and multivariate calculus associated with these objects. In addition, we discuss integral theorems and their generalizations when discontinuous surfaces are present in a continuum region. Here, and throughout the text, we try to make the student comfortable in dealing with three different forms of representing tensors and the associated equations they enter in: by their representation of a coordinate-independent geometrical object, $\mathbf{A}$; by the matrix representing its components, A; and by the specific component elements, a^i_j.

Chapter 3 provides a comprehensive discussion of the kinematics of a continuum body. The deformation and motion of such a body are treated using Lagrangian and Eulerian descriptions. In addition, generalized balance laws are formulated and the important concept of frame-invariance is introduced and utilized.

Chapter 4 is devoted to the fundamental laws of mechanics and thermodynamics. The corresponding global and local forms of the governing equations are developed, and the role that discontinuous surfaces embedded within a continuum region play is discussed. In this chapter, we also consider the effects of the microstructure that underlies the continuum body and subsequently write the balance equations for polar materials. This is done to provide the student interested in this topic a starting point from which to pursue further studies (e.g., the modeling of liquid crystals). In order to focus on major concepts, the text following this chapter only deals with non-polar materials. In this chapter, the stress and couple stress tensors as well as the heat and entropy fluxes are naturally introduced and their properties discussed. In addition, we examine the local equations resulting from Euclidean and Galilean transformations and their implications.

Chapter 5 covers the principles of constitutive theory, where thermodynamics plays an essential role. This chapter provides a unifying theory regardless of the type of matter and forms the centerpiece of the text. The constitutive equations represent macro thermo-mechanical models of real materials. Here, the principles of frame-indifference, causality, equipresence, material smoothness, memory, symmetry, and thermodynamics are systematically utilized to obtain reduced forms of constitutive equations for general materials, and for solids and fluids in particular. Tables are provided for expedient formulations of constitutive equations of isotropic materials. The development of such tables is clearly described and illustrated. Thermodynamics plays an integral part of constitutive theory and many thermodynamic tensor quantities are developed – these reduce to well-known scalar quantities encountered in classical thermodynamics. Formulations are given using the different thermodynamic potentials of internal energy, entropy, Helmholtz free energy, Gibbs free energy, and enthalpy, and the corresponding Maxwell relations provide very useful relations among thermodynamic quantities. Here we also discuss the concepts of thermodynamic equilibrium and stability. In addition, the critical role that the second law of thermodynamics plays in the reduction of constitutive equations is explored using the conventional Coleman–Noll procedure and the more general Müller–Liu procedure that makes use of Lagrange multipliers. Lastly, in this chapter we provide a comprehensive discussion of jump conditions across discontinuous surfaces, and their role in describing material and non-material singular surfaces, including boundary conditions, shocks, and phase-change interfaces.

Chapter 6 is provided to clearly illustrate many of the constitutive theory concepts to the case of spatially uniform material bodies. Here, many of the thermodynamic concepts can easily be applied within a mathematical setting that is not overly burdensome to the student.

This is followed by Chapter 7 where constitutive equations of thermoelastic solids are rigorously developed using the Coleman–Noll procedure. Material symmetries and crystal microscopic structures are fully discussed and linear and nonlinear constitutive equations for non-isotropic and isotropic thermoelastic solids are considered. In addition, we examine a number of fundamental nonlinear equilibrium deformations.

Fluids are discussed in Chapter 8. Here, we provide a rigorous development of constitutive equations using both the Coleman–Noll and the Müller–Liu procedures. General representations of the stress tensor and heat flux are provided. Their simplifications leading to Euler equations, Newtonian equations, the second-order representation, and the Reiner–Rivlin fluid are fully developed. Lastly, comprehensive classifications of fluid motions are provided. The classifications are in the general areas of kinematically restricted types of motions, specialized equations of motion, and specialized constitutive equations.

In Chapter 9, we treat the subject of viscoelasticity. The additional kinematics considerations, aspects of constitutive theory, and general classes of motions of materials having memory are provided. Lastly, the concept of fading memory and application of finite linear viscoelasticity undergoing simple deformations are considered. The treatment of phenomenological constitutive equations, while important, is intentionally left out.

The book includes six appendices that enhance the material presented in the chapters. Of particular note is an appendix that provides a comprehensive discussion of the kinematics, dynamics, and balance laws applicable in Riemannian spaces, such as arbitrary surfaces and curves embedded in the three-dimensional Euclidean space.

Lastly, bibliographies pertinent to material provided in each individual chapter is given at the end of the specific chapter.

I would like to conclude by thanking Jim Jenkins, who first introduced me to continuum mechanics while I was a graduate student in Theoretical and Applied Mechanics at Cornell University, and a number of authors who provided me guidance and inspiration throughout my journey in understanding continuum mechanics and thermodynamics of material bodies. Foremost among them, in alphabetical order, R. Aris, R.M. Bowen, H.B. Callen, D.B. Coleman, D.G.B. Edelen, J.L. Ericksen, A.C. Eringen, I-S. Liu, I. Müeller, W. Noll, R.S. Rivlin, G.F. Smith, A.J.M. Spencer, R. Toupin, and C. Truesdell. In addition, I would like to thank the many students who have taken my course in Continuum Mechanics at the University of Notre Dame; they have provided me useful feedback through multiple versions of the material. In particular, I would like to thank Dr. Gianluca Puliti who drew most of the figures in the text.

Notre Dame, Indiana S. PAOLUCCI

1
Introduction

Classical continuum physics deals with media without a visible microstructure. That is, the scale of observation is large compared to the molecular scale, but small relative to other heterogeneities within the system. More modern continuum theories consider more directly the influence of microstructures; among them are micromorphic, mixture, and nonlocal theories. On the smallest scale, individual molecules are observed. Statistical mechanical theories and some micromorphic field theories may be applicable on this scale. On a slightly larger scale, the material body appears locally uniform with no distinct microstructure. This is the scale of observation on which classical continuum theories apply. On yet a larger scale of observation, large heterogeneities in space and/or time are evident. Such heterogeneities are well characterized by the solution of continuum mechanics problems at these scales.

From the atomic point of view, a macroscopic sample of matter is an agglomerate of an enormous number of nuclei and electrons. A complete mathematical description of a sample consists of the specification of suitable coordinates for each nucleus and electron; the number of such coordinates is enormous considering the magnitude of Avogadro's number of 6.0221×10^{23} mol^{-1} which gives us the number of molecules in one mole of a substance.

In contrast to the atomistic description, only a few parameters are required to describe the system macroscopically. The key to this reduction is the slowness and large scale of macroscopic measurements in comparison to the speed of atomic motions (typically of the order of 10^{-15} s) and atomic distance scales (typically of the order of 10^{-10} m). For example, some of our fastest macroscopic measurements are of the order of 10^{-6} s. Consequently, macroscopic measurements sense only averages of the atomic coordinates. The mathematical process of averaging eliminates coordinates and thus reduces the level of description in going from the atomic to the macroscopic level.

Of the enormous number of atomic coordinates, a very few, with unique symmetry properties, survive the statistical averaging. Certain of these are mechanical in nature (e.g., volume, shape, and components of elastic strain), others are thermal in nature (e.g., temperature and internal energy), or electrical/magnetical in nature (e.g., electric and magnetic dipole moments). The subject of mechanics (e.g., elasticity and fluid mechanics) is the study of one set of surviving coordinates, the subject of thermal sciences (e.g., thermodynamics and heat transfer) is the study of another set of surviving coordinates, and the subject of electricity and magnetism is the study of still another set of coordinates. In general, all these sets of coordinates are coupled, and the study of continuum mechanics provides the framework to study the coupling between these coordinates.

For many materials the behavior of large samples can be studied without recourse to the details of the atomic level structure. We can describe fluids, solids, glasses, bio-materials, mixtures, etc., by making use of the framework provided by continuum mechanics.

1.1 Continuum mechanics

Continuum mechanics is the study of the macroscopic consequences of the large number of atomic coordinates, which, by virtue of statistical averaging, do not appear explicitly in the macroscopic description of a system. It is a branch of physics that deals with materials. The fact that matter is made of atoms and that it commonly has some sort of heterogeneous microstructure is mostly ignored in the simplifying approximation that physical quantities, such as mass, momentum, and energy, can be handled in the infinitesimal limit. For most materials, this is possible as long as the characteristic length scale is far larger than 10^{-9} m and the characteristic speed is much less than the speed of light (3×10^8 m/s). If the length scale is of the order of 10^{-9} m or less, then quantum mechanics applies. If the speed is near the speed of light, then relativistic mechanics applies. If the length scale is of the order of 10^{-9} or less and the speed is near that of light, then quantum field theory applies.

What are the consequences of the existence of the "hidden" atomic motion? Recall that in mechanics, thermal sciences, and electricity and magnetism we are much concerned with the concept of energy. Energy transferred to a mechanical mode of a system is called *mechanical work* $\delta\mathcal{W}$. Similarly, energy can be transferred to an electrical mode of the system. Mechanical work is typified by the term $-p\,dV$ (p is pressure and V is volume), and *electrical work* is typified by the term $-\mathcal{E}\,d\mathcal{P}$ ($\mathcal{E}$ is the electric field and $\mathcal{P}$ is the electric dipole moment). It is equally possible to transfer energy to the hidden atomic modes of motion as well as to those which happen to be macroscopically observable. Energy transfer to the hidden atomic modes is called *heat*. The energy residing in the hidden atomic motions we call *internal energy*. Heat transfer and internal energy are typified by terms such as $\delta\mathcal{Q}$ and dU.

Continuum mechanics is very general; it applies to complicated systems with mechanical, thermal, and electrical/magnetical properties. In this book, we will focus on mechanical and thermal properties of materials, keeping in mind that this is not a limitation of continuum mechanics theory. Differential equations are employed in solving problems in continuum mechanics. Some of these differential equations are specific to the materials being investigated, while others capture fundamental physical laws, such as conservation of mass or conservation of momentum.

The physical laws of a material's response to forces do not depend on the coordinate system in which they are observed. Continuum mechanics is thus described by tensors, which are mathematical objects that are independent of a coordinate system. Such tensors can be expressed in coordinate systems for computational convenience.

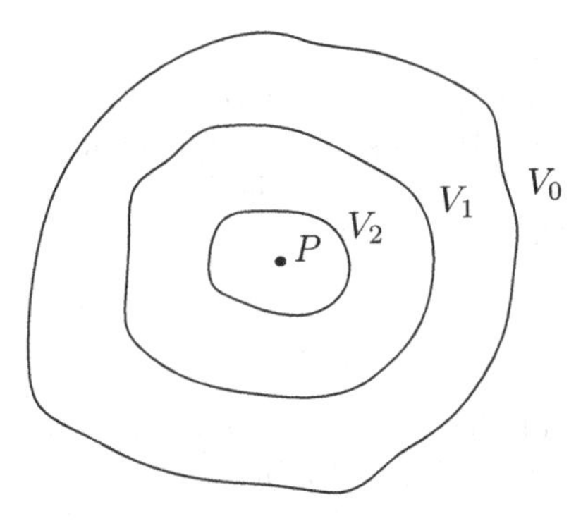

Figure 1.1: Limit at a point P.

1.2 Continuum

A *continuum* is a classical concept derived from mathematics:

a) the real number system is a continuum;

b) *time* can be represented by a real number system;

c) *three-dimensional space* can be represented by three real number systems;

d) time-space together is identified as a four-dimensional continuum.

A material continuum is characterized by quantities such as mass, momentum, energy, and state variables.

Matter, as measured by its mass m, is assumed to have a continuous distribution in space. A certain amount of mass occupies a definite volume V. As illustrated in Fig. 1.1, we define the mass density at an arbitrary point P by

$$\rho(P) = \lim_{\substack{n\to\infty \\ V_n\to 0}} \frac{m_n}{V_n}, \tag{1.1}$$

where m_n is the mass contained in the averaging volume V_n.

Since the averaging volume must be sufficiently larger than molecular scales, to conform to the real world, we take the definition of the density of the material at P with an acceptable variability $\epsilon > 0$ in a defining limit volume $\delta > 0$:

$$\lim_{\substack{n\to\infty \\ V_n\to\delta\ll 1}} \left| \frac{\rho(P)}{m_n/V_n} - 1 \right| < \epsilon \ll 1. \tag{1.2}$$

It is our responsibility to make sure that δ is sufficiently large and ϵ sufficiently small for the concepts of a continuum to make sense. For example, δ should be large enough in the four-dimensional time-space continuum to include a sufficiently large number of molecules so that the number of molecules entering or leaving δ is such as to lead to ϵ sufficiently small. Similarly, we define densities of momentum and energy. For vector quantities, the definition applies to each component individually. Note that in general the size of the limit volume δ for a fixed acceptable variability ϵ is different for different physical quantities. Thus, again, it is our responsibility to understand that the continuum description only makes sense when describing average properties at scales larger than the largest δ among all quantities that we are interested in describing within the acceptable variability ϵ.

Continuum mechanics ignores all the fine detail of atomic and molecular (or particle) level structure and assumes that

- the highly discontinuous structure of real materials can be replaced by a smoothed hypothetical continuum;

- every portion of the continuum, however small, exhibits the macroscopic physical properties of the bulk material.

In any branch of continuum mechanics, the field variables (i.e., density, displacement, and velocity) are conceptual constructs. They are taken to be defined at all points of the imagined continuum and their values are calculated via axiomatic rules of procedure.

The continuum model breaks down over distances comparable to interatomic spacing (in solids about 10^{-10} m). Nonetheless, the *average* of a field variable over a small but *finite* region is meaningful. Such an average can, in principle, be compared directly to its nominal counterpart found by experiment, which will itself represent an average of a kind taken over a region containing many atoms, because of the finite physical size of any measuring probe. For solids, the continuum model is valid in this sense down to a scale of order 10^{-8} m which is the side of a cube containing a million or so atoms. Further, when field variables change slowly with position at a microscopic level $\sim 10^{-6}$ m, their averages over such volumes (10^{-20} m^3 say) differ insignificantly from their centroidal values. In this case, pointwise values can be compared directly to observations. Such behaviors are illustrated in Fig. 1.2 for the mass density at point P as a function of the size of the averaging volume.

Within the continuum we take the behavior to be determined by balance laws for mass, linear momentum, angular momentum, energy, and the second law of thermodynamics. The continuum hypothesis enables us to apply these laws on a local as well as a global scale.

1.3 Mechanics

Classical mechanics is the study of the motion and deformation changes in a body composed of matter due to the action of forces. It is often referred to as *Newtonian mechanics* after Newton and his laws of motion. Classical mechanics is subdivided into statics (which models objects at rest), kinematics (which models objects in motion), and dynamics (which models objects subjected to forces). In continuum mechanics, we deal with all three aspects that are based on the concepts of time, space, and forces. To understand the concept of forces, knowledge is needed from all branches of engineering, physics, chemistry, and biology.

Classical mechanics produces very accurate results within the domain of everyday experience. It is superseded by relativistic mechanics for systems moving at large velocities (near the speed of light), quantum mechanics for systems at small spatial scales (atomic or subatomic scales), and relativistic quantum field theory for systems with both properties. Nevertheless, classical mechanics is still very

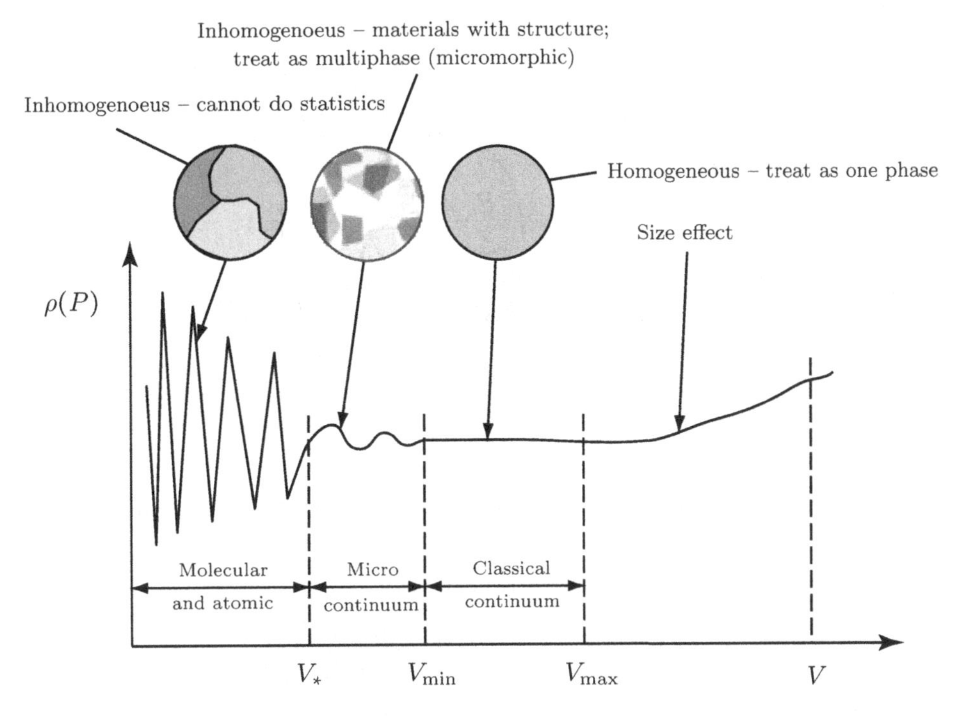

Figure 1.2: Density limit with acceptable variability at a point P as a function of averaging volume.

useful, because (i) it is much simpler and easier to apply than these other theories, and (ii) it has a very large range of approximate validity. Classical mechanics can be used to describe the motion of human-sized objects (i.e., tops and baseballs), many astronomical objects (i.e., planets and galaxies), and certain microscopic objects (i.e., sand grains and organic molecules.)

1.3.1 Deformation and strain

If we take a solid cube and subject it to some deformation, the most obvious change in external characteristics will be a modification of the shape.

The specification of the deformation is thus a geometrical problem and may be carried out from two different viewpoints: relate the deformation

1. with respect to the undeformed state (Lagrangian), or

2. with respect to the deformed state (Eulerian).

Locally, the mapping from the deformed to the undeformed state can be assumed to be linear and described by a differential relation, which is a combination of pure stretch (a rescaling of each coordinate) and a pure rotation.

The mechanical effects of the deformation are confined to the stretch and it is convenient to characterize this by a *strain* measure. For example, for a wire under

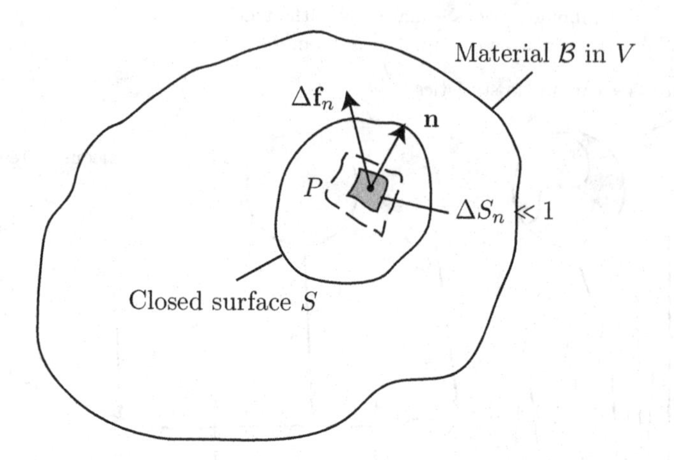

Figure 1.3: Traction with acceptable variability at a point P.

load the strain would be the relative extension, i.e.,

$$\text{linear strain} = \frac{\text{change in length}}{\text{initial length}}.$$

The generalization of this idea requires us to introduce a strain tensor at each point of the continuum.

1.3.2 Stress field

Stress is a measure of force intensity or density. As illustrated in Fig. 1.3, the traction or stress vector $\mathbf{t}$ at an arbitrary point P on a surface with normal vector $\mathbf{n}$ with an acceptable variability $\epsilon > 0$ is defined by

$$\lim_{\substack{n\to\infty \\ \Delta S_n \to \alpha \ll 1}} \left| \frac{\mathbf{t}(P, \mathbf{n})}{\Delta \mathbf{f}_n / \Delta S_n} - 1 \right| < \epsilon \ll 1, \tag{1.3}$$

where $\alpha > 0$ is sufficiently small, or more simply

$$\mathbf{t}(P, \mathbf{n}) = \frac{d\mathbf{f}}{dS}.$$

Within a deformed continuum there will be a force system acting. If we were able to cut the continuum in the neighborhood of a point P as illustrated in Fig. 1.4, we would find a force acting on the cut surface, which would depend on the inclination of the surface and is not necessarily perpendicular to the surface. This force system can be described by introducing a stress tensor $\boldsymbol{\sigma}$ at each point whose components describe the loading characteristics.

1.4 Thermodynamics

Thermodynamics is the physics of energy, heat, work, entropy, and the spontaneity of processes.

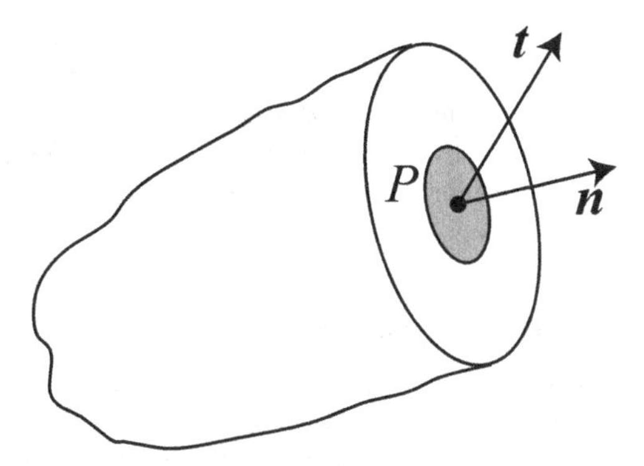

Figure 1.4: Stress vector at a point P.

While dealing with processes in which systems exchange matter or energy, classical thermodynamics is not concerned with the rate at which such processes take place, termed kinetics. For this reason, the use of the term *thermodynamics* usually refers to equilibrium thermodynamics. In this connection, a central concept in thermodynamics is that of quasistatic processes, which are idealized *infinitely slow* processes. Because thermodynamics is not concerned with the concept of time, it has been suggested that a better name for equilibrium thermodynamics would have been *thermostatics*. Time-dependent thermodynamic processes are studied by *nonequilibrium thermodynamics*. In continuum mechanics, we deal with both equilibrium and non-equilibrium thermodynamics.

Thermodynamic laws are of very general validity, and they do not depend on the details of the interactions or the systems being studied. This means they can be applied to systems about which one knows nothing other than the balance of energy and matter transfer between them and the environment.

The quantities that set thermostatics apart from classical particle mechanics are temperature and entropy. The significance of entropy can be illustrated as follows. Consider a flowing fluid. The fluid molecules possess kinetic energy which can be broken into two components, a part which is ordered and contributes to the bulk velocity, and another part which is random. The ordered energy is similar to the macroscopic kinetic energy of particle mechanics, and is mechanical in form. It is capable of being converted to work. Extraction of the ordered kinetic energy would leave only the random (thermal) energy in the fluid. The random component of the energy would contribute nothing to the work, as molecules would impact with such forces so as to cancel each other. Theoretically, one could extract all the ordered energy from the fluid leaving only the random energy. Now suppose that rather than extracting the organized energy, we somehow bring the convective fluid to a stop. The total energy of the fluid would remain unchanged, but there would no longer be any ordered component. All the kinetic energy of the molecules is now coming from random motions, and any attempt to convert this energy to work is fruitless. Entropy is a measure of the randomness, or of the energy's inability to do work – except through transfer of randomness from one body to another.

The random thermal energy will not freely convert back to mechanical form. That is, the likelihood that the molecules will realign to travel in some preferred direction is extremely small. Thus, since entropy is a measure of the randomness, it will not decrease without some external interactions. The only way one can decrease molecular randomness is to transfer some of this randomness to another body, and thereby increasing the randomness (entropy) of the other body. Thus, transfer of thermal energy from one body to another effectively transfers entropy. The transfer of thermal energy (randomness) as heat in this fashion is the only known way by which it is possible to reduce a body's entropy.

1.5 Constitutive theory

The specification of the stress and strain states of a body is insufficient to describe its full behavior; we need in addition to link these two fields.

This is achieved by introducing a *constitutive relation*, which prescribes the response of the continuum to arbitrary loading and thus defines the connection between the stress and strain tensors for the particular material.

At best a mathematical expression provides an approximation to the actual behavior of the material, but as we shall see we can simulate the behavior of a wide class of media.

In general, we think of materials as existing in either a solid or fluid state. The distinction between solid and fluid matter is relative; it depends on time scales over which the material deforms. In turn, we can view a solid as either hard or soft. Hard solids tend to respond elastically to an applied force, they have large acoustic speeds, their energy character is enthalpic, they tend to be anisotropic, they usually rupture upon yielding, and they retain perfect memory of only their initial state. On the other hand, soft solids tend to be dissipative, have a low acoustic speed, their energy behavior is entropic, they tend to be isotropic, they fail through plastic deformation (fluid like), and they behave as viscous on a short time scale and elastic on a long time scale. Fluids, in turn, can be classified as either isotropic or anisotropic. Isotropic fluids can exhibit time scale effects. In general, they respond elastically at short times and viscous at long times. Anisotropic fluids, such as liquid crystals, behave solid-like and exhibit elastic behavior in some directions.

1.5.1 Solids

To get an idea of the behavior of solids, we consider extension of a wire under loading. The tensile stress $\boldsymbol{\sigma}$ and tensile strain $\mathbf{e}$ are then typically related. A typical stress-strain curve is illustrated in Fig. 1.5.

a) Elasticity: If the wire returns to its original configuration when the load is removed, the behavior is said to be elastic.

   i) for linear elasticity $\boldsymbol{\sigma} = E\,\mathbf{e}$ – called Hooke's law and is usually valid for small strains (E is the elastic modulus);

   ii) for nonlinear elasticity $\boldsymbol{\sigma} = \mathbf{f}(\mathbf{e})$ – it is important for rubber-like materials.

b) Plasticity: Once the yield point is exceeded, permanent deformation occurs and there is no unique stress-strain curve, but a unique $d\sigma$-$d\mathbf{e}$ relation. Due to microscopic processes, the yield stress rises with $\boldsymbol{\sigma}$ (work hardening).

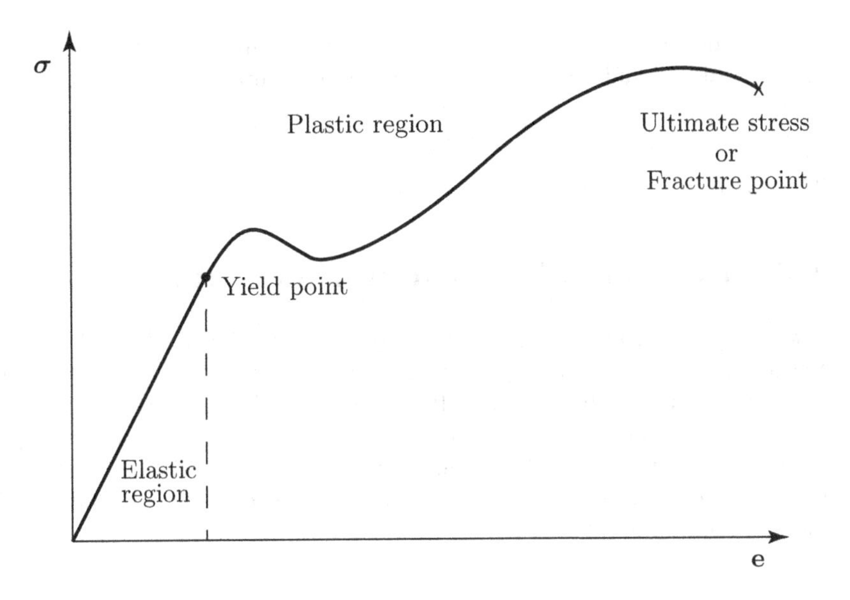

Figure 1.5: A typical stress-strain curve.

c) Viscoelasticity (rate-dependent behavior): Materials may creep and show slow long-term deformation, e.g., plastics and metals at elevated temperatures. Simple models of viscoelasticity are

   i) Maxwell model:

$$\dot{\boldsymbol{\sigma}} + \frac{E}{\mu}\boldsymbol{\sigma} = E\,\dot{\mathbf{e}}$$

   which allows for instantaneous elasticity and represents a crude description of a fluid (μ is the viscosity of the material).

   ii) Kelvin–Voigt model:

$$\boldsymbol{\sigma} = E\,\mathbf{e} + \mu\,\dot{\mathbf{e}}$$

   which displays long-term elasticity.

More complex models can be written down, but all have the same characteristic of depending on the time history of deformation.

1.5.2 Fluids

The simplest constitutive equation encountered in continuum mechanics is that of an ideal fluid:

$$\boldsymbol{\sigma} = -p(\rho, T)\ \mathbf{1},$$

where ρ is the density, T is the absolute temperature, the pressure field p is isotropic and depends on density and temperature, and $\mathbf{1}$ is the unit tensor. If the fluid is incompressible, ρ is a constant. The next level of complication is to allow the stress to depend on the flow of the fluid. The simplest such form, a Newtonian viscous fluid, includes a linear dependence on strain rate

$$\boldsymbol{\sigma} = -p(\rho, T)\ \mathbf{1} + \mu(\rho, T)\ \dot{\mathbf{e}}.$$

The quantity μ is called the *shear viscosity.*

1.6 Pioneers of continuum mechanics

The study of continuum mechanics originated from the works of James and John Bernoulli, Euler, and Cauchy. The field remained stagnant for a very long period of time after them. It was only after World War II that interest in the field was renewed. The modern field of continuum mechanics is the result of pioneering works from Truesdell, Noll, Toupin, Rivlin, Coleman, Ericksen, Müller, Eringen, Gurtin, and Liu, among others. Clifford Truesdell is considered the father of modern continuum mechanics.

Bibliography

B.D. Coleman. Thermodynamics of materials with memory. *Archive for Rational Mechanics and Analysis*, 17:1–46, 1964.

B.D. Coleman and W. Noll. The thermodynamics of elastic materials with heat conduction and viscosity. *Archive for Rational Mechanics and Analysis*, 13(1):167–178, 1963.

A.C. Eringen. Basic principles: Balance laws. In A.C. Eringen, editor, *Continuum Physics*, volume II, pages 69–88. Academic Press, Inc., New York, NY, 1975.

A.C. Eringen. Basic principles: Deformation and motion. In A.C. Eringen, editor, *Continuum Physics*, volume II, pages 3–67. Academic Press, Inc., New York, NY, 1975.

A.C. Eringen. Basic principles: Thermodynamics of continua. In A.C. Eringen, editor, *Continuum Physics*, volume II, pages 89–127. Academic Press, Inc., New York, NY, 1975.

A.C. Eringen. Constitutive equations for simple materials: General theory. In A.C. Eringen, editor, *Continuum Physics*, volume II, pages 131–172. Academic Press, Inc., New York, NY, 1975.

A.C. Eringen. *Mechanics of Continua.* R.E. Krieger Publishing Company, Inc., Melbourne, FL, 1980.

Y.C. Fung. *A First Course in Continuum Mechanics.* Prentice Hall, Inc., Englewood Cliffs, NJ, 3rd edition, 1994.

M.E. Gurtin. Modern continuum thermodynamics. In S. Nemat-Nasser, editor, *Mechanics Today,* volume 1, pages 168–213. Pergamon Press, New York, 1974.

I. Müller. *Thermodynamics.* Pitman Publishing, Inc., Boston, MA, 1985.

W. Noll. Lectures on the foundations of continuum mechanics and thermodynamics. *Archive for Rational Mechanics and Analysis*, 52(1):62–92, 1973.

W. Noll. *The Foundations of Mechanics and Thermodynamics – Selected Papers.* Springer-Verlag, New York, 1974.

C. Truesdell. The mechanical foundations of elasticity and fluid dynamics. *Journal of Rational Mechanics and Analysis*, 1(1):125–300, 1952.

C. Truesdell. *Rational Thermodynamics.* Springer-Verlag, New York, NY, 2nd edition, 1984.

C. Truesdell and W. Noll. The non-linear field theories of mechanics. In S. Flügge, editor, *Handbuch der Physik*, volume III/3. Springer, Berlin-Heidelberg-New York, 1965.

C. Truesdell and R.A. Toupin. The classical field theories. In S. Flügge, editor, *Handbuch der Physik*, volume III/1. Springer, Berlin-Heidelberg-New York, 1960.

2

Tensor analysis

Tensor analysis is the language used to describe continuum mechanics. Physical laws, if they really describe the real world, should be independent of the position and orientation of the observer. Two individuals using two coordinate systems in the same reference frame should observe the same physical event. For this reason, the equations of physical laws, which are tensor equations, should hold in any coordinate system. Invariance of physical laws to two frames of reference in accelerated motion relative to each other is more difficult and requires general relativity theory (tensors in four-dimensional space-time). For simplicity, we limit ourselves to tensors in three-dimensional Euclidean space. We will not consider tensors on manifolds (curved spaces) in the main text. This topic is discussed in Appendix D.

We will discuss general tensors on an arbitrary curvilinear coordinate system, although for the development of continuum mechanics theory, we will use Cartesian tensors. For the solution of specific problems, orthogonal curvilinear coordinates and indeed rectangular coordinates will be used.

In three-dimensional space, a scalar quantity has the same magnitude irrespective of the coordinate system. A vector may be visualized as an arrow that has length equal to the magnitude of the vector and that is pointing in the direction of the vector. It has an existence independent of any coordinate system in which it is observed. Thus a vector is unchanged if it is moved parallel with itself. Two vectors that have the same direction and length are equal. They do not have to have the same origin. In vector operations, we shift the vectors so that they have the same origin. A vector of unit length is called a unit vector. A second-order tensor is a quantity with two directions associated with it. Although it cannot be visualized simply as an arrow, it also has an existence independent of the coordinate system. This remains true for tensors of arbitrary order. It is precisely this independence of the coordinate system that motivates us to study tensors.

2.1 Review of linear algebra

Notation: All scalar quantities will be denoted using the lowercase italic font. Matrices will be denoted by the uppercase (not bold) italic font. Vectors and higher order tensors are denoted by the lowercase and uppercase bold fonts, respectively.

Volumes, surfaces, and curves will be denoted using the script font. Lastly, we will use the fraktur font to denote functionals. In all cases, exceptions are made to respect the notation used historically.

Examples

$$\begin{aligned} \text{Scalars} &: a, b, c \\ \text{Vectors} &: \mathbf{u}, \mathbf{v}, \mathbf{w} \\ \text{Tensors} &: \mathbf{T}, \mathbf{D}, \mathbf{W}, \mathbf{1}, \boldsymbol{\epsilon}, \boldsymbol{\sigma} \\ \text{Matrices} &: A, B, C; \\ \text{Geometric objects} &: \mathscr{V}, \mathscr{S} \\ \text{Functionals} &: \mathfrak{F}, \mathfrak{G} \end{aligned}$$

Definition: The vector $\mathbf{u}$ is said to be a linear combination of the vectors $\mathbf{v}_1, \mathbf{v}_2, \ldots, \mathbf{v}_n$ if

$$\mathbf{u} = c_1\mathbf{v}_1 + c_2\mathbf{v}_2 + \cdots + c_n\mathbf{v}_n,$$

where the c_i's are real numbers.

Definition: A set of vectors $\{\mathbf{v}_1, \mathbf{v}_2, \ldots \mathbf{v}_n\}$ is said to be linearly independent if

$$c_1\mathbf{v}_1 + c_2\mathbf{v}_2 + \cdots + c_n\mathbf{v}_n = \mathbf{0}$$

implies that $c_1 = c_2 = \cdots = c_n = 0$; i.e., the only linear combination that is equal to zero is the trivial linear combination.

Definition: A vector space $\mathcal{V}$ is a collection of vectors, which is closed under linear combinations; i.e., if $(\mathbf{w}_1, \mathbf{w}_2) \in \mathcal{V}$, then $(c_1\mathbf{w}_1 + c_2\mathbf{w}_2) \in \mathcal{V}$ for all real (c_1, c_2).

Definition: A basis of $\mathcal{V}$ is a set $\{\mathbf{u}_1, \mathbf{u}_2, \ldots, \mathbf{u}_n\}$, which

i) is linearly independent, and

ii) spans $\mathcal{V}$ (i.e., if $\mathbf{w} \in \mathcal{V}$, then $\exists$ c_i's such that $\mathbf{w} = c_1\mathbf{u}_1 + c_2\mathbf{u}_2 + \cdots + c_n\mathbf{u}_n$).

Definition: The dimension of a vector space $\mathcal{V}$ corresponds to the number of vectors (unique) in a basis (not unique).

From now on, if the dimension of the vector space is n, we will indicate this by a superscript, i.e., $\mathcal{V}^n$.

To connect some of the later discussions with vectors and matrices, we use the following convention. If A denotes a matrix $[a^i_j]$, then a^i_j denotes the entry in the ith row and jth column of the matrix. Sometimes the notation a_{ij} or a^{ij} is used instead. These symbols also refer to the entry in the ith row and jth column. The transpose of the matrix A is denoted by A^T, and the entries denoted by a_{ji}, while the trace of A is denoted by $\text{tr } A = \text{tr } A^T$ and the entries by a_{ii}. Note that if α is a scalar, then $\text{tr}\,(\alpha A) = \alpha \, \text{tr}\, A$. The transpose of the product of matrices A and B is given by $(AB)^T = B^T A^T$. Also note that $\text{tr}\,(AB) = \text{tr}\,(BA)$. The determinant of matrix A is denoted as $\det A = \det A^T$. We also have that

$\det(\alpha A) = \alpha^n \det A$, where the dimension of A is $n \times n$. The determinant of the product of two matrices A and B is given by the product of the determinant of each matrix, i.e., $\det(AB) = (\det A)(\det B)$. The determinant of the exponential of a matrix is equal to the exponential of the trace of the matrix, i.e., $\det(e^A) = e^{\operatorname{tr} A}$. A matrix is said to be symmetric if $a_{ij} = a_{ji}$ and skew-symmetric if $a_{ij} = -a_{ji}$. Note that a symmetric three-dimensional matrix has six independent components while a skew-symmetric one has only three independent components since it has a zero trace. Any matrix can be decomposed as a sum of its symmetric and skew-symmetric parts. This is denoted symbolically as

$$A = \text{sym } A + \text{skw } A, \tag{2.1}$$

where

$$\text{sym } A = \frac{1}{2}\left(A + A^T\right) \qquad \text{and} \qquad \text{skw } A = \frac{1}{2}\left(A - A^T\right). \tag{2.2}$$

If D is a symmetric matrix and W a skew-symmetric matrix, then it is evident that

$$D^T = D \qquad \text{and} \qquad W^T = -W. \tag{2.3}$$

The inverse of matrix A is defined by

$$AA^{-1} = A^{-1}A = I, \tag{2.4}$$

where I is the identity matrix with entries of unity along the main diagonal and zero for all other entries. The entry in the ith row and jth column of the identity matrix I is denoted by

$$\delta^i_j = \begin{cases} 1, & i = j \\ 0, & i \neq j \end{cases}, \tag{2.5}$$

where δ^i_j is called the Kronecker delta symbol. We note that $(A^{-1})^{-1} = A$ and $(\alpha A^{-1})^{-1} = \alpha^{-1}A$. A matrix is said to be orthogonal if $A^{-1} = A^T$. The inverse of the product of matrices A and B is given by $(AB)^{-1} = B^{-1}A^{-1}$. The determinant of the inverse of matrix A is given by the reciprocal of the determinant of the matrix, i.e., $\det A^{-1} = (\det A)^{-1}$. For an orthogonal matrix $AA^T = A^TA = I$, so $\det(AA^T) = (\det A)(\det A^T) = \det I = 1$ and subsequently $\det A = \pm 1$. Note that for a three-dimensional system, $\operatorname{tr} I = 3$. An orthogonal matrix whose determinant equals $+1$ is called a *proper* orthogonal matrix, and one whose determinant equals -1 is called an *improper* orthogonal matrix. A proper orthogonal matrix, when viewed as a transformation, transforms a right-handed set of axes into a right-handed set of axes, whereas an improper orthogonal matrix transforms a right-handed set of axes into a left-handed set of axes or vice versa. In either case, the transformation preserves vector lengths and angles between vectors. If U is a column matrix, the entry in the ith row is denoted by u^i, where it is understood that the column entry is $j = 1$. Similarly, a matrix V having only one row is called a row matrix and the superscript $i = 1$ is understood, so the jth column is denoted by v_j.

It is straightforward to show that the determinant of a 3×3 matrix A is given by

$$a \equiv \det A = \det[a_{ij}] = \sum_{i=1}^{3}\sum_{j=1}^{3}\sum_{k=1}^{3} \epsilon_{ijk} a^i_1 a^j_2 a^k_3 = \sum_{i=1}^{3}\sum_{j=1}^{3}\sum_{k=1}^{3} \epsilon^{ijk} a^1_i a^2_j a^3_k. \tag{2.6}$$

These equations are obtained by expanding by minors the determinant of A along a column or row, respectively. The entries ϵ_{ijk} and ϵ^{ijk} correspond to the Levi–Civita (or permutation) symbol

$$\epsilon_{ijk} = \epsilon^{ijk} = \begin{cases} 1 & \text{if } (i,j,k) \text{ is an even permutation of } (1,2,3), \\ -1 & \text{if } (i,j,k) \text{ is an odd permutation of } (1,2,3), \\ 0 & \text{if any two labels are the same,} \end{cases} \tag{2.7}$$

i.e., $\epsilon_{ijk} = \epsilon_{jki} = \epsilon_{kij} = -\epsilon_{ikj} = -\epsilon_{kji} = -\epsilon_{jik}$.

Example

The determinant obtained by expanding by minors along the first column is given by

$$\begin{aligned} \det A &= \det \begin{bmatrix} a_1^1 & a_2^1 & a_3^1 \\ a_1^2 & a_2^2 & a_3^2 \\ a_1^3 & a_2^3 & a_3^3 \end{bmatrix} \\ &= a_1^1 \left(a_2^2 a_3^3 - a_2^3 a_3^2\right) - a_1^2 \left(a_2^1 a_3^3 - a_2^3 a_3^1\right) + a_1^3 \left(a_2^1 a_3^2 - a_2^2 a_3^1\right) \\ &= \sum_{j=1}^{3} \sum_{k=1}^{3} a_1^1 \left(\epsilon_{1jk} a_2^j a_3^k\right) + a_1^2 \left(\epsilon_{2jk} a_2^j a_3^k\right) + a_1^3 \left(\epsilon_{3jk} a_2^j a_3^k\right) \\ &= \sum_{i=1}^{3} \sum_{j=1}^{3} \sum_{k=1}^{3} \epsilon_{ijk} a_1^i a_2^j a_3^k. \end{aligned}$$

Some properties of the Kronecker delta and Levi–Civita symbols are given in Appendix A.

2.2 Tensor algebra

Think of curvilinear coordinates as an application. At a point $P \in \mathcal{E}^3$, where $\mathcal{E}^3$ denotes the three-dimensional Euclidean space, one could define three arbitrary coordinates. The length (or norm) of a vector $\mathbf{v}$ in Euclidean space is given by

$$\|\mathbf{v}\| = \sqrt{\mathbf{v} \cdot \mathbf{v}} \geq 0, \tag{2.8}$$

where $\mathbf{u} \cdot \mathbf{v} = \sum_{i=1}^{3} u_i v^i$ denotes the usual scalar inner product. As shown in Fig. 2.1, we could choose a basis, called the natural basis, at a point P as vectors tangent to any three coordinate lines (x^1, x^2, x^3) at P (not unique vectors): $\{\mathbf{e}_1, \mathbf{e}_2, \mathbf{e}_3\} = \{\mathbf{e}_i\}$. The natural basis is a basis as long as the scalar triple product is nonzero, $\mathbf{e}_1 \cdot (\mathbf{e}_2 \times \mathbf{e}_3) \neq 0$. The quantity $\mathbf{u} \times \mathbf{v}$ is a vector that is orthogonal to both $\mathbf{u}$ and $\mathbf{v}$, and is obtained from the vector, or cross, product of two vectors. We will say more about this shortly. From now on, unless stated otherwise, italic subscripts and superscripts are understood to take on values from 1 to 3.

We could have chosen as a basis the normals to the coordinate surfaces: $\{\mathbf{e}^1, \mathbf{e}^2, \mathbf{e}^3\} = \{\mathbf{e}^i\}$. Again, this is a basis as long as their scalar triple product is nonzero, $\mathbf{e}^1 \cdot (\mathbf{e}^2 \times \mathbf{e}^3) \neq 0$. For orthogonal coordinates, $\{\mathbf{e}_i\} = \{\mathbf{e}^i\}$; otherwise, they are not equal. The basis $\{\mathbf{e}^i\}$ is called a reciprocal basis relative to $\{\mathbf{e}_i\}$. Now $\mathbf{e}_1$ and $\mathbf{e}_2$

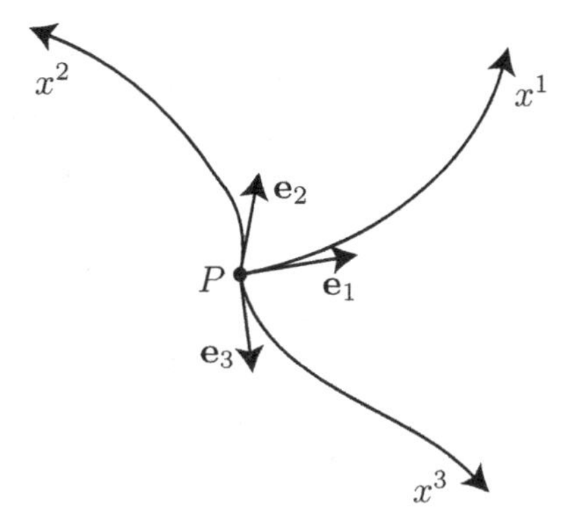

Figure 2.1: Basis at point P.

are tangent to the x^3 surface, and $\mathbf{e}^3$ is normal to the x^3 surface, i.e.,

$$\mathbf{e}_1 \cdot \mathbf{e}^3 = 0 \qquad \text{and} \qquad \mathbf{e}_2 \cdot \mathbf{e}^3 = 0.$$

We require that the length of $\mathbf{e}^3$ be chosen such that

$$\mathbf{e}_3 \cdot \mathbf{e}^3 = 1.$$

Given a basis $\{\mathbf{e}_1, \mathbf{e}_2, \mathbf{e}_3\} \in \mathcal{E}^3$, we can always find the reciprocal basis $\{\mathbf{e}^1, \mathbf{e}^2, \mathbf{e}^3\}$ with the properties

$$\mathbf{e}_i \cdot \mathbf{e}^j = \delta_i^j. \tag{2.9}$$

In $\mathcal{E}^3$, $\mathbf{e}^1 \perp (\mathbf{e}_2, \mathbf{e}_3)$; therefore, $\mathbf{e}^1 = \text{const. } \mathbf{e}_2 \times \mathbf{e}_3$. To find the constant, we require

$$\mathbf{e}_1 \cdot \mathbf{e}^1 = 1 = \text{const. } \mathbf{e}_1 \cdot (\mathbf{e}_2 \times \mathbf{e}_3),$$

or abbreviating the *scalar triple product* by

$$[\mathbf{e}_1, \mathbf{e}_2, \mathbf{e}_3] \equiv \mathbf{e}_1 \cdot (\mathbf{e}_2 \times \mathbf{e}_3), \tag{2.10}$$

$$\text{const.} = \frac{1}{[\mathbf{e}_1, \mathbf{e}_2, \mathbf{e}_3]},$$

thus

$$\mathbf{e}^1 = \frac{\mathbf{e}_2 \times \mathbf{e}_3}{[\mathbf{e}_1, \mathbf{e}_2, \mathbf{e}_3]}. \tag{2.11}$$

Similarly

$$\mathbf{e}^2 = \frac{\mathbf{e}_3 \times \mathbf{e}_1}{[\mathbf{e}_1, \mathbf{e}_2, \mathbf{e}_3]}, \tag{2.12}$$

and

$$\mathbf{e}^3 = \frac{\mathbf{e}_1 \times \mathbf{e}_2}{[\mathbf{e}_1, \mathbf{e}_2, \mathbf{e}_3]}. \tag{2.13}$$

Using the above results, it can be easily shown that

$$[\mathbf{e}^1, \mathbf{e}^2, \mathbf{e}^3] = \frac{1}{[\mathbf{e}_1, \mathbf{e}_2, \mathbf{e}_3]}. \tag{2.14}$$

It is straightforward to show that the scalar triple product can be rewritten in the equivalent forms

$$[\mathbf{u}, \mathbf{v}, \mathbf{w}] = [\mathbf{v}, \mathbf{w}, \mathbf{u}] = [\mathbf{w}, \mathbf{u}, \mathbf{v}]. \tag{2.15}$$

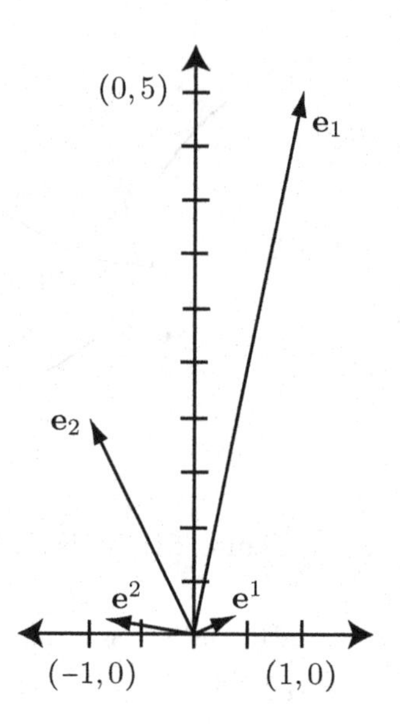

Figure 2.2: Basis in example.

Example

In $\mathcal{E}^2$, given $\mathbf{e}_1 = (1,5)$ and $\mathbf{e}_2 = (-1,2)$, find $\{\mathbf{e}^1, \mathbf{e}^2\}$.
Let $\mathbf{e}^1 = (a,b)$ and $\mathbf{e}^2 = (c,d)$, so that

$$\begin{aligned}
\mathbf{e}_1 \cdot \mathbf{e}^1 &= 1 = a + 5b, \\
\mathbf{e}_1 \cdot \mathbf{e}^2 &= 0 = c + 5d, \\
\mathbf{e}_2 \cdot \mathbf{e}^1 &= 0 = -a + 2b, \\
\mathbf{e}_2 \cdot \mathbf{e}^2 &= 1 = -c + 2d.
\end{aligned}$$

We easily find that $\mathbf{e}^1 = \frac{1}{7}(2,1)$ and $\mathbf{e}^2 = \frac{1}{7}(-5,1)$. The results are displayed in Fig. 2.2.

If we have 2 bases at a point in $\mathcal{E}^3$, $\{\mathbf{e}_i\}$ and $\{\mathbf{e}^i\}$, given a vector $\mathbf{v}$ in the same space, we can write it with respect to either basis:

$$\mathbf{v} = v^1\mathbf{e}_1 + v^2\mathbf{e}_2 + v^3\mathbf{e}_3 = \sum_{i=1}^{3} v^i\mathbf{e}_i = v^i\mathbf{e}_i, \tag{2.16}$$

or

$$\mathbf{v} = v_1\mathbf{e}^1 + v_2\mathbf{e}^2 + v_3\mathbf{e}^3 = \sum_{i=1}^{3} v_i\mathbf{e}^i = v_i\mathbf{e}^i. \tag{2.17}$$

The summation convention used (Einstein's) is that we implicitly sum on repeated indices from 1 to 3, one variable with a superscript index and the other variable with the subscript index. The variables v^i are called contravariant components, and v_i are called covariant components. The summed index, i in this case, is also called a *dummy index* since it can be replaced by any other symbol, say j, without changing the value of the expression. An index appearing in an expression that does not sum is called a *free index.* The indicial notation and summation convention permits us to write equations of continuum mechanics in a much shorter form than would otherwise be possible. This brevity makes the equations easier to remember and understand, once the notation is learned. For example, our previous equations for the reciprocal basis can be combined into the single equation

$$\mathbf{e}^i = \frac{\epsilon^{ijk}\mathbf{e}_j \times \mathbf{e}_k}{[\mathbf{e}_1, \mathbf{e}_2, \mathbf{e}_3]}. \tag{2.18}$$

Note that in the above equation i is a free index, while j and k are dummy indices that can be replaced by other indices. In using Einstein's notation, it is imperative that an index appear no more than twice on either side of an equation, and that the same free row and column indices appear on both sides of the equation.

We now also note that ϵ_{ijk}, and similarly ϵ^{ijk}, are skew-symmetric since the interchange of any two indices changes the sign of the quantity, i.e.,

$$\epsilon_{ijk}a^i_m a^j_n a^k_p = \epsilon_{jik}a^j_m a^i_n a^k_p = -\epsilon_{ijk}a^i_n a^j_m a^k_p.$$

In the same way, we can show that interchanging any two of the indices m, n, and p alters the sign. Furthermore, in view of (2.6), this suggests that we may write more generally, for an expansion by columns,

$$\epsilon_{ijk}a^i_m a^j_n a^k_p = a\ \epsilon_{mnp}.$$

The same type of argument may be used to infer that for an expansion by rows,

$$\epsilon^{ijk}a^m_i a^n_j a^p_k = a\ \epsilon^{mnp}.$$

The number of superscripts or subscripts that the components of a variable has is called the order (or rank) of the tensor. A scalar variable is a zeroth order tensor, a vector variable is a first order tensor. The variable $\boldsymbol{\tau}$ such that

$$\boldsymbol{\tau} = \tau^{ij}\mathbf{e}_i\mathbf{e}_j \qquad \text{or} \qquad \boldsymbol{\tau} = \tau_{lm}\mathbf{e}^l\mathbf{e}^m \tag{2.19}$$

is a second order tensor, also called a dyadic, and τ^{ij} and τ_{lm} are its contravariant and covariant components, respectively. Note that the variables used in the summation are arbitrary, but it is emphasized that there should be no more than two such index variables in each expression.

The main issue of tensor analysis is the requirement that scalars, vectors, and higher order tensor quantities remain invariant when a different curvilinear coordinate system is used, i.e., when a new basis $\{\bar{\mathbf{e}}_1, \bar{\mathbf{e}}_2, \bar{\mathbf{e}}_3\}$ is used instead of $\{\mathbf{e}_1, \mathbf{e}_2, \mathbf{e}_3\}$. Note that corresponding to the new basis $\{\bar{\mathbf{e}}_1, \bar{\mathbf{e}}_2, \bar{\mathbf{e}}_3\}$, there will be a reciprocal basis $\{\bar{\mathbf{e}}^1, \bar{\mathbf{e}}^2, \bar{\mathbf{e}}^3\}$, where we also take $\bar{\mathbf{e}}_i \cdot \bar{\mathbf{e}}^j = \delta^j_i$. Now at a point P we have 4 bases:

$$\{\mathbf{e}_i\} \ : \ \text{the given basis,} \tag{2.20}$$

$$\{\mathbf{e}^i\} \ : \ \text{the reciprocal to the given basis,} \tag{2.21}$$

$$\{\bar{\mathbf{e}}_i\} \ : \ \text{the new basis,} \tag{2.22}$$

$$\{\bar{\mathbf{e}}^i\} \ : \ \text{the reciprocal to the new basis.} \tag{2.23}$$

A change of coordinate systems can be made by a translation of axes without rotation followed by a rotation of the axes keeping the origin fixed (and possibly followed by a reflection in a coordinate plane if we allow transformations between right-handed and left-handed systems). The components of a vector (and similarly higher order tensors) are not changed by translation of axes, since the projections of a vector onto parallel axes are the same in magnitude and sense. Hence we need only to consider transformations with fixed origin.

What conditions do the invariance requirement impose on the tensor components?

A scalar imposes no condition since it has the same value in any coordinate system.

Consider the vector $\mathbf{v}$, which we can write in any of the following forms:

$$\mathbf{v} = v^i \mathbf{e}_i = v_i \mathbf{e}^i = \bar{v}^i \bar{\mathbf{e}}_i = \bar{v}_i \bar{\mathbf{e}}^i. \tag{2.24}$$

Since $\mathbf{v} = v^i \mathbf{e}_i = \bar{v}^j \bar{\mathbf{e}}_j$, how are the components v^i and $\bar{v}^j$ related? First we must ask how are $\mathbf{e}_i$ and $\bar{\mathbf{e}}_j$ related at a point? The answer is by a coordinate transformation. As illustrated in Fig. 2.3, in general we have

$$\bar{\mathbf{e}}_j = a^i_j \mathbf{e}_i, \tag{2.25}$$

or

		New basis		
		$\bar{\mathbf{e}}_1$	$\bar{\mathbf{e}}_2$	$\bar{\mathbf{e}}_3$
Given basis	$\mathbf{e}_1$	a^1_1	a^1_2	a^1_3
	$\mathbf{e}_2$	a^2_1	a^2_2	a^2_3
	$\mathbf{e}_3$	a^3_1	a^3_2	a^3_3

The matrix component a^i_j is the cosine of the angle between the new coordinate vector $\bar{\mathbf{e}}_j$ and the given coordinate vector $\mathbf{e}_i$, i.e., it is the direction cosine. We note that in general $[a^i_j] \neq [a^i_j]^T$. The nine components a^i_j are not all independent if the coordinates are orthogonal; only 6 are. Furthermore, in such case $[a^i_j]^{-1} = [a^i_j]^T = [a^j_i]$ so that $\mathbf{e}_i = a^j_i \bar{\mathbf{e}}_j$. Now, returning to the general case, we have

$$\mathbf{v} = v^i \mathbf{e}_i = \bar{v}^j \bar{\mathbf{e}}_j = \bar{v}^j a^i_j \mathbf{e}_i,$$

so that

$$a^i_j \bar{v}^j = v^i. \tag{2.26}$$

Components with superscript indices are called contravariant because the transformation equation for components is *contrary* (or opposite) to the transformation equation of the basis (2.25). How are v_i and $\bar{v}_i$ related? To see this, we write

$$\mathbf{v} = v_i \mathbf{e}^i = \bar{v}_j \bar{\mathbf{e}}^j.$$

Dotting from the right with $\bar{\mathbf{e}}_k$, we have

$$v_i \mathbf{e}^i \cdot \bar{\mathbf{e}}_k = \bar{v}_j \bar{\mathbf{e}}^j \cdot \bar{\mathbf{e}}_k = \bar{v}_j \delta^j_k = \bar{v}_k.$$

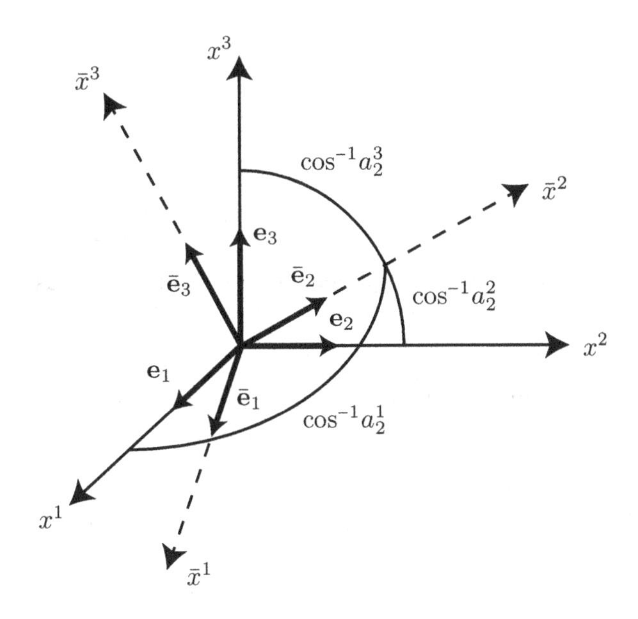

Figure 2.3: Coordinate transformation.

Now using (2.25) on the left-hand side of the above equation, we have

$$\bar{v}_k = v_i \mathbf{e}^i \cdot a_k^j \mathbf{e}_j = v_i a_k^j \delta_j^i,$$

or

$$\bar{v}_k = a_k^i v_i. \tag{2.27}$$

Components with subscript indices are called covariant because the transformation equation for components is *the same* as the transformation equation of the basis (2.25).

For a second-order tensor $\boldsymbol{\tau}$, we have

$$\boldsymbol{\tau} = \tau_{ij} \mathbf{e}^i \mathbf{e}^j = \bar{\tau}_{lm} \bar{\mathbf{e}}^l \bar{\mathbf{e}}^m = \tau^{rs} \mathbf{e}_r \mathbf{e}_s = \bar{\tau}^{pq} \bar{\mathbf{e}}_p \bar{\mathbf{e}}_q. \tag{2.28}$$

A product such as $\mathbf{e}^i \mathbf{e}^j$ is understood to mean a tensor (in this case a dyadic) product. Sometimes the symbol $\otimes$ is used to make this dyadic product explicit so that one would write $\mathbf{e}^i \otimes \mathbf{e}^j$, but we will not use such notation. Products of tensors without any explicit operator between the tensors are always understood to mean tensor products in the present text. Now consider

$$\tau_{ij} \mathbf{e}^i \mathbf{e}^j = \bar{\tau}_{lm} \bar{\mathbf{e}}^l \bar{\mathbf{e}}^m.$$

Dotting the above equation from the right with $\bar{\mathbf{e}}_k$ and $\bar{\mathbf{e}}_p$ consecutively, we have

$$\tau_{ij} (\mathbf{e}^i \cdot \bar{\mathbf{e}}_p)(\mathbf{e}^j \cdot \bar{\mathbf{e}}_k) = \bar{\tau}_{lm} (\bar{\mathbf{e}}^l \cdot \bar{\mathbf{e}}_p)(\bar{\mathbf{e}}^m \cdot \bar{\mathbf{e}}_k) = \bar{\tau}_{lm} \delta_p^l \delta_k^m = \bar{\tau}_{pk}.$$

Now, using (2.25) on the left-hand side of the above equation, we can write

$$\bar{\tau}_{pk} = \tau_{ij} (\mathbf{e}^i \cdot a_p^r \mathbf{e}_r)(\mathbf{e}^j \cdot a_k^q \mathbf{e}_q) = \tau_{ij} a_p^i a_k^j.$$

Thus the covariant components of a second-order tensor transform as

$$\bar{\tau}_{pk} = a^i_p a^j_k \tau_{ij}. \tag{2.29}$$

We also have

$$\tau^{rs}\mathbf{e}_r\mathbf{e}_s = \bar{\tau}^{pq}\bar{\mathbf{e}}_p\bar{\mathbf{e}}_q = \bar{\tau}^{pq}a^r_p\mathbf{e}_r a^s_q\mathbf{e}_s = \bar{\tau}^{pq}a^r_p a^s_q\mathbf{e}_r\mathbf{e}_s,$$

and thus the contravariant components of a second-order tensor transform as

$$a^r_p a^s_q \bar{\tau}^{pq} = \tau^{rs}. \tag{2.30}$$

A similar procedure can be used to obtain relations between covariant and contravariant components of a tensor of any order.

2.3 The metric tensor and its properties

How are v^i and v_j related? To answer this question, we first note that

$$\mathbf{v} = v_i\mathbf{e}^i = v^j\mathbf{e}_j.$$

Dotting the above equation from the right with $\mathbf{e}_k$, we have

$$v_i\mathbf{e}^i\cdot\mathbf{e}_k = v_i\delta^i_k = v_k = v^j\mathbf{e}_j\cdot\mathbf{e}_k,$$

or

$$v_k = g_{jk}v^j, \tag{2.31}$$

where we have defined

$$g_{ij} \equiv \mathbf{e}_i\cdot\mathbf{e}_j = g_{ji}. \tag{2.32}$$

If we think of $G = [g_{ij}]$ as a matrix, then we see that this matrix is symmetric since it is equal to its transpose, i.e., $G = G^T$.

Is the use of g_{ij} as covariant components of some tensor $\mathbf{g}$ justified? That is, do the g_{ij} transform like covariant components of a second-order tensor? To see this, we first define analogously

$$\bar{g}_{lm} \equiv \bar{\mathbf{e}}_l\cdot\bar{\mathbf{e}}_m = \bar{g}_{ml}. \tag{2.33}$$

Subsequently

$$\bar{g}_{lm} = \bar{\mathbf{e}}_l\cdot\bar{\mathbf{e}}_m = a^i_l\mathbf{e}_i\cdot a^j_m\mathbf{e}_j = a^i_l a^j_m\mathbf{e}_i\cdot\mathbf{e}_j = a^i_l a^j_m g_{ij}.$$

We conclude that the second-order tensor

$$\mathbf{g} = g_{ij}\mathbf{e}^i\mathbf{e}^j \tag{2.34}$$

indeed transforms like a second-order tensor with covariant components g_{ij}. The tensor $\mathbf{g}$ is called the *metric* tensor.

We could also write it as

$$\mathbf{g} = g^{lm}\mathbf{e}_l\mathbf{e}_m = g_{ij}\mathbf{e}^i\mathbf{e}^j. \tag{2.35}$$

Now dotting the above equation from the right with $\mathbf{e}_r$ and $\mathbf{e}^p$ consecutively, we have

$$g^{lm}(\mathbf{e}_l \cdot \mathbf{e}^p)(\mathbf{e}_m \cdot \mathbf{e}_r) = g_{ij}(\mathbf{e}^i \cdot \mathbf{e}^p)(\mathbf{e}^j \cdot \mathbf{e}_r),$$

or

$$g^{lm}\delta_l^p g_{mr} = g_{ij}(\mathbf{e}^i \cdot \mathbf{e}^p)\delta_r^j,$$

or

$$g^{pm}g_{mr} = g_{ir}(\mathbf{e}^i \cdot \mathbf{e}^p) = g_{mr}(\mathbf{e}^m \cdot \mathbf{e}^p).$$

Thus

$$g^{pm} = \mathbf{e}^m \cdot \mathbf{e}^p = g^{mp}. \tag{2.36}$$

Now consider the equation

$$\mathbf{v} = v_i\mathbf{e}^i = v^j\mathbf{e}_j,$$

and dot both sides from the right by $\mathbf{e}^k$ to obtain

$$v_i\mathbf{e}^i \cdot \mathbf{e}^k = v^j\mathbf{e}_j \cdot \mathbf{e}^k,$$

or

$$v_i g^{ik} = v^j\delta_j^k,$$

or

$$v^k = g^{ik}v_i. \tag{2.37}$$

We see that the covariant and contravariant components g_{ij} and g^{ij} of the metric tensor provide the means of computing the covariant or contravariant components (v_i or v^j) of vector $\mathbf{v}$ if we know the contravariant or covariant components (v^j or v_i). More simply, the components of the metric tensor provide us the means of raising or lowering indices.

This is also true when operating on the basis vectors:

$$\mathbf{e}_i = g_{ij}\mathbf{e}^j \qquad \text{and} \qquad \mathbf{e}^i = g^{ij}\mathbf{e}_j, \tag{2.38}$$

since we see that if we take the dot product of $\mathbf{e}_k$ from the right with the first equation, we have the identity

$$\mathbf{e}_i \cdot \mathbf{e}_k = g_{ij}\mathbf{e}^j \cdot \mathbf{e}_k = g_{ik},$$

while if we take the dot product of $\mathbf{e}^k$ from the right with the second equation, we have the identity

$$\mathbf{e}^i \cdot \mathbf{e}^k = g^{ij}\mathbf{e}_j \cdot \mathbf{e}^k = g^{ik}.$$

Alternately, if we take the dot product of $\mathbf{e}^k$ from the right with the first equation, we obtain

$$\mathbf{e}_i \cdot \mathbf{e}^k = g_{ij}\mathbf{e}^j \cdot \mathbf{e}^k,$$

or

$$\delta_i^k = g_{ij} g^{jk}. \tag{2.39}$$

In matrix form, this last equation is

$$I = G\,[g^{jk}],$$

thus we see that

$$[g^{jk}] = G^{-1},$$

i.e., the covariant and contravariant components of the metric tensor are inverses of each other. In matrix form, the equations would be written as

$$I = GG^{-1} = G^{-1}G.$$

2.4 General polyadic tensor of order m

In general, a tensor $\tau^{i_1 i_2 \cdots i_r}_{j_1 j_2 \cdots j_n} \mathbf{e}_{i_1} \mathbf{e}_{i_2} \cdots \mathbf{e}_{i_r} \mathbf{e}^{j_1} \mathbf{e}^{j_2} \cdots \mathbf{e}^{j_n}$ with $\binom{r}{n}$ components such that $r + n = m$ is called a polyadic tensor of order m. For example, a tensor $\boldsymbol{\tau}$ of order $m = 5$ with $\binom{5}{0}$ components is written as $\tau^{ijklm} \mathbf{e}_i \mathbf{e}_j \mathbf{e}_k \mathbf{e}_l \mathbf{e}_m$. When r and n are not zero, then the components are said to be mixed.

Consider the dyadic tensor

$$\boldsymbol{\tau} = \tau_{ij} \mathbf{e}^i \mathbf{e}^j = \tau^{kl} \mathbf{e}_k \mathbf{e}_l = \tau^r_s \mathbf{e}_r \mathbf{e}^s. \tag{2.40}$$

We can rewrite the above as

$$\tau_{is} \mathbf{e}^i \mathbf{e}^s = \tau^{rl} \mathbf{e}_r \mathbf{e}_l = \tau^r_s \mathbf{e}_r \mathbf{e}^s.$$

Transforming coordinates by using the metric tensor

$$\tau_{is} g^{ir} \mathbf{e}_r \mathbf{e}^s = \tau^{rl} \mathbf{e}_r g_{ls} \mathbf{e}^s = \tau^r_s \mathbf{e}_r \mathbf{e}^s,$$

we obtain

$$\tau_{is} g^{ir} = \tau^{rl} g_{ls} = \tau^r_s. \tag{2.41}$$

Again we see that the metric tensor provides us with the means of "raising and lowering" component indices for a dyadic tensor. This remains true for tensors of any order.

Example

Suppose we have the equation

$$a_i = \tau_{ij} b^j,$$

and suppose that we wanted to obtain a^r (contravariant components). Then to raise the subscript, we would multiply the equation by g^{ir}:

$$a_i g^{ir} = \tau_{ij} g^{ir} b^j,$$

or

$$a^r = \tau^r_j b^j.$$

We recall from before that the covariant and contravariant components of a second-order tensor under a change of basis transform as

$$\bar{\tau}_{ij} = \tau_{kl} a^k_i a^l_j$$

and

$$\bar{\tau}^{ij} a^k_i a^l_j = \tau^{kl}.$$

How does τ^i_j transform when the basis is changed? To see this, we first write

$$\bar{\mathbf{e}}_i = a^j_i \mathbf{e}_j,$$

and

$$\boldsymbol{\tau} = \tau^i_j \mathbf{e}_i \mathbf{e}^j = \bar{\tau}^l_k \bar{\mathbf{e}}_l \bar{\mathbf{e}}^k.$$

Now if we take the dot product of the above equation with $\bar{\mathbf{e}}_r$ from the right, we have

$$\tau^i_j \mathbf{e}_i (\mathbf{e}^j \cdot \bar{\mathbf{e}}_r) = \bar{\tau}^l_k \bar{\mathbf{e}}_l (\bar{\mathbf{e}}^k \cdot \bar{\mathbf{e}}_r) = \bar{\tau}^l_k \bar{\mathbf{e}}_l \delta^k_r = \bar{\tau}^l_r \bar{\mathbf{e}}_l.$$

Now, changing the basis, we have

$$\tau^i_j \mathbf{e}_i a^p_r \delta^j_p = \bar{\tau}^l_r a^i_l \mathbf{e}_i,$$

or

$$\tau^i_p a^p_r \mathbf{e}_i = \bar{\tau}^l_r a^i_l \mathbf{e}_i.$$

Thus, the mixed components transform as

$$\bar{\tau}^l_r a^i_l = \tau^i_p a^p_r. \tag{2.42}$$

That is, the respective contravariant and covariant components transform as one would expect contravariant and covariant components would. This remains true for a tensor of any order, i.e.,

$$\bar{\tau}^{i_1 i_2 \cdots i_r}_{j_1 j_2 \cdots j_n} a^{k_1}_{i_1} a^{k_2}_{i_2} \cdots a^{k_r}_{i_r} = \tau^{k_1 k_2 \cdots k_r}_{l_1 l_2 \cdots l_n} a^{l_1}_{j_1} a^{l_2}_{j_2} \cdots a^{l_n}_{j_n}. \tag{2.43}$$

All tensors that we have been discussing up to this point are called *absolute* tensors. We now extend the definition and call a tensor whose components transform as

$$\bar{\tau}^{i_1 i_2 \cdots i_r}_{j_1 j_2 \cdots j_n} a^{k_1}_{i_1} a^{k_2}_{i_2} \cdots a^{k_r}_{i_r} = \operatorname{sgn} a^s \; |a|^w \; \tau^{k_1 k_2 \cdots k_r}_{l_1 l_2 \cdots l_n} a^{l_1}_{j_1} a^{l_2}_{j_2} \cdots a^{l_n}_{j_n} \tag{2.44}$$

a *relative* or *weighted* tensor of order $m = r+n$, where $a = \det[a^i_j]$ is the determinant of the transformation, which is assumed to be non-singular. If $s = w = 0$, $\boldsymbol{\tau}$ is called an *absolute tensor.* If $s = 0$ and $w \neq 0$, $\boldsymbol{\tau}$ is called a *relative tensor of weight* w. If $s = 1$ and $w = 0$, $\boldsymbol{\tau}$ is called an *axial tensor.* Lastly, if $s = 1$ and $w \neq 0$, $\boldsymbol{\tau}$ is called an *axial relative tensor of weight* w. When the word tensor is used from now on, it is understood to mean an absolute tensor unless explicitly stated otherwise. Relative or weighted tensors are also referred to as *pseudo-tensors.*

Does the Kronecker delta δ^i_j transform like mixed components of a tensor of order two? To see if it does, we take this to be the components of the unit, or identity, tensor

$$\mathbf{1} = \delta^i_j \mathbf{e}_i \mathbf{e}^j = \bar{\delta}^k_l \bar{\mathbf{e}}_k \bar{\mathbf{e}}^l. \tag{2.45}$$

It is immediately obvious that if we let $\tau^i_j \to \delta^i_j$, the previous derivation applies and thus the Kronecker delta transforms like mixed components of an absolute second-order tensor. Since it has the same value in all coordinate systems, i.e., $\bar{\delta}^i_j = \delta^i_j$, such a tensor is called *isotropic.* More specifically, a tensor is isotropic if its components are the same under arbitrary rotations of the basis vectors. Thus, e.g., if we have

$$\bar{B}_{ij} = a^k_i a^m_j B_{km},$$

then the tensor $\mathbf{B}$ is isotropic if $\bar{B}_{ij} = B_{ij}$.

A simple example of a third-order isotropic tensor is the *Levi–Civita tensor*

$$\boldsymbol{\epsilon} = \epsilon_{ijk} \mathbf{e}^i \mathbf{e}^j \mathbf{e}^k. \tag{2.46}$$

Since the tensor is isotropic, we have that

$$\bar{\epsilon}_{ijk} = \epsilon_{ijk}. \tag{2.47}$$

In addition, as noted earlier (see (2.6)), since

$$a = \epsilon_{rst} a^r_1 a^s_2 a^t_3, \tag{2.48}$$

it then follows that

$$\bar{\epsilon}_{ijk} a = \epsilon_{rst} a^r_i a^s_j a^t_k, \tag{2.49}$$

or

$$\bar{\epsilon}_{ijk} = a^{-1} \epsilon_{rst} a^r_i a^s_j a^t_k. \tag{2.50}$$

We see that ϵ_{ijk} transform like covariant components of a third-order relative tensor with weight $w = -1$. Note that a is nothing more than the determinant of the transformation. In a similar fashion, it can be shown that

$$\bar{\epsilon}^{ijk} a^r_i a^s_j a^t_k = a \epsilon^{rst}, \tag{2.51}$$

so that ϵ^{rst} transforms like contravariant components of a third-order relative tensor with weight $w = 1$.

A general discussion of isotropic tensors is given in Appendix B.

2.5 Scalar product of two vectors

Let

$$\mathbf{u} = u^i \mathbf{e}_i = u_j \mathbf{e}^j,$$

and

$$\mathbf{v} = v^l \mathbf{e}_l = v_m \mathbf{e}^m.$$

Now the scalar (or dot or inner) product of the two vectors is given by

$$\mathbf{u} \cdot \mathbf{v} = (u^i \mathbf{e}_i) \cdot (v^l \mathbf{e}_l) = u^i v^l (\mathbf{e}_i \cdot \mathbf{e}_l),$$

or

$$\mathbf{u} \cdot \mathbf{v} = g_{il} u^i v^l. \tag{2.52}$$

Similarly

$$\mathbf{u} \cdot \mathbf{v} = g^{jm} u_j v_m. \tag{2.53}$$

Also,

$$\mathbf{u} \cdot \mathbf{v} = (u^i \mathbf{e}_i) \cdot (v_m \mathbf{e}^m) = u^i v_m (\mathbf{e}_i \cdot \mathbf{e}^m) = u^i v_m \delta_i^m,$$

so we can see that

$$\mathbf{u} \cdot \mathbf{v} = u^i v_i = u_i v^i. \tag{2.54}$$

2.6 Vector product of two vectors

Consider first the scalar triple product

$$\begin{aligned}
[\mathbf{e}_1, \mathbf{e}_2, \mathbf{e}_3] &= (g_{1i} \mathbf{e}^i) \cdot \left[(g_{2j} \mathbf{e}^j) \times (g_{3k} \mathbf{e}^k) \right], \\
&= g_{1i} g_{2j} g_{3k} \left[\mathbf{e}^i \cdot (\mathbf{e}^j \times \mathbf{e}^k) \right], \\
&= \epsilon^{ijk} g_{1i} g_{2j} g_{3k} \left(\mathbf{e}^1 \cdot \mathbf{e}^2 \times \mathbf{e}^3 \right), \\
&= \frac{\epsilon^{ijk} g_{1i} g_{2j} g_{3k}}{\mathbf{e}_1 \cdot \mathbf{e}_2 \times \mathbf{e}_3}, \\
&= \frac{\det G}{[\mathbf{e}_1, \mathbf{e}_2, \mathbf{e}_3]},
\end{aligned}$$

where we have used (2.18) and (2.6). We now see that

$$[\mathbf{e}_1, \mathbf{e}_2, \mathbf{e}_3] = \sqrt{g}, \tag{2.55}$$

and subsequently

$$\left[\mathbf{e}^1, \mathbf{e}^2, \mathbf{e}^3 \right] = \frac{1}{\sqrt{g}}, \tag{2.56}$$

where

$$g \equiv \det G. \tag{2.57}$$

The vector (or cross) product of two vectors, using the definition of the reciprocal basis, is now given by

$$\mathbf{u} \times \mathbf{v} = u^i v^j \mathbf{e}_i \times \mathbf{e}_j = \sqrt{g}\epsilon_{ijk} u^i v^j \mathbf{e}^k = w_k \mathbf{e}^k = \mathbf{w}, \tag{2.58}$$

thus

$$w_k = \sqrt{g}\epsilon_{ijk} u^i v^j = \sqrt{g}\epsilon_{kij} u^i v^j. \tag{2.59}$$

Similarly, we find that

$$\mathbf{w} = w^k \mathbf{e}_k, \tag{2.60}$$

where

$$w^k = \frac{1}{\sqrt{g}}\epsilon^{ijk} u_i v_j = \frac{1}{\sqrt{g}}\epsilon^{kij} u_i v_j. \tag{2.61}$$

Note that $\mathbf{v} \times \mathbf{u} = -\mathbf{u} \times \mathbf{v}$ since this corresponds to an interchange of two adjacent indices in the Levi–Civita symbol.

We now define

$$\varepsilon_{ijk} \equiv \sqrt{g}\,\epsilon_{ijk} \qquad \text{and} \qquad \bar{\varepsilon}_{ijk} \equiv \sqrt{\bar{g}}\,\bar{\epsilon}_{ijk}. \tag{2.62}$$

Then, using our earlier result, we have

$$\bar{\varepsilon}_{ijk} = \sqrt{\bar{g}}\,\bar{\epsilon}_{ijk} = \sqrt{g}\,\epsilon_{rst} a_i^r a_j^s a_k^t, \tag{2.63}$$

or

$$\bar{\varepsilon}_{ijk} = \varepsilon_{rst} a_i^r a_j^s a_k^t. \tag{2.64}$$

Thus ε_{ijk} transform like the covariant components of a third-order absolute tensor. Similarly, we can show that

$$\bar{\varepsilon}^{ijk} a_i^r a_j^s a_k^t = \varepsilon^{rst}, \tag{2.65}$$

transform like contravariant components of a third-order absolute tensor, where

$$\varepsilon^{rst} \equiv \frac{\epsilon^{rst}}{\sqrt{g}}. \tag{2.66}$$

Subsequently, we define the third-order absolute Levi–Civita tensor

$$\boldsymbol{\varepsilon} = \varepsilon^{ijk} \mathbf{e}_i \mathbf{e}_j \mathbf{e}_k = \varepsilon_{ijk} \mathbf{e}^i \mathbf{e}^j \mathbf{e}^k. \tag{2.67}$$

Note that with the above definitions, the vector, or cross, product of two vectors can now be rewritten as

$$\mathbf{w} = \mathbf{u} \times \mathbf{v} = w_k \mathbf{e}^k = w^k \mathbf{e}_k, \tag{2.68}$$

where

$$w_k = \varepsilon_{kij} u^i v^j \qquad \text{and} \qquad w^k = \varepsilon^{kij} u_i v_j. \tag{2.69}$$

From the above, formally we can also define the following vector operator that is occasionally found to be useful:

$$\times \mathbf{v} = -\mathbf{v}\times. \tag{2.70}$$

The proof is obtained by simply operating on an arbitrary vector from the right and left.

2.7 Tensor product of two vectors

The tensor product of vectors $\mathbf{u}$ and $\mathbf{v}$ is simply $\mathbf{u}\mathbf{v}$. If $\mathbf{w}$ is a vector and $\mathbf{A}$ is a second rank tensor, then it is clear that

$$\operatorname{tr}(\mathbf{u}\mathbf{v}) = \mathbf{u}\cdot\mathbf{v}, \tag{2.71}$$

$$(\mathbf{u}\mathbf{v})\cdot\mathbf{w} = \mathbf{u}(\mathbf{v}\cdot\mathbf{w}) = (\mathbf{v}\cdot\mathbf{w})\,\mathbf{u}, \tag{2.72}$$

$$\mathbf{A}\cdot(\mathbf{u}\mathbf{v}) = (\mathbf{A}\cdot\mathbf{u})\,\mathbf{v}, \tag{2.73}$$

$$(\mathbf{u}\mathbf{v})\cdot\mathbf{A} = \mathbf{u}(\mathbf{v}\cdot\mathbf{A}) = \mathbf{u}\left(\mathbf{A}^T\cdot\mathbf{v}\right). \tag{2.74}$$

Note that while the product of $\mathbf{u}$ and $\mathbf{v}$ yields a second-order tensor, it is important to realize that a general second-order tensor, say $\mathbf{A}$, cannot be represented by the tensor product of two vectors, say $\mathbf{u}\mathbf{v}$. This should be immediately obvious by noting that the components of a general second-order tensor has nine independent elements, while the components of the second-order tensor resulting from the tensor product of two vectors has only six independent entries (the three components of the two vectors). Furthermore, while in general $\det\mathbf{A} \neq 0$, we have that $\det\mathbf{u}\mathbf{v} = 0$. However, if $\mathbf{A}$ is symmetric, then it can be shown that it can be represented as a tensor product of the vectors $\mathbf{u}$ and $\mathbf{v}$, the components of which represents a *quadric surface* (see Section 2.10). Lastly, we note that the scalar and vector products of $\mathbf{u}$ and $\mathbf{v}$ are composed of particular linear combinations of the components of their tensor product.

2.8 Contraction of tensors

Contraction enables us to obtain an $(n-2)$ rank tensor from a given tensor of rank n. This is accomplished by placing a dot product between any polyadic expression. For example, with the tetrad $\mathbf{abcd}$ (where $\mathbf{a}$, $\mathbf{b}$, $\mathbf{c}$ and $\mathbf{d}$ are vectors), contraction can be done in three ways with adjacent vectors to yield

$$\mathbf{a}\cdot\mathbf{bcd} = (\mathbf{a}\cdot\mathbf{b})\mathbf{cd}, \tag{2.75}$$

$$\mathbf{ab}\cdot\mathbf{cd} = (\mathbf{b}\cdot\mathbf{c})\mathbf{ad}, \tag{2.76}$$

$$\mathbf{abc}\cdot\mathbf{d} = (\mathbf{c}\cdot\mathbf{d})\mathbf{ab}, \tag{2.77}$$

where the dot product in parentheses is a scalar function multiplying the resulting dyad. Note that

$$\mathbf{ab}\cdot\mathbf{cd} \neq \mathbf{cd}\cdot\mathbf{ab}. \tag{2.78}$$

There are two possible scalar (double-dot) products of two dyads and they are defined by

$$\mathbf{ab}:\mathbf{cd} \equiv (\mathbf{a}\cdot\mathbf{c})(\mathbf{b}\cdot\mathbf{d}), \tag{2.79}$$

$$\mathbf{ab}\cdot\cdot\,\mathbf{cd} \equiv (\mathbf{b}\cdot\mathbf{c})(\mathbf{a}\cdot\mathbf{d}). \tag{2.80}$$

It is easily shown that the scalar product is commutative as it should be since a scalar is invariant in any coordinate system, although these products would be different if higher order tensors were involved. For example, if $\mathbf{u}$ and $\mathbf{v}$ are vectors and $\mathbf{A}$ is a second-order tensor, then we note that

$$\mathbf{u}\cdot\mathbf{A}\cdot\mathbf{v} \neq \mathbf{v}\cdot\mathbf{A}\cdot\mathbf{u}, \tag{2.81}$$

even though both sides are scalar quantities. The two sides are equal only if $\mathbf{A}$ is symmetric, i.e., $\mathbf{A} = \mathbf{A}^T$. If $\mathbf{A}$ is skew-symmetric, i.e., $\mathbf{A} = -\mathbf{A}^T$, then

$$\mathbf{u}\cdot\mathbf{A}\cdot\mathbf{v} = -\mathbf{v}\cdot\mathbf{A}\cdot\mathbf{u}. \tag{2.82}$$

Note that the contraction of a second rank tensor $\mathbf{A}$ with the unit tensor yields the scalar corresponding to the trace of the tensor:

$$\operatorname{tr}\mathbf{A} = \mathbf{1}:\mathbf{A} = \mathbf{A}:\mathbf{1} \qquad \text{or} \qquad A_{ii} = \delta_{ij}\,A_{ij}. \tag{2.83}$$

Example

Let $\boldsymbol{\tau} = \tau^{ij}_{kl}\mathbf{e}_i\mathbf{e}_j\mathbf{e}^k\mathbf{e}^l$ be a tensor of rank 4. Placing a dot product between any two basis vectors gives a new tensor of rank 2. For example,

$$\boldsymbol{\beta} = \tau^{ij}_{kl}(\mathbf{e}_i\cdot\mathbf{e}_j)\mathbf{e}^k\mathbf{e}^l = \tau^{ij}_{kl}\,g_{ij}\mathbf{e}^k\mathbf{e}^l = \tau^{i}_{ikl}\mathbf{e}^k\mathbf{e}^l.$$

Alternatively, we could have written $\boldsymbol{\tau} = \tau^{i}_{jkl}\mathbf{e}_i\mathbf{e}^j\mathbf{e}^k\mathbf{e}^l$ so that by contracting we have

$$\boldsymbol{\beta} = \tau^{i}_{jkl}(\mathbf{e}_i\cdot\mathbf{e}^j)\mathbf{e}^k\mathbf{e}^l = \tau^{i}_{jkl}\delta^j_i\mathbf{e}^k\mathbf{e}^l = \tau^{i}_{ikl}\mathbf{e}^k\mathbf{e}^l.$$

Since the new tensor $\boldsymbol{\beta}$ should be independent of any coordinate system, we also have

$$\boldsymbol{\beta} = \bar{\tau}^{i}_{ikl}\bar{\mathbf{e}}^k\bar{\mathbf{e}}^l.$$

Example

We show that the cross product of two vectors can also be written as

$$\begin{aligned}\mathbf{w} = \mathbf{u} \times \mathbf{v} = \boldsymbol{\epsilon} : (\mathbf{u}\mathbf{v}) &= \left(\epsilon^{ijk}\mathbf{e}_i\mathbf{e}_j\mathbf{e}_k\right) : \left(u^l\mathbf{e}_l\right)\left(v^m\mathbf{e}_m\right) \\ &= \epsilon^{ijk}u^l v^m \left(\mathbf{e}_j \cdot \mathbf{e}_l\right)\left(\mathbf{e}_k \cdot \mathbf{e}_m\right)\mathbf{e}_i \\ &= \epsilon^{ijk} g_{jl} g_{km} u^l v^m \mathbf{e}_i \\ &= \epsilon^{ijk} g_{in} g_{jl} g_{km} u^l v^m \mathbf{e}^n \\ &= \sqrt{g}\epsilon_{nlm} u^l v^m \mathbf{e}^n \\ &= w_n \mathbf{e}^n,\end{aligned}$$

or alternatively as

$$\begin{aligned}\mathbf{w} = \mathbf{u} \times \mathbf{v} = \boldsymbol{\epsilon} \cdot\cdot (\mathbf{v}\mathbf{u}) &= (\boldsymbol{\epsilon} \cdot \mathbf{v}) \cdot \mathbf{u} \\ &= \left[\left(\epsilon^{ijk}\mathbf{e}_i\mathbf{e}_j\mathbf{e}_k\right) \cdot \left(v^l\mathbf{e}_l\right)\right] \cdot \left(u^m\mathbf{e}_m\right) \\ &= \epsilon^{ijk} g_{kl} g_{jm} v^l u^m \mathbf{e}_i \\ &= \epsilon^{ijk} g_{in} g_{jm} g_{kl} u^m v^l \mathbf{e}^n \\ &= \sqrt{g}\epsilon_{nml} u^m v^l \mathbf{e}^n \\ &= w_n \mathbf{e}^n,\end{aligned}$$

recovering our previous result.

2.9 Transpose of a tensor

The transpose of a tensor of any rank is given by the permutation of any two components. For example, if $\mathbf{A} = a^{ijk}\mathbf{e}_i\mathbf{e}_j\mathbf{e}_k$, then $\mathbf{B} = b^{ijk}\mathbf{e}_i\mathbf{e}_j\mathbf{e}_k$ is a transpose of $\mathbf{A}$ if $b^{ijk} = a^{ikj}$. The superscript of T on a matrix $A = [a_{ij}]$, A^T is understood to denote the permutation of the two indices, i.e., $A^T = [a_{ji}]$. The transpose operation for a second rank tensor can also be viewed as the unique transposition satisfying

$$\left(\mathbf{v} \cdot \mathbf{A}^T\right) \cdot \mathbf{u} = (\mathbf{A} \cdot \mathbf{v}) \cdot \mathbf{u} = \mathbf{u} \cdot (\mathbf{A} \cdot \mathbf{v}) \tag{2.84}$$

for all vector $\mathbf{u}$ and $\mathbf{v}$. Note that $(\mathbf{A}^T)^T = \mathbf{A}$, and if $\mathbf{A}^{-1} = \mathbf{A}^T$, then $\mathbf{A} \cdot \mathbf{A}^T = \mathbf{A}^T \cdot \mathbf{A} = \mathbf{1}$ and $\mathbf{A}$ is an orthogonal tensor. The above rule applies to vectors as well, since a column or row vector is understood to correspondingly have unity in the row or column entry, so that upon transposition it becomes a row or column vector correspondingly. If $\mathbb{A}$ is a tensor of rank 4, generalizing the above definition for tensors of rank 2, the unique transpose of $\mathbb{A}$, which we denote simply as $\mathbb{A}^T$, satisfies

$$\left(\mathbf{B} : \mathbb{A}^T\right) : \mathbf{C} = (\mathbb{A} : \mathbf{B}) : \mathbf{C} = \mathbf{C} : (\mathbb{A} : \mathbf{B}) \tag{2.85}$$

for all second rank tensors $\mathbf{B}$ and $\mathbf{C}$. Note that $(\mathbb{A}^T)_{klij} = [a_{ijkl}]$, $(\mathbb{A}^T)^T = \mathbb{A}$, and $(\mathbf{BC})^T = \mathbf{C}^T\mathbf{B}^T$.

For more general transposition, the superscript $T_{\{m,n\}}$ will denote that the m and n entries are to be transposed. For example, if $A = [a_{ijkl}]$, then $A^{T_{\{2,3\}}} = [a_{ikjl}]$. If multiple transpositions are required, the operator is applied repeatedly, e.g., $A^{T_{\{2,3\}}T_{\{1,4\}}} = [a_{lkji}]$.

2.10 Symmetric and skew-symmetric tensors

A tensor $\mathbf{A}$ is said to be *symmetric* in two indices of *the same type* (both covariant or both contravariant) if the value of any component is not changed by permuting them. The corresponding indices are enclosed in parentheses, the unaffected indices being separated by vertical bars on either side. Thus, if $a_{ijkm} = a_{mjki}$, then we write $a_{(i|jk|m)}$ to denote the symmetry. A tensor $\mathbf{A}$ is said to be *completely symmetric* in any set of upper or lower indices if its components are not altered in value by any permutation of the set, e.g.,

$$a_{ijkm} = a_{jikm} = a_{jimk} = \cdots. \tag{2.86}$$

This is often denoted by sym $\mathbf{A}$ or its indices enclosed in parentheses, e.g., $a_{(ijkm)}$. The validity of this property in one coordinate system ensures it in all coordinate systems.

A tensor $\mathbf{A}$ is *skew-symmetric* (or *anti-symmetric*) if its components are not altered in value by any *even* permutation of the indices and are merely changed in sign by an *odd* permutation of these indices. The corresponding indices are enclosed in brackets, the unaffected indices being separated by vertical bars on either side. Thus, if $a_{ijkm} = -a_{mjki}$, then we write $a_{[i|jk|m]}$. A tensor $\mathbf{A}$ of rank k that is skew-symmetric in all indices is called *completely skew-symmetric*, e.g.,

$$a_{ijkm} = -a_{jikm} = a_{jimk} = \cdots. \tag{2.87}$$

This is often denoted by skw $\mathbf{A}$ or its indices enclosed in brackets, e.g., $a_{[ijkm]}$. In such tensors, all components having two equal indices are zero. A covariant or contravariant completely skew-symmetric tensor of rank k is called a *k-vector.* Note that 0-vectors are scalars and 1-vectors are vectors.

We point out that the generalized Levi–Civita (or permutation) symbol

$$\epsilon^{ijk\ldots} = \epsilon_{ijk\ldots} = \begin{cases} +1 & \text{if } (i,j,k,\ldots) \text{ is an even permutation of } (1,2,3,\ldots), \\ -1 & \text{if } (i,j,k,\ldots) \text{ is an odd permutation of } (1,2,3,\ldots), \\ 0 & \text{if any two labels are the same,} \end{cases} \tag{2.88}$$

is completely skew-symmetric (see Appendix A). Note that $\epsilon^{ijk\ldots}$ and $\epsilon_{ijk\ldots}$ are relative isotropic tensors of weight +1 and −1, respectively. The corresponding generalized absolute contravariant and covariant components of the Levi–Civita tensors are defined by

$$\varepsilon^{ijk\ldots} \equiv \frac{\epsilon^{ijk\ldots}}{\sqrt{g}} \qquad \text{and} \qquad \varepsilon_{ijk\ldots} \equiv \sqrt{g}\,\epsilon_{ijk\ldots}, \tag{2.89}$$

where

$$g \equiv \frac{1}{n!}\,\epsilon^{i_1\ldots i_n}\epsilon^{j_1\ldots j_n} g_{i_1 j_1}\cdots g_{i_n j_n} \tag{2.90}$$

is the contracted product of two epsilons of weight +1 and n absolute dyadics. Hence the discriminant g of the fundamental quadratic form is a relative scalar of weight 2.

Associated with the components of the generalized Levi–Civita tensor is the generalized Kronecker delta

$$\delta^{i_1\ldots i_k}_{j_1\ldots j_k} = \det\begin{bmatrix} \delta^{i_1}_{j_1} & \cdots & \delta^{i_1}_{j_k} \\ \vdots & \vdots & \vdots \\ \delta^{i_k}_{j_1} & \cdots & \delta^{i_k}_{j_k} \end{bmatrix}, \qquad k \le n. \tag{2.91}$$

We note that the k superscripts and subscripts in the component of the generalized Kronecker delta can range from 1 to n. This tensor has the properties that

$$\delta^{i_1\ldots i_k}_{j_1\ldots j_k} = \epsilon^{i_1\ldots i_k}\epsilon_{j_1\ldots j_k}, \tag{2.92}$$

$$(n-k)!\ \delta^{i_1\ldots i_k}_{j_1\ldots j_k} = \epsilon^{i_1\ldots i_k i_{k+1}\ldots i_n}\epsilon_{j_1\ldots j_k i_{k+1}\ldots i_n}, \tag{2.93}$$

$$\delta^{i_1\ldots i_k i_{k+1}\ldots i_n}_{i_1\ldots i_k j_{k+1}\ldots j_n} = k!\ \delta^{i_{k+1}\ldots i_n}_{j_{k+1}\ldots j_n}, \tag{2.94}$$

and

$$\delta^{i_1\ldots i_n}_{i_1\ldots i_n} = n! \tag{2.95}$$

It can be shown that $\delta^{i_1\ldots i_k}_{j_1\ldots j_k}$ is the component of an absolute tensor of rank $2k$, while each Levi–Civita tensor is of rank n. If both the upper and lower indices consist of the same set of distinct numbers, chosen from $1,\ldots,n$, the Kronecker delta is +1 or −1 according to whether the upper indices form an even or odd permutation of the lower indices; in all other cases, it is zero. This rule is a direct consequence of the epsilons. For example, in the case of $n = 3$ and $k = 2$, we have

$$\delta^{12}_{12} = +1, \quad \delta^{12}_{21} = -1, \quad \delta^{32}_{23} = -1, \quad \delta^{23}_{11} = \delta^{13}_{21} = 0, \tag{2.96}$$

while with $k = 3$ we have

$$\delta^{123}_{123} = \delta^{231}_{123} = +1, \quad \delta^{213}_{123} = \delta^{321}_{123} = -1, \quad \delta^{322}_{123} = 0. \tag{2.97}$$

In the case $n = 4$, we can write

$$\delta^{i}_{p} = \frac{1}{3!}\epsilon^{iqrs}\epsilon_{pqrs}, \quad \delta^{ij}_{pq} = \frac{1}{2!}\epsilon^{ijrs}\epsilon_{pqrs}, \quad \delta^{ijk}_{pqr} = \frac{1}{1!}\epsilon^{ijks}\epsilon_{pqrs}, \quad \delta^{ijkl}_{pqrs} = \epsilon^{ijkl}\epsilon_{pqrs}. \tag{2.98}$$

Symmetric and skew-symmetric parts of any tensor can be constructed by an appropriate linear combination of the tensor and its transposes. For example, to construct symmetric parts of the rank-2 tensor $\mathbf{A}$ and rank-3 tensor $\mathbf{B}$, one averages the components of the tensor with all of its transposes:

$$a_{(ij)} = \frac{1}{2!}\left(a_{ij} + a_{ji}\right), \tag{2.99}$$

$$b_{(ijk)} = \frac{1}{3!}\left(b_{ijk} + b_{jki} + b_{kij} + b_{jik} + b_{ikj} + b_{kji}\right). \tag{2.100}$$

The skew-symmetric parts are constructed similarly:

$$a_{[ij]} = \frac{1}{2!}\left(a_{ij} - a_{ji}\right), \tag{2.101}$$

$$b_{[ijk]} = \frac{1}{3!}\left(b_{ijk} + b_{jki} + b_{kij} - b_{jik} - b_{ikj} - b_{kji}\right). \tag{2.102}$$

Note that

$$b_{(ij)k} = \frac{1}{2!}\left(b_{ijk} + b_{jik}\right), \tag{2.103}$$

and

$$b_{[ij]k} = \frac{1}{2!}\left(b_{ijk} - b_{jik}\right). \tag{2.104}$$

It can be shown that in an n-dimensional space, a completely symmetric tensor of rank r can have at most $[(n-1)+r]!/(n-1)!r!$ distinct components. Similarly, a completely skew-symmetric tensor can have at most $n!/r!(n-r)!$ distinct components, and if all particular indices are different, the corresponding components differ from each other only in sign, while if even two such indices are equal, the components vanish. Clearly for $n = 3$ and $r = 2$, the symmetric tensor has 6 components while the skew-symmetric one has 3 components. Also note that (i) if $n = 3$ and $r = 3$, then the symmetric tensor has 10 components while the skew-symmetric one has 1 component; (ii) if $n = 3$ all components of skew-symmetric tensors of rank $r \geq 4$ vanish; and (iii) if $n = r$, then the skew-symmetric tensor has one component. Subsequently, we can write the above completely skew-symmetric third rank tensor in three-dimensional space as

$$b_{[ijk]} = c\,\epsilon_{ijk}, \tag{2.105}$$

where c is a scalar.

As remarked earlier, and as easily verified, any rank-2 tensor $\mathbf{A}$ can be decomposed into symmetric and skew-symmetric parts:

$$\mathbf{A} = \text{sym}\ \mathbf{A} + \text{skw}\ \mathbf{A} \qquad \text{or} \qquad a_{ij} = a_{(ij)} + a_{[ij]}. \tag{2.106}$$

Note that if $\mathbf{D} = \text{sym}\ \mathbf{A}$, $\mathbf{W} = \text{skw}\ \mathbf{A}$, and $\mathbf{C}$ is a general rank-2 tensor, then $\text{tr}\ \mathbf{W} = 0$, $\mathbf{D} = \mathbf{D}^T$, $\mathbf{W} = -\mathbf{W}^T$, $\mathbf{D}:\mathbf{W} = 0$, and

$$\mathbf{D}:\mathbf{C} = \mathbf{D}^T:\mathbf{C} = \mathbf{D}:\frac{1}{2}\left(\mathbf{C}+\mathbf{C}^T\right), \tag{2.107}$$

$$\mathbf{W}:\mathbf{C} = -\mathbf{W}^T:\mathbf{C} = \mathbf{W}:\frac{1}{2}\left(\mathbf{C}-\mathbf{C}^T\right). \tag{2.108}$$

The symmetric part of $\mathbf{A}$ can also be made traceless by subtracting $\frac{1}{3}(\text{tr}\ \mathbf{A})\mathbf{1}$ from it, i.e.,

$$\widehat{\mathbf{A}} = \text{sym}\ \mathbf{A} - \frac{1}{3}(\text{tr}\ \mathbf{A})\mathbf{1} \qquad \text{or} \qquad \widehat{a}_{(ij)} = a_{(ij)} - \frac{1}{3}a_{kk}\delta_{ij}, \tag{2.109}$$

so that $\mathbf{A}$ can be uniquely decomposed as a linear superposition of three contributions

$$\mathbf{A} = \frac{1}{3}(\text{tr}\ \mathbf{A})\mathbf{1} + \widehat{\mathbf{A}} + \text{skw}\ \mathbf{A} \qquad \text{or} \qquad a_{ij} = \frac{1}{3}a_{kk}\delta_{ij} + \widehat{a}_{(ij)} + a_{[ij]} \tag{2.110}$$

each one of which has unique symmetry properties: the first term is isotropic, the second is a traceless symmetric rank-2 tensor, and the third is a skew-symmetric rank-2 tensor. The first term is also called the *spherical* part of $\mathbf{A}$, while the sum of the second and third terms is also called the *deviatoric* part of $\mathbf{A}$ and is denoted as $\mathbf{A}'$. It is clear that in such case $\mathbf{A}$ is decomposed into three orthogonal components since

$$\mathbf{1} : (\text{skw } \mathbf{A}) = 0, \quad \mathbf{1} : \widehat{\mathbf{A}} = 0, \quad \widehat{\mathbf{A}} : (\text{skw } \mathbf{A}) = 0. \tag{2.111}$$

This leads to the following decomposition of the inner product between two arbitrary rank-2 tensors:

$$\mathbf{A} : \mathbf{C} = \frac{1}{3}(\text{tr } \mathbf{A})(\text{tr } \mathbf{C}) + \widehat{\mathbf{A}} : \widehat{\mathbf{C}} + (\text{skw } \mathbf{A}) : (\text{skw } \mathbf{C}). \tag{2.112}$$

We note that any symmetric second rank tensor $[S_{ij}] = [S_{ij}]^T = [S_{ji}]$ can be represented by a quadric surface. Consider the equation

$$\mathbf{x} \cdot \mathbf{S} \cdot \mathbf{x} = 1 \qquad \text{or} \qquad x_i S_{ij} x_j = 1. \tag{2.113}$$

Performing the summations and using the symmetry of the tensor, we obtain

$$S_{11}x_1^2 + S_{22}x_2^2 + S_{33}x_3^2 + 2\left(S_{23}x_2x_3 + S_{31}x_3x_1 + S_{12}x_1x_2\right) = 1. \tag{2.114}$$

This is a general equation of a second degree surface, a quadric, referred to its center as origin. An important property of a symmetric tensor that we shall explore in Sections 3.1.7 and 3.1.8 is the possession of principal axes. These are three directions at right angles to each other such that, when the general quadric (2.113) is referred to them as axes, the equation takes the simpler form

$$S_1x_1^2 + S_2x_2^2 + S_3x_3^2 = 1. \tag{2.115}$$

It is clear that the semi-axes of the representation quadric are of lengths $S_1^{-1/2}$, $S_2^{-1/2}$, and $S_3^{-1/2}$. If S_1, S_2, and S_3 are all positive, the surface is an ellipsoid. If two coefficients are positive and one negative, it is a hyperboloid of one sheet. If one coefficient is positive and two are negative, it is a hyperboloid of two sheets. If all three coefficients are negative, the surface is an imaginary ellipsoid.

Now, it can also be easily shown that a general rank-3 tensor $\mathbf{B}$ with components b_{ijk} cannot be decomposed into its symmetric and skew-symmetric parts $b_{(ijk)}$ and $b_{[ijk]}$, respectively, since they together contain less information than b_{ijk}. However, it can be shown that $\mathbf{B}$ can be written as

$$\mathbf{B} = \mathbf{S} + \mathbf{A}, \tag{2.116}$$

where $\mathbf{S}$ is a completely symmetric tensor and $\mathbf{A}$ is a skew-symmetric tensor defined by the properties

$$a_{\underline{iii}} = 0, \qquad a_{\underline{ii}j} + a_{\underline{i}j\underline{i}} + a_{j\underline{ii}} = 0, \qquad |\epsilon_{ijk}|\, a_{ijk} = 0, \tag{2.117}$$

where there is no summation over the underlined indices. The component of $\mathbf{S}$ and $\mathbf{A}$ are given by

$$s_{ijk} = \frac{1}{6}\left(b_{ijk} + b_{jki} + b_{kij} + b_{jik} + b_{kji} + b_{ikj}\right), \tag{2.118}$$

$$a_{ijk} = \frac{1}{6}\left(5\, b_{ijk} - b_{jki} - b_{kij} - b_{jik} - b_{kji} - b_{ikj}\right). \tag{2.119}$$

It should be noted that $s_{ijk}a_{ijk} = 0$. Now it can also be shown that the skew-symmetrix tensor $\mathbf{A}$ can be represented as

$$a_{ijk} = c_{im}\epsilon_{mjk} - c_{km}\epsilon_{mij} + d_{jm}\epsilon_{mik} - d_{km}\epsilon_{mij} + c\,\epsilon_{ijk}, \quad (2.120)$$

where c_{ij} and d_{ij} are traceless rank-2 tensors, and c is a constant. The tensors c_{ij} and d_{ij}, in turn, may be expressed by the components of a_{ijk} as

$$c_{jn} = \frac{1}{3}\left(a_{ijk} + a_{ikj}\right)\epsilon_{ink}, \quad (2.121)$$

$$d_{jn} = a_{ijk}\epsilon_{ikn} + \frac{1}{3}a_{imk}\epsilon_{imk}\delta_{jn}, \quad (2.122)$$

$$c = \frac{1}{6}a_{ijk}\epsilon_{ijk}. \quad (2.123)$$

Alternatively, the third rank tensor $\mathbf{B}$ can be decomposed as

$$b_{ijk} = b_{(ij)k} + b_{[ij]k} = \delta_{ij}n_k + b'_{(ij)k} + b_{[ij]k}, \quad (2.124)$$

where

$$n_k = \frac{1}{3}b_{llk}, \quad \text{and} \quad b'_{(ij)k} = b_{(ij)k} - \delta_{ij}n_k. \quad (2.125)$$

As is clear, the decomposition of a third rank tensor is not unique even in this case since one can choose any of the pairs (i,j), (j,k), or (i,k) in the indices of b_{ijk} to perform the decomposition. However, if for physical reasons b_{ijk} happens to be symmetric or skew symmetric with respect to a specific pair of indices, then the above decomposition becomes very useful.

While a decomposition may not be unique, in general it is very advantageous to decompose a tensor into *irreducible invariant subspaces.* To make this clear, we note that we could have written (2.106) as

$$\mathbf{A} = \sum_{s=1}^{2}\mathbf{I}_s \cdot \mathbf{A} = \mathbf{I}_1 \cdot \mathbf{A} + \mathbf{I}_2 \cdot \mathbf{A} = \mathbf{A}_1 + \mathbf{A}_2, \quad (2.126)$$

where the components of the symmetrizing and anti-symmetrizing linear operators $\mathbf{I}_1$ and $\mathbf{I}_2$ are given by

$$\left(I_{ij}^{lm}\right)_1 = \frac{1}{2!}\left(\delta_i^l\delta_j^m + \delta_i^m\delta_j^l\right) = \delta_i^{(l}\delta_j^{m)} \text{ and } \left(I_{ij}^{lm}\right)_2 = \frac{1}{2!}\left(\delta_i^l\delta_j^m - \delta_i^m\delta_j^l\right) = \delta_i^{[l}\delta_j^{m]} = \frac{1}{2!}\delta_{ij}^{lm}, \quad (2.127)$$

and they lead to the symmetric and skew-symmetric decompositions $\mathbf{A}_1$ and $\mathbf{A}_2$ with components

$$(a_{ij})_1 = \frac{1}{2!}\left(a_{ij} + a_{ji}\right) = a_{(ij)} \quad \text{and} \quad (a_{ij})_2 = \frac{1}{2!}\left(a_{ij} - a_{ji}\right) = a_{[ij]}. \quad (2.128)$$

Furthermore, it should be noted that $\mathbf{I}_1$ and $\mathbf{I}_2$ have the following properties:

$$\mathbf{I}_p \cdot \mathbf{I}_q = \begin{cases} \mathbf{I}_p & \text{if} \quad p = q \\ \mathbf{0} & \text{if} \quad p \neq q \end{cases} \quad \text{and} \quad \sum_{s=1}^{n}\mathbf{I}_s = \mathbf{I}, \quad (2.129)$$

where $\mathbf{I}$ is the identity operator and $n = 2$ in this case. It turns out that a decomposition satisfying (2.129) can always be accomplished for tensors of any rank. For tensor $\mathbf{B}$ of rank 3, there are four symmetry operators, $n = 4$, and their components are given by

$$\left(I_{ijk}^{lmn}\right)_1 = \delta_i^{(l}\delta_j^m\delta_k^{n)}, \tag{2.130}$$

$$\left(I_{ijk}^{lmn}\right)_2 = \delta_i^{[l}\delta_j^m\delta_k^{n]} = \frac{1}{3!}\delta_{ijk}^{lmn}, \tag{2.131}$$

$$\begin{aligned}\left(I_{ijk}^{lmn}\right)_3 &= \frac{2}{3!}\left[\left(\delta_i^l\delta_j^m + \delta_i^m\delta_j^l\right)\delta_k^n - \left(\delta_i^m\delta_j^n + \delta_i^n\delta_j^m\right)\delta_k^l\right] \\ &= \frac{1}{3!}\left[\delta_i^{(l}\delta_j^{m)}\delta_k^n - \delta_i^{(m}\delta_j^{n)}\delta_k^l\right],\end{aligned} \tag{2.132}$$

$$\begin{aligned}\left(I_{ijk}^{lmn}\right)_4 &= \frac{2}{3!}\left[\left(\delta_i^l\delta_k^n + \delta_i^n\delta_k^l\right)\delta_j^m - \left(\delta_i^m\delta_k^n + \delta_i^n\delta_k^m\right)\delta_j^l\right] \\ &= \frac{1}{3!}\left[\delta_i^{(l}\delta_k^{n)}\delta_j^m - \delta_i^{(m}\delta_k^{n)}\delta_j^l\right].\end{aligned} \tag{2.133}$$

Clearly, $\mathbf{I}_1$ and $\mathbf{I}_2$ are the symmetrizing and anti-symmetrizing operators, while $\mathbf{I}_3$ and $\mathbf{I}_4$ are operators with mixed symmetry. These operators lead to the following decomposition of $\mathbf{B}$:

$$\mathbf{B} = \sum_{s=1}^{4}\mathbf{I}_s \cdot \mathbf{B} = \sum_{s=1}^{4}\mathbf{B}_s, \tag{2.134}$$

where the components of $\mathbf{B}_s$ are given by

$$\left(b_{ijk}\right)_1 = \frac{1}{3!}\left(b_{ijk} + b_{jki} + b_{kij} + b_{jik} + b_{kji} + b_{ikj}\right) = b_{(ijk)}, \tag{2.135}$$

$$\left(b_{ijk}\right)_2 = \frac{1}{3!}\left(b_{ijk} + b_{jki} + b_{kij} - b_{jik} - b_{kji} - b_{ikj}\right) = b_{[ijk]}, \tag{2.136}$$

$$\left(b_{ijk}\right)_3 = \frac{2}{3!}\left[\left(b_{ijk} + b_{jik}\right) - \left(b_{jki} + b_{kji}\right)\right] = \frac{1}{3!}\left[b_{(ij)k} - b_{(jk)i}\right], \tag{2.137}$$

$$\left(b_{ijk}\right)_4 = \frac{2}{3!}\left[\left(b_{ijk} + b_{kji}\right) - \left(b_{jik} + b_{kij}\right)\right] = \frac{1}{3!}\left[b_{(i|j|k)} - b_{(j|i|k)}\right]. \tag{2.138}$$

Note that in three-dimensional space $\mathbf{B}$ has $3^3 = 27$ components, and in conformance with this fact, $\mathbf{B}_1$, the symmetric part, has 10 components, $\mathbf{B}_2$, the anti-symmetric part, has 1 component, and $\mathbf{B}_3$ and $\mathbf{B}_4$, the parts with mixed symmetry, each have 8 components.

2.11 Dual of a tensor

As noted previously, an n-vector is a completely skew-symmetric tensor of rank n. Within the space n, it is possible to associate with any k-vector, $0 \leq k \leq n$, an $(n-k)$-vector, its *dual*, as follows. If $V_{j_1\cdots j_k}$ and $W^{i_1\cdots i_k}$ are k-vectors of weight N, their duals, also called *Hodge duals*, are the $(n-k)$-vectors

$$v^{i_1\cdots i_{n-k}} \equiv (\mathrm{dual}\,\mathbf{V})^{i_1\cdots i_{n-k}} = \frac{1}{k!}\varepsilon^{i_1\cdots i_{n-k}j_1\cdots j_k}V_{j_1\cdots j_k} \tag{2.139}$$

or

$$w_{j_1\cdots j_{n-k}} \equiv (\mathrm{dual}\,\mathbf{W})_{j_1\cdots j_{n-k}} = \frac{1}{k!} W^{i_1\cdots i_k} \varepsilon_{i_1\cdots i_k j_1\cdots j_{n-k}} \tag{2.140}$$

of weights $N+1$ and $N-1$, respectively. Note that the epsilons have n indices, in the contracted products the contravariant tensor is written first, and the summation indices are adjacent. It can be easily shown that these definitions imply that

$$\mathrm{dual}\,(\mathrm{dual}\,\mathbf{W}) = s\,(-1)^{k(n-k)}\,\mathbf{W}, \tag{2.141}$$

so that $\mathbf{w}$ and $\mathbf{W}$ contain exactly the same information (aside from the sign). The signature s is the sign of the determinant of the inner product tensor. For ordinary Euclidean spaces, the signature is always positive, and so $s = +1$.

Note that in $\mathcal{E}^3$ if $\mathbf{w}$ is a 1-vector, $\mathbf{W}$ is a 2-vector, and $\boldsymbol{\varepsilon}$ is the rank-3 absolute Levi–Civita tensor, then

$$w^i = \frac{1}{2}\varepsilon^{ijk}W_{jk} \quad \text{or} \quad w_k = \frac{1}{2}W^{ij}\varepsilon_{ijk}, \tag{2.142}$$

and

$$W_{jk} = w^i\varepsilon_{ijk} \quad \text{or} \quad W^{ij} = \varepsilon^{ijk}w_k. \tag{2.143}$$

As shorthand notation, from now on we will also use the angle bracket notation to denote the dual of a tensor field, i.e.,

$$\mathbf{w} = \langle\mathbf{W}\rangle. \tag{2.144}$$

The components w^i and W_{jk} are related as follows:

$$W = [W_{jk}] = \begin{bmatrix} 0 & w^3 & -w^2 \\ -w^3 & 0 & w^1 \\ w^2 & -w^1 & 0 \end{bmatrix}. \tag{2.145}$$

It is easy to show that if $\mathbf{W}_1$ and $\mathbf{W}_2$ are two skew-symmetric tensors with corresponding axial vectors $\mathbf{w}_1$ and $\mathbf{w}_2$, then

$$\mathbf{W}_1 : \mathbf{W}_2 = 2\,\mathbf{w}_1 \cdot \mathbf{w}_2, \tag{2.146}$$

and subsequently $|\mathbf{W}| = \sqrt{2}\,|\mathbf{w}|$. In addition, if $\mathbf{a}$ is an arbitrary vector, we have

$$\mathbf{W}\cdot\mathbf{w} = \mathbf{0} \qquad \text{and} \qquad \mathbf{W}\cdot\mathbf{a} = -\mathbf{w}\times\mathbf{a}. \tag{2.147}$$

Remark: To avoid the minus sign in the last expression, some authors use the direct relation often used in rigid body dynamics,

$$\mathbf{W}\cdot\mathbf{a} = \mathbf{w}\times\mathbf{a}, \tag{2.148}$$

and subsequently the axial vector $\mathbf{w}$ is chosen to be the negative of that given in (2.142)–(2.145). In this case, one obtains

$$w^i = -\frac{1}{2}\varepsilon^{ijk}W_{jk} \quad \text{or} \quad w_k = -\frac{1}{2}W^{ij}\varepsilon_{ijk}, \tag{2.149}$$

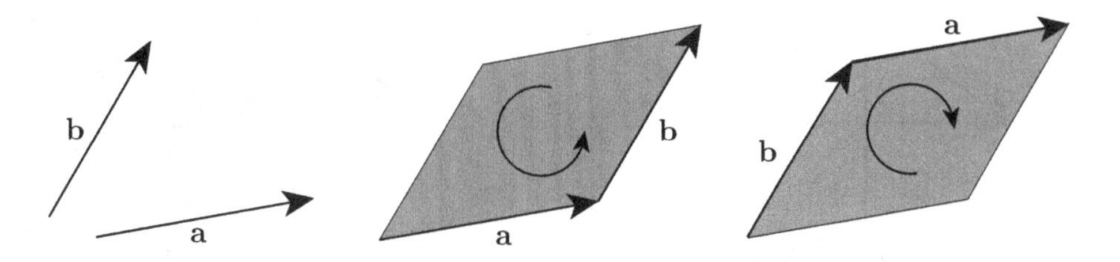

Figure 2.4: Vectors and bivectors.

and

$$W_{jk} = -w^i \varepsilon_{ijk} \quad \text{or} \quad W^{ij} = -\varepsilon^{ijk} w_k, \tag{2.150}$$

where now the components w^i and W_{jk} are related as follows:

$$W = [W_{jk}] = \begin{bmatrix} 0 & -w^3 & w^2 \\ w^3 & 0 & -w^1 \\ -w^2 & w^1 & 0 \end{bmatrix}. \tag{2.151}$$

One should be aware of such sign differences when comparing references.

2.12 Exterior product

The *exterior* or *wedge* product provides a generalization of the standard vector product, which is restricted to three-dimensional vector spaces. For any two scalars a and b and any three multivectors $\mathbf{A}$, $\mathbf{B}$, and $\mathbf{C}$, all familiar rules of addition and multiplication hold, such as

$$(a\,\mathbf{A} + b\,\mathbf{B}) \wedge \mathbf{C} = a\,\mathbf{A} \wedge \mathbf{C} + b\,\mathbf{B} \wedge \mathbf{C}, \tag{2.152}$$

$$(\mathbf{A} \wedge \mathbf{B}) \wedge \mathbf{C} = \mathbf{A} \wedge (\mathbf{B} \wedge \mathbf{C}) = \mathbf{A} \wedge \mathbf{B} \wedge \mathbf{C}, \tag{2.153}$$

except for a modified commutation law between a p-vector $\mathbf{A}$ and a q-vector $\mathbf{B}$:

$$\mathbf{A} \wedge \mathbf{B} = (-1)^{pq}\,\mathbf{B} \wedge \mathbf{A}. \tag{2.154}$$

Application to 1-vectors $\mathbf{a}$ and $\mathbf{b}$ results in a 2-vector, or *bivector*,

$$\mathbf{a} \wedge \mathbf{b} = -\mathbf{b} \wedge \mathbf{a} \quad \text{and} \quad \mathbf{a} \wedge \mathbf{a} = \mathbf{0}, \tag{2.155}$$

which is clearly bilinear and anti-symmetric. A bivector represents an oriented two-dimensional area as indicated in Fig. 2.4. The order of the product matters, as indicated in the figure. In addition,

$$\mathbf{a} \wedge \mathbf{b} = \left(a^j \mathbf{e}_j\right) \wedge \left(b^k \mathbf{e}_k\right) = a^j\, b^k\, \mathbf{e}_j \wedge \mathbf{e}_k = \frac{1}{2}\left(a^j\, b^k - b^j\, a^k\right) \mathbf{e}_j \wedge \mathbf{e}_k. \tag{2.156}$$

If $\mathbf{a}$ and $\mathbf{b}$ are vectors in $\mathcal{E}^3$, then we have

$$\mathbf{a}\wedge\mathbf{b} = \left(a^2\, b^3 - b^2\, a^3\right) \mathbf{e}_2 \wedge \mathbf{e}_3 + \left(a^3\, b^1 - b^3\, a^1\right) \mathbf{e}_3 \wedge \mathbf{e}_1 + \left(a^1\, b^2 - b^1\, a^2\right) \mathbf{e}_1 \wedge \mathbf{e}_2 \tag{2.157}$$

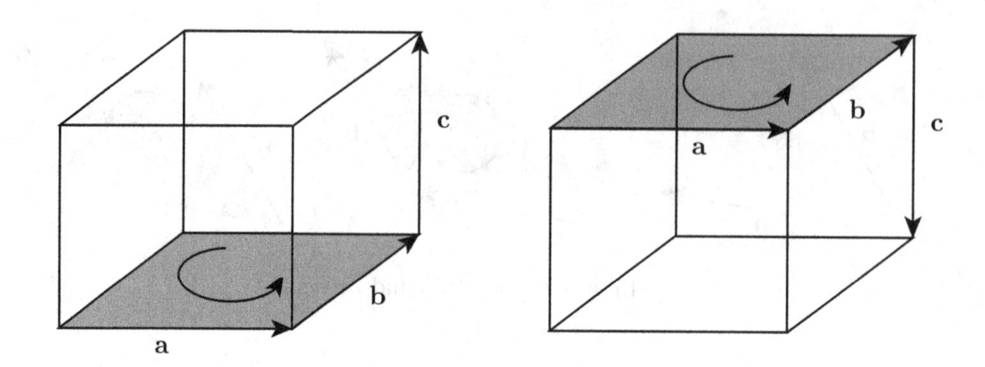

Figure 2.5: Trivectors.

and we note that the components of the bivector are the same as those of the cross product $\mathbf{a} \times \mathbf{b}$. Similarly, the product of three 1-vectors $\mathbf{a}$, $\mathbf{b}$, and $\mathbf{c}$ in $\mathcal{E}^3$, illustrated in Fig. 2.5, yields the 3-vector, or *trivector*,

$$\mathbf{a} \wedge \mathbf{b} \wedge \mathbf{c} = \left(a^1b^2c^3 + a^2b^3c^1 + a^3b^1c^2 - a^1b^3c^2 - a^2b^1c^3 - a^3b^2c^1\right) \mathbf{e}_1 \wedge \mathbf{e}_2 \wedge \mathbf{e}_3, \tag{2.158}$$

noting that the magnitude of the trivector is just $\det(\mathbf{a}\,\mathbf{b}\,\mathbf{c})$ and is equal to the scalar triple product $\mathbf{a} \cdot (\mathbf{b} \times \mathbf{c})$.

In $\mathcal{E}^n$, the number of basis k-vector elements is

$$\begin{pmatrix} n \\ k \end{pmatrix} \tag{2.159}$$

and the total number of elements is 2^n. Thus, in $\mathcal{E}^3$ the basis elements are

$$\{1, \mathbf{e}_1, \mathbf{e}_2, \mathbf{e}_3, \mathbf{e}_{12}, \mathbf{e}_{23}, \mathbf{e}_{31}, \mathbf{e}_{123}\}, \tag{2.160}$$

where we have used the shorthand notation

$$\mathbf{e}_{ij} = \mathbf{e}_i \wedge \mathbf{e}_j \qquad \text{and} \qquad \mathbf{e}_{ijk} = \mathbf{e}_i \wedge \mathbf{e}_j \wedge \mathbf{e}_k. \tag{2.161}$$

Note that in $\mathcal{E}^3$ no higher k-vector exists than $\mathbf{e}_{123}$. The highest vector in the space is usually denoted as I and is a pseudoscalar. Thus in $\mathcal{E}^3$ we have that $I = \mathbf{e}_{123}$ and its inverse is given by $I^{-1} = \mathbf{e}_{321}$.

The wedge product is defined as

$$\mathbf{e}_i \wedge \mathbf{e}_j = \mathbf{e}_{ij} = \begin{cases} \mathbf{0} & \text{for } i = j, \\ \mathbf{e}_i\,\mathbf{e}_j & \text{for } i \neq j, \end{cases} \tag{2.162}$$

so that $\mathbf{e}_i\,\mathbf{e}_j = -\mathbf{e}_j\,\mathbf{e}_i$ if $i \neq j$, with the defining equation of the algebra given by

$$\mathbf{e}_i\,\mathbf{e}_i = 1. \tag{2.163}$$

This implies, e.g., that

$$(\mathbf{e}_i\,\mathbf{e}_j\,\mathbf{e}_k)\,\mathbf{e}_k = (\mathbf{e}_i\,\mathbf{e}_j)(\mathbf{e}_k\,\mathbf{e}_k) = (\mathbf{e}_i\,\mathbf{e}_j)\,1 = \mathbf{e}_i\,\mathbf{e}_j, \tag{2.164}$$

and

$$(\mathbf{e}_i\,\mathbf{e}_j\,\mathbf{e}_k)\,\mathbf{e}_j = (\mathbf{e}_i\,\mathbf{e}_j)\,(\mathbf{e}_k\,\mathbf{e}_j) = -\mathbf{e}_i\,(\mathbf{e}_j\,\mathbf{e}_j)\,\mathbf{e}_k = -\mathbf{e}_i\,\mathbf{e}_k. \tag{2.165}$$

From (2.156), this also implies that

$$\mathbf{a}\wedge\mathbf{b} = \frac{1}{2}\,(\mathbf{a}\,\mathbf{b} - \mathbf{b}\,\mathbf{a})\,. \tag{2.166}$$

Now the dual of a multivector $\mathbf{A}$ is given by

$$(\mathrm{dual}\,\mathbf{A}) = \mathbf{A}\,I^{-1}. \tag{2.167}$$

In $\mathcal{E}^3$, it now follows from (2.157) and (2.158) that

$$[\mathrm{dual}\,(\mathbf{a}\wedge\mathbf{b})] = (\mathbf{a}\wedge\mathbf{b})\,I^{-1} = \mathbf{a}\times\mathbf{b} \tag{2.168}$$

and

$$[\mathrm{dual}\,(\mathbf{a}\wedge\mathbf{b}\wedge\mathbf{c})] = (\mathbf{a}\wedge\mathbf{b}\wedge\mathbf{c})\,I^{-1} = \mathbf{a}\cdot(\mathbf{b}\times\mathbf{c})\,, \tag{2.169}$$

since

$$\begin{aligned}
&(\mathbf{e}_2\,\mathbf{e}_3)\,(\mathbf{e}_3\,\mathbf{e}_2\,\mathbf{e}_1) = \mathbf{e}_1,\\
&(\mathbf{e}_3\,\mathbf{e}_1)\,(\mathbf{e}_3\,\mathbf{e}_2\,\mathbf{e}_1) = \mathbf{e}_2,\\
&(\mathbf{e}_1\,\mathbf{e}_2)\,(\mathbf{e}_3\,\mathbf{e}_2\,\mathbf{e}_1) = \mathbf{e}_3,\\
&(\mathbf{e}_1\,\mathbf{e}_2\,\mathbf{e}_3)\,(\mathbf{e}_3\,\mathbf{e}_2\,\mathbf{e}_1) = 1.
\end{aligned}$$

In summary, and in full accordance with (2.142) and (2.143), the exterior product of two 1-vectors $\mathbf{u}$ and $\mathbf{v}$ in $\mathcal{E}^3$ is a 2-vector or skew-symmetric tensor $\mathbf{W}$ with associated dual (also called *axial vector*) $\mathbf{w}$ corresponding to the vector (or cross product) of the two vectors $\mathbf{u}\times\mathbf{v}$, i.e., if we take

$$\mathbf{W} = \mathbf{u}\wedge\mathbf{v}, \tag{2.170}$$

then it easily follows from (2.168) that

$$\mathbf{w} = \mathbf{u}\times\mathbf{v}. \tag{2.171}$$

2.13 Tensor fields

2.13.1 Cartesian coordinate system

Let (ξ^1,ξ^2,ξ^3) be coordinates in the Euclidean space $\mathcal{E}^3$, and $\mathbf{r}$ the position vector

$$\mathbf{r} = \xi^1\mathbf{i}_1 + \xi^2\mathbf{i}_2 + \xi^3\mathbf{i}_3 = \xi^k\mathbf{i}_k, \tag{2.172}$$

where $(\mathbf{i}_1,\mathbf{i}_2,\mathbf{i}_3) = (\mathbf{i}^1,\mathbf{i}^2,\mathbf{i}^3)$ is the Cartesian (or rectangular) coordinate unit basis, as shown in Fig. 2.6. Quite often, the Cartesian coordinate system is used as a reference coordinate system for general curvilinear coordinate systems. In this system, the basis is taken to be constant, i.e., it is independent of coordinates. The Cartesian coordinates for a vector $\mathbf{u}$ are given by

$$u^j = \mathbf{u}\cdot\mathbf{i}^j = \mathbf{u}\cdot\mathbf{i}_j = u\,l_j, \tag{2.173}$$

where $u = \|\mathbf{u}\|$ and $l_j = \cos\alpha_j$'s are the *direction cosines*. Note that if we have a unit vector ($u = 1$), then l_j is just the component of the unit vector along the ξ^j direction and $l_j l_j = 1$, which is just a standard trigonometric identity.

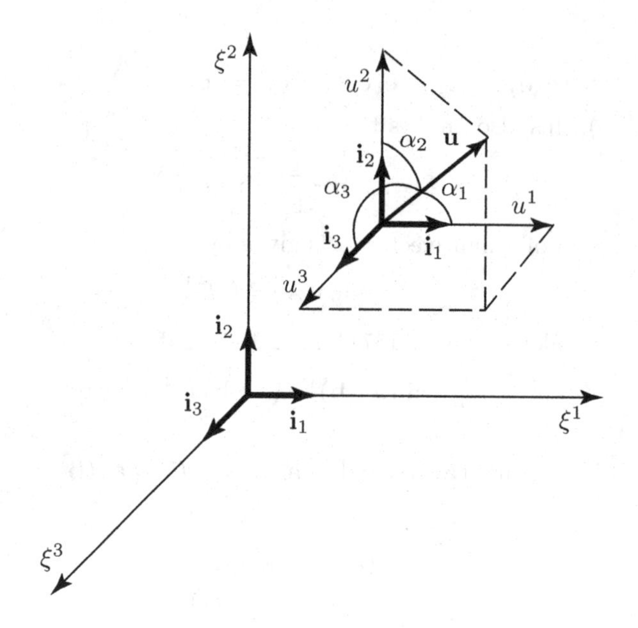

Figure 2.6: Cartesian coordinates and basis.

2.13.2 Curve in space

Let t be some arbitrary parameter and take $\mathbf{r} = \mathbf{r}(t)$ or $\xi^i = \xi^i(t)$. The vector function $\mathbf{r} = \mathbf{r}(t)$ represents a curve in Cartesian coordinates as illustrated in Fig. 2.7. Define the derivative

$$\frac{d\mathbf{r}}{dt} \equiv \lim_{\Delta t \to 0} \frac{\mathbf{r}(t + \Delta t) - \mathbf{r}(t)}{\Delta t} = \lim_{\Delta t \to 0} \frac{\Delta \mathbf{r}}{\Delta t}. \tag{2.174}$$

Now we see that $d\mathbf{r}/dt$ is tangent to the curve $\mathbf{r}(t)$ at t. We could also define the curve in terms of the arc length s of the curve $\mathbf{r} = \mathbf{r}(s)$. Then $d\mathbf{r}/ds$ is the unit vector tangent to the curve since

$$ds^2 = d\mathbf{r} \cdot d\mathbf{r} = \frac{d\mathbf{r}}{ds} \cdot \frac{d\mathbf{r}}{ds}\, ds^2 \left(= \frac{d\mathbf{r}}{dt} \cdot \frac{d\mathbf{r}}{dt}\, dt^2 \right), \tag{2.175}$$

so

$$1 = \frac{d\mathbf{r}}{ds} \cdot \frac{d\mathbf{r}}{ds} \qquad \left(\text{and alternatively } \left(\frac{ds}{dt} \right)^2 = \frac{d\mathbf{r}}{dt} \cdot \frac{d\mathbf{r}}{dt} \right). \tag{2.176}$$

It also follows that $d\mathbf{r} = d\xi^k \mathbf{i}_k$ since the Cartesian basis $\mathbf{i}_k$ is constant.

2.13.3 Derivatives

Below we briefly mention generalizations of the standard derivative of a real-valued function of a single real variable and the directional derivative of such a function to corresponding derivatives of a tensor-valued function of a real tensor variable. To illustrate this, let $f(\mathbf{v})$ be a scalar function of a vector $\mathbf{v}$. The derivative of f

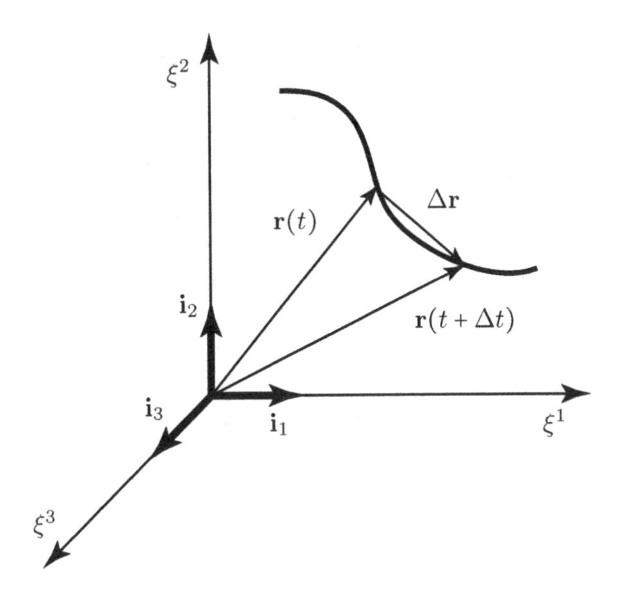

Figure 2.7: Curve in Cartesian coordinate space.

with respect to $\mathbf{v}$ is a vector denoted by $\partial f/\partial \mathbf{v}$ and defined by its scalar product with an arbitrary vector $\mathbf{a}$:

$$\frac{\partial f}{\partial \mathbf{v}} \cdot \mathbf{a} \equiv \lim_{s\to 0} \frac{f(\mathbf{v}+s\,\mathbf{a}) - f(\mathbf{v})}{s}. \tag{2.177}$$

In a Cartesian coordinate system, this expression takes the form

$$\frac{\partial f}{\partial \mathbf{v}} \cdot \mathbf{a} = \frac{\partial f}{\partial v_i} a_i.$$

For the particular case of $\mathbf{a} = \mathbf{e}_j$, we have

$$\frac{\partial f}{\partial \mathbf{v}} \cdot \mathbf{e}_j = \frac{\partial f}{\partial v_j}$$

and we conclude that

$$\frac{\partial f}{\partial \mathbf{v}} = \frac{\partial f}{\partial v_i} \mathbf{e}_i. \tag{2.178}$$

In a similar manner, if $f(\mathbf{D})$ is a scalar function of a tensor $\mathbf{D}$ of rank 2, the derivative of f with respect to $\mathbf{D}$ is a tensor of rank 2 denoted by $\partial f/\partial \mathbf{D}$ and defined by

$$\operatorname{tr}\left(\frac{\partial f}{\partial \mathbf{D}} \cdot \mathbf{A}^T\right) \equiv \lim_{s\to 0} \frac{f(\mathbf{D}+s\,\mathbf{A}) - f(\mathbf{D})}{s}, \tag{2.179}$$

where $\mathbf{A}$ is an arbitrary tensor of rank 2. We conclude that

$$\frac{\partial f}{\partial \mathbf{D}} = \frac{\partial f}{\partial D_{ij}} \mathbf{e}_i \mathbf{e}_j. \tag{2.180}$$

Such generalizations are essential in defining different quantities that we will encounter later on (e.g., the deformation gradient). Nevertheless, we will not delve into mathematical details associated with limits in different functional spaces nor prove the theorem below.

Specifically, the *Fréchet derivative* is commonly used to generalize the standard derivative to the case of differentiating tensor-valued functions of tensor variables.

Definition: A tensor function $\mathbf{A}(\mathbf{X})$ is Fréchet differentiable at $\mathbf{X}_0$ if there is a linear operator $\mathbf{L}$ such that in a neighborhood $\mathbf{V}$ of $\mathbf{X}_0$

$$||\mathbf{A}(\mathbf{X}) - \mathbf{A}(\mathbf{X}_0) - \mathbf{L} \cdot (\mathbf{X} - \mathbf{X}_0)|| = o\,(||\mathbf{X} - \mathbf{X}_0||)\,. \tag{2.181}$$

In this case, we write $\mathbf{L} = D_{\mathbf{X}}\mathbf{A}(\mathbf{X}_0)$, and $D_{\mathbf{X}}\mathbf{A}(\mathbf{X}_0)$ is called the Fréchet derivative of $\mathbf{A}(\mathbf{X})$ at $\mathbf{X}_0$.

That is, if $\mathbf{A}$ is a tensor function of rank n that depends on the scalar x, vector $\mathbf{x}$, and rank-2 tensor $\mathbf{X}$, then the corresponding Fréchet derivatives or ranks n, $n+1$, and $n+2$ are given by

$$D_x\mathbf{A}(x,\mathbf{x},\mathbf{X}) = \frac{\partial \mathbf{A}(x,\mathbf{x},\mathbf{X})}{\partial x}, \tag{2.182}$$

$$D_{\mathbf{x}}\mathbf{A}(x,\mathbf{x},\mathbf{X}) = \frac{\partial \mathbf{A}(x,\mathbf{x},\mathbf{X})}{\partial \mathbf{x}}, \tag{2.183}$$

$$D_{\mathbf{X}}\mathbf{A}(x,\mathbf{x},\mathbf{X}) = \frac{\partial \mathbf{A}(x,\mathbf{x},\mathbf{X})}{\partial \mathbf{X}}. \tag{2.184}$$

The derivative of a quantity represented by a tensor of rank p with respect to a tensor of rank q is defined as a tensor of rank $n = p+q$ whose components are equal to the derivatives of the quantity with respect to the corresponding components of the tensor of rank q. For example, the derivative of a scalar quantity with respect to a vector is given by the vector whose components are equal to the derivatives of the scalar with respect to the corresponding components of the vector.

Analogously, the *Gateaux derivative* provides a generalization of the classical directional derivative.

Definition: A tensor function $\mathbf{A}(\mathbf{X})$ is Gateaux differentiable at $\mathbf{X}_0$ if there is an operator $\mathbf{D}_{\mathbf{X}}\mathbf{A}(\mathbf{X}_0, \mathbf{H})$ such that

$$\lim_{s \to 0} ||\mathbf{A}(\mathbf{X}_0 + s\,\mathbf{H}) - \mathbf{A}(\mathbf{X}_0) - s\,\mathbf{D}_{\mathbf{X}}\mathbf{A}(\mathbf{X}_0, \mathbf{H})|| = \mathbf{0} \tag{2.185}$$

for $(\mathbf{X}_0 + s\,\mathbf{H}) \in \mathbf{V}$, a neighborhood of $\mathbf{X}_0$. Furthermore, $\mathbf{D}_{\mathbf{X}}\mathbf{A}(\mathbf{X}_0, \mathbf{H})$ is called the Gateaux derivative of $\mathbf{A}(\mathbf{X})$ at $\mathbf{X}_0$, and we write

$$\mathbf{D}_{\mathbf{X}}\mathbf{A}(\mathbf{X}_0, \mathbf{H}) = \frac{d}{ds}\mathbf{A}(\mathbf{X}_0 + s\,\mathbf{H})\Big|_{s=0}\,. \tag{2.186}$$

Theorem: If $\mathbf{A}(\mathbf{X})$ is Fréchet differentiable at $\mathbf{X}_0$, it is Gateaux differentiable at $\mathbf{X}_0$. Conversely, if the Gateaux derivative of $\mathbf{A}(\mathbf{X})$ at $\mathbf{X}_0$, $\mathbf{D}_{\mathbf{X}}\mathbf{A}(\mathbf{X}_0, \mathbf{H})$, is linear in $\mathbf{H}$, i.e., $\mathbf{D}_{\mathbf{X}}\mathbf{A}(\mathbf{X}_0, \cdot) \in \mathbf{L}$ and is continuous in $\mathbf{X}$, then $\mathbf{A}(\mathbf{X})$ is Fréchet differentiable at $\mathbf{X}_0$. In either case, we have the formula

$$D_{\mathbf{X}}\mathbf{A}(\mathbf{X}_0) \cdot \mathbf{H} = \mathbf{D}_{\mathbf{X}}\mathbf{A}(\mathbf{X}_0, \mathbf{H}). \tag{2.187}$$

That is,

$$D_x \mathbf{A}(x, \mathbf{x}, \mathbf{X})\ u = \left. \frac{d\mathbf{A}(x + s\,u, \mathbf{x}, \mathbf{X})}{ds} \right|_{s=0}, \tag{2.188}$$

$$D_{\mathbf{x}} \mathbf{A}(x, \mathbf{x}, \mathbf{X}) \cdot \mathbf{u} = \left. \frac{d\mathbf{A}(x, \mathbf{x} + s\,\mathbf{u}, \mathbf{X})}{ds} \right|_{s=0}, \tag{2.189}$$

$$D_{\mathbf{X}} \mathbf{A}(x, \mathbf{x}, \mathbf{X}) : \mathbf{U} = \left. \frac{d\mathbf{A}(x, \mathbf{x}, \mathbf{X} + s\,\mathbf{U})}{ds} \right|_{s=0}, \tag{2.190}$$

where u, $\mathbf{u}$, and $\mathbf{U}$ are arbitrary scalar, vector, and rank-2 tensors.

2.13.4 Surface in space

A surface in space is defined by

$$\phi(\xi^1, \xi^2, \xi^3) = \phi(\xi^k) = \text{const.}, \tag{2.191}$$

or alternatively

$$f(\xi^1, \xi^2, \xi^3) = f(\xi^k) = 0. \tag{2.192}$$

On this surface

$$df = 0 = \frac{\partial f}{\partial \xi^1} d\xi^1 + \frac{\partial f}{\partial \xi^2} d\xi^2 + \frac{\partial f}{\partial \xi^3} d\xi^3 = \frac{\partial f}{\partial \xi^i} d\xi^i = \frac{\partial f}{\partial \xi^i} \mathbf{i}^i \cdot d\xi^k \mathbf{i}_k = \nabla f \cdot d\mathbf{r}, \tag{2.193}$$

where we define the gradient operator in Cartesian coordinates as

$$\nabla \equiv \mathbf{i}^i \frac{\partial}{\partial \xi^i}. \tag{2.194}$$

We note that a superscript index in a denominator should be understood as a subscript for the term; i.e., $F_i = \partial f / \partial \xi^i$. Now we readily see that $d\mathbf{r}$ is a vector that is tangent to the surface while ∇f is a vector normal to the surface.

2.13.5 Curvilinear coordinate system

A general curvilinear coordinate system is given by

$$x^i = x^i(\xi^1, \xi^2, \xi^3) = x^i(\xi^k) \tag{2.195}$$

and

$$\xi^k = \xi^k(x^1, x^2, x^3) = \xi^k(x^i) \tag{2.196}$$

with the transformation being non-singular, i.e.,

$$\det \left[\frac{\partial x^i}{\partial \xi^k} \right] \neq \{0, \pm\infty\}, \tag{2.197}$$

where i and k are the row and column indices, respectively. Now x^i = const. is the x^i coordinate surface (whose normal points in the x^i direction), and the intersection of x^1 = const. and x^2 = const. is the x^3 coordinate curve.

Example

For cylindrical polar coordinates:

$$x^1 = \left[(\xi^1)^2 + (\xi^2)^2\right]^{1/2}, \qquad \xi^1 = x^1 \cos x^2,$$
$$x^2 = \tan^{-1}\left(\frac{\xi^2}{\xi^1}\right), \qquad \xi^2 = x^1 \sin x^2,$$
$$x^3 = \xi^3, \qquad \xi^3 = x^3.$$

At each point in space, we can define two sets of "natural" base vectors:

(i) Let

$$\mathbf{e}_i = \frac{\partial \mathbf{r}}{\partial x^i} = \frac{\partial}{\partial x^i}\left(\xi^k \mathbf{i}_k\right) = \frac{\partial \xi^k}{\partial x^i}\mathbf{i}_k \tag{2.198}$$

which is tangent to the x^i coordinate curve and not necessarily a unit vector. Note that for the Cartesian basis, $\mathbf{i}^k = \mathbf{i}_k$ and $\mathbf{i}_k$ is not a function of the coordinates.

Example

For cylindrical polar coordinates:

$$\begin{aligned}\mathbf{e}_1 &= \cos x^2 \mathbf{i}_1 + \sin x^2 \mathbf{i}_2,\\ \mathbf{e}_2 &= -x^1 \sin x^2 \mathbf{i}_1 + x^1 \cos x^2 \mathbf{i}_2,\\ \mathbf{e}_3 &= \mathbf{i}_3.\end{aligned}$$

Note that $\mathbf{e}_i$ is not necessarily a unit vector; e.g., by inspection $\mathbf{e}_2$ is not.

(ii) Let

$$\mathbf{e}^i = \nabla x^i = \frac{\partial x^i}{\partial \xi^k}\mathbf{i}^k \tag{2.199}$$

which is normal to the x^i coordinate surface, i.e., it is normal to $x^i(\xi^k) =$ const.

Example

For cylindrical polar coordinates:

$$\mathbf{e}^1 = \frac{\xi^1}{[(\xi^1)^2 + (\xi^2)^2]^{1/2}}\mathbf{i}^1 + \frac{\xi^2}{[(\xi^1)^2 + (\xi^2)^2]^{1/2}}\mathbf{i}^2 = \cos x^2 \mathbf{i}^1 + \sin x^2 \mathbf{i}^2.$$

Now since $\tan x^2 = \xi^2/\xi^1$, differentiating with respect to ξ^1 and ξ^2 in turn we have

$$\sec^2 x^2 dx^2 = -\frac{\xi^2}{(\xi^1)^2}d\xi^1 = \frac{1}{\xi^1}d\xi^2,$$

so

$$\frac{dx^2}{d\xi^1} = -\frac{\xi^2 \cos^2 x^2}{(\xi^1)^2} = -\frac{x^1 \sin x^2 \cos^2 x^2}{(x^1)^2 \cos^2 x^2} = -\frac{\sin x^2}{x^1}$$

and

$$\frac{dx^2}{d\xi^2} = \frac{\cos^2 x^2}{\xi^1} = \frac{\cos^2 x^2}{x^1 \cos x^2} = \frac{\cos x^2}{x^1}.$$

Thus

$$\begin{aligned} \mathbf{e}^2 &= -\frac{\sin x^2}{x^1}\mathbf{i}^1 + \frac{\cos x^2}{x^1}\mathbf{i}^2, \\ \mathbf{e}^3 &= \mathbf{i}^3. \end{aligned}$$

Are such bases reciprocal? The answer is yes since

$$\mathbf{e}_i \cdot \mathbf{e}^j = \left(\frac{\partial \xi^k}{\partial x^i}\mathbf{i}_k\right) \cdot \left(\frac{\partial x^j}{\partial \xi^l}\mathbf{i}^l\right) = \frac{\partial \xi^k}{\partial x^i}\frac{\partial x^j}{\partial \xi^k} = \frac{\partial x^j}{\partial x^i} = \delta_i^j$$

and the components of the metric tensor are given by

$$g_{ij} = \mathbf{e}_i \cdot \mathbf{e}_j = \frac{\partial \xi^k}{\partial x^i}\frac{\partial \xi^k}{\partial x^j} = g_{ij}(x^l) \tag{2.200}$$

and

$$g^{ij} = \mathbf{e}^i \cdot \mathbf{e}^j = \frac{\partial x^i}{\partial \xi^k}\frac{\partial x^j}{\partial \xi^k} = g^{ij}(x^l). \tag{2.201}$$

The reason why g_{ij} is called a metric tensor is now evident since

$$ds^2 = d\xi^k \mathbf{i}_k \cdot d\xi^l \mathbf{i}_l = \frac{\partial \xi^k}{\partial x^i} dx^i \mathbf{i}_k \cdot \frac{\partial \xi^l}{\partial x^j} dx^j \mathbf{i}_l = \mathbf{e}_i \cdot \mathbf{e}_j dx^i dx^j = g_{ij} dx^i dx^j.$$

What are the components a_i^j in the basis coordinate transformation

$$\bar{\mathbf{e}}_i = a_i^j \mathbf{e}_j?$$

Suppose we have another curvilinear coordinate system such that

$$\bar{x}^i = \bar{x}^i(\xi^k) \qquad \text{and} \qquad \xi^k = \xi^k(\bar{x}^i).$$

Then

$$\bar{\mathbf{e}}_i = \frac{\partial \mathbf{r}}{\partial \bar{x}^i} = \frac{\partial \xi^k}{\partial \bar{x}^i}\mathbf{i}_k = \frac{\partial x^j}{\partial \bar{x}^i}\frac{\partial \xi^k}{\partial x^j}\mathbf{i}_k = \frac{\partial x^j}{\partial \bar{x}^i}\mathbf{e}_j,$$

so that at a point in space we have

$$a_i^j = \frac{\partial x^j}{\partial \bar{x}^i}. \tag{2.202}$$

This result also allows us to bypass the rectangular coordinate system when transforming between two curvilinear systems, i.e., instead of having

$$x^i = x^i(\xi^k) \qquad \text{and} \qquad \xi^k = \xi^k(\bar{x}^j),$$

we can go directly to

$$x^i = x^i(\bar{x}^j).$$

Example

The mapping from cylindrical polar coordinates

$$\begin{aligned} x^1 &= \left[(\xi^1)^2 + (\xi^2)^2\right]^{1/2}, \\ x^2 &= \tan^{-1}\left(\frac{\xi^2}{\xi^1}\right), \\ x^3 &= \xi^3, \end{aligned}$$

to spherical coordinates is given by

$$\begin{aligned} \bar{x}^1 &= \left[(x^1)^2 + (x^3)^2\right]^{1/2}, \\ \bar{x}^2 &= x^2, \\ \bar{x}^3 &= \tan^{-1}\frac{x^1}{x^3}. \end{aligned}$$

The inverse mapping is

$$\begin{aligned} x^1 &= \bar{x}^1 \sin \bar{x}^3, \\ x^2 &= \bar{x}^2, \\ x^3 &= \bar{x}^1 \cos \bar{x}^3. \end{aligned}$$

To transform basis vectors in the two coordinate systems, the covariant (contravariant) coefficients can be calculated using the above transformations.

Now

$$g = \det G = \det\left[g_{ij}\right] = \det\left[\frac{\partial \xi^k}{\partial x^i}\right]\det\left[\frac{\partial \xi^k}{\partial x^j}\right] = \left(\det\left[\frac{\partial \xi^i}{\partial x^j}\right]\right)^2, \tag{2.203}$$

or

$$\sqrt{g} = \det\left[\frac{\partial \xi^i}{\partial x^j}\right]. \tag{2.204}$$

Similarly

$$\sqrt{\bar{g}} = \det\left[\frac{\partial \xi^i}{\partial \bar{x}^j}\right]. \tag{2.205}$$

Subsequently, we have that

$$\det\left[\frac{\partial x^i}{\partial \bar{x}^j}\right] = \det\left[\frac{\partial x^i}{\partial \xi^k}\right]\det\left[\frac{\partial \xi^k}{\partial \bar{x}^j}\right] = \sqrt{\frac{\bar{g}}{g}} = \det\left[a^i_j\right] = \det A = a. \tag{2.206}$$

Note that subscripts inside determinant quantities are understood not to be free indices. From before, we recall that g_{ij} transform like covariant components of a second-order tensor; thus

$$\bar{g}_{ij} = \frac{\partial x^r}{\partial \bar{x}^i}\frac{\partial x^s}{\partial \bar{x}^j}g_{rs}. \tag{2.207}$$

2.14 Gradient of a scalar field

We now take

$$\mathbf{r} = x^i\mathbf{e}_i \qquad \text{and} \qquad d\mathbf{r} = dx^i\mathbf{e}_i, \tag{2.208}$$

and define the scalar field

$$\phi = \phi(x^k) = \phi(\bar{x}^k) \tag{2.209}$$

which is a tensor of order zero (a scalar), or a relative tensor of order zero and weight zero.

The gradient of a scalar field ϕ is defined by

$$\nabla\phi \equiv \frac{\partial \phi}{\partial x^j}\mathbf{e}^j, \tag{2.210}$$

where now

$$\nabla \equiv \mathbf{e}^j\frac{\partial}{\partial x^j}. \tag{2.211}$$

Note that the Cartesian definition is recovered when we have an identity coordinate transformation, in which case $(x^1, x^2, x^3) = (\xi^1, \xi^2, \xi^3)$ and $(\mathbf{e}^1, \mathbf{e}^2, \mathbf{e}^3) = (\mathbf{i}^1, \mathbf{i}^2, \mathbf{i}^3)$. Does this definition give a tensor (in this case a vector)? In other words, does $\partial\phi/\partial x^j$ transform like a covariant component of a tensor of order one? Recall that vector covariant components transform as

$$\bar{v}_i = a^j_i v_j.$$

Now if we take

$$v_j \to \frac{\partial \phi}{\partial x^j}, \quad \bar{v}_i \to \frac{\partial \phi}{\partial \bar{x}^i}, \quad a^j_i = \frac{\partial x^j}{\partial \bar{x}^i},$$

do we get an equality upon substitution:

$$\frac{\partial \phi}{\partial \bar{x}^i} = \frac{\partial x^j}{\partial \bar{x}^i}\frac{\partial \phi}{\partial x^j}?$$

The answer is yes by the chain rule, and so the gradient of a scalar defines a proper tensor (a vector).

2.15 Gradient of a vector field

Assume that we have the following vector field

$$\mathbf{v} = v^i(x^k)\,\mathbf{e}_i. \tag{2.212}$$

Consider the following definition for the gradient of this vector field (a dyad)

$$\nabla\mathbf{v} = \frac{\partial v^i}{\partial x^j}\,\mathbf{e}_i\mathbf{e}^j. \tag{2.213}$$

Is this a tensor, i.e., do the components transform like $\binom{1}{1}$ tensor components? Now given the dyadic tensor

$$\boldsymbol{\tau} = \tau^i_j \mathbf{e}_i\mathbf{e}^j = \bar{\tau}^l_m \bar{\mathbf{e}}_l\bar{\mathbf{e}}^m$$

recall that $\binom{1}{1}$ components transform as

$$\bar{\tau}^l_m a^i_l = \tau^i_j a^j_m.$$

Now take

$$\tau^i_j \to \frac{\partial v^i}{\partial x^j}, \quad \bar{\tau}^l_m \to \frac{\partial \bar{v}^l}{\partial \bar{x}^m}, \quad a^i_j = \frac{\partial x^i}{\partial \bar{x}^j}.$$

Is the following equality true:

$$\frac{\partial \bar{v}^l}{\partial \bar{x}^m}\frac{\partial x^i}{\partial \bar{x}^l} = \frac{\partial v^i}{\partial x^j}\frac{\partial x^j}{\partial \bar{x}^m}?$$

Since $\mathbf{v}$ is a vector, from before we have that

$$v^i = \bar{v}^l a^i_l = \bar{v}^l \frac{\partial x^i}{\partial \bar{x}^l},$$

so that

$$\frac{\partial v^i}{\partial x^j} = \frac{\partial}{\partial x^j}\left(\bar{v}^l \frac{\partial x^i}{\partial \bar{x}^l}\right) = \frac{\partial}{\partial \bar{x}^m}\left(\bar{v}^l \frac{\partial x^i}{\partial \bar{x}^l}\right)\frac{\partial \bar{x}^m}{\partial x^j},$$

or

$$\frac{\partial v^i}{\partial x^j}\frac{\partial x^j}{\partial \bar{x}^m} = \frac{\partial}{\partial \bar{x}^m}\left(\bar{v}^l \frac{\partial x^i}{\partial \bar{x}^l}\right) = \frac{\partial \bar{v}^l}{\partial \bar{x}^m}\frac{\partial x^i}{\partial \bar{x}^l} + \bar{v}^l \frac{\partial^2 x^i}{\partial \bar{x}^m \partial \bar{x}^l}.$$

So we see that the equality is not true due to the presence of the second term in the above equation. Thus $(\partial v^i/\partial x^j)\mathbf{e}_i\mathbf{e}^j$ is not a tensor since it does not remain invariant under coordinate transformations.

Now let us try to define the gradient of the vector field by

$$\nabla\mathbf{v} \equiv \frac{\partial \mathbf{v}}{\partial x^j}\mathbf{e}^j = \frac{\partial}{\partial x^j}\left(v^i\mathbf{e}_i\right)\mathbf{e}^j = \frac{\partial v^i}{\partial x^j}\mathbf{e}_i\mathbf{e}^j + v^i\frac{\partial \mathbf{e}_i}{\partial x^j}\mathbf{e}^j. \tag{2.214}$$

Now we have just shown that the first term on the right-hand side is not a tensor. It can also be shown that the second term on the right-hand side is not a tensor either. However, the sum of the two terms is indeed a tensor! We note that $\partial \mathbf{e}_i/\partial x^j$ is a vector for fixed j, and thus can be written as a linear combination of $\mathbf{e}_k$, i.e.,

$$\frac{\partial \mathbf{e}_i}{\partial x^j} = \Gamma^k_{ij}\mathbf{e}_k = \left\{ \begin{matrix} k \\ ij \end{matrix} \right\} \mathbf{e}_k, \tag{2.215}$$

where Γ^k_{ij} are not components of a tensor. It is called the Christoffel symbol of the second kind. If we take the dot product of the above equation with $\mathbf{e}^r$, we obtain

$$\frac{\partial \mathbf{e}_i}{\partial x^j} \cdot \mathbf{e}^r = \Gamma^k_{ij}\mathbf{e}_k \cdot \mathbf{e}^r = \Gamma^k_{ij}\delta^r_k = \Gamma^r_{ij}. \tag{2.216}$$

However, we recall that $\mathbf{e}_i = \partial \mathbf{r}/\partial x^i$ so that $\partial \mathbf{e}_i/\partial x^j = \partial^2 \mathbf{r}/\partial x^j \partial x^i$, and

$$\frac{\partial \mathbf{e}_i}{\partial x^j} \cdot \mathbf{e}^r = \frac{\partial^2 \mathbf{r}}{\partial x^j \partial x^i} \cdot \nabla x^r = \frac{\partial^2 \xi^k}{\partial x^j \partial x^i}\mathbf{i}_k \cdot \frac{\partial x^r}{\partial \xi^l}\mathbf{i}^l = \frac{\partial^2 \xi^l}{\partial x^j \partial x^i}\frac{\partial x^r}{\partial \xi^l}.$$

Thus

$$\Gamma^k_{ij} = \frac{\partial^2 \xi^l}{\partial x^j \partial x^i}\frac{\partial x^k}{\partial \xi^l}. \tag{2.217}$$

Note that the Christoffel symbol of the second kind is symmetric with respect to the exchange of lower indices i and j. It can be shown that it can be expressed fully in terms of components of the metric tensor $\mathbf{g}$, i.e.,

$$\Gamma^k_{ij} = \frac{1}{2}g^{kl}\left(\frac{\partial g_{il}}{\partial x^j} + \frac{\partial g_{jl}}{\partial x^i} - \frac{\partial g_{ij}}{\partial x^l}\right). \tag{2.218}$$

Furthermore, it can also be rewritten as

$$\Gamma^k_{ij} = g^{kl}\Gamma_{ijl} = g^{kl}\left[ij, l\right], \tag{2.219}$$

where Γ_{ijl} is the Christoffel symbol of the first kind and are not components of a tensor. Because of the symmetry of the Christoffel symbol of the second kind, we note that the Christoffel symbol of the first kind is also symmetric with respect to the exchange of the indices i and j, and it is given by

$$\Gamma_{ijk} = \frac{1}{2}\left(\frac{\partial g_{ik}}{\partial x^j} + \frac{\partial g_{jk}}{\partial x^i} - \frac{\partial g_{ij}}{\partial x^k}\right). \tag{2.220}$$

Returning to our definition of the gradient of a vector field, we now have

$$\nabla \mathbf{v} = \frac{\partial v^i}{\partial x^j}\mathbf{e}_i\mathbf{e}^j + v^i\Gamma^k_{ij}\mathbf{e}_k\mathbf{e}^j = \left(\frac{\partial v^i}{\partial x^j} + \Gamma^i_{kj}v^k\right)\mathbf{e}_i\mathbf{e}^j$$

or

$$\nabla \mathbf{v} = v^i_{,j}\mathbf{e}_i\mathbf{e}^j, \tag{2.221}$$

where

$$v^i_{,j} \equiv \frac{\partial v^i}{\partial x^j} + \Gamma^i_{kj}v^k \tag{2.222}$$

is called the covariant derivative. In general, all subscripts following a comma denote covariant derivatives with respect to the corresponding components.

We note that the definition of the gradient of a vector as

$$\nabla \mathbf{v} = \frac{\partial \mathbf{v}}{\partial x^j}\mathbf{e}^j$$

is consistent with that of the gradient of a scalar when we take $\mathbf{v} \to \phi$. Indeed, this definition remains true for a tensor $\mathbf{T}$ of arbitrary order, i.e., we can write more generally

$$\nabla \mathbf{T} = \frac{\partial \mathbf{T}}{\partial x^j}\mathbf{e}^j.$$

2.16 Covariant differentiation of a vector

If instead we write our vector field in terms of covariant components

$$\mathbf{v} = v_i \mathbf{e}^i,$$

we then have

$$\nabla \mathbf{v} \equiv \frac{\partial}{\partial x^j}\left(v_i \mathbf{e}^i\right)\mathbf{e}^j = \frac{\partial v_i}{\partial x^j}\mathbf{e}^i\mathbf{e}^j + v_i \frac{\partial \mathbf{e}^i}{\partial x^j}\mathbf{e}^j. \tag{2.223}$$

Now since

$$\mathbf{e}_i \cdot \mathbf{e}^j = \delta_i^j,$$

upon differentiating we have

$$\frac{\partial \mathbf{e}_i}{\partial x^k} \cdot \mathbf{e}^j + \mathbf{e}_i \cdot \frac{\partial \mathbf{e}^j}{\partial x^k} = 0,$$

or

$$\mathbf{e}_i \cdot \frac{\partial \mathbf{e}^j}{\partial x^k} = -\Gamma^r_{ik}\mathbf{e}_r \cdot \mathbf{e}^j = -\Gamma^r_{ik}\delta^j_r = -\Gamma^j_{ik},$$

and

$$\frac{\partial \mathbf{e}^j}{\partial x^k} = -\Gamma^j_{lk}\mathbf{e}^l.$$

Now

$$\nabla \mathbf{v} = \frac{\partial v_i}{\partial x^j}\mathbf{e}^i\mathbf{e}^j - v_i \Gamma^i_{kj}\mathbf{e}^k\mathbf{e}^j = \left(\frac{\partial v_i}{\partial x^j} - v_k \Gamma^k_{ij}\right)\mathbf{e}^i\mathbf{e}^j,$$

or

$$\nabla \mathbf{v} = v_{i,j}\mathbf{e}^i\mathbf{e}^j, \tag{2.224}$$

where

$$v_{i,j} \equiv \frac{\partial v_i}{\partial x^j} - \Gamma^k_{ij}v_k. \tag{2.225}$$

Note that

$$v_{i,j} - v_{j,i} = \frac{\partial v_i}{\partial x^j} - \frac{\partial v_j}{\partial x^i}. \tag{2.226}$$

Furthermore,

$$\nabla \mathbf{v} \cdot d\mathbf{r} = \left(\frac{\partial \mathbf{v}}{\partial x^k} \mathbf{e}^k \right) \cdot \left(dx^i \mathbf{e}_i \right) = \frac{\partial \mathbf{v}}{\partial x^i} dx^i = d\mathbf{v}. \tag{2.227}$$

These results generalize to any $m = r + n$ order tensor with $\binom{r}{n}$ components.

Example

Assume we have the third-order (triadic) tensor

$$\mathbf{T} = T^i_{jk} \mathbf{e}_i \mathbf{e}^j \mathbf{e}^k. \tag{2.228}$$

Then it follows that

$$\nabla \mathbf{T} = T^i_{jk,l} \mathbf{e}_i \mathbf{e}^j \mathbf{e}^k \mathbf{e}^l, \tag{2.229}$$

where

$$T^i_{jk,l} = \frac{\partial T^i_{jk}}{\partial x^l} + \Gamma^i_{rl} T^r_{jk} - \Gamma^r_{jl} T^i_{rk} - \Gamma^r_{kl} T^i_{jr}. \tag{2.230}$$

The rules for covariant differentiation are mostly the same as for ordinary differentiation. That is, (i) the covariant derivative of a sum of two tensors is the sum of the covariant derivatives of the tensors; (ii) the covariant derivative of a product of two tensors is the covariant derivative of the first tensor times the second plus the first tensor times the covariant derivative of the second; (iii) higher covariant derivatives are defined as covariant derivatives of covariant derivatives – however, one should be careful in calculating these higher order derivatives since in general

$$v_{i,jk} \neq v_{i,kj}. \tag{2.231}$$

To see this, we calculate the components of the second covariant derivative of $\mathbf{v}$, i.e., $v_{i,jk}$. As we have shown, the components of the covariant derivative of $\mathbf{v}$ are given by

$$v_{i,j} = \frac{\partial v_i}{\partial x^j} - \Gamma^l_{ij} v_l. \tag{2.232}$$

By definition, the components of the second covariant derivative are those of the covariant derivative of the covariant derivative:

$$v_{i,jk} = \left(v_{i,j} \right)_{,k} = \frac{\partial}{\partial x^k} \left[\frac{\partial}{\partial x^j} - \Gamma^l_{ij} v_l \right] - \Gamma^m_{ik} v_{m,j} - \Gamma^m_{jk} v_{i,m}. \tag{2.233}$$

Expanding the expression, we have

$$v_{i,jk} = \frac{\partial^2 v_i}{\partial x^j \partial x^k} - \Gamma^l_{ij} \frac{\partial v_l}{\partial x^k} - \frac{\partial \Gamma^l_{ij}}{\partial x^k} v_l - \left[\frac{\partial v_m}{\partial x^j} - \Gamma^l_{mj} v_l \right] \Gamma^m_{ik} - \left[\frac{\partial v_i}{\partial x^m} - \Gamma^l_{im} v_l \right] \Gamma^m_{jk}. \tag{2.234}$$

Rearranging terms, the components of the second covariant derivative of the vector $\mathbf{v}$ can be expressed in the form

$$v_{i,jk} = \frac{\partial^2 v_i}{\partial x^j \partial x^k} - \Gamma^l_{ij}\frac{\partial v_l}{\partial x^k} - \Gamma^m_{ik}\frac{\partial v_m}{\partial x^j} - \Gamma^m_{jk}\frac{\partial v_i}{\partial x^m} - \left[\frac{\partial \Gamma^l_{ij}}{\partial x^k} - \Gamma^l_{mj}\Gamma^m_{ik} - \Gamma^l_{im}\Gamma^m_{jk}\right] v_l. \tag{2.235}$$

Now it is easy to see that

$$v_{i,jk} - v_{i,kj} = R^l_{ijk} v_l, \tag{2.236}$$

where

$$R^l_{ijk} \equiv \frac{\partial \Gamma^l_{ik}}{\partial x^j} - \frac{\partial \Gamma^l_{ij}}{\partial x^k} + \Gamma^l_{mj}\Gamma^m_{ik} - \Gamma^l_{mk}\Gamma^m_{ij} \tag{2.237}$$

is called the *Riemann–Christoffel tensor.* The covariant form of this tensor is

$$R_{mijk} = g_{lm} R^l_{ijk}. \tag{2.238}$$

It is an easy exercise to show that the covariant form can be expressed as

$$R_{mijk} = \frac{\partial \Gamma_{ikm}}{\partial x^j} - \frac{\partial \Gamma_{ijm}}{\partial x^k} + \Gamma_{mkn}\Gamma^n_{ij} - \Gamma_{mjn}\Gamma^n_{ik} \tag{2.239}$$

or

$$R_{mijk} = \frac{1}{2}\left(\frac{\partial g_{mk}}{\partial x^i \partial x^j} - \frac{\partial g_{ik}}{\partial x^m \partial x^j} - \frac{\partial g_{mj}}{\partial x^i \partial x^k} + \frac{\partial g_{ij}}{\partial x^m \partial x^k}\right) + g^{pq}\left(\Gamma_{mkp}\Gamma_{ijq} - \Gamma_{mjp}\Gamma_{ikq}\right), \tag{2.240}$$

from which we see that the Riemann–Christoffel tensor is skew-symmetric in the first two indices and the last two indices, and symmetric in the interchange of the first two and last two indices. Consequently,

$$R_{imjk} = -R_{mijk}, \qquad R_{mikj} = -R_{mijk}, \qquad R_{jkmi} = R_{mijk}. \tag{2.241}$$

Now, using these symmetry conditions, it is easy to show that we can write

$$R_{mijk} = \varepsilon_{pmi}\varepsilon_{qjk}S^{pq} \qquad \text{and} \qquad S^{pq} = \frac{1}{4}\varepsilon^{pmi}\varepsilon^{qjk}R_{mijk}, \tag{2.242}$$

where $\mathbf{S}$ is a symmetric rank 2 tensor which clearly has only six components. However, these six components are not all independent as the Riemann–Christoffel tensor satisfies *Bianchi's identities*:

$$R^m_{ijk,l} + R^m_{ikl,j} + R^m_{ilj,k} = 0 \qquad \text{or} \qquad R_{mijk,l} + R_{mikl,j} + R_{milj,k} = 0. \tag{2.243}$$

Bianchi's identities provide three additional equations that restrict the six components of $\mathbf{S}$.

Note that if the metric tensor is constant throughout the space, then $R_{mijk} = 0$, and we have a flat or *Euclidean space* so that $v_{i,jk} = v_{i,kj}$. It can also be shown that if $R_{mijk} = 0$, then we can find a coordinate system in which the components of the metric tensor are constant throughout space. Thus, $R_{mijk} = 0$ is a necessary and sufficient condition for a (flat) Euclidean space. If $R_{mijk} \neq 0$, then the space is curved and is called a *Riemannian space.*

2.17 Divergence of a vector field

Consider the vector field

$$\mathbf{v} = v^i \mathbf{e}_i$$

so that

$$\nabla \mathbf{v} = v^i_{,j} \mathbf{e}_i \mathbf{e}^j.$$

Now contraction of $\nabla \mathbf{v}$ gives $\nabla \cdot \mathbf{v}$, i.e., the divergence of $\mathbf{v}$. Thus we define

$$\nabla \cdot \mathbf{v} \equiv v^i_{,j} \mathbf{e}_i \cdot \mathbf{e}^j = v^i_{,j} \delta^j_i = v^i_{,i}, \tag{2.244}$$

where now

$$v^i_{,i} = \frac{\partial v^i}{\partial x^i} + \Gamma^i_{ki} v^k. \tag{2.245}$$

We recall that

$$\varepsilon_{ijk} = \sqrt{g} \epsilon_{ijk}.$$

Now noting that

$$\frac{\partial \epsilon_{ijk}}{\partial x^p} = 0,$$

we can easily show that ε_{ijk} is also an isotropic tensor, so that we have

$$\varepsilon_{ijk,p} = 0, \tag{2.246}$$

or

$$\epsilon_{ijk} \frac{\partial \sqrt{g}}{\partial x^p} - \sqrt{g}\, \Gamma^l_{ip} \epsilon_{ljk} - \sqrt{g}\, \Gamma^l_{jp} \epsilon_{ilk} - \sqrt{g}\, \Gamma^l_{kp} \epsilon_{ijl} = 0.$$

If we take $i = 1$, $j = 2$, and $k = 3$, then

$$\frac{1}{\sqrt{g}} \frac{\partial \sqrt{g}}{\partial x^p} = \Gamma^1_{1p} + \Gamma^2_{2p} + \Gamma^3_{3p} = \Gamma^k_{kp} = \Gamma^k_{pk}.$$

Subsequently we can rewrite

$$\nabla \cdot \mathbf{v} = v^i_{,i} = \frac{\partial v^i}{\partial x^i} + \frac{1}{\sqrt{g}} \frac{\partial \sqrt{g}}{\partial x^i} v^i = \frac{1}{\sqrt{g}} \frac{\partial}{\partial x^i} \left(\sqrt{g} v^i \right). \tag{2.247}$$

From before we obtained the vector field

$$\nabla \phi = \frac{\partial \phi}{\partial x^i} \mathbf{e}^i = \frac{\partial \phi}{\partial x^i} g^{ij} \mathbf{e}_j.$$

Now the divergence of this vector field is given by

$$\nabla \cdot \nabla \phi = \nabla^2 \phi = \left(\frac{\partial \phi}{\partial x^i} g^{ij} \right)_{,j},$$

where the operator ∇^2 is called the Laplacian. It can be shown that

$$g^{ij}_{,l} = 0 \tag{2.248}$$

(called *Ricci's theorem*), i.e., the metric tensor behaves like a constant under covariant differentiation. Subsequently, we can write

$$\nabla^2 \phi = g^{ij} \left(\frac{\partial \phi}{\partial x^i} \right)_{,j} = g^{ij} \left(\frac{\partial^2 \phi}{\partial x^j \partial x^i} - \Gamma^k_{ij} \frac{\partial \phi}{\partial x^k} \right). \tag{2.249}$$

Using the previous result of the divergence of a vector field, with

$$v^i = g^{ij} \frac{\partial \phi}{\partial x^j}, \tag{2.250}$$

we can rewrite this last result as

$$v^i_{,i} = \nabla^2 \phi = \frac{1}{\sqrt{g}} \frac{\partial}{\partial x^i} \left(\sqrt{g} g^{ij} \frac{\partial \phi}{\partial x^j} \right). \tag{2.251}$$

2.18 Curl of a vector field

The curl of a vector field is given by

$$\begin{aligned}
\mathbf{w} = \text{curl } \mathbf{v} = \nabla \times \mathbf{v} &= \left(\mathbf{e}^i \frac{\partial}{\partial x^i} \right) \times \left(v_j \mathbf{e}^j \right) \\
&= \mathbf{e}^i \times \frac{\partial}{\partial x^i} \left(v_j \mathbf{e}^j \right) \\
&= \mathbf{e}^i \times v_{j,i} \mathbf{e}^j \\
&= v_{j,i} \mathbf{e}^i \times \mathbf{e}^j \\
&= v_{j,i} \frac{\epsilon^{ijk}}{\sqrt{g}} \mathbf{e}_k \\
&= w^k \mathbf{e}_k,
\end{aligned}$$

where we can write

$$w^k = \varepsilon^{kij} v_{j,i}. \tag{2.252}$$

Similarly, in terms of covariant components, we find that

$$\mathbf{w} = \text{curl } \mathbf{v} = w_l \mathbf{e}^l,$$

where

$$w_l = \varepsilon_{kjl} g^{ki} v^j_{,i}. \tag{2.253}$$

Example

The curl $\mathbf{v}$ can be computed by an alternate method that is sometimes convenient. This is done through a double contraction of the absolute Levi–Civita tensor:

$$\begin{aligned} \mathbf{w} = \operatorname{curl}\mathbf{v} = \boldsymbol{\varepsilon} : (\nabla\mathbf{v}) &= \left(\varepsilon^{ijk}\mathbf{e}_i\mathbf{e}_j\mathbf{e}_k\right) : \left(v_{m,l}\mathbf{e}^l\mathbf{e}^m\right) \\ &= \varepsilon^{ijk} v_{m,l}\delta_j^l\delta_k^m\mathbf{e}_i \\ &= \varepsilon^{ijk} v_{k,j}\mathbf{e}_i, \end{aligned}$$

recovering the result above. The analogous result in terms of covariant components can be obtained in a similar fashion.

Vectors (like moment of a force, angular velocity, vorticity, etc.) whose direction is established by convention, and which therefore change direction when the "handedness" of the coordinate system is changed (from right-handed to left-handed, say) are called *axial vectors* (or *pseudo-vectors*). Note that to discuss such vectors we resort to "right-hand-screw direction" or "left-hand-screw direction." Vectors (like force, velocity, etc.) whose direction depends only on their physical meaning, and which therefore do not change direction when the "handedness" of the coordinate system is changed, are called *polar vectors*. Such vectors can be represented without ambiguity by an arrow pointing in a certain direction. Thus, a polar vector is symbolized by a line with an arrow indicating the direction, while an axial vector by a line with a sense of rotation around the line with the line forming its axis (or center of rotation). To determine the nature of a vector, imagine it reflected in a mirror perpendicular to itself as illustrated in Fig. 2.8. If the reflection preserves the direction of the quantity describing a physical phenomenon, then the vector is axial. The vector product of two polar vectors is not a true polar vector but an anti-symmetrical tensor. It happens, as a coincidence, that in three dimensions an anti-symmetrical tensor of the second rank has the same number of independent components as a vector, but this is not true in any other number of dimensions. For instance, in four dimensions an anti-symmetrical tensor of the second rank has 4×4 components with only 6 being independent; a vector, on the other hand, has only four components in this case. Lastly, we note that the components of a skew-symmetric second rank tensor transform like the components of an axial vector, whose components change sign when the "handedness" of the coordinate system is changed. In a similar fashion, any triple scalar product $[\mathbf{u}, \mathbf{v}, \mathbf{w}]$ is an *axial-* or *pseudo-scalar* since basis reversal in any component formulation results in a change of sign. A *polar* or *genuine scalar* has a value completely independent of any basis. For more on this, see Section 2.11.

Example

If $\boldsymbol{\varepsilon}$ is the absolute Levi–Civita tensor and $\mathbf{W}$ is a skew-symmetric second-order tensor, then its dual is given by the axial vector

$$\varpi = \frac{1}{2}\boldsymbol{\varepsilon} : \mathbf{W}. \tag{2.254}$$

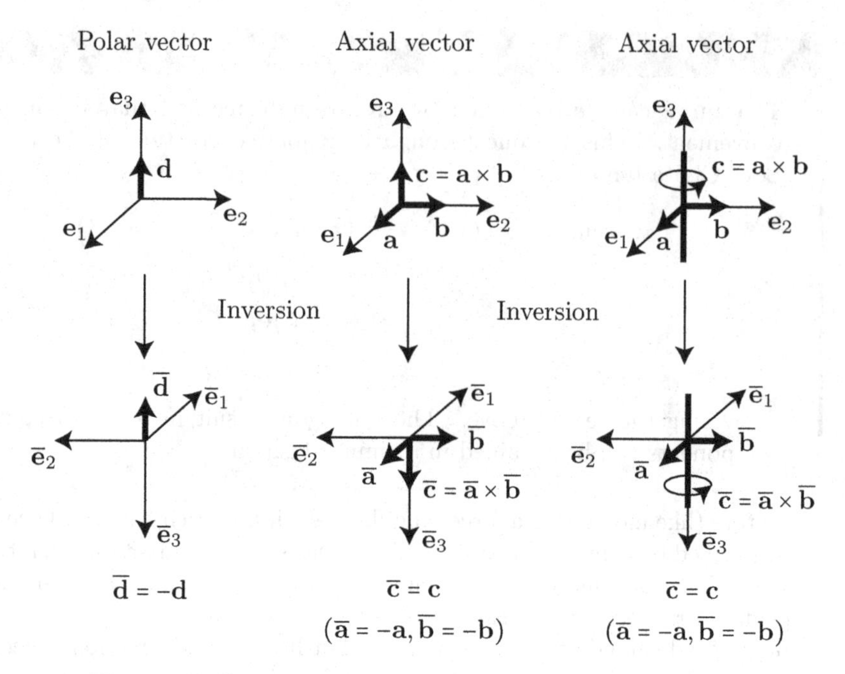

Figure 2.8: Polar vectors **a**, **b**, and **d**, and axial vector **c**.

The vorticity vector $\boldsymbol{\omega}$ will be shown later to be related to the angular velocity $\boldsymbol{\varpi}$, the axial vector corresponding to the second-order skew-symmetric spin tensor $\mathbf{W}$, i.e., $\boldsymbol{\omega} = \text{curl } \mathbf{v} = -2\,\boldsymbol{\varpi}$, where $\mathbf{v}$ is the linear velocity.

2.19 Orthogonal curvilinear coordinate system

In the special case of an orthogonal curvilinear coordinate system, the components of the metric tensor are such that

$$g_{ij} = \mathbf{e}_i \cdot \mathbf{e}_j \begin{cases} = 0 & \text{if } i \neq j, \\ \neq 0 & \text{if } i = j, \end{cases} \tag{2.255}$$

or in matrix notation

$$G = [g_{ij}] = \begin{bmatrix} g_{11} & 0 & 0 \\ 0 & g_{22} & 0 \\ 0 & 0 & g_{33} \end{bmatrix} = \begin{bmatrix} h_1^2 & 0 & 0 \\ 0 & h_2^2 & 0 \\ 0 & 0 & h_3^2 \end{bmatrix}, \tag{2.256}$$

where

$$h_i \equiv \sqrt{g_{\underline{ii}}}, \tag{2.257}$$

and the underline under an index means that that index does not sum. Similarly

$$g^{ij} = \mathbf{e}^i \cdot \mathbf{e}^j \begin{cases} = 0 & \text{if } i \neq j, \\ \neq 0 & \text{if } i = j. \end{cases} \tag{2.258}$$

From earlier results it follows that

$$g^{\underline{ii}} = \frac{1}{g_{\underline{ii}}}, \tag{2.259}$$

and

$$ds^2 = h_1^2 \left(dx^1\right)^2 + h_2^2 \left(dx^2\right)^2 + h_3^2 \left(dx^3\right)^2 = h_i^2 \left(dx^i\right)^2. \tag{2.260}$$

Example

In cylindrical polar coordinates

$$(h_1, h_2, h_3) = \left(1, x^1, 1\right), \tag{2.261}$$

and in spherical coordinates

$$(h_1, h_2, h_3) = \left(1, x^1, x^1 \sin x^2\right). \tag{2.262}$$

For orthogonal coordinates, it can be readily seen that $\Gamma^i_{jk} = \Gamma^i_{kj} = 0$ if $i \neq j \neq k$, e.g., for $i = 1$, $j = 2$, and $k = 3$, we have

$$\Gamma^1_{23} = \frac{1}{2} g^{11} \left(\frac{\partial g_{31}}{\partial x^2} + \frac{\partial g_{21}}{\partial x^3} - \frac{\partial g_{23}}{\partial x^1} \right) + 0 + 0 = 0. \tag{2.263}$$

We also have

$$\Gamma^i_{\underline{jj}} = -\frac{h_{\underline{j}}}{h_{\underline{i}}^2} \frac{\partial h_{\underline{j}}}{\partial x^i} \quad (i \neq j), \tag{2.264}$$

and

$$\Gamma^{\underline{i}}_{j\underline{i}} = \Gamma^{\underline{i}}_{\underline{i}j} = \frac{1}{h_{\underline{i}}} \frac{\partial h_{\underline{i}}}{\partial x^j}. \tag{2.265}$$

Example

In cylindrical polar coordinates the only nonzero components are

$$\Gamma^1_{22} = -\frac{h_2}{h_1^2} \frac{\partial h_2}{\partial x^1} = -x^1, \tag{2.266}$$

$$\Gamma^2_{12} = \Gamma^2_{21} = \frac{1}{h_2} \frac{\partial h_2}{\partial x^1} = \frac{1}{x^1}. \tag{2.267}$$

For rectangular or Cartesian coordinates, $\Gamma^i_{jk} = 0$ for any i, j, and k so that covariant differentiation reduces to standard partial differentiation.

2.19.1 Physical components

We now normalize the basis,

$$\hat{\mathbf{e}}_i \equiv \frac{\mathbf{e}_i}{\|\mathbf{e}_{\underline{i}}\|} = \frac{\mathbf{e}_i}{\sqrt{\mathbf{e}_{\underline{i}} \cdot \mathbf{e}_{\underline{i}}}} = \frac{\mathbf{e}_i}{h_{\underline{i}}} \tag{2.268}$$

and similarly

$$\mathbf{e}^i = g^{ij}\mathbf{e}_j = g^{\underline{ii}}\mathbf{e}_i = \frac{1}{h_{\underline{i}}^2}\mathbf{e}_i = \frac{1}{h_{\underline{i}}}\hat{\mathbf{e}}_i. \tag{2.269}$$

We then see that

$$\mathbf{v} = v^i\mathbf{e}_i = v_i\mathbf{e}^i = v_{<i>}\hat{\mathbf{e}}_i, \tag{2.270}$$

where we have defined the physical components by

$$v_{<i>} = h_{\underline{i}}v^i = \frac{1}{h_{\underline{i}}}v_i. \tag{2.271}$$

2.19.2 Gradient of a scalar field

We now note that

$$\nabla\phi = \frac{\partial\phi}{\partial x^i}\mathbf{e}^i = \frac{\partial\phi}{\partial x^i}g^{ij}\mathbf{e}_j,$$

so

$$\nabla\phi = \frac{\partial\phi}{\partial x^i}\frac{1}{h_{\underline{i}}^2}\mathbf{e}_i,$$

or using physical components

$$\nabla\phi = \frac{1}{h_{\underline{i}}}\frac{\partial\phi}{\partial x^i}\hat{\mathbf{e}}_i. \tag{2.272}$$

Example

In the cylindrical polar coordinate system

$$\nabla\phi = \frac{\partial\phi}{\partial x^1}\hat{\mathbf{e}}_1 + \frac{1}{x^1}\frac{\partial\phi}{\partial x^2}\hat{\mathbf{e}}_2 + \frac{\partial\phi}{\partial x^3}\hat{\mathbf{e}}_3. \tag{2.273}$$

2.19.3 Gradient and divergence of a vector field

In terms of physical components, the covariant derivative of a vector field is now given by

$$\nabla\mathbf{v} = v^i_{,j}\mathbf{e}_i\mathbf{e}^j = \left(\frac{\partial v^i}{\partial x^j} + \Gamma^i_{kj}v^k\right)\mathbf{e}_i\mathbf{e}^j = v_{<i,j>}\hat{\mathbf{e}}_i\hat{\mathbf{e}}_j, \tag{2.274}$$

where we have defined

$$v_{<i,j>} \equiv \frac{h_{\underline{i}}}{h_{\underline{j}}}\left[\frac{\partial}{\partial x^j}\left(\frac{v_{<i>}}{h_{\underline{i}}}\right) + \Gamma^i_{kj}\frac{v_{<k>}}{h_{\underline{k}}}\right]. \tag{2.275}$$

The divergence of a vector field is now easily obtained by contracting the above result

$$\nabla\cdot\mathbf{v} = v^i_{,i} = \frac{1}{\sqrt{g}}\frac{\partial}{\partial x^i}\left(\sqrt{g}v^i\right) = \frac{1}{h_1h_2h_3}\frac{\partial}{\partial x^i}\left(\frac{h_1h_2h_3}{h_{\underline{i}}}v_{<i>}\right). \tag{2.276}$$

Example

In the cylindrical polar coordinate system

$$\nabla \cdot \mathbf{v} = \frac{1}{x^1} \frac{\partial}{\partial x^i} \left(\frac{x^1}{h_{\underline{i}}} v_{<i>} \right) = \frac{1}{x^1} \frac{\partial}{\partial x^1} \left(x^1 v_{<1>} \right) + \frac{1}{x^1} \frac{\partial v_{<2>}}{\partial x^2} + \frac{\partial v_{<3>}}{\partial x^3}. \tag{2.277}$$

2.19.4 Curl of a vector field

Using our earlier results and following a similar procedure, we find that

$$\nabla \times \mathbf{v} = \epsilon^{ijk} \frac{1}{h_{\underline{i}} h_{\underline{j}}} \frac{\partial}{\partial x^i} \left(h_{\underline{j}} v_{<j>} \right) \hat{\mathbf{e}}_k. \tag{2.278}$$

Example

In the cylindrical polar coordinate system

$$\begin{aligned} \nabla \times \mathbf{v} &= \left(\frac{1}{x^1} \frac{\partial v_{<3>}}{\partial x^2} - \frac{\partial v_{<2>}}{\partial x^3} \right) \hat{\mathbf{e}}_1 + \left(\frac{\partial v_{<1>}}{\partial x^3} - \frac{\partial v_{<3>}}{\partial x^1} \right) \hat{\mathbf{e}}_2 + \\ & \frac{1}{x^1} \left[\frac{\partial}{\partial x^1} \left(x^1 v_{<2>} \right) - \frac{\partial v_{<1>}}{\partial x^2} \right] \hat{\mathbf{e}}_3. \end{aligned} \tag{2.279}$$

2.19.5 Laplacian of a scalar field

Using our earlier result

$$\nabla^2 \phi = \frac{1}{\sqrt{g}} \frac{\partial}{\partial x^i} \left(\sqrt{g} g^{ij} \frac{\partial \phi}{\partial x^j} \right) \tag{2.280}$$

in an orthogonal curvilinear system, we have

$$\nabla^2 \phi = \frac{1}{\sqrt{g}} \frac{\partial}{\partial x^i} \left(\sqrt{g} g^{\underline{ii}} \frac{\partial \phi}{\partial x^i} \right) = \frac{1}{h_1 h_2 h_3} \frac{\partial}{\partial x^i} \left(\frac{h_1 h_2 h_3}{h_{\underline{i}}^2} \frac{\partial \phi}{\partial x^i} \right). \tag{2.281}$$

Example

In the cylindrical polar coordinate system

$$\nabla^2 \phi = \frac{1}{x^1} \frac{\partial}{\partial x^i} \left(\frac{x^1}{h_{\underline{i}}^2} \frac{\partial \phi}{\partial x^i} \right) = \frac{1}{x^1} \frac{\partial}{\partial x^1} \left(x^1 \frac{\partial \phi}{\partial x^1} \right) + \frac{1}{(x^1)^2} \frac{\partial^2 \phi}{\partial (x^2)^2} + \frac{\partial^2 \phi}{\partial (x^3)^2}. \tag{2.282}$$

2.19.6 Divergence of a dyadic tensor field

It will also be useful to write the divergence of a second-order tensor in terms of physical components:

$$
\begin{aligned}
\nabla \cdot \boldsymbol{\tau} &= \tau^{ij}_{,j}\mathbf{e}_i, \\
&= \left(\frac{\partial \tau^{ij}}{\partial x^j} + \Gamma^i_{jk}\tau^{kj} + \Gamma^j_{jk}\tau^{ik}\right)\mathbf{e}_i, \\
&= \left[\frac{1}{\sqrt{g}}\frac{\partial}{\partial x^j}\left(\sqrt{g}\tau^{ij}\right) + \Gamma^i_{jk}\tau^{kj}\right]\mathbf{e}_i \quad \left(\text{since } \Gamma^j_{jk} = \frac{1}{\sqrt{g}}\frac{\partial \sqrt{g}}{\partial x^k}\right), \\
&= h_{\underline{i}}\left[\frac{1}{h_1h_2h_3}\frac{\partial}{\partial x^j}\left(h_1h_2h_3\frac{\tau_{<ij>}}{h_{\underline{i}}h_{\underline{j}}}\right) + \Gamma^i_{jk}\frac{\tau_{<kj>}}{h_{\underline{k}}h_{\underline{j}}}\right]\hat{\mathbf{e}}_i,
\end{aligned}
$$

or

$$
\nabla \cdot \boldsymbol{\tau} = \tau_{<ij,j>}\hat{\mathbf{e}}_i, \tag{2.283}
$$

where

$$
\tau_{<ij,j>} \equiv \frac{h_{\underline{i}}}{h_1h_2h_3}\frac{\partial}{\partial x^j}\left(\frac{h_1h_2h_3}{h_{\underline{i}}h_{\underline{j}}}\tau_{<ij>}\right) + \frac{h_{\underline{i}}}{h_{\underline{k}}h_{\underline{j}}}\Gamma^i_{jk}\tau_{<kj>}, \tag{2.284}
$$

and

$$
\tau_{<ij>} \equiv h_{\underline{i}}h_{\underline{j}}\tau^{ij} = \frac{h_{\underline{i}}}{h_{\underline{j}}}\tau^i_j = \frac{1}{h_{\underline{i}}h_{\underline{j}}}\tau_{ij}. \tag{2.285}
$$

Example

In the cylindrical polar coordinate system

$$
\tau_{<ij,j>} = \frac{h_{\underline{i}}}{x^1}\frac{\partial}{\partial x^j}\left(\frac{x^1}{h_{\underline{i}}h_{\underline{j}}}\tau_{<ij>}\right) + \frac{h_{\underline{i}}}{h_{\underline{k}}h_{\underline{j}}}\Gamma^i_{jk}\tau_{<kj>}. \tag{2.286}
$$

2.20 Integral theorems and generalizations

If V is any composite volume with piecewise smooth bounding surface S and outward normal $\mathbf{n} = n_i\mathbf{e}^i$, and F is any continuous function in V whose gradient ∇F is also continuous in V, then

$$
\int_V \nabla F\, dV = \int_S F\, d\mathbf{S} = \int_S F\,\mathbf{n}\, dS \quad \text{or} \quad \int_V F_{,i}\, dV = \int_S F\, dS_i = \int_S F\, n_i\, dS. \tag{2.287}
$$

We will not prove this general form of the Gauss–Green theorem here.

The utility of the theorem stems largely from the observation that the function F may be either a scalar or a tensor of any order. For instance, replacing the function F with an arbitrary vector $\mathbf{v} = v^j\mathbf{e}_j$ gives us

$$
\int_V \nabla\mathbf{v}\, dV = \int_S \mathbf{v}\,\mathbf{n}\, dS \qquad \text{or} \qquad \int_V v^j_{,i}\, dV = \int_S v^j\, n_i\, dS. \tag{2.288}
$$

If we contract the above result, we obtain the usual *divergence theorem*

$$\int_V \nabla \cdot \mathbf{v}\, dV = \int_S \mathbf{v} \cdot \mathbf{n}\, dS \qquad \text{or} \qquad \int_V v^i_{,i}\, dV = \int_S v^i n_i\, dS. \tag{2.289}$$

Replacing the function F with $\times\mathbf{v} = \times(v_j \mathbf{e}^j)$, and recalling that $\times\mathbf{v} = -\mathbf{v}\times$, gives us the result

$$\int_V \nabla \times \mathbf{v}\, dV = \int_S \mathbf{n} \times \mathbf{v}\, dS \qquad \text{or} \qquad \int_V \varepsilon^{kij} v_{j,i}\, dV = \int_S \varepsilon^{kij} n_i v^j\, dS. \tag{2.290}$$

A variety of useful results relating integrals over a closed curve C to integrals over an open surface A bounded by C can be obtained using the two-dimensional form of the theorem, i.e.,

$$\int_A \nabla G\, dA = \int_C G \mathbf{n}'\, dl \qquad \text{or} \qquad \int_A G_{,i}\, dA = \int_C G n'_i\, dl, \tag{2.291}$$

where G is a continuous function defined on an open surface with bounding curve C and $\mathbf{n}' = n'_i \mathbf{e}^i$ is the outward normal to the curve C lying on a plane tangent to A. A result similar to the above can be written when A is a curved surface, but this would require a discussion of tensors on curved, or Riemannian, spaces which is beyond the scope of the present discussion, but see Appendix D.

We now replace G by $\times\mathbf{v} \cdot \mathbf{n} = \times(v_i \mathbf{e}^i) \cdot (n_j \mathbf{e}^j)$, where $\mathbf{n}$ is the normal to the surface A and $\mathbf{v}$ is a continuous vector function. Then we have

$$\begin{aligned}
\int_A \nabla \times \mathbf{v} \cdot \mathbf{n}\, dA &= \int_C \times\mathbf{v} \cdot \mathbf{n}\, \mathbf{n}'\, dl, \\
&= -\int_C \mathbf{n} \cdot \mathbf{v} \times \mathbf{n}'\, dl, \\
&= \int_C \mathbf{n} \cdot \mathbf{n}' \times \mathbf{v}\, dl, \\
&= \int_C \mathbf{v} \cdot \mathbf{n} \times \mathbf{n}'\, dl.
\end{aligned}$$

Now since $\mathbf{n}$ and $\mathbf{n}'$ are normal to both C and to each other, the vector product $\mathbf{n} \times \mathbf{n}'$ in the integral is equal to the vector $\mathbf{t} = \mathbf{n} \times \mathbf{n}' = t^k \mathbf{e}_k$ tangent to C, where $t^k = \varepsilon^{kij} n_i n'_j$. Subsequently, using our previous result of the curl of a vector field, and noting that $\mathbf{t}dl = d\mathbf{r}$, where $d\mathbf{r} = dt^k \mathbf{e}_k$ is the oriented differential length of the line tangent to C, we arrive at the result

$$\int_A \nabla \times \mathbf{v} \cdot \mathbf{n}\, dA = \int_C \mathbf{v} \cdot d\mathbf{r} \qquad \text{or} \qquad \int_A \varepsilon^{kij} n_k v_{j,i}\, dA = \int_C v_k\, dt^k, \tag{2.292}$$

which is known as *Stokes' theorem*.

In addition, setting $\mathbf{v} = \phi\mathbf{a}$ where $\mathbf{a}$ is an arbitrary *constant* vector, and using the property of invariance to cyclic permutations of the triple scalar product, we obtain the following useful identity:

$$\int_A \mathbf{n} \times \nabla\phi\, dA = \int_C \phi\, d\mathbf{r} \qquad \text{or} \qquad \int_A \varepsilon^{kij} n_i \phi_{,j}\, dA = \int_C \phi\, dt^k. \tag{2.293}$$

2.20.1 Regions with discontinuous surfaces, curves, and points

In the above theorems, we have assumed that the tensor field F is continuous in V. When V contains a discontinuous surface across which F undergoes a jump, we can decompose the volume into two subvolumes separated by the discontinuous surface ζ with unit normal $\boldsymbol{\nu}$ as illustrated in Fig. 2.9. Then within each subvolume, the

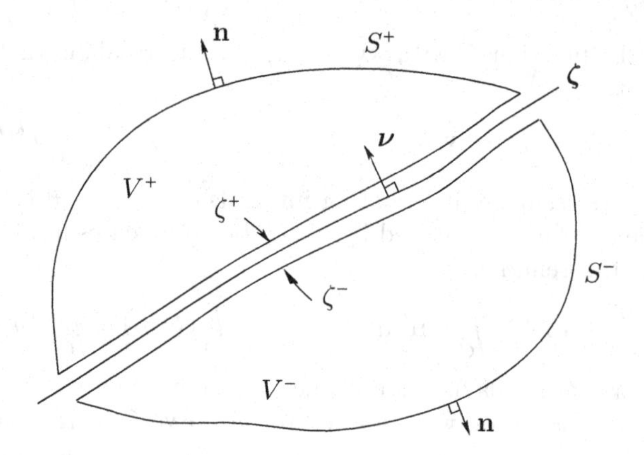

Figure 2.9: Arbitrary volume V containing a discontinuous surface.

field F is continuous and our previous results apply. Subsequently, we can write

$$\int_{V^+} \nabla F \, dV = \int_{S^+} F \, d\mathbf{S} + \int_{\zeta^+} F^+ d\boldsymbol{\zeta}^+ \tag{2.294}$$

and

$$\int_{V^-} \nabla F \, dV = \int_{S^-} F \, d\mathbf{S} + \int_{\zeta^-} F^- d\boldsymbol{\zeta}^-. \tag{2.295}$$

Now adding the above equations, letting $\boldsymbol{\zeta}^+$ and $\boldsymbol{\zeta}^-$ approach $\boldsymbol{\zeta}$, and recognizing that in this limit $d\boldsymbol{\zeta}^+ = -d\boldsymbol{\zeta}^- = -d\boldsymbol{\zeta} = -\boldsymbol{\nu}\, d\zeta$, we obtain

$$\int_{V-\zeta} \nabla F \, dV = \int_{S-\zeta} F \, d\mathbf{S} - \int_{\zeta} [\![F]\!] \, d\boldsymbol{\zeta}, \tag{2.296}$$

where we have defined the *jump* operator

$$[\![F]\!] \equiv F^+ - F^-, \tag{2.297}$$

and

$$F^+(\mathbf{x}) \equiv \lim_{\mathbf{x}\downarrow\zeta} F(\mathbf{x}) \qquad \text{and} \qquad F^-(\mathbf{x}) \equiv \lim_{\mathbf{x}\uparrow\zeta} F(\mathbf{x}). \tag{2.298}$$

For example, if we have a vector field $\mathbf{v}$, the *generalized divergence theorem* is obtained by taking $F \to \mathbf{v}$ and contracting the result to obtain

$$\int_{V-\zeta} \nabla \cdot \mathbf{v} \, dV = \int_{S-\zeta} \mathbf{v} \cdot d\mathbf{S} - \int_{\zeta} [\![\mathbf{v}]\!] \cdot d\boldsymbol{\zeta}. \tag{2.299}$$

Analogously, if area $\mathbf{A}$ with unit normal $\mathbf{n}$ contains a discontinuous line γ with tangential unit vector $\mathbf{t}$, as illustrated in Fig. 2.10, across which $\mathbf{v}$ changes suddenly, then using the same procedure as before that led to Stokes' theorem, and

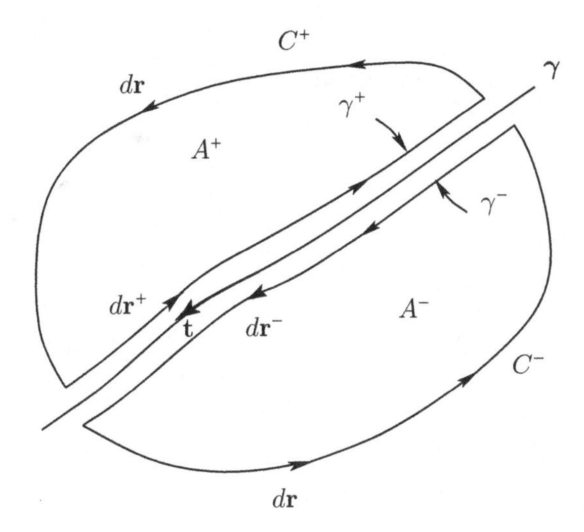

Figure 2.10: Arbitrary surface **A** containing a discontinuous curve.

noting that $d\mathbf{r}^+ = -d\mathbf{r}^- = -d\boldsymbol{\gamma} = -\mathbf{t}\,d\gamma$, we now obtain the *generalized Stokes theorem:*

$$\int_{A-\gamma} \nabla \times \mathbf{v} \cdot d\mathbf{A} = \int_{C-\gamma} \mathbf{v} \cdot d\mathbf{r} - \int_{\gamma} [\![\mathbf{v}]\!] \cdot d\boldsymbol{\gamma}. \tag{2.300}$$

For reference, we also note that in one dimension we have an analogous version of the generalized divergence theorem. In this case, as illustrated in Fig. 2.11, if a function $v(\xi)$ defined on the curve $C : \xi_1 \leq \xi \leq \xi_2$ is discontinuous at the point σ in the interval of C, but continuous in the subintervals $\xi_1 < \xi < \sigma$ and $\sigma < \xi < \xi_2$, then

$$\int_{C(t)-\sigma(t)} \frac{dv}{d\xi}\, d\xi = [v(\xi)]_{\xi_1}^{\xi_2} - [\![v(\sigma)]\!], \tag{2.301}$$

where

$$[\![v(\sigma)]\!] = v^+(\sigma) - v^-(\sigma), \tag{2.302}$$

and

$$v^+(\sigma) \equiv \lim_{\xi \downarrow \sigma} v(\xi) \qquad \text{and} \qquad v^-(\sigma) \equiv \lim_{\xi \uparrow \sigma} v(\xi). \tag{2.303}$$

Problems

1. The scalar triple product of three vectors $[\mathbf{u}, \mathbf{v}, \mathbf{w}]$ is given by $\mathbf{u} \cdot (\mathbf{v} \times \mathbf{w})$. Establish the following property of the scalar triple product:

$$\mathbf{u} \cdot (\mathbf{v} \times \mathbf{w}) = \mathbf{v} \cdot (\mathbf{w} \times \mathbf{u}) = \mathbf{w} \cdot (\mathbf{u} \times \mathbf{v}) = -\mathbf{u} \cdot (\mathbf{w} \times \mathbf{v}) = -\mathbf{v} \cdot (\mathbf{u} \times \mathbf{w}) = -\mathbf{w} \cdot (\mathbf{v} \times \mathbf{u})$$

for all $(\mathbf{u}, \mathbf{v}, \mathbf{w}) \in \mathcal{E}^3$.

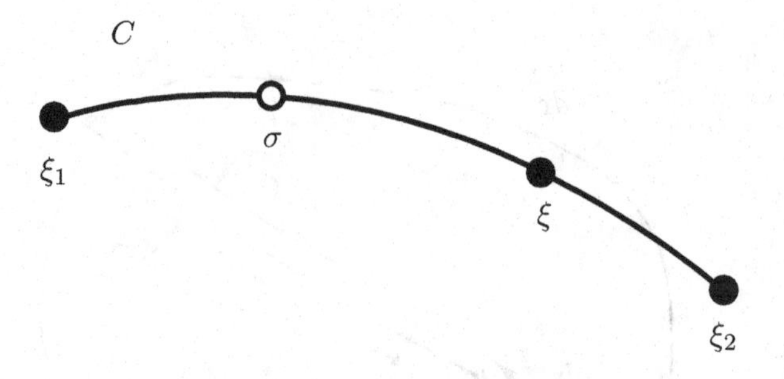

Figure 2.11: Arbitrary curve C containing a discontinuous point.

2. Show that

$$\mathbf{e}_i \times \mathbf{e}_j = \pm\epsilon_{ijk}\mathbf{e}_k,$$

and, subsequently,

$$\mathbf{u} \cdot (\mathbf{v} \times \mathbf{w}) = \pm\epsilon_{ijk}u_i v_j w_k.$$

3. Let $\mathbf{u}$, $\mathbf{v}$, $\mathbf{w}$, and $\mathbf{x}$ be arbitrary vectors, $\mathbf{A}$, $\mathbf{B}$, and $\mathbf{C}$ second rank tensors, $\mathbf{1}$ the second-rank unit tensor, $\operatorname{tr}\mathbf{A}$ the trace of $\mathbf{A}$, and $\det\mathbf{A}$ the determinant of $\mathbf{A}$. Show that

    a) $(\mathbf{uv})^T = \mathbf{vu}$,
    b) $\operatorname{tr}(\mathbf{uv}) = \mathbf{1} : \mathbf{uv} = \mathbf{u} \cdot \mathbf{v}$,
    c) $\det(\mathbf{uv}) = 0$,
    d) $(\mathbf{uv}) : (\mathbf{wx}) = (\mathbf{u} \cdot \mathbf{w})(\mathbf{v} \cdot \mathbf{x})$,
    e) $\mathbf{v} \cdot \mathbf{A}^T \cdot \mathbf{u} = (\mathbf{A} \cdot \mathbf{v}) \cdot \mathbf{u} = \mathbf{u} \cdot \mathbf{A} \cdot \mathbf{v}$,
    f) $\mathbf{A} : \mathbf{uv} = \mathbf{uv} : \mathbf{A} = \mathbf{u} \cdot \mathbf{A} \cdot \mathbf{v}$,
    g) $||\mathbf{A}|| = (\mathbf{A} : \mathbf{A})^{1/2} \geq 0$,
    h) $\operatorname{tr}\mathbf{A} = \mathbf{1} : \mathbf{A} = \mathbf{A} : \mathbf{1}$,
    i) $\det(\mathbf{A}^{-1}) = (\det\mathbf{A})^{-1}$,
    j) $(\mathbf{A}^{-1})^T = (\mathbf{A}^T)^{-1}$,
    k) $\det(\mathbf{A} \cdot \mathbf{B}) = \det\mathbf{A}\det\mathbf{B}$,
    l) $(\mathbf{A} \cdot \mathbf{B})^T = \mathbf{B}^T \cdot \mathbf{A}^T$,
    m) $\mathbf{A} : \mathbf{B} = \mathbf{B} : \mathbf{A} = \operatorname{tr}(\mathbf{A}^T \cdot \mathbf{B}) = \operatorname{tr}(\mathbf{A} \cdot \mathbf{B}^T) = \operatorname{tr}(\mathbf{B}^T \cdot \mathbf{A}) = \operatorname{tr}(\mathbf{B} \cdot \mathbf{A}^T)$,
    n) $\mathbf{A} : (\mathbf{B} \cdot \mathbf{C}) = (\mathbf{B}^T \cdot \mathbf{A}) : \mathbf{C} = (\mathbf{A} \cdot \mathbf{C}^T) : \mathbf{B}$.

4. Show that $\mathbf{B} = \mathbf{F}^{-1} \cdot \mathbf{A} \cdot \left(\mathbf{F}^{-1}\right)^T$ is symmetric if $\mathbf{A}$ is symmetric.

5. Under the assumption that $\mathbf{A}$ is symmetric, construct the partial derivative of $||\mathbf{A}||$ with respect to $\mathbf{A}$.

6. Let $\mathbf{A}$ be a tensor of rank 4, and $\mathbf{B}$ and $\mathbf{C}$ tensors of rank 2. Show that $(\mathbf{A}^T)^T = \mathbf{A}$, and

$$\mathbf{B} : \mathbf{A}^T : \mathbf{C} = \mathbf{C} : \mathbf{A} : \mathbf{B} = (\mathbf{A} : \mathbf{B}) : \mathbf{C}.$$

7. Let $\mathbf{A}$ be a tensor of rank 2, the deviatoric part of $\mathbf{A}$ be defined by $\mathbf{A}' \equiv \mathbf{A} - \frac{1}{3}(\mathbf{1} : \mathbf{A})\mathbf{1}$, $\mathbb{1}$ be the unit tensor of rank 4 such that $(\mathbb{1})_{ijkl} = \delta_{ik}\delta_{jl}$, and $\bar{\mathbb{1}}$ be the unit tensor of rank 4 such that $(\bar{\mathbb{1}})_{ijkl} = \delta_{il}\delta_{jk}$. Show that

   a) $\operatorname{tr} \mathbf{A}' = 0$,
   b) $\bar{\mathbb{1}} \neq \mathbb{1}^T$,
   c) $\mathbf{A} = \mathbb{1} : \mathbf{A}$ and $\mathbf{A}^T = \bar{\mathbb{1}} : \mathbf{A}$,
   d) $\frac{1}{2}(\mathbf{A} + \mathbf{A}^T) = \mathbb{D} : \mathbf{A}$ and $\frac{1}{2}(\mathbf{A} - \mathbf{A}^T) = \mathbb{W} : \mathbf{A}$, where $\mathbb{D} \equiv \frac{1}{2}(\mathbb{1} + \bar{\mathbb{1}})$ and $\mathbb{W} \equiv \frac{1}{2}(\mathbb{1} - \bar{\mathbb{1}})$ are symmetric and skew-symmetric tensors of rank 4.
   e) $\mathbf{A}' = \mathbb{P} : \mathbf{A}$, where the projector $\mathbb{P}$ is given by $\mathbb{P} = \mathbb{1} - \frac{1}{3}\mathbf{1}\mathbf{1}$ with components $P_{ijkl} = \delta_{ik}\delta_{jl} - \frac{1}{3}\delta_{ij}\delta_{kl}$,
   f) $\mathbb{1}$, $\bar{\mathbb{1}}$, and $\mathbf{1}\mathbf{1}$ are isotropic tensors, and the most general isotropic tensor of rank 4 is of the form $\alpha\mathbf{1}\mathbf{1} + \beta\mathbb{1} + \gamma\bar{\mathbb{1}}$ with components

$$(\alpha\mathbf{1}\mathbf{1} + \beta\mathbb{1} + \gamma\bar{\mathbb{1}})_{ijkl} = \alpha\delta_{ij}\delta_{kl} + \beta\delta_{ik}\delta_{jl} + \gamma\delta_{il}\delta_{jk}.$$

8. How many distinct components are there in the completely symmetric tensor of rank 3 in $\mathcal{E}^3$?

9. Show that a completely skew-symmetric tensor of rank 3 in $\mathcal{E}^3$ has only one nonzero distinct component.

10. Assuming that $\mathbf{u}$, $\mathbf{v}$, and $\mathbf{a}$ are arbitrary vectors, and $\mathbf{w}$ is the axial vector corresponding to $\mathbf{W} = \mathbf{u} \wedge \mathbf{v}$. By using (2.147), show that

$$\mathbf{w} = \mathbf{u} \times \mathbf{v}. \tag{2.304}$$

11. Let $\mathbf{W}$ be an arbitrary rank-2 three-dimensional skew-symmetric tensor, $\mathbf{w}$ the corresponding axial vector, and $\mathbf{v}$ an arbitrary three-dimensional vector satisfying (2.147). Using the identity $(\mathbf{a} \times \mathbf{b}) \times \mathbf{c} = (\mathbf{a} \cdot \mathbf{c})\,\mathbf{b} - (\mathbf{b} \cdot \mathbf{c})\,\mathbf{a}$, show that

$$\mathbf{W}^2 = \mathbf{w}\,\mathbf{w} - (\mathbf{w} \cdot \mathbf{w})\,\mathbf{1} \tag{2.305}$$

and hence that

$$|\mathbf{w}|^2 = -\frac{1}{2}\operatorname{tr} \mathbf{W}^2. \tag{2.306}$$

12. Let w_{ij} be the components of a skew-symmetric tensor and s^{ij} the components of a symmetric tensor.

   a) Provide justifications for each of the following equal signs:

$$w_{ij}s^{ij} = -w_{ji}s^{ij} = -w_{ji}s^{ji} = -w_{kl}s^{kl} = -w_{ij}s^{ij} = 0.$$

b) Establish the following two identities for any arbitrary tensor of rank 2 with components v_{ij}:

$$v^{ij} w_{ij} = \frac{1}{2}\left(v^{ij} - v^{ji}\right) w_{ij}, \qquad v^{ij} s_{ij} = \frac{1}{2}\left(v^{ij} + v^{ji}\right) s_{ij}.$$

13. Let Q_{ijr} be the components of a tensor of rank 3 that is skew-symmetric in its first two indices, i.e., $Q_{[ij]r}$. Show that the tensor can be decomposed into three orthogonal parts

$$Q_{[ij]r} = Q^{(1)}_{ijr} + Q^{(2)}_{ijr} + Q^{(3)}_{ijr},$$

where

$$\begin{aligned} Q^{(1)}_{ijr} &= \frac{1}{3}\left(Q_{ijr} + Q_{rij} + Q_{jri}\right), \\ Q^{(2)}_{ijr} &= \frac{1}{6}\left(4\, Q_{ijr} - 2\, Q_{rij} - 2\, Q_{jri} - 3\, Q_{ikk}\, \delta_{jr} - 3\, Q_{jkk}\, \delta_{ir}\right), \\ Q^{(3)}_{ijr} &= \frac{1}{2}\left(Q_{ikk}\, \delta_{jr} + Q_{kjk}\, \delta_{ir}\right). \end{aligned}$$

14. Let $\mathbf{e}_1 = (1,0)$ and $\mathbf{e}_2 = (1,1)$ be a basis for $\mathcal{E}^2$.

    a) Find the reciprocal basis $(\mathbf{e}^1, \mathbf{e}^2)$.

    b) Compute $[g_{ij}]$ and $[g^{ij}]$ directly from inner products involving basis vectors, and show that $[g^{ij}] = [g_{ij}]^{-1}$.

    c) Let $\mathbf{v}$ be a vector in $\mathcal{E}^2$ with contravariant components (relative to the above basis) $v^1 = 1$ and $v^2 = 2$. Find the covariant components of $\mathbf{v}$ relative to this basis.

    d) Compute $\|\mathbf{v}\|$.

15. In the treatment of the metric tensor, both covariant g_{ij} and contravariant g^{ij} components are discussed. The mixed components g^i_j are, however, not mentioned. Explain why. (Hint: Compute g^i_j.)

16. a) Show that the triadic tensor $t^{ijk}\mathbf{e}_i\mathbf{e}_j\mathbf{e}_k$ can be contracted to a vector by the formula $t^{ijk}\mathbf{e}_i \cdot \mathbf{e}_j\mathbf{e}_k$ and that the result is independent of the basis used (i.e., show that $t^{ijk}g_{ij}\mathbf{e}_k = \bar{t}^{ijk}\bar{g}_{ij}\bar{\mathbf{e}}_k$).

    b) Generalize the result for any order tensor. Notice that the result depends on where the dot is placed.

17. Show that in cylindrical polar coordinates

$$\begin{aligned} \mathbf{e}^1 &= \cos x^2 \mathbf{i}^1 + \sin x^2 \mathbf{i}^2, \\ \mathbf{e}^2 &= -\frac{\sin x^2}{x^1}\mathbf{i}^1 + \frac{\cos x^2}{x^1}\mathbf{i}^2, \\ \mathbf{e}^3 &= \mathbf{i}^3. \end{aligned}$$

18. Calculate g_{ik} for the cylindrical polar coordinate system.

19. Show that

$$\bar{v}^i = \frac{\partial \bar{x}^i}{\partial x^j} v^j.$$

20. Show that

$$\mathbf{v} \cdot \mathbf{v} = v_i v_j g^{ij}$$

and

$$v^i = v_k g^{ik}.$$

21. Show that

$$\epsilon_{ijk}\epsilon_{rst} = \delta_{ir}\delta_{js}\delta_{kt} + \delta_{is}\delta_{jt}\delta_{kr} + \delta_{it}\delta_{jr}\delta_{ks} - \delta_{is}\delta_{jr}\delta_{kt} - \delta_{ir}\delta_{jt}\delta_{ks} - \delta_{it}\delta_{js}\delta_{kr},$$

and that then

a) $\epsilon_{ijk}\epsilon_{ist} = \delta_{js}\delta_{kt} - \delta_{jt}\delta_{ks}$,

b) $\epsilon_{ijk}\epsilon_{ijt} = 2\delta_{kt}$,

c) $\epsilon_{ijk}\epsilon_{ijk} = 6$,

d) and compute $\nabla \times \nabla \times \mathbf{v}$.

22. From the definition

$$g_{ij} = \frac{\partial \mathbf{r}}{\partial x^i} \cdot \frac{\partial \mathbf{r}}{\partial x^j},$$

verify the formula

$$\Gamma^i_{jk} = \frac{1}{2} g^{ir} \left(\frac{\partial g_{rj}}{\partial x^k} + \frac{\partial g_{rk}}{\partial x^j} - \frac{\partial g_{jk}}{\partial x^r} \right).$$

23. Show that the Christoffel symbol Γ^i_{jk} follows the transformation law

$$\bar{\Gamma}^i_{jk} = \frac{\partial \bar{x}^i}{\partial x^r} \frac{\partial x^s}{\partial \bar{x}^k} \frac{\partial x^t}{\partial \bar{x}^j} \Gamma^r_{st} + \frac{\partial \bar{x}^i}{\partial x^r} \frac{\partial^2 x^r}{\partial \bar{x}^k \partial \bar{x}^j},$$

which, in view of the second term, shows that Γ^i_{jk} is not of the form of components of a tensor with respect to an arbitrary coordinate system.

24. Prove that

$$\bar{\epsilon}^{ijk} \left| \frac{\partial \bar{x}}{\partial x} \right| = \epsilon^{rst} \frac{\partial \bar{x}^i}{\partial x^r} \frac{\partial \bar{x}^j}{\partial x^s} \frac{\partial \bar{x}^k}{\partial x^t}.$$

25. Show that $T^i_{k,l}$ transforms like a third-order tensor.

26. Show that

$$v_{i,j} - v_{j,i} = \frac{\partial v_i}{\partial x^j} - \frac{\partial v_j}{\partial x^i}.$$

27. a) Show that $g_{ij,l} = 0$ by taking the covariant derivative of the $\binom{0}{2}$ component form of a second-order tensor and using the definition of Γ^i_{jk}.

    b) Take the covariant derivative of $g_{ij}g^{jk} = \delta_i^k$ and use part a) to show that $g^{ij}_{,l} = 0$.

28. Show that

$$\varepsilon_{ijk,p} = 0.$$

29. Given a set of curvilinear coordinates x^i defined by

$$\xi^1 = x^1 x^2, \quad \xi^2 = x^1 + x^3, \quad \xi^3 = x^3.$$

    a) Find the inverse transformation $x^i = x^i(\xi^j)$.
    b) Find the $\mathbf{e}_i$'s.
    c) Find the $\mathbf{e}^i$'s.
    d) Find the g_{ij}'s.
    e) Find the g^{ij}'s.
    f) Find the Γ^i_{jk}'s (there are 27 of them).
    g) Show that $g_{ij,l} = 0$ by direct computation.
    h) Write out the equation $\nabla^2\phi = 0$ in x^i coordinates, where $\phi = \phi(x^i)$ is a scalar field.

30. Calculate Γ^i_{jk} for cylindrical polar coordinates.

31. Assume that in cylindrical polar coordinates a rank-2 tensor $\mathbf{T}$ has mixed components

$$\left[T^i_j\right] = \begin{bmatrix} 2 & -1 & 1 \\ 0 & 1 & 2 \\ 3 & 0 & -2 \end{bmatrix}$$

at the point $(x^1, x^2, x^3) = (1, \pi/4, -\sqrt{3})$. Find the component $\overline{T}^1_2$ of $\mathbf{T}$ in spherical coordinates.

Bibliography

R. Aris. *Vectors, Tensors and the Basic Equations of Fluid Mechanics.* Dover Publications, Inc., Mineola, NY, 1962.

H.D. Block. *Introduction to Tensor Analysis.* Charles E. Merrill Books, Inc., Columbus, Ohio, 1978.

A.I. Borisenko and I.E. Tarapov. *Vector and Tensor Analysis with Applications.* Dover Publications, Inc., New York, NY, 1968.

R.M. Bowen and C.-C. Wang. *Introduction to Vectors and Tensors – Linear and Multilinear Algebra*, volume 1. Plenum Press, New York, NY, 1976.

R.M. Bowen and C.-C. Wang. *Introduction to Vectors and Tensors – Vector and Tensor Analysis*, volume 2. Plenum Press, New York, NY, 1976.

L. Brand. *Vector and Tensor Analysis.* John Wiley & Sons, Inc., New York, NY, 1955.

L. Brand. *Vector Analysis.* John Wiley & Sons, Inc., New York, NY, 1957.

J.H. Heinbockel. *Introduction to Tensor Calculus and Continuum Mechanics.* Trafford Publishing, Victoria, B.C., Canada, 2001.

M. Itskov. *Tensor Algebra and Tensor Analysis for Engineers.* Springer-Verlag, Berlin, 2nd edition, 2009.

H. Jeffreys. *Cartesian Tensors.* Cambridge University Press, London, 1969.

J.K. Knowles. *Linear Vector Spaces and Cartesian Tensors.* Oxford University Press, New York, 1998.

A.J. McConnell. *Applications of Tensor Analysis.* Dover Publications, Inc., New York, NY, 1957.

C. Perwass. *Geometric Algebra with Applications in Engineering.* Springer-Verlag, Berlin, 2009.

J.G. Simmonds. *A Brief on Tensor Analysis.* Springer-Verlag, New York, NY, 1994.

G. Temple. *Cartesian Tensors.* Methuen & Co. Ltd., London, 1960.

J. Vince. *Rotation Transforms for Computer Graphics.* Springer-Verlag, London, 2011.

T.L. Wade. Tensor algebra and Young's symmetry operators. *American Journal of Mathematics*, 63(3):645–657, 1941.

T.L. Wade and R.H. Bruck. Types of symmetries. *The American Mathematical Monthly*, 51(3):123–129, 1944.

S. Winitzki. *Linear Algebra via Exterior Products.* lulu.com, 2010.

3

Kinematics

Kinematics is the study of deformation and motion of material bodies. The relationship between the initial position of *material points* or *material particles* of a body and their subsequent places is essential in the description of the local length and angle changes and translations and rotations of elements of the body. We are concerned with such changes and their measures irrespective of the type of material and the external effects.

To describe the positions of material points, we introduce two sets of coordinate systems, one for the undeformed body and one for the deformed body. The deformation of a point is then described by the relation of the coordinates of the same material point in the undeformed and deformed states.

The material points of a continuum medium, at a certain time, occupy a region $\mathcal{B}$ in space. In order to describe the body in space, we will identify it with a region in a three-dimensional Euclidean space $\mathcal{E}^3$ relative to a frame of reference. We call a one-to-one mapping from $\mathcal{B}$ into $\mathcal{E}^3$, or a complete specification of the positions of particles of a body, a *configuration* of $\mathcal{B}$. It is usually convenient to choose a particular configuration of $\mathcal{B}$, say $\boldsymbol{\kappa}$, as a reference:

$$\boldsymbol{\kappa} : \mathcal{B} \to \mathcal{E}^3 \qquad \text{or} \qquad \boldsymbol{\kappa}(X) = \mathbf{X}, \tag{3.1}$$

where $X \in \mathcal{B}$ labels a material point. We call $\boldsymbol{\kappa}$ a *reference configuration* of $\mathcal{B}$. In general, the initial or undeformed configuration need not be chosen to be the same as the reference configuration. The coordinate $\mathbf{X}$, with components X^K ($K = 1, 2, 3$), is called the *reference coordinate* or *material coordinate* since the point $\mathbf{X}$ in the reference configuration is identified with the material point labeled X of the body $\mathcal{B}$. We note that the set of all points in the material body defines the volume in the reference configuration, i.e.,

$$V \equiv \boldsymbol{\kappa}(\mathcal{B}), \tag{3.2}$$

so that $\mathbf{X} \in V$.

Let $\boldsymbol{\kappa}$ be a reference configuration and $\boldsymbol{\chi}$ an arbitrary deformed configuration of $\mathcal{B}$. Then the mapping of the material point labeled X

$$\mathbf{x} = \boldsymbol{\chi}_{\boldsymbol{\kappa}}(X) = \boldsymbol{\chi}(\boldsymbol{\kappa}^{-1}(\mathbf{X})) = \boldsymbol{\chi}_{\boldsymbol{\kappa}}(\mathbf{X}) \tag{3.3}$$

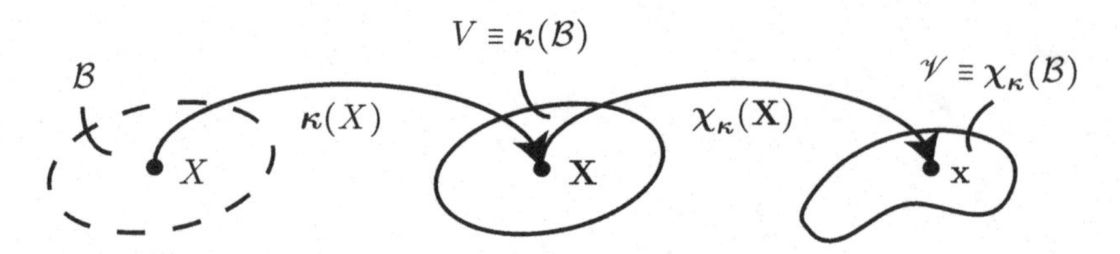

Figure 3.1: Material body, reference configuration, and deformed configuration.

is called the *deformation* of $\mathcal{B}$ from $\boldsymbol{\kappa}$ to $\boldsymbol{\chi}$. In terms of coordinate systems in the deformed, x^i $(i = 1,2,3)$, and the reference, X^K $(K = 1,2,3)$, configurations, the deformation $\boldsymbol{\chi}_{\boldsymbol{\kappa}}$ can be expressed as

$$x^i = \chi^i_{\boldsymbol{\kappa}}(X^K), \tag{3.4}$$

where $\boldsymbol{\chi}_{\boldsymbol{\kappa}}$ is called the *deformation function.* After deformation takes place, the volume V, with surface S, occupies a region consisting of the deformed volume $\mathscr{V} \equiv \boldsymbol{\chi}_{\boldsymbol{\kappa}}(\mathcal{B})$ with surface $\mathscr{S}$. In this deformed state, the material point X occupies the spatial location $\mathbf{x} \in \mathscr{V}$ in the deformed configuration, with components x^i $(i = 1,2,3)$. The configurations are illustrated in Fig. 3.1. We call X^K the *material* or *Lagrangian* coordinates of a *particle* or *material point* and x^i the *spatial* or *Eulerian* coordinates. They both have dimensions of length, $[L]$. The brackets denote the dimensions of the quantity enclosed inside of them. The deformation of the body carries various materials points through various spatial points expressed by the deformation function. *The aim of continuum mechanics is the determination of the explicit form of the deformation function when the external effects and the initial and boundary conditions of a prescribed body are known.*

From now on, for notational simplicity, we shall write the deformation function as $\boldsymbol{\chi}$ where it is understood that this is a deformation relative to the configuration $\boldsymbol{\kappa}$, unless otherwise noted. We also note that we will use quite interchangeably the following nomenclature to denote a tensor. Say that $\mathbf{T}$ is a second-order tensor. Then, using the contravariant component form, we write that $\mathbf{T} = T^{ij}\mathbf{e}_i\mathbf{e}_j$, where T^{ij} is understood to be the component in the directions $\mathbf{e}_i\mathbf{e}_j$. We shall also write the components as the matrix $T = [T^{ij}]$. Lastly, we shall take the reference as well as deformed coordinate systems to be Cartesian so that $\mathbf{e}_k \to \mathbf{i}_k$ and $T^{ij} \to T_{ij}$. However, to distinguish between quantities in the reference and deformed states, we will use uppercase letters to denote quantities associated with the reference configuration and lowercase letters with those associated with the deformed configuration, so that, e.g., $\mathbf{I}_K$ and $\mathbf{i}_k$ are the respective basis vectors in the reference and deformed coordinate systems. Sometimes it is advantageous to select two different reference frames for the reference and deformed configurations, particularly when curvilinear coordinates are used. In this case, the general picture is described as in Fig. 3.2. Note that

$$\mathbf{I}_K \cdot \mathbf{I}_L = \delta_{KL}, \quad \mathbf{i}_k \cdot \mathbf{i}_l = \delta_{kl}, \quad \mathbf{I}_K \cdot \mathbf{i}_k = g_{Kk} = g_{kK}. \tag{3.5}$$

Subsequently, we write

$$\mathbf{x} = \boldsymbol{\chi}(\mathbf{X}) \qquad \text{or} \qquad x_k = \chi_k(X_K) \tag{3.6}$$

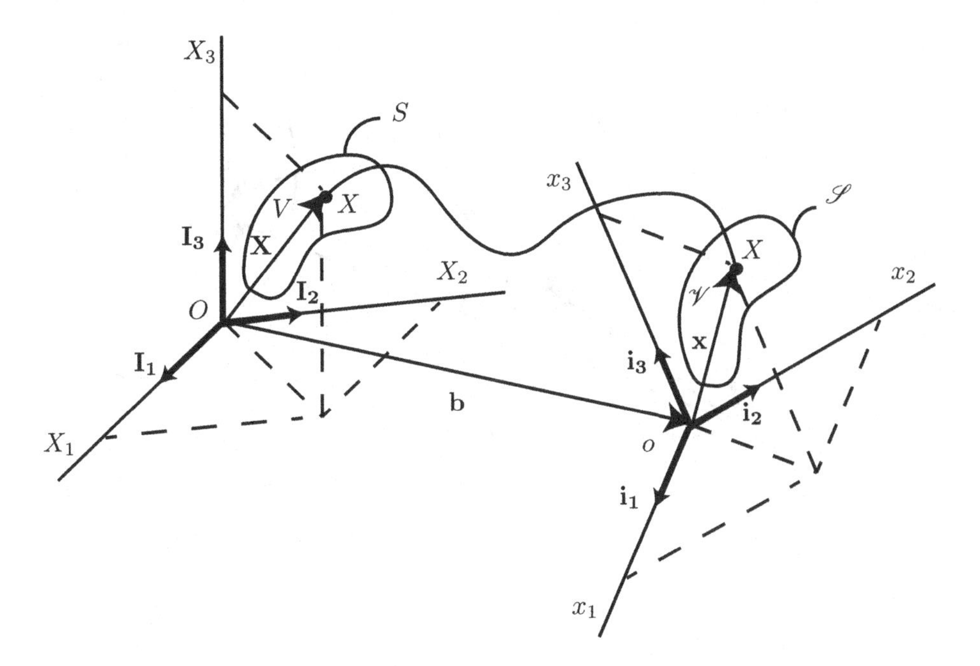

Figure 3.2: Two reference frames.

for every point in the reference configuration. Conversely, we write

$$\mathbf{X} = \chi^{-1}(\mathbf{x}) \qquad \text{or} \qquad X_K = \chi_K^{-1}(x_k). \tag{3.7}$$

We assume that the mappings are single valued and possess continuous derivatives with respect to their arguments of whatever order is desired, except possibly at some singular points, curves, and surfaces. Furthermore, we assume that the above are unique inverses of each other in a neighborhood of any material point. This assumption is known as the *axiom of continuity.* It expresses the fact that matter is *indestructible*, i.e., no region of a finite volume of matter can be deformed into one of zero volume. Furthermore, it implies that matter is *impenetrable*, i.e., the motion carries every region into a region, every surface into a surface, and every curve into a curve. This assumption is embodied into the requirement that

$$J \equiv \det\left[\frac{\partial x_k}{\partial X_K}\right] \neq (0, \pm\infty) \tag{3.8}$$

is satisfied for all material points $X \in \mathcal{B}$, except possibly at some singular points, curves, and surfaces. Note that as long as the handedness of the coordinate system in the reference and deformed configurations are the same, then $J > 0$. Unless specifically noted otherwise, we will assume this to be the case. If $J = 1$, the deformation is said to be *isochoric.*

Components of a tensor field quantity of any order can be written using either the material or the spatial descriptions

$$\psi(\mathbf{X}) = \psi[\chi^{-1}(\mathbf{x})] = \widehat{\psi}(\mathbf{x}), \tag{3.9}$$

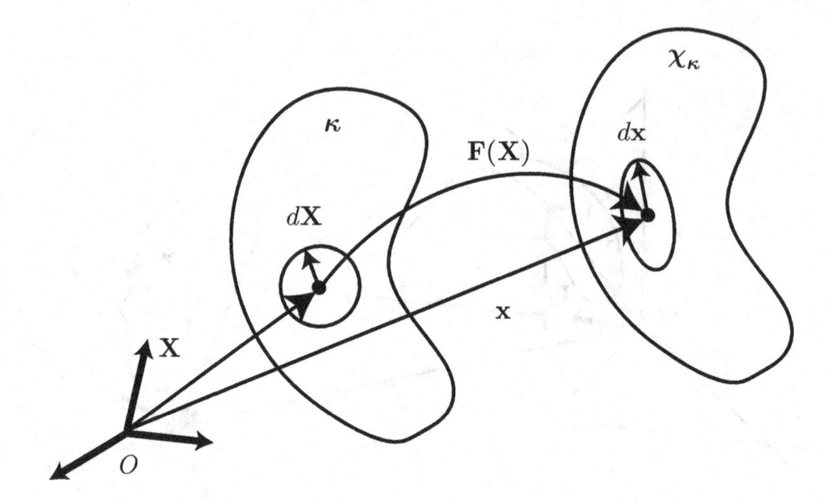

Figure 3.3: Mapping of neighborhood by the deformation gradient.

or

$$\psi_{\dots}(X_K) = \psi_{\dots}[\chi_K^{-1}(x_k)] = \widehat{\psi}_{\dots}(x_k). \tag{3.10}$$

For notational simplicity we drop the hat on the function since the functional dependence should be obvious in applications or will be explicitly displayed when necessary.

3.1 Deformation

3.1.1 Deformation gradient

The *material deformation gradient* at $\mathbf{X}$, as illustrated in Fig. 3.3, is a linear transformation defined by

$$\mathbf{F} = \mathbf{F}(\mathbf{X}) \equiv (\text{Grad } \mathbf{x})^T = (\nabla_{\mathbf{X}}\mathbf{x})^T \qquad \text{or} \qquad F_{kK} \equiv x_{k,K} = \frac{\partial x_k}{\partial X_K}, \tag{3.11}$$

where it is obvious that $J = \det[F_{kK}] \neq (0, \pm\infty)$. Above we have denoted the gradient with respect to the material coordinates by "Grad." Similarly, we will denote the divergence with respect to the material coordinates by "Div" and the curl by "Curl." The analogous gradient, divergence, and curl operators with respect to spatial coordinates will be denoted by "grad", "div", and "curl" respectively. Note that $\mathbf{F}$ depends on the reference configuration, but the $\boldsymbol{\kappa}$ subscript is dropped since this leads to no confusion at the moment. The deformation gradient is not a true tensor since it relates two points in different coordinate systems, the current to the reference. Subsequently, it is often referred to as a *two-point tensor* or a *double vector.*

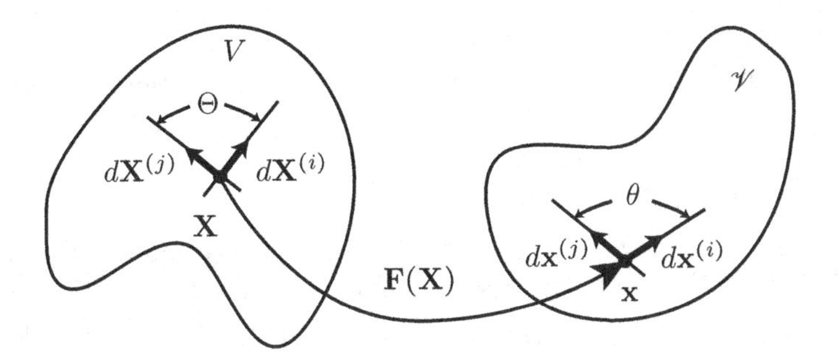

Figure 3.4: Transformation of a vector element.

3.1.2 Transformation of linear elements

It follows from the definition of the deformation gradient that

$$d\mathbf{x} = \mathbf{F} \cdot d\mathbf{X} \qquad \text{or} \qquad dx_k = F_{kK}(\mathbf{X})\; dX_K. \tag{3.12}$$

This equation represents the transformation for infinitesimal linear elements of material under the deformation $\mathbf{x} = \boldsymbol{\chi}(\mathbf{X})$ illustrated in Fig. 3.4. Clearly the inverse transformation is given by

$$d\mathbf{X} = \mathbf{F}^{-1} \cdot d\mathbf{x} \qquad \text{or} \qquad dX_K = F^{-1}_{Kk}(\mathbf{x})\; dx_k. \tag{3.13}$$

In the deformed configuration, the length of a differential vector element is

$$(dx_k dx_k)^{1/2} = (F_{kK} F_{kL} dX_K dX_L)^{1/2}. \tag{3.14}$$

Now introducing the direction unit vectors

$$d\mathbf{x} = \mathbf{t}\, dx \qquad \text{or} \qquad dx_k = t_k dx \tag{3.15}$$

and

$$d\mathbf{X} = \mathbf{T}\, dX \qquad \text{or} \qquad dX_K = T_K dX, \tag{3.16}$$

we have

$$(t_k dx\; t_k dx)^{1/2} = (F_{kK} F_{kL} T_K dX\, T_L dX)^{1/2}, \tag{3.17}$$

or, noting that $t_k t_k = 1$,

$$dx = (C_{KL} T_K T_L)^{1/2}\, dX, \tag{3.18}$$

which relates the length of a differential element before and after deformation. Above we have defined the symmetric second-order tensor

$$\mathbf{C} \equiv \mathbf{F}^T \cdot \mathbf{F} \qquad \text{or} \qquad C_{KL} \equiv F_{kK} F_{kL}, \tag{3.19}$$

which is called the *right Cauchy–Green strain tensor*. The quantity

$$\lambda \equiv \frac{dx}{dX} = (C_{KL} T_K T_L)^{1/2} = (\mathbf{C} : \mathbf{TT})^{1/2} \tag{3.20}$$

is called the *length stretch ratio.*

To examine the change in orientation of the differential line element, we note that

$$t_k = \frac{dx_k}{dx} = \frac{F_{kK}T_K dX}{(C_{LM}T_L T_M)^{1/2}\,dX} = \frac{F_{kK}T_K}{\lambda} \qquad \text{or} \qquad \mathbf{t} = \frac{1}{\lambda}\mathbf{F}\cdot\mathbf{T}. \tag{3.21}$$

Example

We recall that $dx_i = F_{iI}dX_I$ where $F_{iI} = \partial x_i/\partial X_I$, and the deformation function is given by $x_i = \chi_i(X_I)$. Let us assume that we have the deformation

$$\begin{aligned} x_1 &= X_1 - AX_1, \\ x_2 &= X_2 - AX_2, \\ x_3 &= X_3 + BX_3, \end{aligned}$$

where A and B are positive constants. The deformation gradient is then given by

$$F_{iI} = \begin{bmatrix} 1-A & 0 & 0 \\ 0 & 1-A & 0 \\ 0 & 0 & 1+B \end{bmatrix}.$$

In this example F_{iI} is independent of X_K since the deformation function is linear. When this is so, it is called a *homogeneous deformation.* The right Cauchy–Green strain tensor is then given by

$$C_{KL} = F_{iK}F_{iL},$$

so

$$C_{KL} = \begin{bmatrix} (1-A)^2 & 0 & 0 \\ 0 & (1-A)^2 & 0 \\ 0 & 0 & (1+B)^2 \end{bmatrix}.$$

Now let us examine how a differential material element with direction cosine $T_K^{(1)} = (0,0,1)$ in the reference configuration stretches when deformed. Since

$$dx^{(1)} = (C_{KL}T_K T_L)^{1/2} dX^{(1)},$$

then $\lambda^{(1)} = (1+B)$. If we look in the direction $T_K^{(2)} = (0,1,0)$, then $\lambda^{(2)} = (1-A)$. Let us examine the stretch and change in orientation for a material differential element with direction cosine $T_K^{(3)} = (1/\sqrt{2}, 1/\sqrt{2}, 0)$. The material is stretched by

$$\lambda^{(3)} = \left(C_{KL}T_K^{(3)}T_L^{(3)}\right)^{1/2} = (1-A),$$

and the new orientation is given by

$$t_k^{(3)} = \frac{F_{kK}T_K^{(3)}}{\lambda^{(3)}} = \frac{[(1-A)/\sqrt{2}, (1-A)/\sqrt{2}, 0]}{(1-A)} = \left(\frac{1}{\sqrt{2}}, \frac{1}{\sqrt{2}}, 0\right);$$

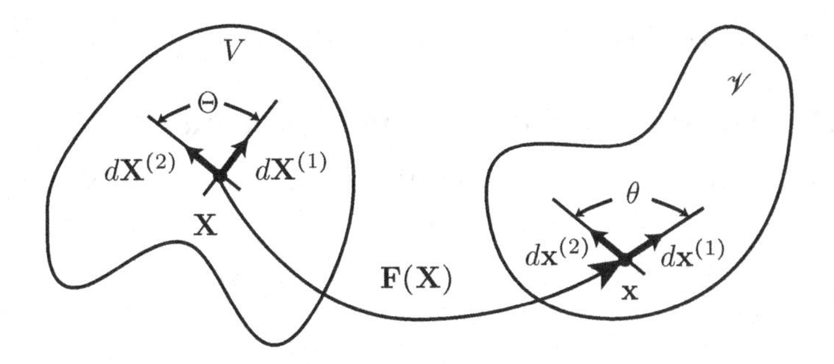

Figure 3.5: Reference and current configurations of elementary arcs on intersecting material curves.

thus we see that there is no change in orientation for a vector in that direction. Lastly, we look at a material line element in the direction $T_K^{(4)} = (1/\sqrt{3}, 1/\sqrt{3}, 1/\sqrt{3})$, then

$$\lambda^{(4)} = \left(C_{KL}T_K^{(4)}T_L^{(4)}\right)^{1/2} = \left(\frac{C_{11}+C_{22}+C_{33}}{3}\right)^{1/2} = \left[\frac{2}{3}(1-A)^2 + \frac{1}{3}(1+B)^2\right]^{1/2}$$

and

$$t_k^{(4)} = \frac{F_{kK}T_K^{(4)}}{\lambda^{(4)}} = \frac{1}{\lambda^{(4)}}\left(\frac{1-A}{\sqrt{3}}, \frac{1-A}{\sqrt{3}}, \frac{1+B}{\sqrt{3}}\right).$$

Now we take two oriented differential line elements at two different orientations, as illustrated in Fig. 3.5, and examine the change in angle between the two line elements before and after deformation. First we take

$$dx_i^{(1)} = t_i^{(1)} dx^{(1)}, \tag{3.22}$$

$$dx_j^{(2)} = t_j^{(2)} dx^{(2)}, \tag{3.23}$$

$$dX_I^{(1)} = T_I^{(1)} dX^{(1)}, \tag{3.24}$$

$$dX_J^{(2)} = T_J^{(2)} dX^{(2)}, \tag{3.25}$$

and since

$$t_i^{(\alpha)} = \frac{F_{iI}T_I^{(\alpha)}}{\lambda^{(\alpha)}}, \tag{3.26}$$

we then have

$$\cos\theta_{(ii)} = t_i^{(1)} t_i^{(2)} = \frac{C_{IJ}}{\lambda^{(1)}\lambda^{(2)}} \cos\Theta_{(IJ)}. \tag{3.27}$$

Note that $\cos\Theta_{(IJ)} = T_I^{(1)} T_J^{(2)}$.

At this point, we emphasize that F_{kK} contains all the information about any deformation. Additionally, we note that the integral of any arbitrary tensor field $G(\mathbf{x})$ along an arbitrary curve is easily obtained from

$$\int_C G(x_i)\, dx_k = \int_C G(X_I)\, F_{kK}(X_J)\, dX_K, \tag{3.28}$$

since in the second integral C is independent of the deformation.

3.1.3 Transformation of a surface element

Oriented differential area elements in the reference and deformed coordinate frames, as illustrated in Fig. 3.6, are given by

$$d\mathbf{S} = d\mathbf{X}^{(1)} \times d\mathbf{X}^{(2)} \qquad \text{and} \qquad d\mathbf{s} = d\mathbf{x}^{(1)} \times d\mathbf{x}^{(2)}, \tag{3.29}$$

or

$$dS_I = \epsilon_{IJK} dX_J^{(1)} dX_K^{(2)} \qquad \text{and} \qquad ds_i = \epsilon_{ijk} dx_j^{(1)} dx_k^{(2)}. \tag{3.30}$$

Since $dx_j = F_{jJ} dX_J = x_{j,J} dX_J$, we have

$$ds_i = \epsilon_{ijk} F_{jJ} F_{kK} dX_J^{(1)} dX_K^{(2)} = \epsilon_{ijk} x_{j,J} x_{k,K} dX_J^{(1)} dX_K^{(2)}. \tag{3.31}$$

Now recalling that $J = \det[F_{jJ}]$, using (2.49), we can write

$$\epsilon_{LJK} J = \epsilon_{ljk} x_{l,L} x_{j,J} x_{k,K} \tag{3.32}$$

or, since $\epsilon_{LJK}\epsilon_{LJK} = 6$,

$$J = \frac{1}{6} \epsilon_{LJK} \epsilon_{ljk} x_{l,L} x_{j,J} x_{k,K}, \tag{3.33}$$

and

$$\epsilon_{LJK} X_{L,i} J = \epsilon_{ljk} \left(X_{L,i} x_{l,L} \right) x_{j,J} x_{k,K} = \epsilon_{ljk} \delta_{li} x_{j,J} x_{k,K} = \epsilon_{ijk} x_{j,J} x_{k,K}. \tag{3.34}$$

Thus

$$ds_i = J X_{L,i} \epsilon_{LJK} dX_J^{(1)} dX_K^{(2)}, \tag{3.35}$$

or finally

$$d\mathbf{s} = J \left(\mathbf{F}^{-1} \right)^T \cdot d\mathbf{S} \qquad \text{or} \qquad ds_i = J X_{L,i} dS_L, \tag{3.36}$$

and since we can also write

$$x_{i,L} ds_i = J dS_L, \tag{3.37}$$

we additionally have

$$d\mathbf{S} = J^{-1} \mathbf{F}^T \cdot d\mathbf{s} \qquad \text{or} \qquad dS_L = J^{-1} x_{i,L} ds_i. \tag{3.38}$$

From above we also see that $(\mathbf{F}^{-1})^T = (\mathbf{F}^T)^{-1}$ (also written quite often as $\mathbf{F}^{-T}$), and we have used the fact that $J^{-1} = 1/J$ since

$$\delta_{ik} = \frac{\partial x_i}{\partial X_J} \frac{\partial X_J}{\partial x_k}, \tag{3.39}$$

and taking the determinant of both sides,

$$1 = \det\left[\frac{\partial x_i}{\partial X_J} \frac{\partial X_J}{\partial x_k} \right] = \det\left[\frac{\partial x_i}{\partial X_J} \right] \det\left[\frac{\partial X_J}{\partial x_k} \right] = J J^{-1}. \tag{3.40}$$

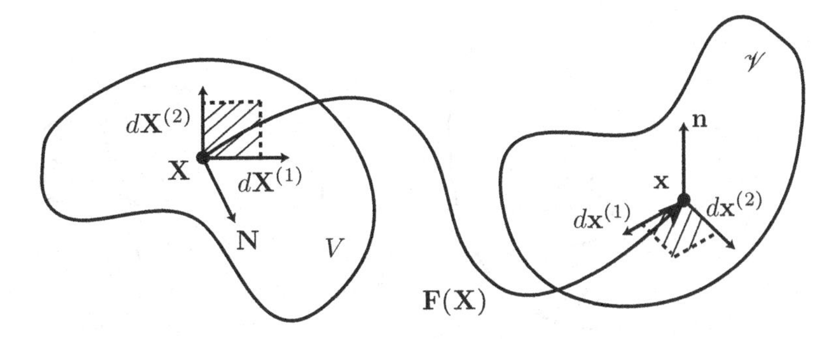

Figure 3.6: Reference and current configurations of an element of a material surface.

Lastly, since

$$d\mathbf{s} = \mathbf{n}\, ds \qquad \text{or} \qquad ds_i = n_i\, ds \tag{3.41}$$

and

$$d\mathbf{S} = \mathbf{N}\, dS \qquad \text{or} \qquad dS_I = N_I\, dS, \tag{3.42}$$

and using the fact that $n_i n_i = 1$, we can also write

$$ds = J\left(X_{J,l} X_{K,l} N_J N_K\right)^{1/2} dS. \tag{3.43}$$

The quantity

$$\eta = \frac{ds}{dS} = J\left(X_{J,l} X_{K,l} N_J N_K\right)^{1/2} = J\left(\mathbf{C}^{-1} : \mathbf{NN}\right)^{1/2} \tag{3.44}$$

is called the *area stretch ratio*, and using (3.36), (3.41), (3.42), and (3.44), it is easy to see that the area normals are related by

$$\mathbf{n} = \frac{J}{\eta}\left(\mathbf{F}^{-1}\right)^T \cdot \mathbf{N}. \tag{3.45}$$

As we can see from (3.21) and (3.45), the linear and area stretches are related for $\mathbf{t} = \mathbf{n}$ and $\mathbf{T} = \mathbf{N}$ by

$$\lambda \eta = J, \tag{3.46}$$

or, equivalently,

$$dx\, ds = dv, \tag{3.47}$$

since $dX\, dS = dV$. For isochoric deformations, we see that $\lambda \eta = 1$.

We also see that the surface integral of an arbitrary tensor field $G(\mathbf{x})$ can be written as

$$\int_{\mathscr{S}} G(x_i)\; ds = \int_S G(X_I)\, J(X_L)\left(X_{J,l} X_{K,l} N_J N_K\right)^{1/2} dS, \tag{3.48}$$

where the second integral is easier to compute since S is independent of the deformation.

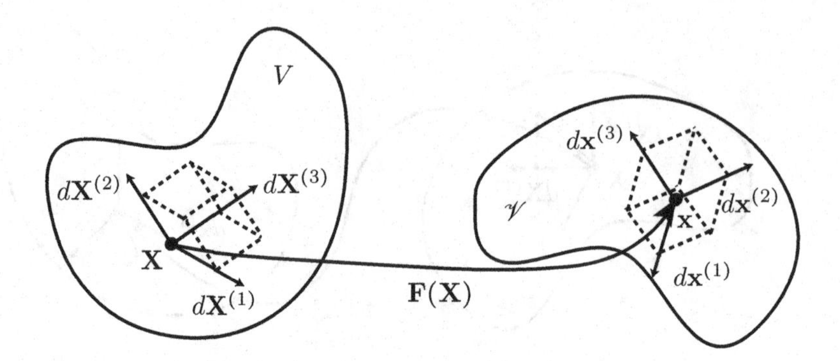

Figure 3.7: Reference and current configurations of an element of a material volume.

3.1.4 Transformation of a volume element

By definition, the differential volume elements in the reference and deformed coordinate frames illustrated in Fig. 3.7 are respectively given by

$$dV = \left|\left[d\mathbf{X}^{(1)}, d\mathbf{X}^{(2)}, d\mathbf{X}^{(3)}\right]\right| \tag{3.49}$$

and

$$dv = \left|\left[d\mathbf{x}^{(1)}, d\mathbf{x}^{(2)}, d\mathbf{x}^{(3)}\right]\right|. \tag{3.50}$$

Now using our previous results, we have

$$\begin{aligned} dv &= \left|\epsilon_{ijk} dx_i^{(1)} dx_j^{(2)} dx_k^{(3)}\right|, \\ &= \left|\epsilon_{ijk} x_{i,I}^{(1)} x_{j,J}^{(2)} x_{k,K}^{(3)} dX_I^{(1)} dX_J^{(2)} dX_K^{(3)}\right|, \\ &= \left|\epsilon_{IJK} J dX_I^{(1)} dX_J^{(2)} dX_K^{(3)}\right|, \end{aligned}$$

where we used (3.32) in the last step, or since $J > 0$, we finally have

$$dv = J\, dV. \tag{3.51}$$

We see that the quantity

$$J = \frac{dv}{dV} \tag{3.52}$$

also represents the *volume stretch ratio*. For this reason, it is sometimes convenient to perform a multiplicative decomposition of the deformation gradient tensor,

$$\mathbf{F} = J^{1/3}\,\overline{\mathbf{F}} \qquad \text{or} \qquad F_{ij} = J^{1/3}\,\overline{F}_{ij}, \tag{3.53}$$

into the dilatational part, $J^{1/3}\,\mathbf{1}$, and the isochoric part, $\overline{\mathbf{F}}$, since $\det \overline{\mathbf{F}} = \mathbf{1}$.

Lastly, we note that a volume integral of a tensor field $G(\mathbf{x})$ can now be written as

$$\int_{\mathcal{V}} G(\mathbf{x})\, dv = \int_V G(\mathbf{X})\, J(\mathbf{X})\, dV, \tag{3.54}$$

where V is independent of the deformation.

3.1.5 Relations between deformation and inverse deformation gradients

If we have the inverse deformation function $X_{I,i} = X_{I,i}(x_{j,J})$, we would like the ability to express it and its gradient as explicit functions of $x_{j,J}$ or of $X_{K,k}$. Similarly, if we are given the deformation function $x_{i,I} = x_{i,I}(X_{J,j})$, we would like the ability to express it and its gradient as explicit functions of $X_{J,j}$ or of $x_{k,K}$. To do this, we start with our previous result (3.34):

$$\epsilon_{LJK} J X_{L,i} = \epsilon_{ijk} x_{j,J} x_{k,K}. \tag{3.55}$$

Multiplying both sides by ϵ_{IJK} and noting that $\epsilon_{IJK}\epsilon_{LJK} = 2\delta_{IL}$, we have

$$2\delta_{IL} J X_{L,i} = \epsilon_{ijk}\epsilon_{IJK} x_{j,J} x_{k,K}, \tag{3.56}$$

or

$$X_{I,i} = \frac{1}{2} J^{-1} \epsilon_{ijk}\epsilon_{IJK} x_{j,J} x_{k,K}. \tag{3.57}$$

In a similar fashion we can show that

$$x_{i,I} = \frac{1}{2} J \epsilon_{ijk}\epsilon_{IJK} X_{J,j} X_{K,k}. \tag{3.58}$$

Also from (3.33), we have

$$J = \frac{1}{6} \epsilon_{ijk}\epsilon_{IJK} x_{i,I} x_{j,J} x_{k,K}. \tag{3.59}$$

Now differentiating with respect to $x_{l,L}$, we have

$$\begin{aligned} 6\frac{\partial J}{\partial x_{l,L}} &= \epsilon_{ljk}\epsilon_{LJK} x_{j,J} x_{k,K} + \epsilon_{ilk}\epsilon_{ILK} x_{i,I} x_{k,K} + \epsilon_{ijl}\epsilon_{IJL} x_{i,I} x_{j,J}, \\ &= 3\epsilon_{ljk}\epsilon_{LJK} x_{j,J} x_{k,K}, \end{aligned}$$

or, using (3.56),

$$\frac{\partial J}{\partial x_{l,L}} = \frac{1}{2}\epsilon_{ljk}\epsilon_{LJK} x_{j,J} x_{k,K} = J X_{L,l}. \tag{3.60}$$

Similarly, it can be shown that

$$\frac{\partial J^{-1}}{\partial X_{L,l}} = \frac{1}{2}\epsilon_{LJK}\epsilon_{ljk} X_{J,j} X_{K,k} = J^{-1} x_{l,L}. \tag{3.61}$$

Lastly, since

$$x_{i,J} X_{J,k} = \delta_{ik}, \tag{3.62}$$

differentiating both sides with respect to $x_{l,L}$,

$$\delta_{il}\delta_{JL} X_{J,k} + x_{i,J}\frac{\partial X_{J,k}}{\partial x_{l,L}} = 0. \tag{3.63}$$

Now multiplying both sides by $X_{K,i}$,

$$\frac{\partial X_{K,k}}{\partial x_{l,L}} = -X_{K,l} X_{L,k}. \tag{3.64}$$

Similarly, it can be shown that

$$\frac{\partial x_{l,L}}{\partial X_{K,k}} = -x_{k,L} x_{l,K}. \tag{3.65}$$

3.1.6 Identities of Euler–Piola–Jacobi

We want to show that

$$\left(JX_{I,i}\right)_{,I} = 0 \tag{3.66}$$

and

$$\left(J^{-1}x_{i,I}\right)_{,i} = 0, \tag{3.67}$$

which are known as Euler–Piola–Jacobi identities.

Now

$$\begin{aligned}
\left(JX_{I,i}\right)_{,I} &= \frac{\partial J}{\partial X_I}X_{I,i} + J\frac{\partial X_{I,i}}{\partial X_I}, \\
&= \frac{\partial J}{\partial x_{j,J}}\frac{\partial x_{j,J}}{\partial X_I}X_{I,i} + J\frac{\partial X_{I,i}}{\partial x_{j,J}}\frac{\partial x_{j,J}}{\partial X_I}, \\
&= \frac{\partial J}{\partial x_{j,J}}\frac{\partial^2 x_j}{\partial X_I \partial X_J}X_{I,i} + J\frac{\partial X_{I,i}}{\partial x_{j,J}}\frac{\partial^2 x_j}{\partial X_I \partial X_J},
\end{aligned} \tag{3.68}$$

but from (3.60)

$$\frac{\partial J}{\partial x_{j,J}} = JX_{J,j} \tag{3.69}$$

and from (3.64)

$$\frac{\partial X_{I,i}}{\partial x_{j,J}} = -X_{I,j}X_{J,i}, \tag{3.70}$$

so

$$\left(JX_{I,i}\right)_{,I} = J\left(X_{J,j}X_{I,i} - X_{I,j}X_{J,i}\right)x_{j,IJ}, \tag{3.71}$$

or

$$\left(JX_{I,i}\right)_{,I} = JS_{IJ,ij}D_{j,IJ}, \tag{3.72}$$

where we have defined

$$S_{IJ,ij} = X_{J,j}X_{I,i} - X_{I,j}X_{J,i} = -S_{JI,ij} \tag{3.73}$$

and

$$D_{j,IJ} = x_{j,IJ} = D_{j,JI}. \tag{3.74}$$

Now since

$$S_{IJ,ij}D_{j,IJ} = -S_{JI,ij}D_{j,IJ} = -S_{JI,ij}D_{j,JI} = -S_{IJ,ij}D_{j,IJ} = 0, \tag{3.75}$$

then

$$\left(JX_{I,i}\right)_{,I} = 0. \tag{3.76}$$

In a similar fashion, it can be shown that

$$\left(J^{-1}x_{i,I}\right)_{,i} = 0. \tag{3.77}$$

3.1.7 Cayley–Hamilton theorem

A scalar λ is called an *eigenvalue* of the matrix A of size $n \times n$ if there exists a nonzero vector $\mathbf{v}$ such that it satisfies the following *eigenvalue problem*:

$$(\mathbf{A} - \lambda \mathbf{1}) \cdot \mathbf{v} = \mathbf{0} \qquad \text{or} \qquad (A - \lambda I)\, \mathbf{v} = \mathbf{0} \qquad \text{or} \qquad (a_{ik} - \lambda \delta_{ik})\, v_k = 0. \quad (3.78)$$

It follows that λ is an eigenvalue if and only if

$$\det (A - \lambda I) = 0 \qquad \text{or} \qquad \det (a_{ik} - \lambda \delta_{ik}) = 0, \tag{3.79}$$

or more explicitly, if λ is a root of the following *characteristic polynomial* equation:

$$f(\lambda) = (-\lambda)^n + A_{(1)} (-\lambda)^{n-1} + \cdots + A_{(n-1)} (-\lambda) + A_{(n)} = 0. \tag{3.80}$$

The coefficients $A_{(1)}, \ldots, A_{(n)}$ are scalar functions of A, called the *principal invariants* of A, and are given by (see (2.91)–(2.95))

$$\begin{aligned}
A_{(1)} &= \frac{1}{1!} \delta_{i_1 j_1} a_{i_1 j_1} = \mathrm{tr}\ A, && (3.81)\\
A_{(2)} &= \frac{1}{2!} \delta_{i_1 j_1 i_2 j_2} a_{i_1 j_1} a_{i_2 j_2} = \frac{1}{2} \left(A_{(1)} \mathrm{tr}\ A - \mathrm{tr}\ A^2 \right), && (3.82)\\
A_{(3)} &= \frac{1}{3!} \delta_{i_1 j_1 i_2 j_2 i_3 j_3} a_{i_1 j_1} a_{i_2 j_2} a_{i_3 j_3} = \frac{1}{3} \left(A_{(2)} \mathrm{tr}\ A - A_{(1)} \mathrm{tr}\ A^2 + \mathrm{tr}\ A^3 \right), && (3.83)\\
&\ \vdots \\
A_{(k)} &= \frac{1}{k!} \delta_{i_1 j_1 \cdots i_k j_k} a_{i_1 j_1} \cdots a_{i_k j_k}, \\
&= \frac{1}{k} \left(A_{(k-1)} \mathrm{tr}\ A - \cdots + (-1)^{k-1} \mathrm{tr}\ A^k \right), \qquad 1 < k \le n, && (3.84)\\
&\ \vdots \\
A_{(n)} &= \frac{1}{n!} \delta_{i_1 j_1 \cdots i_n j_n} a_{i_1 j_1} \cdots a_{i_n j_n}, \\
&= \frac{1}{n} \left(A_{(n-1)} \mathrm{tr}\ A - A_{(n-2)} \mathrm{tr}\ A^2 + A_{(n-3)} \mathrm{tr}\ A^3 - \cdots + (-1)^{n-1} \mathrm{tr}\ A^n \right), \\
&= \det A. && (3.85)
\end{aligned}$$

To determine the gradients of the principal invariants with respect to A, it is useful to rewrite the above relation

$$\det (A - \lambda I) = \sum_{k=0}^{n} (-\lambda)^{n-k} A_{(k)}, \tag{3.86}$$

where $A_{(0)} = 1$, and it can be shown that

$$\frac{\partial A_{(n)}}{\partial A} = \frac{\partial \det A}{\partial A} = \det A \left(A^{-1} \right)^T. \tag{3.87}$$

Using this result in conjunction with (3.86), the following recursion can be obtained:

$$\frac{\partial A_{(k+1)}}{\partial A} = A_{(k)} I - A^T \frac{\partial A_{(k)}}{\partial A}, \qquad k = 0, 1, \ldots, n, \tag{3.88}$$

where $A_{(n+1)} = 0$. By induction, the above recursion can also be written in the form

$$\frac{\partial A_{(k)}}{\partial A} = \left[\sum_{j=0}^{k-1} (-1)^j \, A_{(k-j-1)} A^j \right]^T . \tag{3.89}$$

The *Cayley–Hamilton theorem* states that the matrix A satisfies its own characteristic polynomial equation, i.e.,

$$f(A) = (-A)^n + A_{(1)} (-A)^{n-1} + \cdots + A_{(n-1)} (-A) + A_{(n)} I = 0. \tag{3.90}$$

The proof is straightforward. First rewrite the eigenvalue problem as

$$A\mathbf{v} = \lambda \mathbf{v}.$$

We note that for any $r = 1, \ldots, n$

$$A^r \mathbf{v} = A^{r-1}(A\mathbf{v}) = A^{r-1}\lambda\mathbf{v} = \lambda A^{r-1}\mathbf{v} = \cdots = \lambda^r \mathbf{v}.$$

Subsequently, since the actions of λ^r and A^r on the nonzero vector $\mathbf{v}$ are the same, replacing λ^r by A^r in the characteristic polynomial equation (3.80) establishes the theorem.

The case with $n = 3$ is most relevant to our discussions. In this case, we write

$$-\lambda^3 + A_{(1)}\lambda^2 - A_{(2)}\lambda + A_{(3)} = 0, \tag{3.91}$$

where

$$\begin{aligned}
A_{(1)} &= a_{ii} = \operatorname{tr} A, \\
&= \lambda^{(1)} + \lambda^{(2)} + \lambda^{(3)}, && (3.92) \\
A_{(2)} &= \frac{1}{2}\left(a_{ii}a_{kk} - a_{ik}a_{ki}\right) = \frac{1}{2}\left[(\operatorname{tr} A)^2 - \operatorname{tr} A^2\right], \\
&= \lambda^{(1)}\lambda^{(2)} + \lambda^{(2)}\lambda^{(3)} + \lambda^{(3)}\lambda^{(1)}, && (3.93) \\
A_{(3)} &= \frac{1}{6}\epsilon_{ijk}\epsilon_{rst}a_{ir}a_{js}a_{kt} = \frac{1}{6}\left[(\operatorname{tr} A)^3 - 3\operatorname{tr} A \operatorname{tr} A^2 + 2\operatorname{tr} A^3\right] = \det A, \\
&= \lambda^{(1)}\lambda^{(2)}\lambda^{(3)}. && (3.94)
\end{aligned}$$

The eigenvalues $\lambda^{(k)}$ $(k = 1, 2, 3)$ are the zeros of the characteristic polynomial equation (3.91), and for each eigenvalue, there corresponds an associated *eigenvector* $\mathbf{v}^{(k)} = v_i^{(k)}\mathbf{i}_i$. Use of the Cayley–Hamilton theorem provides

$$-A^3 + A_{(1)}A^2 - A_{(2)}A + A_{(3)}I = 0 \tag{3.95}$$

and

$$-A^2 + A_{(1)}A - A_{(2)}I + A_{(3)}A^{-1} = 0. \tag{3.96}$$

It should be noted that by taking the trace of (3.95), we obtain

$$-\operatorname{tr} A^3 + A_{(1)}\operatorname{tr} A^2 - A_{(2)}\operatorname{tr} A + 3\,A_{(3)} = 0, \tag{3.97}$$

so that subsequently, using (3.92)–(3.94), we have

$$\operatorname{tr} A = A_{(1)}, \tag{3.98}$$

$$\operatorname{tr} A^2 = A_{(1)}^2 - 2\,A_{(2)}, \tag{3.99}$$

$$\operatorname{tr} A^3 = A_{(1)}^3 - 3\,A_{(1)}A_{(2)} + 3\,A_{(3)}. \tag{3.100}$$

In addition, by taking the trace of (3.96) and using (3.93), it is easy to show that we can alternatively write

$$A_{(2)} = \det A \, \operatorname{tr} A^{-1}. \tag{3.101}$$

Lastly, from (3.89), we also have that

$$\frac{\partial A_{(1)}}{\partial A} = I, \quad \frac{\partial A_{(2)}}{\partial A} = A_{(1)}I - A^T, \quad \frac{\partial A_{(3)}}{\partial A} = \left(A^2 - A_{(1)}A + A_{(2)}I\right)^T = A_{(3)}\left(A^{-1}\right)^T. \tag{3.102}$$

In addition, any matrix A can be decomposed into a spherical or mean part, $\frac{1}{3}A_{(1)}I$, and a deviatoric part, A':

$$A = \frac{1}{3}A_{(1)}I + A'. \tag{3.103}$$

Note that any tensor of the form $\alpha\,\mathbf{1}$, where α is a scalar, is known as a spherical tensor. Now it is easy to show that

$$\operatorname{tr} A' = 0, \tag{3.104}$$

$$\operatorname{tr} A'^2 = \frac{2}{3}\,A_{(1)}^2 - 2\,A_{(2)}, \tag{3.105}$$

$$\operatorname{tr} A'^3 = \frac{2}{9}\,A_{(1)}^3 - A_{(1)}A_{(2)} + 3\,A_{(3)}, \tag{3.106}$$

and the invariants of the deviatoric part A' are related[1] to those of A as follows:

$$A'_{(1)} = \operatorname{tr} A' = 0, \tag{3.107}$$

$$A'_{(2)} = -\frac{1}{2}\operatorname{tr} A'^2 = -\frac{1}{3}\,A_{(1)}^2 + A_{(2)}, \tag{3.108}$$

$$A'_{(3)} = \frac{1}{3}\operatorname{tr} A'^3 = \frac{2}{27}\,A_{(1)}^3 - \frac{1}{3}\,A_{(1)}A_{(2)} + A_{(3)}. \tag{3.109}$$

We note that the eigenvectors or principal directions of A' are the same as those of A, while, if we denote the eigenvalues or principal values of A' as $s^{(i)}$, then it is easy to show that

$$s^{(i)} = \lambda^{(i)} - \frac{1}{3}\left(\lambda^{(1)} + \lambda^{(2)} + \lambda^{(3)}\right). \tag{3.110}$$

Subsequently, we can also write the principal scalar invariants of A' as

$$A'_{(1)} = s^{(1)} + s^{(2)} + s^{(3)} = 0, \tag{3.111}$$

$$A'_{(2)} = \left(s^{(1)}s^{(2)} + s^{(2)}s^{(3)} + s^{(3)}s^{(1)}\right) = -\frac{1}{2}\left({s^{(1)}}^2 + {s^{(2)}}^2 + {s^{(3)}}^2\right), \tag{3.112}$$

$$A'_{(3)} = s^{(1)}s^{(2)}s^{(3)}. \tag{3.113}$$

[1] Some authors define $A'_{(2)}$ as the negative of our definition. Our convention is consistent with the definitions of the invariants of A given in (3.91)–(3.94).

We note that we can also write $A'_{(2)}$ in terms of the eigenvalues of A:

$$A'_{(2)} = -\frac{1}{6}\left[\left(\lambda^{(1)} - \lambda^{(2)}\right)^2 + \left(\lambda^{(2)} - \lambda^{(3)}\right)^2 + \left(\lambda^{(3)} - \lambda^{(1)}\right)^2\right]. \tag{3.114}$$

3.1.8 Real symmetric matrices

Let $A = [a_{ik}] = [a_{ki}]$ be a real symmetric matrix, and

$$\mathbf{A} = a_{ik}\mathbf{i}_i\mathbf{i}_k. \tag{3.115}$$

We would like to transform it to

$$\bar{\mathbf{A}} = \bar{a}_{pq}\bar{\mathbf{i}}_p\bar{\mathbf{i}}_q, \tag{3.116}$$

such that

$$[\bar{a}_{pq}] = [\bar{a}_{qp}] = \begin{bmatrix} \lambda^{(1)} & 0 & 0 \\ 0 & \lambda^{(2)} & 0 \\ 0 & 0 & \lambda^{(3)} \end{bmatrix}. \tag{3.117}$$

Noting that $d\bar{\xi}_i = (\partial\bar{\xi}_i/\partial\xi_k)d\xi_k$ and $ds^2 = d\bar{\xi}_i d\bar{\xi}_i = d\xi_k d\xi_k$, and since

$$\frac{\partial\bar{\xi}_i}{\partial\xi_k}\frac{\partial\bar{\xi}_i}{\partial\xi_l}d\xi_k d\xi_l = d\xi_i d\xi_i, \tag{3.118}$$

we have that

$$\frac{\partial\bar{\xi}_i}{\partial\xi_k}\frac{\partial\bar{\xi}_i}{\partial\xi_l} = \delta_{kl}. \tag{3.119}$$

If we let

$$R_{ik} \equiv \frac{\partial\bar{\xi}_i}{\partial\xi_k}, \tag{3.120}$$

and $R = [R_{ik}]$, then we can write

$$R^T R = I \qquad \text{or} \qquad R_{ik}R_{il} = \delta_{kl}. \tag{3.121}$$

This shows that the matrix R is orthogonal since $R^{-1} = R^T$. The matrix R is said to be a *proper orthogonal matrix* if $\det R = 1$ and an *improper orthogonal matrix* if $\det R = -1$. In the following we will always assume that R is a proper orthogonal matrix.

According to the transformation rule for components of a second-order tensor, we can now write that

$$\bar{A} = R\,A\,R^T \qquad \text{or} \qquad \bar{a}_{pq} = R_{pi}R_{qk}a_{ik}. \tag{3.122}$$

We now establish the following results for our matrix A:

1) If A is real and symmetric, then the eigenvalues $\lambda^{(k)}$ are real. To see this, we write

$$a_{ik}v_k = \lambda v_i, \tag{3.123}$$
$$a^*_{ik}v^*_k = \lambda^* v^*_i, \tag{3.124}$$

where the star superscripts denote complex conjugates. Now multiplying the first equation by v^*_i and the second equation by v_i, subtracting, and noting that a_{ik} is symmetric and real and $v_i v^*_i$ is real, we obtain

$$(\lambda - \lambda^*)\, v_i v^*_i = 0. \tag{3.125}$$

Since $v_i \neq 0$ (and hence $v^*_i \neq 0$), then $\lambda^* = \lambda$ and thus λ is real.

2) Eigenvectors corresponding to distinct eigenvalues are orthogonal. To see this, we write

$$a_{ik}v^{(\alpha)}_k = \lambda^{(\alpha)} v^{(\alpha)}_i, \tag{3.126}$$
$$a_{ik}v^{(\beta)}_k = \lambda^{(\beta)} v^{(\beta)}_i. \tag{3.127}$$

Now multiplying the first equation by $v^{(\beta)}_i$ and the second by $v^{(\alpha)}_i$ and proceeding as before, we obtain

$$\left(\lambda^{(\alpha)} - \lambda^{(\beta)}\right) v^{(\alpha)}_i v^{(\beta)}_i = 0. \tag{3.128}$$

If $\lambda^{(\alpha)} \neq \lambda^{(\beta)}$, then $v^{(\alpha)}_i$ and $v^{(\beta)}_i$ are orthogonal, i.e., $\mathbf{v}^{(\alpha)} \cdot \mathbf{v}^{(\beta)} = 0$.

Note that if the eigenvalues are not distinct, say $\lambda^{(1)} = \lambda^{(2)} = \lambda$, then

$$a_{ik}v^{(1)}_k = \lambda v^{(1)}_i, \tag{3.129}$$
$$a_{ik}v^{(2)}_k = \lambda v^{(2)}_i, \tag{3.130}$$

and in this case the linear combination $\alpha\mathbf{v}^{(1)} + \beta\mathbf{v}^{(2)}$ is also an eigenvector of λ for arbitrary α and β.

The matrix A is said to be *positive definite* if

$$a_{ik}v^{(\alpha)}_i v^{(\alpha)}_k > 0 \qquad \text{or} \qquad \mathbf{v}^{(\alpha)} \cdot \mathbf{A} \cdot \mathbf{v}^{(\alpha)} > 0. \tag{3.131}$$

Subsequently, since $\mathbf{A} \cdot \mathbf{v}^{(\alpha)} = A\mathbf{v}^{(\alpha)} = \lambda^{(\alpha)}\mathbf{v}^{(\alpha)}$, and since for any $\mathbf{v}^{(\alpha)} \neq \mathbf{0}$ we have that $\mathbf{v}^{(\alpha)} \cdot \mathbf{v}^{(\alpha)} > 0$, it then follows that A is positive definite if $\lambda^{(\alpha)} > 0$. It is *semi-definite* if and only if $\lambda^{(\alpha)} \geq 0$, in which case

$$a_{ik}v^{(\alpha)}_i v^{(\alpha)}_k \geq 0 \qquad \text{or} \qquad \mathbf{v}^{(\alpha)} \cdot \mathbf{A} \cdot \mathbf{v}^{(\alpha)} \geq 0. \tag{3.132}$$

The terms *negative definite* and *negative semi-definite* apply to tensors $\mathbf{A}$ whose eigenvalues are negative definite and negative semi-definite, respectively. It can be shown that if A is an arbitrary square matrix, then $AA^T = A^T A$ are positive semi-definite, and if A is invertible, then they are positive definite.

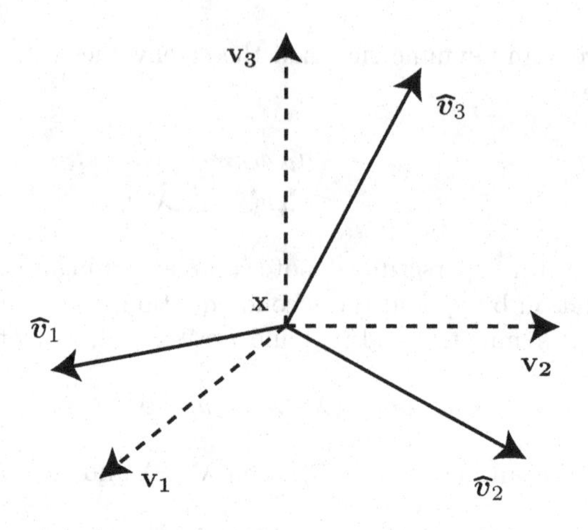

Figure 3.8: Rotation of principal axes.

If we now return to our transformation of matrix A, take $\widehat{v}_k^{(i)}$ to be the normalized eigenvectors of A, and define the orthogonal matrix as

$$R_{ik} \equiv \widehat{v}_k^{(i)}, \tag{3.133}$$

we then have

$$\begin{aligned}
\bar{a}_{pq} &= R_{pi}R_{qk}a_{ik} \\
&= R_{pi}\widehat{v}_k^{(q)}a_{ik} \\
&= R_{pi}\lambda^{(q)}\widehat{v}_i^{(q)} \\
&= \widehat{v}_i^{(p)}\lambda^{(q)}\widehat{v}_i^{(q)} \\
&= \lambda^{(q)}\widehat{v}_i^{(p)}\widehat{v}_i^{(q)} \\
&= \begin{bmatrix} \lambda^{(1)} & 0 & 0 \\ 0 & \lambda^{(2)} & 0 \\ 0 & 0 & \lambda^{(3)} \end{bmatrix}.
\end{aligned} \tag{3.134}$$

The normalized eigenvectors correspond to the *principal directions* in the new coordinate system as illustrated in Fig. 3.8, while the eigenvalues are denoted as *principal values.* Furthermore, in the new coordinate system, the invariants of $\bar{A}$, given by (3.92)–(3.94), can now be written as

$$\bar{A}_{(1)} = \lambda^{(1)} + \lambda^{(2)} + \lambda^{(3)}, \tag{3.135}$$

$$\bar{A}_{(2)} = \lambda^{(1)}\lambda^{(2)} + \lambda^{(2)}\lambda^{(3)} + \lambda^{(3)}\lambda^{(1)}, \tag{3.136}$$

$$\bar{A}_{(3)} = \lambda^{(1)}\lambda^{(2)}\lambda^{(3)}. \tag{3.137}$$

We note that a symmetric tensor obviously has six independent components or degrees of freedom. When referred to principal axes, it still has 6 degrees of freedom: 3 are the directions of the principal axes and the other 3 are the magnitudes of the principal components.

It then follows that any symmetric tensor $\mathbf{S}$ of rank 2 can be written in terms of its principal values and directions:

$$\mathbf{S} = \sum_{i=1}^{3} \lambda^{(i)} \, \widehat{\mathbf{v}}_i \widehat{\mathbf{v}}_i, \tag{3.138}$$

and subsequently we can write

$$\sqrt{\mathbf{S}} = \sum_{i=1}^{3} \sqrt{\lambda^{(i)}} \, \widehat{\mathbf{v}}_i \widehat{\mathbf{v}}_i, \tag{3.139}$$

$$\mathbf{S}^{-1} = \sum_{i=1}^{3} \lambda^{(i)^{-1}} \, \widehat{\mathbf{v}}_i \widehat{\mathbf{v}}_i, \tag{3.140}$$

$$e^{\mathbf{S}} = \sum_{i=1}^{3} e^{\lambda^{(i)}} \, \widehat{\mathbf{v}}_i \widehat{\mathbf{v}}_i, \tag{3.141}$$

$$\log \mathbf{S} = \sum_{i=1}^{3} \log \lambda^{(i)} \, \widehat{\mathbf{v}}_i \widehat{\mathbf{v}}_i. \tag{3.142}$$

3.1.9 Polar decomposition theorem

It is clear that any second-order tensor $\mathbf{F}$ can be decomposed into the sum of symmetric and skew-symmetric tensors:

$$\mathbf{F} = \text{sym } \mathbf{F} + \text{skw } \mathbf{F} = \frac{1}{2}\left(\mathbf{F} + \mathbf{F}^T\right) + \frac{1}{2}\left(\mathbf{F} - \mathbf{F}^T\right). \tag{3.143}$$

This is sometimes called the *Cartesian decomposition* of a tensor.

Another useful decomposition is given by the *polar decomposition theorem* which states that for any real non-singular second-order tensor $\mathbf{F}$, which we take here to be the material deformation gradient, there exist real symmetric positive-definite transformations U and V and a real orthogonal transformation R such that

$$\mathbf{F} = \mathbf{R} \cdot \mathbf{U} = \mathbf{V} \cdot \mathbf{R} \qquad \text{or} \qquad F = R\,U = VR. \tag{3.144}$$

Note that a positive-definite symmetric tensor represents a pure stretch deformation along three mutually orthogonal axes, i.e., in the directions of the eigenvectors, while an orthogonal tensor represents a rotation. Therefore, the above states that any local deformation is a combination of a pure stretch and a rotation: first stretch U and then rotate R, or first rotate R and then stretch V. The two decompositions of the deformation gradient are illustrated in Fig. 3.9. We call $\mathbf{R}$ the *rotation tensor*, while $\mathbf{U}$ and $\mathbf{V}$ are called the *right* and the *left stretch tensors*. From this decomposition, we see that no deformation corresponds to $\mathbf{U} = \mathbf{V} = \mathbf{1}$ in which case the most general deformation gradient that does not lead to a deformation is given by $\mathbf{F} = \mathbf{R}$, i.e., a pure rotation.

The above theorem is proved as follows. Clearly we have, assuming that R is a proper orthogonal transformation,

$$U^2 = F^T F, \qquad V^2 = FF^T, \qquad \det U = \det V = \det F. \tag{3.145}$$

Let the eigenvalues and eigenvectors of U be $\lambda^{(k)}$ and $\mathbf{e}_k$, respectively, so that

$$U\mathbf{e}_k = \lambda^{(k)}\mathbf{e}_k. \tag{3.146}$$

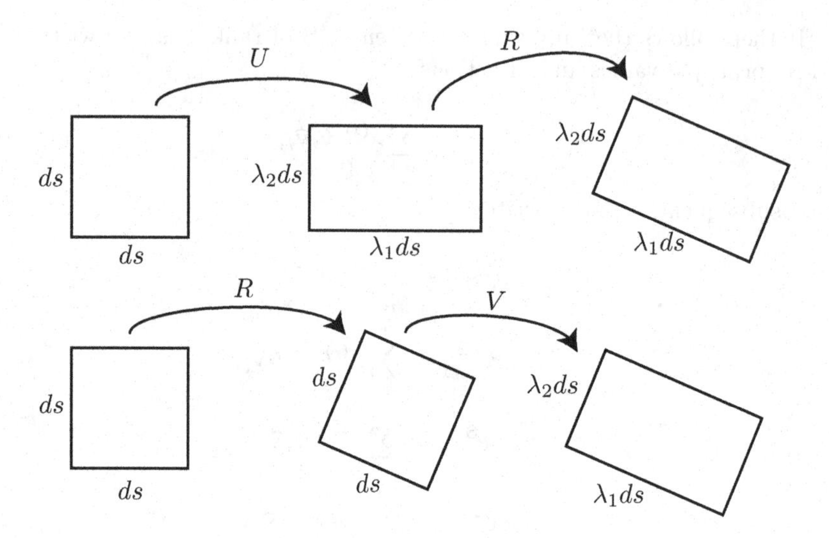

Figure 3.9: Polar decomposition of deformation gradient.

Then, since $V = RUR^T$, we have that

$$V(R\mathbf{e}_k) = RUR^T(R\mathbf{e}_k) = RU\mathbf{e}_k = \lambda^{(k)}(R\mathbf{e}_k). \tag{3.147}$$

In other words, V and U have the same eigenvalues and their eigenvectors differ only by the rotation R. The eigenvalues $\lambda^{(k)}$ are called the *principal stretches*, and the corresponding mutually orthogonal eigenvectors $\mathbf{e}_k$ are called the *principal directions of stretch*.

Given a non-singular deformation gradient $\mathbf{F}$, it is convenient to introduce the *right* and *left Cauchy–Green strain tensors* respectively defined by

$$C = U^2 = F^T F \qquad \text{and} \qquad B = V^2 = FF^T. \tag{3.148}$$

Now it is clear that C (analogously B) is symmetric since

$$C^T = \left(F^T F\right)^T = F^T F = C, \tag{3.149}$$

and positive definite since

$$C_{IK} = F_{lI} F_{lK} = \frac{\partial x_l}{\partial X_I}\frac{\partial x_l}{\partial X_K}, \tag{3.150}$$

and for any $v_I \neq 0$

$$C_{IK} v_I v_K = F_{lI} v_I F_{lK} v_K = W_l W_l > 0, \tag{3.151}$$

where $W_l = F_{lI} v_I$. Since C is a symmetric positive-definite matrix, then it has positive eigenvalues, and the corresponding eigenvectors form an orthogonal basis such that C can be written in the form

$$C = Q^T \Lambda^2 Q, \tag{3.152}$$

where

$$\Lambda^2 = \left[\left(\lambda^{(k)}\right)^2\right], \tag{3.153}$$

$$Q = [\mathbf{e}_k] = [e_{kj}]. \tag{3.154}$$

The eigenvalues of U are the positive square roots of those of C associated with the same eigenvectors. Subsequently we write

$$U \equiv C^{1/2} \tag{3.155}$$

and call U the "square root" of C. In other words,

$$U = C^{1/2} = Q^T \Lambda Q. \tag{3.156}$$

Lastly, we define

$$R \equiv FU^{-1}. \tag{3.157}$$

Clearly R is orthogonal, since

$$R^T = \left(FU^{-1}\right)^T = \left(U^{-1}\right)^T F^T = \left(U^T\right)^{-1} F^T = \left(FU^{-1}\right)^{-1} = R^{-1}, \tag{3.158}$$

where the equality in the next to the last step follows since

$$\left[\left(U^T\right)^{-1} F^T\right]\left(FU^{-1}\right) = U^{-1}CU^{-1} = U^{-1}UUU^{-1} = I, \tag{3.159}$$

and we have also used the fact that

$$\left(U^{-1}\right)^T = \left(U^T\right)^{-1} = U^{-1}, \tag{3.160}$$

which can be easily proved. Using a similar procedure, we can also demonstrate the decomposition

$$F = VR. \tag{3.161}$$

Thus, if $\det F \neq 0$, then $F = RU = VR$, where $R^{-1} = R^T$, and $U = U^T$, and $V = V^T$ are positive definite. We can also prove that the decompositions are unique, i.e., if $F = \overline{V}\,\overline{R}$, then $\overline{V} = V$, and $\overline{R} = R$.

The above decomposition is a consequence of a more general theorem which states that every non-singular complex matrix can be uniquely written as a product of a positive-definite Hermitian matrix A and a unitary matrix B. To clarify the above statement, we recall that a matrix A is Hermitian (or self-adjoint) if it is equal to its adjoint matrix $A^\dagger$, i.e., $A = A^\dagger$, where the adjoint matrix $A^\dagger$ is defined as the transpose of the complex conjugate of A, i.e., $A^\dagger = (A^\star)^T$, and $A^\star$ denotes the complex conjugate of A. Furthermore, a matrix B is normal if $B B^\dagger = B^\dagger B$ and is unitary if $B B^\dagger = B^\dagger B = I$. Clearly, if such matrices are real, then we obtain the above enunciated polar decomposition theorem. We also note that if the complex matrix is of unit dimension, we then obtain the polar representation of a complex number: $z = re^{i\theta}$. The number r is a positive real number, which is the magnitude of z, and the quantity $e^{i\theta}$ is a complex number of magnitude unity representing an angular rotation.

Example

Let C be given by

$$C = \begin{bmatrix} 3 & \sqrt{2} \\ \sqrt{2} & 2 \end{bmatrix},$$

which has eigenvalues $(\lambda^{(1)})^2 = 4$ and $(\lambda^{(2)})^2 = 1$ and corresponding eigenvectors $\mathbf{e}_1 = (\sqrt{2/3}, \sqrt{1/3})$ and $\mathbf{e}_2 = (-\sqrt{1/3}, \sqrt{2/3})$. Therefore, we have

$$\Lambda = \begin{bmatrix} 2 & 0 \\ 0 & 1 \end{bmatrix},$$

$$Q = \frac{1}{\sqrt{3}} \begin{bmatrix} \sqrt{2} & 1 \\ -1 & \sqrt{2} \end{bmatrix},$$

and subsequently

$$U = C^{1/2} = Q^T \Lambda Q = \frac{1}{3} \begin{bmatrix} 5 & \sqrt{2} \\ \sqrt{2} & 4 \end{bmatrix}.$$

One can easily verify that $U^2 = C$.

Example

Consider the deformation $\mathbf{x} = \boldsymbol{\chi}(\mathbf{X})$ given in Cartesian coordinates, in both the reference and the deformed configurations, by

$$x_1 = X_1 + \kappa X_2, \tag{3.162}$$

$$x_2 = X_2, \tag{3.163}$$

$$x_3 = X_3. \tag{3.164}$$

This deformation, illustrated in Fig. 3.10, is called a *simple shear* and $\kappa > 0$ is called the *amount of shear*. Now we have the deformation gradient

$$F = \begin{bmatrix} 1 & \kappa & 0 \\ 0 & 1 & 0 \\ 0 & 0 & 1 \end{bmatrix}, \tag{3.165}$$

and since $J = \det F = 1$, simple shear is a volume-preserving, or isochoric, deformation. The right Cauchy–Green tensor is given by

$$C = \begin{bmatrix} 1 & \kappa & 0 \\ \kappa & 1 + \kappa^2 & 0 \\ 0 & 0 & 1 \end{bmatrix}. \tag{3.166}$$

From the eigenvalues and eigenvectors of C, we find the principal stretches

$$\left(\lambda^{(1,2)}\right)^2 = 1 + \frac{1}{2}\kappa^2 \pm \kappa\sqrt{1 + \frac{1}{4}\kappa^2}, \quad \left(\lambda^{(3)}\right)^2 = 1, \tag{3.167}$$

with corresponding principal directions of stretches

$$\mathbf{e}_{1,2} = \left(\frac{1}{2}\kappa \pm \frac{1}{2}\sqrt{4+\kappa^2}\right)\mathbf{i}_1 + \mathbf{i}_2, \quad \mathbf{e}_3 = \mathbf{i}_3. \tag{3.168}$$

We note that $\left(\lambda^{(2)}\right)^2 = 1/\left(\lambda^{(1)}\right)^2$. From the square root of C, we obtain the right stretch tensor

$$U = \begin{bmatrix} \dfrac{2}{\sqrt{4+\kappa^2}} & \dfrac{\kappa}{\sqrt{4+\kappa^2}} & 0 \\ \dfrac{\kappa}{\sqrt{4+\kappa^2}} & \dfrac{2+\kappa^2}{\sqrt{4+\kappa^2}} & 0 \\ 0 & 0 & 1 \end{bmatrix}. \tag{3.169}$$

Note that in the principal direction $\mathbf{e}_1$, the principal stretch, $\lambda^{(1)} > 1$, is an extension, while in the direction $\mathbf{e}_2$, the stretch, $\lambda^{(2)} < 1$, is a contraction. Similarly, we have

$$B = \begin{bmatrix} 1+\kappa^2 & \kappa & 0 \\ \kappa & 1 & 0 \\ 0 & 0 & 1 \end{bmatrix} \quad \text{and} \quad V = \begin{bmatrix} \dfrac{2+\kappa^2}{\sqrt{4+\kappa^2}} & \dfrac{\kappa}{\sqrt{4+\kappa^2}} & 0 \\ \dfrac{\kappa}{\sqrt{4+\kappa^2}} & \dfrac{2}{\sqrt{4+\kappa^2}} & 0 \\ 0 & 0 & 1 \end{bmatrix}. \tag{3.170}$$

The rotation tensor can be calculated from $R = FU^{-1}$:

$$R = \begin{bmatrix} \dfrac{2}{\sqrt{4+\kappa^2}} & \dfrac{\kappa}{\sqrt{4+\kappa^2}} & 0 \\ \dfrac{-\kappa}{\sqrt{4+\kappa^2}} & \dfrac{2}{\sqrt{4+\kappa^2}} & 0 \\ 0 & 0 & 1 \end{bmatrix}. \tag{3.171}$$

If we denote $\theta = \tan^{-1}(\kappa/2)$, then R becomes

$$R = \begin{bmatrix} \cos\theta & \sin\theta & 0 \\ -\sin\theta & \cos\theta & 0 \\ 0 & 0 & 1 \end{bmatrix}, \tag{3.172}$$

which is a clockwise rotation about the x_3-axis by the angle θ.

3.1.10 Strain kinematics

From before we recall that from the deformation of a material line segment, we have

$$d\mathbf{x}^{(1)} \cdot d\mathbf{x}^{(2)} = (\mathbf{F} \cdot d\mathbf{X}^{(1)}) \cdot (\mathbf{F} \cdot d\mathbf{X}^{(2)}) = \left(\mathbf{F}^T \cdot \mathbf{F}\right) \cdot d\mathbf{X}^{(1)} \cdot d\mathbf{X}^{(2)} = \mathbf{C} \cdot d\mathbf{X}^{(1)} \cdot d\mathbf{X}^{(2)}. \tag{3.173}$$

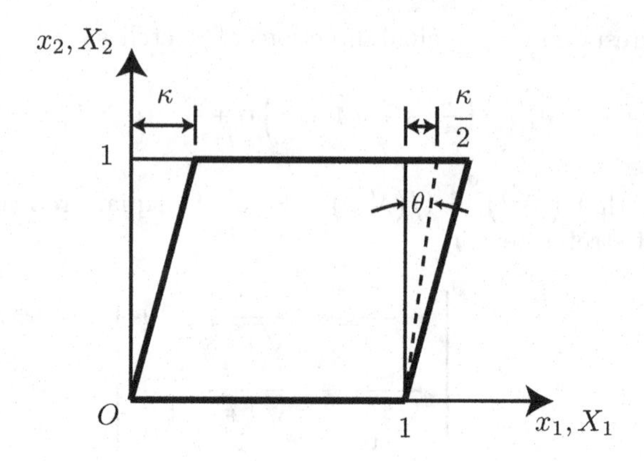

Figure 3.10: Simple shear.

Now the change in length and orientation between the current and reference configurations is

$$d\mathbf{x}^{(1)} \cdot d\mathbf{x}^{(2)} - d\mathbf{X}^{(1)} \cdot d\mathbf{X}^{(2)} = 2\mathbf{E} \cdot d\mathbf{X}^{(1)} \cdot d\mathbf{X}^{(2)}, \tag{3.174}$$

where

$$\mathbf{E} = \frac{1}{2}(\mathbf{C} - \mathbf{1}) \qquad \text{or} \qquad E = \frac{1}{2}(C - I) \tag{3.175}$$

is called the *Green–St. Venant strain tensor*, or the *finite strain tensor in the reference configuration*. Similarly, since $d\mathbf{X} = \mathbf{F}^{-1} \cdot d\mathbf{x}$, we also have

$$d\mathbf{X}^{(1)} \cdot d\mathbf{X}^{(2)} = (\mathbf{F}^{-1} \cdot d\mathbf{x}^{(1)}) \cdot (\mathbf{F}^{-1} \cdot d\mathbf{x}^{(2)}) = \left(\mathbf{F} \cdot \mathbf{F}^T\right)^{-1} \cdot d\mathbf{x}^{(1)} \cdot d\mathbf{x}^{(2)} = \mathbf{B}^{-1} \cdot d\mathbf{x}^{(1)} \cdot d\mathbf{x}^{(2)}, \tag{3.176}$$

and then

$$d\mathbf{x}^{(1)} \cdot d\mathbf{x}^{(2)} - d\mathbf{X}^{(1)} \cdot d\mathbf{X}^{(2)} = 2\,\mathbf{e} \cdot d\mathbf{x}^{(1)} \cdot d\mathbf{x}^{(2)}, \tag{3.177}$$

where

$$\mathbf{e} = \frac{1}{2}\left(\mathbf{1} - \mathbf{B}^{-1}\right) \qquad \text{or} \qquad e = \frac{1}{2}\left(I - B^{-1}\right) \tag{3.178}$$

is called the *Almansi–Hamel strain tensor*, or the *finite strain tensor in the current configuration.*

We note that the mapping from the reference configuration frame to the deformed configuration frame is given by the rigid frame transformation

$$\mathbf{x} = \mathbf{b} + \mathbf{Q} \cdot \mathbf{X}, \tag{3.179}$$

where $\mathbf{b}$ and $\mathbf{Q}$ are the constant frame translation vector and orthogonal rotation tensor, respectively. Note that $\mathbf{Q}^{-1} = \mathbf{Q}^T$. Subsequently, the deformation in the current configuration by *displacement* of a material point from $\mathbf{X}$ to $\mathbf{X} + \mathbf{u}(\mathbf{X})$ is given by

$$\mathbf{x} = \mathbf{b} + \mathbf{Q} \cdot [\mathbf{X} + \mathbf{u}(\mathbf{X})], \tag{3.180}$$

where $\mathbf{u}(\mathbf{X})$ is the displacement in the reference configuration:

$$\mathbf{u}(\mathbf{X}) = \mathbf{Q}^T \cdot (\mathbf{x} - \mathbf{b}) - \mathbf{X}. \tag{3.181}$$

The *material displacement gradient* at $\mathbf{X}$ is a linear transformation defined by

$$\mathbf{H} = \mathbf{H}(\mathbf{X}) \equiv (\text{Grad } \mathbf{u})^T \qquad \text{or} \qquad H_{IJ} = u_{I,J}. \tag{3.182}$$

It easily follows that

$$\mathbf{H} = \mathbf{Q}^T \cdot \mathbf{F} - \mathbf{1}. \tag{3.183}$$

Analogously, the deformation in the reference configuration by *displacement* of a material point in the current configuration from $\mathbf{x}$ to $\mathbf{x} - \mathbf{u}(\mathbf{x})$ is given by

$$[\mathbf{x} - \mathbf{u}(\mathbf{x})] = \mathbf{b} + \mathbf{Q} \cdot \mathbf{X}, \tag{3.184}$$

where $\mathbf{u}(\mathbf{x})$ is the displacement in the current configuration:

$$\mathbf{u}(\mathbf{x}) = \mathbf{x} - (\mathbf{b} + \mathbf{Q} \cdot \mathbf{X}). \tag{3.185}$$

Now, the *spatial displacement gradient* at $\mathbf{x}$ is a linear transformation defined by

$$\mathbf{h} = \mathbf{h}(\mathbf{x}) \equiv (\text{grad } \mathbf{u})^T \qquad \text{or} \qquad h_{ij} = u_{i,j}. \tag{3.186}$$

It easily follows that

$$\mathbf{h} = \mathbf{1} - \mathbf{Q} \cdot \mathbf{F}^{-1}. \tag{3.187}$$

Both strain tensors E and e vanish when there is no deformation, i.e., when $U = V = I$, in which case we also see that $F = R = Q$ and $H = h = 0$. For small deformations, these strains are, therefore, expected to be small. Since from above we have that the deformation gradient and its inverse, in terms of the material and spatial displacement gradients, are given by

$$F = Q\,(I + H) \qquad \text{and} \qquad F^{-1} = Q^T(I - h), \tag{3.188}$$

the finite strain tensors (3.175) and (3.178) can be rewritten in terms of the displacement gradients as

$$E = \frac{1}{2}\left(H + H^T + H^T H\right) \qquad \text{and} \qquad e = \frac{1}{2}\left(h + h^T - h^T h\right). \tag{3.189}$$

Note that, using (3.175), (3.178), and (3.189), we can also write

$$C = I + H + H^T + H^T H \qquad \text{and} \qquad B^{-1} = I - h - h^T + h^T h. \tag{3.190}$$

For small deformations, $H^T H$ and $h^T h$ are second-order quantities; thus neglecting these terms, we obtain the *infinitesimal strain tensors*

$$\widetilde{\mathbf{E}} \equiv \frac{1}{2}\left(\mathbf{H} + \mathbf{H}^T\right) = \frac{1}{2}\left[(\text{Grad } \mathbf{u}) + (\text{Grad } \mathbf{u})^T\right] \tag{3.191}$$

and

$$\widetilde{\mathbf{e}} \equiv \frac{1}{2}\left(\mathbf{h} + \mathbf{h}^T\right) = \frac{1}{2}\left[(\text{grad } \mathbf{u}) + (\text{grad } \mathbf{u})^T\right]. \tag{3.192}$$

The linear strain tensor $\widetilde{\mathbf{E}}$ was introduced by Cauchy in the classical theory of elasticity.

For small displacement gradients, the right stretch tensor and the rotation tensor can be approximated by

$$U = (F^T F)^{1/2} \approx I + \frac{1}{2}(H + H^T) = I + \widetilde{E}, \tag{3.193}$$

$$R = FU^{-1} \approx Q\left[I + \frac{1}{2}(H - H^T)\right] = Q(I + \widetilde{R}), \tag{3.194}$$

where

$$\widetilde{R} \equiv \frac{1}{2}(H - H^T) = \frac{1}{2}\left[(\text{Grad } \mathbf{u}) - (\text{Grad } \mathbf{u})^T\right] \tag{3.195}$$

is called the *infinitesimal rotation tensor.* Note that the infinitesimal strain and rotation tensors correspond to the symmetric and skew-symmetric parts of the finite displacement gradient! Also note that when we have no deformation, we have that $\widetilde{E} = \widetilde{e} = \widetilde{R} = 0$ and $R = Q$.

3.1.11 Compatibility conditions

In three dimensions the right and left Cauchy–Green tensors $\mathbf{C}$ and $\mathbf{B}$ and the finite strain tensors $\mathbf{E}$ and $\mathbf{e}$ are all symmetric and hence have six independent components, which can be written in terms of the gradient of the displacement $\mathbf{u}$, which has three components:

$$2E_{IJ} = C_{IJ} - \delta_{IJ} = H_{IJ} + H_{JI} + H_{KI}H_{KJ} = u_{I,J} + u_{J,I} + u_{K,I}u_{K,J}, \tag{3.196}$$

$$2e_{ij} = \delta_{ij} - B_{ij}^{-1} = h_{ij} + h_{ji} - h_{ki}h_{kj} = u_{i,j} + u_{j,i} - u_{k,i}u_{k,j}. \tag{3.197}$$

If we are given a displacement vector $\mathbf{u}$ which is differentiable, by differentiation and substitution in the previous equations we can obtain $\mathbf{C}$, $\mathbf{B}$, $\mathbf{E}$, and $\mathbf{e}$. If, on the other hand, $\mathbf{C}$, $\mathbf{B}$, $\mathbf{E}$, or $\mathbf{e}$ is given, then it's not clear that we can obtain the corresponding single-valued continuous displacement field $\mathbf{u}$. From this standpoint, (3.196) or (3.197) correspond to an over determined system of six partial differential equations that may not possess a unique solution for $\mathbf{u}$ unless specific integrability conditions are satisfied. We emphasize that if the displacement field is given, then the compatibility conditions are not needed. On the other hand, if the problem is formulated in terms of strain tensors, then such conditions are required to ensure compatibility with a single-valued differentiable displacement field. We note that if the compatibility conditions are violated, the corresponding displacement field in the body is not unique and then the body may possess *dislocations.*

To find such conditions, we first note that both the undeformed and deformed bodies are embedded in a three-dimensional Euclidean space. If we consider that the deformation and inverse deformation

$$x_i = \chi_i(X_I) \qquad \text{and} \qquad X_I = \chi_I^{-1}(x_i) \tag{3.198}$$

are nothing more than coordinate transformations from $\mathbf{X}$ to $\mathbf{x}$ and from $\mathbf{x}$ to $\mathbf{X}$, then it is clear that the right Cauchy–Green tensor $\mathbf{C}$ plays the role of the metric tensor in the curvilinear coordinate $\mathbf{X}$ while the inverse of the left Cauchy–Green tensor $\mathbf{B}^{-1}$ plays the role of the metric tensor in the curvilinear coordinate $\mathbf{x}$. In Section 2.16 we have shown that in a Euclidean space the Riemann–Christoffel tensor vanishes. Thus, if we now correspondingly replace $\mathbf{g}$ and $\mathbf{x}$ with $\mathbf{C}$ and $\mathbf{X}$ or $\mathbf{g}$ with $\mathbf{B}^{-1}$ in the Riemann–Christoffel tensor (2.240) and subsequently in the Christoffel symbol (2.220), we have

$$R^{(C)}_{MIJK} = 0 \qquad \text{and} \qquad R^{(B)}_{mijk} = 0. \tag{3.199}$$

These correspond to $3^4 = 81$ equations, respectively. If we now note the symmetry conditions (2.241), we can use (2.242) to write

$$S^{(C)}_{PQ} = \frac{1}{4}\epsilon_{PMI}\epsilon_{QJK}R^{(C)}_{MIJK} = 0 \qquad \text{and} \qquad S^{(B)}_{pq} = \frac{1}{4}\epsilon_{pmi}\epsilon_{qjk}R^{(B)}_{mijk} = 0, \tag{3.200}$$

where $\mathbf{S}^{(C)}$ and $\mathbf{S}^{(B)}$ are easily shown to be symmetric tensors; subsequently, these yield six equations. However, these six equations are not all independent since the Riemann–Christoffel tensor satisfies Bianchi's identities (2.243):

$$R^{(C)}_{MIJK,L} + R^{(C)}_{MIKL,J} + R^{(C)}_{MILJ,K} = 0, \tag{3.201}$$

with analogous identities for $R^{(B)}_{mijk}$. Bianchi's identities provide three additional equations which restrict the six components of $\mathbf{S}^{(C)}$ and $\mathbf{S}^{(B)}$ to three degrees of freedom. If we subsequently use (3.175) or (3.178), the six equations are given explicitly by

$$\begin{aligned} &E_{MI,JK} + E_{JK,MI} - E_{MJ,IK} - E_{IK,MJ} + \\ &\quad C^{-1}_{PQ}\left[\left(E_{IP,M} + E_{MP,I} - E_{MI,P}\right)\left(E_{JQ,K} + E_{KQ,J} - E_{JK,Q}\right) - \right. \\ &\quad\quad \left.\left(E_{IP,K} + E_{KP,I} - E_{IK,P}\right)\left(E_{JQ,M} + E_{MQ,J} - E_{JM,Q}\right)\right] = 0 \end{aligned} \tag{3.202}$$

and

$$\begin{aligned} &e_{mi,jk} + e_{jk,mi} - e_{mj,ik} - e_{ik,mj} - \\ &\quad B_{pq}\left[\left(e_{ip,m} + e_{mp,i} - e_{mi,p}\right)\left(e_{jq,k} + e_{kq,j} - e_{jk,q}\right) - \right. \\ &\quad\quad \left.\left(e_{ip,k} + e_{kp,i} - e_{ik,p}\right)\left(e_{jq,m} + e_{mq,j} - e_{jm,q}\right)\right] = 0. \end{aligned} \tag{3.203}$$

For the case of infinitesimal strains, all quadratic terms are small and thus we obtain, say,

$$\tilde{e}_{mi,jk} + \tilde{e}_{kj,mi} - \tilde{e}_{mj,ik} - \tilde{e}_{ki,mj} = 0. \tag{3.204}$$

Note that if we set

$$j^{(\tilde{e})}_{mkji} \equiv \tilde{e}_{mi,jk} + \tilde{e}_{kj,mi} - \tilde{e}_{mj,ik} - \tilde{e}_{ki,mj} = 0, \tag{3.205}$$

then we have identically

$$j^{(\tilde{e})}_{mkji,l} + j^{(\tilde{e})}_{mkil,j} + j^{(\tilde{e})}_{mklj,i} = 0, \tag{3.206}$$

a relation formally analogous to Bianchi's identities (2.243) in flat Euclidean space. The compatibility conditions (3.204) may be rewritten in the form $j^{(\tilde{e})}_{mkji} = 0$. Subsequently, the conditions (3.204) may be divided into two sets of three conditions, $j^{(\tilde{e})}_{1212} = j^{(\tilde{e})}_{2323} = j^{(\tilde{e})}_{3131} = 0$ and $j^{(\tilde{e})}_{1213} = j^{(\tilde{e})}_{2321} = j^{(\tilde{e})}_{3132} = 0$, or, more explicitly,

$$\tilde{e}_{11,22} + \tilde{e}_{22,11} - 2\tilde{e}_{12,12} = 0, \tag{3.207}$$

$$\tilde{e}_{22,33} + \tilde{e}_{33,22} - 2\tilde{e}_{23,23} = 0, \tag{3.208}$$

$$\tilde{e}_{33,11} + \tilde{e}_{11,33} - 2\tilde{e}_{31,31} = 0, \tag{3.209}$$

and

$$\tilde{e}_{12,23} + \tilde{e}_{23,12} - \tilde{e}_{22,31} - \tilde{e}_{31,22} = 0, \tag{3.210}$$

$$\tilde{e}_{23,31} + \tilde{e}_{31,23} - \tilde{e}_{33,12} - \tilde{e}_{12,33} = 0, \tag{3.211}$$

$$\tilde{e}_{31,12} + \tilde{e}_{12,31} - \tilde{e}_{11,23} - \tilde{e}_{23,11} = 0, \tag{3.212}$$

such that if both sets are satisfied upon the boundary of a region, then the vanishing of either set in the interior implies the vanishing of both sets.

When these conditions are satisfied, then the single-valued integral of the linear form of (3.197),

$$\tilde{e}_{ij} = \frac{1}{2}\left(u_{i,j} + u_{j,i}\right), \tag{3.213}$$

exists and is given by

$$u_i = u_i^0 + \widetilde{Q}_{ji}x_j + b_i, \tag{3.214}$$

where u_i^0 is any solution of (3.213), $\widetilde{Q}_{ij}$ is a skew-symmetric rotation tensor independent of x_i, and b_i is an arbitrary vector independent of x_i. This means that the displacement field $\mathbf{u}$ is single valued and uniquely determined to within a rigid motion.

3.2 Motion

The *motion* of a body $\mathcal{B}$ is regarded as a continuous sequence of configurations in time. Thus the motion χ_κ can be expressed as the map

$$\chi_\kappa : \mathcal{B}_\kappa \times \mathcal{R} \to \mathcal{E}^3 \qquad \text{or} \qquad \mathbf{x} = \chi_\kappa(\mathbf{X}, t). \tag{3.215}$$

It represents a one-parameter family of deformations. The quantity of time has physical dimension of $[T]$, which is independent of $[L]$. More simply, by dropping the subscript κ referring to the reference configuration for the time being, we write

$$\mathbf{x} = \chi(\mathbf{X}, t) \qquad \text{or} \qquad x_k = \chi_k(X_K, t) \tag{3.216}$$

for every point in the reference configuration. Conversely, we write

$$\mathbf{X} = \chi^{-1}(\mathbf{x}, t) \qquad \text{or} \qquad X_K = \chi_K^{-1}(x_k, t). \tag{3.217}$$

For a fixed material point X with coordinate $\mathbf{X}$, $\chi : \mathcal{R} \to \mathcal{E}^3$ is a curve called a *path* or *trajectory* of the material point.

The coordinates $\mathbf{X}$ are assumed to be assigned once and for all to given particles in the material. Since they are the coordinates of the particles at an arbitrary initial time t_0, they serve for all time as names for the particles of the material. The coordinates $\mathbf{x}$, on the other hand, are thought of as assigned once and for all to a point in the Euclidean space where material body resides. They are the names of places. The motion $\mathbf{x} = \chi(\mathbf{X}, t)$ chronicles the places that $\mathbf{x}$ is occupied by the particles $\mathbf{X}$ in the course of time. Problems for which $\mathbf{X}$ and t are taken as independent variables are said to be set in the *material description*; those in $\mathbf{x}$ and t, the *spatial description*. For purposes of interpretation, $\mathbf{x} = \chi(\mathbf{X}, t)$ should be thought of purely as a continuous coordinate transformation. The material description is an immediate extension of the scheme used in the mechanics of mass-points, where the paths of the several distinct masses are traced, while the spatial description has no counterpart in elementary mechanics. While the material description is more fundamental, it leads to mathematical difficulties; this is the reason why the spatial description is usually preferred. A fully general spatial description was first given by Euler; that is why it's referred to as the *Eulerian description*. The material description is called the *Lagrangian description*, even though it was also Euler who first formulated such description.

Analogously, components of a tensor of any order can now be written using either the material or the spatial descriptions

$$\psi(\mathbf{X}, t) = \psi[\chi^{-1}(\mathbf{x}, t), t] = \widehat{\psi}(\mathbf{x}, t), \tag{3.218}$$

or

$$\psi_{\dots}(X_K, t) = \psi_{\dots}[\chi_K^{-1}(x_k, t), t] = \widehat{\psi}_{\dots}(x_k, t). \tag{3.219}$$

Again, for notational simplicity, we drop the hat on the function and thus do not differentiate between a function and its values since the functional dependence should be obvious in applications or will be explicitly displayed when necessary.

Since $J = \det F \neq (0, \pm\infty)$, when we have motion, J has to remain of the same sign. Thus, as before, and without loss of generality, we assume that $0 < J < \infty$.

3.2.1 Velocity and acceleration

Given the motion χ, we can calculate the velocity of the particle at $\mathbf{X}$ in the reference configuration by

$$\mathbf{v} \equiv \dot{\mathbf{x}} \equiv \left.\frac{\partial \mathbf{x}}{\partial t}\right|_{\mathbf{X}} = \left.\frac{\partial \chi(\mathbf{X}, t)}{\partial t}\right|_{\mathbf{X}} \quad \text{or} \quad v_k \equiv \dot{x}_k \equiv \left.\frac{\partial x_k}{\partial t}\right|_{X_K} = \left.\frac{\partial \chi_k(X_K, t)}{\partial t}\right|_{X_K}, \tag{3.220}$$

whose components can be written as

$$v_i = \begin{cases} v_i(X_K, t), & \text{in the material (Lagrangian) description,} \\ v_i(x_k, t), & \text{in the spatial (Eulerian) description,} \end{cases} \tag{3.221}$$

and we have introduced the material derivative of a tensor quantity z as

$$\dot{z} \equiv \frac{dz}{dt}. \tag{3.222}$$

To make these descriptions clear, if ψ is a tensor field such that $\psi = \psi(\mathbf{X}, t)$, then

$$\dot{\psi}(\mathbf{X}, t) = \left.\frac{\partial \psi(\mathbf{X}, t)}{\partial t}\right|_{\mathbf{X}} + \dot{\mathbf{X}} \cdot \text{Grad}\, \psi(\mathbf{X}, t) = \left.\frac{\partial \psi(\mathbf{X}, t)}{\partial t}\right|_{\mathbf{X}} \tag{3.223}$$

since $\dot{\mathbf{X}} = \mathbf{0}$ because $\mathbf{X}$ moves with the particle associated with $\mathbf{X}$. On the other hand, if we write $\psi = \psi(\mathbf{x}, t)$, then

$$\dot{\psi}(\mathbf{x}, t) = \left.\frac{\partial \psi(\mathbf{x}, t)}{\partial t}\right|_{\mathbf{x}} + \dot{\mathbf{x}} \cdot \text{grad}\, \psi(\mathbf{x}, t) = \left.\frac{\partial \psi(\mathbf{x}, t)}{\partial t}\right|_{\mathbf{x}} + \mathbf{v} \cdot \text{grad}\, \psi(\mathbf{x}, t) \tag{3.224}$$

since $\mathbf{x} = \boldsymbol{\chi}(\mathbf{X}, t)$ and $\dot{\mathbf{x}} = \partial \boldsymbol{\chi}/\partial t|_{\mathbf{X}} = \mathbf{v}$.

Now, more generally, we define the *material derivative* for the components of a tensor field of any order as

$$\dot{\psi}_{\dots} \equiv \left.\frac{\partial \psi_{\dots}(X_K, t)}{\partial t}\right|_{X_K} = \left.\frac{\partial \psi_{\dots}(x_k, t)}{\partial t}\right|_{x_k} + \left.\frac{\partial \psi_{\dots}(x_k, t)}{\partial x_l}\right|_t \left.\frac{\partial x_l}{\partial t}\right|_{X_K},$$

or, dropping the explicit indication of what is kept fixed in the differentiations,

$$\dot{\psi} \equiv \frac{\partial \psi(\mathbf{X}, t)}{\partial t} = \frac{\partial \psi(\mathbf{x}, t)}{\partial t} + \mathbf{v}(\mathbf{x}, t) \cdot \text{grad}\, \psi(\mathbf{x}, t) \tag{3.225}$$

or

$$\dot{\psi}_{\dots} \equiv \frac{\partial \psi_{\dots}(X_K, t)}{\partial t} = \frac{\partial \psi_{\dots}(x_l, t)}{\partial t} + v_k(x_l, t)\, \frac{\partial \psi_{\dots}(x_l, t)}{\partial x_k}, \tag{3.226}$$

or using general tensor notation

$$\dot{\psi}^{\dots}_{\dots} \equiv \frac{\partial \psi^{\dots}_{\dots}}{\partial t} + v^k\, \psi^{\dots}_{\dots,k} = \frac{\partial \psi^{\dots}_{\dots}}{\partial t} + v^k \left[\frac{\partial \psi^{\dots}_{\dots}}{\partial x^k} + \Gamma^{\cdot}_{k\cdot} \psi^{\dots}_{\dots} + \cdots - \Gamma^{\cdot}_{\cdot k} \psi^{\dots}_{\dots} - \cdots\right]. \tag{3.227}$$

We note that sometimes the operator

$$\frac{D}{Dt} \equiv \frac{\partial}{\partial t} + \mathbf{v}(\mathbf{x}, t) \cdot \text{grad} \tag{3.228}$$

is defined to indicate the material derivative when written in spatial coordinates. We will not use such convention here. We will just use the overdot accent to denote the material derivative; the form that it will take in the material or spatial description will be understood as indicated in (3.225). A motion is *homogeneous* if it is of the form

$$\mathbf{v} = \dot{\mathbf{x}} = \mathbf{b}(t) + \mathbf{Q}(t) \cdot \mathbf{x}. \tag{3.229}$$

We note that when $\mathbf{v}$ is given, we can also determine the identity of the material point at $\mathbf{x}$, but this requires solving the ordinary differential equations

$$\frac{d\mathbf{x}}{dt} = \mathbf{v}(\mathbf{x}, t), \tag{3.230}$$

subject to the initial conditions

$$\mathbf{x}(t_0) = \boldsymbol{\chi}(\mathbf{X}, t_0) = \mathbf{X}. \tag{3.231}$$

A surface $f(\mathbf{x},t) = 0$ consisting of a set of particles is said to be a *material surface* if

$$\dot{f} = \frac{\partial f}{\partial t} + \mathbf{v}\cdot\text{grad } f = 0. \tag{3.232}$$

A material *boundary* is a surface which the material does not cross. The material inside a material surface is called a *body*. At a boundary, the normal component of velocity $\dot{x}_n$ is equal to the normal velocity of the boundary c_n,

$$\dot{x}_n = \dot{\mathbf{x}}\cdot\mathbf{n} = \mathbf{c}\cdot\mathbf{n} = c_n \qquad \text{on} \qquad \mathscr{S}. \tag{3.233}$$

We will denote by $\mathbf{N}$ the normal component to the boundary $\mathbf{S}$ in the material configuration. If $c_n = 0$, the boundary is said to be *stationary*. If the velocity vector $\dot{\mathbf{x}}$ is equal to the velocity of the surface $\mathbf{c}$, i.e.,

$$\dot{\mathbf{x}} = \mathbf{c} \qquad \text{on} \qquad \mathscr{S}, \tag{3.234}$$

then the material is said to *adhere* to the boundary, or is referred to as a *no-slip/no-penetration* condition. If in addition the boundary is stationary, then this condition becomes

$$\dot{\mathbf{x}} = \mathbf{0} \qquad \text{on} \qquad \mathscr{S}. \tag{3.235}$$

A more extensive discussion of boundary conditions is given later in Section 5.12.

If we take $\psi \to \mathbf{v}$ in (3.225), we obtain the acceleration

$$\mathbf{a} \equiv \ddot{\mathbf{x}} \equiv \left.\frac{\partial^2\mathbf{x}}{\partial t^2}\right|_{\mathbf{X}} \equiv \dot{\mathbf{v}} = \frac{\partial\mathbf{v}}{\partial t} + (\mathbf{v}\cdot\text{grad})\mathbf{v} \tag{3.236}$$

or

$$a_k \equiv \ddot{x}_k \equiv \left.\frac{\partial^2 x_k}{\partial t^2}\right|_{X_K} \equiv \dot{v}_k = \frac{\partial v_k}{\partial t} + v_j\frac{\partial v_k}{\partial x_j}. \tag{3.237}$$

The *spatial velocity gradient* at $(\mathbf{x},t)$ is a linear transformation given by

$$\mathbf{L} = \mathbf{L}(\mathbf{x},t) \equiv (\text{grad }\dot{\mathbf{x}})^T = (\text{grad }\mathbf{v})^T \qquad \text{or} \qquad L_{ij} = v_{i,j}. \tag{3.238}$$

We now note that

$$\begin{aligned}
\dot{\mathbf{F}} &= \overline{(\text{Grad }\mathbf{x})^T}^{\cdot}, \\
&= \overline{(\text{Grad }\boldsymbol{\chi}(\mathbf{X},t))^T}^{\cdot}, \\
&= \left(\text{Grad }\frac{\partial\boldsymbol{\chi}(\mathbf{X},t)}{\partial t}\right)^T, \\
&= \left(\text{Grad }\mathbf{x}\cdot\text{grad }\frac{\partial\boldsymbol{\chi}(\mathbf{X},t)}{\partial t}\right)^T, \\
&= (\text{grad }\dot{\mathbf{x}})^T\cdot(\text{Grad }\mathbf{x})^T,
\end{aligned}$$

or

$$\dot{\mathbf{F}} = \mathbf{L} \cdot \mathbf{F}. \tag{3.239}$$

Above, we have used the overbar to span the terms affected by the dot operation. We now see that the spatial gradient of velocity is related to the rate of deformation by

$$(\text{grad}\,\mathbf{v})^T = \mathbf{L} = \dot{\mathbf{F}} \cdot \mathbf{F}^{-1}. \tag{3.240}$$

Using (3.239), it is easy to obtain the following interesting result

$$\dot{J} = J\,\text{div}\,\mathbf{v}. \tag{3.241}$$

A motion such that the volume occupied by any material region is unaltered, i.e., $J = 1$, is called *isochoric*. It immediately follows that a motion is isochoric if and only if its velocity is solenoidal, i.e., $\text{div}\,\mathbf{v} = 0$.

3.2.2 Path lines, stream lines, and streak lines

A point where $\dot{\mathbf{x}} = \mathbf{0}$ is called a *stagnation point*. A motion such that the velocity field does not change with time, i.e., $\dot{\mathbf{x}} = \dot{\chi}(\mathbf{x})$ is said to be *steady*. More generally, any quantity which is independent of time is said to be steady. For example, it is sometimes convenient to have the observer move at a constant speed V in the x-direction, in which case a quantity f can be rewritten as

$$f(x,y,z,t) = g(\xi,y,z), \qquad \xi = x - Vt. \tag{3.242}$$

Subsequently, we have that

$$\frac{\partial f}{\partial t} = -V\frac{\partial g}{\partial \xi}. \tag{3.243}$$

The curve in space traversed by $\mathbf{X}$ as t varies is the *path line* of $\mathbf{X}$:

$$\mathbf{x} = \chi(\mathbf{X},t) \quad \text{for} \quad \mathbf{X}\ \text{fixed} \quad \text{and} \quad -\infty < t < \infty. \tag{3.244}$$

It also corresponds to the integral curve of the system

$$d\mathbf{x} = \mathbf{v}\ dt \tag{3.245}$$

which passes through $\mathbf{X}$ at $t = t_0$.

Vector lines of the field $\mathbf{v}$ at time t are the *stream lines*. They correspond to the integral curves

$$f_1(\mathbf{x},t) = 0 \quad \text{and} \quad f_2(\mathbf{x},t) = 0 \tag{3.246}$$

of the system

$$dx_1 : dx_2 : dx_3 = v_1 : v_2 : v_3 \quad \text{for} \quad t = \text{const.} \tag{3.247}$$

To see this, we note that we can rewrite (3.245) or (3.247) as

$$\frac{dx_1}{v_1} = \frac{dx_2}{v_2} = \frac{dx_3}{v_3}, \tag{3.248}$$

or

$$v_2 dx_1 - v_1 dx_2 = 0, \quad (3.249)$$
$$v_3 dx_2 - v_2 dx_3 = 0, \quad (3.250)$$
$$v_1 dx_3 - v_3 dx_1 = 0, \quad (3.251)$$

or more succinctly

$$\epsilon_{ijk} v_j dx_k = 0. \quad (3.252)$$

Because only two of the equations (3.252) can be independent, we can write

$$g_{ij}(x_k) dx_j = 0, \qquad i = 1,2, \quad j = 1,2,3. \quad (3.253)$$

The solution of these equations are given by the integral curves (3.246).

The *streak line* through $\mathbf{x}$ at time t is the locus at time t of all particles that at any time, past or future, will occupy or have occupied the place $\mathbf{x}$. If we write the motion in the forms $\mathbf{x} = \boldsymbol{\chi}(\mathbf{X}, t)$ and $\mathbf{X} = \boldsymbol{\chi}^{-1}(\mathbf{x}, t)$, then the streak line through $\mathbf{x}$ at time t is given parametrically by the locus of x, where

$$\mathrm{x} = \boldsymbol{\chi}(\boldsymbol{\chi}^{-1}(\mathbf{x}, t'), t) \quad \text{for} \quad -\infty < t' < \infty. \quad (3.254)$$

At a given place $\mathbf{x}$ and time t, the stream line through $\mathbf{x}$, the path line of the particle occupying $\mathbf{x}$, and the streak line through $\mathbf{x}$ all have a common tangent. When the motion is steady, all three curves coincide, but in general for unsteady motion, they are distinct. A stream line never crosses itself nor ends, except possibly at a stagnation point. Since the stream lines are determined at a fixed instant, singularities such as stagnation points may be generated or destroyed in the course of time. The path lines and streak lines of an unsteady motion may cross themselves or double back upon themselves.

Example

Consider the plane motion whose spatial description is given by

$$\mathbf{v} = \dot{\mathbf{x}} = (\dot{x}_1, \dot{x}_2) = \left(\frac{x_1}{1+t}, 1 \right). \quad (3.255)$$

By integrating the equivalent of (3.247) for this planar case, we get the equation for the stream lines, which at $t_0 = 0$ go through $\mathbf{x}_0 = (x_{10}, x_{20})$:

$$(x_2 - x_{20}) - (1+t) \ln \left| \frac{x_1}{x_{10}} \right| = 0, \quad (3.256)$$

which corresponds to (3.246) for the planar case. Equivalently, a parametric equation for the stream line at time t is obtained by integrating

$$\frac{d\mathbf{x}}{d\tau} = \mathbf{v}(\mathbf{x}, t), \quad (3.257)$$

the solution of which is given by

$$\mathbf{x} = (x_1, x_2) = \left(x_{10} e^{\tau/(1+t)}, x_{20} + \tau \right). \quad (3.258)$$

Also note that if we eliminate the parameter τ in (3.258), we obtain (3.256). By integrating (3.245) with $t_0 = 0$, we get the material description by taking $\mathbf{x}(0) = \mathbf{X}$:

$$\mathbf{x} = (x_1, x_2) = (X_1(1+t), X_2 + t) \,. \tag{3.259}$$

Note that $\dot{\mathbf{x}} = (\dot{x}_1, \dot{x}_2) = (X_1, 1)$, which upon substituting for X_1 from (3.259), we recover (3.255). By eliminating t from the above, we get the path line of the particle $\mathbf{X}$:

$$x_1 - X_1 x_2 = X_1 (1 - X_2) \,. \tag{3.260}$$

Thus, each particle moves in a straight line at constant speed, but the stream lines change in time according to (3.256). To get the streak line through $\mathbf{x}$ when $t = 0$, we need only hold $\mathbf{x}$ fixed in (3.260):

$$X_1 X_2 - X_1 (1 + x_2) + x_1 = 0. \tag{3.261}$$

This is a hyperbola. To get the streak line through $\mathbf{x}$ at time t, we first invert (3.259) at time t':

$$\mathbf{X} = (X_1, X_2) = \left(\frac{x_1}{1+t'}, x_2 - t' \right). \tag{3.262}$$

This gives the particle that occupies the place $\mathbf{x}$ at time t'. The place x occupied by this particle at time t follows from (3.259):

$$\mathrm{x} = (\mathrm{x}_1, \mathrm{x}_2) = \left(x_1 \frac{1+t}{1+t'}, x_2 + t - t' \right), \tag{3.263}$$

which, by eliminating t', corresponds to the hyperbolic curve

$$\mathrm{x}_1 \mathrm{x}_2 - \mathrm{x}_1 (1 + x_2 + t) + x_1 (1+t) = 0 \tag{3.264}$$

that includes (3.261) as the special case with $t = 0$. The stream, path, and streak lines are illustrated in Fig. 3.11.

3.2.3 Relative deformation

In practice, for a given motion, the reference configuration is often chosen as the configuration at some instant $t = t_0$. For some media (e.g., fluids) this choice is not only unnecessary but also inconvenient. The configuration can be chosen independently of any motion. It is more convenient to choose the *current configuration* at time t as the reference configuration and measure changes from an earlier time $\tau \leq t$ to this configuration.

Thus we denote the position of the material point $\mathbf{X}$ at time τ by $\boldsymbol{\xi}$:

$$\boldsymbol{\xi} = \boldsymbol{\chi}(\mathbf{X}, \tau). \tag{3.265}$$

Subsequently we can write, with a little abuse of functional notation,

$$\boldsymbol{\xi} = \boldsymbol{\chi}(\boldsymbol{\chi}^{-1}(\mathbf{x}, t), \tau) \equiv {}_{(t)}\boldsymbol{\chi}(\mathbf{x}, \tau) \quad \text{or} \quad \xi_\alpha = \chi_\alpha(\chi_K^{-1}(x_k, t), \tau) \equiv {}_{(t)}\chi_\alpha(x_k, \tau) \,, \tag{3.266}$$

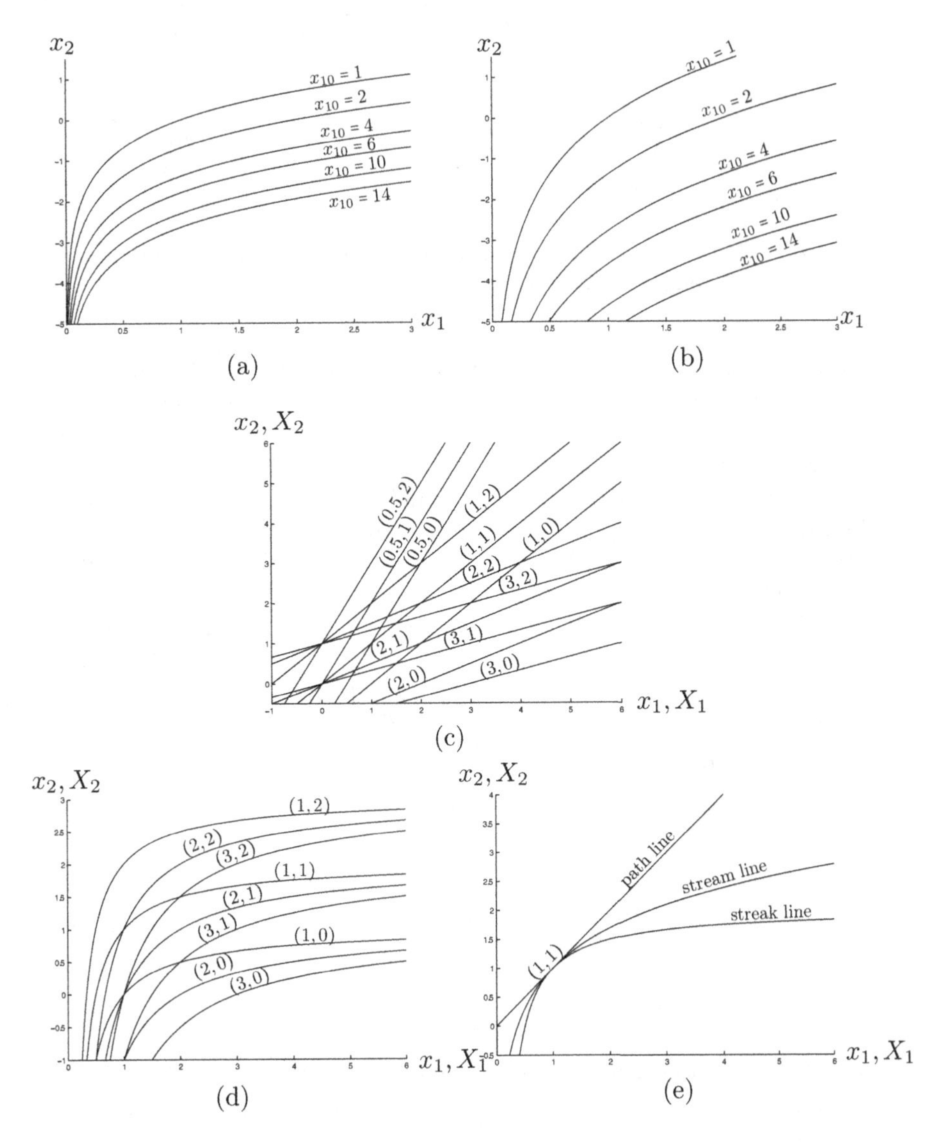

Figure 3.11: Stream, path, and streak lines corresponding to the velocity field (3.255): (a) stream lines at $t = 0$, with x_{10} as indicated and $x_{20} = 0$; (b) stream lines at $t = 1$, with x_{10} as indicated and $x_{20} = 0$; (c) path lines of the particles $\mathbf{X}$ given; (d) streak lines at $t = 0$ of all particles ever occupying $\mathbf{x}$; (e) path line of the particle $(1, 1)$, stream line at $t = 0$, and initial streak line through $(1, 1)$.

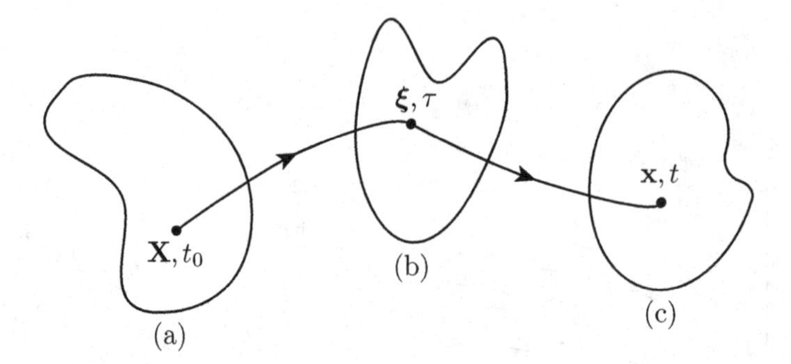

Figure 3.12: (a) Initial configuration at $t = t_0$, (b) current configuration at time $\tau \leq t$, and (c) reference configuration at time t.

where ${}_{(t)}\boldsymbol{\chi}$ is called the *relative motion.* The different configurations are illustrated in Fig. 3.12. Note that we use lowercase Greek subscripts for indices associated with the current configuration at time τ, and these indices become Latin lowercase letters when we take the limit $\tau \to t$. Thus $\boldsymbol{\xi}$ is the place at time τ of the material point, which at time t is located at $\mathbf{x}$ and at time t_0 is located at $\mathbf{X}$. In the following, unless explicitly given, it will be understood that the relative deformation gradient is to be evaluated as $\mathbf{x}$, but for notational simplicity, we will not display this functional dependence explicitly.

As before, we assume that the mapping from τ to t is one-to-one, i.e.,

$$ {}_{(t)}J(\tau) = \det\left[\frac{\partial \xi_\alpha}{\partial x_l}\right] > 0. \tag{3.267}$$

Note that ${}_{(t)}J(t) = 1$.

The *relative deformation gradient* at τ relative to time t, ${}_{(t)}\mathbf{F}(\tau)$, is defined by

$$ {}_{(t)}\mathbf{F}(\tau) \equiv (\text{grad}\ \boldsymbol{\xi})^T \qquad \text{or} \qquad {}_{(t)}F_{\alpha k}(\tau) \equiv \frac{\partial \xi_\alpha}{\partial x_k}. \tag{3.268}$$

We note that

$$ {}_{(t)}\mathbf{F}(t) = \mathbf{1} \qquad \text{or} \qquad {}_{(t)}F_{ik}(t) = \delta_{ik}, \tag{3.269}$$

and if ${}_{(t)}\mathbf{F}(\tau)$ and ${}_{(t')}\mathbf{F}(\tau)$ are two deformation gradients, where $\tau \leq t' \leq t$, then they are related by

$$ {}_{(t)}\mathbf{F}(\tau) = {}_{(t')}\mathbf{F}(\tau) \cdot {}_{(t)}\mathbf{F}(t'). \tag{3.270}$$

This follows since, by definition, we have

$$ \boldsymbol{\xi} = {}_{(t)}\boldsymbol{\chi}(\mathbf{x}, \tau) = {}_{(t')}\boldsymbol{\chi}(\mathbf{x}', \tau), \tag{3.271}$$

and if $\mathbf{x}$ at time t is a reference configuration relative to $\mathbf{x}'$ at time t', then we also have

$$ \mathbf{x} = {}_{(t')}\boldsymbol{\chi}(\mathbf{x}', t). \tag{3.272}$$

The chain rule now gives

$$\frac{\partial \xi_\alpha}{\partial x_l} = \frac{\partial \xi_\alpha}{\partial x'_k}\frac{\partial x'_k}{\partial x_l}, \tag{3.273}$$

which corresponds to (3.270). Note that if $t \to t_0$ corresponding to the original configuration, and $t' \to t$, then we have

$$_{(t_0)}\mathbf{F}(\tau) = {}_{(t)}\mathbf{F}(\tau) \cdot {}_{(t_0)}\mathbf{F}(t), \tag{3.274}$$

or more explicitly

$$F(\mathbf{X},\tau) = {}_{(t)}F(\mathbf{x},\tau)\; F(\mathbf{X},t). \tag{3.275}$$

Example

Consider the following motion in the x_1-x_2 plane relative to the reference configuration $\boldsymbol{\kappa}$ in the Cartesian coordinate system:

$$\mathbf{x} = \boldsymbol{\chi}(\mathbf{X},t) = \left(X_1 e^t, X_2(t+1)\right), \tag{3.276}$$

with inverse

$$\mathbf{X} = \boldsymbol{\chi}^{-1}(\mathbf{x},t) = \left(x_1 e^{-t}, \frac{x_2}{t+1}\right). \tag{3.277}$$

Note that for this motion the reference configuration is the configuration of the body at the instant $t = t_0 = 0$.
Then, the relative deformation is given by

$$\begin{aligned}\boldsymbol{\xi} = {}_{(t)}\boldsymbol{\chi}(\mathbf{x},\tau) = \boldsymbol{\chi}\left(\boldsymbol{\chi}^{-1}(\mathbf{x},t),\tau\right) &= \boldsymbol{\chi}\left(\left(x_1 e^{-t}, \frac{x_2}{t+1}\right),\tau\right),\\ &= \left(x_1 e^{\tau-t}, x_2\left(\frac{\tau+1}{t+1}\right)\right).\end{aligned} \tag{3.278}$$

From the above, we now calculate the deformation gradients

$$\mathbf{F}(t) = e^t\, \mathbf{i}_1\mathbf{i}_1 + (t+1)\, \mathbf{i}_2\mathbf{i}_2, \tag{3.279}$$

$$_{(t)}\mathbf{F}(\tau) = e^{\tau-t}\, \mathbf{i}_1\mathbf{i}_1 + \left(\frac{\tau+1}{t+1}\right)\mathbf{i}_2\mathbf{i}_2. \tag{3.280}$$

One can also obtain the path of an arbitrary material point $\mathbf{X}_0$ in this motion, by eliminating time t from the deformation function,

$$\begin{aligned} x_2 &= X_{20}\left(1+\ln\frac{x_1}{X_{10}}\right) && \text{for} \quad X_{10} \neq 0,\\ x_1 &= 0 && \text{for} \quad X_{10} = 0.\end{aligned} \tag{3.281}$$

Note that by choosing four points $\mathbf{X}_0$ which would correspond to corners of a square in the reference configuration, one can examine the image of this square at any instant in the motion.

Since the mapping is invertible, $\det[_{(t)}F_{ik}(\tau)] \neq (0, \pm\infty)$, and using the polar decomposition theorem,

$$_{(t)}F(\tau) = {}_{(t)}R(\tau)_{(t)}U(\tau) = {}_{(t)}V(\tau)_{(t)}R(\tau). \tag{3.282}$$

We note that $_{(t)}R(t) = {}_{(t)}U(t) = {}_{(t)}V(t) = I$.

We can also introduce the *relative right* and *left Cauchy–Green strain tensors* by

$$_{(t)}\mathbf{C}(\tau) = {}_{(t)}\mathbf{F}^T(\tau) \cdot {}_{(t)}\mathbf{F}(\tau) \quad \text{or} \quad {}_{(t)}C_{kl} = \xi_{\alpha,k}\xi_{\beta,l}\delta_{\alpha\beta}, \tag{3.283}$$

$$_{(t)}\mathbf{B}(\tau) = {}_{(t)}\mathbf{F}(\tau) \cdot {}_{(t)}\mathbf{F}^T(\tau) \quad \text{or} \quad {}_{(t)}B_{\alpha\beta} = \xi_{\alpha,k}\xi_{\beta,l}\delta_{kl}, \tag{3.284}$$

and recognize that

$$_{(t)}\mathbf{C}(t) = {}_{(t)}\mathbf{B}(t) = \mathbf{1}. \tag{3.285}$$

We also note that since

$$C_{KL}(\tau) = \xi_{\alpha,K}\xi_{\beta,L}\delta_{\alpha\beta} = \xi_{\alpha,k}\xi_{\beta,l}x_{k,K}x_{l,L}\delta_{\alpha\beta} = x_{k,K}\,{}_{(t)}C_{kl}\,x_{l,L}, \tag{3.286}$$

$$B_{\alpha\beta}(\tau) = \xi_{\alpha,K}\xi_{\beta,L}\delta_{KL} = \xi_{\alpha,k}\xi_{\beta,l}x_{k,K}x_{l,L}\delta_{KL} = {}_{(t)}F_{\alpha,k}\,B_{kl}\,{}_{(t)}F_{\beta,l}, \tag{3.287}$$

the relationships between the absolute and relative Cauchy–Green tensors are given by

$$\mathbf{C}(\tau) = \mathbf{F}^T(t) \cdot {}_{(t)}\mathbf{C}(\tau) \cdot \mathbf{F}(t) \quad \text{and} \quad \mathbf{B}(\tau) = {}_{(t)}\mathbf{F}(\tau) \cdot \mathbf{B}(t) \cdot {}_{(t)}\mathbf{F}^T(\tau), \tag{3.288}$$

which reduce to the standard definitions when $\tau = t$.

3.2.4 Stretch and spin

While the deformation gradient measures the local deformation, the material time derivative of the deformation gradient measures the rate at which such changes occur. Another measure for the rate of deformation is the *spatial gradient of velocity*. From before, they are related by $\mathbf{L} = (\text{grad } \mathbf{v})^T = \dot{\mathbf{F}} \cdot \mathbf{F}^{-1}$, where $\dot{\mathbf{F}}$ is the rate of change of deformation relative to the reference configuration. Similarly, we can define the rate of change of deformation relative to the current configuration by

$$L_{ik}(t) \equiv \frac{\partial v_i}{\partial x_k} = \left.\frac{\partial \dot{\xi}_\alpha}{\partial x_k}\right|_{\tau=t} = \frac{\partial}{\partial x_k}\left.\left(\frac{\partial \xi_\alpha}{\partial \tau}\right)\right|_{\tau=t} = \frac{\partial}{\partial \tau}\left.\left(\frac{\partial \xi_\alpha}{\partial x_k}\right)\right|_{\tau=t} = {}_{(t)}\dot{F}_{\alpha k}(\tau)\Big|_{\tau=t} = {}_{(t)}\dot{F}_{ik}(t), \tag{3.289}$$

or

$$\mathbf{L}(\mathbf{x}, t) = {}_{(t)}\dot{\mathbf{F}}(t). \tag{3.290}$$

In other words, the velocity gradient can also be interpreted as the rate of change of deformation relative to the current configuration. This can also be seen by using the derivative of (3.275) with respect to τ.

If we hold $\mathbf{x}$ and t fixed and take the derivative of the relative deformation gradient with respect to τ, using the polar decomposition, we obtain

$$ {}_{(t)}\dot{\mathbf{F}}(\tau) = {}_{(t)}\mathbf{R}(\tau) \cdot {}_{(t)}\dot{\mathbf{U}}(\tau) + {}_{(t)}\dot{\mathbf{R}}(\tau) \cdot {}_{(t)}\mathbf{U}(\tau), \tag{3.291} $$

and setting $\tau = t$, we have

$$ \mathbf{L}(t) = {}_{(t)}\dot{\mathbf{U}}(t) + {}_{(t)}\dot{\mathbf{R}}(t) = \mathbf{D}(t) + \mathbf{W}(t). \tag{3.292} $$

where we define

$$ D_{ik}(t) \equiv {}_{(t)}\dot{U}_{ik}(t) = \frac{\partial}{\partial \tau}\, {}_{(t)}U_{\alpha k}(\tau)\big|_{\tau=t} \tag{3.293} $$

and

$$ W_{ik}(t) \equiv {}_{(t)}\dot{R}_{ik}(t) = \frac{\partial}{\partial \tau}\, {}_{(t)}R_{\alpha k}(\tau)\big|_{\tau=t}. \tag{3.294} $$

The quantity $\mathbf{D}$ is called the *stretch* or *rate of strain tensor*, while the quantity $\mathbf{W}$ is called the *spin* or *rate of rotation tensor.*

Now we can easily see that since ${}_{(t)}U_{\alpha k}(\tau)|_{\tau=t}$ is symmetric, so is ${}_{(t)}\dot{U}_{ik}(t)$, thus the stretching tensor is symmetric, i.e.,

$$ D^T(t) = D(t) \qquad \text{or} \qquad D_{ki}(t) = D_{ik}(t). \tag{3.295} $$

Furthermore, to see that the spin tensor is skew-symmetric, we first note that since $R^{-1} = R^T$, we can write

$$ {}_{(t)}R_{\alpha k}(\tau)\ {}_{(t)}R_{\alpha l}(\tau) = \delta_{kl}, \tag{3.296} $$

and differentiating with respect to τ, we have

$$ {}_{(t)}\dot{R}_{\alpha k}(\tau)\ {}_{(t)}R_{\alpha l}(\tau) + {}_{(t)}R_{\alpha k}(\tau)\ {}_{(t)}\dot{R}_{\alpha l}(\tau) = 0. \tag{3.297} $$

Now evaluating at $\tau = t$, we obtain

$$ W_{ik}(t)\delta_{il} + \delta_{ik}W_{il}(t) = 0. \tag{3.298} $$

Thus

$$ W^T(t) = -W(t) \qquad \text{or} \qquad W_{ki}(t) = -W_{ik}(t). \tag{3.299} $$

Therefore, the decomposition of $\mathbf{L}(t)$,

$$ \mathbf{L}(t) = \mathbf{D}(t) + \mathbf{W}(t) \qquad \text{or} \qquad L_{ik} = D_{ik} + W_{ik}, \tag{3.300} $$

corresponds to the Cartesian decomposition into symmetric and skew-symmetric parts of $(\operatorname{grad}\mathbf{v})^T$ where we see that

$$ \mathbf{D}(t) = \frac{1}{2}\left(\mathbf{L} + \mathbf{L}^T\right) = \frac{1}{2}\left((\text{grad }\mathbf{v})^T + \text{grad }\mathbf{v}\right) \text{ or } D_{ik} = \frac{1}{2}\left(\frac{\partial v_i}{\partial x_k} + \frac{\partial v_k}{\partial x_i}\right), \tag{3.301} $$

$$ \mathbf{W}(t) = \frac{1}{2}\left(\mathbf{L} - \mathbf{L}^T\right) = \frac{1}{2}\left((\text{grad }\mathbf{v})^T - \text{grad }\mathbf{v}\right) \text{ or } W_{ik} = \frac{1}{2}\left(\frac{\partial v_i}{\partial x_k} - \frac{\partial v_k}{\partial x_i}\right). \tag{3.302} $$

3.2.5 Kinematical significance of D and W

We can write

$$d\xi_\alpha = {}_{(t)}F_{\alpha k}(\tau)dx_k \tag{3.303}$$

or differentiating with respect to τ,

$$\dot{\overline{d\xi_\alpha}} = {}_{(t)}\dot{F}_{\alpha k}(\tau)dx_k. \tag{3.304}$$

Evaluating the above at $\tau = t$, we have

$$\dot{\overline{dx_i}} = L_{ik}(t)dx_k. \tag{3.305}$$

If we let $d\xi_\alpha = l_\alpha d\xi$ and $dx_k = l_k dx$, then

$$d\xi = \left[{}_{(t)}C_{\alpha k}(\tau)l_\alpha l_k\right]^{1/2} dx, \tag{3.306}$$

where ${}_{(t)}C_{\alpha k}(\tau)$ is the relative right Cauchy–Green tensor. Now

$$\dot{\overline{d\xi}} = \frac{1}{2}\frac{{}_{(t)}\dot{C}_{\alpha k}(\tau)l_\alpha l_k}{\left[{}_{(t)}C_{\beta n}(\tau)l_\beta l_n\right]^{1/2}}dx, \tag{3.307}$$

and evaluating at $\tau = t$, we obtain

$$\dot{\overline{dx}} = \frac{1}{2}{}_{(t)}\dot{C}_{ik}(t)l_i l_k dx. \tag{3.308}$$

Since

$${}_{(t)}C(\tau) = {}_{(t)}F^T(\tau)\ {}_{(t)}F(\tau), \tag{3.309}$$

differentiating with respect to τ, we have

$${}_{(t)}\dot{C}(\tau) = {}_{(t)}\dot{F}^T(\tau)\ {}_{(t)}F(\tau) + {}_{(t)}F^T(\tau)\ {}_{(t)}\dot{F}(\tau), \tag{3.310}$$

which, when evaluated at $\tau = t$, gives

$${}_{(t)}\dot{C}(t) = L^T(t) + L(t) = 2D(t) \quad \text{or} \quad {}_{(t)}\dot{C}_{ik}(t) = \frac{\partial v_k}{\partial x_i} + \frac{\partial v_i}{\partial x_k} = 2D_{ik}(t). \tag{3.311}$$

Thus

$$\dot{\overline{dx}} = D_{ik}(t)l_i l_k dx. \tag{3.312}$$

Example

Suppose that we have $l_i = (1,0,0)$. Then

$$D_{11} = \frac{\dot{\overline{dx}}}{dx} = \dot{\overline{\ln|dx|}}. \tag{3.313}$$

The quantity D_{11} is the rate of change in length per unit length of an element parallel to the x_1 axis.

The relation between differential volumes at t and at τ is given by

$$dv(\tau) = {}_{(t)}J(\tau)\ dv(t), \tag{3.314}$$

and

$$\dot{\overline{dv(\tau)}} = \left[\frac{\partial}{\partial\tau}{}_{(t)}J(\tau)\right] dv(t). \tag{3.315}$$

Now using the chain rule and (3.60), we can write

$$\frac{\partial_{(t)}J(\tau)}{\partial\tau} = \frac{\partial_{(t)}J(\tau)}{\partial\xi_{\alpha,k}}\frac{\partial^2\xi_\alpha}{\partial\tau\partial x_k} = {}_{(t)}J(\tau)\frac{\partial x_k}{\partial\xi_\alpha}\frac{\partial^2\xi_\alpha}{\partial x_k\partial\tau}, \tag{3.316}$$

and at $\tau = t$,

$$\dot{\overline{dv(t)}} = \delta_{ki}\frac{\partial v_i}{\partial x_k}dv(t), \tag{3.317}$$

so

$$\dot{\overline{dv(t)}} = \frac{\partial v_i}{\partial x_i}dv(t) = D_{ii}(t)\ dv(t), \tag{3.318}$$

or

$$\frac{\dot{\overline{dv(t)}}}{dv(t)} = D_{ii}(t). \tag{3.319}$$

Thus, the trace of the stretch tensor corresponds to the local time rate of change of the volume per unit volume.

In addition,

$$d\xi_\alpha^{(1)}d\xi_\alpha^{(2)} = {}_{(t)}F_{\alpha k}(\tau)\,{}_{(t)}F_{\alpha l}(\tau)dx_k^{(1)}dx_l^{(2)}, \tag{3.320}$$

so

$$\dot{\overline{d\xi_\alpha^{(1)}d\xi_\alpha^{(2)}}} = \left({}_{(t)}\dot{F}_{\alpha k}(\tau)\,{}_{(t)}F_{\alpha l}(\tau) + {}_{(t)}F_{\alpha k}(\tau)\,{}_{(t)}\dot{F}_{\alpha l}(\tau)\right)dx_k^{(1)}dx_l^{(2)}, \tag{3.321}$$

and at $\tau = t$,

$$\dot{\overline{dx_i^{(1)}dx_i^{(2)}}} = \left(\frac{\partial v_l}{\partial x_k} + \frac{\partial v_k}{\partial x_l}\right)dx_k^{(1)}dx_l^{(2)} = 2D_{kl}\ dx_k^{(1)}dx_l^{(2)}. \tag{3.322}$$

Also at $\tau = t$, and since $l_\alpha^{(1)}l_\alpha^{(2)} = \cos\theta$,

$$\dot{\overline{dx_i^{(1)}dx_i^{(2)}}} = \left(\dot{\overline{dx^{(1)}}}dx^{(2)} + dx^{(1)}\dot{\overline{dx^{(2)}}}\right)\cos\theta - dx^{(1)}dx^{(2)}\dot{\theta}\sin\theta, \tag{3.323}$$

and if we let $\theta = \pi/2$, then we get

$$|D_{ij}| = -\frac{\dot{\theta}}{2}, \quad i \neq j, \tag{3.324}$$

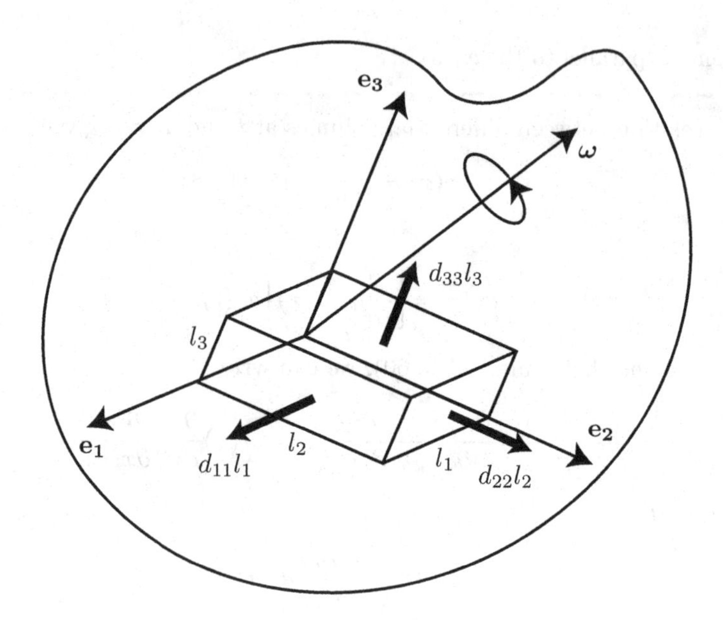

Figure 3.13: Decomposition of relative rate of deformation into rates of stretch and spin.

indicating that the off-diagonal terms of the stretch tensor provide information on the local rate of relative rotation between material line elements.

Lastly, we note that

$$d\xi_\alpha = {}_{(t)}F_{\alpha k}(\tau) dx_k \tag{3.325}$$

and

$$\overline{d\xi_\alpha}^{\boldsymbol{\cdot}} = {}_{(t)}\dot{F}_{\alpha k}(\tau) dx_k, \tag{3.326}$$

so that at $\tau = t$,

$$\overline{dx_i}^{\boldsymbol{\cdot}} = \frac{\partial v_i}{\partial x_k} dx_k = (D_{ik} + W_{ik})\, dx_k, \tag{3.327}$$

or

$$\frac{\overline{dx_i}^{\boldsymbol{\cdot}}}{dx_k} = D_{ik} + W_{ik}. \tag{3.328}$$

The decomposition is illustrated in Fig. 3.13.

If the body is rigid (in this case, it is also isochoric), $D_{ik} = 0$, and using (2.143) we have

$$\overline{dx_i}^{\boldsymbol{\cdot}} = W_{ik} dx_k = \epsilon_{ikl} \varpi_l dx_k = -\epsilon_{ilk} \varpi_l dx_k = -(\boldsymbol{\varpi} \times d\mathbf{x})_i, \tag{3.329}$$

so

$$\overline{d\mathbf{x}}^{\boldsymbol{\cdot}} = -\boldsymbol{\varpi} \times d\mathbf{x}, \tag{3.330}$$

where $\boldsymbol{\varpi}$ is the rigid body *angular velocity* (see (2.147)). Above, we have used the fact that it is always possible to associate an axial vector, in our case the angular velocity $\boldsymbol{\varpi}$, with any second-order skew-symmetric tensor, in our case the spin tensor $\mathbf{W}$, by the relation

$$\mathbf{W} = \boldsymbol{\varpi} \cdot \boldsymbol{\epsilon} \qquad \text{or} \qquad W_{ik} = \varpi_l \epsilon_{lik}. \tag{3.331}$$

Subsequently, since

$$\epsilon_{pik} W_{ik} = \epsilon_{pik}\epsilon_{lik}\varpi_l = 2\delta_{pl}\varpi_l = 2\varpi_p, \tag{3.332}$$

we also have, as in (2.142),

$$\boldsymbol{\varpi} = \frac{1}{2}\boldsymbol{\epsilon} : \mathbf{W} \qquad \text{or} \qquad \varpi_p = \frac{1}{2}\epsilon_{pik} W_{ik}. \tag{3.333}$$

Note that in this case stretching occurs in a direction perpendicular to the plane containing $\boldsymbol{\varpi}$ and $d\mathbf{x}$ and forms a right-handed system.

From above, we have that

$$\begin{aligned}
2\varpi_p &= \epsilon_{pik} W_{ik}, \\
&= -\epsilon_{pki}\left[\frac{1}{2}\left(\frac{\partial v_i}{\partial x_k} - \frac{\partial v_k}{\partial x_i}\right)\right], \\
&= -\frac{1}{2}\left(\epsilon_{pki} v_{i,k} - \epsilon_{pki} v_{k,i}\right), \\
&= -\epsilon_{pki} v_{i,k}, \\
&= -\left(\nabla \times \mathbf{v}\right)_p,
\end{aligned}$$

or, taking $\boldsymbol{\omega} = -2\boldsymbol{\varpi}$,

$$\boldsymbol{\omega} = \nabla \times \mathbf{v}. \tag{3.334}$$

The quantity $\boldsymbol{\omega}$ is called the *vorticity*, and we see that its magnitude is twice the angular velocity $\boldsymbol{\varpi}$. A motion for which the spin vanishes, i.e., $\mathbf{W} = \mathbf{0}$, is called *irrotational*.

Lastly, it is easy to see that

$$\mathbf{W}_{(1)} = \operatorname{tr}\mathbf{W} = 0, \qquad \mathbf{W}_{(2)} = \boldsymbol{\varpi} \cdot \boldsymbol{\varpi} = \frac{1}{4}\boldsymbol{\omega} \cdot \boldsymbol{\omega}, \quad \text{and} \quad \mathbf{W}_{(3)} = \det \mathbf{W} = 0. \tag{3.335}$$

The vorticity squared is called *enstrophy* and is a quantity directly related to the contribution of spin in the local kinetic energy of the associated motion.

3.2.6 Kinematics and dynamical systems

We will presently extend the kinematical tools that we have developed and apply them to the study of dynamical systems. As we will see, this leads to an interesting approach to understanding the dynamics of such systems.

The dynamical system

Consider the autonomous dynamical system

$$\frac{d\mathbf{x}}{dt} = \mathbf{v}(\mathbf{x}(t)), \qquad \mathbf{x}(t_0) = \mathbf{X}, \tag{3.336}$$

where $\mathbf{x}$ is considered a spatial curvilinear coordinate in an n-dimensional Euclidean space $\mathcal{E}^n$, and $\mathbf{v}(\mathbf{x})$ is the velocity at $\mathbf{x}(t)$. Under very general conditions, the solution of (3.336) is given by

$$\mathbf{x} = \chi(\mathbf{X}, t), \tag{3.337}$$

where we now think of $\mathbf{X}$ as coordinates from a reference frame in $\mathcal{E}^n$, which we take to be the same as that for $\mathbf{x}(t)$. We consider $\mathbf{x}(t)$ to be a field generated by the reference field $\mathbf{X}$ of all possible initial conditions. Thus, our phase space represents a continuum of all possible fields $\mathbf{x} \in \mathcal{E}^n$ emanating from the continuum of all possible initial conditions $\mathbf{X} \in \mathcal{E}^n$.

Now note from (3.336) that the velocity is tangential to the trajectory. Differentiating the velocity (taking the material derivative), we have

$$\frac{d\mathbf{v}}{dt} = (\operatorname{grad}\mathbf{v})^T \cdot \mathbf{v} = \mathbf{L} \cdot \mathbf{v}, \tag{3.338}$$

where $\mathbf{L}$ denotes the velocity gradient. Using the Cartesian decomposition, the velocity gradient can be rewritten as

$$\mathbf{L} = \mathbf{D} + \mathbf{W}. \tag{3.339}$$

Locally, $\mathbf{D}$ and $\mathbf{W}$ represent the stretching and twisting of the *tangent bundle* of trajectories in the neighborhood of the specific trajectory $\mathbf{x}(t)$ evolving from any point $\mathbf{X}$ and points near it.

Local tangent and normal spaces

As noted, since $\mathbf{v}(x)$ is tangential to the trajectory $\mathbf{x}(t)$, we can define the unit tangential vector $\mathbf{t}$ as

$$\mathbf{t} \equiv \frac{\mathbf{v}}{|\mathbf{v}|}. \tag{3.340}$$

Subsequently, dividing (3.338) by $|\mathbf{v}|$ and manipulating the equation, we arrive at the evolution equation for the tangent vector:

$$\frac{d\mathbf{t}}{dt} = \mathbf{L} \cdot \mathbf{t} - (\mathbf{t} \cdot \mathbf{L} \cdot \mathbf{t})\,\mathbf{t} = \mathbf{L} \cdot \mathbf{t} - (\mathbf{t} \cdot \mathbf{D} \cdot \mathbf{t})\,\mathbf{t}, \tag{3.341}$$

since $\mathbf{t} \cdot \mathbf{W} \cdot \mathbf{t} = 0$. Note that since $\mathbf{t} \cdot \mathbf{t} = 1$, and thus $\mathbf{t} \cdot d\mathbf{t}/dt = 0$, we see that taking the inner product of $\mathbf{t}$ with (3.341), the above equation is identically satisfied, as it should be.

Now, any unit vector normal to $\mathbf{t}$, say $\mathbf{n}_i$ for $i = 1, \ldots, n-1$, can be obtained by requiring that

$$\mathbf{n}_i \cdot \mathbf{t} = 0 \qquad \text{and} \qquad \mathbf{n}_i \cdot \mathbf{n}_j = \begin{cases} 1 & \text{if } i = j, \\ 0 & \text{if } i \neq j. \end{cases} \tag{3.342}$$

Note that $d\mathbf{n}_i/dt \cdot \mathbf{n}_j = 0$ for any $i, j = 1, \dots, n-1$. Differentiating (taking the material derivative of) $(3.342)_1$, and using (3.341) and $(3.342)_1$, we obtain

$$\frac{d\mathbf{n}_i}{dt} \cdot \mathbf{t} = -\mathbf{n}_i \cdot \frac{d\mathbf{t}}{dt} = -\mathbf{n}_i \cdot [\mathbf{L} \cdot \mathbf{t} - (\mathbf{t} \cdot \mathbf{D} \cdot \mathbf{t})\,\mathbf{t}] = -\mathbf{n}_i \cdot \mathbf{L} \cdot \mathbf{t}. \tag{3.343}$$

Subsequently, it follows that

$$\frac{d\mathbf{n}_i}{dt} = -\mathbf{n}_i \cdot \mathbf{L} + G\,\mathbf{n}_i, \tag{3.344}$$

where G is an arbitrary scalar function. To determine it, we take the inner product of this last equation with $\mathbf{n}_j$, from which we obtain

$$\frac{d\mathbf{n}_i}{dt} \cdot \mathbf{n}_j = -\mathbf{n}_i \cdot \mathbf{L} \cdot \mathbf{n}_j + G\,\mathbf{n}_i \cdot \mathbf{n}_j. \tag{3.345}$$

Now, using $(3.342)_2$ and its derivative (see above), we see that $G = \mathbf{n}_j \cdot \mathbf{L} \cdot \mathbf{n}_j = \mathbf{n}_j \cdot \mathbf{D} \cdot \mathbf{n}_j$, $\mathbf{n}_j \cdot \mathbf{W} \cdot \mathbf{n}_j = 0$, and $\mathbf{n}_i \cdot \mathbf{L} \cdot \mathbf{n}_j = 0$ when $i \neq j$. Subsequently, the evolution equation for the normal vector (3.344) becomes

$$\frac{d\mathbf{n}_i}{dt} = -\mathbf{n}_i \cdot \mathbf{L} + (\mathbf{n}_j \cdot \mathbf{D} \cdot \mathbf{n}_j)\,\mathbf{n}_i. \tag{3.346}$$

We note from (3.341) and (3.346) that the unit tangent and normal vectors can be taken as their negatives ($\mathbf{t} \to -\mathbf{t}$ and $\mathbf{n}_i \to -\mathbf{n}_i$) without changing the respective evolution equations.

Local deformations along the trajectory

We now examine the local deformations in the neighborhood of a specified trajectory $\mathbf{x}(t)$ by examining neighboring trajectories.

Three useful measures of deformation are the stretch in the tangential direction, the deformation of an area whose normal is in the tangential direction (say, how an initial circular area deforms into an ellipsoidal area as we move along the trajectory), and how a local volume deforms (say, from an initial spherical volume into an ellipsoidal volume as we move along the trajectory).

To examine the stretch in the tangential direction, we look at the square of the arc length $dx(t)$ (local differential element in $\mathcal{E}^n$):

$$(dx)^2 = d\mathbf{x} \cdot d\mathbf{x}. \tag{3.347}$$

Taking the material derivative, we find that

$$\frac{d}{dt}(dx)^2 = 2\,d\mathbf{x} \cdot \frac{d}{dt}(d\mathbf{x}) = 2\,d\mathbf{x} \cdot d\mathbf{v} = 2\,d\mathbf{x} \cdot \left(\frac{\partial \mathbf{v}}{\partial \mathbf{x}}\right)^T \cdot d\mathbf{x} = 2\,d\mathbf{x} \cdot \mathbf{L} \cdot d\mathbf{x} = 2\,d\mathbf{x} \cdot \mathbf{D} \cdot d\mathbf{x}. \tag{3.348}$$

Now, dividing through by $(dx)^2$, and noting that

$$\mathbf{t} = \frac{d\mathbf{x}}{dx}, \tag{3.349}$$

we obtain the equation for the *local relative tangent stretch rate*:

$$\omega_{\mathbf{t}} \equiv \frac{d}{dt}(\ln dx) = \frac{d}{dt}(\ln \lambda) = \mathbf{t} \cdot \mathbf{D} \cdot \mathbf{t}, \tag{3.350}$$

where we have noted from (3.20) that the arc length is related to the length stretch ratio by $\lambda = dx/dX$. In addition, we should recognize that $|\mathbf{t}\cdot\mathbf{D}\cdot\mathbf{t}| = |\mathbf{D}:\mathbf{tt}| \leq |\mathbf{D}|$, since $|\mathbf{tt}| = 1$, so that

$$\omega_{\mathbf{t}} \leq |\mathbf{D}|\,. \tag{3.351}$$

It is noted that this equation is also readily obtained from (3.305) or (3.312). Also, it is obvious that $\omega_{\mathbf{t}} = \lambda_{\mathbf{D}}$ is an extremum (or principal stretch) if $\mathbf{t} = \mathbf{t}_{\mathbf{D}}$, where $\mathbf{t}_{\mathbf{D}}$ is a principal direction of $\mathbf{D}$:

$$(\mathbf{D} - \lambda_{\mathbf{D}}\,\mathbf{1})\cdot\mathbf{t}_{\mathbf{D}} = \mathbf{0}, \tag{3.352}$$

since (3.350) corresponds to the necessary and sufficient conditions for the solution of (3.352) when $\lambda_{\mathbf{D}} = \omega_{\mathbf{t}}$. In such case, the corresponding instantaneous rate of rotation, from (3.352) and (3.341), is given by

$$\frac{d\mathbf{t}_{\mathbf{D}}}{dt} = \mathbf{W}\cdot\mathbf{t}_{\mathbf{D}} = \mathbf{w}\times\mathbf{t}_{\mathbf{D}}, \tag{3.353}$$

where $\mathbf{w} = \langle\mathbf{W}\rangle$ is the axial vector associated with the spin tensor $\mathbf{W}$ (see (2.148)).

To examine the local relative change in the differential volume $dv(t)$, one follows a similar procedure as above, or alternatively, using (3.319), to obtain the *local relative volume stretch rate*

$$\omega_V \equiv \frac{d}{dt}(\ln dv) = \frac{d}{dt}(\ln J) = \operatorname{tr}\mathbf{D}, \tag{3.354}$$

where we have recognized from (3.52) that the volume stretch ratio is given by $J = dv/dV$. Note that $|\operatorname{tr}\mathbf{D}| = |\mathbf{D}:\mathbf{1}| \leq |\mathbf{D}|$, since $|\mathbf{1}| = 1$, so that

$$\omega_V \leq |\mathbf{D}|\,. \tag{3.355}$$

Now, the easiest way to obtain the local relative change in the differential area $ds(t)$ whose normal is in the tangential direction is to first use (3.47):

$$dv = dx\,ds \qquad \text{so that} \qquad \ln dv = \ln dx + \ln ds. \tag{3.356}$$

Then,

$$\frac{d}{dt}(\ln ds) = \frac{d}{dt}(\ln dv) - \frac{d}{dt}(\ln dx) = \operatorname{tr}\mathbf{D} - \mathbf{t}\cdot\mathbf{D}\cdot\mathbf{t}. \tag{3.357}$$

But

$$\operatorname{tr}\mathbf{D} - \mathbf{t}\cdot\mathbf{D}\cdot\mathbf{t} = \mathbf{D}:\mathbf{1} - \mathbf{D}:\mathbf{t}\,\mathbf{t} = \mathbf{D}:(\mathbf{1} - \mathbf{t}\,\mathbf{t}) = \mathbf{D}:\mathbf{n}\,\mathbf{n} = \mathbf{n}\cdot\mathbf{D}\cdot\mathbf{n} = \sum_{i=1}^{n-1}\mathbf{n}_i\cdot\mathbf{D}\cdot\mathbf{n}_i. \tag{3.358}$$

Subsequently, we have that the *local relative area stretch rate*

$$\omega_{\mathbf{n}} \equiv \frac{d}{dt}(\ln ds) = \frac{d}{dt}(\ln \eta) = \mathbf{n}\cdot\mathbf{D}\cdot\mathbf{n} = \sum_{i=1}^{n-1}\mathbf{n}_i\cdot\mathbf{D}\cdot\mathbf{n}_i, \tag{3.359}$$

where we have recognized from (3.44) that the area stretch ratio is given by $\eta = ds/dS$. It is noted that $|\mathbf{n}\cdot\mathbf{D}\cdot\mathbf{n}| = |\mathbf{D}:\mathbf{nn}| \leq |\mathbf{D}|$, since $|\mathbf{nn}| = 1$, so that

$$\omega_{\mathbf{n}} \leq |\mathbf{D}|\,. \tag{3.360}$$

3.2.7 Internal angular velocity and acceleration

In the next chapter, we will formulate balance laws for a *polar material.* Such material is characterized kinematically by an *internal angular velocity* (an axial vector field), $\boldsymbol{\nu}$, that is independent of the translational velocity field. As noted earlier, for an ordinary continuum, the angular velocity field is equal to one-half of the vorticity (or the curl of the velocity field):

$$\boldsymbol{\varpi} = -\frac{1}{2}\boldsymbol{\omega} = -\frac{1}{2}\nabla \times \mathbf{v} \qquad \text{or} \qquad \varpi_i = -\frac{1}{2}\omega_i = \frac{1}{2}\epsilon_{ijk}\, v_{j,k}. \tag{3.361}$$

We interpret the usual angular velocity $\boldsymbol{\varpi}$ as an average angular velocity at location $\mathbf{x}$ and time t, while the internal angular velocity $\boldsymbol{\nu}$ represents the angular velocity of the polar-material particle at the same location and time. The internal angular velocity can also be represented by the second rank skew-symmetric rotation tensor

$$\boldsymbol{\Upsilon} = \boldsymbol{\nu}\cdot\boldsymbol{\epsilon} \qquad \text{or} \qquad \Upsilon_{jk} = \Upsilon_{[jk]} = \nu_i\,\epsilon_{ijk}. \tag{3.362}$$

The internal or particle angular velocity relative to the average local angular velocity is given by

$$\boldsymbol{\Theta} \equiv \boldsymbol{\nu} - \boldsymbol{\varpi} \qquad \text{or} \qquad \Theta_i = \nu_i - \varpi_i. \tag{3.363}$$

The quantity $\boldsymbol{\Theta}$ is called the *relative angular velocity.* In the case of irrotational internal motion, the particle angular velocity, $\boldsymbol{\nu}$, vanishes, and we note that in this case $\boldsymbol{\Theta} = -\boldsymbol{\varpi}$. This type of motion is more restrictive than ordinary irrotational motion, which requires that $\boldsymbol{\varpi}$ vanish. Stationary motion requires that the velocity field, and subsequently the average angular velocity, $\boldsymbol{\varpi}$, vanish. In this case, $\boldsymbol{\Theta} = \boldsymbol{\nu}$, so that the material particle is stationary but rotating. The necessary and sufficient conditions for rigid-body motion are that the stretch tensor $\mathbf{D}$ and the relative angular velocity $\boldsymbol{\Theta}$ both vanish.

Analogous to the translational acceleration, $\dot{\mathbf{v}}$, and the velocity gradient, $\mathbf{L} = (\text{grad}\,\mathbf{v})^T$, we define the *internal angular acceleration* by $\dot{\boldsymbol{\nu}}$, and the *internal angular velocity gradient* by

$$\boldsymbol{\Xi} \equiv (\text{grad}\,\boldsymbol{\nu})^T \qquad \text{or} \qquad \Xi_{kl} \equiv \nu_{k,l}. \tag{3.364}$$

Note that Ξ_{kl} are components of an axial second rank tensor and associated with it are the components of a third-rank tensor that is anti-symmetric with respect to its first pair of indices,

$$\widehat{\Xi}_{ijl} = \Xi_{kl}\,\epsilon_{kij} \qquad \text{and} \qquad \widehat{\Xi}_{ijl} = -\widehat{\Xi}_{jil}, \tag{3.365}$$

and thus does not correspond to a general tensor of rank 3. Its irreducible symmetry parts can be found by substituting $\widehat{\boldsymbol{\Xi}}$ in the relations (2.135)–(2.138) to find that

$$\left(\widehat{\Xi}_{ijl}\right)_1 = \left(\widehat{\Xi}_{(ijl)}\right) = 0, \tag{3.366}$$

$$\left(\widehat{\Xi}_{ijl}\right)_2 = \left(\widehat{\Xi}_{[ijl]}\right) = \xi\,\epsilon_{ijl}, \tag{3.367}$$

$$\left(\widehat{\Xi}_{ijl}\right)_3 = \frac{1}{3!}\left[\widehat{\Xi}_{(ij)l} - \widehat{\Xi}_{(jl)i}\right] = 0, \tag{3.368}$$

$$\left(\widehat{\Xi}_{ijl}\right)_4 = \frac{1}{3!}\left[\widehat{\Xi}_{(i|j|l)} - \widehat{\Xi}_{(j|i|l)}\right] = \Xi'_{kl}\,\epsilon_{kij}, \tag{3.369}$$

where we see that $(\widehat{\Xi}_{ijl})_2$ has one component and $(\widehat{\Xi}_{ijl})_4$ has eight components. Furthermore, we note that

$$\nu^{(0)} = \frac{1}{3}\operatorname{tr}\boldsymbol{\Xi} = \frac{1}{3}\operatorname{div}\boldsymbol{\nu} \qquad \text{and} \qquad \boldsymbol{\Xi}' = \boldsymbol{\Xi} - \frac{1}{3}(\operatorname{tr}\boldsymbol{\Xi})\,\mathbf{1}, \tag{3.370}$$

so that $\nu^{(0)}\,\mathbf{1}$ and $\boldsymbol{\Xi}'$ are the spherical and deviatoric parts of $\boldsymbol{\Xi}$.

3.3 Objective tensors

We use concepts associated with space and time to describe the motion of material objects based on our experience. Based on this experience, we ascribe certain fundamental properties to space and time. Specifically, we consider the properties of homogeneity and isotropy of space and time.

Space homogeneity means that a location in space is identical to any other location of space. That is, any physical process will occur the same way no matter where it occurs; i.e., under identical initial conditions, an experiment will yield the same result no matter where it is conducted. Shifting the origin means displacing the system. The implication of this is that we can choose the origin of our coordinate system anywhere we wish without affecting processes.

Space isotropy means that one direction in space is equivalent to any other direction. A particular experiment will yield the same result whether the laboratory is pointing north or east. That is, an arbitrary rotation of our coordinate system should not change the internal state of an isolated system, and thus, we should be able to orient our coordinate system any way we wish.

Time homogeneity means that one instant (or duration) of time is identical to any other instant (or duration) of time. An experiment should yield the same result when performed under the same conditions independent of the time of the day or the day of the year it is performed. Homogeneity of time implies that we can choose the initial time for an observation of a physical process to be any instant of time we desire.

Time isotropy means the equivalence of time directions. That is, the future direction is equivalent to the past direction. While such isotropy applies to the dynamics of Newtonian particles, processes consisting of a large number of particles do not show time reversibility and thus do not happen in nature. For complex macroscopic systems, this concept is replaced by the concept of entropy and the second law of thermodynamics, which are discussed in Chapters 4 and 5.

A *frame of reference* can be interpreted as an *observer* who observes an *event* in terms of positions and time with a ruler, a protractor, and a clock. Different observers may use different rulers, protractors, and clocks and come up with different results for the same event. However, if the same units of measure for their rulers, protractors, and clocks are used, they should obtain the same *distance*, *angle*, and *time lapse* between any two events under observation, even though the values of their observations may still be different. We shall impose these requirements on a change of frame from one to another.

In the formulation of physical laws, it is desirable to use quantities that are independent of the motion of the observer. Such quantities are called *objective* or *frame indifferent* or *frame invariant* since they reflect the objective properties of the object they embody. Such quantities are represented by tensors of various orders and they should be independent of the choice of the coordinate system in which they are expressed. Within the realm of classical mechanics, the most general transformation which represents the homogeneity and isotropy of space and homogeneity of time is a time-dependent rigid transformation, referred to as the *Euclidean transformation.* Under such transformation, lengths, angles, and time lapses are preserved. Since we are staying within the realm of classical mechanics, we require that physical quantities must be objective with respect to time-dependent rigid motions of the spatial frame of reference. This is known as the *principle of objectivity.* Let a Cartesian frame $\mathcal{F}$ be in relative rigid motion with respect to another frame, $\mathcal{F}'$. A point with Cartesian coordinates $\mathbf{x}$ at time t in $\mathcal{F}$ will have the Cartesian coordinates $\mathbf{x}'$ at time t' in $\mathcal{F}'$. Since the frames are in rigid motion with respect to one another, the two motions $\mathbf{x}(\mathbf{X},t)$ and $\mathbf{x}'(\mathbf{X},t')$ are equivalent if

$$\mathbf{x}'(\mathbf{X},t') = \mathbf{b}(t) + \mathbf{Q}(t)\cdot\mathbf{x}(\mathbf{X},t) \quad \text{or} \quad x_i'(X_j,t') = b_i(t) + Q_{ik}(t)x_k(X_j,t) \tag{3.371}$$

with

$$t' = a + t, \tag{3.372}$$

where

$$\mathbf{b}(t_0) = \mathbf{x}_0' - \mathbf{Q}(t_0)\cdot\mathbf{x}_0 \qquad \text{and} \qquad a = t_0' - t_0 \tag{3.373}$$

represent the position vector between the origins of the two frames and the initial time shift, respectively, $\mathbf{x}_0 \in \mathcal{E}^3$ and $\mathbf{x}_0' \in \mathcal{E}^3$ correspond to the absolute coordinates of the origins of $\mathcal{F}$ and $\mathcal{F}'$, $t_0 \in \mathcal{R}$ and $t_0' \in \mathcal{R}$ represent the absolute time coordinates in the two frames, and $\mathbf{Q}(t) = \text{grad}\,\mathbf{x}'$ is the orthogonal rotation tensor, so that

$$\mathbf{Q}\cdot\mathbf{Q}^T = \mathbf{Q}^T\cdot\mathbf{Q} = \mathbf{1} \qquad \text{or} \qquad Q_{ij}Q_{kj} = Q_{ji}Q_{jk} = \delta_{ik}. \tag{3.374}$$

The transformation between the two frames is illustrated in Fig. 3.14. Note that $\mathbf{b}$ and $\mathbf{Q}$ are functions of time only, and $\det Q(t) = \pm 1$. We recognize that the matrix $Q(t)$ has components Q_{ij} corresponding to direction cosines between the axes x_i' and x_j. We note that under this transformation, lengths, angles, and time lapses remain invariant; e.g.,

$$ds'^2 = dx_k'dx_k' = Q_{ki}dx_iQ_{kj}dx_j = \delta_{ij}dx_idx_j = dx_idx_i = ds^2. \tag{3.375}$$

Now any tensorial quantity is said to be objective, or frame indifferent, if in any two objectively equivalent rigid motions, it obeys the appropriate tensor transformation law for all times. In such case, we say that objective quantities are *invariant* under a change of observers. Thus, we recall that if $\mathbf{u}$ is a vector and $\mathbf{T}$ is a second-order tensor, then for them to be objective, they must satisfy the relations

$$\mathbf{u}'(\mathbf{X},t') = \mathbf{Q}(t)\cdot\mathbf{u}(\mathbf{X},t) \qquad \text{or} \qquad u_k'(\mathbf{X},t') = Q_{kl}(t)u_l(\mathbf{X},t) \tag{3.376}$$

and

$$\mathbf{T}'(\mathbf{X},t') = \mathbf{Q}(t)\cdot\mathbf{T}(\mathbf{X},t)\cdot\mathbf{Q}^T(t) \quad \text{or} \quad T_{kl}'(\mathbf{X},t') = Q_{km}(t)Q_{ln}(t)T_{mn}(\mathbf{X},t). \tag{3.377}$$

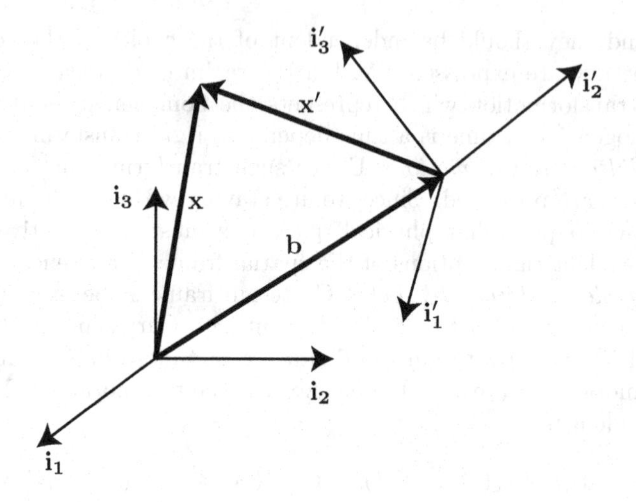

Figure 3.14: Rigid translation and rotation of Cartesian frames.

More generally, a tensor ψ of order n must satisfy the relation

$$\psi'_{i_1 i_2 \cdots i_n}(\mathbf{X}, t') = Q_{i_1 j_1}(t) Q_{i_2 j_2}(t) \cdots Q_{i_n j_n}(t) \psi_{j_1 j_2 \cdots j_n}(\mathbf{X}, t). \tag{3.378}$$

Example

For any pair of coordinates $\mathbf{x}_1$ and $\mathbf{x}_2$ in $\mathcal{F}$, we can associate the distance vector $\mathbf{u}$ such that

$$\mathbf{u} = \mathbf{x}_2 - \mathbf{x}_1. \tag{3.379}$$

Alternately, in $\mathcal{F}'$ we have

$$\mathbf{u}' = \mathbf{x}_2' - \mathbf{x}_1'. \tag{3.380}$$

Subsequently, using the Euclidean transformation (3.371), we have

$$\mathbf{u}' = \mathbf{Q}(t) \cdot (\mathbf{x}_2 - \mathbf{x}_1) = \mathbf{Q}(t) \cdot \mathbf{u}, \tag{3.381}$$

verifying that indeed distance vectors are objective quantities.

3.3.1 Apparent velocity

The velocity in $\mathcal{F}'$ is related to that in $\mathcal{F}$ by

$$\dot{x}_i' = \dot{b}_i(t) + \dot{Q}_{ik}(t) x_k + Q_{ik}(t) \dot{x}_k. \tag{3.382}$$

We immediately recognize that the velocity does not transform like a tensor, so it is not objective. Now we can rewrite

$$\dot{x}_i' = \dot{b}_i + \dot{Q}_{im} Q_{lm} Q_{lk} x_k + Q_{ik} \dot{x}_k = \dot{b}_i + \Omega_{il} (x_l' - b_l) + Q_{ik} \dot{x}_k, \tag{3.383}$$

where we have used (3.371) and (3.374), and have defined the *relative frame rotation tensor*, or *frame spin tensor*, by

$$\boldsymbol{\Omega}(t) \equiv \dot{\mathbf{Q}}(t) \cdot \mathbf{Q}^T(t) \qquad \text{or} \qquad \Omega_{il} \equiv \dot{Q}_{ik} Q_{lk}. \tag{3.384}$$

We want to show that $\boldsymbol{\Omega}^T = -\boldsymbol{\Omega}$ or $\Omega_{il} = -\Omega_{li}$. Now

$$Q_{ik} Q_{lk} = \delta_{il}, \tag{3.385}$$

and differentiating,

$$\dot{Q}_{ik} Q_{lk} + Q_{ik} \dot{Q}_{lk} = 0, \tag{3.386}$$

or

$$\Omega_{il} = -\Omega_{li}. \tag{3.387}$$

Subsequently, using (2.142) and (2.143), we can write

$$\Omega_{il} = w_p \epsilon_{pil} \qquad \text{and} \qquad w_p = \frac{1}{2} \epsilon_{plm} \Omega_{lm}, \tag{3.388}$$

where $\mathbf{w}$ is the *frame angular velocity* of $\mathcal{F}'$ with respect to frame $\mathcal{F}$. Thus, we write

$$\dot{\mathbf{x}}' = \dot{\widehat{\mathbf{x}}} + \mathbf{v}' \qquad \text{or} \qquad \dot{x}_i' = \dot{\widehat{x}}_i + v_i', \tag{3.389}$$

where

$$(\dot{\mathbf{x}}' - \mathbf{v}') \equiv \dot{\widehat{\mathbf{x}}} = \mathbf{Q} \cdot \dot{\mathbf{x}} \qquad \text{or} \qquad (\dot{x}_i' - v_i') \equiv \dot{\widehat{x}}_i = Q_{ik} \dot{x}_k, \tag{3.390}$$

and

$$\mathbf{v}' \equiv \dot{\mathbf{b}} - \mathbf{w} \times (\mathbf{x}' - \mathbf{b}) \qquad \text{or} \qquad v_i' \equiv \dot{b}_i - \epsilon_{ipl} w_p (x_l' - b_l). \tag{3.391}$$

We note that $\dot{\mathbf{x}}'$ is the *true velocity* while $\dot{\widehat{\mathbf{x}}}$ is called the *apparent velocity.* The difference between these two velocities is given by the *inertial velocity* $\mathbf{v}'$ of frame $\mathcal{F}'$ relative to frame $\mathcal{F}$. Note that while the true velocity is not an objective quantity, the apparent velocity is.

Furthermore, the internal angular velocity $\boldsymbol{\nu}$ (an axial vector) at a material point in frame $\mathcal{F}$ is related to that in frame $\mathcal{F}'$ by

$$(\boldsymbol{\nu}' - \mathbf{w}) \equiv \widehat{\boldsymbol{\nu}} = (\det \mathbf{Q})\, \mathbf{Q} \cdot \boldsymbol{\nu}. \tag{3.392}$$

It is pointed out that $\boldsymbol{\nu}'$ is the *true internal angular velocity* while $\widehat{\boldsymbol{\nu}}$ is called the *apparent internal angular velocity.* The difference between these two velocities is given by the *frame angular velocity* $\mathbf{w}$ of frame $\mathcal{F}'$ relative to frame $\mathcal{F}$. Note that while the true internal angular velocity is not an objective quantity, the apparent internal angular velocity is.

In general, the true velocity and angular velocity are objective if and only if $\mathbf{v}' = \mathbf{0}$ and $\mathbf{w} = \mathbf{0}$, which can be easily shown to lead to the requirements that $\mathbf{b} = \mathbf{b}_0$ = const. and $\mathbf{Q} = \mathbf{Q}_0$ = const. Subsequently,

$$\mathbf{x}'(\mathbf{X}, t') = \mathbf{b}_0 + \mathbf{Q}_0 \cdot \mathbf{x}(\mathbf{X}, t'), \tag{3.393}$$

$$t' = a + t, \tag{3.394}$$

which is called a *time-independent rigid transformation.*

3.3.2 Apparent acceleration

In transforming the acceleration from frame $\mathcal{F}$ to frame $\mathcal{F}'$, we obtain

$$\ddot{x}'_i = \ddot{b}_i + \ddot{Q}_{ik}x_k + 2\dot{Q}_{ik}\dot{x}_k + Q_{ik}\ddot{x}_k, \tag{3.395}$$

which again does not transform like a tensor. Now, using (3.374), we have

$$\ddot{x}'_i = \ddot{b}_i + \ddot{Q}_{im}Q_{lm}Q_{lk}x_k + 2\dot{Q}_{im}Q_{lm}Q_{lk}\dot{x}_k + Q_{ik}\ddot{x}_k. \tag{3.396}$$

Differentiating the frame spin tensor, we also have

$$\begin{aligned}\dot{\Omega}_{il} &= \ddot{Q}_{ik}Q_{lk} + \dot{Q}_{ik}\dot{Q}_{lk},\\ &= \ddot{Q}_{ik}Q_{lk} + \dot{Q}_{ip}Q_{sp}Q_{sk}\dot{Q}_{lk},\\ &= \ddot{Q}_{ik}Q_{lk} + \Omega_{is}\Omega_{ls},\end{aligned}$$

so that, also using (3.371) and (3.384), (3.396) becomes

$$\begin{aligned}\ddot{x}'_i &= \ddot{b}_i + (\dot{\Omega}_{il} - \Omega_{is}\Omega_{ls})(x'_l - b_l) + 2\Omega_{il}(\dot{x}'_l - \dot{b}_l) + 2\Omega_{is}\Omega_{ls}(x'_l - b_l) + Q_{ik}\ddot{x}_k,\\ &= \ddot{b}_i + \dot{\Omega}_{il}(x'_l - b_l) + 2\Omega_{il}(\dot{x}'_l - \dot{b}_l) - \Omega_{is}\Omega_{sl}(x'_l - b_l) + Q_{ik}\ddot{x}_k.\end{aligned} \tag{3.397}$$

Thus, in terms of the frame's angular velocity, we have

$$\ddot{\mathbf{x}}' = \ddot{\bar{\mathbf{x}}} + \mathbf{i}', \tag{3.398}$$

where

$$(\ddot{\mathbf{x}}' - \mathbf{i}') \equiv \ddot{\bar{\mathbf{x}}} = \mathbf{Q}(t)\cdot\ddot{\mathbf{x}}, \tag{3.399}$$

and

$$\mathbf{i}' \equiv \ddot{\mathbf{b}} - \dot{\mathbf{w}} \times (\mathbf{x}' - \mathbf{b}) - 2\mathbf{w} \times (\dot{\mathbf{x}}' - \dot{\mathbf{b}}) - \mathbf{w} \times [\mathbf{w} \times (\mathbf{x}' - \mathbf{b})]. \tag{3.400}$$

We note that $\ddot{\mathbf{x}}'$ is the *true acceleration* while $\ddot{\bar{\mathbf{x}}}$ is called the *apparent acceleration*. The difference between these two accelerations is given by the *inertial acceleration* $\mathbf{i}'$, which consists respectively of the inertial acceleration of relative translation of the frames, the *Euler acceleration*, the *Coriolis acceleration*, and the *centripetal acceleration* (its negative is also called the *centrifugal acceleration*).

We also note that the internal angular acceleration $\dot{\boldsymbol{\nu}}$ at a material point in frame $\mathcal{F}$ is related to that in frame $\mathcal{F}'$ by

$$(\dot{\boldsymbol{\nu}}' - \boldsymbol{\upsilon}') \equiv \dot{\bar{\boldsymbol{\nu}}} = (\det\mathbf{Q})\,\mathbf{Q}\cdot\dot{\boldsymbol{\nu}}, \tag{3.401}$$

where

$$\boldsymbol{\upsilon}' \equiv \dot{\mathbf{w}} + \boldsymbol{\Omega}\cdot(\boldsymbol{\nu}' - \mathbf{w}) = \dot{\mathbf{w}} - \mathbf{w} \times (\boldsymbol{\nu}' - \mathbf{w}). \tag{3.402}$$

It is pointed out that $\dot{\boldsymbol{\nu}}'$ is the *true internal angular acceleration* while $\dot{\bar{\boldsymbol{\nu}}}$ is called the *apparent internal angular acceleration*. The difference between these two angular accelerations is given by the *inertial internal angular acceleration* $\boldsymbol{\upsilon}'$, which consists respectively of the angular acceleration of relative rotation of the frames and the *Coriolis angular acceleration*.

Now we know that the location of a point will appear different to observers located at different places. Similarly, as we have seen, the velocity of a point is dependent upon the velocity of the observer. Therefore, these quantities are not objective (however, we note that the apparent quantities are). On the other hand, the distance between two points and the angles between two directions are independent of the rigid motion of the frame of reference (observer).

Note that in general the translational and internal angular accelerations are objective if and only if $\mathbf{i}' = \mathbf{0}$ and $\boldsymbol{v}' = \mathbf{0}$, which can be shown to lead to $\mathbf{Q} = \mathbf{Q}_0 =$ const. and $\mathbf{b}(t) = \mathbf{b}_0 + \mathbf{V}t$, where $\mathbf{b}_0 =$ const. and $\mathbf{V} =$ const. Subsequently, we have

$$\mathbf{x}'(\mathbf{X}, t') = \mathbf{b}_0 + \mathbf{V}t + \mathbf{Q}_0 \cdot \mathbf{x}(\mathbf{X}, t), \tag{3.403}$$

$$t' = a + t, \tag{3.404}$$

which is called a *Galilean transformation.* A Galilean frame differs from a time-independent rigid frame by the constant translation velocity $\mathbf{V}$. Newton's second law of motion is known to be valid only in this special frame of reference, also known as the *inertial frame.* Note that the velocity is not frame indifferent with respect to the Galilean transformation, but is so only under the more restrictive time-independent rigid transformation.

Einstein removed these restrictions by examining frame-invariance properties of arbitrary four-dimensional space-time transformations in his work on general relativity.

3.3.3 Properties of kinematic quantities

In classical mechanics, physical properties of materials should not depend on the coordinate frame selected. Properties should be the same whether or not the observer is in motion. Thus, the evolution equations as well as the constitutive equations that we address later must be objective, or frame invariant, with respect to rigid motions of the spatial, or Euclidean, frame of reference. We refer to this as *objectivity* or *frame indifference.*

In order to see how a reference configuration may be affected by a change of frame, we choose the reference configuration as the configuration occupied by some body at time t_0, so that for some arbitrary point X_K

$$x_i = \chi_i(X_K, t).$$

By noting that at $t = t_0$ we have that $\mathbf{x} = \mathbf{X}$, it now follows that in the new frame

$$X'_L = \chi'_L(X_K, t'_0) = b_L(t_0) + Q_{LK}(t_0)X_K.$$

On the other hand, the motion relative to the change of frame is given by

$$x'_i(\mathbf{X}', t') = b_i(t) + Q_{ik}(t)x_k(\mathbf{X}, t). \tag{3.405}$$

The deformation gradient in the new frame is then given by

$$F'_{iK}(\mathbf{X}', t') = \frac{\partial x'_i}{\partial X'_K} = \frac{\partial x'_i}{\partial x_j}\frac{\partial x_j}{\partial X_L}\frac{\partial X_L}{\partial X'_K}.$$

Subsequently, and more simply, we see that the deformation gradient in the new frame is given by

$$\mathbf{F}'(t) = \mathbf{Q}(t)\cdot\mathbf{F}\cdot\mathbf{Q}^T(t_0) \qquad \text{or} \qquad F'_{iK} = Q_{ij}(t)F_{jL}Q_{KL}(t_0), \tag{3.406}$$

and is thus not objective. The deformation gradient is not an absolute tensor; it is referred to as a *two-point tensor* or a *double vector*, since it is a quantity that transforms as a vector with respect to each of the indices.

With polar decompositions of $\mathbf{F}$ and $\mathbf{F}'$, we also have

$$\mathbf{R}'\cdot\mathbf{U}' = \mathbf{Q}(t)\cdot\mathbf{R}\cdot\mathbf{U}\cdot\mathbf{Q}^T(t_0) \qquad \text{and} \qquad \mathbf{V}'\cdot\mathbf{R}' = \mathbf{Q}(t)\cdot\mathbf{V}\cdot\mathbf{R}\cdot\mathbf{Q}^T(t_0).$$

By the uniqueness of such decompositions, we find that

$$\mathbf{U}' = \mathbf{Q}(t_0)\cdot\mathbf{U}\cdot\mathbf{Q}^T(t_0), \quad \mathbf{V}' = \mathbf{Q}(t)\cdot\mathbf{V}\cdot\mathbf{Q}^T(t), \quad \mathbf{R}' = \mathbf{Q}(t)\cdot\mathbf{R}\cdot\mathbf{Q}^T(t_0), \tag{3.407}$$

and subsequently

$$\mathbf{C}' = \mathbf{Q}(t_0)\cdot\mathbf{C}\cdot\mathbf{Q}^T(t_0), \qquad \mathbf{B}' = \mathbf{Q}(t)\cdot\mathbf{B}\cdot\mathbf{Q}^T(t). \tag{3.408}$$

Therefore, we conclude that $\mathbf{V}$ and $\mathbf{B}$ are objective tensors, while $\mathbf{R}$, $\mathbf{U}$, and $\mathbf{C}$ are not objective tensors.

Moreover, if we take the material derivative of the deformation gradient in the moving frame, we have

$$\dot{\mathbf{F}}' = \mathbf{Q}(t)\cdot\dot{\mathbf{F}}\cdot\mathbf{Q}^T(t_0) + \dot{\mathbf{Q}}(t)\cdot\mathbf{F}\cdot\mathbf{Q}^T(t_0),$$

and since $\dot{\mathbf{F}} = \mathbf{L}\cdot\mathbf{F}$, we have

$$\begin{aligned}\mathbf{L}'\cdot\mathbf{F}' &= \mathbf{Q}(t)\cdot\mathbf{L}\cdot\mathbf{F}\cdot\mathbf{Q}^T(t_0) + \dot{\mathbf{Q}}(t)\cdot\mathbf{F}\cdot\mathbf{Q}^T(t_0),\\ &= \mathbf{Q}(t)\cdot\mathbf{L}\cdot\mathbf{Q}^T(t)\cdot\mathbf{F}' + \dot{\mathbf{Q}}(t)\cdot\mathbf{Q}^T(t)\cdot\mathbf{F}',\end{aligned}$$

or, using (3.384) and since $\mathbf{F}'$ is non-singular,

$$\mathbf{L}' = \mathbf{Q}(t)\cdot\mathbf{L}\cdot\mathbf{Q}^T(t) + \boldsymbol{\Omega}(t). \tag{3.409}$$

Lastly, since $\mathbf{L} = \mathbf{D} + \mathbf{W}$, the above becomes

$$\mathbf{D}' + \mathbf{W}' = \mathbf{Q}(t)\cdot(\mathbf{D} + \mathbf{W})\cdot\mathbf{Q}^T(t) + \boldsymbol{\Omega}(t).$$

By separating symmetric and skew-symmetric parts, we obtain

$$\mathbf{D}' = \mathbf{Q}(t)\cdot\mathbf{D}\cdot\mathbf{Q}^T(t) \qquad \text{or} \qquad D'_{ik} = Q_{ip}Q_{kq}D_{pq}, \tag{3.410}$$

and

$$\mathbf{W}' = \mathbf{Q}(t)\cdot\mathbf{W}\cdot\mathbf{Q}^T(t) + \boldsymbol{\Omega}(t) \qquad \text{or} \qquad W'_{ik} = Q_{ip}Q_{kq}W_{pq} + \Omega_{ik}. \tag{3.411}$$

Therefore, the rate of strain tensor $\mathbf{D}$ is an objective quantity, while the velocity gradient $\mathbf{L}$ and the rate of rotation tensor $\mathbf{W}$ are not objective quantities.

Note that since the average angular velocity and the internal angular velocity transform as

$$\boldsymbol{\varpi}' = (\det\mathbf{Q})\,\mathbf{Q}\cdot\boldsymbol{\varpi} + \mathbf{w} \qquad \text{and} \qquad \boldsymbol{\nu}' = (\det\mathbf{Q})\,\mathbf{Q}\cdot\boldsymbol{\nu} + \mathbf{w}, \tag{3.412}$$

the relative angular velocity and internal angular velocity gradient transform as objective axial quantities (see (3.363) and (3.364)):

$$\mathbf{\Theta}' = (\det \mathbf{Q})\, \mathbf{Q} \cdot \mathbf{\Theta} \qquad \text{and} \qquad \mathbf{\Xi}' = (\det \mathbf{Q})\, \mathbf{Q} \cdot \mathbf{\Xi} \cdot \mathbf{Q}^T. \tag{3.413}$$

Now assume that we have an objective vector field $\mathbf{u}(\mathbf{x}, t)$, so that

$$\mathbf{u}'(\mathbf{x}', t') = \mathbf{Q}(t) \cdot \mathbf{u}(\mathbf{x}, t).$$

Taking the gradient with respect to $\mathbf{x}$, we have

$$\text{grad } \mathbf{u}'(\mathbf{x}', t') = (\text{grad}' \ \mathbf{u}') \cdot (\text{grad } \mathbf{x}') = \mathbf{Q}(t) \cdot \text{grad } \mathbf{u}(\mathbf{x}, t).$$

But since $\mathbf{Q}(t) = \text{grad } \mathbf{x}'$, we have

$$(\text{grad } \mathbf{u})' = \mathbf{Q}(t) \cdot (\text{grad } \mathbf{u}(\mathbf{x}, t)) \cdot \mathbf{Q}^T(t). \tag{3.414}$$

On the other hand, if we express this vector field in the material coordinate

$$\mathbf{u}'(\mathbf{X}', t') = \mathbf{Q}(t) \cdot \mathbf{u}(\mathbf{X}, t),$$

then by taking the gradient with respect to $\mathbf{X}$, we easily find that

$$(\text{Grad } \mathbf{u})' = \mathbf{Q}(t) \cdot (\text{Grad } \mathbf{u}) \cdot \mathbf{Q}^T(t_0). \tag{3.415}$$

Furthermore, if we take the material derivative of the vector field, we have

$$\begin{aligned} \dot{\mathbf{u}}' &= \mathbf{Q}(t) \cdot \dot{\mathbf{u}} + \dot{\mathbf{Q}}(t) \cdot \mathbf{u}, \\ &= \mathbf{Q}(t) \cdot \dot{\mathbf{u}} + \dot{\mathbf{Q}}(t) \cdot \mathbf{Q}^T(t) \cdot \mathbf{u}', \\ &= \mathbf{Q}(t) \cdot \dot{\mathbf{u}} + \mathbf{\Omega}(t) \cdot \mathbf{u}'. \end{aligned} \tag{3.416}$$

Therefore, if $\mathbf{u}$ is an objective vector field, then its spatial gradient, grad $\mathbf{u}$, is an objective quantity, while its material gradient, Grad $\mathbf{u}$, and its material time derivative, $\dot{\mathbf{u}}$, are not objective quantities.

If ϕ is an objective scalar field, then we easily find that

$$\dot{\phi}' = \dot{\phi}, \quad (\text{grad } \phi)' = \mathbf{Q}(t) \cdot (\text{grad } \phi), \quad (\text{Grad } \phi)' = \mathbf{Q}(t_0) \cdot (\text{Grad } \phi), \tag{3.417}$$

so that the material derivative and the spatial gradient are objective, while the material gradient is not.

Similarly, we can show that if ψ is an objective tensor field of order n, then the material derivative $\dot{\psi}$ is not objective for $n > 0$, the spatial gradient grad ψ is an objective tensor field of order $n+1$, while the material gradient Grad ψ is not an objective tensor quantity.

It should be noted that our analysis of frame invariance is based on the Euclidean transformation (3.371)–(3.372) since it is expected that physical quantities should be invariant under such transformation. We observe that if the reference configuration is unaffected by the change of frame, then $\mathbf{Q}(t_0) = \mathbf{1}$. Lastly, it is easy to see that a number of quantities (e.g., $\mathbf{F}$ and $\mathbf{C}$) that are not invariant under the Euclidean transformation are invariant under the more restrictive Galilean transformation (3.403)–(3.404), since then $\mathbf{Q}(t) = \mathbf{Q}(t_0) = \mathbf{Q}_0$.

3.3.4 Corotational and convected derivatives

Suppose we have the objective vector field $\mathbf{u}$ so that

$$u_i' = Q_{ik}(t)u_k. \tag{3.418}$$

As we have seen, the material derivative of an objective vector field $\mathbf{u}$ is not objective. However, let's look at the vector

$$\mathring{\mathbf{u}} \equiv \dot{\mathbf{u}} - \mathbf{W} \cdot \mathbf{u} \qquad \text{or} \qquad \mathring{u}_i \equiv \dot{u}_i - W_{ik}u_k, \tag{3.419}$$

and see how it transforms. Using (3.411), we have

$$\begin{aligned}
\mathring{u}_i' = \dot{u}_i' - W_{ik}'u_k' &= \dot{Q}_{ik}u_k + Q_{ik}\dot{u}_k - \Omega_{ik}Q_{kl}u_l - Q_{ip}Q_{kl}Q_{ks}W_{pl}u_s, \\
&= \dot{Q}_{ik}u_k + Q_{ik}\dot{u}_k - \dot{Q}_{ip}Q_{kp}Q_{kl}u_l - Q_{ip}\delta_{ls}W_{pl}u_s, \\
&= \dot{Q}_{ik}u_k + Q_{ik}\dot{u}_k - \dot{Q}_{ip}u_p - Q_{ip}W_{pl}u_l, \\
&= Q_{ik}\left(\dot{u}_k - W_{kl}u_l\right), \\
&= Q_{ik}\mathring{u}_k.
\end{aligned} \tag{3.420}$$

Thus we find that the quantity $\mathring{\mathbf{u}}$, called the *corotational time derivative*, does transform like a tensor. Such derivative of an objective vector field is not unique. For example, one can easily verify that the *convected time derivative*

$$\overset{\odot}{\mathbf{u}} \equiv \dot{\mathbf{u}} - \mathbf{L} \cdot \mathbf{u} \qquad \text{or} \qquad \overset{\odot}{u}_i \equiv \dot{u}_i - L_{ik}u_k \tag{3.421}$$

is also objective.

Analogously, for a second-order objective tensor field $\mathbf{T}$, using the same procedure as above, one can show that the quantity

$$\mathring{\mathbf{T}} \equiv \dot{\mathbf{T}} - \mathbf{W} \cdot \mathbf{T} + \mathbf{T} \cdot \mathbf{W} \qquad \text{or} \qquad \mathring{T}_{ij} \equiv \dot{T}_{ij} - W_{ik}T_{kj} + T_{ik}W_{kj}, \tag{3.422}$$

called the corotational or *Jaumann derivative*, transforms like an objective second-order tensor. Again we note that the definitions of such derivatives are not unique. For example, it can also be shown that the convected rates given by the *Oldroyd* tensor

$$\overset{\star}{\mathbf{T}} \equiv \dot{\mathbf{T}} - \mathbf{T} \cdot \mathbf{L}^T - \mathbf{L} \cdot \mathbf{T} \qquad \text{or} \qquad \overset{\star}{T}_{ij} \equiv \dot{T}_{ij} - T_{ik}L_{jk} - L_{ik}T_{kj}, \tag{3.423}$$

the *Truesdell* tensor

$$\overset{\diamond}{\mathbf{T}} \equiv \dot{\mathbf{T}} - \mathbf{L}^T \cdot \mathbf{T} - \mathbf{T} \cdot \mathbf{L} + (\mathrm{tr}\ \mathbf{L})\mathbf{T} \quad \text{or} \quad \overset{\diamond}{T}_{ij} \equiv \dot{T}_{ij} - L_{ki}T_{kj} - T_{ik}L_{kj} + L_{kk}T_{ij}, \tag{3.424}$$

as well as the *Cotter–Rivlin* tensor

$$\overset{\triangle}{\mathbf{T}} \equiv \dot{\mathbf{T}} + \mathbf{L}^T \cdot \mathbf{T} + \mathbf{T} \cdot \mathbf{L} \quad \text{or} \quad \overset{\triangle}{T}_{ij} \equiv \dot{T}_{ij} + L_{ki}T_{kj} + T_{ik}L_{kj} \tag{3.425}$$

are also objective.

3.3.5 Push-forward and pull-back operations

Transformations between material and spatial descriptions are sometimes called *push-forward* and *pull-back operations*. A push-forward operation transforms a tensor-valued quantity based on the reference configuration to the current configuration. A pull-back operation transforms a tensor-valued quantity based in the current configuration to the reference configuration. A pull-back operation is an inverse of the push-forward operation.

Consider the Green–St. Venant strain tensor $\mathbf{E}$, which is defined in the reference configuration. From it, it is possible to compute the corresponding Almansi–Hamel strain tensor $\mathbf{e}$ in the current configuration by a push-forward operation. To affect this, we rewrite (3.178) as follows:

$$\begin{aligned}
\mathbf{e} &= \frac{1}{2}\left(\mathbf{1} - \mathbf{B}^{-1}\right) \\
&= \frac{1}{2}\left[\mathbf{1} - \left(\mathbf{F}\cdot\mathbf{F}^T\right)^{-1}\right] \\
&= \frac{1}{2}\left[\mathbf{1} - \mathbf{F}^{-T}\cdot\mathbf{F}^{-1}\right] \\
&= \frac{1}{2}\mathbf{F}^{-T}\cdot\left[\mathbf{F}^T\cdot\left(\mathbf{1} - \mathbf{F}^{-T}\cdot\mathbf{F}^{-1}\right)\cdot\mathbf{F}\right]\cdot\mathbf{F}^{-1} \\
&= \frac{1}{2}\mathbf{F}^{-T}\cdot\left(\mathbf{F}^T\cdot\mathbf{F} - \mathbf{1}\right)\cdot\mathbf{F}^{-1} \\
&= \frac{1}{2}\mathbf{F}^{-T}\cdot\left(\mathbf{C} - \mathbf{1}\right)\cdot\mathbf{F}^{-1} \\
&= \mathbf{F}^{-T}\cdot\mathbf{E}\cdot\mathbf{F}^{-1} \\
&\equiv \chi_\star\left(\mathbf{E}\right),
\end{aligned} \tag{3.426}$$

where we have used (3.175). Note that $\mathbf{F}^{-1}$ maps the current configuration into the reference configuration, $\mathbf{E}$ maps the reference configuration into the reference configuration, and $\mathbf{F}^{-T}$ maps the reference configuration into the current configuration. The operator $\chi_\star\left(\bullet\right)$ is the push-forward operator.

The inverse, or pull-back operation of $\mathbf{e}$, from (3.175), is given from

$$\begin{aligned}
\mathbf{E} &= \frac{1}{2}\left(\mathbf{C} - \mathbf{1}\right) \\
&= \frac{1}{2}\left(\mathbf{F}^T\cdot\mathbf{F} - \mathbf{1}\right) \\
&= \frac{1}{2}\mathbf{F}^T\cdot\left[\mathbf{F}^{-T}\cdot\left(\mathbf{F}^T\cdot\mathbf{F} - \mathbf{1}\right)\cdot\mathbf{F}^{-1}\right]\cdot\mathbf{F} \\
&= \frac{1}{2}\mathbf{F}^T\cdot\left(\mathbf{1} - \mathbf{F}^{-T}\cdot\mathbf{F}^{-1}\right)\cdot\mathbf{F} \\
&= \frac{1}{2}\mathbf{F}^T\cdot\left[\mathbf{1} - \left(\mathbf{F}\cdot\mathbf{F}^T\right)^{-1}\right]\cdot\mathbf{F} \\
&= \frac{1}{2}\mathbf{F}^T\cdot\left(\mathbf{1} - \mathbf{B}^{-1}\right)\cdot\mathbf{F} \\
&= \mathbf{F}^T\cdot\mathbf{e}\cdot\mathbf{F} \\
&\equiv \chi_\star^{-1}\left(\mathbf{e}\right),
\end{aligned} \tag{3.427}$$

where we have used (3.178). Note that $\mathbf{F}$ maps the reference configuration into the current configuration, $\mathbf{e}$ maps the current configuration into the current configuration, and $\mathbf{F}^T$ maps the current configuration into the reference configuration. The operator $\chi_\star^{-1}(\bullet)$ is the pull-back operator.

The above push-forward and pull-back operators can be applied to push-forward or pull-back other corresponding rank 2 quantities. Analogous push-forward and pull-back operators can be obtained for tensors of other ranks. The push-forward and pull-back operators provide relationships between the same type of components (contravariant, covariant, or mixed) between the current and reference configurations. Thus, for a vector $\mathbf{v}$ in the reference configuration, there are two possible push-forward operators ($\mathbf{F}^{-T}\cdot\mathbf{v}$ for covariant components and $\mathbf{F}\cdot\mathbf{v}$ for contravariant components) and two corresponding push-back operators ($\mathbf{F}^T\cdot\mathbf{v}$ for covariant components and $\mathbf{F}^{-1}\cdot\mathbf{v}$ for contravariant components). For relations involving a second rank tensor $\mathbf{A}$ in the reference configuration, there are four possible push-forward operators ($\mathbf{F}^{-T}\cdot\mathbf{A}\cdot\mathbf{F}^{-1}$ for covariant components, $\mathbf{F}\cdot\mathbf{A}\cdot\mathbf{F}^T$ for contravariant components, and $\mathbf{F}\cdot\mathbf{A}\cdot\mathbf{F}^{-1}$ and $\mathbf{F}^{-T}\cdot\mathbf{A}\cdot\mathbf{F}^T$ between the two mixed components) and four corresponding push-back operators ($\mathbf{F}^T\cdot\mathbf{A}\cdot\mathbf{F}$ for covariant components, $\mathbf{F}^{-1}\cdot\mathbf{A}\cdot\mathbf{F}^{-T}$ for contravariant components, and $\mathbf{F}^{-1}\cdot\mathbf{A}\cdot\mathbf{F}$ and $\mathbf{F}^T\cdot\mathbf{A}\cdot\mathbf{F}^{-T}$ between the two mixed components). Note that the definitions (3.426) and (3.427) provide the corresponding relations for $\mathbf{E}$ and $\mathbf{e}$ written using covariant components.

Lastly, we point out that the *Lie time derivative* can be defined using pull-back and push-forward operations. For example, if $\mathbf{v} = v^i\,\mathbf{e}_i$ is a spatial vector written using contravariant components, the material time derivative is given by

$$\dot{\mathbf{v}} = \dot{v}^i\,\mathbf{e}_i + v^i\,\dot{\mathbf{e}}_i. \tag{3.428}$$

The Lie time derivative is a material derivative holding the deformed basis constant, i.e., it corresponds to the first term on the right hand side of the above equation:

$$\mathcal{L}\mathbf{v} = \dot{v}^i\,\mathbf{e}_i. \tag{3.429}$$

In terms of pull-back and push-forward operations, it is given by

$$\mathcal{L}\mathbf{v} = \chi_\star\left(\frac{d}{dt}\left[\chi_\star^{-1}(\mathbf{v})\right]\right). \tag{3.430}$$

The spatial vector is first pulled back to the reference configuration, there the differentiation is carried out, where the base vectors are constant, and then the vector is pushed forward again to the current configuration.

One of the most important uses of the Lie time derivative is that Lie time derivatives of objective spatial tensors are objective spatial tensors. For example, it can be easily shown that the Lie time derivative of a rank 2 spatial tensor $\mathbf{A}$ written using covariant components is given by

$$\mathcal{L}\mathbf{A} = \dot{\mathbf{A}} + \mathbf{L}^T\cdot\mathbf{A} + \mathbf{A}\cdot\mathbf{L}, \tag{3.431}$$

which we recognize as the Cotter–Rivlin tensor (3.425).

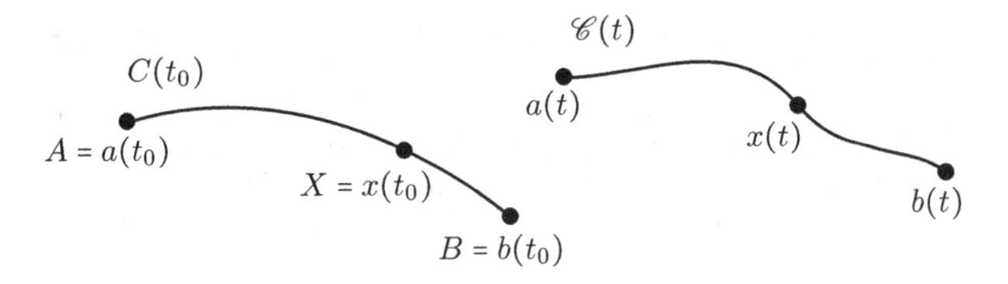

Figure 3.15: Material curve segment.

3.4 Transport theorems

3.4.1 Material derivative of a line integral

Let $\psi(\mathbf{x}, t)$ be a tensor field of arbitrary order that is continuous on the arbitrary oriented curve $\mathscr{C}(t)$ illustrated in Fig. 3.15. Then, using (3.12), the material derivative of the line integral of field ψ over the material curve is given by

$$\begin{aligned}
\frac{d}{dt}\int_{\mathscr{C}(t)} \psi(x_k,t)\, dx_i &= \frac{d}{dt}\int_C \psi(X_L,t)\frac{\partial x_i}{\partial X_K}\, dX_K, \\
&= \int_C \frac{\partial}{\partial t}\left[\psi(X_L,t)\frac{\partial x_i}{\partial X_K}\right] dX_K, \\
&= \int_C \left(\dot{\psi}(X_L,t)\frac{\partial x_i}{\partial X_K} + \psi(X_L,t)\frac{\partial^2 x_i}{\partial t \partial X_K}\right) dX_K, \\
&= \int_C \left(\dot{\psi}(X_L,t)\frac{\partial x_i}{\partial X_K} + \psi(X_L,t) v_{i,l}\frac{\partial x_l}{\partial X_K}\right) dX_K.
\end{aligned}$$

Now we can rewrite

$$\frac{d}{dt}\int_{\mathscr{C}(t)} \psi(x_k,t)\, dx_i = \int_{\mathscr{C}(t)} \dot{\psi}\, dx_i + \int_{\mathscr{C}(t)} \psi\, v_{i,l}\, dx_l, \tag{3.432}$$

or

$$\frac{d}{dt}\int_{\mathscr{C}(t)} \psi(\mathbf{x},t)\, d\mathbf{x} = \int_{\mathscr{C}(t)} \dot{\psi}\, d\mathbf{x} + \int_{\mathscr{C}(t)} \psi\, (\operatorname{grad}\mathbf{v})^T \cdot d\mathbf{x}, \tag{3.433}$$

and note that

$$\dot{\psi} = \frac{\partial \psi(\mathbf{X},t)}{\partial t} = \frac{\partial \psi(\mathbf{x},t)}{\partial t} + \mathbf{v}(\mathbf{x},t)\cdot \operatorname{grad}\psi(\mathbf{x},t), \tag{3.434}$$

where $\mathbf{v} = \dot{\mathbf{x}}$. If ψ is a vector field, i.e., $\psi \to u_l$, then if we project it on the curve, we have

$$\frac{d}{dt}\int_{\mathscr{C}(t)} u_i(x_k,t)\, dx_i = \int_{\mathscr{C}(t)} (\dot{u}_i + u_l\, v_{l,i})\, dx_i, \tag{3.435}$$

or

$$\frac{d}{dt}\int_{\mathscr{C}(t)} \mathbf{u}(\mathbf{x},t)\cdot d\mathbf{x} = \int_{\mathscr{C}(t)} \left[\dot{\mathbf{u}} + \mathbf{u}\cdot(\operatorname{grad}\mathbf{v})^T\right]\cdot d\mathbf{x}. \tag{3.436}$$

Example

As an illustration on the use of (3.436), we examine the evolution of the length of the curve between the points $\mathbf{x}_1$ and $\mathbf{x}_2$ by taking the vector $\mathbf{u}$ to be the unit tangent vector $\mathbf{t}$ along the curve. In this case, if we call the length of the material curve L, using (3.436) and the fact that $\dot{\mathbf{t}} \cdot \mathbf{t} = 0$, we have

$$\frac{dL}{dt} = \frac{d}{dt}\int_{\mathscr{C}(t)} dx = \frac{d}{dt}\int_{\mathscr{C}(t)} \mathbf{t}\cdot d\mathbf{x} = \int_{\mathscr{C}(t)} \left(\dot{\mathbf{t}} + \mathbf{t}\cdot\mathbf{L}\right)\cdot d\mathbf{x} = \int_{\mathscr{C}(t)} \mathbf{t}\cdot\mathbf{L}\cdot\mathbf{t}\,dx. \tag{3.437}$$

But note from (3.15) that

$$\mathbf{L}\cdot\mathbf{t} = (\text{grad }\mathbf{v})^T \cdot \frac{d\mathbf{x}}{dx} = \frac{d\mathbf{v}}{dx}; \tag{3.438}$$

thus we obtain

$$\frac{dL}{dt} = \int_{\mathscr{C}(t)} \mathbf{t}\cdot d\mathbf{v} = \int_{\mathscr{C}(t)} d(\mathbf{v}\cdot\mathbf{t}) - \int_{\mathscr{C}(t)} \mathbf{v}\cdot d\mathbf{t} = \mathbf{v}\cdot\mathbf{t}\Big|_{\mathbf{x}_1}^{\mathbf{x}_2} - \int_{\mathscr{C}(t)} \kappa_L\,\mathbf{v}\cdot\mathbf{n}\,dx, \tag{3.439}$$

where $d\mathbf{t}/dx = \kappa_L\,\mathbf{n}$, $\mathbf{n}$ is the principal normal to $\mathbf{t}$, and κ_L is the principal curvature. Here we note that since $d\mathbf{t}/dx\cdot\mathbf{t} = 0$, then $d\mathbf{t}/dx$ is orthogonal to $\mathbf{t}$, and thus one chooses $d\mathbf{t}/dx = \kappa_L\,\mathbf{n}$ with $d\mathbf{t}/dx\cdot d\mathbf{t}/dx = \kappa_L^2$ so that $\mathbf{n}$ is a unit vector.

We note that if $\mathscr{C}(t)$ is a closed curve ($\mathbf{x}_1 = \mathbf{x}_2$) or $\mathbf{v}\cdot\mathbf{t} = \mathbf{0}$ at $\mathbf{x}_1$ and $\mathbf{x}_2$, we then have that

$$\frac{dL}{dt} = -\int_{\mathscr{C}(t)} \kappa_L\,\mathbf{v}\cdot\mathbf{n}\,dx. \tag{3.440}$$

Note that if the tensor field ψ depends on one spatial dimension, i.e., $\psi = \psi(\xi,t)$, where ξ is measured along the curve, and $\{\mathscr{C}(t) : \xi_1(t) \le \xi \le \xi_2(t)\}$, then taking $v(\xi,t) = \dot{\xi}(t)$, the above result leads to the Leibniz rule since

$$\frac{d}{dt}\int_{\xi_1(t)}^{\xi_2(t)} \psi(\xi,t)\,d\xi = \int_{\xi_1(t)}^{\xi_2(t)} \left(\frac{\partial\psi}{\partial t} + v\frac{\partial\psi}{\partial\xi}\right)d\xi + \int_{\xi_1(t)}^{\xi_2(t)} \psi\frac{\partial v}{\partial\xi}\,d\xi, \tag{3.441}$$

$$= \int_{\xi_1(t)}^{\xi_2(t)} \frac{\partial\psi}{\partial t}\,d\xi + \int_{\xi_1(t)}^{\xi_2(t)} \frac{\partial(\psi v)}{\partial\xi}\,d\xi, \tag{3.442}$$

$$= \int_{\xi_1(t)}^{\xi_2(t)} \frac{\partial\psi}{\partial t}\,d\xi + \psi(\xi_2,t)v(\xi_2,t) - \psi(\xi_1,t)v(\xi_1,t), \tag{3.443}$$

$$= \int_{\xi_1(t)}^{\xi_2(t)} \frac{\partial\psi}{\partial t}\,d\xi + \psi(\xi_2,t)\dot{\xi}_2(t) - \psi(\xi_1,t)\dot{\xi}_1(t). \tag{3.444}$$

In the above steps, we have assumed that ψ is differentiable on $\mathscr{C}(t)$. To remove this assumption, and thus allow for ψ to be discontinuous, as indicated in Fig. 3.16,

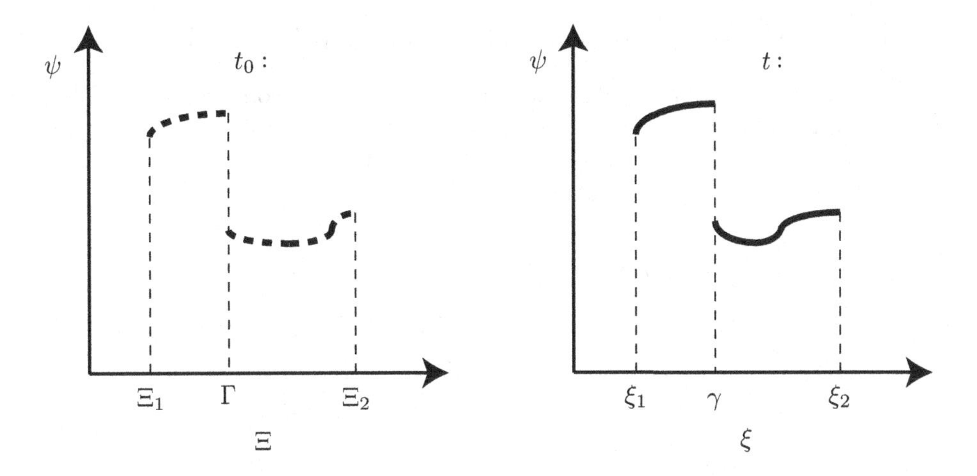

Figure 3.16: Material line segment with discontinuity of ψ at $\gamma(t)$.

assume that it is discontinuous at the point $\gamma(t)$ in the interval $\mathscr{C}(t)$. Now ψ is continuous in the subintervals $\xi_1(t) \leq \xi < \gamma(t)$ and $\gamma(t) < \xi \leq \xi_2(t)$, so we can apply the above result in these subintervals to obtain

$$\frac{d}{dt}\int_{\xi_1(t)}^{\gamma(t)} \psi(\xi,t)\,d\xi = \int_{\xi_1(t)}^{\gamma(t)} \frac{\partial\psi}{\partial t}\,d\xi + \psi^-(\gamma,t)\dot{\gamma}(t) - \psi(\xi_1,t)\dot{\xi}_1(t) \tag{3.445}$$

and

$$\frac{d}{dt}\int_{\gamma(t)}^{\xi_2(t)} \psi(\xi,t)\,d\xi = \int_{\gamma(t)}^{\xi_2(t)} \frac{\partial\psi}{\partial t}\,d\xi + \psi(\xi_2,t)\dot{\xi}_2(t) - \psi^+(\gamma,t)\dot{\gamma}(t), \tag{3.446}$$

where we have defined

$$\psi^+(\gamma,t) \equiv \lim_{\xi\downarrow\gamma(t)} \psi(\xi,t) \qquad \text{and} \qquad \psi^-(\gamma,t) \equiv \lim_{\xi\uparrow\gamma(t)} \psi(\xi,t). \tag{3.447}$$

Now adding the above results, we obtain the generalized Leibniz rule

$$\frac{d}{dt}\int_{\mathscr{C}(t)-\gamma(t)} \psi(\xi,t)\,d\xi = \int_{\mathscr{C}(t)-\gamma(t)} \frac{\partial\psi}{\partial t}\,d\xi + \left[\psi(\xi,t)v(\xi,t)\right]_{\xi_1}^{\xi_2} - [\![\psi(\gamma,t)]\!]\dot{\gamma}(t), \tag{3.448}$$

where we have denoted the jump in ψ by

$$[\![\psi(\gamma,t)]\!] \equiv \psi^+(\gamma,t) - \psi^-(\gamma,t). \tag{3.449}$$

Lastly, using the generalized divergence theorem (2.301) on a line, we can rewrite the generalized Leibniz rule in the form

$$\frac{d}{dt}\int_{\mathscr{C}(t)-\gamma(t)} \psi(\xi,t)\,d\xi = \int_{\mathscr{C}(t)-\gamma(t)} \left[\frac{\partial\psi}{\partial t} + \frac{\partial(\psi v)}{\partial\xi}\right] d\xi + [\![\psi(\gamma,t)\left[v(\gamma,t) - \dot{\gamma}(t)\right]]\!]. \tag{3.450}$$

3.4.2 Material derivative of a surface integral

Let $\psi(\mathbf{x},t)$ be a tensor field of arbitrary order that is continuous on an arbitrary material surface $\mathscr{S}(t)$, which is moving with velocity $\mathbf{v}(t)$ and is bounded by a closed curve $\mathscr{C}(t)$. Then, using (3.36), the material derivative of the differential element of surface area is given by

$$\begin{aligned}\frac{d}{dt}\int_{\mathscr{S}(t)}\psi(x_k,t)\,ds_i &= \frac{d}{dt}\int_{\mathbf{S}}\psi(X_K,t)J(X_K,t)X_{L,i}\,dS_L,\\ &= \int_{\mathbf{S}}\frac{\partial}{\partial t}\left[\psi(X_K,t)J(X_K,t)X_{L,i}\right]dS_L,\\ &= \int_{\mathbf{S}}\left[\frac{\partial\psi(X_K,t)}{\partial t}J(X_K,t)X_{L,i}+\right.\\ &\qquad \psi(X_K,t)\frac{\partial J(X_K,t)}{\partial t}X_{L,i}+\\ &\qquad \left.\psi(X_K,t)J(X_K,t)\frac{\partial X_{L,i}}{\partial t}\right]dS_L.\end{aligned}$$

However, using (3.60), we have

$$\left.\frac{\partial J}{\partial t}\right|_{X_K} = \frac{\partial J}{\partial x_{i,K}}\frac{\partial^2 x_i}{\partial t\partial X_K} = JX_{K,i}\frac{\partial v_i}{\partial X_K} = J\frac{\partial v_i}{\partial x_i}, \tag{3.451}$$

and it can be shown that

$$\dot{\overline{\mathbf{F}^{-1}}} = -\mathbf{F}^{-1}\cdot\dot{\mathbf{F}}\cdot\mathbf{F}^{-1} = -\mathbf{F}^{-1}\cdot\mathbf{L}. \tag{3.452}$$

Subsequently, we have

$$\begin{aligned}\frac{d}{dt}\int_{\mathscr{S}(t)}\psi(x_k,t)\,ds_i = \int_{\mathbf{S}}&\left[\frac{\partial\psi(X_K,t)}{\partial t}J(X_K,t)X_{L,i}+\right.\\ \psi(X_K,t)J(X_K,t)\frac{\partial v_j}{\partial x_j}X_{L,i} &- \left.\psi(X_K,t)J(X_K,t)X_{L,j}\frac{\partial v_j}{\partial x_i}\right]dS_L,\end{aligned}$$

and, using (3.36), we arrive at

$$\frac{d}{dt}\int_{\mathscr{S}(t)}\psi(x_k,t)\,ds_i = \int_{\mathscr{S}(t)}(\dot{\psi}+\psi v_{j,j})\,ds_i - \int_{\mathscr{S}(t)}\psi v_{j,i}\,ds_j, \tag{3.453}$$

or

$$\frac{d}{dt}\int_{\mathscr{S}(t)}\psi(\mathbf{x},t)\,d\mathbf{s} = \int_{\mathscr{S}(t)}(\dot{\psi}+\psi\,\mathrm{div}\,\mathbf{v})\,d\mathbf{s} - \int_{\mathscr{S}(t)}\psi\,(\mathrm{grad}\,\mathbf{v})^T\cdot d\mathbf{s}. \tag{3.454}$$

Note that if ψ is a vector field projected on the surface, i.e., $\psi\to u_i$, then we have

$$\frac{d}{dt}\int_{\mathscr{S}(t)}u_i(x_k,t)\,ds_i = \int_{\mathscr{S}(t)}(\dot{u}_i+u_iv_{k,k}-u_kv_{i,k})\,ds_i, \tag{3.455}$$

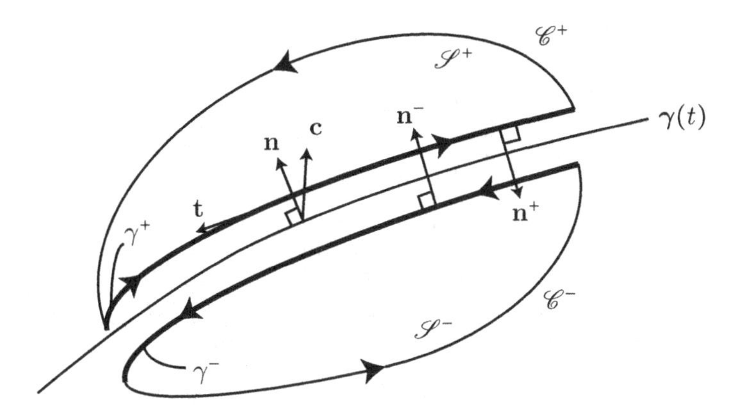

Figure 3.17: Material surface $\mathscr{S}(t)$ with discontinuity along curve $\gamma(t)$.

or

$$\begin{aligned}
\frac{d}{dt}\int_{\mathscr{S}(t)} \mathbf{u}(\mathbf{x},t)\cdot d\mathbf{s} &= \int_{\mathscr{S}(t)}\left[\frac{\partial \mathbf{u}}{\partial t} + (\mathbf{v}\cdot \operatorname{grad})\,\mathbf{u} + \mathbf{u}\,(\operatorname{div}\mathbf{u}) - (\mathbf{u}\cdot\operatorname{grad})\,\mathbf{v}\right]\cdot d\mathbf{s},\\
&= \int_{\mathscr{S}(t)}\left[\frac{\partial \mathbf{u}}{\partial t} + \mathbf{v}\ \operatorname{div}\mathbf{u} + \operatorname{curl}(\mathbf{u}\times\mathbf{v})\right]\cdot d\mathbf{s}, \qquad (3.456)\\
&= \int_{\mathscr{S}(t)}\left(\frac{\partial \mathbf{u}}{\partial t} + \mathbf{v}\ \operatorname{div}\mathbf{u}\right)\cdot d\mathbf{s} + \int_{\mathscr{C}(t)} \mathbf{u}\times\mathbf{v}\cdot d\mathbf{x}, \qquad (3.457)
\end{aligned}$$

where we have used Stokes' theorem (2.292) in the last step. We note from (3.456) that in order that the flux of the vector field $\mathbf{u}(\mathbf{x},t)$ across every material surface remain constant in time, it is necessary and sufficient that *Zorawski's criterion* be satisfied:

$$\frac{\partial \mathbf{u}}{\partial t} + \mathbf{v}\ \operatorname{div}\mathbf{u} + \operatorname{curl}(\mathbf{u}\times\mathbf{v}) = \mathbf{0}. \qquad (3.458)$$

A similar argument can be extended to a surface $\mathscr{S}(t)$ intersected by a discontinuity line $\gamma(t)$ moving with a velocity $\mathbf{c}(t)$ on the surface, as illustrated in Fig. 3.17. Applying our result (3.457) of the transport of vector quantity $\mathbf{u}$ to the two subsurfaces separated by $\gamma(t)$, adding the results, letting $\gamma^+(t)$ and $\gamma^-(t)$ approach $\gamma(t)$, noting that $d\gamma^+ = -d\gamma^- = -d\gamma$ and $\mathbf{c}^+ = \mathbf{c}^- = \mathbf{c}$, and using the generalized Stokes theorem (2.300), we obtain

$$\begin{aligned}
\frac{d}{dt}\int_{\mathscr{S}(t)-\gamma(t)} \mathbf{u}(\mathbf{x},t)\cdot d\mathbf{s} &= \int_{\mathscr{S}(t)-\gamma(t)}\left[\frac{\partial \mathbf{u}}{\partial t} + \mathbf{v}\ \operatorname{div}\mathbf{u} + \operatorname{curl}(\mathbf{u}\times\mathbf{v})\right]\cdot d\mathbf{s} + \\
&\quad \int_{\gamma(t)} [\![\mathbf{u}\times(\mathbf{v}-\mathbf{c})]\!]\cdot d\boldsymbol{\gamma}. \qquad (3.459)
\end{aligned}$$

Example

As an illustration on the use of (3.457), we examine the rate of increase of a surface area by taking the vector $\mathbf{u}$ to be the unit vector $\mathbf{n}$ normal to the surface. In this case, if we call the area of the material surface A, using

(3.457) and the fact that $\partial\mathbf{n}/\partial t\cdot\mathbf{n}=0$, we have

$$\frac{dA}{dt} = \frac{d}{dt}\int_{\mathscr{S}(t)} ds \tag{3.460}$$

$$= \frac{d}{dt}\int_{\mathscr{S}(t)} \mathbf{n}\cdot d\mathbf{s} \tag{3.461}$$

$$= \int_{\mathscr{S}(t)}\left[\frac{\partial\mathbf{n}}{\partial t}+\mathbf{v}\,(\operatorname{div}\mathbf{n})\right]\cdot d\mathbf{s}+\int_{\mathscr{C}(t)}\mathbf{n}\times\mathbf{v}\cdot d\mathbf{x} \tag{3.462}$$

$$= \int_{\mathscr{S}(t)}(\operatorname{div}\mathbf{n})\,\mathbf{v}\cdot d\mathbf{s}+\int_{\mathscr{C}(t)}\mathbf{n}\times\mathbf{v}\cdot d\mathbf{x}. \tag{3.463}$$

We now note from (D.217) that $\operatorname{div}\mathbf{n}=-2K_M$, where K_M is the mean surface curvature. Subsequently, we obtain

$$\frac{dA}{dt} = -\int_{\mathscr{S}(t)} K_M\,\mathbf{v}\cdot\mathbf{n}\,ds+\int_{\mathscr{C}(t)}\mathbf{n}\times\mathbf{v}\cdot\mathbf{t}\,dx. \tag{3.464}$$

3.4.3 Material derivative of a volume integral

Let $\psi(\mathbf{x},t)$ be a tensor field of arbitrary order that is continuous in $\mathscr{V}(t)$, an arbitrary volume bounded by the closed surface $\mathscr{S}(t)$, and let a material point in $\mathscr{V}(t)$ move with velocity $\mathbf{v}(t)$. Then, using (3.51) the material derivative of the field ψ over the volume is given by

$$\begin{aligned}\frac{d}{dt}\int_{\mathscr{V}(t)}\psi(x_k,t)\;dv &= \frac{d}{dt}\int_V\psi(X_K,t)J(X_K,t)\;dV,\\ &= \int_V\frac{\partial}{\partial t}\left[\psi(X_K,t)J(X_K,t)\right]\;dV,\\ &= \int_V\left[\frac{\partial\psi(X_K,t)}{\partial t}J(X_K,t)+\psi(X_K,t)\frac{\partial J(X_K,t)}{\partial t}\right]dV.\end{aligned}$$

However, using (3.451), we have

$$\begin{aligned}\frac{d}{dt}\int_{\mathscr{V}(t)}\psi(x_k,t)\;dv &= \int_V\left[\dot\psi(X_K,t)+\psi(X_K,t)\frac{\partial X_K}{\partial x_i}\frac{\partial v_i(X_K,t)}{\partial X_K}\right]J(X_K,t)\;dV,\\ &= \int_{\mathscr{V}(t)}\left(\dot\psi+\psi\frac{\partial v_i}{\partial x_i}\right)dv,\\ &= \int_{\mathscr{V}(t)}\left(\frac{\partial\psi}{\partial t}+v_i\frac{\partial\psi}{\partial x_i}+\psi\frac{\partial v_i}{\partial x_i}\right)dv.\end{aligned}$$

Thus we arrive at

$$\frac{d}{dt}\int_{\mathscr{V}(t)}\psi(x_k,t)\;dv=\int_{\mathscr{V}(t)}\left[\frac{\partial\psi}{\partial t}+\frac{\partial}{\partial x_i}(v_i\psi)\right]dv= \int_{\mathscr{V}(t)}\frac{\partial\psi}{\partial t}\;dv+\int_{\mathscr{S}(t)}\psi v_i\;ds_i, \tag{3.465}$$

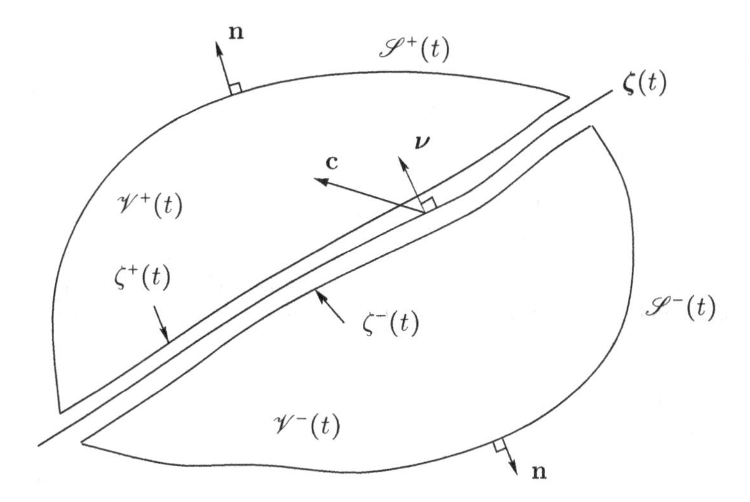

Figure 3.18: Material volume $\mathscr{V}(t)$ with discontinuity along surface $\boldsymbol{\zeta}(t)$.

or

$$\frac{d}{dt}\int_{\mathscr{V}(t)} \psi(\mathbf{x},t)\, dv = \int_{\mathscr{V}(t)} \left[\frac{\partial\psi}{\partial t} + \operatorname{div}(\mathbf{v}\psi)\right] dv = \int_{\mathscr{V}(t)} \frac{\partial\psi}{\partial t}\, dv + \int_{\mathscr{S}(t)} \psi\mathbf{v}\cdot d\mathbf{s}, \tag{3.466}$$

where the last step is obtained by the use of the divergence theorem (2.289). This result is known as *Reynolds' transport theorem.*

Example

Let $\psi = 1$ and $V(t)$ be an arbitrary subvolume of a continuous material body which is in motion. It then follows from (3.466) that

$$\frac{dV}{dt} = \frac{d}{dt}\int_{\mathscr{V}(t)} dv = \int_{\mathscr{V}(t)} \operatorname{div}\mathbf{v}\, dv = \int_{\mathscr{S}(t)} \mathbf{v}\cdot d\mathbf{s} = \int_{\mathscr{S}(t)} \mathbf{v}\cdot\mathbf{n}\, ds. \tag{3.467}$$

We see that the volume remains constant, i.e., the motion is *incompressible,* if and only if $\operatorname{div}\mathbf{v} = 0$.

Note that in the last step, the conventional divergence theorem is used, which assumes that the quantity $\psi\mathbf{v}$ is continuous in $\mathscr{V}(t)$. However, time rates of integrals over regions containing a discontinuity surface are common occurrences. Thus, we generalize the above result by assuming that the volume $\mathscr{V}(t)$ is intersected by a surface of discontinuity $\zeta(t)$ with unit normal $\boldsymbol{\nu}$ moving with velocity $\mathbf{c}(t)$, as illustrated in Fig. 3.18. Now applying the above result to the two subvolumes in which the quantity $\psi\mathbf{v}$ is continuous, we have

$$\frac{d}{dt}\int_{\mathscr{V}^+(t)} \psi(\mathbf{x},t)\, dv = \int_{\mathscr{V}^+(t)} \frac{\partial\psi}{\partial t}\, dv + \int_{\mathscr{S}^+(t)} \psi\mathbf{v}\cdot\mathbf{n}\, ds - \int_{\zeta^+(t)} \psi^+\mathbf{c}^+\cdot\boldsymbol{\nu}^+ d\zeta \tag{3.468}$$

and

$$\frac{d}{dt}\int_{\mathscr{V}^-(t)} \psi(\mathbf{x},t)\, dv = \int_{\mathscr{V}^-(t)} \frac{\partial \psi}{\partial t}\, dv + \int_{\mathscr{S}^-(t)} \psi \mathbf{v}\cdot\mathbf{n}\, ds + \int_{\varsigma^-(t)} \psi^- \mathbf{c}^- \cdot \boldsymbol{\nu}^- d\varsigma, \quad (3.469)$$

where we have defined

$$\psi^+(\mathbf{x},t) \equiv \lim_{\mathbf{x}\downarrow\varsigma(t)} \psi(\mathbf{x},t) \qquad \text{and} \qquad \psi^-(\mathbf{x},t) \equiv \lim_{\mathbf{x}\uparrow\varsigma(t)} \psi(\mathbf{x},t), \quad (3.470)$$

and analogously for $\mathbf{c}$ and $\boldsymbol{\nu}$. Now upon adding the two equations, letting ς^+ and ς^- approach ς, and noting that $d\boldsymbol{\varsigma}^+ = -d\boldsymbol{\varsigma}^- = -d\boldsymbol{\varsigma} = -\boldsymbol{\nu}d\varsigma$ and $\mathbf{c}^+ = \mathbf{c}^- = \mathbf{c}$, we obtain

$$\frac{d}{dt}\int_{\mathscr{V}(t)-\varsigma(t)} \psi(\mathbf{x},t)\, dv = \int_{\mathscr{V}(t)-\varsigma(t)} \frac{\partial \psi}{\partial t}\, dv + \int_{\mathscr{S}(t)-\varsigma(t)} \psi \mathbf{v}\cdot d\mathbf{s} - \int_{\varsigma(t)} [\![\psi \mathbf{c}]\!]\cdot d\boldsymbol{\varsigma}, \quad (3.471)$$

where we define

$$[\![\mathbf{A}]\!] \equiv \mathbf{A}^+ - \mathbf{A}^-. \quad (3.472)$$

If we use the generalized divergence theorem (2.299) to replace the second term on the right hand side, we obtain the *generalized Reynolds' transport theorem*

$$\frac{d}{dt}\int_{\mathscr{V}(t)-\varsigma(t)} \psi(\mathbf{x},t)\, dv = \int_{\mathscr{V}(t)-\varsigma(t)} \left[\frac{\partial \psi}{\partial t} + \operatorname{div}(\psi \mathbf{v})\right] dv + \int_{\varsigma(t)} [\![\psi(\mathbf{v}-\mathbf{c})]\!]\cdot d\boldsymbol{\varsigma}. \quad (3.473)$$

Problems

1. Given the deformation function

$$\begin{aligned} x_1 &= X_1 + \kappa X_2, \\ x_2 &= X_2, \\ x_3 &= X_3, \end{aligned}$$

where κ is a positive constant, obtain the stretch ratio in the directions $L_K^{(1)} = (1,0,0)$, $L_K^{(2)} = (0,1,0)$, $L_K^{(3)} = (0,0,1)$, $L_K^{(4)} = (1/\sqrt{2}, 1/\sqrt{2}, 0)$, $L_K^{(5)} = 1/\sqrt{3}(1,1,1)$, and the changes in orientation of line elements in the last two directions.

2. Show that

$$x_{i,I} = \frac{1}{2} J \epsilon_{ijk} \epsilon_{IJK} X_{J,j} X_{K,k}.$$

3. Show that

$$\frac{\partial J^{-1}}{\partial X_{I,i}} = \frac{1}{2} \epsilon_{IJK} \epsilon_{ijk} X_{J,j} X_{K,k}.$$

4. Show that

$$\frac{\partial x_{i,K}}{\partial X_{I,k}} = -x_{k,K} x_{i,I}.$$

5. Show that

$$\left(J^{-1} x_{i,I}\right)_{,i} = 0.$$

6. If **A** is a second rank tensor, show that

$$\dot{\overline{A^{-1}}} = -A^{-1} \dot{A} A^{-1}, \tag{3.474}$$

and

$$\frac{\partial \det A}{\partial A} = \det A \left(A^{-1}\right)^T. \tag{3.475}$$

7. If **A** is a second rank tensor, show that

$$\dot{\overline{\det A}} = \det A \operatorname{tr}\left(\dot{A} A^{-1}\right). \tag{3.476}$$

8. If **A** is a second rank tensor, show that

$$\frac{\partial \operatorname{tr} A}{\partial A} = I, \qquad \frac{\partial \operatorname{tr} A^2}{\partial A} = 2 A^T, \quad \text{and} \quad \frac{\partial \operatorname{tr} A^3}{\partial A} = 3 \left(A^2\right)^T, \tag{3.477}$$

or in general

$$\frac{\partial \operatorname{tr} A^k}{\partial A} = k \left(A^{k-1}\right)^T. \tag{3.478}$$

9. If **A** is a second rank tensor, use (3.86) and (3.87) to obtain the following recursion relation:

$$\frac{\partial A_{(k+1)}}{\partial A} = A_{(k)} I - A^T \frac{\partial A_{(k)}}{\partial A}, \quad k = 0, 1, \ldots, n, \tag{3.479}$$

where $A_{(0)} = I$ and $A_{(n+1)} = 0$. By induction, show that the above recursion can also be written in the following form:

$$\frac{\partial A_{(k)}}{\partial A} = \left[\sum_{j=0}^{k-1} (-1)^j A_{(k-j-1)} A^j\right]^T. \tag{3.480}$$

10. Evaluate the invariants of a three-dimensional skew-symmetric second rank tensor **A**.

11. Show that the characteristic polynomial for $A = [a_{ik}]$ is given by

$$\lambda^3 - A_{(1)} \lambda^2 + A_{(2)} \lambda - A_{(3)} = 0, \tag{3.481}$$

where

$$A_{(1)} = a_{ii}, \tag{3.482}$$

$$A_{(2)} = \frac{1}{2}\left[\left(a_{ii}\right)^2 - a_{ik} a_{ki}\right], \tag{3.483}$$

$$A_{(3)} = \det[a_{ik}]. \tag{3.484}$$

a) Show that the eigenvalues of A are given by

$$\lambda^{(k)} = \frac{1}{3}\left\{A_{(1)} + 2\left(A_{(1)}^2 - 3\,A_{(2)}\right)^{1/2} \cos\left[\frac{1}{3}\,(\theta + 2\,\pi\,k)\right]\right\}, \qquad k = 1,2,3, \tag{3.485}$$

where

$$\cos\theta = \frac{2\,A_{(1)}^3 - 9\,A_{(1)}A_{(2)} + 27\,A_{(3)}}{2\left(A_{(1)}^2 - 3\,A_{(2)}\right)^{3/2}} \tag{3.486}$$

b) As noted in (3.138), the spectral representation of a symmetric matrix A is given by

$$A = \sum_1^3 \lambda^{(i)} A_i, \qquad A_i = \widehat{\mathbf{v}}_i \widehat{\mathbf{v}}_i, \tag{3.487}$$

where $\widehat{\mathbf{v}}_i$ are the normalized eigenvectors of A. If the eigenvalues are all distinct, show that

$$A_k = \frac{\left(A - \lambda^{(l)} I\right)\left(A - \lambda^{(m)} I\right)}{\left(\lambda^{(k)} - \lambda^{(l)}\right)\left(\lambda^{(l)} - \lambda^{(m)}\right)}, \tag{3.488}$$

where (k, l, m) represents a cyclic permutation of $(1,2,3)$.

c) In the case of coalescence of two eigenvalues ($\lambda^{(1)} \neq \lambda^{(2)} = \lambda^{(3)} = \lambda$), show that

$$A_1 = \frac{(A - \lambda\,I)}{\left(\lambda^{(1)} - \lambda\right)}. \tag{3.489}$$

d) For the case of coalescence of all eigenvalues ($\lambda^{(1)} = \lambda^{(2)} = \lambda^{(3)} = \lambda$), show that

$$A = \lambda\,I. \tag{3.490}$$

12. Show that if W is a 3×3 skew-symmetric matrix, then $\operatorname{tr} A^4 = \frac{1}{2}(\operatorname{tr} A^2)^2$. [Hint: Use Cayley–Hamilton theorem.]

13. Given the right Cauchy–Green tensor $\mathbf{C}$, show that the necessary and sufficient conditions for a *rigid deformation* are $C_{(1)} = C_{(2)} = 3$ and $C_{(3)} = 1$.

14. Show that $(U^{-1})^T = (U^T)^{-1}$.

15. Show that $\mathbf{F} = \mathbf{V} \cdot \mathbf{R}$.

16. Prove that the decomposition $\mathbf{F} = \mathbf{V} \cdot \mathbf{R}$ is unique.

17. Determine the eigenvalues and normalized eigenvectors of A given by

$$A = \begin{bmatrix} \frac{11}{6} & -\frac{2}{3} & -\frac{1}{6} \\ -\frac{2}{3} & \frac{7}{3} & -\frac{2}{3} \\ -\frac{1}{6} & -\frac{2}{3} & \frac{11}{6} \end{bmatrix}.$$

18. For the two-dimensional small strain theory, the strains in a beam are given by

$$\tilde{e}_{11} = a\,X_1\,X_2, \qquad \tilde{e}_{22} = -a\,b\,X_1\,X_2, \qquad \tilde{e}_{12} = \frac{1}{2}\,a\,(1+b)\left(c^2 - X_2^2\right),$$

where a, b, and c are positive constants, $a \ll 1$, and $b \leq 1/2$. Assume that displacements u_1 and u_2 along the X_1 and X_2 axes are functions of X_1 and X_2.

a) Using compatibility conditions, show that continuous single-valued displacements u_1 and u_2 are possible.

b) Subsequently, derive expressions for u_1 and u_2 in terms of X_1 and X_2 with the conditions $u_1 = u_2 = u_{1,2} = 0$ at the point $X_1 = L$, $X_2 = 0$.

19. Show that the stretching tensor $\mathbf{D}(t)$ and the spin tensor $\mathbf{W}(t)$ may be expressed in terms of the polar decomposition $\mathbf{F}(t) = \mathbf{R}(t) \cdot \mathbf{U}(t)$ by

$$\mathbf{D}(t) = \frac{1}{2}\mathbf{R}(t) \cdot \left[\dot{\mathbf{U}}(t) \cdot \mathbf{U}^{-1}(t) + \mathbf{U}^{-1}(t) \cdot \dot{\mathbf{U}}(t)\right] \cdot \mathbf{R}^T(t), \tag{3.491}$$

$$\mathbf{W}(t) = \dot{\mathbf{R}}(t) \cdot \mathbf{R}^T(t) + \frac{1}{2}\mathbf{R}(t) \cdot \left[\dot{\mathbf{U}}(t) \cdot \mathbf{U}^{-1}(t) - \mathbf{U}^{-1}(t) \cdot \dot{\mathbf{U}}(t)\right] \cdot \mathbf{R}^T(t). \tag{3.492}$$

20. Derive (3.202) from (3.200).

21. Derive (3.203) from (3.200).

22. Derive (3.207)–(3.212) from (3.204).

23. Show that the material derivative of the differential area (3.36) is given by

$$\dot{\overline{d\mathbf{s}}} = \operatorname{tr}\mathbf{D}\, d\mathbf{s} - \mathbf{L}^T \cdot d\mathbf{s}. \tag{3.493}$$

24. Using (3.239), show that the evolution equation for $\mathbf{f} \equiv \mathbf{F}^{-1}$ is given by

$$\frac{\partial \mathbf{f}}{\partial t} + \nabla\left(\mathbf{f} \cdot \mathbf{v}\right) = \mathbf{0}. \tag{3.494}$$

Is this equation objective?

25. Using (3.239), show that the evolution equation for right Cauchy–Green tensor $\mathbf{C}$ is given by

$$\dot{\mathbf{C}} = 2\,\mathbf{F}^T \cdot \mathbf{D} \cdot \mathbf{F}. \tag{3.495}$$

Is this equation objective?

26. Using (3.239), show that the evolution equation for left Cauchy–Green tensor $\mathbf{B}$ is given by

$$\overset{\star}{\mathbf{B}} = \mathbf{0}, \tag{3.496}$$

where $\overset{\star}{\mathbf{B}}$ is the Oldroyd tensor defined in (3.423). Is this equation objective?

27. A deformation of the form

$$\begin{aligned} x_1 &= f_1(X_1, X_2), \\ x_2 &= f_2(X_1, X_2), \\ x_3 &= X_3, \end{aligned}$$

where f_1 and f_2 are smooth functions, is called *plane strain*. Show that for such a deformation, the principal stretch λ_3 (in the X_3 direction) is unity. Show further that the deformation is isochoric if and only if the other principal stretches, λ_1 and λ_2, satisfy

$$\lambda_1 = \frac{1}{\lambda_2}.$$

28. A motion is *plane* if the velocity field has the form

$$\mathbf{v}(\mathbf{x},t) = v_1(x_1,x_2,t)\mathbf{i}_1 + v_2(x_1,x_2,t)\mathbf{i}_2,$$

in some Cartesian frame. Show that in a plane motion

$$\mathbf{W}\cdot\mathbf{D} + \mathbf{D}\cdot\mathbf{W} = (\operatorname{div}\mathbf{v})\ \mathbf{W},$$

where $\mathbf{D}$ and $\mathbf{W}$ are the stretch and spin tensors, respectively.

29. Later we will be making use of the convected tensor $\overline{T}$ defined by

$$\overline{T} = \mathbf{F}^T\cdot T\cdot\mathbf{F}. \tag{3.497}$$

a) Show that the spin tensor satisfies the differential equation

$$\dot{\mathbf{W}} + \mathbf{D}\cdot\mathbf{W} + \mathbf{W}\cdot\mathbf{D} = \mathbf{J}, \tag{3.498}$$

where

$$\mathbf{J} = \frac{1}{2}\left[(\operatorname{grad}\dot{\mathbf{v}}) - (\operatorname{grad}\dot{\mathbf{v}})^T\right] \tag{3.499}$$

is the skew-symmetric part of the acceleration gradient.

b) Show that convected spin tensor satisfies the differential equation

$$\dot{\overline{\mathbf{W}}} = \overline{\mathbf{J}}. \tag{3.500}$$

30. Prove that

$$\mathbf{D}(t) = {}_{(t)}\dot{\mathbf{U}}(t) = {}_{(t)}\dot{\mathbf{V}}(t). \tag{3.501}$$

31. Consider the motion

$$\dot{x}_1 = 0, \quad \dot{x}_2 = \kappa x_1, \quad \dot{x}_3 = 0.$$

Calculate $\boldsymbol{\xi}(\mathbf{x},\tau;t)$, ${}_{(t)}\dot{\mathbf{R}}(\tau)$, and $\mathbf{W}(t)$.

32. A motion of a body is given by

$$\begin{aligned}
x_1 &= e^{-at}\left[X_1\cos(bX_3t) - X_2\sin(bX_3t)\right],\\
x_2 &= e^{-at}\left[X_1\sin(bX_3t) + X_2\cos(bX_3t)\right],\\
x_3 &= \phi(t)X_3,
\end{aligned}$$

where $a,b>0$ are constants. Find the form of $\phi(t)$ for which the motion is isochoric. Determine, for the isochoric motion, the components of velocity and acceleration in the spatial description, and show that both the particle paths and the stream lines lie on the surfaces

$$\left(x_1^2 + x_2^2\right)x_3 = \text{constant}.$$

33. A plane circular shearing motion of a body is given by

$$\begin{aligned} x_1 &= X_1 + \phi(X_3)\cos(\omega t) + \psi(X_3)\sin(\omega t), \\ x_2 &= X_2 + \phi(X_3)\sin(\omega t) - \psi(X_3)\cos(\omega t), \\ x_3 &= X_3, \end{aligned}$$

where $\omega > 0$ is a constant, the functions ϕ and ψ are differentiable, and the referential and spatial coordinates refer to a common Cartesian system.

a) Show that the motion is isochoric and the particle paths and stream lines are circles.

b) Discuss the stretch and rotation undergone by material line elements, which, in the reference configuration, lie parallel and orthogonal to the X_3 direction.

34. In a plane motion of a continuum, the velocity field is given by

$$\dot{x}_1 = -V\left[1 - \frac{a^2(x_1^2 - x_2^2)}{(x_1^2 + x_2^2)^2}\right], \qquad \dot{x}_2 = 2Va^2\frac{x_1 x_2}{r^4}, \qquad \dot{x}_3 = 0,$$

where a and V are constants, (x_1, x_2, x_3) are Cartesian coordinates, and $r^2 = x_1^2 + x_2^2 + x_3^2$. Determine the stream lines.

35. The velocity field of a continuum is given by

$$\dot{x}_1 = Va^2\frac{(x_1^2 - x_2^2)}{(x_1^2 + x_2^2)^2}, \qquad \dot{x}_2 = 2Va^2\frac{x_1 x_2}{(x_1^2 + x_2^2)^2}, \qquad \dot{x}_3 = 0,$$

where a and V are constants. Determine the path lines.

36. In a plane motion of a continuum, the velocity field is given by

$$\dot{x}_1 = -\frac{k}{4\pi}\frac{x_1^2}{x_2(x_1^2 + x_2^2)}, \qquad \dot{x}_2 = \frac{k}{4\pi}\frac{x_2^2}{x_1(x_1^2 + x_2^2)}, \qquad \dot{x}_3 = 0,$$

where k is a constant. Determine the stream lines and vortex lines.

37. Give a geometrical description of the deformation

$$\begin{aligned} x_1 &= X_1 - \tau X_2 X_3, \\ x_2 &= X_2 + \tau X_1 X_3, \\ x_3 &= X_3, \end{aligned}$$

where τ is a constant, and calculate the components of $\mathbf{F}$ and $\mathbf{C}$. Is the deformation isochoric? Find the surface into which the cylinder $X_1^2 + X_2^2 = a^2$ (a = const.) deforms.

38. A steady two-dimensional flow (pure straining) is given by

$$v_1 = \alpha x_1 \qquad \text{and} \qquad v_2 = -\alpha x_2,$$

with α = const.

a) Find the equation for a general stream line of the flow, and sketch some of them.

b) At $t = 0$ the fluid on the curve $x_1^2 + x_2^2 = a^2$ is marked by some technique. Find the equation for this material fluid curve for $t > 0$.

c) Does the area within the curve change in time, and why?

39. Do Problem 38, but for the two-dimensional flow (simple shear) given by

$$v_1 = \gamma x_2 \qquad \text{and} \qquad v_2 = 0,$$

with γ = const. Which of the two flows stretches the curve faster at long times?

40. Verify that the two-dimensional flow given by

$$v_1 = \frac{x_2 - c_2}{(x_1 - c_1)^2 + (x_2 - c_2)^2} \qquad \text{and} \qquad v_2 = \frac{c_1 - x_1}{(x_1 - c_1)^2 + (x_2 - c_2)^2},$$

where c_1 and c_2 are constants, satisfies div $\mathbf{v} = 0$, and then find the stream function $\psi(x_1, x_2)$ such that

$$v_1 = \frac{\partial \psi}{\partial x_2} \qquad \text{and} \qquad v_2 = -\frac{\partial \psi}{\partial x_1}.$$

Sketch the stream lines.

41. Verify that the two-dimensional flow given in cylindrical polar coordinates by

$$v_r = U\left(1 - \frac{a^2}{r^2}\right)\cos\theta \qquad \text{and} \qquad v_\theta = -U\left(1 + \frac{a^2}{r^2}\right)\sin\theta$$

satisfies div $\mathbf{v} = 0$, and find the stream function $\psi(r, \theta)$ such that

$$v_r = \frac{1}{r}\frac{\partial \psi}{\partial \theta} \qquad \text{and} \qquad v_\theta = -\frac{\partial \psi}{\partial r}.$$

Sketch the stream lines.

42. Verify that the axisymmetrical flow (uniaxial straining) given in cylindrical polar coordinates by

$$v_r = -\frac{1}{2}\alpha r \qquad \text{and} \qquad v_z = \alpha z$$

satisfies div $\mathbf{v} = 0$, and find the Stokes stream function $\psi(r, z)$ such that

$$v_r = -\frac{1}{r}\frac{\partial \psi}{\partial z} \qquad \text{and} \qquad v_z = \frac{1}{r}\frac{\partial \psi}{\partial r}.$$

Sketch the stream lines.

43. Show that if the first invariant of the stretch tensor is zero, i.e., $D_{(1)} = 0$, then the motion is isochoric.

44. When the stretching tensor is spherical, i.e., all three principal stretches are equal, the motion is purely dilatational. Show that in this case

$$\left(\frac{1}{3}D_{(1)}\right)^3 = \left(\frac{1}{3}D_{(2)}\right)^{3/2} = D_{(3)}.$$

45. Show that for the simple shearing motion $\dot{x}_1 = \kappa x_2, \dot{x}_2 = \dot{x}_3 = 0$,

    a) $D_{(1)} = D_{(3)} = 0$;
    b) the principal stretches are such that $d_1 = -d_3$ and $d_2 = 0$;
    c) vorticity is parallel to the principal axis of stretch along which the stretch is zero;
    d) the amount of shear is given by $\kappa = \omega = \sqrt{-4D_{(2)}}$, where $\omega = |\boldsymbol{\omega}|$.

46. The *kinematical vorticity number* A is the dimensionless ratio of the magnitudes of the spin and stretch tensors. Show that

$$A = \sqrt{\frac{\mathbf{W}:\mathbf{W}}{\mathbf{D}:\mathbf{D}}} = \sqrt{-\frac{W_{(2)}}{D_{(2)}}} = \frac{\omega}{\sqrt{2D_{(2)}}} = \frac{\omega}{\sqrt{2(d_1^2 + d_2^2 + d_3^2)}}, \tag{3.502}$$

where d_i's are the principal stretches and $\omega = |\boldsymbol{\omega}|$. Note that if $A \ll 1$, then the motion is nearly irrotational.

47. Let the exponential of a matrix A be defined by the series

$$\exp(A) = 1 + A + \frac{A^2}{2!} + \frac{A^3}{3!} + \cdots.$$

Show that

    i) if $AB = BA$, then $\exp(A)\exp(B) = \exp(A + B)$;
    ii) if A is skew-symmetric, then $\exp(A)$ is orthogonal;
    iii) for any skew-symmetric matrix W, the matrix function $Q(t) = \exp[-(t-t_0)W]$ satisfies the following differential equation

$$\dot{Q} + QW = 0 \qquad \text{and} \qquad Q(t_0) = 1.$$

48. Suppose that $\psi(\mathbf{x}, t)$ transforms like a scalar field under a change of frame. Is $\partial\psi/\partial t$ objective? Is $\dot{\psi}$ objective?

49. Suppose that $\mathbf{q}$ is a frame-indifferent vector, i.e., $\mathbf{q}' = \mathbf{Q}\cdot\mathbf{q}$. Is the quantity $\dot{\mathbf{q}} - \mathbf{L}\cdot\mathbf{q}$ frame indifferent?

50. Suppose that $\boldsymbol{\sigma}$ is a symmetric frame indifferent second rank tensor, i.e., $\boldsymbol{\sigma} = \boldsymbol{\sigma}^T$ and $\boldsymbol{\sigma}' = \mathbf{Q}\cdot\boldsymbol{\sigma}\cdot\mathbf{Q}^T$. Is the quantity $\dot{\boldsymbol{\sigma}} - 2\mathbf{D}\cdot\boldsymbol{\sigma}$ frame indifferent?

51. Show that (3.422) is an objective tensor.

52. Show that (3.423) is an objective tensor.

53. Show that (3.424) is an objective tensor.

54. Show that (3.425) is an objective tensor.

Bibliography

R.C. Batra. *Elements of Continuum Mechanics.* AIAA, Reston, VA, 2006.

B.S. Berger and M. Rokni. Lyapunov exponents and continuum kinematics. *International Journal of Engineering Science*, 25(10):1251–1257, 1987.

R.M. Bowen. *Introduction to Continuum Mechanics for Engineers.* Plenum Press, New York, NY, 1989.

A.C. Eringen. *Nonlinear Theory of Continuous Media.* McGraw-Hill Book Company, Inc., New York, NY, 1962.

A.C. Eringen. Basic principles: Deformation and motion. In A.C. Eringen, editor, *Continuum Physics*, volume II, pages 3–67. Academic Press, Inc., New York, NY, 1975.

A.C. Eringen. *Mechanics of Continua.* R.E. Krieger Publishing Company, Inc., Melbourne, FL, 1980.

M.E. Gurtin. *An Introduction to Continuum Mechanics.* Academic Press, San Diego, CA, 2003.

M.E. Gurtin, E. Fried, and L. Anand. *The Mechanics and Thermodynamics of Continua.* Cambridge University Press, Cambridge, UK, 2010.

P. Haupt. *Continuum Mechanics and Theory of Materials.* Springer-Verlag, Berlin, 2000.

G.A. Holzapfel. *Nonlinear Solid Mechanics.* John Wiley & Sons, Ltd., Chichester, England, 2005.

K. Hutter and K. Jöhnk. *Continuum Methods of Physical Modeling.* Springer-Verlag, Berlin, 1981.

W.M. Lai, D. Rubin, and E. Krempl. *Introduction to Continuum Mechanics.* Butterworth-Heinemann, Burlington, MA, 2010.

I.-S. Liu. *Continuum Mechanics.* Springer-Verlag, Berlin, 2002.

L.E. Malvern. *Introduction to the Mechanics of a Continuous Medium.* Prentice-Hall, Inc., Upper Saddle River, NJ, 1969.

A.G. McLellan. *The Classical Thermodynamics of Deformable Materials.* Cambridge University Press, 1980.

W. Noll. On the continuity of the solid and fluid states. *Journal of Rational Mechanics and Analysis*, 4(1):3–81, 1955.

W. Noll. A mathematical theory of the mechanical behavior of continuous media. *Archive for Rational Mechanics and Analysis*, 2(1):197–226, 1958.

J.M. Ottino. *The Kinematics of Mixing: Stretching, Chaos, and Transport.* Cambridge University Press, 1989.

C. Truesdell and W. Noll. The non-linear field theories of mechanics. In S. Flügge, editor, *Handbuch der Physik*, volume III/3. Springer, Berlin-Heidelberg-New York, 1965.

C. Truesdell and R.A. Toupin. The classical field theories. In S. Flügge, editor, *Handbuch der Physik*, volume III/1. Springer, Berlin-Heidelberg-New York, 1960.

K. Washizu. A note on the conditions of compatibility. *Journal of Mathematics and Physics*, 36:306–312, 1958.

A. Wintner and F.D. Murnaghan. On a polar representation of non singular square matrices. *Proceedings of the National Academy of Sciences of the United States of America*, 17:676–678, 1931.

4

Mechanics and thermodynamics

4.1 Balance law

Consider an arbitrary material body having volume $\mathscr{V}(t)$ and surface $\mathscr{S}(t)$ which is separated into two parts $\mathscr{V}^+(t)$ and $\mathscr{V}^-(t)$, or $\mathscr{V}(t) - \zeta(t)$, by a singular surface whose intersection with $\mathscr{V}$ is denoted by the surface $\zeta(t)$, as shown in Fig. 4.1. The surface $\zeta(t)$ has a unit normal vector $\mathbf{n}$ pointing into $\mathscr{V}^+(t)$ and velocity $\mathbf{c}$. The line of intersection of $\mathscr{S}(t)$ and $\zeta(t)$ will be denoted by $\mathscr{C}(t)$ and the unit vector normal to $\mathscr{C}(t)$ which is tangential to $\zeta(t)$ and pointing externally to $\mathscr{V}(t)$ is called $\widetilde{\boldsymbol{\mu}}$. The parts of $\mathscr{S}(t)$ not on $\mathscr{C}(t)$ will be denoted as $\mathscr{S}^+(t)$ and $\mathscr{S}^-(t)$, or $\mathscr{S}(t) - \mathscr{C}(t)$. Note that $\mathscr{S}^+(t)$ and $\mathscr{S}^-(t)$ also include the surface areas of $\mathscr{V}^+(t)$ and $\mathscr{V}^-(t)$ adjacent to $\zeta(t)$, and there they have exterior normals $\mathbf{n}^+$ and $\mathbf{n}^-$, respectively. The symbol $\mathbf{n}$ also denotes the unit normal on $\mathscr{S}(t) - \mathscr{C}(t)$, which points out of $\mathscr{V}(t)$. The material in $\mathscr{V}(t) - \zeta(t)$ moves with particle velocity $\mathbf{v}$.

Generally, from the physical point of view, tensor fields or one of their derivatives would change very fast within a small layer near $\zeta(t)$. We idealize such situation by considering this layer to be of zero thickness and subsequently allow the tensor fields to be singular there. The singularity will appear as a discontinuity in functions or their derivatives. A surface that is singular with respect to some quantity and that has a nonzero speed of propagation is said to be a *propagating singular surface* or *wave.* Singularities of the first type are usually associated with material singular surfaces which are formed by the same material particles at all times. Examples of the second type are nonmaterial singular surfaces, or *shocks*, or *acceleration waves.* We note that in addition to singular surfaces, singular lines and singular points are also idealizations of common occurrences.

The general balance statement of a tensor quantity Ψ associated with the body is given in the form

$$\frac{d\Psi}{dt} = \mathcal{T}(\Psi) + \mathcal{G}(\Psi), \tag{4.1}$$

where $\mathcal{T}(\Psi)$ denotes the flux of Ψ through the surface of the body, and $\mathcal{G}(\Psi)$ the combined external supply of Ψ to the body and internal production of Ψ within the body. One should note that while we represent symbolically the sum of supply and production by a single term in the present derivation, physically supply is different from production because it may be controlled from the exterior of the

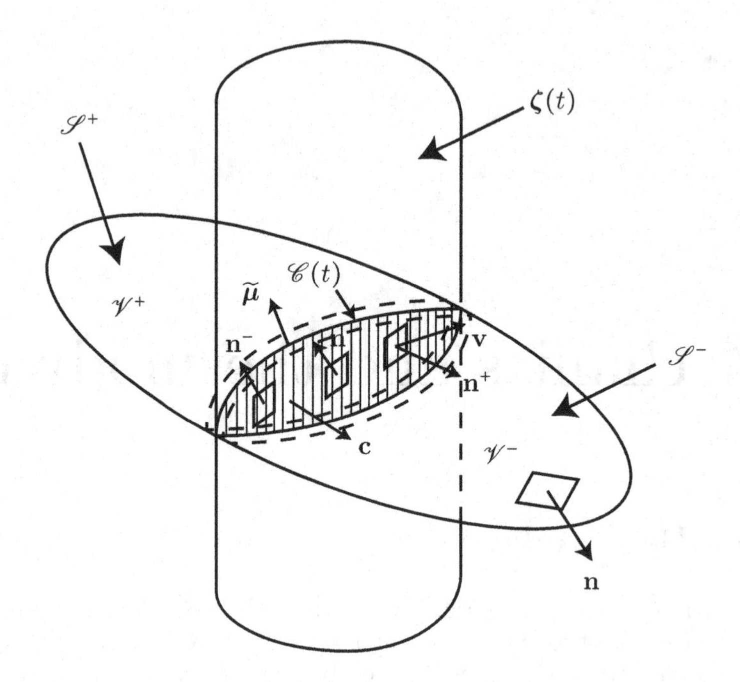

Figure 4.1: Arbitrary volume $\mathscr{V}$ intersected by a discontinuous surface.

body. Subsequently, in later applications, we will recognize this difference by writing their contributions separately. By the Radón–Nikodym theorem, we deduce that additive densities of Ψ, $\mathcal{T}(\Psi)$, and $\mathcal{G}(\Psi)$ exist, and denote the corresponding quantities that are defined within $\mathscr{V}(t) - \zeta(t)$ by ψ, $\mathbf{t}$, and g, respectively. Analogously, we also infer the existence of those quantities that are defined only on the singular surface $\zeta(t)$ by a tilde superscript, i.e., $\widetilde{\psi}$, $\widetilde{\mathbf{t}}$, and $\widetilde{g}$. Subsequently, (4.1) can be rewritten more explicitly as

$$\frac{d}{dt}\left(\int_{\mathscr{V}(t)-\zeta(t)} \psi\, dv + \int_{\zeta(t)} \widetilde{\psi}\, ds\right) = \left(\int_{\mathscr{S}(t)-\mathscr{C}(t)} \mathbf{t}\cdot\mathbf{n}\, ds + \int_{\mathscr{C}(t)} \widetilde{\mathbf{t}}\cdot\widetilde{\boldsymbol{\mu}}\, dl\right) + \left(\int_{\mathscr{V}(t)-\zeta(t)} g\, dv + \int_{\zeta(t)} \widetilde{g}\, ds\right). \quad (4.2)$$

Now using the generalized Reynolds and surface transport theorems (3.473) and (D.224), and the generalized divergence and surface divergence theorems (2.299) and (D.216), we have

$$\int_{\mathscr{V}(t)-\zeta(t)} \left[\frac{\partial\psi}{\partial t} + \operatorname{div}(\psi\mathbf{v}) - \operatorname{div}\mathbf{t} - g\right] dv + \int_{\zeta(t)} \left[\frac{\partial\widetilde{\psi}}{\partial t} + \widetilde{\nabla}\cdot(\widetilde{\psi}\mathbf{v}) - 2K_M\widetilde{\psi}v_{(n)} - \widetilde{\nabla}\cdot\widetilde{\mathbf{t}} - \widetilde{g} + [\![\psi(\mathbf{v}-\mathbf{c}) - \mathbf{t}]\!]\cdot\mathbf{n}\right] ds = 0, \quad (4.3)$$

where $\widetilde{\nabla}$ is the surface gradient operator, $v_{(n)}$ is the velocity component normal to $\zeta(t)$, and K_M is the mean curvature of the surface $\zeta(t)$ (see Appendix D and (D.183)).

Now evaluating the balance law over an arbitrary volume in the continuous region $\mathscr{V}(t) - \zeta(t)$, we obtain the local form of the balance law for the volumetric tensor quantity ψ over the region

$$\frac{\partial \psi}{\partial t} + \operatorname{div}(\psi \mathbf{v}) - \operatorname{div} \mathbf{t} - g = 0 \tag{4.4}$$

since this holds for all sufficiently regular volumes $\mathscr{V}(t) - \zeta(t)$, however small. Alternately, taking the limit by shrinking a volume down to $\zeta(t)$ in such a way that the volume tends to zero, while the area of $\zeta(t)$ remains unchanged, and assuming that $\partial\psi/\partial t$ and g remain bounded, then the volume integral vanishes in the limit. In addition, since the integrand of the remaining surface integral is smooth on $\zeta(t)$, and thus holds for any surface area no matter how small, then the integrand of the surface integral must vanish. Subsequently, we obtain the balance law for the areal tensor quantity $\widetilde{\psi}$ defined on the singular surface:

$$\frac{\partial \widetilde{\psi}}{\partial t} + \widetilde{\nabla} \cdot (\widetilde{\psi} \mathbf{v}) - 2K_M \widetilde{\psi} v_{(n)} - \widetilde{\nabla} \cdot \widetilde{\mathbf{t}} - \widetilde{g} + [\![\psi (\mathbf{v} - \mathbf{c}) - \mathbf{t}]\!] \cdot \mathbf{n} = 0. \tag{4.5}$$

This surface balance law is very important in the continuum mechanics of membranes or thin shells. However, from here on, we shall assume that the surface of discontinuity does not possess any properties of its own, i.e., $\widetilde{\psi} = 0$, $\widetilde{\mathbf{t}} = \mathbf{0}$, and $\widetilde{g} = 0$. Subsequently, the integral balance law (4.3) becomes

$$\int_{\mathscr{V}(t)-\zeta(t)} \left[\frac{\partial \psi}{\partial t} + \operatorname{div}(\psi \mathbf{v}) - \operatorname{div} \mathbf{t} - g \right] dv + \int_{\zeta(t)} [\![\psi (\mathbf{v} - \mathbf{c}) - \mathbf{t}]\!] \cdot \mathbf{n} \, ds = 0, \tag{4.6}$$

and the local balance equation for the surface of discontinuity (4.5) reduces to a jump condition on volumetric quantities that has to be satisfied across the surface $\zeta(t)$:

$$[\![\psi (\mathbf{v} - \mathbf{c}) - \mathbf{t}]\!] \cdot \mathbf{n} = 0. \tag{4.7}$$

In the rest of this chapter, we shall only make use of the integral balance laws (4.1) and (4.6), and the local balance laws (4.4) and (4.7), applied to an arbitrary material volume to obtain field equations and jump conditions in the spatial description. The corresponding balance laws in material coordinates are given in Appendix C, while the balance laws pertaining to material surfaces and material lines are discussed in Appendix D.

4.2 Fundamental axioms of mechanics

Associated with each material body, there is a quantity called *mass*, which has a dimension $[M]$ independent of the dimensions of length, $[L]$, and time, $[T]$. This quantity is positive definite and additive, i.e., it is an extensive property, and is absolutely continuous in the space variables. Subsequently, there exists a density ρ called the *mass density* with dimensions $[\rho] = [M]/[L]^3$. The quantity of mass is assigned to a set of particles having positive volume. Every finite body has finite mass, and zero volume implies zero mass. However, to deal with singular surfaces, we allow $0 \le \rho < \infty$. The *specific volume* is given by the reciprocal of the mass density, $\mathsf{v} = 1/\rho$, and the volume of the body is given by

$$\mathscr{V} = \int_{\mathscr{V}} dv. \tag{4.8}$$

Definition: The total mass $\mathcal{M}$ of a material body having volume $\mathscr{V}$ is determined by

$$\mathcal{M} \equiv \int_{\mathscr{V}} \rho \, dv. \tag{4.9}$$

Definition: The center of mass $\mathcal{M}$ of a material body having volume $\mathscr{V}$ is given by

$$\mathbf{x}_c \equiv \mathbf{x}_P + \frac{1}{\mathcal{M}} \int_{\mathscr{V}} \rho \left(\mathbf{x} - \mathbf{x}_P \right) dv, \tag{4.10}$$

where $\mathbf{x}_P$ is an arbitrary point fixed in the reference frame. It is easily verified that $\mathbf{x}_c$ is independent of the choice of $\mathbf{x}_P$. In rigid bodies, the center of mass is fixed relative to the body's geometry. This is not the case when deformations occur. In a deformable body, the center of mass moves about within (or possibly outside) the body in the course of time.

Definition: The linear momentum $\mathcal{P}$ of a continuous mass medium contained in $\mathscr{V}$ is given by

$$\mathcal{P} \equiv \int_{\mathscr{V}} \rho \mathbf{v} \, dv, \tag{4.11}$$

where $\mathbf{v} = \dot{\mathbf{x}}$ is the velocity of the material particle.

Definition: The barycentric velocity $\mathbf{v}_c = \dot{\mathbf{x}}_c$ of a continuous mass medium contained in $\mathscr{V}$ is given by

$$\mathbf{v}_c \equiv \frac{1}{\mathcal{M}} \int_{\mathscr{V}} \rho \mathbf{v} \, dv. \tag{4.12}$$

In the following, we will formulate the balance laws for a *polar material.* Polar materials arise from a statistical mechanics model that assumes noncentral forces of interaction between particles. When such forces are noncentral, an interparticle couple, in addition to an interparticle force, manifests itself. Under the action of the couple, the material particle will have a tendency to rotate relative to its neighbors. The essential idea of a polar material is obtained by introducing a kinematic variable to model the rotation of the particle relative to its neighbors and a skew-symmetric tensor to model the forces that balance the action of the couple. Subsequently, an internal (particle) angular velocity and an associated internal spin are defined independently of the velocity field, along with other pertinent quantities.

Definition: The internal spin $\boldsymbol{s}$ of a continuous medium contained in $\mathscr{V}$ is given by

$$\boldsymbol{s} \equiv \int_{\mathscr{V}} \rho \boldsymbol{\varphi} \, dv, \tag{4.13}$$

where the internal spin per unit mass $\boldsymbol{\varphi}$ is assumed to exist and to be linearly related to the internal angular velocity $\boldsymbol{\nu}$ by

$$\boldsymbol{\varphi} = \boldsymbol{i} \cdot \boldsymbol{\nu}, \tag{4.14}$$

where $\boldsymbol{i}$ is the symmetric positive-definite rank-2 *internal inertia tensor* per unit mass. The internal inertia tensor describes the average inertia of material particles relative to the position vector $\mathbf{x}$, where $\mathbf{x}$ is viewed as the center of mass of particles near $\mathbf{x}$. The internal spin and the internal spin per unit mass can be taken to correspond to axial vectors of the respective skew-symmetric *internal spin tensors* $\boldsymbol{\mathcal{S}} = \boldsymbol{s} \cdot \boldsymbol{\epsilon}$ and $\boldsymbol{\Phi} = \boldsymbol{\varphi} \cdot \boldsymbol{\epsilon}$ so that the internal spin tensors are related by

$$\boldsymbol{\mathcal{S}} \equiv \int_{\mathscr{V}} \rho \boldsymbol{\Phi} \, dv. \tag{4.15}$$

Definition: The moment of momentum $\boldsymbol{h}$, about point $\mathbf{x}_P$, of a continuous mass medium contained in $\mathscr{V}$ is defined by

$$\boldsymbol{h} \equiv \int_{\mathscr{V}} \mathbf{r} \times \rho \mathbf{v} \, dv, \tag{4.16}$$

where $\mathbf{r} = \mathbf{x} - \mathbf{x}_P$, $\mathbf{x}$ is the position vector, and $\mathbf{r}$ is the vector from the moment-center $\mathbf{x}_P$ to the point $\mathbf{x}$ on the line of action of the force. Note that if we take the moment-center to be the origin, then $\mathbf{r} = \mathbf{x}$; most often the moment-center is taken to be the center of mass, in which case $\mathbf{r} = \mathbf{x} - \mathbf{x}_c$. Since $\boldsymbol{h}$ can be taken to correspond to the axial vector of the skew-symmetric tensor $\boldsymbol{\mathcal{H}}$, i.e., $\boldsymbol{\mathcal{H}} = \boldsymbol{h} \cdot \boldsymbol{\epsilon}$, we can also write (see (2.170) and (2.171))

$$\boldsymbol{\mathcal{H}} \equiv \int_{\mathscr{V}} (\mathbf{r} \wedge \rho \mathbf{v}) \, dv. \tag{4.17}$$

Definition: The kinetic energy $\mathcal{K}$ of the continuous mass medium in $\mathscr{V}$ is given by the sum of the internal spin and translational kinetic energies:

$$\mathcal{K} \equiv \frac{1}{2} \int_{\mathscr{V}} \rho \left(\boldsymbol{\nu} \cdot \boldsymbol{\varphi} + \mathbf{v} \cdot \mathbf{v} \right) dv. \tag{4.18}$$

We now state the three fundamental laws of mechanics.

Axiom 1 – Conservation of mass: *The total mass of a body is unchanged during motion. When this is valid for an arbitrarily small neighborhood of each material point, we say that the mass is conserved locally.*

The total or global mass conservation may be expressed by

$$\mathcal{M} = \int_{\mathscr{V}} \rho \, dv = \int_{V} \rho_R \, dV \equiv \mathcal{M}_R, \tag{4.19}$$

where $\rho_R(\mathbf{X}, t_0)$ is the mass density in the reference configuration. Note that the above volume integrals can both be expressed in either the reference material coordinate system or the spatial coordinate system:

$$\int_{V} (\rho J - \rho_R) \, dV \qquad \text{or} \qquad \int_{\mathscr{V}} \left(\rho - \rho_R J^{-1} \right) dv.$$

As a result, global mass conservation can be rewritten in the reference configuration as

$$\mathcal{M} - \mathcal{M}_R = \int_{V} (\rho J - \rho_R) \, dV = 0. \tag{4.20}$$

If the volume is taken to correspond to an arbitrary volume element within the body, then the integrand must be identically zero to satisfy the equation. Setting the integrand to zero, we obtain the local mass balance

$$\rho(\mathbf{X}, t) J(\mathbf{X}, t) = \rho_R(\mathbf{X}, t_0) \qquad \text{or} \qquad \rho(\mathbf{x}, t) = \rho_R(\mathbf{X}, t_0) J^{-1}(\mathbf{x}, t). \tag{4.21}$$

If $J = 1$, then $\rho(\mathbf{X}, t) = \rho_R(\mathbf{X}, t_0)$ and the motion is *isochoric*. If in addition ρ_R = const., then the motion is *homochoric*. The equation above is the general solution of the conservation of mass equation. However, it is expressed within the material description, and in order to render it explicit, we must know the motion. A spatial form, for application of which the motion itself need not be given, is preferable. This will be given later. The material derivative of the global mass conservation equation (4.19) in the current configuration is

$$\frac{d\mathcal{M}}{dt} = \frac{d}{dt}\int_{\mathcal{V}} \rho \, dv = 0. \tag{4.22}$$

Axiom 2 – Balance of linear momentum: *The time rate of change of linear momentum $\mathcal{P}$ is equal to the resultant force $\mathcal{F}$ acting on the body. We postulate that this is valid for an arbitrarily small neighborhood of each material point, thus giving rise to the local form of the linear momentum balance.*

This statement is expressed by the following equation:

$$\frac{d\mathcal{P}}{dt} = \mathcal{F} \qquad \text{or} \qquad \frac{d}{dt}\int_{\mathcal{V}} \rho \mathbf{v} \, dv = \mathcal{F}. \tag{4.23}$$

Axiom 3 – Balance of angular momentum: *The time rate of angular momentum is given by the change of the sum of the internal spin and the moment of momentum of a body about a fixed point, and is equal to the resultant moment $\boldsymbol{m}$ about the point. Analogously, we postulate that this is valid for an arbitrarily small neighborhood of each material point in the body, thus giving rise to the local form of the angular momentum balance.*

This statement is expressed by the following equation:

$$\frac{d}{dt}(\mathbf{s} + \boldsymbol{h}) = \boldsymbol{m} \qquad \text{or} \qquad \frac{d}{dt}\int_{\mathcal{V}} \rho\,(\boldsymbol{\varphi} + \mathbf{r} \times \mathbf{v}) \, dv = \boldsymbol{m}. \tag{4.24}$$

In terms of an equivalent resultant-moment skew-symmetric tensor $\boldsymbol{\mathcal{M}} = \boldsymbol{m} \cdot \boldsymbol{\epsilon}$, we have

$$\frac{d}{dt}(\boldsymbol{\mathcal{S}} + \boldsymbol{\mathcal{H}}) = \boldsymbol{\mathcal{M}} \qquad \text{or} \qquad \frac{d}{dt}\int_{\mathcal{V}} \rho\,(\boldsymbol{\Phi} + \mathbf{r} \wedge \mathbf{v}) \, dv = \boldsymbol{\mathcal{M}}. \tag{4.25}$$

The equations for linear and angular momenta are called *Euler's equations of motion* and are considered extensions of Newton's second and third laws of motion of a particle.

4.3 Fundamental axioms of thermodynamics

Internal energy is defined to be the energy of material particles due to internal mechanisms not explicitly modeled. From correspondence between kinetic theory and continuum mechanics, this energy includes classically the kinetic energy due to the motion of molecules (translational, rotational, vibrational) and the potential energy associated with the vibrational and electric energy of atoms within molecules or crystals (it includes the energy in all the chemical bonds, and the energy of the free conduction electrons in metals). More generally, it is understood to include all forms of energies of material particles due to mechanisms whose scales are not modeled. It is an extensive property that, in classical thermodynamics, is a *state function*, i.e., it is independent of the *process* followed in changing the state of the body.

Definition: The internal energy, $\mathcal{E}$, of a continuous mass medium contained in $\mathscr{V}$ is given by

$$\mathcal{E} \equiv \int_{\mathscr{V}} \rho e \, dv, \tag{4.26}$$

where a local internal energy density per unit mass, e, has been assumed to exist.

Axiom 4 – Balance of energy: *The time rate of change of the internal energy plus kinetic energy is equal to the heat energy that enters or leaves the body per unit time plus the sum of the rates of work of the external forces per unit time. We postulate that this statement remains true for an arbitrarily small neighborhood of a material point.*

The above statement is expressed by the equation

$$\frac{d}{dt}(\mathcal{E} + \mathcal{K}) = \dot{\mathcal{Q}} + \sum_{\alpha} \dot{\mathcal{W}}_{\alpha} \quad \text{or} \quad \frac{d}{dt} \int_{\mathscr{V}} \rho \left[e + \frac{1}{2} (\boldsymbol{\nu} \cdot \boldsymbol{\varphi} + \mathbf{v} \cdot \mathbf{v}) \right] dv = \dot{\mathcal{Q}} + \sum_{\alpha} \dot{\mathcal{W}}_{\alpha}, \tag{4.27}$$

where $\dot{\mathcal{Q}}$ is the *heat energy* per unit time, and $\dot{\mathcal{W}}_{\alpha}$ is the αth kind of *work* of external forces per unit time (mechanical, chemical, electrical, magnetical, etc.).

The above axiom implies that the energies are additive, and that if proper accounting is made of all the energies due to external effects, what is left over to balance is the rate of the internal energy.

In particle mechanics, the conservation of energy is obtained as the first integral of Newton's second law of motion when the forces are not explicit functions of velocity and time. In thermodynamics, it is stated as *the first principle of thermostatics*, applicable to systems in equilibrium. The axiom stated above is an extension of both that of classical mechanics and that of thermostatics. It is valid for every system, including dissipative ones, in which the energy principle of both classical mechanics and thermostatics fails to apply.

The first law of thermodynamics states that energy is conserved. However, there are many thermodynamic processes that conserve energy but that actually never occur. Furthermore, the first law of thermodynamics does not restrict our ability to convert work into heat or heat into work, except that energy must be conserved in the process. And yet in practice, although we can convert a given quantity of work completely into heat, we have never been able to find a scheme that converts a given amount of heat completely into work. A quantity that we call *entropy* provides a measure of a system's thermal energy that is unavailable for doing useful work. In terms of statistical mechanics, the entropy of a system describes the number of possible microscopic configurations the particles in the system can have. The statistical definition of entropy is generally thought to be the more fundamental definition, from which all other properties of entropy follow. Entropy is an extensive quantity that represents a state function of the system.

Definition: The entropy, $\mathcal{S}$, of a continuous mass medium contained in $\mathscr{V}$ is given by

$$\mathcal{S} \equiv \int_{\mathscr{V}} \rho \eta \, dv, \tag{4.28}$$

where a local entropy density per unit mass, η, has been assumed to exist.

Axiom 5 – Second law of thermodynamics: *The time rate of change of the total entropy $\mathcal{S}$ in an arbitrary material body having volume $\mathscr{V}$ enclosed by surface $\mathscr{S}$ with exterior normal* $\mathbf{n}$ *is never less than the sum of the contact entropy supply Π through the surface of the body and the entropy $\mathcal{B}$ produced by external sources. It is postulated that this is true for all parts of the body and for all independent processes.*

According to the first part of the above axiom, also referred to as the *entropy inequality*, we have

$$\Gamma \equiv \frac{d\mathcal{S}}{dt} - \mathcal{B} - \Pi \geq 0 \qquad \text{or} \qquad \Gamma \equiv \frac{d}{dt}\int_{\mathscr{V}} \rho\eta \, dv - \mathcal{B} - \Pi \geq 0, \tag{4.29}$$

where Γ is the *total entropy production.*

4.4 Forces and moments

Generically, we will refer to forces and moments (or couples) as loads. In continuum mechanics, as opposed to particle mechanics, loads may depend on spatial gradients of various orders, their various time rates and integrals, as well as other variables pertaining to various loads of mechanical, electrical, or some other origin. (While they may be included, we assume that there are no concentrated loads acting at points.) A body will undergo a deformation when subjected to loads that may be either external (acting on the body) or internal (acting between two parts of the same body) in character. By a suitable choice of a free body imagined to be cut out of the complete body, any internal load in the original body may become an external load on the isolated body. The term *free body* denotes a portion of the complete body instantaneously bounded by an arbitrary closed surface.

External loads arise from external effects and are classified as being either *volumetric (or body) loads* or *surface (or contact) loads*:

i) *Body loads* act on elements of the volume or mass inside the body (e.g., gravity). These are "action-at-a-distance" forces. A body load density per unit mass is assumed to exist. External body loads are assumed to be objective.

ii) *Contact loads* arise from the action of one body upon another through the bounding surface contact. A surface force density per unit area, called the *surface traction*, and a surface couple density per unit area called the *surface couple*, are assumed to exist. Surface tractions and couples depend on the orientation of the surface on which they act.

Internal (or mutual) loads are the result of the mutual interaction of pairs of material particles that are located in the interior of the body. According to Newton's third law, the mutual action of a pair of particles consists of two forces acting along the line connecting the particles (central forces), equal in magnitude, and opposite in direction to one another. *Therefore, the resultant internal force and couple are zero.* Subsequently, such mutual loads are objective. However, we shall not make this assumption for the moment and thus allow the existence of a body couple and a couple stress vector. The effect of interparticle forces in a continuum appears in the form of a resultant effect of one part of the body on another part of the body through the latter's bounding surface. This concept gives rise to the *stress hypothesis*, and the existence of quantities called *stress* and *couple stress*, which will be discussed later.

In mechanics, real forces are always exerted by one body on another body (possibly by one part of a body acting on another part), regardless of whether they are body forces or surface forces. Now let $\mathbf{f}$ be the body force per unit mass acting on an infinitesimal volume dv of the body, so that body force on the volume is $\rho\mathbf{f}\,dv$. In general, the vector $\mathbf{f}$ varies from point to point in the body at any given time and may also vary with time at any given point, thus $\mathbf{f}(\mathbf{x},t)$. The vector sum of the body forces acting on an arbitrary finite volume $\mathscr{V}$ is then given by the spacial integral over the volume

$$\int_{\mathscr{V}} \rho\mathbf{f}\;dv. \tag{4.30}$$

Let $\mathbf{t}(\mathbf{n})$ be the surface traction per unit area acting on the differential surface ds of the body with exterior normal $\mathbf{n}$. In general, in addition to being dependent on the surface normal $\mathbf{n}$, the vector $\mathbf{t}$ varies from point to point on the body surface at any given time and may also vary with time at any given point, thus $\mathbf{t}(\mathbf{n},\mathbf{x},t)$. However, here we only display the dependence on $\mathbf{n}$, while the other dependences are suppressed for the moment. The force exerted across the differential area element is then $\mathbf{t}(\mathbf{n})\,ds$, and the vector sum of the forces across the bounding surface $\mathscr{S}$ of the arbitrary volume $\mathscr{V}$ is given by the vector surface integral

$$\int_{\mathscr{S}} \mathbf{t}(\mathbf{n})\,ds. \tag{4.31}$$

Subsequently, the resultant force acting on a body is given by

$$\mathcal{F} = \int_{\mathscr{S}} \mathbf{t}(\mathbf{n})\,ds + \int_{\mathscr{V}} \rho\mathbf{f}\,dv. \tag{4.32}$$

The moment of a force $\mathbf{g}$ about a point $\mathbf{x}_P$ is given by the vector product $\mathbf{r}\times\mathbf{g}$. The total moment on a given body of volume $\mathscr{V}$ bounded by a closed surface $\mathscr{S}$ is the sum of the total moments due to external and internal forces and couples. If the force is the body force $\rho\mathbf{f}$, then the moments about the three axes are given by $\mathbf{r}\times\rho\mathbf{f}\,dv$, and the total moment of all forces acting on a finite volume is given by

$$\int_{\mathscr{V}} \mathbf{r}\times\rho\mathbf{f}\,dv. \tag{4.33}$$

The moment of surface traction $\mathbf{t}(\mathbf{n})$ on an element of area ds can be expressed in a similar fashion, and the total of the distributed force on a finite surface $\mathscr{S}$ is obtained by means of the surface integral

$$\int_{\mathscr{S}} \mathbf{r}\times\mathbf{t}(\mathbf{n})\,ds. \tag{4.34}$$

To account for the internal spin, we let $\mathbf{l}$ and $\mathbf{m}(\mathbf{n})$, respectively, denote the external supply of spin per unit mass and the contact couple stress vector per unit surface area. In general, the axial vector $\mathbf{l}$ (or its associated skew-symmetric second-rank external supply tensor $\mathbf{L} = \mathbf{l}\cdot\boldsymbol{\epsilon}$) varies from point to point in the body at any given time and may also vary with time at any given point, so that $\mathbf{l}(\mathbf{x}, t)$, and in general, the axial vector $\mathbf{m}$ (or its associated skew-symmetric second-rank couple stress tensor $\mathsf{M} = \mathbf{m} \cdot \boldsymbol{\epsilon}$), in addition to being dependent on the surface normal $\mathbf{n}$, also varies from point to point on the body surface at any given time and may also vary with time at any given point, so that $\mathbf{m}(\mathbf{n}, \mathbf{x}, t)$; however, we will suppress the other dependences for the moment. The vector sum of all the body couples $\mathbf{l}$ acting on an arbitrary finite volume $\mathscr{V}$ is given by the space integral over the volume

$$\int_{\mathscr{V}} \rho \mathbf{l} \, dv, \tag{4.35}$$

and the vector sum of all surface couples $\mathbf{m}(\mathbf{n})$ acting on an arbitrary finite surface $\mathscr{S}$ is given by the space integral over the surface

$$\int_{\mathscr{S}} \mathbf{m}(\mathbf{n}) \, ds. \tag{4.36}$$

Subsequently, the resultant moment about $\mathbf{x}_c$ is given by

$$m = \int_{\mathscr{S}} [\mathbf{m}(\mathbf{n}) + \mathbf{r} \times \mathbf{t}(\mathbf{n})] \, ds + \int_{\mathscr{V}} \rho (\mathbf{l} + \mathbf{r} \times \mathbf{f}) \, dv \tag{4.37}$$

or

$$M = \int_{\mathscr{S}} [\mathsf{M}(\mathbf{n}) + \mathbf{r} \wedge \mathbf{t}(\mathbf{n})] \, ds + \int_{\mathscr{V}} \rho (\mathsf{L} + \mathbf{r} \wedge \mathbf{f}) \, dv. \tag{4.38}$$

Now, using (4.32), (4.37), and (4.38), the basic Axioms 2 and 3 (see (4.23)–(4.25)) become

$$\frac{d}{dt} \int_{\mathscr{V}} \rho \mathbf{v} \, dv = \int_{\mathscr{S}} \mathbf{t}(\mathbf{n}) \, ds + \int_{\mathscr{V}} \rho \mathbf{f} \, dv \tag{4.39}$$

and

$$\frac{d}{dt} \int_{\mathscr{V}} \rho (\boldsymbol{\varphi} + \mathbf{r} \times \mathbf{v}) \, dv = \int_{\mathscr{S}} [\mathbf{m}(\mathbf{n}) + \mathbf{r} \times \mathbf{t}(\mathbf{n})] \, ds + \int_{\mathscr{V}} \rho (\mathbf{l} + \mathbf{r} \times \mathbf{f}) \, dv \tag{4.40}$$

or

$$\frac{d}{dt} \int_{\mathscr{V}} \rho (\boldsymbol{\Phi} + \mathbf{r} \wedge \mathbf{v}) \, dv = \int_{\mathscr{S}} [\mathsf{M}(\mathbf{n}) + \mathbf{r} \wedge \mathbf{t}(\mathbf{n})] \, ds + \int_{\mathscr{V}} \rho (\mathsf{L} + \mathbf{r} \wedge \mathbf{f}) \, dv. \tag{4.41}$$

These are Euler's equations that govern the global motion of the body. Note that the volume and surface integrals can be taken over the space occupied instantaneously by the deformed configuration of the body, as is usually done in solid mechanics, or applied to a given fixed volume of space (with the closed surface called a control surface), which at different times does not usually contain the same material, as is usually done in fluid mechanics.

4.5 Rigid body dynamics

A deformation is rigid if the distance between any two material points remains constant during the motion, i.e.,

$$|\mathbf{x}(\mathbf{X}, t) - \mathbf{x}(\mathbf{Y}, t)|^2 = |\mathbf{X} - \mathbf{Y}|^2, \tag{4.42}$$

where $\mathbf{X}$ and $\mathbf{Y}$ denote two points in the reference configuration. By differentiating with respect to $\mathbf{X}$, we obtain

$$\mathbf{F}^T(\mathbf{X},t)\cdot[\mathbf{x}(\mathbf{X},t)-\mathbf{x}(\mathbf{Y},t)]=\mathbf{X}-\mathbf{Y}, \tag{4.43}$$

where $\mathbf{F}(\mathbf{X},t)$ is the usual deformation gradient. Repeating the procedure by differentiating with respect to $\mathbf{Y}$, we obtain

$$\mathbf{F}^T(\mathbf{Y},t)\cdot[\mathbf{x}(\mathbf{X},t)-\mathbf{x}(\mathbf{Y},t)]=\mathbf{X}-\mathbf{Y}. \tag{4.44}$$

Subsequently, it follows that

$$\mathbf{F}(\mathbf{X},t)\cdot\mathbf{F}^T(\mathbf{Y},t)=\mathbf{1}. \tag{4.45}$$

Since this result is valid for any arbitrary point $\mathbf{Y}$, then we must have that the deformation gradient is independent of the spacial location, i.e., $\mathbf{F}(\mathbf{X},t)=\mathbf{F}(t)$, so

$$\mathbf{F}(t)\cdot\mathbf{F}^T(t)=\mathbf{1}, \tag{4.46}$$

and thus the deformation gradient is orthogonal. From the polar decomposition, we subsequently have that $\mathbf{F}(t)=\mathbf{R}(t)$, $\mathbf{U}=\mathbf{V}=\mathbf{1}$, and we can rewrite (4.43) as the rigid motion

$$\mathbf{x}(\mathbf{X},t)=\mathbf{b}(t)+\mathbf{R}(t)\cdot\mathbf{X}, \tag{4.47}$$

where $\mathbf{b}(t)=\mathbf{x}(\mathbf{Y},t)-\mathbf{R}(t)\cdot\mathbf{Y}$. It is easily verified that the motion (4.47) satisfies (4.42). In particular, if $\mathbf{X}_c$ is the center of mass in the reference configuration, i.e.,

$$\mathbf{X}_c=\mathbf{X}_P+\int_V\rho(\mathbf{X})\,(\mathbf{X}-\mathbf{X}_P)\;dV, \tag{4.48}$$

and $\mathbf{X}_P$ is an arbitrary point, then

$$\mathbf{x}_c(t)=\mathbf{b}(t)+\mathbf{R}(t)\cdot\mathbf{X}_c, \tag{4.49}$$

where $\mathbf{x}_c(t)\equiv\mathbf{x}(\mathbf{X}_c,t)$. Now subtracting (4.49) from (4.47), we obtain

$$[\mathbf{x}(\mathbf{X},t)-\mathbf{x}_c(t)]=\mathbf{R}(t)\cdot(\mathbf{X}-\mathbf{X}_c). \tag{4.50}$$

Differentiating with respect to time and substituting (4.50), we obtain

$$\boldsymbol{\upsilon}(\mathbf{x},t)=\mathbf{W}(t)\cdot\mathbf{r}(t)=-\boldsymbol{\varpi}(t)\times\mathbf{r}(t), \tag{4.51}$$

where $\boldsymbol{\upsilon}(\mathbf{x},t)=\mathbf{v}(\mathbf{x},t)-\mathbf{v}_c(t)$ is the particle velocity relative to the velocity of the center of mass (barycentric velocity), $\mathbf{W}(t)=\dot{\mathbf{R}}(t)\cdot\mathbf{R}^T(t)$ is the skew-symmetric rigid spin tensor, and $\boldsymbol{\varpi}(t)=\langle\mathbf{W}(t)\rangle$ is the angular velocity, which is the axial vector corresponding to the spin tensor (see (2.147)). From (3.491) and (3.492), we also see that $\mathbf{D}(\mathbf{x},t)=\mathbf{0}$, $\mathbf{W}(\mathbf{x},t)=\mathbf{W}(t)$, and then it follows that $\mathbf{L}(\mathbf{x},t)=\mathbf{W}(t)$, or from (4.51) that

$$\mathbf{W}(t)=\frac{1}{2}\left((\operatorname{grad}\mathbf{v})^T-\operatorname{grad}\mathbf{v}\right)\qquad\text{and}\qquad\boldsymbol{\varpi}(t)=-\frac{1}{2}\operatorname{curl}\mathbf{v}. \tag{4.52}$$

We also note from (4.12) that

$$\int_{\mathcal{V}}\rho\,\boldsymbol{\upsilon}\,dv=\mathbf{0}. \tag{4.53}$$

Now, using (4.12) and (4.22), we can write the linear momentum equation (4.23) in the form

$$\mathcal{M}\,\mathbf{a}_c = \mathcal{F}, \tag{4.54}$$

where $\mathbf{a}_c = \dot{\mathbf{v}}_c = \ddot{\mathbf{x}}_c$ is the acceleration of the center of mass. Equation (4.54) is just Newton's equation of motion of the center of mass of the material body.

Below we will take the material to be nonpolar. The moment of momentum equation (4.16) can then be rewritten by separating the contributions of the center of mass and that relative to the center of mass:

$$\boldsymbol{h} = \mathbf{r}_c \times \mathcal{M}\,\mathbf{v}_c + \int_{\mathscr{V}} (\mathbf{x} - \mathbf{x}_c) \times \rho\,(\mathbf{v} - \mathbf{v}_c)\,dv, \tag{4.55}$$

where $\mathbf{r}_c = \mathbf{x}_c - \mathbf{x}_P$ is the moment arm about the fixed position $\mathbf{x}_P$. In the following development, we select the point $\mathbf{x}_P$ to correspond to the center of mass, i.e., $\mathbf{r} = \mathbf{x} - \mathbf{x}_c$ and $\mathbf{r}_c = \mathbf{0}$. Subsequently, we then have

$$\boldsymbol{h} = \int_{\mathscr{V}} \mathbf{r} \times \rho\,\boldsymbol{v}\,dv \qquad \text{and} \qquad \mathcal{H} = \int_{\mathscr{V}} \mathbf{r} \wedge \rho\,\boldsymbol{v}\,dv. \tag{4.56}$$

Substituting (4.51) into (4.56), we obtain

$$\boldsymbol{h} = -\mathbf{I}_c \cdot \boldsymbol{\varpi} \qquad \text{and} \qquad \mathcal{H} = -\left(\mathbf{I} \cdot \mathbf{W} + \mathbf{W} \cdot \mathbf{I}\right), \tag{4.57}$$

where we have defined the inertia tensors

$$\mathbf{I}(t) \equiv \int_{\mathscr{V}} \rho\,\mathbf{r}\,\mathbf{r}\,dv \qquad \text{and} \qquad \mathbf{I}_c(t) \equiv (\operatorname{tr}\mathbf{I}(t))\,\mathbf{1} - \mathbf{I}(t). \tag{4.58}$$

Principal moments of inertia, $I_c^{(i)}$ for $i = 1, 2, 3$, can be readily obtained from

$$\det\,(\mathbf{I}_c - I_c \mathbf{1}) = 0. \tag{4.59}$$

Similarly, we can rewrite the kinetic energy by separating the contributions of the center of mass and that relative to it:

$$\mathcal{K} = \frac{1}{2}\mathcal{M}\,\mathbf{v}_c \cdot \mathbf{v}_c + \frac{1}{2}\int_{\mathscr{V}} \rho\,\boldsymbol{v} \cdot \boldsymbol{v}\,dv \tag{4.60}$$

and subsequently it can be rewritten in the following forms:

$$\mathcal{K} = \frac{1}{2}\mathcal{M}\,\mathbf{v}_c \cdot \mathbf{v}_c + \frac{1}{2}\boldsymbol{\varpi} \cdot \mathbf{I}_c \cdot \boldsymbol{\varpi} = \frac{1}{2}\mathcal{M}\,\mathbf{v}_c \cdot \mathbf{v}_c + \frac{1}{2}\mathbf{W} \cdot \mathbf{I} \cdot \mathbf{W}. \tag{4.61}$$

We now note that

$$\mathring{\mathbf{I}} \equiv \dot{\mathbf{I}} - \mathbf{W} \cdot \mathbf{I} + \mathbf{I} \cdot \mathbf{W} = \mathbf{0} \qquad \text{and} \qquad \dot{\mathbf{I}}_c = -\dot{\mathbf{I}}, \tag{4.62}$$

where $\mathring{\mathbf{I}}$ is the corotational or Jaumann derivative (see (3.422)), so that subsequently, by taking the time derivative of equations (4.56), we can write the equations for the moment of momentum of a rigid body as

$$\dot{\boldsymbol{h}} = -\mathbf{I}_c \cdot \dot{\boldsymbol{\varpi}} + \boldsymbol{\varpi} \times (\mathbf{I}_c \cdot \boldsymbol{\varpi}) = \boldsymbol{m} \quad \text{and} \quad \dot{\mathcal{H}} = -\left(\mathbf{I} \cdot \dot{\mathbf{W}} + \dot{\mathbf{W}} \cdot \mathbf{I}\right) + \left(\mathbf{I} \cdot \mathbf{W}^2 - \mathbf{W}^2 \cdot \mathbf{I}\right) = \boldsymbol{\mathcal{M}}. \tag{4.63}$$

In terms of the principal moments of inertia, we can rewrite the above equation as

$$I_c^{(1)}\dot{\varpi}_1 + \varpi_2\,\varpi_3\left(I_c^{(2)} - I_c^{(3)}\right) = -m_1, \tag{4.64}$$

$$I_c^{(2)}\dot{\varpi}_2 + \varpi_3\,\varpi_1\left(I_c^{(3)} - I_c^{(1)}\right) = -m_2, \tag{4.65}$$

$$I_c^{(3)}\dot{\varpi}_3 + \varpi_1\,\varpi_2\left(I_c^{(1)} - I_c^{(2)}\right) = -m_3. \tag{4.66}$$

These correspond to the Euler equations of motion for a rigid body about the center of mass.

Lastly, taking the time derivative of the kinetic energy (4.61), and using (4.54), (4.58), and (4.62), we readily see that

$$\dot{\mathcal{K}} = \mathbf{v}_c \cdot \boldsymbol{\mathcal{F}} + \boldsymbol{\varpi} \cdot \boldsymbol{m} = \mathbf{v}_c \cdot \boldsymbol{\mathcal{F}} + \mathbf{W} : \boldsymbol{\mathcal{M}}. \tag{4.67}$$

That is, the time rate of change of the kinetic energy is equal to the total power expended by the external forces and couples.

4.6 Stress and couple stress hypotheses

Internal loads and their connection to surface loads may be understood by the application of the balance of momenta on a small continuous region of volume Δv bounded by the closed surface Δs fully contained in a body. At a point on the surface, the effect of the other part of the body is equivalent to a system of forces $\mathbf{t}(\mathbf{n})$ called *stress vectors* and surface couples $\mathbf{m}(\mathbf{n})$ called *couple stress vectors*. Both stress and couple stress vectors are objective. On surfaces passing through the same point, but oriented differently, the stress and couple stress vectors are in general different. Thus these loads depend not only on their location on the surface but also on the exterior normal vector of the surface. This is why we are indicating this dependence explicitly. To determine the dependence of the stress and couple stress vectors on $\mathbf{n}$, we apply the balance of momenta to a volume in the shape of a small tetrahedron with its vertex fixed at $\mathbf{r}$ and having three of its faces on the Cartesian coordinate surfaces and the fourth face being a curved surface, as illustrated in Fig. 4.2. The stress vector on the coordinate surface x_k = const. (whose exterior normal is in the $-\mathbf{i}_k$ direction) is denoted by $\mathbf{t}(-\mathbf{i}_k)$.

Now from the *mean value theorem* we can define the mean density

$$\rho^\star \equiv \frac{1}{\Delta v}\int_{\Delta v} \rho\, dv, \tag{4.68}$$

the mean linear momentum

$$\rho^\star \mathbf{v}^\star \equiv \frac{1}{\Delta v}\int_{\Delta v} \rho\mathbf{v}\, dv, \tag{4.69}$$

the mean body force

$$\rho^\star \mathbf{f}^\star \equiv \frac{1}{\Delta v}\int_{\Delta v} \rho\mathbf{f}\, dv, \tag{4.70}$$

and the mean stress vector

$$\mathbf{t}^\star(\mathbf{n}) \equiv \frac{1}{\Delta s}\int_{\Delta s} \mathbf{t}(\mathbf{n})\, ds. \tag{4.71}$$

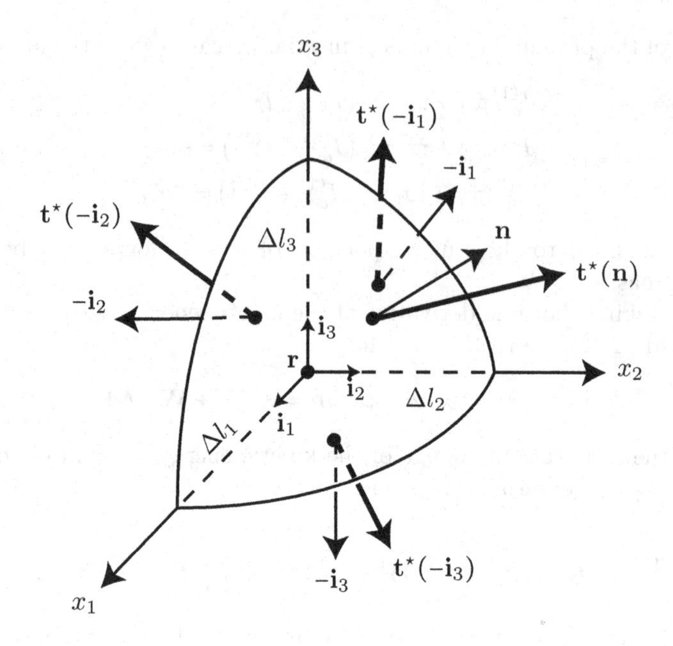

Figure 4.2: Infinitesimal tetrahedral volume Δv with surface Δs.

It is stressed that from the mean value theorem, there exist points within Δv where volumetric mean quantities are defined, and points within the curved surface Δs or the coordinate surface Δs_k where corresponding mean surface quantities are defined. Subsequently, the balances of mass (4.22) and linear momentum (4.39) applied to the tetrahedron, by making use of the mean quantities, become

$$\frac{d}{dt}(\rho^\star \Delta v) = 0, \tag{4.72}$$

and

$$\frac{d}{dt}(\rho^\star \mathbf{v}^\star \Delta v) = \mathbf{t}^\star(\mathbf{n})\Delta s + \mathbf{t}^\star(-\mathbf{i}_k)\Delta s_k + \rho^\star \mathbf{f}^\star \Delta v. \tag{4.73}$$

We note that by using the mass balance, the balance of linear momentum can be rewritten in the form

$$\rho^\star (\dot{\mathbf{v}}^\star - \mathbf{f}^\star)\Delta v = \mathbf{t}^\star(\mathbf{n})\Delta s + \mathbf{t}^\star(-\mathbf{i}_k)\Delta s_k. \tag{4.74}$$

Now dividing both sides of the equation by Δs, letting $\Delta v \to 0$ and $\Delta s \to 0$ so that the curved surface approaches $\mathbf{r}$, noting that $\Delta v/\Delta s \to 0$, and assuming that the quantities $\rho^\star$, $\dot{\mathbf{v}}^\star$, $\mathbf{f}^\star$, and $\Delta s_k/\Delta s$ remain bounded, we obtain

$$\mathbf{t}(\mathbf{n})\, ds = -\mathbf{t}(-\mathbf{i}_k)\, ds_k. \tag{4.75}$$

If we denote by Δl_k a typical dimension of our tetrahedron in the x_k direction (see Fig. 4.2), and if in the above limiting process we take $\Delta v \to 0$ and $\Delta l_k \to 0$, we obtain

$$\mathbf{t}(\mathbf{i}_k) = -\mathbf{t}(-\mathbf{i}_k), \tag{4.76}$$

which demonstrates that the stress vector acting on opposite sides of the same surface at a given point is equal in magnitude and opposite in sign. This result is known as *Cauchy's lemma*. The four faces of the tetrahedron form a closed surface. Therefore, in the limit, the vector sum of the coordinate surfaces must add up to the area vector $d\mathbf{s}$, i.e.,

$$d\mathbf{s} = \mathbf{n}\, ds \qquad \text{or} \qquad ds_k = n_k\, ds. \tag{4.77}$$

Substituting the last two expressions in our limiting result, we obtain

$$\mathbf{t}(\mathbf{n}) = \mathbf{t}(\mathbf{i}_k) n_k. \tag{4.78}$$

This proves that the stress vector at a point on a surface with an exterior normal $\mathbf{n}$ is a linear function of the stress vectors acting on the coordinate surfaces through the same point, the coefficients being the direction cosines of $\mathbf{n}$.

Analogously, from the mean value theorem, we can define the mean internal spin

$$\rho^\star \varphi^\star \equiv \frac{1}{\Delta v} \int_{\Delta v} \rho\varphi \, dv, \tag{4.79}$$

the mean body couple

$$\rho^\star \mathbf{l}^\star \equiv \frac{1}{\Delta v} \int_{\Delta v} \rho\mathbf{l} \, dv, \tag{4.80}$$

and the mean surface couple

$$\mathbf{m}^\star(\mathbf{n}) \equiv \frac{1}{\Delta s} \int_{\Delta s} \mathbf{m}(\mathbf{n}) \, ds. \tag{4.81}$$

Subsequently, using the mass balance (4.72) and the linear momentum balance (4.74) on the tetrahedron, the angular momentum balance (4.40) becomes a balance for the internal spin:

$$\rho^\star \left(\dot{\varphi}^\star - \mathbf{l}^\star\right) \Delta v = \mathbf{m}^\star(\mathbf{n})\Delta s + \mathbf{m}^\star(-\mathbf{i}_k)\Delta s_k. \tag{4.82}$$

Now, assuming that the quantities $\rho^\star$, $\dot{\varphi}^\star$, and $\mathbf{l}^\star$ remain bounded in the limit $\Delta v \to 0$, it is immediately clear that application of the same procedure to the balance of internal spin leads to the equivalent statements for the couple stress axial vector:

$$\mathbf{m}(\mathbf{i}_k) = -\mathbf{m}(-\mathbf{i}_k), \tag{4.83}$$

and

$$\mathbf{m}(\mathbf{n}) = \mathbf{m}(\mathbf{i}_k) n_k. \tag{4.84}$$

4.6.1 Stress and couple stress tensors

We now define the lth component of the stress vector $\mathbf{t}(\mathbf{i}_k)$ acting on the positive side of the kth coordinate surface as

$$\mathbf{t}(\mathbf{i}_k) \equiv \mathbf{t}_k \equiv \sigma_{lk} \mathbf{i}_l. \tag{4.85}$$

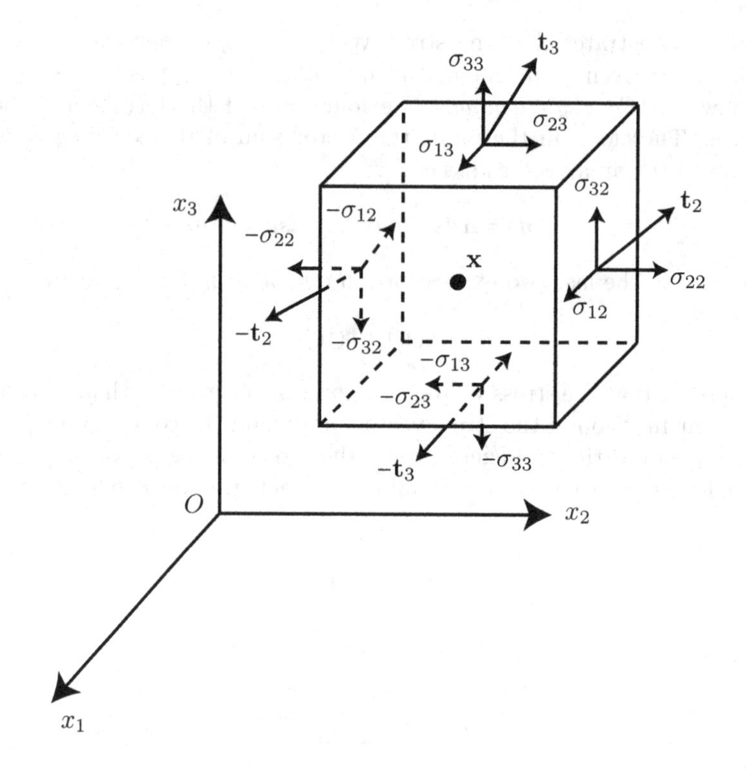

Figure 4.3: Components of stress tensor.

The components σ_{lk} correspond to the components of a second-order tensor $\boldsymbol{\sigma}$ that we call the *stress tensor*:

$$\boldsymbol{\sigma} = \sigma_{lk}\mathbf{i}_l\mathbf{i}_k. \tag{4.86}$$

It is a simple exercise to show that the stress tensor is objective. As we see, now writing complete explicit dependencies, the stress vector $\mathbf{t}(\mathbf{n}, \mathbf{x}, t)$ is related to the stress tensor $\boldsymbol{\sigma}(\mathbf{x}, t)$ by the linear transformation

$$\mathbf{t}(\mathbf{n}, \mathbf{x}, t) = \boldsymbol{\sigma}(\mathbf{x}, t) \cdot \mathbf{n} \qquad \text{or} \qquad t_l(\mathbf{i}_k, \mathbf{x}, t) = \sigma_{lk}(\mathbf{x}, t). \tag{4.87}$$

This result is known as *Cauchy's theorem*, and it asserts that $\mathbf{t}(\mathbf{n}, \mathbf{x}, t)$ depends upon the surface orientation only in a *linear* fashion. Note that the second subscript in σ_{lk} indicates the coordinate surface x_k = const. on which the stress vector $\mathbf{t}_k$ acts, and the first subscript indicates the direction of the component of $\mathbf{t}_k$, as illustrated in Fig. 4.3. Some authors use the opposite convention for the subscripts.

The components σ_{kk} are called *normal stresses* and the mixed components σ_{lk} with $l \neq k$ are called *shearing stresses*. In matrix form they are given by

$$[\sigma_{lk}] = \begin{bmatrix} \sigma_{11} & \sigma_{12} & \sigma_{13} \\ \sigma_{21} & \sigma_{22} & \sigma_{23} \\ \sigma_{31} & \sigma_{32} & \sigma_{33} \end{bmatrix}. \tag{4.88}$$

A stress is called *hydrostatic* if and only if

$$\mathbf{t}(\mathbf{n},\mathbf{x},t) = \boldsymbol{\sigma}(\mathbf{x},t)\cdot\mathbf{n} = -p(\mathbf{x},t)\,\mathbf{n} \tag{4.89}$$

for all $\mathbf{n}$, where p is a scalar independent of $\mathbf{n}$ (the negative sign is used by convention). Alternately, the stress is hydrostatic if and only if every component of $\boldsymbol{\sigma}$ transforms like a scalar.

It can be shown in a similar manner that the axial couple stress vector $\mathbf{m}(\mathbf{n},\mathbf{x},t)$ is related to an axial tensor $\boldsymbol{\Sigma}(\mathbf{x},t)$ by the linear transformation

$$\mathbf{m}(\mathbf{n},\mathbf{x},t) = \boldsymbol{\Sigma}(\mathbf{x},t)\cdot\mathbf{n} \qquad \text{or} \qquad m_l(\mathbf{i}_k,\mathbf{x},t) = \Sigma_{lk}(\mathbf{x},t), \tag{4.90}$$

where

$$\boldsymbol{\Sigma} = \Sigma_{lk}\mathbf{i}_l\mathbf{i}_k \tag{4.91}$$

is an objective second-order axial tensor called the *couple stress tensor*. This tensor is dual to the third-order tensor

$$\widehat{\boldsymbol{\Sigma}} = \boldsymbol{\Sigma}\cdot\boldsymbol{\epsilon} \tag{4.92}$$

that is skew-symmetric with respect to the first pair of indices, i.e., $\widehat{\Sigma}_{jkl} = -\widehat{\Sigma}_{kjl}$. Its irreducible symmetry is analogous to that of $\widehat{\boldsymbol{\Xi}}$ given by (3.366)–(3.370).

With the introductions of the stress and the couple stress tensors through (4.87) and (4.90), the balances of linear and angular momenta (4.39) and (4.40) give rise to *Cauchy's first and second laws of motion*:

$$\frac{d}{dt}\int_{\mathscr{V}} \rho\mathbf{v}\,dv = \int_{\mathscr{S}} \boldsymbol{\sigma}\cdot d\mathbf{s} + \int_{\mathscr{V}} \rho\mathbf{f}\,dv \tag{4.93}$$

and

$$\frac{d}{dt}\int_{\mathscr{V}} \rho(\boldsymbol{\varphi} + \mathbf{r}\times\mathbf{v})\,dv = \int_{\mathscr{S}} (\boldsymbol{\Sigma} + \mathbf{r}\times\boldsymbol{\sigma})\cdot d\mathbf{s} + \int_{\mathscr{V}} \rho(\mathbf{l} + \mathbf{r}\times\mathbf{f})\,dv \tag{4.94}$$

or

$$\frac{d}{dt}\int_{\mathscr{V}} \rho(\boldsymbol{\Phi} + \mathbf{r}\wedge\mathbf{v})\,dv = \int_{\mathscr{S}} (\widehat{\boldsymbol{\Sigma}} + \mathbf{r}\wedge\boldsymbol{\sigma})\cdot d\mathbf{s} + \int_{\mathscr{V}} \rho(\mathbf{L} + \mathbf{r}\wedge\mathbf{f})\,dv. \tag{4.95}$$

4.7 Local forms of axioms of mechanics

The local form of the mass conservation equation is obtained by comparing the integral form of the equation (4.22), which reflects the statement of Axiom 1, with the general balance law (4.1). In this comparison, we note that $\psi \to \rho$, $\mathbf{t} \to \mathbf{0}$, and $g \to 0$. Subsequently, from (4.6) we have

$$\int_{\mathscr{V}(t)-\zeta(t)} \left[\frac{\partial\rho}{\partial t} + \operatorname{div}(\rho\mathbf{v})\right] dv + \int_{\zeta(t)} [\![\rho(\mathbf{v}-\mathbf{c})]\!]\cdot\mathbf{n}\,d\zeta = 0. \tag{4.96}$$

Setting the integrand of the first integral to zero, we obtain the local mass conservation which applies at a regular point in a continuous region

$$\frac{\partial\rho}{\partial t} + \operatorname{div}(\rho\mathbf{v}) = 0, \tag{4.97}$$

which can also be rewritten in the form

$$\dot{\rho} + \rho \operatorname{div} \mathbf{v} = 0 \qquad \text{or} \qquad \dot{\rho} + \rho v_{k,k} = 0, \tag{4.98}$$

where we use the definition of the material derivative:

$$\dot{\rho} = \frac{\partial \rho}{\partial t} + (\mathbf{v} \cdot \operatorname{grad})\rho. \tag{4.99}$$

It should be noted that, in contrast to $(4.21)_2$, in general ρ can be determined from $\mathbf{v}$ without knowing the motion $\boldsymbol{\chi}$ explicitly. If we set the integrand of the second integral to zero, we obtain the jump condition across the singular moving surface $\boldsymbol{\zeta}$:

$$[\![\rho(\mathbf{v} - \mathbf{c})]\!] \cdot \mathbf{n} = 0 \qquad \text{or} \qquad [\![\rho(v_k - c_k)]\!] n_k = 0. \tag{4.100}$$

The jump condition allows us to define the mass flux (per unit area)

$$m = \rho(\mathbf{v} - \mathbf{c}) \cdot \mathbf{n}, \tag{4.101}$$

so we see that the mass flux is continuous across the singular surface:

$$[\![m]\!] = 0. \tag{4.102}$$

Note that m is continuous at the jump, so $m^+ = m^- = m$, where m is the value at the singular surface. Furthermore, since

$$[\![\mathbf{v}]\!] \cdot \mathbf{n} = [\![\mathbf{v} - \mathbf{c}]\!] \cdot \mathbf{n} = \left[\!\!\left[\frac{1}{\rho} \rho(\mathbf{v} - \mathbf{c}) \right]\!\!\right] \cdot \mathbf{n} = m \, [\![\mathsf{v}]\!], \tag{4.103}$$

where $\mathsf{v} = 1/\rho$ is the specific volume, we can decompose the jump in velocity into tangential and normal components,

$$[\![\mathbf{v}]\!] = [\![\mathbf{v}_{(n)}]\!] + [\![\mathbf{v}_{(s)}]\!] = m \, [\![\mathsf{v}]\!] \, \mathbf{n} + [\![v_{(s)}]\!] \, \mathbf{s}, \tag{4.104}$$

where

$$\mathbf{v}_{(n)} = v_{(n)} \, \mathbf{n} = (\mathbf{v} \cdot \mathbf{n}) \, \mathbf{n} \qquad \text{and} \qquad \mathbf{v}_{(s)} = v_{(s)} \, \mathbf{s} = (\mathbf{v} \cdot \mathbf{s}) \, \mathbf{s} = \mathbf{v} \cdot (\mathbf{1} - \mathbf{n}\mathbf{n}) \tag{4.105}$$

are the normal and tangential velocity components, $[\![v_{(s)}]\!]$ is the *slip*, and $\mathbf{s}$ is a vector tangent to the singular surface.

It will be particularly useful to take advantage of the local mass balances (4.97) and (4.102) in conjunction with taking $\psi \to \rho\psi$ and $g \to \rho g$ in the general balance laws (4.1) and (4.6) for an arbitrary volume. In such case, the balance laws reduce to the more suggestive forms

$$\frac{d}{dt} \int_{\mathscr{V}} \rho \psi \, dv = \int_{\mathscr{S}} \mathbf{t} \cdot d\mathbf{s} + \int_{\mathscr{V}} \rho g \, dv \tag{4.106}$$

and

$$\int_{\mathscr{V}(t) - \zeta(t)} \left[\rho(\dot{\psi} - g) - \operatorname{div} \mathbf{t} \right] dv + \int_{\zeta(t)} (m [\![\psi]\!] - [\![\mathbf{t}]\!] \cdot \mathbf{n}) \, d\zeta = 0. \tag{4.107}$$

The local form of the balance of linear momentum is obtained by comparing the integral form of equation (4.93), which reflects the statement of Axiom 2, with the reformulated general balance laws given above. In this comparison, we note that $\psi \to \mathbf{v}$, $\mathbf{t} \to \boldsymbol{\sigma}$, and $g \to \mathbf{f}$. Subsequently, we write

$$\int_{\mathscr{V}(t)-\zeta(t)} [\rho(\mathbf{a}-\mathbf{f}) - \operatorname{div}\boldsymbol{\sigma}]\, dv + \int_{\zeta(t)} (m[\![\mathbf{v}]\!] - [\![\boldsymbol{\sigma}]\!]\cdot\mathbf{n})\, d\zeta = \mathbf{0}, \tag{4.108}$$

where $\mathbf{a} = \dot{\mathbf{v}}$ is the material particle acceleration.

Setting the integrand of the first integral to zero, we obtain the local balance of linear momentum, which applies at a regular point in a continuous region

$$\rho(\mathbf{a}-\mathbf{f}) = \operatorname{div}\boldsymbol{\sigma} \qquad \text{or} \qquad \rho(a_l - f_l) = \sigma_{lk,k}, \tag{4.109}$$

and setting the integrand of the second integral to zero, we obtain the jump condition across the singular moving surface $\boldsymbol{\zeta}$

$$m[\![\mathbf{v}]\!] - [\![\boldsymbol{\sigma}]\!]\cdot\mathbf{n} = \mathbf{0} \qquad \text{or} \qquad m[\![v_l]\!] - [\![\sigma_{lk}]\!]n_k = 0. \tag{4.110}$$

The above equation valid at a regular point is sometimes called *Cauchy's first law of motion.* Taking the inner product of (4.110) with $\mathbf{n}$ and $\mathbf{s}$, and using (4.104) and (4.105), we have

$$m^2[\![\mathrm{v}]\!] - \mathbf{n}\cdot[\![\boldsymbol{\sigma}]\!]\cdot\mathbf{n} = 0 \qquad \text{and} \qquad m[\![v_{(s)}]\!] - \mathbf{s}\cdot[\![\boldsymbol{\sigma}]\!]\cdot\mathbf{n} = 0. \tag{4.111}$$

Lastly, it should be emphasized that in general $\mathbf{a}$ can be determined from $\mathbf{v}$ without knowing the motion $\boldsymbol{\chi}$ explicitly. In arriving at (4.109) from (4.108), we have assumed that the body force $\mathbf{f}$ is continuous.

Lastly, the local form of the equation stating the balance of angular momentum is obtained by comparing the integral form of equation (4.94), that reflects the statement of Axiom 3, with the reformulated balance laws (4.106) and (4.107). In this comparison, we note that $\psi \to (\boldsymbol{\varphi} + \mathbf{r}\times\mathbf{v})$, $\mathbf{t} \to (\boldsymbol{\Sigma} + \mathbf{r}\times\boldsymbol{\sigma})$, and $g \to (\mathbf{l} + \mathbf{r}\times\mathbf{f})$. Subsequently, we have

$$\begin{aligned}\int_{\mathscr{V}(t)-\zeta(t)} \{\rho[(\dot{\boldsymbol{\varphi}} + \mathbf{r}\times\mathbf{a}) - (\mathbf{l} + \mathbf{r}\times\mathbf{f})] - \operatorname{div}(\boldsymbol{\Sigma} + \mathbf{r}\times\boldsymbol{\sigma})\}\, dv \,+ \\ \int_{\zeta(t)} (m[\![\boldsymbol{\varphi} + \mathbf{r}\times\mathbf{v}]\!] - [\![\boldsymbol{\Sigma} + \mathbf{r}\times\boldsymbol{\sigma}]\!]\cdot\mathbf{n})\, d\zeta = \mathbf{0}.\end{aligned} \tag{4.112}$$

We now recognize that

$$\operatorname{div}(\mathbf{r}\times\boldsymbol{\sigma}) = \mathbf{r}\times(\operatorname{div}\boldsymbol{\sigma}) - \boldsymbol{\varepsilon} : \boldsymbol{\sigma}, \tag{4.113}$$

so that, by rearranging the above integrals, we have

$$\begin{aligned}\int_{\mathscr{V}(t)-\zeta(t)} \{[\rho(\dot{\boldsymbol{\varphi}} - \mathbf{l}) - \operatorname{div}\boldsymbol{\Sigma} + \boldsymbol{\varepsilon}:\boldsymbol{\sigma}] + \mathbf{r}\times[\rho(\mathbf{a}-\mathbf{f}) - \operatorname{div}\boldsymbol{\sigma}]\}\, dv \,+ \\ \int_{\zeta(t)} [(m[\![\boldsymbol{\varphi}]\!] - [\![\boldsymbol{\Sigma}]\!]\cdot\mathbf{n}) + \mathbf{r}\times(m[\![\mathbf{v}]\!] - [\![\boldsymbol{\sigma}]\!]\cdot\mathbf{n})]\, d\zeta = \mathbf{0}.\end{aligned} \tag{4.114}$$

Using the balances of linear momentum at regular and singular points (4.109) and (4.110), we finally obtain

$$\int_{\mathscr{V}(t)-\zeta(t)} [\rho(\dot{\boldsymbol{\varphi}} - \mathbf{l}) - \operatorname{div}\boldsymbol{\Sigma} + \boldsymbol{\varepsilon}:\boldsymbol{\sigma}]\, dv + \int_{\zeta(t)} (m[\![\boldsymbol{\varphi}]\!] - [\![\boldsymbol{\Sigma}]\!]\cdot\mathbf{n})\, d\zeta = \mathbf{0}. \tag{4.115}$$

Now setting the integrand of the first integral to zero, we obtain the local balance of angular momentum, which applies at a regular point in a continuous region

$$\rho(\dot{\varphi} - \mathbf{l}) = \operatorname{div} \boldsymbol{\Sigma} - \boldsymbol{\varepsilon} : \boldsymbol{\sigma} \qquad \text{or} \qquad \rho(\dot{\varphi}_l - l_l) = \Sigma_{lk,k} - \varepsilon_{lkm}\sigma_{km}, \tag{4.116}$$

and setting the integrand of the second integral to zero, we obtain the jump condition across the singular moving surface ζ

$$m[\![\varphi]\!] - [\![\boldsymbol{\Sigma}]\!] \cdot \mathbf{n} = \mathbf{0} \qquad \text{or} \qquad m[\![\varphi_l]\!] - [\![\Sigma_{lk}]\!] n_k = 0. \tag{4.117}$$

The above equation valid at a regular point is sometimes called *Cauchy's second law of motion.* In arriving at (4.116) from (4.115), we have assumed that the internal body couple $\mathbf{l}$ is continuous. Note that we can also write (4.116) in the form

$$\rho\left(\dot{\boldsymbol{\Phi}} - \mathsf{L}\right) = \operatorname{div} \widehat{\boldsymbol{\Sigma}} - \operatorname{skw} \boldsymbol{\sigma} \qquad \text{or} \qquad \rho\left(\dot{\Phi}_{[ij]} - \mathsf{L}_{[ij]}\right) = \widehat{\Sigma}_{[ij]k,k} - \sigma_{[ij]}. \tag{4.118}$$

We see that if $\varphi = \mathbf{0}$ (from (4.14) when $\boldsymbol{i} = \mathbf{0}$), $\mathbf{l} = \mathbf{0}$, and $\boldsymbol{\Sigma} = \mathbf{0}$, then (4.116) becomes

$$\boldsymbol{\varepsilon} : \boldsymbol{\sigma} = \mathbf{0}, \tag{4.119}$$

or

$$\boldsymbol{\sigma} = \boldsymbol{\sigma}^T \qquad \text{or} \qquad \sigma_{lk} = \sigma_{kl}, \tag{4.120}$$

i.e., the stress tensor is symmetric. Materials in which internal spin, body couples, and couple stresses occur are called *polar materials*, and those in which they do not occur are called *nonpolar materials.* A couple, in classical mechanics, is viewed as a pair of parallel (bounded) forces having equal magnitude and opposite direction, separated by a moment arm. If we let the moment arm approach zero, which occurs when we previously took $\Delta v \to 0$ and $\Delta s \to 0$, the moment of the couple approaches zero. Thus, in the classical continuum mechanics picture of central forces, one cannot have a body couple or a couple stress vector without having some other forces acting on the body (e.g., due to a magnetic field). Nevertheless, if the infinitesimals dv and ds are considered to represent some physical volume and surface small enough, but with an acceptable variability, as noted in the introduction, then the existence of an internal spin density, a body couple, and a couple stress vector may be admitted even without the need of additional forces.

By taking the inner product of $\mathbf{v}$ with (4.109) and $\boldsymbol{\nu}$ with (4.116), we find that the local balance of kinetic energy is given by

$$\frac{1}{2}\rho\left(\overline{\boldsymbol{\nu}\cdot\varphi} + \overline{\mathbf{v}\cdot\mathbf{v}}^{\,\cdot}\right) = \operatorname{div}(\boldsymbol{\nu}\cdot\boldsymbol{\Sigma} + \mathbf{v}\cdot\boldsymbol{\sigma}) + \rho(\boldsymbol{\nu}\cdot\mathbf{l} + \mathbf{v}\cdot\mathbf{f}) - \Phi \tag{4.121}$$

or

$$\frac{1}{2}\rho\left(\overline{\nu_l\varphi_l}^{\,\cdot} + \overline{v_l v_l}^{\,\cdot}\right) = (\nu_l\Sigma_{lk} + v_l\sigma_{lk})_{,k} + \rho(\nu_l l_l + v_l f_l) - \Phi, \tag{4.122}$$

where, using (3.362), we have

$$\Phi \equiv \frac{1}{2}\rho\boldsymbol{\nu}\cdot\dot{\boldsymbol{i}}\cdot\boldsymbol{\nu} + \boldsymbol{\Xi} : \boldsymbol{\Sigma} + (\boldsymbol{\Upsilon} + \mathbf{L}) : \boldsymbol{\sigma} \tag{4.123}$$

or

$$\Phi \equiv \frac{1}{2}\rho\nu_k \dot{i}_{kl}\nu_l + \Xi_{kl}\Sigma_{kl} + (\Upsilon_{kl} + L_{kl})\,\sigma_{kl} \tag{4.124}$$

which is called the *mechanical energy*, also called the *stress power*, that may be regarded as an *internal production* of energy, and we recall from (4.14), (3.238), and (3.364) that $\boldsymbol{\varphi} = \boldsymbol{i}\cdot\boldsymbol{\nu}$, $\boldsymbol{\Xi} \equiv (\text{grad}\ \boldsymbol{\nu})^T$, and $\mathbf{L} = (\text{grad}\ \mathbf{v})^T$. The term $\frac{1}{2}\rho\boldsymbol{\nu}\cdot\dot{\boldsymbol{i}}\cdot\boldsymbol{\nu}$ represents the *internal spin production.* If the internal specific inertia tensor is such that $\dot{\boldsymbol{i}} = \mathbf{0}$, the material is called *micropolar*, and if $\dot{\boldsymbol{i}} \neq \mathbf{0}$, it is called *micromorphic*; it is called a *microstretch* material if $\dot{\boldsymbol{i}} = \dot{i}\,\mathbf{1}$.

The kinetic energy equation (4.121) can also be rewritten in integral form if it is integrated over an arbitrary continuous volume and use is made of (4.106), (4.107), (2.289), (4.87), and (4.90):

$$\frac{d}{dt}\int_{\mathscr{V}} \frac{1}{2}\rho\,(\boldsymbol{\nu}\cdot\boldsymbol{\varphi} + \mathbf{v}\cdot\mathbf{v})\,dv = \int_{\mathscr{S}} [\boldsymbol{\nu}\cdot\mathbf{m}(\mathbf{n}) + \mathbf{v}\cdot\mathbf{t}(\mathbf{n})]\,ds + \int_{\mathscr{V}} \rho\,(\boldsymbol{\nu}\cdot\mathbf{l} + \mathbf{v}\cdot\mathbf{f})\,dv - \int_{\mathscr{V}} \Phi\,dv. \tag{4.125}$$

4.8 Properties of stress vector and tensor

The discussions in this section apply to nonpolar materials for which the Cauchy stress tensor $\boldsymbol{\sigma}$ is symmetric.

4.8.1 Principal stresses and principal stress directions

The stress vector $\mathbf{t}(\mathbf{n})$ acting at a point on a surface with exterior normal $\mathbf{n}$ can be decomposed into a component normal to the surface and a component that is tangent to it:

$$\mathbf{t}(\mathbf{n}) = \mathbf{t}_{(n)}(\mathbf{n}) + \mathbf{t}_{(s)}(\mathbf{n}), \tag{4.126}$$

where

$$\mathbf{t}_{(n)}(\mathbf{n}) = \sigma_n\,\mathbf{n} \qquad \text{and} \qquad \mathbf{t}_{(s)}(\mathbf{n}) = \tau\,\mathbf{s}, \tag{4.127}$$

and the normal and tangential vectors are such that

$$\mathbf{n}\cdot\mathbf{n} = 1, \qquad \mathbf{s}\cdot\mathbf{s} = 1, \quad \text{and} \quad \mathbf{n}\cdot\mathbf{s} = 0. \tag{4.128}$$

Note that, using (4.87), we have that

$$\sigma_n = \mathbf{n}\cdot\mathbf{t}(\mathbf{n}) = \mathbf{n}\cdot\boldsymbol{\sigma}\cdot\mathbf{n}, \qquad \tau = \mathbf{s}\cdot\mathbf{t}(\mathbf{n}) = \mathbf{s}\cdot\boldsymbol{\sigma}\cdot\mathbf{n}, \quad \text{and} \quad \tau^2 = \mathbf{t}(\mathbf{n})\cdot\mathbf{t}(\mathbf{n}) - \sigma_n^2. \tag{4.129}$$

Furthermore, the stress tensor can be written in terms of the normal and shear components as follows:

$$\boldsymbol{\sigma} = \sigma_n\,\mathbf{n}\,\mathbf{n} + \tau\,(\mathbf{n}\,\mathbf{s} + \mathbf{s}\,\mathbf{n}). \tag{4.130}$$

In analyzing internal (mutual) tractions at a material point, it is useful to find the direction for which the traction vector is purely normal. The traction vector is purely normal in the direction $\widehat{\mathbf{n}}$ when $\mathbf{t}(\widehat{\mathbf{n}}) = \boldsymbol{\sigma}\cdot\widehat{\mathbf{n}} = \sigma\widehat{\mathbf{n}}$ for some σ. Subsequently, we can write

$$(\boldsymbol{\sigma} - \sigma\mathbf{1})\cdot\widehat{\mathbf{n}} = \mathbf{0} \tag{4.131}$$

to find σ and $\widehat{\mathbf{n}}$. This is just an eigenvalue problem. The normalized eigenvectors $\pm\widehat{\mathbf{n}}^{(l)}$ are called the *principal axes of stress* and the eigenvalues $\sigma^{(l)}$, given by the roots of $\det[\sigma_{ij} - \sigma\delta_{ij}] = 0$, are called the *principal stresses.* Since for nonpolar materials σ_{ij} is symmetric, the principal stresses $\sigma^{(l)}$ are real and the principal axes are orthogonal (see Section 3.1.7). The characteristic equation is given by the cubic

$$\sigma^3 - \sigma_{(1)}\sigma^2 + \sigma_{(2)}\sigma - \sigma_{(3)} = 0, \tag{4.132}$$

where the *stress invariants* are given by (see (3.92)–(3.94) and (3.135)–(3.137))

$$\sigma_{(1)} = \sigma^{(1)} + \sigma^{(2)} + \sigma^{(3)} = \sigma_{ii}, \tag{4.133}$$

$$\sigma_{(2)} = \sigma^{(1)}\sigma^{(2)} + \sigma^{(2)}\sigma^{(3)} + \sigma^{(3)}\sigma^{(1)} = \frac{1}{2}\left(\sigma_{ii}\sigma_{jj} - \sigma_{ij}\sigma_{ji}\right), \tag{4.134}$$

$$\sigma_{(3)} = \sigma^{(1)}\sigma^{(2)}\sigma^{(3)} = \det\left[\sigma_{ij}\right]. \tag{4.135}$$

The above invariants provide formal expressions for the principal stresses in terms of stress tensor components. Here we follow the convention of ordering the principal stresses so that $\sigma^{(1)} \geq \sigma^{(2)} \geq \sigma^{(3)}$. If $\sigma^{(k)}$ is positive, it is said to be a *tension,* and if it is negative, it is said to be a *compression.* If $\sigma^{(1)} \neq 0$ and $\sigma^{(2)} = \sigma^{(3)} = 0$, then we have a *simple tension* or *compression.* If one and only one principal stress vanishes, the state of stress is called *biaxial.* We have a *triaxial stress* otherwise. When the principal axes and stresses are known, stress components are then given by the following spectral decomposition:

$$\boldsymbol{\sigma} = \mathbf{N}^T \cdot \boldsymbol{\Lambda}_\sigma \cdot \mathbf{N}, \tag{4.136}$$

where $N_{jl} = \widehat{n}_j^{(l)}$ (note that $\mathbf{N}^{-1} = \mathbf{N}^T$) and $\boldsymbol{\Lambda}_\sigma$ is the diagonal matrix composed of principal stresses.

Suppose that σ_n is kept fixed and the orientation $\mathbf{n}$ of the surface at a point is varied. This is accomplished by defining different volumes with different surfaces all going through the same point. Then

$$\sigma_n = \mathbf{t}(\mathbf{n}) \cdot \mathbf{n} = \sigma_{lk} n_l n_k = \text{const.} \tag{4.137}$$

represents a quadric surface (see Section 2.10) called the *stress quadric of Cauchy.* We would like to find the extremal values of σ_n, which are in the directions $\widehat{\mathbf{n}}$. Accordingly, we can write

$$f(\widehat{n}_k) = \sigma_{ij}\widehat{n}_i\widehat{n}_j - \sigma(\widehat{n}_i\widehat{n}_i - 1), \tag{4.138}$$

where σ serves the role of a Lagrange multiplier to impose the normalization constraint. Now differentiating $f(\widehat{n}_k)$ with respect to $\widehat{n}_k$, noting that $\partial\widehat{n}_i/\partial\widehat{n}_k = \delta_{ik}$, and taking advantage of the symmetry of σ_{ij}, the extremal problem reduces to the solution of the following problem:

$$(\sigma_{ij} - \sigma\delta_{ij})\widehat{n}_j = 0, \tag{4.139}$$

which is identical to the eigenvalue problem for the principal stresses. Thus, the Lagrange multiplier σ is the same as a principal stress. Furthermore, the principal stresses $\sigma^{(l)}$ include both the maximum and minimum values of normal stress, and the vectors $\widehat{\mathbf{n}}^{(l)}$ can also be thought of geometrically as the principal axes of the stress quadric surface. Note that referring to the principal axes with $\widehat{\mathbf{n}} = \widehat{n}_k \mathbf{i}_k$, where $\widehat{n}_k$ are direction cosines, the stress vector can be rewritten as

$$\mathbf{t}(\widehat{\mathbf{n}}) = \boldsymbol{\sigma} \cdot \widehat{\mathbf{n}} = \sigma^{(1)} \widehat{n}_1 \mathbf{i}_1 + \sigma^{(2)} \widehat{n}_2 \mathbf{i}_2 + \sigma^{(3)} \widehat{n}_3 \mathbf{i}_3, \tag{4.140}$$

and, subsequently, the normal component takes the form

$$\sigma_n = \mathbf{t}(\widehat{\mathbf{n}}) \cdot \widehat{\mathbf{n}} = \sigma^{(1)} \widehat{n}_1^2 + \sigma^{(2)} \widehat{n}_2^2 + \sigma^{(3)} \widehat{n}_3^2. \tag{4.141}$$

With regard to the extreme values of the shear component τ, it is easiest to express it in terms of the principal stresses:

$$\begin{aligned} \tau^2 &= \mathbf{t}(\widehat{\mathbf{n}}) \cdot \mathbf{t}(\widehat{\mathbf{n}}) - \sigma_n^2, \\ &= {\sigma^{(1)}}^2 \widehat{n}_1^2 + {\sigma^{(2)}}^2 \widehat{n}_2^2 + {\sigma^{(3)}}^2 \widehat{n}_3^2 - \left(\sigma^{(1)} \widehat{n}_1^2 + \sigma^{(2)} \widehat{n}_2^2 + \sigma^{(3)} \widehat{n}_3^2\right)^2. \end{aligned} \tag{4.142}$$

Now since $\widehat{n}_k \widehat{n}_k = 1$, we can eliminate one of the direction cosines from the above expression, say $\widehat{n}_3$, to obtain an expression in terms of $\widehat{n}_1$ and $\widehat{n}_2$:

$$\begin{aligned} \tau^2 = {} & \left({\sigma^{(1)}}^2 - {\sigma^{(3)}}^2\right) \widehat{n}_1^2 + \left({\sigma^{(2)}}^2 - {\sigma^{(3)}}^2\right) \widehat{n}_2^2 + {\sigma^{(3)}}^2 \\ & - \left[\left(\sigma^{(1)} - \sigma^{(3)}\right) \widehat{n}_1^2 + \left(\sigma^{(2)} - \sigma^{(3)}\right) \widehat{n}_2^2 + \sigma^{(3)}\right]^2. \end{aligned} \tag{4.143}$$

To find extreme values of $\tau(\widehat{n}_1, \widehat{n}_2)$, we differentiate the above expression with respect to $\widehat{n}_1$ and $\widehat{n}_2$ and set the results equal to zero. After some algebra, we have

$$\begin{aligned} \frac{\partial \tau^2}{\partial \widehat{n}_1} = {} & \widehat{n}_1 \left(\sigma^{(1)} - \sigma^{(3)}\right) \left\{\sigma^{(1)} - \sigma^{(3)} - 2\left[\left(\sigma^{(1)} - \sigma^{(3)}\right) \widehat{n}_1^2 + \right.\right. \\ & \left.\left.\left(\sigma^{(2)} - \sigma^{(3)}\right) \widehat{n}_2^2\right]\right\} = 0, \end{aligned} \tag{4.144}$$

$$\begin{aligned} \frac{\partial \tau^2}{\partial \widehat{n}_2} = {} & \widehat{n}_2 \left(\sigma^{(2)} - \sigma^{(3)}\right) \left\{\sigma^{(2)} - \sigma^{(3)} - 2\left[\left(\sigma^{(1)} - \sigma^{(3)}\right) \widehat{n}_1^2 + \right.\right. \\ & \left.\left.\left(\sigma^{(2)} - \sigma^{(3)}\right) \widehat{n}_2^2\right]\right\} = 0. \end{aligned} \tag{4.145}$$

An obvious solution is $\widehat{n}_1 = \widehat{n}_2 = 0$, for which $\widehat{n}_3 = \pm 1$ and then $\tau = 0$. This is the expected result since $\widehat{n}_3 = \pm 1$ is the principal plane where σ_n is an extreme value for which the shear component is zero. If we had eliminated $\widehat{n}_1$ or $\widehat{n}_2$ instead on $\widehat{n}_3$, similar calculations would lead to the other two principal planes for which the shear component is also zero.

A second solution is obtained by taking $\widehat{n}_1 = 0$ and solving the resulting second quadratic equation for $\widehat{n}_2$. The result is $\widehat{n}_2 = \pm 1/\sqrt{2}$ and, from the normalization, we find $\widehat{n}_3 = \pm 1/\sqrt{2}$. For this solution, the shear component is given by

$$\tau = \pm \frac{1}{2} \left(\sigma^{(2)} - \sigma^{(3)}\right). \tag{4.146}$$

As before, if we had taken $\widehat{n}_2 = 0$ or $\widehat{n}_3 = 0$, we would obtain equivalent solutions. The complete solutions are

$$\tau = \begin{cases} \pm\frac{1}{2}\left(\sigma^{(2)} - \sigma^{(3)}\right) & \text{when} \quad \widehat{n}_1 = 0,\ \widehat{n}_2 = \pm\frac{1}{\sqrt{2}},\ \widehat{n}_3 = \pm\frac{1}{\sqrt{2}}, \\ \pm\frac{1}{2}\left(\sigma^{(3)} - \sigma^{(1)}\right) & \text{when} \quad \widehat{n}_1 = \pm\frac{1}{\sqrt{2}},\ \widehat{n}_2 = 0,\ \widehat{n}_3 = \pm\frac{1}{\sqrt{2}}, \\ \pm\frac{1}{2}\left(\sigma^{(1)} - \sigma^{(2)}\right) & \text{when} \quad \widehat{n}_1 = \pm\frac{1}{\sqrt{2}},\ \widehat{n}_2 = \pm\frac{1}{\sqrt{2}},\ \widehat{n}_3 = 0. \end{cases} \tag{4.147}$$

Because of our ordering of the principal stresses, it is clear that the maximum shear stress is subsequently given by

$$\tau_{\max} = \frac{1}{2}\left|\sigma^{(3)} - \sigma^{(1)}\right|. \tag{4.148}$$

It may be shown that for distinctive principal stresses, the above are the only two possible maximal solutions.

4.8.2 Mean stress and deviatoric stress tensor

It is often convenient to define the *spherical (or mean) stress* as

$$\sigma^{(0)} = \frac{1}{3}\left(\sigma^{(1)} + \sigma^{(2)} + \sigma^{(3)}\right) = \frac{1}{3}\sigma_{kk} = \frac{1}{3}\sigma_{(1)}. \tag{4.149}$$

Subsequently, using the relation (3.103), the *deviatoric stress* is defined as

$$\sigma'_{ij} = \sigma_{ij} - \frac{1}{3}\sigma_{(1)}\delta_{ij}. \tag{4.150}$$

Now, recalling (3.107)–(3.109), we see that the invariants of the deviatoric stress tensor are related to those of the stress tensor by

$$\sigma'_{(1)} = 0, \tag{4.151}$$

$$\sigma'_{(2)} = -\frac{1}{3}\sigma^2_{(1)} + \sigma_{(2)}, \tag{4.152}$$

$$\sigma'_{(3)} = \frac{2}{27}\sigma_{(1)} - \frac{1}{3}\sigma_{(1)}\sigma_{(2)} + \sigma_{(3)}. \tag{4.153}$$

4.8.3 Lamé's stress ellipsoid

Let the principal axes of the stress tensor be chosen as the coordinate axes ($\widehat{n}_1, \widehat{n}_2$, $\widehat{n}_3$) in the directions $(\mathbf{i}_1, \mathbf{i}_2, \mathbf{i}_3)$, respectively. Then the components of stress vector $\mathbf{t}(\widehat{\mathbf{n}})$ are given by

$$t_1(\widehat{\mathbf{n}}) = \sigma^{(1)}\widehat{n}_1, \qquad t_2(\widehat{\mathbf{n}}) = \sigma^{(2)}\widehat{n}_2, \qquad t_3(\widehat{\mathbf{n}}) = \sigma^{(3)}\widehat{n}_3. \tag{4.154}$$

Now solving the above equations for $\widehat{n}_k$ and substituting them into the unit vector normalization condition

$$\widehat{n}_1^2 + \widehat{n}_2^2 + \widehat{n}_3^2 = 1, \tag{4.155}$$

we obtain the following equation:

$$\left(\frac{t_1(\widehat{\mathbf{n}})}{\sigma^{(1)}}\right)^2 + \left(\frac{t_2(\widehat{\mathbf{n}})}{\sigma^{(2)}}\right)^2 + \left(\frac{t_3(\widehat{\mathbf{n}})}{\sigma^{(3)}}\right)^2 = 1. \tag{4.156}$$

This is just the equation for an ellipsoid, called the *Lamé's stress ellipsoid*, with reference to a system of rectangular coordinates with axes $(t_1(\widehat{\mathbf{n}}), t_2(\widehat{\mathbf{n}}), t_3(\widehat{\mathbf{n}}))$ having semi-axes $(\sigma^{(1)}, \sigma^{(2)}, \sigma^{(3)})$ with the stress vector $\mathbf{t}(\widehat{\mathbf{n}})$ issuing from the origin.

4.8.4 Mohr's circles

We note that the equations for given normal and shear components of the stress vector along with the normalization condition referred to the principal axes make up a system of equations for the direction cosines $\widehat{n}_k$:

$$\sigma_n = \sigma^{(1)}\widehat{n}_1^2 + \sigma^{(2)}\widehat{n}_2^2 + \sigma^{(3)}\widehat{n}_3^2, \tag{4.157}$$

$$\sigma_n^2 + \tau^2 = \sigma^{(1)^2}\widehat{n}_1^2 + \sigma^{(2)^2}\widehat{n}_2^2 + \sigma^{(3)^2}\widehat{n}_3^2, \tag{4.158}$$

$$\widehat{n}_1^2 + \widehat{n}_2^2 + \widehat{n}_3^2 = 1. \tag{4.159}$$

The solution of this system, assuming that the principal stresses are distinct, is given by

$$\widehat{n}_1^2 = \frac{\left(\sigma_n - \sigma^{(2)}\right)\left(\sigma_n - \sigma^{(3)}\right) + \tau^2}{\left(\sigma^{(1)} - \sigma^{(2)}\right)\left(\sigma^{(1)} - \sigma^{(3)}\right)}, \tag{4.160}$$

$$\widehat{n}_2^2 = \frac{\left(\sigma_n - \sigma^{(3)}\right)\left(\sigma_n - \sigma^{(1)}\right) + \tau^2}{\left(\sigma^{(2)} - \sigma^{(3)}\right)\left(\sigma^{(2)} - \sigma^{(1)}\right)}, \tag{4.161}$$

$$\widehat{n}_3^2 = \frac{\left(\sigma_n - \sigma^{(1)}\right)\left(\sigma_n - \sigma^{(2)}\right) + \tau^2}{\left(\sigma^{(3)} - \sigma^{(1)}\right)\left(\sigma^{(3)} - \sigma^{(2)}\right)}. \tag{4.162}$$

These solutions can be interpreted graphically by using σ_n as the abscissa and τ as the ordinate as illustrated in Fig. 4.4. First note that, due to the ordering of the principal stresses, the denominators in the first and third solutions are positive, while the denominator in the second solution is negative. Subsequently, the numerators must be such that

$$\left(\sigma_n - \sigma^{(2)}\right)\left(\sigma_n - \sigma^{(3)}\right) + \tau^2 \geq 0, \tag{4.163}$$

$$\left(\sigma_n - \sigma^{(3)}\right)\left(\sigma_n - \sigma^{(1)}\right) + \tau^2 \leq 0, \tag{4.164}$$

$$\left(\sigma_n - \sigma^{(1)}\right)\left(\sigma_n - \sigma^{(2)}\right) + \tau^2 \geq 0, \tag{4.165}$$

which can be rearranged in the following geometric quadratic forms:

$$\left[\sigma_n - \frac{1}{2}\left(\sigma^{(2)} + \sigma^{(3)}\right)\right]^2 + \tau^2 \geq \left[\frac{1}{2}\left(\sigma^{(2)} - \sigma^{(3)}\right)\right]^2, \tag{4.166}$$

$$\left[\sigma_n - \frac{1}{2}\left(\sigma^{(1)} + \sigma^{(3)}\right)\right]^2 + \tau^2 \leq \left[\frac{1}{2}\left(\sigma^{(1)} - \sigma^{(3)}\right)\right]^2, \tag{4.167}$$

$$\left[\sigma_n - \frac{1}{2}\left(\sigma^{(1)} + \sigma^{(2)}\right)\right]^2 + \tau^2 \geq \left[\frac{1}{2}\left(\sigma^{(1)} - \sigma^{(2)}\right)\right]^2. \tag{4.168}$$

Clearly, admissible stress vectors in the (σ_n, τ) plane lie in the region exterior to the circles defined by the first and third equations and interior to the circle defined by the second equation. Furthermore, since the radius of the circle defined by the second equation is $\tau_{\max}$, it must be such that this circle encloses the other two circles. These circles are known as *Mohr's circles.*

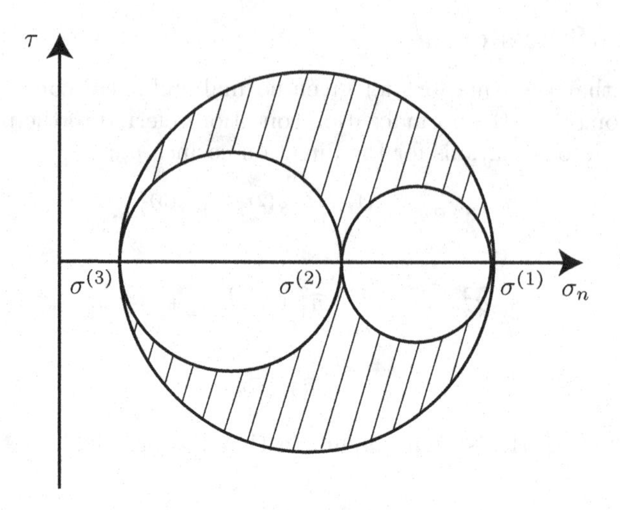

Figure 4.4: Mohr's circles.

4.9 Work and heat

For the sake of simplicity, we consider thermomechanical systems in which only the mechanical work $\mathcal{W}$ is present, i.e., $\sum_\alpha \dot{\mathcal{W}}_\alpha \to \dot{\mathcal{W}}$. Thus, the conservation of energy statement (4.27) takes the form

$$\dot{\mathcal{E}} + \dot{\mathcal{K}} = \dot{\mathcal{Q}} + \dot{\mathcal{W}}. \tag{4.169}$$

In *thermostatics*, we have $\dot{\mathcal{K}} = 0$, and if we write $d\mathcal{E} \equiv \dot{\mathcal{E}}dt$, $đ\mathcal{Q} \equiv \dot{\mathcal{Q}}dt$, and $đ\mathcal{W} \equiv \dot{\mathcal{W}}dt$, we arrive at the *first law of thermodynamics*:

$$d\mathcal{E} = đ\mathcal{Q} + đ\mathcal{W}, \tag{4.170}$$

where we make a distinction between exact differentials between states that are path independent such as $d\mathcal{E}$, and inexact differentials that are path dependent such as $đ\mathcal{Q}$ and $đ\mathcal{W}$. Quantities that are path independent denote properties of the material body. Here we recognize that (4.170) should more accurately be called the first law of thermostatics.

The mechanical energy consists of the work done by the surface and body forces per unit time (see (4.125)):

$$\dot{\mathcal{W}} = \int_{\mathcal{S}} [\boldsymbol{\nu} \cdot \mathbf{m}(\mathbf{n}) + \mathbf{v} \cdot \mathbf{t}(\mathbf{n})]\, ds + \int_{\mathcal{V}} \rho(\boldsymbol{\nu} \cdot \mathbf{l} + \mathbf{v} \cdot \mathbf{f})\, dv, \tag{4.171}$$

where

$$\boldsymbol{\nu} \cdot \mathbf{m}(\mathbf{n}) = (\boldsymbol{\nu} \cdot \boldsymbol{\Sigma}) \cdot \mathbf{n} \qquad \text{and} \qquad \mathbf{v} \cdot \mathbf{t}(\mathbf{n}) = (\mathbf{v} \cdot \boldsymbol{\sigma}) \cdot \mathbf{n}. \tag{4.172}$$

In a continuous medium, heat may enter an arbitrary region through contact heat supply per unit area $H(\mathbf{n})$ through the surface with exterior normal $\mathbf{n}$, or it may be supplied volumetrically from external sources (e.g., thermal radiation) per unit mass of the body so that r denotes the energy supply density. In general, r

varies from point to point in the body at any given time and may also vary with time at any given point, so that $r(\mathbf{x}, t)$, and in general, H, in addition to being dependent on the surface normal $\mathbf{n}$, also varies from point to point on the body surface at any given time and may also vary with time at any given point, so that $H(\mathbf{n}, \mathbf{x}, t)$; however, we will suppress the other dependences for the moment. Then the total heat input per unit time is given by

$$\dot{\mathcal{Q}} = \int_{\mathscr{S}} H(\mathbf{n})\, ds + \int_{\mathscr{V}} \rho r\, dv. \tag{4.173}$$

Subsequently, the balance of energy statement (4.169) becomes

$$\frac{d}{dt}\int_{\mathscr{V}} \rho\left(e + \frac{1}{2}\boldsymbol{\nu}\cdot\boldsymbol{\varphi} + \frac{1}{2}\mathbf{v}\cdot\mathbf{v}\right) dv = \int_{\mathscr{S}} [\boldsymbol{\nu}\cdot\mathbf{m}(\mathbf{n}) + \mathbf{v}\cdot\mathbf{t}(\mathbf{n}) + H(\mathbf{n})]\, ds + \int_{\mathscr{V}} \rho\,(\boldsymbol{\nu}\cdot\mathbf{l} + \mathbf{v}\cdot\mathbf{f} + r)\, dv. \tag{4.174}$$

Note that by subtracting (4.125), we can also write the integral balance of internal energy:

$$\frac{d}{dt}\int_{\mathscr{V}} \rho e\, dv = \int_{\mathscr{S}} H(\mathbf{n}) ds + \int_{\mathscr{V}} (\Phi + \rho r)\, dv. \tag{4.175}$$

4.10 Heat flux hypothesis

To determine the dependence of the contact heat supply on $\mathbf{n}$, we apply the internal energy balance (4.175) to the same infinitesimal tetrahedral volume shown in Fig. 4.2. From the mean value theorem, we define the mean internal energy

$$\rho^\star e^\star \equiv \frac{1}{\Delta v}\int_{\Delta v} \rho e\, dv, \tag{4.176}$$

the mean external body energy source

$$\rho^\star r^\star \equiv \frac{1}{\Delta v}\int_{\Delta v} \rho r\, dv, \tag{4.177}$$

the mean contact heat supply

$$H^\star(\mathbf{n}) \equiv \frac{1}{\Delta s}\int_{\Delta s} H(\mathbf{n})\, ds, \tag{4.178}$$

and the mean mechanical energy

$$\Phi^\star \equiv \frac{1}{\Delta v}\int_{\Delta v} \Phi\, dv. \tag{4.179}$$

Subsequently, the internal energy balance (4.175) applied to the tetrahedron, by making use of mean quantities and the mass balance (4.72), becomes

$$\rho^\star\,(\dot{e}^\star - r^\star)\,\Delta v = \Phi^\star \Delta v + H^\star(\mathbf{n})\Delta s + H^\star(-\mathbf{i}_k)\Delta s_k. \tag{4.180}$$

Now dividing both sides of the equation by Δs, letting $\Delta v \to 0$ and $\Delta s \to 0$ so that a point on the curved surface approaches $\mathbf{r}$, noting that $\Delta v/\Delta s \to 0$, and assuming that the quantities $\rho^\star$, $\dot{e}^\star$, $r^\star$, $\Phi^\star$, and $\Delta s_k/\Delta s$ remain bounded, we obtain

$$H(\mathbf{n})\, ds = -H(-\mathbf{i}_k)\, ds_k. \tag{4.181}$$

If we denote by Δl_k a typical dimension of our tetrahedron in the x_k direction, and if in the above limiting process we take $\Delta v \to 0$ and $\Delta l_k \to 0$, we obtain

$$H(\mathbf{i}_k) = -H(-\mathbf{i}_k), \tag{4.182}$$

which demonstrates that the contact heat supply acting on opposite sides of the same surfaces at a given point are equal in magnitude and opposite in sign. The four faces of the tetrahedron form a closed surface. As before, in the limit, the vector sum of the coordinate surfaces must add up to the area vector $d\mathbf{s}$, i.e.,

$$d\mathbf{s} = \mathbf{n}\, ds \qquad \text{or} \qquad ds_k = n_k\, ds. \tag{4.183}$$

Substituting the last two expressions in our limiting result, we obtain

$$H(\mathbf{n}) = H(\mathbf{i}_k) n_k. \tag{4.184}$$

This proves that the contact heat supply at a point on a surface with an exterior normal $\mathbf{n}$ is a linear function of the heat supply acting on the coordinate surface through the same point, the coefficient being the direction cosine of $\mathbf{n}$.

We now define the component $-q_k$ as the component of the contact heat supply $H(\mathbf{i}_k)$ acting on the positive side of the kth coordinate surface (the negative sign is used by convention):

$$H(\mathbf{i}_k) \equiv -q_k. \tag{4.185}$$

As we see, now writing the complete explicit dependence, the heat supply $H(\mathbf{n}, \mathbf{x}, t)$ is related to the *heat flux* vector $\mathbf{q}(\mathbf{x}, t)$ by the linear transformation

$$H(\mathbf{n}, \mathbf{x}, t) = -\mathbf{q}(\mathbf{x}, t) \cdot \mathbf{n} \qquad \text{or} \qquad H(\mathbf{i}_k, \mathbf{x}, t) = -q_k(\mathbf{x}, t). \tag{4.186}$$

This result is the counterpart of Cauchy's postulate and Cauchy's theorem for contact heat supply and is called the *Fourier–Stokes heat flux theorem.* It asserts that $H(\mathbf{n}, \mathbf{x}, t)$ depends upon the surface orientation only in a *linear* fashion.

Using the above result with (4.172), the global balance of energy (4.174) becomes

$$\frac{d}{dt}\int_{\mathscr{V}} \rho\left(e + \frac{1}{2}\mathbf{v}\cdot\mathbf{v} + \frac{1}{2}\boldsymbol{\nu}\cdot\boldsymbol{\varphi}\right) dv = \int_{\mathscr{S}} (\mathbf{v}\cdot\boldsymbol{\sigma} + \boldsymbol{\nu}\cdot\boldsymbol{\Sigma} - \mathbf{q})\cdot d\mathbf{s} + \int_{\mathscr{V}} \rho(\mathbf{v}\cdot\mathbf{f} + \boldsymbol{\nu}\cdot\mathbf{l} + r)\, dv. \tag{4.187}$$

4.11 Entropy flux hypothesis

The entropy inequality, which is reflected in Axiom 5, applied to an arbitrary body is given by

$$\int_{\mathscr{V}} \gamma\, dv \equiv \frac{d}{dt}\int_{\mathscr{V}} \rho\eta\, dv - \int_{\mathscr{V}} \rho b\, dv - \int_{\mathscr{S}} \pi(\mathbf{n})\, ds \geq 0, \tag{4.188}$$

where, for a continuous material mass, local entropy supply per unit mass, contact entropy supply per unit area, and entropy production per unit volume have been assumed to exist and given by

$$\mathcal{B} \equiv \int_{\mathscr{V}} \rho b \, dv, \tag{4.189}$$

$$\Pi \equiv \int_{\mathscr{S}} \pi(\mathbf{n}) \, ds, \tag{4.190}$$

$$\Gamma \equiv \int_{\mathscr{V}} \gamma \, dv. \tag{4.191}$$

In addition, we have assumed that the contact entropy supply is a function of the surface exterior normal direction.

As before, from the mean value theorem, we can define the mean entropy density

$$\rho^\star \eta^\star = \frac{1}{\Delta v} \int_{\Delta v} \rho \eta \, dv, \tag{4.192}$$

the mean external entropy supply

$$\rho^\star b^\star = \frac{1}{\Delta v} \int_{\Delta v} \rho b \, dv, \tag{4.193}$$

and the mean contact entropy supply

$$\pi^\star(\mathbf{n}) = \frac{1}{\Delta s} \int_{\Delta s} \pi(\mathbf{n}) \, ds. \tag{4.194}$$

Subsequently, using the mass balance (4.72) applied to the infinitesimal tetrahedron in the positive quadrant shown in Fig. 4.5, we obtain

$$\rho^\star \left(\dot{\eta}^\star - b^\star \right) \Delta v - \pi^\star(\mathbf{n}) \Delta s - \pi^\star(-\mathbf{i}_k) \Delta s_k \geq 0. \tag{4.195}$$

Now proceeding as before by taking $\Delta v \to 0$ and $\Delta s \to 0$ with $\Delta v / \Delta s \to 0$, we obtain

$$\pi(\mathbf{n}) ds + \pi(-\mathbf{i}_k) ds_k \leq 0, \tag{4.196}$$

and upon taking $\Delta v \to 0$ and $\Delta l_k \to 0$, we obtain that

$$\pi(\mathbf{i}_k) = -\pi(-\mathbf{i}_k). \tag{4.197}$$

Using the fact that $ds_k = n_k ds$ leads to the result that

$$\pi(\mathbf{n}) - \pi(\mathbf{i}_k) n_k \leq 0. \tag{4.198}$$

Repeating the above procedure over the tetrahedron in the negative quadrant shown in Fig. 4.5, we also obtain

$$\pi(-\mathbf{n}) ds + \pi(\mathbf{i}_k) ds_k \leq 0, \tag{4.199}$$

or

$$-\pi(\mathbf{n}) + \pi(\mathbf{i}_k) n_k \leq 0. \tag{4.200}$$

Now the only way that the quantity on the left-hand side of (4.198) and its negative on the left-hand side of (4.200) be both less than or equal to zero is only if the quantity is equal to zero, i.e.,

$$\pi(\mathbf{n}) = \pi(\mathbf{i}_k) n_k. \tag{4.201}$$

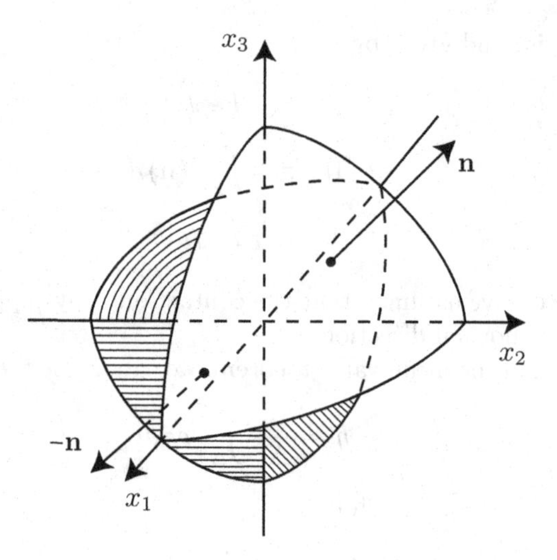

Figure 4.5: Tetrahedral infinitesimal volumes.

This proves that the contact entropy supply at a point on a surface with an exterior normal $\mathbf{n}$ is a linear function of the entropy supply acting on the coordinate surface through the same point, the coefficient being the direction cosine of $\mathbf{n}$.

We now define the component $-h_k$ as the component of the contact entropy supply $P(\mathbf{i}_k)$ acting on the positive side of the kth coordinate surface (the negative sign is used by convention):

$$\pi(\mathbf{i}_k) \equiv -h_k. \tag{4.202}$$

As we see, now writing the complete explicit dependence, the entropy supply $P(\mathbf{n}, \mathbf{x}, t)$ is related to the *entropy flux* vector $\mathbf{h}(\mathbf{x}, t)$ by the linear transformation

$$\pi(\mathbf{n}, \mathbf{x}, t) = -\mathbf{h}(\mathbf{x}, t) \cdot \mathbf{n} \qquad \text{or} \qquad \pi(\mathbf{i}_k, \mathbf{x}, t) = -h_k(\mathbf{x}, t). \tag{4.203}$$

This result is the counterpart of Cauchy's theorem, but for contact entropy supply. It asserts that $\pi(\mathbf{n}, \mathbf{x}, t)$ depends upon the surface orientation only in a *linear* fashion.

Using the above result in (4.188), we obtain the *entropy inequality*

$$\int_{\mathscr{V}} \gamma \, dv \equiv \frac{d}{dt} \int_{\mathscr{V}} \rho\eta \, dv - \int_{\mathscr{V}} \rho b \, dv + \int_{\mathscr{S}} \mathbf{h} \cdot d\mathbf{s} \geq 0. \tag{4.204}$$

4.12 Local forms of axioms of thermodynamics

The local form of conservation of energy is obtained by comparing (4.187) with the reformulated general balance laws (4.106). In this comparison, we note that $\psi \to (e + \frac{1}{2}\mathbf{v}\cdot\mathbf{v} + \frac{1}{2}\boldsymbol{\nu}\cdot\boldsymbol{\varphi})$, $\mathbf{t} \to (\mathbf{v}\cdot\boldsymbol{\sigma} + \boldsymbol{\nu}\cdot\boldsymbol{\Sigma} - \mathbf{q})$, and $g \to (\mathbf{v}\cdot\mathbf{f} + \boldsymbol{\nu}\cdot\mathbf{l} + r)$. Subsequently, from (4.107), we have

$$\int_{\mathscr{V}(t)-\varsigma(t)} \left\{\rho\left[\left(\dot{e}+\frac{1}{2}\overline{\mathbf{v}\cdot\mathbf{v}}^{\cdot}+\frac{1}{2}\overline{\boldsymbol{\nu}\cdot\boldsymbol{\varphi}}^{\cdot}\right)-(\mathbf{v}\cdot\mathbf{f}+\boldsymbol{\nu}\cdot\mathbf{l}+r)\right]- \operatorname{div}(\mathbf{v}\cdot\boldsymbol{\sigma}+\boldsymbol{\nu}\cdot\boldsymbol{\Sigma}-\mathbf{q})\right\}dv+$$
$$\int_{\varsigma(t)}\left(m\left[\!\!\left[e+\frac{1}{2}\mathbf{v}\cdot\mathbf{v}+\frac{1}{2}\boldsymbol{\nu}\cdot\boldsymbol{\varphi}\right]\!\!\right]-[\![\mathbf{v}\cdot\boldsymbol{\sigma}+\boldsymbol{\nu}\cdot\boldsymbol{\Sigma}-\mathbf{q}]\!]\cdot\mathbf{n}\right)d\varsigma=0. \tag{4.205}$$

The above equation, upon subtracting the local equation of kinetic energy (4.121), can be rewritten in the form

$$\int_{\mathscr{V}(t)-\varsigma(t)}\{\rho(\dot{e}-r)-\Phi+\operatorname{div}\mathbf{q}\}\,dv+$$
$$\int_{\varsigma(t)}\left(m\left[\!\!\left[e+\frac{1}{2}\mathbf{v}\cdot\mathbf{v}+\frac{1}{2}\boldsymbol{\nu}\cdot\boldsymbol{\varphi}\right]\!\!\right]-[\![\mathbf{v}\cdot\boldsymbol{\sigma}+\boldsymbol{\nu}\cdot\boldsymbol{\Sigma}-\mathbf{q}]\!]\cdot\mathbf{n}\right)d\varsigma=0. \tag{4.206}$$

Now setting the integrand of the first integral to zero, we obtain the local internal energy balance, which applies at a regular point in a continuous region

$$\rho(\dot{e}-r)=\Phi-\operatorname{div}\mathbf{q} \qquad \text{or} \qquad \rho(\dot{e}-r)=\Phi-q_{k,k}, \tag{4.207}$$

and setting the integrand of the second integral to zero, we obtain the energy jump condition across the singular moving surface $\boldsymbol{\varsigma}$

$$m\left[\!\!\left[e+\frac{1}{2}\mathbf{v}\cdot\mathbf{v}+\frac{1}{2}\boldsymbol{\nu}\cdot\boldsymbol{\varphi}\right]\!\!\right]-[\![\mathbf{v}\cdot\boldsymbol{\sigma}+\boldsymbol{\nu}\cdot\boldsymbol{\Sigma}-\mathbf{q}]\!]\cdot\mathbf{n}=0 \tag{4.208}$$

or

$$m\left[\!\!\left[e+\frac{1}{2}v_l v_l+\frac{1}{2}\nu_l\varphi_l\right]\!\!\right]-[\![v_l\sigma_{lk}+\nu_l\Sigma_{lk}-q_k]\!]\,n_k=0. \tag{4.209}$$

In arriving at (4.207) from (4.206), we have assumed that the external heat supply r is continuous. Equation (4.207), which applies at a regular point in a continuous medium, corresponds to the local first law of thermodynamics stating that the internal energy change per unit time is due to the stress power, the heat flux, and the external heat supply. By decomposing the tensors appearing in the stress power term (given in (4.124)) into spherical, deviatoric (indicated by primes), and skew-symmetric parts (see (2.110)), the stress power can be rewritten in the following illuminating form:

$$\begin{aligned}\Phi = {} & \underbrace{\left[\frac{1}{2}\rho\,\dot{i}^{(0)}\,\nu_k\nu_k+m^{(0)}\nu_{k,k}+\sigma^{(0)}v_{k,k}\right]}_{(1)}+\\ & \underbrace{\left[\frac{1}{2}\rho\,\dot{i}'_{(kl)}\nu_k\nu_l+\Xi'_{(kl)}\Sigma'_{(kl)}+D'_{(kl)}\sigma'_{(kl)}\right]}_{(2)}+\\ & \underbrace{\left[\Xi_{[kl]}\Sigma_{[kl]}+\left(\Upsilon_{[kl]}+W_{[kl]}\right)\sigma_{[kl]}\right]}_{(3)},\end{aligned} \tag{4.210}$$

where $\sigma^{(0)}$ is defined in (4.149), and we have defined the spherical components of the internal inertia and couple stress tensors by

$$i^{(0)} = \frac{1}{3}\operatorname{tr}\boldsymbol{i} \qquad \text{and} \qquad m^{(0)} = \frac{1}{3}\operatorname{tr}\boldsymbol{\Sigma}. \tag{4.211}$$

We see that the stress power is given by three terms which in order represent (1) the local work of changing the volume (dilatation), (2) changing the shape (distortion), and (3) rotating the material element. Furthermore, for $\sigma^{(0)} = -\bar{p}$, with $\bar{p}$ a *mechanical pressure*, the third term in (1) represents the $-\bar{p}\,dV$ work term, and the third term in (2) is just the product of the rate of distortion in shape and the shearing stresses. These are the only terms that apply for nonpolar materials since $\sigma_{[kl]} = 0$ in this case.

The above jump condition can be rewritten in a more convenient and simpler form if we first note that, for any tensors A and B, we have that

$$[\![AB]\!] = [\![A]\!]\langle\!\langle B\rangle\!\rangle + \langle\!\langle A\rangle\!\rangle[\![B]\!], \tag{4.212}$$

where we have defined the average operator

$$\langle\!\langle A\rangle\!\rangle \equiv \frac{1}{2}\left(A^{+} + A^{-}\right). \tag{4.213}$$

Subsequently, we first note that, using (4.212) and (4.110), we have

$$\frac{1}{2} m\,[\![\mathbf{v}\cdot\mathbf{v}]\!] - [\![\mathbf{v}\cdot\boldsymbol{\sigma}]\!]\cdot\mathbf{n} = -[\![\mathbf{v}]\!]\cdot\langle\!\langle\boldsymbol{\sigma}\rangle\!\rangle\cdot\mathbf{n} \tag{4.214}$$

and, using (4.212) and (4.117), we have

$$\frac{1}{2} m\,[\![\boldsymbol{\nu}\cdot\boldsymbol{\varphi}]\!] - [\![\boldsymbol{\nu}\cdot\boldsymbol{\Sigma}]\!]\cdot\mathbf{n} = -[\![\boldsymbol{\nu}]\!]\cdot\langle\!\langle\boldsymbol{\Sigma}\rangle\!\rangle\cdot\mathbf{n}. \tag{4.215}$$

In obtaining this last equality, we have used the fact that $\boldsymbol{\varphi} = \boldsymbol{i}\cdot\boldsymbol{\nu}$ (see (4.14)), and noted that the internal inertia tensor $\boldsymbol{i}$ is symmetric and continuous across a jump discontinuity since it is only a function of the local coordinates. Subsequently, we can rewrite (4.208) as

$$m\,[\![e]\!] - \left([\![\boldsymbol{\nu}]\!]\cdot\langle\!\langle\boldsymbol{\Sigma}\rangle\!\rangle + [\![\mathbf{v}]\!]\cdot\langle\!\langle\boldsymbol{\sigma}\rangle\!\rangle - [\![\mathbf{q}]\!]\right)\cdot\mathbf{n} = 0. \tag{4.216}$$

Using (4.104) and (4.129), we can also rewrite the above equation as

$$m\left([\![e]\!] - [\![v]\!]\,\langle\!\langle\sigma_n\rangle\!\rangle\right) - \left([\![\boldsymbol{\nu}]\!]\cdot\langle\!\langle\boldsymbol{\Sigma}\rangle\!\rangle\cdot\mathbf{n} + [\![v_{(s)}]\!]\,\langle\!\langle\tau\rangle\!\rangle - [\![\mathbf{q}]\!]\cdot\mathbf{n}\right) = 0. \tag{4.217}$$

The local form of the entropy inequality is obtained by comparing (4.204) with the reformulated balance law (4.106). In this comparison we note that $\psi \to \eta$, $\mathbf{t} \to -\mathbf{h}$, and $g \to b$. Subsequently, from (4.107), we have

$$\int_{\mathscr{V}(t)} \gamma\,dv \equiv \int_{\mathscr{V}(t)-\varsigma(t)} \left[\rho(\dot{\eta} - b) + \operatorname{div}\mathbf{h}\right] dv + \int_{\varsigma(t)} \left(m[\![\eta]\!] + [\![\mathbf{h}]\!]\cdot\mathbf{n}\right) d\varsigma \geq 0. \tag{4.218}$$

Since the volume is arbitrary, choosing a volume in the regular region leads to the local entropy inequality, which applies at a regular point in the continuous region

$$\gamma_v \equiv \rho(\dot{\eta} - b) + \operatorname{div}\mathbf{h} \geq 0 \qquad \text{or} \qquad \gamma_v \equiv \rho(\dot{\eta} - b) + h_{k,k} \geq 0, \tag{4.219}$$

and setting the integrand of the last integral to zero, we obtain the jump condition across the singular moving surface $\boldsymbol{\zeta}$:

$$\gamma_s \equiv m[\![\eta]\!] + [\![\mathbf{h}]\!] \cdot \mathbf{n} \geq 0 \qquad \text{or} \qquad \gamma_s \equiv m[\![\eta]\!] + [\![h_k]\!] n_k \geq 0. \tag{4.220}$$

In arriving at (4.219) from (4.218), we have assumed that the external entropy supply b is continuous. Note that γ_v and γ_s represent the entropy productions per unit volume in the continuous region and per unit area on the singular surface, respectively. Equation (4.219) corresponds to the local form of the second law of thermodynamics.

4.13 Field equations in Euclidean frames

In the previous sections we have obtained the local field equations in an inertial frame. Nevertheless, we require that the density ρ, the forces $\mathbf{f}$ and $\mathbf{t}$ (and therefore the stress tensor $\boldsymbol{\sigma}$), the internal inertia tensor $\boldsymbol{i}$, the couples $\mathbf{l}$ and $\mathbf{m}$ (and therefore the couple stress tensor $\boldsymbol{\Sigma}$), the internal energy e, the energy supplies r and h (and therefore the heat flux $\mathbf{q}$), the entropy η, and the entropy productions $\gamma_{v,s}$ and supplies b and P (and therefore the entropy flux $\mathbf{h}$) be all frame indifferent under a Euclidean transformation, i.e.,

$$\begin{gathered} \rho' = \rho, \quad e' = e, \quad \eta' = \eta, \quad r' = r, \quad b' = b, \quad \gamma'_{v,s} = \gamma_{v,s}, \\ \mathbf{f}' = \mathbf{Q}\cdot\mathbf{f}, \quad \mathbf{l}' = (\det\mathbf{Q})\,\mathbf{Q}\cdot\mathbf{l}, \quad \mathbf{q}' = \mathbf{Q}\cdot\mathbf{q}, \quad \mathbf{h}' = \mathbf{Q}\cdot\mathbf{h}, \\ \boldsymbol{\sigma}' = \mathbf{Q}\cdot\boldsymbol{\sigma}\cdot\mathbf{Q}^T, \quad \boldsymbol{i}' = \mathbf{Q}\cdot\boldsymbol{i}\cdot\mathbf{Q}^T, \quad \boldsymbol{\Sigma}' = (\det\mathbf{Q})\,\mathbf{Q}\cdot\boldsymbol{\Sigma}\cdot\mathbf{Q}^T. \end{gathered} \tag{4.221}$$

We note that the frame indifference requirement of the internal spin, $\boldsymbol{\varphi}'$, is connected (see (4.14)) with the frame-indifference requirements of the internal inertia tensor, $\boldsymbol{i}'$, and the internal angular velocity, $\boldsymbol{\nu}'$. Furthermore, as with the internal angular velocity, the internal spin rate $\dot{\boldsymbol{\varphi}}$ is not frame invariant since

$$\dot{\boldsymbol{\varphi}}' = \boldsymbol{i}'\cdot\dot{\boldsymbol{\nu}}' + \dot{\boldsymbol{i}}'\cdot\boldsymbol{\nu}' = (\det\mathbf{Q})\,\mathbf{Q}\cdot\dot{\boldsymbol{\varphi}} + \mathbf{j}', \tag{4.222}$$

or

$$(\dot{\boldsymbol{\varphi}}' - \mathbf{j}') \equiv \bar{\dot{\boldsymbol{\varphi}}} = (\det\mathbf{Q})\,\mathbf{Q}\cdot\dot{\boldsymbol{\varphi}}, \tag{4.223}$$

where we have defined the *inertial internal spin* (see (2.147))

$$\mathbf{j}' \equiv \dot{\boldsymbol{\phi}}' + \boldsymbol{\Omega}\cdot(\boldsymbol{\varphi}' - \boldsymbol{\phi}') = \dot{\boldsymbol{\phi}}' - \mathbf{w}\times(\boldsymbol{\varphi}' - \boldsymbol{\phi}'), \tag{4.224}$$

where

$$\boldsymbol{\phi}' = \boldsymbol{i}'\cdot\mathbf{w} \tag{4.225}$$

is the spin rate of the second frame relative to the first. We note that $\dot{\boldsymbol{\varphi}}'$ is the *true internal spin rate* while $\bar{\dot{\boldsymbol{\varphi}}}$ is called the *apparent internal spin rate*. The difference between these two internal spin rates is given by the inertial angular spin rate $\mathbf{j}'$.

We recall that the material derivative of an objective scalar field is objective (not true for tensor fields in general); therefore,

$$(\dot{\rho})' = \dot{\rho}, \qquad (\dot{e})' = \dot{e}, \qquad (\dot{\eta})' = \dot{\eta}. \tag{4.226}$$

In addition, it is easy to show that the spatial divergence of velocity, heat flux, and entropy flux are frame invariant, i.e.,

$$(\operatorname{div}\mathbf{v})' = \operatorname{div}\mathbf{v}, \qquad (\operatorname{div}\mathbf{q})' = \operatorname{div}\mathbf{q}, \qquad (\operatorname{div}\mathbf{h})' = \operatorname{div}\mathbf{h}, \tag{4.227}$$

and so is the stress power (4.123), i.e.,

$$\Phi' = \Phi. \tag{4.228}$$

Moreover, we can show that the spatial divergence of an objective second rank tensor field is also objective, and in particular

$$(\operatorname{div}\boldsymbol{\sigma})' = \mathbf{Q}\cdot(\operatorname{div}\boldsymbol{\sigma}), \qquad (\operatorname{div}\boldsymbol{\Sigma})' = (\det\mathbf{Q})\,\mathbf{Q}\cdot(\operatorname{div}\boldsymbol{\Sigma}). \tag{4.229}$$

Subsequently, from the above transformation properties, it follows immediately that the balances of mass, energy, and entropy are objective scalar equations and are thus valid in arbitrary Euclidean frames. On the other hand, the balances of linear and angular momenta are only Galilean invariant since they contain the acceleration and the internal spin rate. In an arbitrary Euclidean frame, the linear and angular momenta take the forms

$$\rho'\left[(\mathbf{a}-\mathbf{i})' - \mathbf{f}'\right] = (\operatorname{div}\boldsymbol{\sigma})' \tag{4.230}$$

and

$$\rho'\left[(\dot{\boldsymbol{\varphi}}-\mathbf{j})' - \mathbf{l}'\right] = (\operatorname{div}\boldsymbol{\Sigma})' - \boldsymbol{\varepsilon}:\boldsymbol{\sigma}', \tag{4.231}$$

where $\mathbf{i}'$ is given by (3.400), $\mathbf{j}'$ is given by (4.224), and we have used the fact that $\boldsymbol{\varepsilon}$ is an isotropic axial tensor. However, we note that the *apparent acceleration* and *internal spin rate* are given by (see (3.398) and (4.223))

$$(\mathbf{a}-\mathbf{i})' = \mathbf{Q}\cdot\mathbf{a} \qquad \text{and} \qquad (\dot{\boldsymbol{\varphi}}-\mathbf{j})' = (\det\mathbf{Q})\,\mathbf{Q}\cdot\dot{\boldsymbol{\varphi}}. \tag{4.232}$$

Subsequently, it can be easily shown that the apparent acceleration and the apparent internal spin rate in two non-inertial frames are related as

$$(\mathbf{a}-\mathbf{i})' = \mathbf{Q}\cdot(\mathbf{a}-\mathbf{i}) \qquad \text{and} \qquad (\dot{\boldsymbol{\varphi}}-\mathbf{j})' = (\det\mathbf{Q})\,\mathbf{Q}\cdot(\dot{\boldsymbol{\varphi}}-\mathbf{j}), \tag{4.233}$$

where

$$\mathbf{i} \equiv \ddot{\mathbf{b}} - \dot{\mathbf{w}}\times(\mathbf{x}-\mathbf{b}) - 2\,\mathbf{w}\times(\dot{\mathbf{x}}-\dot{\mathbf{b}}) - \mathbf{w}\times\left[\mathbf{w}\times(\mathbf{x}-\mathbf{b})\right], \tag{4.234}$$

$$\mathbf{j} \equiv \dot{\boldsymbol{\phi}} - \mathbf{w}\times(\boldsymbol{\varphi}-\boldsymbol{\phi}), \tag{4.235}$$

and $\mathbf{w}$ and $\boldsymbol{\phi} = \boldsymbol{i}\cdot\mathbf{w}$ are the frame's angular velocity and spin rate relative to the second frame.

We can interpret the vector fields $\mathbf{i}$ and $\mathbf{j}$ as *inertial (apparent) body force* and *inertial (apparent) body couple.* Now the individual terms retain the same names, but are interpreted as forces and couples. If we combine the inertial force and couple with the real body forces $\mathbf{f}$ and couples $\mathbf{l}$, so that we now think of the body forces and couples as including these apparent forces and couples, then the balance of linear and angular momenta are also seen to be invariant under a Euclidean transformation. From now on, we shall assume this to be the case and interpret the body forces $\mathbf{f}$ to mean $(\mathbf{f}+\mathbf{i})$ and the body couples $\mathbf{l}$ to mean $(\mathbf{l}+\mathbf{j})$.

Subsequently, we take $\ddot{\mathbf{b}} = \mathbf{0}$ and $\mathbf{w} = \mathbf{0}$ in an inertial frame and $\ddot{\mathbf{b}}$ and $\mathbf{w}$ to be nonzero in a non-inertial frame so that $\mathbf{i}$ and $\mathbf{j}$ will be zero or nonzero correspondingly. We remark that, with this understanding, the corresponding integral balance laws are also invariant under a general Euclidean transformation.

4.14 Jump conditions in Euclidean frames

The jump conditions should be invariant with respect to Euclidean transformations. We shall examine whether their invariance places any restrictions on the motion in Cartesian frames.

We first recall that the velocity in a Euclidean frame is given by (3.382),

$$\mathbf{v}' = \dot{\mathbf{b}}(t) + \dot{\mathbf{Q}}(t) \cdot \mathbf{x} + \mathbf{Q}(t) \cdot \mathbf{v}, \tag{4.236}$$

where $\mathbf{b}$ is the position vector between the two Cartesian frames $\mathcal{F}$ and $\mathcal{F}'$, and $\mathbf{Q}$ is the orthogonal rotation tensor between the two frames. To examine the jump in velocity in a new frame, $[\![\mathbf{v}']\!] = (\mathbf{v}')^+ - (\mathbf{v}')^-$, we need to examine the velocity differences at the same location $\mathbf{x}$ on a discontinuous surface. Under such transformation, we have

$$(\mathbf{v}')^+ = \dot{\mathbf{b}}(t) + \dot{\mathbf{Q}}(t) \cdot \mathbf{x} + \mathbf{Q}(t) \cdot \mathbf{v}^+ \quad \text{and} \quad (\mathbf{v}')^- = \dot{\mathbf{b}}(t) + \dot{\mathbf{Q}}(t) \cdot \mathbf{x} + \mathbf{Q}(t) \cdot \mathbf{v}^-. \tag{4.237}$$

We clearly see that velocity differences, and subsequently jumps in velocities, are always objective:

$$[\![\mathbf{v}']\!] = (\mathbf{v}')^+ - (\mathbf{v}')^- = \mathbf{Q}(t) \cdot (\mathbf{v}^+ - \mathbf{v}^-) = \mathbf{Q}(t) \cdot [\![\mathbf{v}]\!]. \tag{4.238}$$

Now, since from (4.221) we have that $\rho' = \rho$, and since $\mathbf{n}' = \mathbf{Q} \cdot \mathbf{n}$, the transformation of the mass balance jump condition (4.102) across the singular moving surface $\boldsymbol{\zeta}$ is given by

$$[\![m']\!] = [\![\rho'(\mathbf{v}' - \mathbf{c}')]\!] \cdot \mathbf{n}' = [\![\rho \mathbf{Q} \cdot (\mathbf{v} - \mathbf{c})]\!] \cdot \mathbf{Q} \cdot \mathbf{n} = [\![\rho (\mathbf{v} - \mathbf{c})]\!] \cdot \mathbf{Q}^T \cdot \mathbf{Q} \cdot \mathbf{n} = [\![\rho (\mathbf{v} - \mathbf{c})]\!] \cdot \mathbf{n} = [\![m]\!] = 0. \tag{4.239}$$

Subsequently, the mass balance jump condition is frame indifferent.

Using the transformations (4.221), the linear momentum jump condition (4.110) takes the form:

$$m'[\![\mathbf{v}']\!] - [\![\boldsymbol{\sigma}']\!] \cdot \mathbf{n}' = \mathbf{Q} \cdot \left(m [\![\mathbf{v}]\!] - [\![\boldsymbol{\sigma}]\!] \cdot \mathbf{Q}^T \cdot \mathbf{Q} \cdot \mathbf{n} \right) = m [\![\mathbf{v}]\!] - [\![\boldsymbol{\sigma}]\!] \cdot \mathbf{n} = \mathbf{0}. \tag{4.240}$$

Thus, the linear momentum balance jump condition is frame indifferent.

Similarly, using the transformations (4.221), the angular momentum jump condition (4.117) takes the form:

$$m'[\![\boldsymbol{\varphi}']\!] - [\![\boldsymbol{\Sigma}']\!] \cdot \mathbf{n}' = (\det \mathbf{Q})\, \mathbf{Q} \cdot \left(m [\![\boldsymbol{\varphi}]\!] - [\![\boldsymbol{\Sigma}]\!] \cdot \mathbf{Q}^T \cdot \mathbf{Q} \cdot \mathbf{n} \right) = m [\![\boldsymbol{\varphi}]\!] - [\![\boldsymbol{\Sigma}]\!] \cdot \mathbf{n} = \mathbf{0}, \tag{4.241}$$

and thus, the angular momentum balance jump condition is frame indifferent as well.

Now, using the transformations (4.221), for the transformation of the energy jump condition, we use the form (4.216):

$$\begin{aligned} m' [\![e']\!] - \left([\![\mathbf{v}']\!] \cdot \langle\!\langle \boldsymbol{\sigma}' \rangle\!\rangle + [\![\boldsymbol{\nu}']\!] \cdot \langle\!\langle \boldsymbol{\Sigma}' \rangle\!\rangle - [\![\mathbf{q}']\!] \right) \cdot \mathbf{n}' = \\ m [\![e]\!] - \left([\![\mathbf{v}]\!] \cdot \mathbf{Q}^T \cdot \mathbf{Q} \cdot \langle\!\langle \boldsymbol{\sigma} \rangle\!\rangle + (\det \mathbf{Q})^2 [\![\boldsymbol{\nu}]\!] \cdot \mathbf{Q}^T \cdot \mathbf{Q} \cdot \langle\!\langle \boldsymbol{\Sigma} \rangle\!\rangle - [\![\mathbf{q}]\!] \right) \cdot \mathbf{Q}^T \cdot \mathbf{Q} \cdot \mathbf{n} = \\ m [\![e]\!] - ([\![\mathbf{v}]\!] \cdot \langle\!\langle \boldsymbol{\sigma} \rangle\!\rangle + [\![\boldsymbol{\nu}]\!] \cdot \langle\!\langle \boldsymbol{\Sigma} \rangle\!\rangle - [\![\mathbf{q}]\!]) \cdot \mathbf{n} = 0. \end{aligned} \tag{4.242}$$

Thus, the energy balance jump condition is frame indifferent.

Lastly, using the transformations (4.221), the transformation of the entropy jump condition (4.220) is given by

$$\gamma_s = m'[\![\eta']\!] + [\![\mathbf{h}']\!] \cdot \mathbf{n}' = m[\![\eta]\!] + [\![\mathbf{h}]\!] \cdot \mathbf{Q}^T \cdot \mathbf{Q} \cdot \mathbf{n} = m[\![\eta]\!] + [\![\mathbf{h}]\!] \cdot \mathbf{n} \geq 0. \quad (4.243)$$

Thus, the entropy inequality jump condition is also frame indifferent.

We close this chapter by noting that while we have arrived at the appropriate global and local balance laws and jump conditions for a polar material, from now on we shall focus all further discussions on nonpolar materials. Specifically, we take $\boldsymbol{\varphi} = \mathbf{0}$ (or $\boldsymbol{i} = \mathbf{0}$), $\mathbf{l} = \mathbf{0}$, and $\boldsymbol{\Sigma} = \mathbf{0}$ in the balance equation and jump condition of angular momentum so that $\boldsymbol{\sigma} = \boldsymbol{\sigma}^T$. The energy equation and its jump condition subsequently simplify accordingly.

Problems

1. Show that
$$\frac{d(\ln J)}{dt} = \operatorname{div} \mathbf{v}. \quad (4.244)$$

2. Obtain the local mass balance equation (4.98) by using (4.21) and each of the following evolution equations for the

   ii) deformation gradient (3.239),

   ii) inverse deformation gradient (3.494),

   iii) right Cauchy–Green tensor (3.495), and

   iv) left Cauchy–Green tensor (3.496).

3. Show that the integral of the evolution equation for the deformation gradient (3.239) is given by
$$\mathbf{F}(\mathbf{x},t) = \exp\left[\nabla \int_{t_0}^{t} \mathbf{v}(\mathbf{x},t')\,dt'\right]^T \cdot \mathbf{F}(\mathbf{x},t_0). \quad (4.245)$$

4. Show that (4.244) can also be obtained from (4.245).

5. Show that for steady motion ($\partial \mathbf{v}/\partial t = \mathbf{0}$) of a continuum, the stream lines and path lines coincide.

6. Show that $\operatorname{div} \mathbf{v}$, $\operatorname{div} \mathbf{q}$, and $\operatorname{div} \mathbf{h}$ are frame indifferent under a Euclidean transformation.

7. Show that $\operatorname{div} \boldsymbol{\sigma}$ and $\operatorname{div} \boldsymbol{\Sigma}$ are frame indifferent under a Euclidean transformation.

8. Show that the stress power Φ given in (4.123) is frame indifferent under a Euclidean transformation.

9. If A and B are two tensors, show that

   i)
$$[\![AB]\!] = [\![A]\!]\langle\!\langle B\rangle\!\rangle + \langle\!\langle A\rangle\!\rangle[\![B]\!], \quad (4.246)$$

ii)

$$\langle\!\langle A \rangle\!\rangle = A^{\mp} \pm \frac{1}{2} [\![A]\!], \tag{4.247}$$

where

$$\langle\!\langle A \rangle\!\rangle \equiv \frac{1}{2} \left(A^{+} + A^{-} \right). \tag{4.248}$$

10. Starting from (4.216), derive (4.217).

11. Show that for any tensor quantity ψ, the mass conservation statement (4.98) implies that

$$\rho \dot{\psi} = \frac{\partial \rho \psi}{\partial t} + \operatorname{div} \left(\rho \psi \mathbf{v} \right). \tag{4.249}$$

12. Show that the acceleration vector $\mathbf{a} = \dot{\mathbf{v}}$ may be expressed in the form

$$\mathbf{a} = \frac{\partial \mathbf{v}}{\partial t} + \boldsymbol{\omega} \times \mathbf{v} + \frac{1}{2} \operatorname{grad} \left(\mathbf{v} \cdot \mathbf{v} \right), \tag{4.250}$$

where $\boldsymbol{\omega} = \operatorname{curl} \mathbf{v}$ is the vorticity vector. Show that the above form is valid for any arbitrary curvilinear coordinates.

13. Consider a flow with the hydrostatic stress tensor $\boldsymbol{\sigma} = -p\,\mathbf{1}$ and the conservative body force $\mathbf{f} = -\operatorname{grad} \phi$.

    i) Show that if the flow is steady ($\partial \mathbf{v} / \partial t = \mathbf{0}$), then

    $$\rho \mathbf{v} \cdot \operatorname{grad} \left(\frac{1}{2} \mathbf{v} \cdot \mathbf{v} + \phi \right) + \mathbf{v} \cdot \operatorname{grad} p = 0. \tag{4.251}$$

    ii) Show that if the flow is steady and irrotational ($\partial \mathbf{v} / \partial t = \mathbf{0}$ and $\operatorname{curl} \mathbf{v} = \mathbf{0}$), then

    $$\rho \operatorname{grad} \left(\frac{1}{2} \mathbf{v} \cdot \mathbf{v} + \phi \right) + \operatorname{grad} p = 0. \tag{4.252}$$

14. Consider a cubic block with its sides parallel to the Cartesian coordinate axes in the reference configuration. Suppose that the stress tensor is given by $\boldsymbol{\sigma} = -p\,\mathbf{1} + \mu \mathbf{B}$, where p and μ are scalar quantities and $\mathbf{B}$ is the left Cauchy–Green strain tensor. Determine the normal and shear components of the tractions on the surface of the block in the deformed configuration, under the simple shear deformation given by (3.162)–(3.164).

15. At a specific point in a deformable body, the components of the Cauchy stress tensor with respect to the Cartesian coordinate system are given by

$$[\boldsymbol{\sigma}] = \begin{bmatrix} 1 & 4 & -2 \\ 4 & 0 & 0 \\ -2 & 0 & 3 \end{bmatrix}. \tag{4.253}$$

    i) Find the components of the traction vector $\mathbf{t}_n$ on the normal to the plane that passes through the point and that is parallel to the plane $2x_1 + 3x_2 + x_3 = 5$.

ii) Find the length of $\mathbf{t}_n$ and the angle that $\mathbf{t}_n$ makes with the normal to the plane.

iii) Find the stress components of $\overline{\boldsymbol{\sigma}}$ along the new normal basis

$$\bar{\mathbf{e}}_1 = \mathbf{e}_1, \qquad \bar{\mathbf{e}}_2 = \frac{1}{\sqrt{2}}(\mathbf{e}_1 - \mathbf{e}_3), \qquad \bar{\mathbf{e}}_3 = \frac{1}{3}(2\,\mathbf{e}_1 - \mathbf{e}_2 + 2\,\mathbf{e}_3). \quad (4.254)$$

16. The components of the symmetric Cauchy stress tensor referred to the Cartesian coordinate system are given by

$$[\boldsymbol{\sigma}] = \begin{bmatrix} 0 & 0 & \alpha x_2 \\ 0 & 0 & -\beta x_3 \\ \alpha x_2 & -\beta x_3 & 0 \end{bmatrix}, \quad (4.255)$$

where α and β are constants. For the point $\mathbf{x} = (0, \beta^2, \alpha)$,

i) find the three principal stress invariants of tensor $\boldsymbol{\sigma}$;

ii) compute the principal stress components and their associated principal directions;

iii) compute the maximum magnitude of shear stress and the plane on which it acts.

17. Assume a plane stress state in a parallelipiped bounded by the planes $x_1 = \pm a$, $x_2 = \pm b$, and $x_3 = \pm c$, so that the components of the symmetric Cauchy stress tensor referred to the Cartesian coordinate system are given by

$$[\boldsymbol{\sigma}] = \begin{bmatrix} \alpha(x_1 - x_2) & \beta x_1^2 x_2 & 0 \\ \beta x_1^2 x_2 & -\alpha(x_1 - x_2) & 0 \\ 0 & 0 & 0 \end{bmatrix}, \quad (4.256)$$

where α and β are constants. For the point $\mathbf{x} = (a/2, -b/2, 0)$, find

i) the principal normal stresses and the associated principal directions;

ii) the planes, characterized by the unit normal $\mathbf{n}$, that give the maximum and minimum shear stresses and the magnitude of the extremal shear stress;

iii) the total Cauchy traction vector on each face of this parallelipiped.

18. A dynamical process is described by the motion

$$x_1 = e^t X_1 - e^{-t} X_2, \quad (4.257)$$

$$x_2 = e^t X_1 + e^{-t} X_2, \quad (4.258)$$

$$x_3 = X_3, \quad (4.259)$$

for $t > 0$, and the symmetric Cauchy stress tensor with components

$$[\sigma_{ij}] = \begin{bmatrix} x_1^2 & \alpha x_2 x_3^2 & 0 \\ \alpha x_2 x_3^2 & x_2^2 & 0 \\ 0 & 0 & \beta x_1^3 \end{bmatrix}, \quad (4.260)$$

where α and β are scalar constants. Find the system of forces so that the mass conservation and linear momentum balance equations are satisfied. The Cauchy traction vector $\mathbf{t}$ is assumed to act at a point $\mathbf{x}$ of a plane tangential to the sphere given by $\phi = x_1^2 + x_2^2 + x_3^2$.

19. Suppose that $\boldsymbol{\sigma} = -p\,\mathbf{1}$, $\boldsymbol{\Sigma} = -m\,\mathbf{1}$, and $\mathbf{q} = \mathbf{0}$. Obtain the simplified jump conditions.

20. If in a motion of a body the material points crossing a surface $\zeta(t)$ gain or lose mass, what would the jump condition at $\zeta(t)$ be?

21. Across a moving surface $\zeta(t)$, mass and linear momentum of the material points of a continuum are seen to undergo jumps because of the creation or destruction of mass. Find the jump conditions at $\zeta(t)$.

22. Across a moving surface $\zeta(t)$, radiation of heat energy is causing sudden energy loss. Express the jump condition across a moving surface $\zeta(t)$.

23. For the velocity field given in Problem 3.34, determine

    i) components of the deformation rate tensor;
    ii) components of the spin tensor and vorticity vector;
    iii) invariants of the deformation rate tensor.

24. For the velocity field given in Problem 3.35, determine

    i) components of the deformation rate tensor;
    ii) components of the spin tensor and vorticity vector;
    iii) invariants of the deformation rate tensor.

25. For the velocity field given in Problem 3.36, determine

    i) components of the deformation rate tensor;
    ii) components of the spin tensor and vorticity vector;
    iii) invariants of the deformation rate tensor.

26. For the isochoric and irrotational motion of a body, show that the kinetic energy of a nonpolar material is given by

$$\mathcal{K} = \frac{1}{2}\rho \int_{\mathscr{S}} \phi \operatorname{grad}\phi \cdot \mathbf{n}\, ds, \tag{4.261}$$

where ϕ is the velocity potential, i.e., $\mathbf{v} = -\operatorname{grad}\phi$, and $\mathbf{n}$ is the exterior normal of the closed surface $\mathscr{S}$.

27. If we have a nonpolar ideal fluid, so that $\boldsymbol{\sigma} = -p\,\mathbf{1}$, show that the stress power may be expressed by the equation

$$\Phi = \frac{p}{\rho}\frac{d\rho}{dt}. \tag{4.262}$$

28. Show that the balance equation for the vorticity vector $\boldsymbol{\omega} = \operatorname{curl}\mathbf{v}$, valid for an ideal fluid subject to a conservative body force ($\boldsymbol{\sigma} = -p\,\mathbf{1}$ and $\mathbf{f} = -\operatorname{grad}\phi$), is given by

$$\frac{d}{dt}\left(\frac{\boldsymbol{\omega}}{\rho}\right) = (\operatorname{grad}\mathbf{v})\cdot\left(\frac{\boldsymbol{\omega}}{\rho}\right) + \frac{\operatorname{grad}\rho \times \operatorname{grad}p}{\rho^3}, \tag{4.263}$$

where $\mathbf{L} = (\operatorname{grad}\mathbf{v})^T$.

29. Show that the balance equation for the vorticity vector $\boldsymbol{\omega} = \operatorname{curl}\mathbf{v}$, valid for a constant density ideal fluid ($\rho = \text{const.}$ and $\boldsymbol{\sigma} = -p\mathbf{1}$), is given by

$$\dot{\boldsymbol{\omega}} = \mathbf{L} \cdot \boldsymbol{\omega}, \tag{4.264}$$

where $\mathbf{L} = (\operatorname{grad}\mathbf{v})^T$.

30. Show that the balance equation for a barotropic ideal fluid subject to a conservative body force ($\rho = \rho(p)$, $\boldsymbol{\sigma} = -p\mathbf{1}$, and $\mathbf{f} = -\operatorname{grad}\phi$) is given by

$$\dot{\boldsymbol{\xi}} = \mathbf{L} \cdot \boldsymbol{\xi}, \tag{4.265}$$

where $\boldsymbol{\xi} = \boldsymbol{\omega}/\rho$ is the specific vorticity (vorticity per unit mass), $\boldsymbol{\omega} = \operatorname{curl}\mathbf{v}$, and $\mathbf{L} = (\operatorname{grad}\mathbf{v})^T$.

   i) Now if $\boldsymbol{\xi}_R$ is the specific vorticity in the reference configuration, and $\mathbf{F}$ is the deformation gradient, show that

   $$\boldsymbol{\xi} = \mathbf{F} \cdot \boldsymbol{\xi}_R \tag{4.266}$$

   integrates the equation.

   ii) Use the polar decomposition $\mathbf{F} = \mathbf{R} \cdot \mathbf{U}$ and the decomposition $\mathbf{L} = \mathbf{D} + \mathbf{W}$ to show that a vortex filament is stretched and rotated during its motion.

   iii) What can you say about the dynamics of vortex filaments in plane motion?

Bibliography

R.G. Bartle. *The Elements of Integration.* John Wiley & Sons, New York, NY, 1966.

R.M. Bowen. *Introduction to Continuum Mechanics for Engineers.* Plenum Press, New York, NY, 1989.

H.B. Callen. *Thermodynamics.* John Wiley & Sons, Inc., New York, NY, 1962.

P. Chadwick. *Continuum Mechanics – Concise Theory and Problems.* Dover Publications, Inc., Mineola, NY, 2nd edition, 1999.

K. Denbigh. *The Principles of Chemical Equilibrium.* Cambridge University Press, Cambridge, England, 1981.

A.C. Eringen. *Nonlinear Theory of Continuous Media.* McGraw-Hill Book Company, Inc., New York, NY, 1962.

A.C. Eringen. Basic principles: Balance laws. In A.C. Eringen, editor, *Continuum Physics*, volume II, pages 69–88. Academic Press, Inc., New York, NY, 1975.

A.C. Eringen. Basic principles: Thermodynamics of continua. In A.C. Eringen, editor, *Continuum Physics*, volume II, pages 89–127. Academic Press, Inc., New York, NY, 1975.

A.C. Eringen. *Mechanics of Continua.* R.E. Krieger Publishing Company, Inc., Melbourne, FL, 1980.

M.E. Gurtin. *An Introduction to Continuum Mechanics.* Academic Press, San Diego, CA, 2003.

M.E. Gurtin, E. Fried, and L. Anand. *The Mechanics and Thermodynamics of Continua.* Cambridge University Press, Cambridge, UK, 2010.

P. Haupt. *Continuum Mechanics and Theory of Materials.* Springer-Verlag, Berlin, 2000.

G.A. Holzapfel. *Nonlinear Solid Mechanics.* John Wiley & Sons, Ltd., Chichester, England, 2005.

K. Hutter and K. Jöhnk. *Continuum Methods of Physical Modeling.* Springer-Verlag, Berlin, 1981.

W. Jaunzemis. *Continuum Mechanics.* The Macmillan Company, New York, NY, 1967.

J. Kestin. *A Course in Thermodynamics*, volume 1. McGraw-Hill Book Company, New York, NY, 1979.

J. Kestin. *A Course in Thermodynamics*, volume 2. McGraw-Hill Book Company, New York, NY, 1979.

I.-S. Liu. *Continuum Mechanics.* Springer-Verlag, Berlin, 2002.

I. Müller. *Thermodynamics.* Pitman Publishing, Inc., Boston, MA, 1985.

W. Noll. On the continuity of the solid and fluid states. *Journal of Rational Mechanics and Analysis*, 4(1):3–81, 1955.

R.S. Rivlin. The fundamental equations of nonlinear continuum mechanics. In S.I. Pai, A.J. Faller, T.L. Lincoln, D.A. Tidman, G.N. Trytten, and T.D. Wilkerson, editors, *Dynamics of Fluids in Porous Media*, pages 83–126, Academic Press, New York, 1966.

M. Silhavy. *The Mechanics and Thermodynamics of Continuous Media.* Springer-Verlag, Berlin, 1997.

C. Truesdell. Thermodynamics for beginners. In M. Parkus and L.I. Sedov, editors, *Irreversible Aspects of Continuum Mechanics and Transfer of Physical Characteristics of Moving Fluids*, pages 373–389. Springer, Wien, 1968.

C. Truesdell. *A First Course in Rational Continuum Mechanics*, volume 1. Academic Press, New York, NY, 1977.

C. Truesdell and W. Noll. The non-linear field theories of mechanics. In S. Flügge, editor, *Handbuch der Physik*, volume III/3. Springer, Berlin-Heidelberg-New York, 1965.

C. Truesdell and R.A. Toupin. The classical field theories. In S. Flügge, editor, *Handbuch der Physik*, volume III/1. Springer, Berlin-Heidelberg-New York, 1960.

5

Principles of constitutive theory

Conservation of mass, the balance of linear momentum, the balance of angular momentum, the balance of energy, and the entropy inequality are valid for all continuous media. However, different material bodies having the same mass and geometry, when subjected to identical external effects, respond differently. The internal constitution of matter is responsible for the different responses. To understand these differences, we need equations that reflect structural differences between materials. This is the subject of constitutive theory.

Constitutive relations can be regarded as mathematical models for material bodies, and as such, they define ideal materials. Since real materials always contain irregularities and defects, the validity of a model should be verified through experiments on the results it predicts. On the contrary, some experiments may suggest certain functional dependence of the constitutive relations on its variables to within a reasonable satisfaction for certain materials. However, experiments alone are rarely, if ever, sufficient to determine constitutive relations of a material body.

There are some universal requirements that a model should obey lest its consequences be contradictory to some well-known physical experience. Therefore, in search of a correct formulation of a mathematical model, in general, we shall first impose these requirements on the proposed model. The most important universal requirements of this kind are:

- *Principle of causality*: causality describes the relationship between causes and effects as governed by the laws of nature. In classical physics, we assume that a cause should always precede its effect; i.e., the cause and its effect are separated by a time interval, and the effect belongs to the future of its cause.

- *Principle of equipresence*: all constitutive functions should be expressed in terms of the same set of independent constitutive variables until the contrary is deduced.

- *Principle of frame indifference*: the constitutive equations must be objective with respect to Euclidean motions of the spatial frame of reference – they must be the same as seen from inertial and non-inertial frames of reference.

- *Principles of material smoothness and memory*: the constitutive functions at regular points should be spatially and temporally smooth so that material gradients and time derivatives up to some orders exist.

- *Principle of material symmetry*: a body subjected to the same thermomechanical history at two different configurations in general have different results. However, it may happen that the results are exactly the same if the material possesses a certain symmetry that makes it indistinguishable in the two configurations.

- *Thermodynamic principles*: the constitutive equations must be consistent with thermodynamic concepts such as thermodynamic states and processes, and must obey the second law of thermodynamics.

These requirements impose severe restrictions on the model and hence lead to a great simplification for general constitutive relations. The reduction of constitutive relations from very general to more specific and mathematically simpler ones for a given class of materials is the main objective of constitutive theories in continuum mechanics.

5.1 General constitutive equation

The principle of causality amounts to the selection of physical independent variables on which the response of a material depends on. For nonpolar thermomechanical materials, in addition to motion, we include temperature in this set of fields. Its existence is postulated through the *zeroth law of thermodynamics*, which speaks to thermal equilibrium between bodies in contact, and also implies the existence of an *empirical temperature* or *coldness function*, which, without loss of generality, can always be related to the *absolute temperature* θ with dimension $[\Theta]$ that is different than the dimensions of length $[L]$, mass $[M]$, and time $[T]$. We shall suppose that it is always possible to assign a positive-definite temperature $\theta > 0$ to each material point X of a body. As will be noted in Section 5.10, the absolute positive-definite thermodynamic temperature and entropy are inextricably connected with each other. If the existence of a quantity we call entropy is postulated, then temperature appears naturally as a dual quantity of entropy through an equation of state. In such case, it is not necessary to invoke the zeroth law of thermodynamics. The choice of fields limits the class of physical and chemical phenomena observable in the material. Mechanics, electricity and magnetism, and thermodynamics are three parallel divisions of classical macroscopic physics which continuum mechanics speaks to. In this book we have not dealt with the balance laws of electricity and magnetism, and thus we will not consider how materials respond to such fields. Here we focus exclusively on the thermomechanical response of materials. Subsequently, the behavior of the material particle located at $\mathbf{X}$ at time t is characterized by a description of the set of independent fields

$$\mathcal{I}(\mathbf{X},t) \equiv \{\mathbf{x}(\mathbf{X},t), \theta(\mathbf{X},t)\} \tag{5.1}$$

called the *basic fields*. It would appear that density should be included in the set of basic fields. However, from the mass balance, $\rho(\mathbf{X},t) = \rho_R(\mathbf{X},t_0)J^{-1}(\mathbf{X},t)$, we see that ρ can be obtained if we know the density in the reference configuration,

$\rho_R(\mathbf{X}, t_0)$, and the motion, $\mathbf{x}(\mathbf{X}, t)$. We note that $\rho_R(\mathbf{X}, t_0)$ is initially given as an explicit function of $\mathbf{X}$.

Given a reference configuration, the conservation of mass and the balances of momenta and energy alone are not sufficient to determine the basic fields since these contain other, unknown, field quantities: the stress tensor $\boldsymbol{\sigma}$, the heat flux $\mathbf{q}$, the entropy flux $\mathbf{h}$, the body force $\mathbf{f}$, the energy supply r, and the entropy supply b. The external supplies $\mathbf{f}$ and r are regarded as known functions which are provided by the environment that the body encounters. It is assumed that material properties are independent of external supplies. Furthermore, while not necessary, it is convenient to reformulate the entropy inequality (4.219) in terms of the Helmholtz free energy density ψ instead of the internal energy density e (i.e., $\psi = e - \theta\eta$). Subsequently, the unknown quantities consist of the stress tensor, the heat flux, the entropy flux, the Helmholtz free energy density, and the entropy density, and they will depend not only on the behavior of the body but also on the kind of material that constitutes the body. Equations for this set of dependent fields,

$$\mathcal{C}(\mathbf{X}, t) \equiv \{\boldsymbol{\sigma}(\mathbf{X}, t), \mathbf{q}(\mathbf{X}, t), \mathbf{h}(\mathbf{X}, t), \psi(\mathbf{X}, t), \eta(\mathbf{X}, t)\}, \tag{5.2}$$

called *constitutive quantities*, must depend on the basic fields, in addition to the location of a material point and the current time $\{\mathbf{X}, t\}$, to characterize thermo-mechanical responses of a particular material body.

The principle of causality also requires that the response of a material is not influenced by future values of the basic fields or by material particles outside of (not in contact with) the body except for their effects as included in external body supplies. Subsequently, we postulate that, in general, the history of the behavior up to the present time determines the present response of the body at any material point located at $\mathbf{X} \in V = \boldsymbol{\kappa}(\mathcal{B})$ and time t, i.e.,

$$\mathop{\mathfrak{F}}_{\substack{\mathbf{Y}\in V \\ -\infty<\tau\leq t}} \{\mathcal{C}(\mathbf{Y}, \tau); \mathcal{I}(\mathbf{Y}, \tau), \mathbf{X}, t\} = 0, \tag{5.3}$$

where $\mathfrak{F}$ is a functional. A functional is simply a function whose arguments are functions and whose values are tensors. It should be noted that the *response functional* $\mathfrak{F}$ depends on the reference configuration $\boldsymbol{\kappa}$ since both the motion $\boldsymbol{\chi}$ and the material particle coordinate $\mathbf{X}$ depend on the reference configuration. Thus, we should write $\mathfrak{F}_{\boldsymbol{\kappa}}$ for the response functional relative to reference configuration $\boldsymbol{\kappa}$. For simplicity, we continue to drop the subscript denoting this dependence. It will be necessary to make this dependence explicit later in the discussion of material symmetries.

Let a tensor field φ be a function of time. If we take $\tau = t - s$, then the history of φ up to time t is defined by

$$\varphi^{(t)}(s) \equiv \varphi(t - s) = \varphi(\tau), \tag{5.4}$$

where $0 \leq s < \infty$ denotes the time coordinate pointing into the past from the present time t. Clearly, when a material has no memory, $\varphi^{(t)}(0) = \varphi(t)$. Now we can rewrite

$$\mathop{\mathfrak{F}}_{\substack{\mathbf{Y}\in V \\ 0\leq s<\infty}} \{\mathcal{C}(\mathbf{Y}, t - s); \mathcal{I}(\mathbf{Y}, t - s), \mathbf{X}, t\} = \mathop{\mathfrak{F}}_{\substack{\mathbf{Y}\in V \\ 0\leq s<\infty}} \{\mathcal{C}^{(t)}(\mathbf{Y}, s); \mathcal{I}^{(t)}(\mathbf{Y}, s), \mathbf{X}, t\} = 0. \tag{5.5}$$

Such response functionals are sufficiently general so that, e.g., the stress could depend on the histories of stress, heat flux, entropy flux, free energy, entropy, and their gradients and rates of varying orders as well as on the motion and temperature at all other points in the body. Indeed, to produce a theory of heat conduction for which thermal disturbances propagate with finite, rather than infinite, speed, it is necessary to consider such formulation. The functional can be solved for the constitutive quantities if it is non-singular and single valued. We assume that this is the case for the class of materials that we wish to consider. So, if we let the tensor $T(\mathbf{X},t) \in \mathcal{C}$, of arbitrary rank, represent a *constitutive function* of any of the unknown fields, then, with some abuse of functional notation, T is given by

$$T(\mathbf{X},t) = \underset{\substack{\mathbf{Y}\in V\\ 0\le s<\infty}}{\mathfrak{F}} \{\mathbf{x}^{(t)}(\mathbf{Y},s), \theta^{(t)}(\mathbf{Y},s), \mathbf{X}, t\}. \tag{5.6}$$

While we call $\mathfrak{F}$ the constitutive function of T, we recognize that in reality $\mathfrak{F}$ is a functional. Such a functional allows the description of arbitrary nonlocal effects of any inhomogeneous material body with a perfect memory of the past. Note that (obviously) the functionals in (5.5) and (5.6) are different. Here, and below, we re-use some of the symbols, such as $\mathfrak{F}$, so as not to unnecessarily cloud the presentation with too many symbols.

We note that for given $\mathbf{x}$ and θ at all material points $\mathbf{X} \in V$ and for all time t, the above response functionals provide all $T \in \mathcal{C}$. Subsequently, for given $\rho_R(\mathbf{X}, t_0)$ and boundary conditions consistent with the given motion and temperature, the mass balance provides the density ρ, the linear momentum balance provides the body force density $\mathbf{f}$, the angular momentum balance is identically satisfied for nonpolar materials with $\boldsymbol{\sigma} = \boldsymbol{\sigma}^T$, and the energy balance provides the energy supply density r. In some sense, the above procedure assures us of the existence of a solution satisfying all the balance laws. The use of the resulting body force and energy supply densities obtained from such procedure can be effectively utilized to verify the correct implementation of a numerical algorithm intended to produce an approximation to the given motion and temperature fields. This procedure is called the method of manufactured solutions. In reality, the program for us is more challenging since, in general, we are given the body force $\mathbf{f}$ and energy supply r along with $\rho_R(\mathbf{X}, t_0)$ and boundary conditions, and are asked to obtain the motion $\mathbf{x}$ and temperature θ. Nevertheless, in either case not all such solutions are physically realized. It is postulated that all solutions that are physically realizable are such that the response functionals satisfy the entropy inequality as well as the other previously noted accepted principles of continuum mechanics.

5.2 Frame indifference

Assuming that the constitutive function T is a scalar field, and recognizing that θ is a scalar as well, then under an arbitrary Euclidean transformation (see Section 3.3) with arbitrary scalar a, vector $\mathbf{b}(t)$, and orthogonal second-order rotation tensor $\mathbf{Q}(t)$, in order for the field to be frame indifferent or objective we must have

$$\underset{\substack{\mathbf{Y}\in V\\ 0\le s<\infty}}{\mathfrak{F}} \{\mathbf{b}(t-s) + \mathbf{Q}(t-s)\cdot\mathbf{x}(\mathbf{Y},t-s), \theta(\mathbf{Y},t-s), \mathbf{X}, t+a\} = \underset{\substack{\mathbf{Y}\in V\\ 0\le s<\infty}}{\mathfrak{F}} \{\mathbf{x}(\mathbf{Y},t-s), \theta(\mathbf{Y},t-s), \mathbf{X}, t\}. \tag{5.7}$$

Now if we choose the special frame with $\mathbf{b}(t-s) = \mathbf{0}$ and $\mathbf{Q}(t-s) = \mathbf{1}$, corresponding to a time shift, the restriction becomes

$$\underset{\substack{\mathbf{Y}\in V\\0\le s<\infty}}{\mathfrak{F}}\{\mathbf{x}(\mathbf{Y},t-s),\theta(\mathbf{Y},t-s),\mathbf{X},t+a\} = \underset{\substack{\mathbf{Y}\in V\\0\le s<\infty}}{\mathfrak{F}}\{\mathbf{x}(\mathbf{Y},t-s),\theta(\mathbf{Y},t-s),\mathbf{X},t\}. \quad (5.8)$$

We now see that if we take $a = -t$, then the scalar field is seen to be objective only if it contains no explicit dependence on time, i.e.,

$$T(\mathbf{X},t) = \underset{\substack{\mathbf{Y}\in V\\0\le s<\infty}}{\mathfrak{F}}\{\mathbf{x}^{(t)}(\mathbf{Y},s),\theta^{(t)}(\mathbf{Y},s),\mathbf{X}\}. \quad (5.9)$$

Accounting for the independence on t, if we now choose the special frame with $\mathbf{b}(t-s) = -\mathbf{x}(\mathbf{X},t-s)$ and $\mathbf{Q}(t-s) = \mathbf{1}$, corresponding to a rigid translation of the frame, the restriction becomes

$$\underset{\substack{\mathbf{Y}\in V\\0\le s<\infty}}{\mathfrak{F}}\{[\mathbf{x}(\mathbf{Y},t-s)-\mathbf{x}(\mathbf{X},t-s)],\theta(\mathbf{Y},t-s),\mathbf{X}\} = \underset{\substack{\mathbf{Y}\in V\\0\le s<\infty}}{\mathfrak{F}}\{\mathbf{x}(\mathbf{Y},t-s),\theta(\mathbf{Y},t-s),\mathbf{X}\}. \quad (5.10)$$

Subsequently, in order for the constitutive function to be objective, it must be of the form

$$T(\mathbf{X},t) = \underset{\substack{\mathbf{Y}\in V\\0\le s<\infty}}{\mathfrak{F}}\{[\mathbf{x}(\mathbf{Y},t-s)-\mathbf{x}(\mathbf{X},t-s)],\theta(\mathbf{Y},t-s),\mathbf{X}\}. \quad (5.11)$$

Lastly, again accounting for the independence on t, if we now choose the special frame with $\mathbf{b}(t-s) = \mathbf{0}$ and $\mathbf{Q}(t-s)$ arbitrary, corresponding to a rigid rotation of the frame, the restriction for a second-order tensor, e.g., becomes

$$\underset{\substack{\mathbf{Y}\in V\\0\le s<\infty}}{\mathfrak{F}}\{\mathbf{Q}(t-s)\cdot[\mathbf{x}(\mathbf{Y},t-s)-\mathbf{x}(\mathbf{X},t-s)],\theta(\mathbf{Y},t-s),\mathbf{X}\} = \mathbf{Q}(t)\cdot \underset{\substack{\mathbf{Y}\in V\\0\le s<\infty}}{\mathfrak{F}}\{[\mathbf{x}(\mathbf{Y},t-s)-\mathbf{x}(\mathbf{X},t-s)],\theta(\mathbf{Y},t-s),\mathbf{X}\}\cdot\mathbf{Q}^T(t), \quad (5.12)$$

or

$$\underset{\substack{\mathbf{Y}\in V\\0\le s<\infty}}{\mathfrak{F}}\{\mathbf{Q}^{(t)}(s)\cdot[\mathbf{x}^{(t)}(\mathbf{Y},s)-\mathbf{x}^{(t)}(\mathbf{X},s)],\theta^{(t)}(\mathbf{Y},s),\mathbf{X}\} = \mathbf{Q}(t)\cdot \underset{\substack{\mathbf{Y}\in V\\0\le s<\infty}}{\mathfrak{F}}\{[\mathbf{x}^{(t)}(\mathbf{Y},s)-\mathbf{x}^{(t)}(\mathbf{X},s)],\theta^{(t)}(\mathbf{Y},s),\mathbf{X}\}\cdot\mathbf{Q}^T(t). \quad (5.13)$$

Since any general rigid motion of a Euclidean frame and time shift can be obtained by a sequence of the above three transformations, then the *general constitutive equation* is of the form

$$T(\mathbf{X},t) = \underset{\substack{\mathbf{Y}\in V\\0\le s<\infty}}{\mathfrak{F}}\{[\mathbf{x}^{(t)}(\mathbf{Y},s)-\mathbf{x}^{(t)}(\mathbf{X},s)],\theta^{(t)}(\mathbf{Y},s),\mathbf{X}\}, \quad (5.14)$$

subject to the restriction arising from the rigid rotation of the frame, such as (5.13) for a second-order tensor.

5.3 Temporal material smoothness

Suppose that there exists a time $\tau < t$ such that the histories $\mathbf{x}^{(t)}(\mathbf{Y},s)$, $\mathbf{x}^{(t)}(\mathbf{X},s)$, and $\theta^{(t)}(\mathbf{Y},s)$ for $s > 0$ possess Taylor series expansions about $s = 0$ for all $\mathbf{Y} \in \mathcal{B}$:

$$x^{(t)}(\mathbf{Y},s) = x(\mathbf{Y},t) - \dot{x}(\mathbf{Y},t)s + \frac{1}{2!}\ddot{x}(\mathbf{Y},t)s^2 + \cdots, \tag{5.15}$$

$$x^{(t)}(\mathbf{X},s) = x(\mathbf{X},t) - \dot{x}(\mathbf{X},t)s + \frac{1}{2!}\ddot{x}(\mathbf{X},t)s^2 + \cdots, \tag{5.16}$$

$$\theta^{(t)}(\mathbf{Y},s) = \theta(\mathbf{Y},t) - \dot{\theta}(\mathbf{Y},t)s + \frac{1}{2!}\ddot{\theta}(\mathbf{Y},t)s^2 + \cdots. \tag{5.17}$$

Now, assume that derivatives up to orders p and q exist for $\mathbf{x}(\mathbf{Y},t)$ and $\theta(\mathbf{Y},t)$. In such case, materials satisfy the constitutive equations of the form

$$T(\mathbf{X},t) = \underset{\mathbf{Y}\in V}{\mathfrak{G}}\{[\overset{(k)}{\mathbf{x}}(\mathbf{Y},t) - \overset{(k)}{\mathbf{x}}(\mathbf{X},t)], \overset{(l)}{\theta}(\mathbf{Y},t), \mathbf{X}\}, \quad k = 0,1,\ldots,p, \quad l = 0,1,\ldots,q \tag{5.18}$$

and are called *materials of rate type.*

Definition: A material is said to be of mechanical rate p and thermal rate q if and only if the constitutive functional depends on the mechanical and thermal rates up to order p and q, respectively.

The condition of objectivity restricts the constitutive functional (5.18). For example, if T is a second-order tensor, we must require that

$$\underset{\mathbf{Y}\in V}{\mathfrak{G}}\{\mathbf{Q}(t)\cdot[\overset{(k)}{\mathbf{x}}(\mathbf{Y},t) - \overset{(k)}{\mathbf{x}}(\mathbf{X},t)], \overset{(l)}{\theta}(\mathbf{Y},t), \mathbf{X}\} = \mathbf{Q}(t)\cdot \underset{\mathbf{Y}\in V}{\mathfrak{G}}\{[\overset{(k)}{\mathbf{x}}(\mathbf{Y},t) - \overset{(k)}{\mathbf{x}}(\mathbf{X},t)], \overset{(l)}{\theta}(\mathbf{Y},t), \mathbf{X}\}\cdot\mathbf{Q}^T(t), \tag{5.19}$$

where $k = 0,1,\ldots,p$ and $l = 0,1,\ldots,q$.

5.4 Spatial material smoothness

Histories of any part of a material body can affect the response at any other point of the body. In most applications, such nonlocal effect is rarely important. It is usually assumed that only gradients up to some order in an arbitrarily small neighborhood of $\mathbf{X}$ affect the material response at the point $\mathbf{X}$. Hence, the motion and temperature can be approximated to some order by a Taylor series expansion about point $\mathbf{X}$:

$$x_k^{(t)}(\mathbf{Y},s) = x_k^{(t)}(\mathbf{X},s) + x_{k,K_1}^{(t)}(\mathbf{X},s)(Y_{K_1} - X_{K_1}) + \frac{1}{2!}x_{k,K_1K_2}^{(t)}(\mathbf{X},s)(Y_{K_1} - X_{K_1})(Y_{K_2} - X_{K_2}) + \cdots, \tag{5.20}$$

$$\theta^{(t)}(\mathbf{Y},s) = \theta^{(t)}(\mathbf{X},s) + \theta_{,K_1}^{(t)}(\mathbf{X},s)(Y_{K_1} - X_{K_1}) + \frac{1}{2!}\theta_{,K_1K_2}^{(t)}(\mathbf{X},s)(Y_{K_1} - X_{K_1})(Y_{K_2} - X_{K_2}) + \cdots, \tag{5.21}$$

so that the constitutive functional (5.14), up to gradient orders P and Q, for $\mathbf{x}^{(t)}$ and $\theta^{(t)}$ respectively, can be rewritten as

$$T(\mathbf{X},t) = \underset{0\le s<\infty}{\mathfrak{F}} \{{}^{i}\mathbf{F}^{(t)}(\mathbf{X},s), \theta^{(t)}(\mathbf{X},s), {}^{j}\mathbf{G}^{(t)}(\mathbf{X},s), \mathbf{D}_K, \mathbf{X}\}, \tag{5.22}$$

where $i = 1,\ldots,P$, $j = 1,\ldots,Q$ and we define deformation gradients of grade i and temperature gradients of grade j by

$${}^{i}F^{(t)}_{kK_1K_2\cdots K_i} \equiv x^{(t)}_{k,K_1K_2\cdots K_i} \qquad \text{and} \qquad {}^{j}G^{(t)}_{K_1K_2\cdots K_j} \equiv \theta^{(t)}_{,K_1K_2\cdots K_j}, \tag{5.23}$$

$K = K_n = \{1,2,3\}$, and the three vectors $\mathbf{D}_K$ represent the decomposition of all directional vectors $(\mathbf{Y}-\mathbf{X})$ originating from $\mathbf{X}$. Note that

$${}^{j}G^{(t)}_{K_1K_2\cdots K_j} = F^{(t)}_{k_1K_1}F^{(t)}_{k_2K_2}\cdots F^{(t)}_{k_jK_j}\,{}^{j}g^{(t)}_{k_1k_2\cdots k_j}, \tag{5.24}$$

where

$${}^{j}g^{(t)}_{k_1k_2\cdots k_j} \equiv \theta^{(t)}_{,k_1k_2\cdots k_j}. \tag{5.25}$$

Furthermore, we observe that

$${}^{1}\mathbf{F}^{(t)}(\mathbf{X},s) = \mathbf{F}^{(t)}(\mathbf{X},s) = \operatorname{Grad}\mathbf{x}^{(t)}(\mathbf{X},s) \tag{5.26}$$

is the conventional deformation gradient,

$${}^{1}\mathbf{G}^{(t)}(\mathbf{X},s) = \mathbf{G}^{(t)}(\mathbf{X},s) = \operatorname{Grad}\theta^{(t)}(\mathbf{X},s) \tag{5.27}$$

is the conventional temperature gradient, and

$$\mathbf{G}^{(t)}(\mathbf{X},s) = \mathbf{F}^{T}(\mathbf{X},t)\cdot\mathbf{g}^{(t)}(\mathbf{x},s), \tag{5.28}$$

where

$${}^{1}\mathbf{g}^{(t)}(\mathbf{x},s) = \mathbf{g}^{(t)}(\mathbf{x},s) = \operatorname{grad}\theta^{(t)}(\mathbf{x},s) \tag{5.29}$$

is the conventional temperature gradient in the spatial coordinates.

The presence of $\mathbf{D}_K$ and $\mathbf{X}$ in the response functional indicate *material anisotropy* and *material inhomogeneity*, respectively. The vectors $\mathbf{D}_K$ are called *material descriptors* and express the directional dependence of the material properties at material point $\mathbf{X}$. Without loss of generality, one may replace $\mathbf{D}_K$ by $\mathbf{1}_K$, which are the unit vectors of coordinates X_k. The presence of $\mathbf{D}_K$ is an indication that the form of the response functional depends on the choice of the material reference configuration. Since, as already noted earlier, we recognize that the response functional depends on the reference configuration, then there is no loss of generality in dropping the dependence on $\mathbf{D}_K$ from the arguments of the response functional. Changes in the material reference configuration are discussed later. Subsequently, we write

$$T(\mathbf{X},t) = \underset{0\le s<\infty}{\mathfrak{F}} \{{}^{i}\mathbf{F}^{(t)}(\mathbf{X},s), \theta^{(t)}(\mathbf{X},s), {}^{j}\mathbf{G}^{(t)}(\mathbf{X},s), \mathbf{X}\}, \tag{5.30}$$

where $i = 1,\ldots,P$ and $j = 1,\ldots,Q$.

Definition: A material is said to be of mechanical grade P and thermal grade Q if and only if the constitutive functional depends on the deformation gradients up to order P and temperature gradients up to order Q.

Definition: Thermomechanical materials of grade one ($P = Q = 1$) are called *simple materials* and their constitutive function takes the form

$$T(\mathbf{X},t) = \underset{0\leq s<\infty}{\mathfrak{F}} \{\mathbf{F}^{(t)}(\mathbf{X},s), \theta^{(t)}(\mathbf{X},s), \mathbf{G}^{(t)}(\mathbf{X},s), \mathbf{X}\}. \tag{5.31}$$

This class of materials has perfect temporal memory but only very limited non-local sensitivity since we only retain first-order gradients. Nevertheless, this class is general enough to include most material bodies of practical interest. *Non-simple material* bodies, corresponding to $P > 1$ and/or $Q > 1$, are by no means unimportant. As an example, in theories of mixtures and porous media, the density gradient, which is related to the second gradient of deformation, must be taken into account to obtain a consistent theory. Note from (5.20) and (5.21) that $P = 0$ corresponds to simple ($Q = 1$) or non-simple ($Q > 1$) rigid materials and $Q = 0$ to simple ($P = 1$) or non-simple ($P > 1$) non-heat-conducting materials.

The condition of objectivity restricts the constitutive functional (5.30). For example, if T is a second-order tensor, we must require that

$$\begin{aligned}\underset{0\leq s<\infty}{\mathfrak{F}} &\{\mathbf{Q}^{(t)}(s)\cdot {}^{i}\mathbf{F}^{(t)}(\mathbf{X},s), \theta^{(t)}(\mathbf{X},s), {}^{j}\mathbf{G}^{(t)}(\mathbf{X},s), \mathbf{X}\} = \\ &\mathbf{Q}(t)\cdot \underset{0\leq s<\infty}{\mathfrak{F}} \{{}^{i}\mathbf{F}^{(t)}(\mathbf{X},s), \theta^{(t)}(\mathbf{X},s), {}^{j}\mathbf{G}^{(t)}(\mathbf{X},s), \mathbf{X}\}\cdot \mathbf{Q}^{T}(t),\end{aligned} \tag{5.32}$$

where $i = 1,\ldots,P$ and $j = 1,\ldots,Q$.

As noted earlier, a body is called homogeneous if the constitutive function does not depend on $\mathbf{X}$ explicitly. Therefore, for a *homogeneous simple material* body, the above constitutive function reduces to

$$T(\mathbf{X},t) = \underset{0\leq s<\infty}{\mathfrak{F}} \{\mathbf{F}^{(t)}(\mathbf{X},s), \theta^{(t)}(\mathbf{X},s), \mathbf{G}^{(t)}(\mathbf{X},s)\}. \tag{5.33}$$

5.5 Spatial and temporal material smoothness

As in Section 5.3, if the constitutive functional in addition possesses continuous material derivatives with respect to s at $s = 0$, then the gradients of motion and temperature can be approximated up to some order by a Taylor series expansion:

$$x^{(t)}_{k,K_1}(\mathbf{X},s) = x_{k,K_1}(\mathbf{X},t) - \dot{x}_{k,K_1}(\mathbf{X},t)s + \frac{1}{2!}\ddot{x}_{k,K_1}(\mathbf{X},t)s^2 + \cdots, \tag{5.34}$$

$$\vdots$$

$$\theta^{(t)}(\mathbf{X},s) = \theta(\mathbf{X},t) - \dot{\theta}(\mathbf{X},t)s + \frac{1}{2!}\ddot{\theta}(\mathbf{X},t)s^2 + \cdots, \tag{5.35}$$

$$\theta^{(t)}_{,K_1}(\mathbf{X},s) = \theta_{,K_1}(\mathbf{X},t) - \dot{\theta}_{,K_1}(\mathbf{X},t)s + \frac{1}{2!}\ddot{\theta}_{,K_1}(\mathbf{X},t)s^2 + \cdots, \tag{5.36}$$

$$\vdots$$

Subsequently, the constitutive functional (5.30) becomes

$$T(\mathbf{X},t) = T\{{}^{i}\overset{(k)}{\mathbf{F}}(\mathbf{X},t), \overset{(l)}{\theta}(\mathbf{X},t), {}^{j}\overset{(m)}{\mathbf{G}}(\mathbf{X},t), \mathbf{X}\}, \tag{5.37}$$

where $i = 1,2,\ldots,P$, $j = 1,2,\ldots,Q$, $k = 0,1,\ldots,p$, $l = 0,1,\ldots,q$, and $m = 0,1,\ldots,r$. Note that now we no longer have a functional, but a tensor-valued function. A simple material ($P = Q = 1$) involving time rates of the deformation gradients up to order p and temperature and its gradients up to order q and r, respectively, is described by

$$T(\mathbf{X},t) = T\{\overset{(k)}{\mathbf{F}}(\mathbf{X},t),\ \overset{(l)}{\theta}(\mathbf{X},t),\ \overset{(m)}{\mathbf{G}}(\mathbf{X},t),\mathbf{X}\} \tag{5.38}$$

with $k = 0,1,\ldots,p$, $l = 0,1,\ldots,q$, and $m = 0,1,\ldots,r$. Clearly, for a simple material with no memory ($p = q = r = 0$), we have

$$T(\mathbf{X},t) = T\{\mathbf{F}(\mathbf{X},t),\theta(\mathbf{X},t),\mathbf{G}(\mathbf{X},t),\mathbf{X}\}. \tag{5.39}$$

For a simple material that includes heat conduction and viscous dissipation, one should take $p = 1$ and $q = r = 0$, in which case the constitutive function is

$$T(\mathbf{X},t) = T\{\mathbf{F}(\mathbf{X},t),\dot{\mathbf{F}}(\mathbf{X},t),\theta(\mathbf{X},t),\mathbf{G}(\mathbf{X},t),\mathbf{X}\}. \tag{5.40}$$

In general, the set of variables defines the class of viscous heat-conducting material for the purpose of the theory, and the form of the constitutive functions defines a particular material within that class.

5.6 Material symmetry

At this stage, we wish to make a distinction between *property tensors* and *field tensors*. A property tensor depends on the structure of the material. A field tensor depends on the field applied to the material; it can have any arbitrary form and *does not* depend on the structure of the material. In constitutive representations, we usually find that two specific field tensors are related (linearly, quadratically, etc.) through property tensors. In this relation, the property tensor expresses the response to a generalized force by yielding a generalized displacement. For example, mass and volume of an object are field tensors of rank 0; they are linearly related through density, which is a rank 0 property tensor. The heat flux and the temperature gradient are field tensors of rank 1. As we will see, for a solid the temperature gradient is linearly related to the heat flux through the rank 2 property tensor of thermal conductivity. The stress and strain tensors are rank 2 field tensors. Again, for a solid we will see that they are linearly related through the rank 4 stiffness or compliance tensors. Clearly, all property and field tensors must transform appropriately (as discussed in Chapter 2) to be frame indifferent.

The rank of a tensor determines the number of components. However, symmetries reduce the number of independent components considerably. Symmetries inherent in field tensors are usually obtained from application of physical principles, such as application of angular momentum (for nonpolar materials) yielding the symmetry of the stress tensor, or from definitions involving second or higher derivatives, where the tensor is not affected from interchange of the order that such derivatives are taken, or from application of thermodynamical reasoning, such as the requirement of reversible changes at equilibrium or application of Onsager's principle near equilibrium. On the other hand, symmetries of property tensors are discovered from examining the intrinsic symmetries of the material at equilibrium conditions. The study of symmetries of materials is the subject of this section.

The constitutive functional of a simple material relative to a reference configuration $\boldsymbol{\kappa}$ can be written in the form

$$T(\mathbf{X},t) = \underset{0\leq s<\infty}{\mathfrak{F}_{\boldsymbol{\kappa}}} \{\mathbf{F}^{(t)}(\mathbf{X},s), \theta^{(t)}(\mathbf{X},s), \mathbf{G}^{(t)}(\mathbf{X},s), \mathbf{X}\}. \tag{5.41}$$

As the notation indicates, the response functional depends on the reference configuration $\boldsymbol{\kappa}$ because $\mathbf{X}$ depends on $\boldsymbol{\kappa}$ and $\mathbf{F}$ depends on the motion $\chi_{\boldsymbol{\kappa}}$, which in turn depends on the reference configuration $\boldsymbol{\kappa}$.

The concept of material symmetry arises when one attempts to determine in what fashion the response functional depends on the choice of reference configuration. Note that

$$\mathbf{X} = \boldsymbol{\kappa}(X) \tag{5.42}$$

indicates the position occupied by the material particle labeled X in the reference configuration $\boldsymbol{\kappa}$. Suppose that $\widehat{\boldsymbol{\kappa}}$ is another reference configuration of the same material point. Then

$$\widehat{\mathbf{X}} = \widehat{\boldsymbol{\kappa}}(X) \tag{5.43}$$

is the position occupied by X in the reference configuration $\widehat{\boldsymbol{\kappa}}$. Note that

$$\widehat{\mathbf{X}} = \widehat{\boldsymbol{\kappa}}(\boldsymbol{\kappa}^{-1}(\mathbf{X})) = \widehat{\boldsymbol{\kappa}}_{\boldsymbol{\kappa}}(\mathbf{X}) \tag{5.44}$$

represents a change of reference configuration. Now define

$$\mathbf{P} \equiv \text{Grad}\, \widehat{\boldsymbol{\kappa}}_{\boldsymbol{\kappa}}(\mathbf{X}) = \frac{\partial \widehat{\mathbf{X}}}{\partial \mathbf{X}}, \tag{5.45}$$

and recall that the motion is defined by

$$\mathbf{x} = \chi_{\boldsymbol{\kappa}}(\mathbf{X},t). \tag{5.46}$$

Likewise, we can define the motion $\widehat{\chi}_{\widehat{\boldsymbol{\kappa}}}$ by

$$\widehat{\mathbf{x}} = \widehat{\chi}_{\widehat{\boldsymbol{\kappa}}}(X,t) = \widehat{\chi}(\widehat{\boldsymbol{\kappa}}^{-1}(\widehat{\mathbf{X}}),t) = \widehat{\chi}_{\widehat{\boldsymbol{\kappa}}}(\widehat{\mathbf{X}},t) \tag{5.47}$$

$$= \widehat{\chi}(\boldsymbol{\kappa}^{-1}(\mathbf{X}),t) = \widehat{\chi}_{\boldsymbol{\kappa}}(\mathbf{X},t), \tag{5.48}$$

where $\widehat{\chi}$ is the motion of $\mathcal{B}$. The general concept is illustrated in Fig. 5.1. Of course if the motion $\widehat{\chi}$ is the same as χ, the spatial regions $\widehat{\chi}(\mathcal{B},t)$ and $\chi(\mathcal{B},t)$ for all $X \in \mathcal{B}$ are the same. Then, it easily follows that

$$\mathbf{F} = \text{Grad}\, \chi_{\boldsymbol{\kappa}}(\mathbf{X},t) = \text{Grad}\, \widehat{\chi}_{\widehat{\boldsymbol{\kappa}}}(\widehat{\mathbf{X}},t) = \widehat{\mathbf{F}} \cdot \mathbf{P} \quad \text{or} \quad F_{kK} = \widehat{F}_{k\widehat{L}} P_{\widehat{L}K}. \tag{5.49}$$

This equation relates the deformation gradients constructed from viewing the motion from two different reference configurations. Analogously, if the temperature at a material point is the same in two different reference configurations, i.e., $\widehat{\theta}_{\widehat{\boldsymbol{\kappa}}}(\widehat{\mathbf{X}},t) = \theta_{\boldsymbol{\kappa}}(\mathbf{X},t)$, then it is easy to see that the relation between temperature gradients between the two different reference configurations is given by

$$\mathbf{G} = \mathbf{P}^T \cdot \widehat{\mathbf{G}} \quad \text{or} \quad \theta_{,K} = P_{\widehat{L}K} \widehat{\theta}_{,\widehat{L}}. \tag{5.50}$$

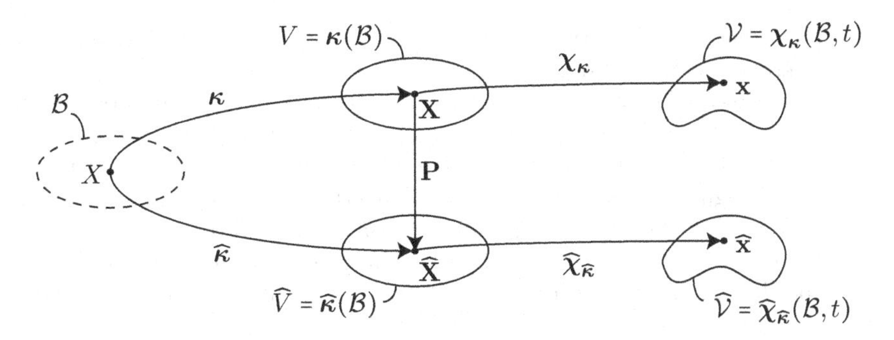

Figure 5.1: Material body, and reference and deformed configurations.

The response functional of particle X in a simple material in reference configuration $\boldsymbol{\kappa}$ is given by

$$T(X,t) = \underset{0\le s<\infty}{\mathfrak{F}_{\boldsymbol{\kappa}}} \{\mathbf{F}^{(t)}, \theta^{(t)}, \mathbf{G}^{(t)}, X\}. \tag{5.51}$$

The response functional of the same particle in the reference configuration $\widehat{\boldsymbol{\kappa}}$ is given by

$$\widehat{T}(X,t) = \underset{0\le s<\infty}{\mathfrak{F}_{\widehat{\boldsymbol{\kappa}}}} \{\widehat{\mathbf{F}}^{(t)}, \theta^{(t)}, \widehat{\mathbf{G}}^{(t)}, X\}, \tag{5.52}$$

where we have used the fact that, for a fixed material point, $\widehat{\theta}^{(t)}(\widehat{\mathbf{X}},t) = \theta^{(t)}(\mathbf{X},t) = \theta^{(t)}(X,t)$.

From the above discussion, we can see that the response functional only depends on $\boldsymbol{\kappa}$ in a neighborhood of X. This neighborhood is called a *local reference configuration*. Our response functional applies to the local reference configuration at material point X and we look for the set of all reference configurations that are equivalent at X. Subsequently, from now on, our reference configuration will be understood to be the local reference configuration and so from (5.44) we take the specific linear transformation

$$\widehat{\mathbf{X}} = \mathbf{A} + \mathbf{H}\cdot\mathbf{X}, \tag{5.53}$$

where $\mathbf{A}$ now represents a constant translation vector and $\mathbf{H}$ a constant rotation or inversion tensor. Note that $\mathbf{A}$ and $\mathbf{H}$ are time independent since the change of local reference configuration is time independent, and in this case, $\mathbf{P} = \mathbf{H}$.

The concept of material symmetry arises when one tries to characterize those changes of reference configuration which do not affect the response of the material. We now want to characterize those linear transformations which produce the *same* value of the response functional at material point X independent of the reference configuration, i.e., $\widehat{T}(X,t) = T(X,t)$ or

$$\underset{0\le s<\infty}{\mathfrak{F}_{\widehat{\boldsymbol{\kappa}}}} \{\widehat{\mathbf{F}}^{(t)}, \theta^{(t)}, \widehat{\mathbf{G}}^{(t)}, \widehat{\mathbf{X}}\} = \underset{0\le s<\infty}{\mathfrak{F}_{\boldsymbol{\kappa}}} \{\widehat{\mathbf{F}}^{(t)}\cdot\mathbf{H}, \theta^{(t)}, \mathbf{H}^T\cdot\widehat{\mathbf{G}}^{(t)}, \mathbf{H}^{-1}\cdot(\widehat{\mathbf{X}}-\mathbf{A})\}, \tag{5.54}$$

where we have used (5.49) to relate $\mathbf{F}$ to $\widehat{\mathbf{F}}$, (5.50) to relate $\mathbf{G}$ to $\widehat{\mathbf{G}}$, and (5.53) to relate $\mathbf{X}$ to $\widehat{\mathbf{X}}$. From above, it follows that a response functional relative to a reference configuration determines the response functional relative to any other reference configuration.

A material body subjected to the same history using two different configurations, in general, yields different results. However, it may happen that the results are exactly the same if the material possesses a certain symmetry that makes it unable to distinguish between the two configurations. The reference configurations are then said to be *materially indistinguishable* and $\widehat{\kappa} = \kappa$. The consequences of these conditions are obtained by requiring objectivity with respect to material coordinates in any reference configuration. Thus, for a tensor of any order, we have from (5.54) that

$$\underset{0 \le s < \infty}{\mathfrak{F}_{\kappa}} \{\mathbf{F}^{(t)}, \theta^{(t)}, \mathbf{G}^{(t)}, \mathbf{X}\} = \underset{0 \le s < \infty}{\mathfrak{F}_{\kappa}} \{\mathbf{F}^{(t)} \cdot \mathbf{H}, \theta^{(t)}, \mathbf{H}^T \cdot \mathbf{G}^{(t)}, \mathbf{H}^{-1} \cdot (\mathbf{X} - \mathbf{A})\}$$
$$\text{for all } (\mathbf{F}^{(t)}, \theta^{(t)}, \mathbf{G}^{(t)}), \quad (5.55)$$

where $\mathbf{H}$ now indicates a *material symmetry transformation.* If a material is such that if we take $\mathbf{A} = \mathbf{X}$ for all material points in the body, the above remains valid, then we see that the response functional (5.55) becomes independent of $\mathbf{X}$. Such materials are called *homogeneous.* From now on, we will only consider homogeneous materials. Furthermore, we will consider only the class of transformations $\mathbf{H}$ such that the mass density of the material remains the same. Thus, since from (4.21) we have that

$$\rho = \rho_R J^{-1} = \rho_R [\det F]^{-1} = \rho_R [\det(FH)]^{-1}, \quad (5.56)$$

and since $\det(FH) = (\det F)(\det H)$, we require that changes in the reference configuration satisfy $\det H = \pm 1$ (recall that following (3.8) we assumed that $J = \det F > 0$; more generally J could be of either sign). Note that $\mathbf{F}$ and $\mathbf{H}$ are both non-singular so that their inverses always exist. A material transformation $\mathbf{H}$ satisfying these properties for the constitutive quantity T is called a *unimodular transformation.* The set of all unimodular transformations forms a group called the *unimodular group* $\mathscr{U}(\mathscr{V})$. Subsequently, we have that $\mathbf{H} \in \mathscr{U}(\mathscr{V})$.

Definition: A collection of all symmetry transformations with respect to reference configuration κ forms the *symmetry group* of a specific material, denoted $\mathscr{G}_{\kappa}$, if

i) for every $(\mathbf{H}_1, \mathbf{H}_2) \in \mathscr{G}_{\kappa}$, $\mathbf{H}_1 \cdot \mathbf{H}_2 \in \mathscr{G}_{\kappa}$ (closure under multiplication);

ii) for every $(\mathbf{H}_1, \mathbf{H}_2, \mathbf{H}_3) \in \mathscr{G}_{\kappa}$, $(\mathbf{H}_1 \cdot \mathbf{H}_2) \cdot \mathbf{H}_3 = \mathbf{H}_1 \cdot (\mathbf{H}_2 \cdot \mathbf{H}_3)$ (associative law for products);

iii) the set $\mathscr{G}_{\kappa}$ contains a unit element $\mathbf{1}$ such that for all $\mathbf{H} \in \mathscr{G}_{\kappa}$, $\mathbf{H} \cdot \mathbf{1} = \mathbf{1} \cdot \mathbf{H} = \mathbf{H}$;

iv) for every $\mathbf{H} \in \mathscr{G}_{\kappa}$, there exists an element $\mathbf{H}^{-1} \in \mathscr{G}_{\kappa}$ called the inverse of $\mathbf{H}$ such that $\mathbf{H} \cdot \mathbf{H}^{-1} = \mathbf{H}^{-1} \cdot \mathbf{H} = \mathbf{1}$.

If $\mathbf{H}_1$ and $\mathbf{H}_2$ are symmetry transformations, then $\mathbf{H}_1 \cdot \mathbf{H}_2$ is also a symmetry transformation, since by hypothesis

$$\underset{0 \le s < \infty}{\mathfrak{F}_{\kappa}} \{\overline{\mathbf{F}}^{(t)}, \theta^{(t)}, \overline{\mathbf{G}}^{(t)}\} = \underset{0 \le s < \infty}{\mathfrak{F}_{\kappa}} \{\overline{\mathbf{F}}^{(t)} \cdot \mathbf{H}_2, \theta^{(t)}, \mathbf{H}_2^T \cdot \overline{\mathbf{G}}^{(t)}\}$$
$$\text{for all } (\overline{\mathbf{F}}^{(t)}, \theta^{(t)}, \overline{\mathbf{G}}^{(t)}), \quad (5.57)$$

and in particular for $\overline{\mathbf{F}}^{(t)} = \mathbf{F}^{(t)} \cdot \mathbf{H}_1$ and $\overline{\mathbf{G}}^{(t)} = \mathbf{H}_1^T \cdot \mathbf{G}^{(t)}$, so

$$\underset{0\le s<\infty}{\mathfrak{F}_{\boldsymbol{\kappa}}}\{\mathbf{F}^{(t)} \cdot \mathbf{H}_1, \theta^{(t)}, \mathbf{H}_1^T \cdot \mathbf{G}^{(t)}\} = \underset{0\le s<\infty}{\mathfrak{F}_{\boldsymbol{\kappa}}}\{\mathbf{F}^{(t)} \cdot \mathbf{H}_1 \cdot \mathbf{H}_2, \theta^{(t)}, \mathbf{H}_2^T \cdot \mathbf{H}_1^T \cdot \mathbf{G}^{(t)}\},$$

and since by assumption $\mathbf{H}_1$ is a symmetry transformation,

$$\underset{0\le s<\infty}{\mathfrak{F}_{\boldsymbol{\kappa}}}\{\mathbf{F}^{(t)}, \theta^{(t)}, \mathbf{G}^{(t)}\} = \underset{0\le s<\infty}{\mathfrak{F}_{\boldsymbol{\kappa}}}\{\mathbf{F}^{(t)} \cdot (\mathbf{H}_1 \cdot \mathbf{H}_2), \theta^{(t)}, (\mathbf{H}_1 \cdot \mathbf{H}_2)^T \cdot \mathbf{G}^{(t)}\} \quad \text{for all} \quad (\mathbf{F}^{(t)}, \theta^{(t)}, \mathbf{G}^{(t)}). \tag{5.58}$$

If $\mathbf{H}$ is a symmetry transformation, then $\mathbf{H}^{-1}$, where $\mathbf{H} \cdot \mathbf{H}^{-1} = \mathbf{H}^{-1} \cdot \mathbf{H} = \mathbf{1}$, is also a symmetry transformation, since if

$$\underset{0\le s<\infty}{\mathfrak{F}_{\boldsymbol{\kappa}}}\{\overline{\mathbf{F}}^{(t)}, \theta^{(t)}, \overline{\mathbf{G}}^{(t)}\} = \underset{0\le s<\infty}{\mathfrak{F}_{\boldsymbol{\kappa}}}\{\overline{\mathbf{F}}^{(t)} \cdot \mathbf{H}, \theta^{(t)}, \mathbf{H}^T \cdot \overline{\mathbf{G}}^{(t)}\} \quad \text{for all} \quad (\overline{\mathbf{F}}^{(t)}, \theta^{(t)}, \overline{\mathbf{G}}^{(t)}), \tag{5.59}$$

then in particular for $\overline{\mathbf{F}}^{(t)} = \mathbf{F}^{(t)} \cdot \mathbf{H}^{-1}$ and $\overline{\mathbf{G}}^{(t)} = (\mathbf{H}^{-1})^T \cdot \mathbf{G}^{(t)}$, we have

$$\underset{0\le s<\infty}{\mathfrak{F}_{\boldsymbol{\kappa}}}\{\mathbf{F}^{(t)} \cdot \mathbf{H}^{-1}, \theta^{(t)}, (\mathbf{H}^{-1})^T \cdot \mathbf{G}^{(t)}\} = \underset{0\le s<\infty}{\mathfrak{F}_{\boldsymbol{\kappa}}}\{\mathbf{F}^{(t)}, \theta^{(t)}, \mathbf{G}^{(t)}\} \quad \text{for all } (\mathbf{F}^{(t)}, \theta^{(t)}, \mathbf{G}^{(t)}). \tag{5.60}$$

A group is called *Abelian* (or *commutative*) if for every pair of elements $(\mathbf{H}_1, \mathbf{H}_2) \in \mathscr{G}_{\boldsymbol{\kappa}}$, $\mathbf{H}_1 \cdot \mathbf{H}_2 = \mathbf{H}_2 \cdot \mathbf{H}_1$.

Any finite set of elements satisfying the four group axioms is said to form a *finite group*, the order of the group being equal to the number of elements in the set. If the group does not have a finite number of elements, it is called an *infinite group*.

A *subgroup* of a group is a subset of a group such that the subset itself is a group. The term *proper subgroup* is defined to be consistent with the terms of subgroup and proper subset.

Example

The set of four numbers $\{1, i, -1, -i\}$, where i is the imaginary number, forms a group of order 4 under multiplication. Group property (i) is clearly satisfied since the product of any two scalar elements (and square of each element) are elements of the set (e.g., $1i = i$, $i(-i) = 1$, $i^2 = -1$, $(-i)^2 = -1$, etc.). The associative law (ii) also holds for the multiplication of numbers. The unit element is taken as the number 1. Finally, if the inverse of every element is taken as its reciprocal (e.g., $1/i = -i$, $1/(-1) = -1$, etc.), then group property iv) is satisfied.

Example

An example of a finite group of matrices which is Abelian is given by

$$\begin{pmatrix} 1 & 0 \\ 0 & 1 \end{pmatrix}, \begin{pmatrix} 0 & 1 \\ -1 & 0 \end{pmatrix}, \begin{pmatrix} -1 & 0 \\ 0 & -1 \end{pmatrix}, \begin{pmatrix} 0 & -1 \\ 1 & 0 \end{pmatrix}, \tag{5.61}$$

which form a group of order 4 under matrix multiplication.

Sets of matrices which form groups with respect to matrix multiplication are usually called *matrix groups*, and are of extreme importance for us.

Example

Another example of a matrix group of order 6 is given by

$$H_1 = \begin{pmatrix} 1 & 0 \\ 0 & 1 \end{pmatrix}, \quad H_2 = \begin{pmatrix} -\frac{1}{2} & \frac{\sqrt{3}}{2} \\ -\frac{\sqrt{3}}{2} & -\frac{1}{2} \end{pmatrix}, \quad H_3 = \begin{pmatrix} -\frac{1}{2} & -\frac{\sqrt{3}}{2} \\ \frac{\sqrt{3}}{2} & -\frac{1}{2} \end{pmatrix},$$

$$H_4 = \begin{pmatrix} 1 & 0 \\ 0 & -1 \end{pmatrix}, \quad H_5 = \begin{pmatrix} -\frac{1}{2} & \frac{\sqrt{3}}{2} \\ \frac{\sqrt{3}}{2} & \frac{1}{2} \end{pmatrix}, \quad H_6 = \begin{pmatrix} -\frac{1}{2} & -\frac{\sqrt{3}}{2} \\ -\frac{\sqrt{3}}{2} & \frac{1}{2} \end{pmatrix}. \quad (5.62)$$

It may be easily verified by examining their products, etc. For example,

$$H_4 H_2 = \begin{pmatrix} 1 & 0 \\ 0 & -1 \end{pmatrix} \begin{pmatrix} -\frac{1}{2} & \frac{\sqrt{3}}{2} \\ -\frac{\sqrt{3}}{2} & -\frac{1}{2} \end{pmatrix} = \begin{pmatrix} -\frac{1}{2} & -\frac{\sqrt{3}}{2} \\ -\frac{\sqrt{3}}{2} & \frac{1}{2} \end{pmatrix} = H_6, \qquad (5.63)$$

and so on.

For a given material, the set of all material symmetry transformations for the constitutive quantity $T(\mathbf{X}, t)$ with respect to $\boldsymbol{\kappa}$, denoted by $\mathscr{G}_{\boldsymbol{\kappa}}(T)$, is a subgroup of the unimodular group,

$$\mathscr{G}_{\boldsymbol{\kappa}}(T) \subseteq \mathscr{U}(\mathscr{V}). \qquad (5.64)$$

We call $\mathscr{G}_{\boldsymbol{\kappa}}$ the *material symmetry group* of T with respect to the reference configuration $\boldsymbol{\kappa}$. It is important to note that the symmetry group depends on the reference configuration $\boldsymbol{\kappa}$. It is clear that $\mathscr{G}_{\boldsymbol{\kappa}}$ depends as well on the constitutive quantity T. In other words, we may have different symmetry groups for different constitutive quantities of the same material body. The largest group contained in the symmetry groups of all constitutive quantities of the material body $\mathcal{B}$ is called the *material symmetry group* of $\mathcal{B}$ and is denoted by $\mathscr{G}_{\boldsymbol{\kappa}}(\mathscr{V})$.

For any $\boldsymbol{\kappa}$ and $\widehat{\boldsymbol{\kappa}}$ such that

$$\mathbf{P} = \frac{\partial \widehat{\mathbf{X}}}{\partial \mathbf{X}}, \qquad (5.65)$$

the following relation, known as *Noll's rule*, between the symmetry groups with respect to the two different reference configurations, holds:

$$\mathscr{G}_{\widehat{\boldsymbol{\kappa}}} = \mathbf{P} \cdot \mathscr{G}_{\boldsymbol{\kappa}} \cdot \mathbf{P}^{-1}. \qquad (5.66)$$

To prove the above result, we note that since $\widehat{\boldsymbol{\kappa}}$ is a different configuration than

$\boldsymbol{\kappa}$, from (5.49) and (5.50), for any $\mathbf{H} \in \mathscr{G}_{\boldsymbol{\kappa}}$, we have

$$\underset{0\leq s<\infty}{\mathfrak{F}_{\widehat{\boldsymbol{\kappa}}}}\{\widehat{\mathbf{F}}^{(t)},\theta^{(t)},\widehat{\mathbf{G}}^{(t)}\} = \underset{0\leq s<\infty}{\mathfrak{F}_{\boldsymbol{\kappa}}}\{\widehat{\mathbf{F}}^{(t)}\cdot\mathbf{P},\theta^{(t)},\mathbf{P}^T\cdot\widehat{\mathbf{G}}^{(t)}\} =$$

$$\underset{0\leq s<\infty}{\mathfrak{F}_{\boldsymbol{\kappa}}}\{(\widehat{\mathbf{F}}^{(t)}\cdot\mathbf{P})\cdot\mathbf{H},\theta^{(t)},\mathbf{H}^T\cdot(\mathbf{P}^T\cdot\widehat{\mathbf{G}}^{(t)})\} =$$

$$\underset{0\leq s<\infty}{\mathfrak{F}_{\boldsymbol{\kappa}}}\{\widehat{\mathbf{F}}^{(t)}\cdot(\mathbf{P}\cdot\mathbf{H}\cdot\mathbf{P}^{-1})\cdot\mathbf{P},\theta^{(t)},\mathbf{P}^T\cdot(\mathbf{P}\cdot\mathbf{H}\cdot\mathbf{P}^{-1})^T\cdot\widehat{\mathbf{G}}^{(t)}\} =$$

$$\underset{0\leq s<\infty}{\mathfrak{F}_{\widehat{\boldsymbol{\kappa}}}}\{\widehat{\mathbf{F}}^{(t)}\cdot(\mathbf{P}\cdot\mathbf{H}\cdot\mathbf{P}^{-1}),\theta^{(t)},(\mathbf{P}\cdot\mathbf{H}\cdot\mathbf{P}^{-1})^T\cdot\widehat{\mathbf{G}}^{(t)}\},$$

which implies that $\widehat{\mathbf{H}} = (\mathbf{P}\cdot\mathbf{H}\cdot\mathbf{P}^{-1}) \in \mathscr{G}_{\widehat{\boldsymbol{\kappa}}}$.

Suppose that $\mathbf{P} = \alpha\mathbf{1}$, where $\alpha \neq 0$ is a scalar constant. If $\alpha > 1$, this transformation represents a dilatation; if $0 < \alpha < 1$, it is a contraction; and if $\alpha = -1$, it is a central inversion. Then $\mathbf{P}^{-1} = \alpha^{-1}\mathbf{1}$ and so

$$\mathscr{G}_{\widehat{\boldsymbol{\kappa}}} = (\alpha\mathbf{1})\cdot\mathscr{G}_{\boldsymbol{\kappa}}\cdot(\alpha\mathbf{1})^{-1} = \mathscr{G}_{\boldsymbol{\kappa}}, \tag{5.67}$$

so that the material symmetry group is invariant under a uniform transformation.

Example

We want to find the generators of the symmetry group for the geometry shown in Fig. 5.2. By inspection, we have

$$H^{(1)} = \begin{bmatrix} 1 & 0 & 0 \\ 0 & 1 & 0 \\ 0 & 0 & 1 \end{bmatrix}, H^{(2)} = \begin{bmatrix} 0 & 1 & 0 \\ -1 & 0 & 0 \\ 0 & 0 & 1 \end{bmatrix}, H^{(3)} = \begin{bmatrix} 1 & 0 & 0 \\ 0 & -1 & 0 \\ 0 & 0 & -1 \end{bmatrix},$$

$$H^{(4)} = \begin{bmatrix} -1 & 0 & 0 \\ 0 & 1 & 0 \\ 0 & 0 & -1 \end{bmatrix}, H^{(5)} = \begin{bmatrix} 0 & 1 & 0 \\ 1 & 0 & 0 \\ 0 & 0 & -1 \end{bmatrix}, \tag{5.68}$$

where

$$X_I = H_{IK}X_K. \tag{5.69}$$

We note that the above generators are not all the generators that form the complete discrete symmetry group for this cubic geometry. A more complete discussion of specific symmetries is given in Chapter 7.
If we call this group $\mathscr{G}_0 = \{H^{(1)}, H^{(2)}, H^{(3)}, H^{(4)}, H^{(5)}\}$, we want to find the group $\mathscr{G}_{\widehat{0}}$ for the transformation shown in Fig. 5.3 and given by

$$\widehat{X}_1 = X_1 + KX_2, \tag{5.70}$$
$$\widehat{X}_2 = X_2, \tag{5.71}$$
$$\widehat{X}_3 = X_3. \tag{5.72}$$

Now

$$P = \left[\frac{\partial \widehat{X}_I}{\partial X_J}\right] = \begin{bmatrix} 1 & K & 0 \\ 0 & 1 & 0 \\ 0 & 0 & 1 \end{bmatrix} \quad \text{and} \quad P^{-1} = \begin{bmatrix} 1 & -K & 0 \\ 0 & 1 & 0 \\ 0 & 0 & 1 \end{bmatrix}. \tag{5.73}$$

Note that P is not orthogonal since $P^{-1} \neq P^T$. We can obtain the new group by the transformation given by Noll's rule:

$$\mathscr{G}_{\widehat{0}} = P\mathscr{G}_0 P^{-1} = \{\widehat{H}^{(1)}, \widehat{H}^{(2)}, \widehat{H}^{(3)}, \widehat{H}^{(4)}, \widehat{H}^{(5)}\}, \tag{5.74}$$

where the $\widehat{H}^{(i)}$, $i = 1, \ldots, 5$, satisfy the symmetry transformation

$$\widehat{X}_I = \widehat{H}_{IK}\widehat{X}_K. \tag{5.75}$$

It is easy to verify that $\widehat{H}^{(2)}$ is a symmetry transformation of $\mathscr{G}_{\widehat{0}}$. For example, for $H^{(2)} \in \mathscr{G}_0$, we have

$$\begin{aligned}\widehat{H}^{(2)} &= \begin{bmatrix} 1 & K & 0 \\ 0 & 1 & 0 \\ 0 & 0 & 1 \end{bmatrix}\begin{bmatrix} 0 & 1 & 0 \\ -1 & 0 & 0 \\ 0 & 0 & 1 \end{bmatrix}\begin{bmatrix} 1 & -K & 0 \\ 0 & 1 & 0 \\ 0 & 0 & 1 \end{bmatrix} \\ &= \begin{bmatrix} -K & 1+K^2 & 0 \\ -1 & K & 0 \\ 0 & 0 & 1 \end{bmatrix}.\end{aligned} \tag{5.76}$$

Now, for $\widehat{\mathbf{H}}^{(2)}$ to be a symmetry transformation of (5.70)–(5.72), it must reproduce the same geometry shown in Fig. 5.3 after transforming the indicated material vector elements. Transforming the vectors $\mathbf{a} = (1, 0, 0)$ and $\mathbf{b} = (K, 1, 0)$, we obtain

$$\widehat{\mathbf{H}}^{(2)} \cdot \mathbf{a} = \begin{bmatrix} -K & 1+K^2 & 0 \\ -1 & K & 0 \\ 0 & 0 & 1 \end{bmatrix}\begin{bmatrix} 1 \\ 0 \\ 0 \end{bmatrix} = -\begin{bmatrix} K \\ 1 \\ 0 \end{bmatrix} = -\mathbf{b}, \tag{5.77}$$

$$\widehat{\mathbf{H}}^{(2)} \cdot \mathbf{b} = \begin{bmatrix} -K & 1+K^2 & 0 \\ -1 & K & 0 \\ 0 & 0 & 1 \end{bmatrix}\begin{bmatrix} K \\ 1 \\ 0 \end{bmatrix} = \begin{bmatrix} 1 \\ 0 \\ 0 \end{bmatrix} = \mathbf{a}, \tag{5.78}$$

thus obtaining the same geometry. Subsequently, $\widehat{\mathbf{H}}^{(2)}$ is a symmetry transformation for (5.70)–(5.72).

The constitutive equations can be specialized to specific classes of homogeneous materials, each class different in the invariance properties assigned to the constitutive functional. According to the admitted symmetry group, i.e., according to the set of all second-order invertible tensors $\mathbf{H}$ for which

$$\underset{0 \le s < \infty}{\mathfrak{F}_{\kappa}} \{\mathbf{F}^{(t)}, \theta^{(t)}, \mathbf{G}^{(t)}\} = \underset{0 \le s < \infty}{\mathfrak{F}_{\kappa}} \{\mathbf{F}^{(t)} \cdot \mathbf{H}, \theta^{(t)}, \mathbf{H}^T \cdot \mathbf{G}^{(t)}\} \quad \text{for all} \quad (\mathbf{F}^{(t)}, \theta^{(t)}, \mathbf{G}^{(t)}), \tag{5.79}$$

we distinguish the different *classes of materials*. In the sequel, $\mathscr{G}_{\kappa}$ will denote the corresponding class of materials. This classification is important since it leads to mathematical definitions for various types of real materials, and once $\mathscr{G}_{\kappa}$ is specified, i.e., the material symmetry is selected, particular restrictive conditions on the form of the constitutive functional result.

The choice of the type of simple material, along with the imposed material sym-

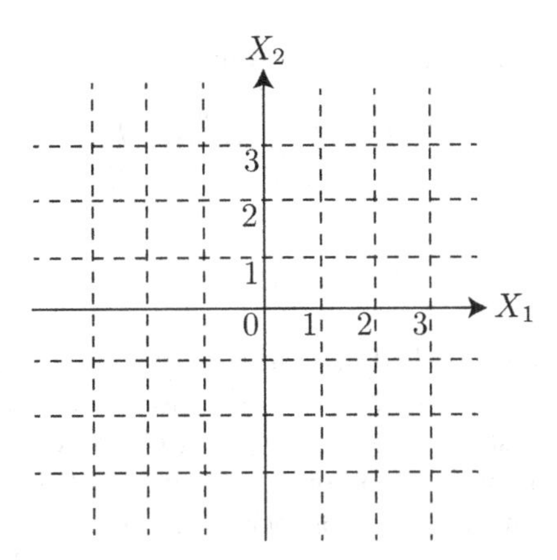

Figure 5.2: Simple material geometry with the X_3 axis pointing out of the page.

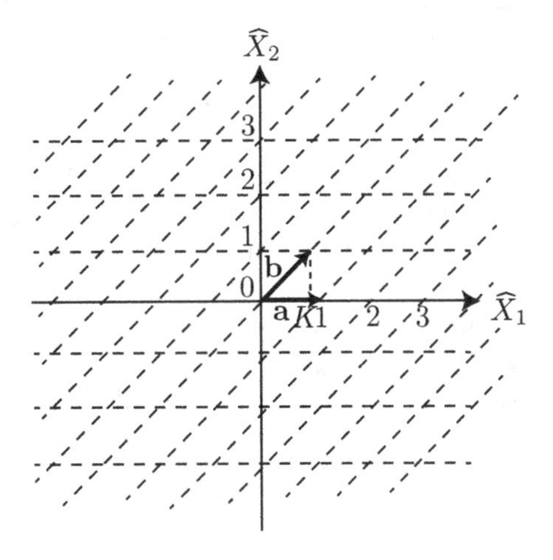

Figure 5.3: Transformed material geometry.

metry, leads to very useful representation theorems for the tensor T. It is then possible to specify the constitutive functional of the material by means of a certain number of independent scalar functions of the history of deformation, called the *response coefficients*, which characterize the different materials. These quantities are mainly to be observed in suitable experiments. The mathematical results sought by experimentalists to guide their design of tests for a practical evaluation of the response coefficients are mainly those concerning problems investigated without further specifications on the constitutive functional. These analytical results are known as the *universal solutions* and the *universal relations*. A *controllable solution* (one satisfying the balance laws) which is the same for all materials in a given class $\mathscr{G}_{\boldsymbol{\kappa}}$ is a universal solution. In correspondence with a given deformation or motion, a universal relation is an equation between, say, the stress components and the position vector components which holds for all $\mathbf{X}$ and t and which is the same for any material in an assigned class.

Definition: A material is called *solid* if there exists a reference configuration $\boldsymbol{\kappa}$

such that $\mathscr{G}_{\boldsymbol{\kappa}}$ is the full orthogonal group

$$\mathscr{O} = \{H \mid H^{-1} = H^T, \det H = \pm 1\}, \tag{5.80}$$

or a subgroup of it, i.e.,

$$\mathscr{G}_{\boldsymbol{\kappa}} \subseteq \mathscr{O}(\mathscr{V}). \tag{5.81}$$

Such a configuration is called an *undistorted configuration* for the solid.

In general, only certain particular reference configurations are undistorted for a solid body. For a distorted configuration of the solid, the symmetry group neither contains the orthogonal group nor is contained within it.

The orthogonal group $\mathscr{O}$ is a proper subgroup of the unimodular group $\mathscr{U}$, i.e., $\mathscr{O} \subset \mathscr{U}$. One can easily construct examples of unimodular linear transformations which are not orthogonal.

Definition: A material is called a *fluid* if for a reference configuration $\boldsymbol{\kappa}$, the symmetry group is the full unimodular group, i.e.,

$$\mathscr{G}_{\boldsymbol{\kappa}} = \mathscr{U}(\mathscr{V}). \tag{5.82}$$

For a fluid, Noll's rule implies that $\mathscr{G}_{\widehat{\boldsymbol{\kappa}}} = \mathbf{P} \cdot \mathscr{U} \cdot \mathbf{P}^{-1} = \mathscr{U} = \mathscr{G}_{\boldsymbol{\kappa}}$ for all $\boldsymbol{\kappa}$ and $\widehat{\boldsymbol{\kappa}}$. Therefore, a fluid has the same symmetry group with respect to any configuration. In other words, a fluid does not have a preferred configuration.

Definition: A material that is neither a fluid nor a solid is called a *fluid crystal*. In other words, for a fluid crystal, there does not exist a reference configuration $\boldsymbol{\kappa}$ for which either $\mathscr{G}_{\boldsymbol{\kappa}} \subseteq \mathscr{O}(\mathscr{V})$ or $\mathscr{G}_{\boldsymbol{\kappa}} = \mathscr{U}(\mathscr{V})$.

Definition: A material is called *hemitropic* if there exists a reference configuration $\boldsymbol{\kappa}$ such that $\mathscr{G}_{\boldsymbol{\kappa}} = \mathscr{O}^+(\mathscr{V})$, where $\mathscr{O}^+$ is the proper orthogonal group, i.e.,

$$\mathscr{O}^+ = \{H \mid H^{-1} = H^T, \det H = 1\}. \tag{5.83}$$

Definition: A material is called *isotropic* if there exists a configuration $\boldsymbol{\kappa}$ such that

$$\mathscr{G}_{\boldsymbol{\kappa}} = \mathscr{O}(\mathscr{V}) \qquad \text{or} \qquad \mathscr{G}_{\boldsymbol{\kappa}} = \mathscr{U}(\mathscr{V}). \tag{5.84}$$

Such a configuration is called *undistorted* for the isotropic material body, or simply an *isotropic* configuration.

The unimodular, orthogonal, and proper orthogonal groups are clearly all infinite groups.

It can be readily seen that the only isotropic materials are isotropic solids, $\mathscr{G}_{\boldsymbol{\kappa}} = \mathscr{O}(\mathscr{V})$, and fluids, $\mathscr{G}_{\boldsymbol{\kappa}} = \mathscr{U}(\mathscr{V})$. Any other materials are *anisotropic*. Anisotropic materials include anisotropic solids, $\mathscr{G}_{\boldsymbol{\kappa}} \subset \mathscr{O}(\mathscr{V})$, and fluid crystals, $\mathscr{O}(\mathscr{V}) \subset \mathscr{G}_{\boldsymbol{\kappa}} \subset \mathscr{U}(\mathscr{V})$. The relationship between the different materials is illustrated in Fig. 5.4.

5.7 Reduced constitutive equations

The frame-indifference requirement allows a reduction of the form of the constitutive equations to a simpler form. In the discussion of this section, we consider

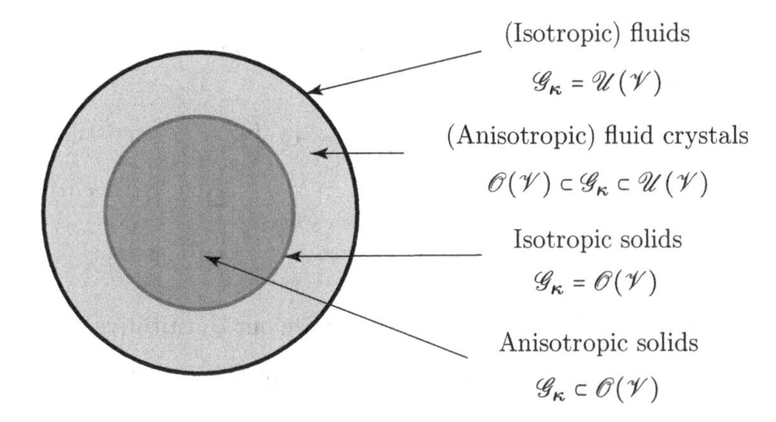

Figure 5.4: Illustration of the classification of material symmetry groups.

simple homogeneous materials. In addition, dependencies of the constitutive functional on $\mathbf{X}$ and s are understood; therefore, we shall omit them from the argument list and, for simplicity, rewrite (5.33) as

$$T = \underset{0\le s<\infty}{\mathfrak{F}} \{\mathbf{F}^{(t)}, \theta^{(t)}, \mathbf{G}^{(t)}\}. \tag{5.85}$$

We note that $\theta^{(t)}$ and $\mathbf{G}^{(t)}$ impose no restrictions on the constitutive functional since they are not affected by a Euclidean transformation. We now examine the restrictions imposed by the functional dependence on $\mathbf{F}^{(t)}$. In particular, we take T to represent a second-order tensor, so that objectivity requires that

$$\underset{0\le s<\infty}{\mathfrak{F}} \{\mathbf{Q}^{(t)} \cdot \mathbf{F}^{(t)}, \theta^{(t)}, \mathbf{G}^{(t)}\} = \mathbf{Q} \cdot \underset{0\le s<\infty}{\mathfrak{F}} \{\mathbf{F}^{(t)}, \theta^{(t)}, \mathbf{G}^{(t)}\} \cdot \mathbf{Q}^T \quad \text{for all} \quad \mathbf{Q} \in \mathscr{O}(\mathscr{V}), \tag{5.86}$$

and material symmetry requires that

$$\underset{0\le s<\infty}{\mathfrak{F}} \{\mathbf{F}^{(t)}, \theta^{(t)}, \mathbf{G}^{(t)}\} = \underset{0\le s<\infty}{\mathfrak{F}} \{\mathbf{F}^{(t)} \cdot \mathbf{H}, \theta^{(t)}, \mathbf{H}^T \cdot \mathbf{G}^{(t)}\} \quad \text{for all} \quad \mathbf{H} \in \mathscr{G}_{\kappa}(T). \tag{5.87}$$

For ease of discussion, we shall temporarily omit the arguments that are not affected by the frame-indifference transformation and rewrite (5.86) as

$$\underset{0\le s<\infty}{\mathfrak{F}} \{\mathbf{Q}^{(t)} \cdot \mathbf{F}^{(t)}\} = \mathbf{Q} \cdot \underset{0\le s<\infty}{\mathfrak{F}} \{\mathbf{F}^{(t)}\} \cdot \mathbf{Q}^T. \tag{5.88}$$

Now using the polar decomposition $\mathbf{F}^{(t)} = \mathbf{R}^{(t)} \cdot \mathbf{U}^{(t)}$ on the left-hand side, we have

$$\underset{0\le s<\infty}{\mathfrak{F}} \{\mathbf{Q}^{(t)} \cdot \mathbf{R}^{(t)} \cdot \mathbf{U}^{(t)}\} = \mathbf{Q} \cdot \underset{0\le s<\infty}{\mathfrak{F}} \{\mathbf{F}^{(t)}\} \cdot \mathbf{Q}^T, \tag{5.89}$$

where $\mathbf{R} \in \mathscr{O}(\mathscr{V})$. If we choose $\mathbf{Q}^{(t)} = \mathbf{R}^{(t)^T}$, then we obtain

$$\underset{0\le s<\infty}{\mathfrak{F}} \{\mathbf{U}^{(t)}\} = \mathbf{R}^T \cdot \underset{0\le s<\infty}{\mathfrak{F}} \{\mathbf{F}^{(t)}\} \cdot \mathbf{R}. \tag{5.90}$$

Conversely, if we originally take

$$\underset{0\le s<\infty}{\mathfrak{F}} \{\mathbf{F}^{(t)}\} = \mathbf{R} \cdot \underset{0\le s<\infty}{\mathfrak{F}} \{\mathbf{U}^{(t)}\} \cdot \mathbf{R}^T, \tag{5.91}$$

then the transformed quantity is seen to be objective since, if we take $\mathbf{R}^{(t)} = \mathbf{Q}^{(t)^T}$, we have

$$\begin{aligned}\mathbf{Q}\cdot \underset{0\le s<\infty}{\mathfrak{F}}\{\mathbf{F}^{(t)}\}\cdot\mathbf{Q}^T &= \mathbf{Q}\cdot\left(\mathbf{R}\cdot \underset{0\le s<\infty}{\mathfrak{F}}\{\mathbf{U}^{(t)}\}\cdot\mathbf{R}^T\right)\cdot\mathbf{Q}^T\\ &= \underset{0\le s<\infty}{\mathfrak{F}}\{\mathbf{Q}^{(t)}\cdot\mathbf{R}^{(t)}\cdot\mathbf{U}^{(t)}\}\\ &= \underset{0\le s<\infty}{\mathfrak{F}}\{\mathbf{Q}^{(t)}\cdot\mathbf{F}^{(t)}\}.\end{aligned}$$

As a result of the above, we can rewrite our constitutive functional in the form

$$T(\mathbf{X},t) = \mathbf{R}(t)\cdot \underset{0\le s<\infty}{\mathfrak{F}}\{\mathbf{U}^{(t)}(\mathbf{X},s),\theta^{(t)}(\mathbf{X},s),\mathbf{G}^{(t)}(\mathbf{X},s)\}\cdot\mathbf{R}^T(t). \tag{5.92}$$

This equation represents the general solution to the frame-indifference restriction requirement. Note that the stress at time t is only affected by the rotation at time t, $\mathbf{R}(t)$, and not by the history of rotation, $\mathbf{R}^{(t)}(s)$ for $s>0$.

The representation (5.92) is not convenient since for practical applications the polar decomposition of $\mathbf{F}$ would have to be worked out. To derive a more practical representation, we now note that, using once again the polar decomposition $\mathbf{F} = \mathbf{R}\cdot\mathbf{U}$, we can rewrite the constitutive function in the form

$$\begin{aligned}\mathbf{F}^T\cdot \underset{0\le s<\infty}{\mathfrak{F}}\{\mathbf{F}^{(t)}\}\cdot\mathbf{F} &= \mathbf{U}^T\cdot\mathbf{R}^T\cdot\left(\mathbf{R}\cdot \underset{0\le s<\infty}{\mathfrak{F}}\{\mathbf{U}^{(t)}\}\cdot\mathbf{R}^T\right)\cdot\mathbf{R}\cdot\mathbf{U}\\ &= \mathbf{U}\cdot \underset{0\le s<\infty}{\mathfrak{F}}\{\mathbf{U}^{(t)}\}\cdot\mathbf{U}.\end{aligned}$$

Subsequently, since $\mathbf{C} = \mathbf{U}^2$, we can write

$$\underset{0\le s<\infty}{\mathfrak{G}}\{\mathbf{C}^{(t)}\} \equiv \mathbf{U}\cdot \underset{0\le s<\infty}{\mathfrak{F}}\{\mathbf{U}^{(t)}\}\cdot\mathbf{U} = \mathbf{F}^T\cdot \underset{0\le s<\infty}{\mathfrak{F}}\{\mathbf{F}^{(t)}\}\cdot\mathbf{F}. \tag{5.93}$$

Thus we finally arrive at the more convenient result that the constitutive functional for a second-order tensor quantity is objective if and only if $\mathbf{Q}^{(t)} = \mathbf{R}^{(t)^T}$ and

$$\overline{T}(\mathbf{X},t) = \underset{0\le s<\infty}{\mathfrak{G}}\{\mathbf{C}^{(t)}(\mathbf{X},s),\theta^{(t)}(\mathbf{X},s),\mathbf{G}^{(t)}(\mathbf{X},s)\}, \tag{5.94}$$

where

$$\overline{T} \equiv \mathbf{F}^T\cdot T\cdot\mathbf{F} \tag{5.95}$$

is called the *convected tensor*. Constitutive functionals of this form, which are not subject to any further restrictions from the objectivity condition, are said to be in *reduced form.*

We shall next develop a form of the constitutive equation in which the reference configuration is taken to be the current configuration. This form is most useful for the discussion of fluids. First, we recall, upon using (3.288), (5.4), and (5.28), that

$$\begin{aligned}\mathbf{C}^{(t)}(\mathbf{X},s) &= \mathbf{C}(\mathbf{X},\tau) = \mathbf{F}^T(\mathbf{X},t)\cdot{}_{(t)}\mathbf{C}(\mathbf{x},\tau)\cdot\mathbf{F}(\mathbf{X},t)\\ &= \mathbf{F}^T(\mathbf{X},t)\cdot{}_{(t)}\mathbf{C}^{(t)}(\mathbf{x},s)\cdot\mathbf{F}(\mathbf{X},t), \qquad (5.96)\\ \mathbf{G}^{(t)}(\mathbf{X},s) &= \mathbf{G}(\mathbf{X},\tau) = \mathbf{F}^T(\mathbf{X},t)\cdot{}_{(t)}\mathbf{g}(\mathbf{x},\tau) = \mathbf{F}^T(\mathbf{X},t)\cdot{}_{(t)}\mathbf{g}^{(t)}(\mathbf{x},s), \quad (5.97)\end{aligned}$$

where we have taken $\tau = t - s$, and note that

$$ {}_{(t)}\mathbf{C}(\mathbf{x},t) = \mathbf{1} \qquad \text{and} \qquad {}_{(t)}\mathbf{g}(\mathbf{x},t) = \mathbf{g}(\mathbf{x},t) = \text{grad}\ \theta(\mathbf{x},t). \tag{5.98} $$

Subsequently, from (5.94) and (5.95), the constitutive functional becomes

$$ \begin{aligned} T(\mathbf{X},t) &= \left(\mathbf{F}^T(\mathbf{X},t)\right)^{-1} \cdot \underset{0\leq s<\infty}{\mathfrak{S}} \{\mathbf{F}^T(\mathbf{X},t)\cdot {}_{(t)}\mathbf{C}^{(t)}(\mathbf{x},s)\cdot \mathbf{F}(\mathbf{X},t), \theta^{(t)}(\mathbf{X},s), \\ &\qquad \mathbf{F}^T(\mathbf{X},t)\cdot {}_{(t)}\mathbf{g}^{(t)}(\mathbf{x},s)\}\cdot \mathbf{F}^{-1}(\mathbf{X},t), \end{aligned} $$

or, with a little abuse of notation, we have

$$ T(\mathbf{x},t) = \underset{0\leq s<\infty}{\mathfrak{F}} \{\mathbf{F}(\mathbf{x},t), {}_{(t)}\mathbf{C}^{(t)}(\mathbf{x},s), \theta^{(t)}(\mathbf{x},s), {}_{(t)}\mathbf{g}^{(t)}(\mathbf{x},s)\}. \tag{5.99} $$

Now, assuming that T is a second-order tensor, for frame indifference we require that

$$ \underset{0\leq s<\infty}{\mathfrak{F}} \{\overline{\mathbf{F}}, {}_{(t)}\overline{\mathbf{C}}^{(t)}, \overline{\theta}^{(t)}, {}_{(t)}\overline{\mathbf{g}}^{(t)}\} = \mathbf{Q}\cdot \underset{0\leq s<\infty}{\mathfrak{F}} \{\mathbf{F}, {}_{(t)}\mathbf{C}^{(t)}, \theta^{(t)}, {}_{(t)}\mathbf{g}^{(t)}\}\cdot \mathbf{Q}^T. \tag{5.100} $$

Since

$$ {}_{(t)}\overline{\mathbf{C}}^{(t)} = {}_{(t)}\overline{\mathbf{F}}^{(t)T}\cdot {}_{(t)}\overline{\mathbf{F}}^{(t)} \qquad \text{and} \qquad {}_{(t)}\overline{\mathbf{g}}^{(t)} = \left(\overline{\mathbf{F}}^T\right)^{-1}\cdot \overline{\mathbf{G}}^{(t)}, \tag{5.101} $$

and since $\overline{\theta}^{(t)} = \theta^{(t)}$ and $\overline{\mathbf{G}}^{(t)} = \mathbf{G}^{(t)}$, if we take $\overline{\mathbf{F}}^{(t)} = \mathbf{Q}^{(t)}\cdot \mathbf{F}^{(t)}$ and use (3.275), we have

$$ \begin{aligned} {}_{(t)}\overline{\mathbf{F}}^{(t)} &= \mathbf{Q}^{(t)}\cdot \mathbf{F}^{(t)}\cdot \mathbf{F}^{-1}\cdot \mathbf{Q}^T = \mathbf{Q}^{(t)}\cdot {}_{(t)}\mathbf{F}^{(t)}\cdot \mathbf{Q}^T, \qquad (5.102) \\ {}_{(t)}\overline{\mathbf{C}}^{(t)} &= \left(\mathbf{Q}^{(t)}\cdot {}_{(t)}\mathbf{F}^{(t)}\cdot \mathbf{Q}^T\right)^T\cdot \left(\mathbf{Q}^{(t)}\cdot {}_{(t)}\mathbf{F}^{(t)}\cdot \mathbf{Q}^T\right) \\ &= \mathbf{Q}\cdot {}_{(t)}\mathbf{F}^{(t)T}\cdot \mathbf{Q}^{(t)T}\cdot \mathbf{Q}^{(t)}\cdot {}_{(t)}\mathbf{F}^{(t)}\cdot \mathbf{Q}^T \\ &= \mathbf{Q}\cdot {}_{(t)}\mathbf{C}^{(t)}\cdot \mathbf{Q}^T, \qquad (5.103) \\ {}_{(t)}\overline{\mathbf{g}}^{(t)} &= \mathbf{Q}\cdot \left(\mathbf{F}^T\right)^{-1}\cdot \mathbf{G}^{(t)} = \mathbf{Q}\cdot {}_{(t)}\mathbf{g}^{(t)}. \qquad (5.104) \end{aligned} $$

Subsequently, the frame indifference condition becomes

$$ \begin{aligned} &\underset{0\leq s<\infty}{\mathfrak{F}} \{\mathbf{Q}\cdot\mathbf{F}, \mathbf{Q}\cdot {}_{(t)}\mathbf{C}^{(t)}\cdot \mathbf{Q}^T, \theta^{(t)}, \mathbf{Q}\cdot {}_{(t)}\mathbf{g}^{(t)}\} = \\ &\qquad \mathbf{Q}\cdot \underset{0\leq s<\infty}{\mathfrak{F}} \{\mathbf{F}, {}_{(t)}\mathbf{C}^{(t)}, \theta^{(t)}, {}_{(t)}\mathbf{g}^{(t)}\}\cdot \mathbf{Q}^T \quad \text{for all} \quad \mathbf{Q} = \mathbf{R}^T \in \mathscr{O}(\mathscr{V}). \end{aligned} \tag{5.105} $$

For material symmetry, with T being an arbitrary tensor, we have

$$ \underset{0\leq s<\infty}{\mathfrak{F}} \{\mathbf{F}, {}_{(t)}\mathbf{C}^{(t)}, \theta^{(t)}, {}_{(t)}\mathbf{g}^{(t)}\} = \underset{0\leq s<\infty}{\mathfrak{F}} \{\overline{\mathbf{F}}, {}_{(t)}\overline{\mathbf{C}}^{(t)}, \theta^{(t)}, {}_{(t)}\overline{\mathbf{g}}^{(t)}\}, \tag{5.106} $$

for $\mathbf{H} \in \mathscr{G}_{\boldsymbol{\kappa}}(T)$. We now note that if we take $\overline{\mathbf{F}}^{(t)} = \mathbf{F}^{(t)}\cdot \mathbf{H}$ and $\overline{\mathbf{G}}^{(t)} = \mathbf{H}^T\cdot \mathbf{G}^{(t)}$, and use (3.275) and (5.101), we have

$$ \begin{aligned} {}_{(t)}\overline{\mathbf{F}}^{(t)} &= \mathbf{F}^{(t)}\cdot \mathbf{H}\cdot \mathbf{H}^{-1}\cdot \mathbf{F}^{-1} = \mathbf{F}^{(t)}\cdot \mathbf{F}^{-1} = {}_{(t)}\mathbf{F}^{(t)}, \\ {}_{(t)}\overline{\mathbf{C}}^{(t)} &= {}_{(t)}\mathbf{C}^{(t)}, \\ {}_{(t)}\overline{\mathbf{g}}^{(t)} &= \left(\mathbf{F}^T\right)^{-1}\cdot \left(\mathbf{H}^T\right)^{-1}\cdot \mathbf{H}^T\cdot \mathbf{G}^{(t)} = \left(\mathbf{F}^T\right)^{-1}\cdot \mathbf{G}^{(t)} = {}_{(t)}\mathbf{g}^{(t)}. \end{aligned} $$

Subsequently, the symmetry condition becomes

$$\underset{0\le s<\infty}{\mathfrak{F}}\{\mathbf{F}, {}_{(t)}\mathbf{C}^{(t)}, \theta^{(t)}, {}_{(t)}\mathbf{g}^{(t)}\} = \underset{0\le s<\infty}{\mathfrak{F}}\{\mathbf{F}\cdot\mathbf{H}, {}_{(t)}\mathbf{C}^{(t)}, \theta^{(t)}, {}_{(t)}\mathbf{g}^{(t)}\}$$
$$\text{for all}\quad \mathbf{H}\in\mathscr{G}_{\boldsymbol{\kappa}}(T). \tag{5.107}$$

5.7.1 Constitutive equation for a simple isotropic solid

Different symmetry groups correspond to solids having different material symmetries. For a simple *isotropic solid*, $\mathscr{G}_{\boldsymbol{\kappa}}(T) = \mathscr{O}(\mathscr{V})$, so that by using the polar decomposition $\mathbf{F} = \mathbf{R}\cdot\mathbf{U}$ and taking $\mathbf{H} = \mathbf{R}^T \in \mathscr{O}(\mathscr{V})$, the symmetry condition (5.107), using the current reference configuration, becomes

$$\underset{0\le s<\infty}{\mathfrak{F}}\{\mathbf{F}, {}_{(t)}\mathbf{C}^{(t)}, \theta^{(t)}, {}_{(t)}\mathbf{g}^{(t)}\} = \underset{0\le s<\infty}{\mathfrak{F}}\{\mathbf{R}\cdot\mathbf{U}\cdot\mathbf{R}^T, {}_{(t)}\mathbf{C}^{(t)}, \theta^{(t)}, {}_{(t)}\mathbf{g}^{(t)}\}. \tag{5.108}$$

Now since $\mathbf{V} = \mathbf{R}\cdot\mathbf{U}\cdot\mathbf{R}^T$ and $\mathbf{B} = \mathbf{V}^2$, with a little abuse of notation, we can rewrite (5.99) as

$$T(\mathbf{x}, t) = \underset{0\le s<\infty}{\mathfrak{F}}\{\mathbf{B}, {}_{(t)}\mathbf{C}^{(t)}, \theta^{(t)}, {}_{(t)}\mathbf{g}^{(t)}\}, \tag{5.109}$$

which is required to satisfy the appropriate invariance condition depending on the order of tensor T. For a second-order tensor, using (5.103) and (5.104), such condition is

$$\underset{0\le s<\infty}{\mathfrak{F}}\{\mathbf{Q}\cdot\mathbf{B}\cdot\mathbf{Q}^T, \mathbf{Q}\cdot{}_{(t)}\mathbf{C}^{(t)}\cdot\mathbf{Q}^T, \theta^{(t)}, \mathbf{Q}\cdot{}_{(t)}\mathbf{g}^{(t)}\} =$$
$$\mathbf{Q}\cdot\underset{0\le s<\infty}{\mathfrak{F}}\{\mathbf{B}, {}_{(t)}\mathbf{C}^{(t)}, \theta^{(t)}, {}_{(t)}\mathbf{g}^{(t)}\}\cdot\mathbf{Q}^T, \tag{5.110}$$

for all $\mathbf{Q} = \mathbf{R}^T \in \mathscr{O}(\mathscr{V})$. The above equation corresponds to the reduced form of the constitutive function for a simple isotropic solid. From this representation, we see that in the current configuration the constitutive dependence on the deformation gradient $\mathbf{F}$ reduces to the dependence on the left Cauchy–Green tensor $\mathbf{B}$. It is noted that (5.109) remains valid for hemitropic solids as long as now $\mathbf{H} = \mathbf{R}^T \in \mathscr{O}^+(\mathscr{V})$.

5.7.2 Constitutive equation for a simple (isotropic) fluid

Now, from the requirement that the mass density remain the same, we must have that $\det F = \det(FH)$. Subsequently, assume that $H = aA$, where $a > 0$ is a scalar and A is a matrix. Now since $\det(aA) = a^3 \det A$, the above requirement becomes $\det F = \det F\ a^3 \det A$. Thus we see that if we take $A = F^{-1}$ and $a = (\det F)^{1/3}$, the requirement is satisfied identically, and the transformation

$$\mathbf{H} = (\det F)^{1/3}\mathbf{F}^{-1} \tag{5.111}$$

is such that $\mathscr{G}_{\boldsymbol{\kappa}} = \mathscr{U}(\mathscr{V})$. Subsequently, using (5.107), for a simple fluid the above symmetry condition becomes

$$\underset{0\le s<\infty}{\mathfrak{F}}\{\mathbf{F}, {}_{(t)}\mathbf{C}^{(t)}, \theta^{(t)}, {}_{(t)}\mathbf{g}^{(t)}\} = \underset{0\le s<\infty}{\mathfrak{F}}\{(\det F)^{1/3}\mathbf{1}, {}_{(t)}\mathbf{C}^{(t)}, \theta^{(t)}, {}_{(t)}\mathbf{g}^{(t)}\} \tag{5.112}$$

or, since $\det F = J = \rho_R/\rho$, with abuse of notation, we have

$$\underset{0\le s<\infty}{\mathfrak{F}}\{\mathbf{F}, {}_{(t)}\mathbf{C}^{(t)}, \theta^{(t)}, {}_{(t)}\mathbf{g}^{(t)}\} = \underset{0\le s<\infty}{\mathfrak{F}}\{\rho, {}_{(t)}\mathbf{C}^{(t)}, \theta^{(t)}, {}_{(t)}\mathbf{g}^{(t)}\}. \tag{5.113}$$

Thus, the constitutive functional (5.99) reduces to

$$T(\mathbf{x},t) = \underset{0\le s<\infty}{\mathfrak{F}}\{\rho, {}_{(t)}\mathbf{C}^{(t)}, \theta^{(t)}, {}_{(t)}\mathbf{g}^{(t)}\}, \tag{5.114}$$

which is required to satisfy the appropriate invariance condition depending on the order of tensor T. For a second-order tensor, using (5.103) and (5.104), such condition is

$$\begin{aligned}\underset{0\le s<\infty}{\mathfrak{F}}\{\rho, \mathbf{Q}\cdot{}_{(t)}\mathbf{C}^{(t)}\cdot\mathbf{Q}^T, \theta^{(t)}, \mathbf{Q}\cdot{}_{(t)}\mathbf{g}^{(t)}\} &= \\ \mathbf{Q}\cdot\underset{0\le s<\infty}{\mathfrak{F}}\{\rho, {}_{(t)}\mathbf{C}^{(t)}, \theta^{(t)}, {}_{(t)}\mathbf{g}^{(t)}\}\cdot\mathbf{Q}^T&,\end{aligned} \tag{5.115}$$

for all $\mathbf{Q} = \mathbf{R}^T \in \mathscr{O}(\mathscr{V})$. Equation (5.114) corresponds to the reduced form of the constitutive function for a simple fluid. From this representation, we see that the constitutive dependence on the deformation gradient $\mathbf{F}$ reduces to the dependence on its determinant only, or equivalently ρ.

5.8 Isotropic and hemitropic representations

Below we consider constitutive equations of different rates and grades as given by (5.37). Furthermore, we consider the representation of isotropic and hemitropic functions. Since representation of more general *non-isotropic functions* apply specifically to solids, we shall discuss this topic in Chapter 7, which deals with thermoelastic solids.

The main problem of invariant theory is the representation of tensor-valued functions. Let ϕ_α, $\mathbf{v}_\beta$, $\mathbf{A}_\gamma$, and $\mathbf{W}_\delta$ denote scalars, vectors, symmetric tensors, and skew-symmetric tensors, respectively. Let ψ, $\mathbf{h}$, and $\mathbf{T}$ be scalar-, vector-, and tensor-valued functions of ϕ_α, $\mathbf{v}_\beta$, $\mathbf{A}_\gamma$, and $\mathbf{W}_\delta$, respectively. An orthogonal tensor $\mathbf{Q}$ is said to be a *symmetry transformation* of the functions ψ, $\mathbf{h}$, and $\mathbf{T}$, respectively, if

$$\psi(\phi'_\alpha, \mathbf{v}'_\beta, \mathbf{A}'_\gamma, \mathbf{W}'_\delta) = \psi(\phi_\alpha, \mathbf{v}_\beta, \mathbf{A}_\gamma, \mathbf{W}_\delta), \tag{5.116}$$

$$\mathbf{h}(\phi'_\alpha, \mathbf{v}'_\beta, \mathbf{A}'_\gamma, \mathbf{W}'_\delta) = \mathbf{Q}\cdot\mathbf{h}(\phi_\alpha, \mathbf{v}_\beta, \mathbf{A}_\gamma, \mathbf{W}_\delta), \tag{5.117}$$

$$\mathbf{T}(\phi'_\alpha, \mathbf{v}'_\beta, \mathbf{A}'_\gamma, \mathbf{W}'_\delta) = \mathbf{Q}\cdot\mathbf{T}(\phi_\alpha, \mathbf{v}_\beta, \mathbf{A}_\gamma, \mathbf{W}_\delta)\cdot\mathbf{Q}^T, \tag{5.118}$$

where

$$\phi'_\alpha = \phi_\alpha, \tag{5.119}$$

$$\mathbf{v}'_\beta = \mathbf{Q}\cdot\mathbf{v}_\beta, \tag{5.120}$$

$$\mathbf{A}'_\gamma = \mathbf{Q}\cdot\mathbf{A}_\gamma\cdot\mathbf{Q}^T, \tag{5.121}$$

$$\mathbf{W}'_\delta = (\det Q)\,\mathbf{Q}\cdot\mathbf{W}_\delta\cdot\mathbf{Q}^T. \tag{5.122}$$

The skew-symmetric tensors $\mathbf{W}_\delta = \mathbf{w}_\delta \cdot \boldsymbol{\varepsilon}$ may be replaced by the axial vectors $\mathbf{w}_\delta = \frac{1}{2}\boldsymbol{\varepsilon} : \mathbf{W}_\delta \equiv \langle \mathbf{W}_\delta \rangle$, where the axial vectors transform according to

$$\mathbf{w}'_\delta = (\det Q)\,\mathbf{Q} \cdot \mathbf{w}_\delta, \tag{5.123}$$

and note that in general $\det Q = \pm 1$. Here we also note that the axial scalars (pseudoscalars) w_δ are frame indifferent if the following holds:

$$w'_\delta = (\det Q)\, w_\delta. \tag{5.124}$$

The *symmetry groups* of the functions ψ, $\mathbf{h}$, and $\mathbf{T}$ consist of the sets of all symmetry transformations of ψ, $\mathbf{h}$, and $\mathbf{T}$, respectively.

Definition: We say that ψ, $\mathbf{h}$, and $\mathbf{T}$ are scalar-, vector-, and tensor-valued *isotropic functions*, respectively, if for any number of scalars ϕ_α, vectors $\mathbf{v}_\beta$, symmetric tensors $\mathbf{A}_\gamma$, and skew-symmetric tensors $\mathbf{W}_\delta$, they are invariant with respect to all $\mathbf{Q} \in \mathscr{O}(\mathscr{V})$.

Definition: We say that ψ, $\mathbf{h}$, and $\mathbf{T}$ are scalar-, vector-, and tensor-valued *hemitropic functions*, respectively, if for any number of scalars ϕ_α, vectors $\mathbf{v}_\beta$, symmetric tensors $\mathbf{A}_\gamma$, and skew-symmetric tensors $\mathbf{W}_\delta$, they are invariant with respect to all $\mathbf{Q} \in \mathscr{O}^+(\mathscr{V})$.

A material property is isotropic if that property at a point is the same in all directions, and transversely isotropic if that property is the same in all directions in a plane. It is called *anisotropic* (or non-isotropic) otherwise. Isotropic functions are also called *isotropic invariants.* From the definition, the conditions impose no restrictions on the scalar variables which a function depends on. Hence, scalar variables are irrelevant as far as the representation of isotropic invariants are concerned. A scalar-valued *function basis* for isotropic or hemitropic functions of $\mathbf{v}_\beta$, $\mathbf{A}_\gamma$, and $\mathbf{W}_\delta$ consists of a set of isotropic or hemitropic scalar-valued functions $I_1, \ldots, I_n$ of $\mathbf{v}_\beta$, $\mathbf{A}_\gamma$, and $\mathbf{W}_\delta$ such that any isotropic or hemitropic scalar-valued function of $\mathbf{v}_\beta$, $\mathbf{A}_\gamma$, and $\mathbf{W}_\delta$ can be expressed as a single-valued function of $I_1, \ldots, I_n$, i.e.,

$$\psi = \psi(\phi_\alpha, \mathbf{v}_\beta, \mathbf{A}_\gamma, \mathbf{W}_\delta) = \psi(\phi_\alpha, I_1, \ldots, I_n). \tag{5.125}$$

Let $\mathbf{h}_0, \ldots, \mathbf{h}_R$ and $\mathbf{T}_0, \ldots, \mathbf{T}_S$ denote respectively vector- and tensor-valued isotropic or hemitropic functions of $\mathbf{v}_\beta$, $\mathbf{A}_\gamma$, and $\mathbf{W}_\delta$. Then, if any vector-valued and tensor-valued isotropic or hemitropic functions $\mathbf{h}(\phi_\alpha, \mathbf{v}_\beta, \mathbf{A}_\gamma, \mathbf{W}_\delta)$ and $\mathbf{T}(\phi_\alpha, \mathbf{v}_\beta, \mathbf{A}_\gamma, \mathbf{W}_\delta)$ may be expressed respectively in the forms

$$\mathbf{h} = \sum_{r=0}^{R} a_r \mathbf{h}_r \qquad \text{and} \qquad \mathbf{T} = \sum_{s=0}^{S} b_s \mathbf{T}_s, \tag{5.126}$$

where a_r and b_s are functions of ϕ_α and the isotropic or hemitropic scalar invariants of $\mathbf{v}_\beta$, $\mathbf{A}_\gamma$, and $\mathbf{W}_\delta$ (i.e., $I_1, \ldots, I_n$), we say that $\{\mathbf{h}_0, \ldots, \mathbf{h}_R\}$ and $\{\mathbf{T}_0, \ldots, \mathbf{T}_S\}$ are *vector* and *tensor generators* of $\mathbf{v}_\beta$, $\mathbf{A}_\gamma$, and $\mathbf{W}_\delta$.

The general representation for isotropic and hemitropic scalar-, vector-, and tensor-valued functions are given by Zheng (1994) and are reproduced in Tables 5.1–5.8. To clarify the nomenclature in the tables, we note that A and W are component matrices of the second-order tensors $\mathbf{A}$ and $\mathbf{W}$, respectively, AW is the component matrix of $\mathbf{A} \cdot \mathbf{W} = A_{ij} W_{jk}$, $\mathbf{u} \cdot A\mathbf{v} = u_i A_{ij} v_j$, etc.

As an illustration on the use of the tables, we examine the isotropic representation of $\mathbf{T} = \mathbf{T}(\mathbf{A})$, where $\mathbf{T}$ and $\mathbf{A}$ are symmetric second-order tensors, and it obeys the invariance condition

$$\mathbf{T}(\mathbf{Q}\cdot\mathbf{A}\cdot\mathbf{Q}^T) = \mathbf{Q}\cdot\mathbf{T}(\mathbf{A})\cdot\mathbf{Q}^T \tag{5.127}$$

for $\mathbf{Q} \in \mathscr{O}(\mathscr{V})$. Then its irreducible representation from Table 5.3 is given by

$$\mathbf{T} = b_0\mathbf{1} + b_1\mathbf{A} + b_2\mathbf{A}^2, \tag{5.128}$$

where b_0, b_1, and b_2 are functions of the scalar invariants $I_1 = \operatorname{tr} A$, $I_2 = \operatorname{tr} A^2$, and $I_3 = \operatorname{tr} A^3$ obtained from Table 5.1. The tensors $\mathbf{1}$, $\mathbf{A}$, and $\mathbf{A}^2$ are the generators of the tensor $\mathbf{T}$.

As a second illustration, the hemitropic representation of $\mathbf{h} = \mathbf{h}(\mathbf{v}, \mathbf{A})$, where $\mathbf{v}$ is a vector and $\mathbf{A}$ is a symmetric second-order tensor, that satisfies the frame-invariance condition

$$\mathbf{h}(\mathbf{Q}\cdot\mathbf{v}, \mathbf{Q}\cdot\mathbf{A}\cdot\mathbf{Q}^T) = \mathbf{Q}\cdot\mathbf{h}(\mathbf{v}, \mathbf{A}), \tag{5.129}$$

for $\mathbf{Q} \in \mathscr{O}^+(\mathscr{V})$, is obtained from Table 5.6:

$$\mathbf{h} = a_0\mathbf{v} + a_1 A\mathbf{v} + a_2\mathbf{v}\times A\mathbf{v}, \tag{5.130}$$

where a_0, a_1, and a_2 are functions of the scalar invariants $I_1, \ldots, I_7$ given by $\mathbf{v}\cdot\mathbf{v}$, tr A, tr A^2, tr A^3, $\mathbf{v}\cdot A\mathbf{v}$, $\mathbf{v}\cdot A^2\mathbf{v}$, and the scalar triple product $[\mathbf{v}, A\mathbf{v}, A^2\mathbf{v}]$ obtained from Table 5.5.

Appendix E provides details illustrating a procedure for obtaining representations of isotropic scalar, vector, and symmetric second-order tensor functions of a vector and a symmetric second-order tensor. While such a procedure can be expanded to obtain more general results, the results presented in the tables have been obtained by a more systematic procedure.

5.9 Expansions of constitutive equations

Let us assume that we have a constitutive function for an arbitrary rank tensor quantity T that is a function of scalar fields ϕ_α, vector fields $\mathbf{v}_\beta$, rank-2 tensor fields $\mathbf{A}_\gamma$, etc.:

$$T = T\left(\phi_\alpha, \mathbf{v}_\beta, \mathbf{A}_\gamma, \ldots\right). \tag{5.131}$$

Let us also assume that at $(\phi_\alpha^0, \mathbf{v}_\beta^0, \mathbf{A}_\gamma^0, \ldots)$ the value of the function T is defined and derivatives of T to all orders exist at this point. Then, in the neighborhood of such point, we can write the Taylor series expansion

$$T\left(\phi_\alpha, \mathbf{v}_\beta, \mathbf{A}_\gamma, \ldots\right) = \sum_{n=0}^{\infty}\frac{1}{n!}\left\{\left(\phi_\alpha - \phi_\alpha^0\right)\frac{\partial}{\partial\phi_\alpha} + \left(\mathbf{v}_\beta - \mathbf{v}_\beta^0\right)\cdot\frac{\partial}{\partial\mathbf{v}_\beta} + \left(\mathbf{A}_\gamma - \mathbf{A}_\gamma^0\right):\frac{\partial}{\partial\mathbf{A}_\gamma} + \cdots\right\}^n T^0, \tag{5.132}$$

where $T^0 \equiv T\left(\phi_\alpha^0, \mathbf{v}_\beta^0, \mathbf{A}_\gamma^0, \ldots\right)$ signifies that after the function is operated on, the resulting terms are evaluated at the reference values of $\phi_\alpha = \phi_\alpha^0$, $\mathbf{v}_\beta = \mathbf{v}_\beta^0$, $\mathbf{A}_\gamma = \mathbf{A}_\gamma^0$, etc., and the resulting terms are understood as

$$\left.\frac{\partial T^0}{\partial \phi_\alpha}\right|_{\phi'_\alpha,\mathbf{v}_\beta,\mathbf{A}_\gamma}, \quad \left.\frac{\partial T^0}{\partial \mathbf{v}_\beta}\right|_{\phi_\alpha,\mathbf{v}'_\beta,\mathbf{A}_\gamma}, \quad \left.\frac{\partial T^0}{\partial \mathbf{A}_\gamma}\right|_{\phi_\alpha,\mathbf{v}_\beta,\mathbf{A}'_\gamma}, \quad \ldots,$$

which correspond to tensors of appropriate orders, and ϕ'_α indicates that all ϕ_1, ϕ_2, ... are kept constant with the exception of ϕ_α; the meaning for the other variables is analogous. It is noted that typically the reference state corresponds to one of thermodynamic equilibrium (see below), in which case the above terms are related to equilibrium or near-equilibrium tensor properties.

5.10 Thermodynamic considerations

A system is a region containing matter and energy that is separated from its surroundings by arbitrarily imposed walls or boundaries. In a thermodynamic analysis, the system is the subject of the investigation. A boundary is a closed surface surrounding the system through which mass and/or energy may enter or leave the system. Everything external to the system is the surroundings.

If no mass can cross the complete boundary of a system (but work and heat can), then we have a *closed system*. If in addition energy does not cross the complete boundary, then we have a mechanically and thermally *isolated system*. An *open system* is one in which both mass and energy can cross the system's boundary, and such boundary is a nonmaterial surface.

The condition of the system at any instant of time is called its state. The state at a given instant of time is described by the properties of the system.

5.10.1 Thermodynamic states

Definition: A *thermodynamic state* corresponds to a set of values of property variables of a system that must be specified to reproduce the system uniquely. The number of values required to specify the state depends on the system. Once a sufficient number has been specified, the values of all other variables are uniquely determined.

Property variables are classified as being extensive or intensive. *Extensive properties* are additive and thus depend on the amount of matter in the system (e.g., internal energy, volume). *Intensive properties* are independent of the amount of matter in the system (e.g., temperature, pressure).

We make a special note that, quite often, in a homogeneous system we write an extensive property, say Φ, as $\Phi = m\,\phi$, where m is the mass of the system, thus referring to ϕ as a property density or specific property. For example, we write the internal energy as $m\,e$, where e is the internal energy density or specific internal energy. Similarly, we write the extensive properties of volume V as $m\,\mathsf{v}$, where v is the volume density or specific volume, and number of moles N_i of a chemical component i as $m\,\mathsf{n}_i$, where n_i is the mole number density of component i. Note that $\mathsf{y}_i = \mathsf{M}_i\,\mathsf{n}_i$ is the mass fraction of component i, where M_i is the molecular mass of the component. Alternately, specific quantities are referred to one mole rather than to a unit of mass of the system. In such case, $\Phi = \mathsf{N}\,\overline{\phi}$, where $\mathsf{N} = \sum_i \mathsf{N}_i$ denotes the total number of moles of matter in the system and $\overline{\phi}$ is referred as a molar density or specific molar quantity. For example, corresponding to the number of moles N_i of chemical component i, one writes $\mathsf{N}\,\bar{\mathsf{n}}_i$, where $\bar{\mathsf{n}}_i$ is the mole fraction of component i. Property densities also do not depend on the system size. But such quantities should not be confused with intensive quantities. In applying a physical statement, it is always best to recall the extensive quantity Φ before applying the statement.

It is postulated that there exists an extensive quantity, called entropy, that is a function of a number of extensive quantities that describe any composite system, which is defined for all states and has the following properties. The values assumed by the extensive variables in the absence of an internal constraint are those that maximize the entropy at an *equilibrium state.* Furthermore, it is assumed that entropy is continuous, is differentiable, and is a monotonically increasing function of internal energy.

Writing the above quantities per unit mass, we assume that the specific entropy η is a function of the set of specific state variables $(e, \boldsymbol{\nu}_\alpha)$, with $\alpha = 1, \ldots, n$, and these quantities specify a thermodynamic state. In addition, we postulate that such state characterizes completely the entropy density η of a material point X and occupying coordinate $\mathbf{x}$ at time t. The choice of $\boldsymbol{\nu}_\alpha$, which are in general tensors of different ranks, and which we will generically call *specific thermostatic volumes*, depend on the system under consideration and define the thermodynamic character of the system. For example, for a single-component simple system, $n = 2$, $\boldsymbol{\nu}_1 \to \mathsf{v}$, and $\boldsymbol{\nu}_2 \to \mathsf{n}$. Here we recognize $\boldsymbol{\nu}_1$ as the *specific volume*, $\mathsf{v} = 1/\rho$, and $\boldsymbol{\nu}_2$ as the *mole number density*, n (note that $\mathsf{y} = \mathsf{M}\,\mathsf{n} = 1$ and $\bar{\mathsf{n}} = 1$ in this case). More generally, and as will become clear later, $\boldsymbol{\nu}_\alpha$ are related to $\mathbf{F}$, $\mathbf{C}$, and $\mathbf{g}$ for a simple solid (see (5.99)), and ρ, $\mathbf{C}$, and $\mathbf{g}$ for a simple fluid (see (5.114)). Once e and $\boldsymbol{\nu}_\alpha$ are selected, we will have the thermodynamic state of the system defined. *Equilibrium states* are, macroscopically, characterized completely by η, e, and $\boldsymbol{\nu}_\alpha$. Note that microscopically there will be fluctuations about an equilibrium state, but macroscopic measurements do not see them.

A basic problem is the determination of an equilibrium state. The problem of thermodynamic equilibrium can be completely solved with the aid of the above extremum principle if the entropy of the system is known as a function of the state variables, i.e.,

$$\eta = \eta(e, \boldsymbol{\nu}_\alpha, X), \qquad \alpha = 1, \ldots, n. \tag{5.133}$$

Such constitutive relation is called a *fundamental relation* and η is considered a *thermodynamic potential.* Thus, if the fundamental relation is known for a particular system, then all thermodynamic information about the system can be obtained. In addition, the continuity, differentiability, and monotonic property of $\partial\eta/\partial e|_{\boldsymbol{\nu}_\alpha, X} > 0$ imply that the entropy function can be inverted with respect to internal energy and that also the internal energy is a single-valued, continuous, and differentiable function of $(\eta, \boldsymbol{\nu}_\alpha)$ for fixed X. Thus (5.133) can be solved uniquely for e in the form

$$e = e(\eta, \boldsymbol{\nu}_\alpha, X), \qquad \alpha = 1, \ldots, n, \tag{5.134}$$

where now e is considered the thermodynamic potential. The set $(\eta, \boldsymbol{\nu}_\alpha)$ describes the thermodynamic state at the material point X. Constitutive equations (5.133) and (5.134) are alternative forms of the fundamental relation, and each contains *all* thermodynamic information about the system. Specific choices of functions define different thermodynamic substances.

For a thermodynamic state to represent a physical one, it must not contradict the basic axioms of mechanics, thermodynamics, and constitutive representations. Such restrictions provide conditions of admissibility on the thermodynamic state of the system. The specific functional form defines different thermodynamic substances.

Definition: The changes that occur in e due to changes of η and $\boldsymbol{\nu}_\alpha$ are called a *thermodynamic process.*

Definition: A thermodynamic process is called *thermodynamically homogeneous* if e is independent of X.

For a thermodynamically homogeneous state, the functional form of the internal energy density is the same at all points of the material body. Such equation is a *thermodynamical constitutive equation* for the internal energy density. It is subject to the restrictions of the the second law of thermodynamics as well as other constitutive principles to be discussed. The selection of $\boldsymbol{\nu}_\alpha$ depends on the thermodynamic state of the body. A certain class of variables cannot be admitted into the class of $\boldsymbol{\nu}_\alpha$ since e is a thermodynamic property of a material. For example, it cannot be an explicit function of t, $\mathbf{x}$, $\mathbf{v}$, and $\mathbf{a}$. On the other hand, η and $\boldsymbol{\nu}_\alpha$ do depend on time and coordinate, i.e.,

$$\eta = \eta(\mathbf{x}, t) \qquad \text{and} \qquad \boldsymbol{\nu}_\alpha = \boldsymbol{\nu}_\alpha(\mathbf{x}, t). \tag{5.135}$$

Since for a given motion we have $\mathbf{x} = \boldsymbol{\chi}(\mathbf{X}, t)$, we see that $\eta = \widehat{\eta}(\mathbf{X}, t)$ and $\boldsymbol{\nu}_\alpha = \widehat{\boldsymbol{\nu}}_\alpha(\mathbf{X}, t)$, and therefore, $e = \widehat{e}(\mathbf{X}, t)$, where $\mathbf{X}$ is the coordinate in the reference configuration of the material point X. Nevertheless, we again note that e is not an explicit function of $\mathbf{x}$ and t, but depends on the values of η and $\boldsymbol{\nu}_\alpha$ at $\mathbf{x}$ and t.

Definition: The *thermostatic temperature* θ and *thermostatic tensions* $\boldsymbol{\tau}_\alpha$ are defined by

$$\theta \equiv \left.\frac{\partial e}{\partial \eta}\right|_{\boldsymbol{\nu}_\gamma, X} \qquad \text{and} \qquad \boldsymbol{\tau}_\alpha \equiv \left.\frac{\partial e}{\partial \boldsymbol{\nu}_\alpha}\right|_{\eta, \boldsymbol{\nu}'_\alpha, X}. \tag{5.136}$$

It is noted that the postulation of the existence of entropy satisfying the previously mentioned properties leads to the above definition of temperature that is independent of that introduced by the zeroth law of thermodynamics. Furthermore, the requirements that the entropy function can be inverted with respect to internal energy and that also the internal energy can be inverted with respect to the entropy requires that $0 < \theta < \infty$.

From now on, since all partial derivatives are for fixed material point X, we will suppress the subscript designations of fixed X in all such derivatives. We also define the additional thermodynamic quantities

$$\boldsymbol{\varphi}_\alpha \equiv \left.\frac{\partial \theta}{\partial \boldsymbol{\nu}_\alpha}\right|_{\eta, \boldsymbol{\nu}'_\alpha} \qquad \text{and} \qquad \boldsymbol{\phi}_{\alpha\beta} \equiv \left.\frac{\partial \boldsymbol{\tau}_\alpha}{\partial \boldsymbol{\nu}_\beta}\right|_{\eta, \boldsymbol{\nu}'_\beta}, \tag{5.137}$$

where $\boldsymbol{\varphi}_\alpha$ is the *isentropic thermal stiffness tensor* and $\boldsymbol{\phi}_{\alpha\beta}$ is the *isentropic elastic stiffness tensor.* When $\boldsymbol{\nu}_1$ is the specific volume, v, then $-\boldsymbol{\tau}_1$ is the *thermodynamic pressure, p,* and $\mathsf{v}^2\,\boldsymbol{\phi}_{11}$ corresponds to the square of the isentropic speed of sound.

For a mixture, if we take $\boldsymbol{\nu}_2, \ldots, \boldsymbol{\nu}_n$ as the mole number densities of the constituents, then $\boldsymbol{\tau}_2, \ldots, \boldsymbol{\tau}_n$ are known as the *chemical potentials.* If $\boldsymbol{\nu}_1$ is related to the strain tensor, then $\boldsymbol{\tau}_1$ will be seen to be the stress tensor.

Taking the first differential of the first and second forms of the fundamental relations (5.133) and (5.134), we see that for an arbitrary change of the thermodynamic state at a given material point X, we have

$$d\eta = \frac{1}{\theta}\, de - \frac{\boldsymbol{\tau}_\alpha}{\theta} \cdot d\boldsymbol{\nu}_\alpha \tag{5.138}$$

and

$$de = \theta\, d\eta + \boldsymbol{\tau}_\alpha \cdot d\boldsymbol{\nu}_\alpha, \tag{5.139}$$

where the inner products denote full contractions between tensors $\boldsymbol{\tau}_\alpha$ and $\boldsymbol{\nu}_\alpha$. Equation (5.139) is the local form of the *Gibbs equation.* The variables $(1/\theta, e)$ and $(-\boldsymbol{\tau}_\alpha/\theta, \boldsymbol{\nu}_\alpha)$, and (θ, η) and $(\boldsymbol{\tau}_\alpha, \boldsymbol{\nu}_\alpha)$ are considered to be conjugate variable pairs of intensive variables and specific extensive variables in the corresponding fundamental relations.

From (5.136), it is clear that

$$\theta = \theta(\eta, \boldsymbol{\nu}_\beta, X) \qquad \text{and} \qquad \boldsymbol{\tau}_\alpha = \boldsymbol{\tau}_\alpha(\eta, \boldsymbol{\nu}_\beta, X). \tag{5.140}$$

Such relations expressing intensive properties in terms of the state variables (the *thermodynamic state*) are called *equations of state.* Knowledge of a single equation of state does *not* give complete knowledge of thermodynamic properties of a system; knowledge of *all* the equations of state is equivalent to knowledge of the fundamental equation and thus, is thermodynamically complete. Equations of state can be derived from the fundamental relation.

If the equation for thermostatic tensions in $(5.140)_2$ is solvable for $\boldsymbol{\nu}_\beta$, then we can select $\boldsymbol{\tau}_\alpha$ as a new state variable to replace $\boldsymbol{\nu}_\beta$. A sufficient condition for this to be possible is that $\boldsymbol{\phi}_{\alpha\beta}$ is continuous and does not vanish in some neighborhood of $\boldsymbol{\nu}_\beta$. In this case, we have

$$\boldsymbol{\nu}_\alpha = \widehat{\boldsymbol{\nu}}_\alpha(\eta, \boldsymbol{\tau}_\beta, X) \tag{5.141}$$

and subsequently, we can write

$$\eta = \eta(e, \widehat{\boldsymbol{\nu}}_\alpha(\eta, \boldsymbol{\tau}_\beta, X), X) = \eta(e, \boldsymbol{\tau}_\beta, X) \tag{5.142}$$

or

$$e = e(\eta, \widehat{\boldsymbol{\nu}}_\alpha(\eta, \boldsymbol{\tau}_\beta, X), X) = e(\eta, \boldsymbol{\tau}_\beta, X). \tag{5.143}$$

Effectively, we have eliminated $\boldsymbol{\nu}_\alpha$ using equations of state $(5.140)_2$. We also define the additional thermodynamic quantities

$$\boldsymbol{\zeta}_\alpha \equiv \left.\frac{\partial \boldsymbol{\nu}_\alpha}{\partial \eta}\right|_{\boldsymbol{\tau}_\gamma} \qquad \text{and} \qquad \boldsymbol{\chi}_{\alpha\beta} \equiv \left.\frac{\partial \boldsymbol{\nu}_\alpha}{\partial \boldsymbol{\tau}_\beta}\right|_{\eta, \boldsymbol{\tau}'_\beta}, \tag{5.144}$$

where $\boldsymbol{\zeta}_\alpha$ is the *isopiestic thermal expansion tensor* and is related to the thermal expansion tensor and specific heat at constant tension (see below) and $\boldsymbol{\chi}_{\alpha\beta}$ is the *isentropic elastic compliance tensor.*

Alternately, if the equation for temperature in $(5.140)_1$ is solvable for η, then we can select θ as a new state variable to replace η. A sufficient condition for this to be possible is that $\partial\theta/\partial\eta|_{\boldsymbol{\nu}_\beta}$ is continuous and does not vanish in some neighborhood of η. In this case, we have

$$\eta = \widehat{\eta}(\theta, \boldsymbol{\nu}_\alpha, X) \qquad \text{and} \qquad e = e(\widehat{\eta}(\theta, \boldsymbol{\nu}_\beta, X), \boldsymbol{\nu}_\alpha, X) = e(\theta, \boldsymbol{\nu}_\alpha, X), \quad (5.145)$$

and subsequently,

$$\boldsymbol{\tau}_\alpha = \boldsymbol{\tau}_\alpha(\theta, \boldsymbol{\nu}_\beta, X) \quad (5.146)$$

or

$$\boldsymbol{\nu}_\alpha = \boldsymbol{\nu}_\alpha(\theta, \boldsymbol{\tau}_\beta, X). \quad (5.147)$$

Effectively, we have eliminated η between the equations of state (5.140). Equations (5.145) are referred to as *caloric equations of state*, while equations (5.146) and (5.147) are referred to as *thermal equations of state.* Note that if $\boldsymbol{\nu}_1$ is the specific volume and $-\boldsymbol{\tau}_1$ the thermodynamic pressure, (5.146) is nothing more than $p = p(\theta, \mathsf{v})$ at material point X, which reduces to the ideal gas equation, or van der Waal's equation of state, etc., depending on the specific fundamental relation. Furthermore, if equation (5.147) is substituted in the relations (5.145), we have

$$\eta = \eta(\theta, \boldsymbol{\tau}_\alpha, X) \qquad \text{and} \qquad e = e(\theta, \boldsymbol{\tau}_\alpha, X). \quad (5.148)$$

Differentiating (5.146) and (5.147) at an arbitrary material point X, we have

$$d\boldsymbol{\tau}_\alpha = \boldsymbol{\beta}_\alpha d\theta + \boldsymbol{\xi}_{\alpha\beta} \cdot d\boldsymbol{\nu}_\beta \qquad \text{and} \qquad d\boldsymbol{\nu}_\alpha = \boldsymbol{\alpha}_\alpha d\theta + \boldsymbol{\upsilon}_{\alpha\beta} \cdot d\boldsymbol{\tau}_\beta, \quad (5.149)$$

where

$$\boldsymbol{\beta}_\alpha \equiv \left.\frac{\partial \boldsymbol{\tau}_\alpha}{\partial \theta}\right|_{\boldsymbol{\nu}_\gamma}, \quad \boldsymbol{\alpha}_\alpha \equiv \left.\frac{\partial \boldsymbol{\nu}_\alpha}{\partial \theta}\right|_{\boldsymbol{\tau}_\gamma}, \quad \boldsymbol{\xi}_{\alpha\beta} \equiv \left.\frac{\partial \boldsymbol{\tau}_\alpha}{\partial \boldsymbol{\nu}_\beta}\right|_{\theta, \boldsymbol{\nu}'_\beta}, \quad \boldsymbol{\upsilon}_{\alpha\beta} \equiv \left.\frac{\partial \boldsymbol{\nu}_\alpha}{\partial \boldsymbol{\tau}_\beta}\right|_{\theta, \boldsymbol{\tau}'_\beta}, \quad (5.150)$$

and $\boldsymbol{\beta}_\alpha$ is the *isochoric thermal tension tensor*, $\boldsymbol{\alpha}_\alpha$ is the *thermal strain tensor* (also called the *piezocaloric tensor*), $\boldsymbol{\xi}_{\alpha\beta}$ is the *isothermal elastic stiffness tensor*, and $\boldsymbol{\upsilon}_{\alpha\beta}$ is the *isothermal elastic compliance tensor.* The quantities (5.150) are related by the identities

$$\boldsymbol{\beta}_\alpha + \boldsymbol{\xi}_{\alpha\beta} \cdot \boldsymbol{\alpha}_\beta = \mathbf{0}, \quad \boldsymbol{\alpha}_\alpha + \boldsymbol{\upsilon}_{\alpha\beta} \cdot \boldsymbol{\beta}_\beta = \mathbf{0}, \quad \boldsymbol{\upsilon}_{\alpha\gamma} \cdot \boldsymbol{\xi}_{\gamma\beta} = \mathbf{1}_{\alpha\beta}, \quad \boldsymbol{\xi}_{\alpha\gamma} \cdot \boldsymbol{\upsilon}_{\gamma\beta} = \mathbf{1}_{\alpha\beta}, \quad (5.151)$$

where $\mathbf{1}_{\alpha\beta}$ is the unit tensor. Since the tensors (5.150) provide changes of measurable quantities, they are useful for inferring the forms of the thermal equations of state from experiments. For example, when $\boldsymbol{\nu}_1$ is the specific volume, v, and $-\boldsymbol{\tau}_1$ the thermodynamic pressure, p, then $\boldsymbol{\beta}_1/p$ is the *isochoric pressure coefficient*, $\boldsymbol{\alpha}_1/\mathsf{v}$ is the *coefficient of thermal* or *volume expansion*, and $\boldsymbol{\upsilon}_{11}/\mathsf{v}$ is the *isothermal compressibility*.

When compared with the fundamental relations, equations of state offer the advantage of connecting easily measurable quantities, but the disadvantage of being insufficient in providing all thermodynamic properties of a material. We elaborate on this aspect below.

Definition: We say that $f(x_1, \ldots, x_n)$ is a *homogeneous function of degree n* for all $x_i > 0$ and $\lambda > 0$ if

$$f(\lambda x_1, \ldots, \lambda x_n) = \lambda^n f(x_1, \ldots, x_n). \tag{5.152}$$

In thermodynamics, intensive functions are homogeneous of degree 0 and extensive functions are homogeneous of degree 1.

Euler's theorem: Let $f(x_1, \ldots, x_n)$, with $x_i > 0$, $i = 1, \ldots, n$, be a continuous and differentiable function. Then f is a homogeneous function of degree n if and only if

$$n f(x_1, \ldots, x_n) = \left.\frac{\partial f(x_1, \ldots, x_n)}{\partial x_i}\right|_{x_i'} x_i. \tag{5.153}$$

Proof: Simply differentiate the homogeneity condition (5.152) with respect to λ:

$$\frac{d}{d\lambda} f(\lambda x_1, \ldots, \lambda x_n) = \frac{d}{d\lambda}\left[\lambda^n f(x_1, \ldots, x_n)\right]$$

to obtain

$$\left.\frac{\partial f(\lambda x_1, \ldots, \lambda x_n)}{\partial(\lambda x_i)}\right|_{x_i'} x_i = n\,\lambda^{n-1} f(x_1, \ldots, x_n).$$

Now setting $\lambda = 1$, we have our proof.

Since the internal energy is an extensive quantity, which, through the fundamental relation (5.134), is a function of extensive quantities of entropy and thermostatic volumes, and since extensive quantities are homogeneous functions of degree 1, from (5.134), Euler's theorem (5.153), and the definitions (5.136), we obtain

$$e = \theta\,\eta + \boldsymbol{\tau}_\alpha \cdot \boldsymbol{\nu}_\alpha. \tag{5.154}$$

This result is known as *Euler's equation.* Now, taking the differential of Euler's equation (5.154) and subtracting the Gibbs equation (5.139), we obtain the *Gibbs–Duhem equation*:

$$\eta\, d\theta + \boldsymbol{\nu}_\alpha \cdot d\boldsymbol{\tau}_\alpha = 0. \tag{5.155}$$

Again we note that the inner products in the above equations denote full contractions.

Before continuing, it is worthwhile to summarize the formal structure of thermodynamics. The fundamental equation (5.133) or (5.134) contains *all* thermodynamic information about a system. To be specific, we use the internal energy representation (5.134) as the fundamental relation. With the definitions of the thermostatic temperature and thermostatic tensions (5.136), the fundamental relation implies the equations of state (5.140). If *all* equations of state are known, they may be substituted into Euler's equation (5.154) to recover the fundamental equation (5.134). Thus, the totality of the equations of state is equivalent to the fundamental equation. Any single equation of state contains less thermodynamic information. If all equations of state minus one are known, the Gibbs–Duhem equation (5.155) may be integrated to obtain the missing one, but such equation will contain an undetermined integration constant. Thus, just one missing equation of state suffices to determine the fundamental equation except for an undetermined constant. Note that, as shown, it is always possible to express e as a function of variables other than η and $\boldsymbol{\nu}_\alpha$. Thus, we could eliminate η between the fundamental relation and the equation of state for θ in (5.140) to obtain an equation of the form $(5.145)_2$. However, this *is not* a fundamental relation since it does not contain all possible thermodynamic information about a system. In fact, recalling the definition of θ in $(5.136)_1$, we see that $e = e(\theta, \boldsymbol{\nu}_\alpha, X)$ is a partial differential equation. Even if this equation was integrable, it would yield a fundamental relation with undetermined functions.

Definition: The *heat* and *work increments* are defined by

$$đq \equiv \theta \, d\eta \qquad \text{and} \qquad đw \equiv \boldsymbol{\tau}_\alpha \cdot d\boldsymbol{\nu}_\alpha. \tag{5.156}$$

We note that the Gibbs equation (5.139) can now be rewritten in the familiar form

$$de = đq + đw, \tag{5.157}$$

which is the local form of the first law of thermodynamics (4.170).

Using the Gibbs equation (5.139), the *specific heat* c, and the *caloric stiffness* $\boldsymbol{\lambda}_{\boldsymbol{\nu}_\alpha}$ and *caloric compliance* $\boldsymbol{\lambda}_{\boldsymbol{\tau}_\alpha}$ for fixed X are subsequently defined by

$$c \equiv \frac{đq}{d\theta} = \theta \frac{d\eta}{d\theta} = \frac{1}{d\theta}\left(de - \boldsymbol{\tau}_\alpha \cdot d\boldsymbol{\nu}_\alpha\right), \tag{5.158}$$

$$\boldsymbol{\lambda}_{\boldsymbol{\nu}_\alpha} \equiv \frac{đq}{d\boldsymbol{\nu}_\alpha} = \theta \frac{d\eta}{d\boldsymbol{\nu}_\alpha} = \frac{1}{d\boldsymbol{\nu}_\alpha}\left(de - \boldsymbol{\tau}_\beta \cdot d\boldsymbol{\nu}_\beta\right), \tag{5.159}$$

$$\boldsymbol{\lambda}_{\boldsymbol{\tau}_\alpha} \equiv \frac{đq}{d\boldsymbol{\tau}_\alpha} = \theta \frac{d\eta}{d\boldsymbol{\tau}_\alpha} = \frac{1}{d\boldsymbol{\tau}_\alpha}\left(de - \boldsymbol{\tau}_\beta \cdot d\boldsymbol{\nu}_\beta\right). \tag{5.160}$$

In practical applications, experimental measurements frequently dictate that a partial derivative be evaluated. For example, we may be concerned with the analysis of the temperature change, which is required to maintain the volume of a single component system constant if the pressure is increased slightly. In such case, we require $\partial\theta/\partial p|_{\mathsf{v}}$. A general feature of the derivatives that arise is that they generally involve both intensive and extensive properties. Of all such derivatives, only $N = (n+1)(n+2)/2$ can be independent, and this number can be shown to correspond to the number of unique second derivatives. As second derivative quantities are associated with special material properties, then one can choose N such properties as conventional and then any other property can be written in terms of these conventional properties. For example, in the case of a single-component simple system where the mole number density is constant, we have $n = 1$ and $\boldsymbol{\nu}_1$ is the specific volume, the number of independent properties are $N = 3$. The three conventional properties, which are defined in terms of the three unique second derivatives, are the specific heat at constant pressure c_p, the coefficient of thermal expansion α, and the isothermal compressibility κ_θ:

$$c_p = \theta \left.\frac{\partial\eta}{\partial\theta}\right|_p, \qquad \alpha = \frac{1}{\mathsf{v}}\left.\frac{\partial \mathsf{v}}{\partial\theta}\right|_p, \qquad \kappa_\theta = -\frac{1}{\mathsf{v}}\left.\frac{\partial \mathsf{v}}{\partial p}\right|_\theta. \tag{5.161}$$

Any other property can then be subsequently written in terms of these three. For example, the specific heat at constant volume and the adiabatic compressibility,

$$c_{\mathsf{v}} = \theta \left.\frac{\partial\eta}{\partial\theta}\right|_{\mathsf{v}} \qquad \text{and} \qquad \kappa_\eta = -\frac{1}{\mathsf{v}}\left.\frac{\partial \mathsf{v}}{\partial p}\right|_\eta, \tag{5.162}$$

can then be obtained from the difference and ratio of specific heats relations:

$$c_p - c_v = \frac{\theta \mathrm{v} \alpha^2}{\kappa_\theta} \tag{5.163}$$

and

$$\gamma = \frac{c_p}{c_v} = \frac{\kappa_\theta}{\kappa_\eta}. \tag{5.164}$$

More generally, for experimental measurements, it is convenient to keep $\boldsymbol{\nu}_\alpha$ = const. or $\boldsymbol{\tau}_\alpha$ = const. Thus, the *specific heats at constant thermostatic volumes* and the *specific heats at constant thermostatic tensions* (using $(5.150)_2$) follow from

$$c_{\boldsymbol{\nu}_\alpha} = \theta \left.\frac{\partial \eta}{\partial \theta}\right|_{\boldsymbol{\nu}_\alpha} = \left.\frac{\partial e}{\partial \theta}\right|_{\boldsymbol{\nu}_\alpha} \qquad \text{and} \qquad c_{\boldsymbol{\tau}_\alpha} = \theta \left.\frac{\partial \eta}{\partial \theta}\right|_{\boldsymbol{\tau}_\alpha} = \left.\frac{\partial e}{\partial \theta}\right|_{\boldsymbol{\tau}_\alpha} - \boldsymbol{\tau}_\alpha \cdot \boldsymbol{\alpha}_\alpha. \tag{5.165}$$

To determine the specific heats at constant tensions in a form more amenable to experimental measurements, we regard e as a function of θ and $\boldsymbol{\nu}_\beta$ at fixed X in (5.158) (see $(5.145)_2$). Then, from (5.158), we have

$$c_{\boldsymbol{\tau}_\alpha} = \frac{1}{d\theta}\left[\left.\frac{\partial e}{\partial \theta}\right|_{\boldsymbol{\nu}_\alpha} d\theta + \left(\left.\frac{\partial e}{\partial \boldsymbol{\nu}_\alpha}\right|_{\theta, \boldsymbol{\nu}'_\alpha} - \boldsymbol{\tau}_\alpha \right) \cdot d\boldsymbol{\nu}_\alpha \right]. \tag{5.166}$$

If we now regard $\boldsymbol{\nu}_\alpha$ as a function of θ and $\boldsymbol{\tau}_\beta$ at fixed X, as in (5.147), and hold the tensions $\boldsymbol{\tau}_\beta$ = const., using $(5.149)_2$, (5.150), and $(5.165)_1$, we get

$$c_{\boldsymbol{\tau}_\alpha} - c_{\boldsymbol{\nu}_\alpha} = \left(\left.\frac{\partial e}{\partial \boldsymbol{\nu}_\alpha}\right|_{\theta, \boldsymbol{\nu}'_\alpha} - \boldsymbol{\tau}_\alpha \right) \cdot \boldsymbol{\alpha}_\alpha, \tag{5.167}$$

which, upon using (5.159), becomes

$$c_{\boldsymbol{\tau}_\alpha} - c_{\boldsymbol{\nu}_\alpha} = \boldsymbol{\lambda}_{\boldsymbol{\nu}_\alpha} \cdot \boldsymbol{\alpha}_\alpha. \tag{5.168}$$

Analogously, by regarding e as a function of θ and $\boldsymbol{\tau}_\alpha$ at fixed X, it is easy to show that also

$$c_{\boldsymbol{\tau}_\alpha} - c_{\boldsymbol{\nu}_\alpha} = -\boldsymbol{\lambda}_{\boldsymbol{\tau}_\alpha} \cdot \boldsymbol{\beta}_\alpha. \tag{5.169}$$

Subsequently, we see that

$$\boldsymbol{\lambda}_{\boldsymbol{\nu}_\alpha} \cdot \boldsymbol{\alpha}_\alpha = -\boldsymbol{\lambda}_{\boldsymbol{\tau}_\alpha} \cdot \boldsymbol{\beta}_\alpha, \tag{5.170}$$

or, using $(5.151)_{1,2}$,

$$\boldsymbol{\lambda}_{\boldsymbol{\nu}_\beta} = \boldsymbol{\lambda}_{\boldsymbol{\tau}_\alpha} \cdot \boldsymbol{\xi}_{\alpha\beta} \qquad \text{or} \qquad \boldsymbol{\lambda}_{\boldsymbol{\tau}_\beta} = \boldsymbol{\lambda}_{\boldsymbol{\nu}_\alpha} \cdot \boldsymbol{v}_{\alpha\beta}. \tag{5.171}$$

We also note from (5.159) and (5.160) (using $(5.149)_2$) that when $\eta = \eta(\theta, \boldsymbol{\nu}_\alpha)$ and $\eta = \eta(\theta, \boldsymbol{\tau}_\alpha)$, we respectively have

$$\boldsymbol{\lambda}_{\boldsymbol{\nu}_\alpha} = \theta \left.\frac{\partial \eta}{\partial \boldsymbol{\nu}_\alpha}\right|_{\theta, \boldsymbol{\nu}'_\alpha} = \left.\frac{\partial e}{\partial \boldsymbol{\nu}_\alpha}\right|_{\theta, \boldsymbol{\nu}'_\alpha} - \boldsymbol{\tau}_\alpha, \tag{5.172}$$

$$\boldsymbol{\lambda}_{\boldsymbol{\tau}_\alpha} = \theta \left.\frac{\partial \eta}{\partial \boldsymbol{\tau}_\alpha}\right|_{\theta, \boldsymbol{\tau}'_\alpha} = \left.\frac{\partial e}{\partial \boldsymbol{\tau}_\alpha}\right|_{\theta, \boldsymbol{\tau}'_\alpha} - \boldsymbol{\tau}_\beta \cdot \boldsymbol{v}_{\beta\alpha}, \tag{5.173}$$

and

$$đq \equiv \theta\, d\eta = c_{\nu_\alpha}\, d\theta + \boldsymbol{\lambda}_{\nu_\alpha} \cdot d\boldsymbol{\nu}_\alpha \qquad \text{and} \qquad đq \equiv \theta\, d\eta = c_{\tau_\alpha}\, d\theta + \boldsymbol{\lambda}_{\tau_\alpha} \cdot d\boldsymbol{\tau}_\alpha. \tag{5.174}$$

The *ratio of specific heats*

$$\gamma_\alpha \equiv \frac{c_{\tau_\alpha}}{c_{\nu_\alpha}}, \tag{5.175}$$

which represents a more general quantity than (5.164), is important and it arises in many branches of physics. It, as well as the specific heats, is connected with other quantities using the following identities (which can be easily proved):

$$\boldsymbol{\phi}_{\alpha\gamma} \cdot \boldsymbol{\upsilon}_{\gamma\beta} + \boldsymbol{\varphi}_\alpha \boldsymbol{\alpha}_\beta = \mathbf{1}_{\alpha\beta}, \quad \boldsymbol{\varphi}_\beta \cdot \boldsymbol{\upsilon}_{\beta\alpha} + \gamma_\alpha \boldsymbol{\zeta}_\alpha = \mathbf{0}, \quad \boldsymbol{\zeta}_\beta \cdot \boldsymbol{\phi}_{\beta\alpha} + \boldsymbol{\varphi}_\alpha = \mathbf{0},$$
$$\gamma_\alpha + \boldsymbol{\varphi}_\alpha \cdot \boldsymbol{\alpha}_\alpha = 1, \tag{5.176}$$

where we have used (5.175) and

$$\boldsymbol{\zeta}_\alpha = (\theta / c_{\tau_\alpha})\, \boldsymbol{\alpha}_\alpha. \tag{5.177}$$

Results (5.168) and (5.169) and similar ones may be obtained easily by considering the following. Assume that we have two specific extensive tensor quantities A and B given in terms of two intensive tensor quantities a and b:

$$\mathrm{A} = \mathrm{A}(a, b) \qquad \text{and} \qquad \mathrm{B} = \mathrm{B}(a, b). \tag{5.178}$$

Then

$$d\mathrm{A} = \left.\frac{\partial \mathrm{A}}{\partial a}\right|_b \cdot da + \left.\frac{\partial \mathrm{A}}{\partial b}\right|_a \cdot db \qquad \text{and} \qquad d\mathrm{B} = \left.\frac{\partial \mathrm{B}}{\partial a}\right|_b \cdot da + \left.\frac{\partial \mathrm{B}}{\partial b}\right|_a \cdot db. \tag{5.179}$$

Now, holding B constant and taking the inner product of the second equation with $\partial b / \partial \mathrm{B}|_a$, we obtain

$$0 = \left[\left.\frac{\partial b}{\partial \mathrm{B}}\right|_a \cdot \left.\frac{\partial \mathrm{B}}{\partial a}\right|_b \cdot da + \left.\frac{\partial b}{\partial \mathrm{B}}\right|_a \cdot \left.\frac{\partial \mathrm{B}}{\partial b}\right|_a \cdot db \right] \tag{5.180}$$

or

$$db = - \left.\frac{\partial b}{\partial \mathrm{B}}\right|_a \cdot \left.\frac{\partial \mathrm{B}}{\partial a}\right|_b \cdot da. \tag{5.181}$$

Introducing this result into the first equation in (5.179) gives

$$d\mathrm{A} = \left.\frac{\partial \mathrm{A}}{\partial a}\right|_b \cdot da - \left.\frac{\partial \mathrm{A}}{\partial b}\right|_a \cdot \left.\frac{\partial b}{\partial \mathrm{B}}\right|_a \cdot \left.\frac{\partial \mathrm{B}}{\partial a}\right|_b \cdot da. \tag{5.182}$$

Now dividing this equation by da at constant B, leads to the final result

$$\left.\frac{\partial \mathrm{A}}{\partial a}\right|_{\mathrm{B}} - \left.\frac{\partial \mathrm{A}}{\partial a}\right|_b = - \left.\frac{\partial \mathrm{A}}{\partial \mathrm{B}}\right|_a \cdot \left.\frac{\partial \mathrm{B}}{\partial a}\right|_b. \tag{5.183}$$

To illustrate the use of (5.183), if we take $\mathrm{A} \to e$, $\mathrm{B} \to \boldsymbol{\nu}_\alpha$, $a \to \theta$, and $b \to \boldsymbol{\tau}_\alpha$, we easily obtain (5.167) and subsequently (5.168). If we take $\mathrm{A} \to \boldsymbol{\nu}_\alpha$, $\mathrm{B} \to \eta$,

$a \to \boldsymbol{\tau}_\alpha$, and $b \to \theta$, we obtain an equation that provides the difference between the isentropic and isothermal elastic compliance tensors:

$$\left.\frac{\partial \boldsymbol{\nu}_\alpha}{\partial \boldsymbol{\tau}_\beta}\right|_{\eta,\boldsymbol{\tau}'_\beta} - \left.\frac{\partial \boldsymbol{\nu}_\alpha}{\partial \boldsymbol{\tau}_\beta}\right|_{\theta,\boldsymbol{\tau}'_\beta} = -\left.\frac{\partial \boldsymbol{\nu}_\alpha}{\partial \eta}\right|_{\boldsymbol{\tau}_\gamma} \left.\frac{\partial \eta}{\partial \boldsymbol{\tau}_\beta}\right|_{\theta,\boldsymbol{\tau}'_\beta}, \tag{5.184}$$

or, using (5.144), $(5.150)_4$, and (5.173),

$$\boldsymbol{\chi}_{\alpha\beta} - \boldsymbol{\upsilon}_{\alpha\beta} = -\frac{1}{\theta}\boldsymbol{\zeta}_\alpha \boldsymbol{\lambda}_{\boldsymbol{\tau}_\beta}. \tag{5.185}$$

We note that in the case where $\boldsymbol{\nu}_1$ is the specific volume, this relationship leads to (5.163).

The Gibbs–Duhem relation (5.155) expresses the existence of a relationship among the first derivatives of the fundamental relation (5.134). Similarly, there are relations among the second derivatives. They are associated with the equality of the various mixed second derivatives. Such identities are known as *Maxwell relations*. Specifically, using the fundamental relation (5.134) and the definitions (5.136), we observe that

$$\left.\frac{\partial}{\partial \boldsymbol{\nu}_\alpha}\left(\left.\frac{\partial e}{\partial \eta}\right|_{\boldsymbol{\nu}_\gamma}\right)\right|_{\eta,\boldsymbol{\nu}'_\alpha} = \left.\frac{\partial}{\partial \eta}\left(\left.\frac{\partial e}{\partial \boldsymbol{\nu}_\alpha}\right|_{\eta,\boldsymbol{\nu}'_\alpha}\right)\right|_{\boldsymbol{\nu}_\gamma} \quad \text{and}$$

$$\left.\frac{\partial}{\partial \boldsymbol{\nu}_\beta}\left(\left.\frac{\partial e}{\partial \boldsymbol{\nu}_\alpha}\right|_{\eta,\boldsymbol{\nu}'_\alpha}\right)\right|_{\eta,\boldsymbol{\nu}'_\beta} = \left.\frac{\partial}{\partial \boldsymbol{\nu}_\alpha}\left(\left.\frac{\partial e}{\partial \boldsymbol{\nu}_\beta}\right|_{\eta,\boldsymbol{\nu}'_\beta}\right)\right|_{\eta,\boldsymbol{\nu}'_\alpha}, \tag{5.186}$$

or

$$\left.\frac{\partial \theta}{\partial \boldsymbol{\nu}_\alpha}\right|_{\eta,\boldsymbol{\nu}'_\alpha} = \left.\frac{\partial \boldsymbol{\tau}_\alpha}{\partial \eta}\right|_{\boldsymbol{\nu}_\gamma} \quad \text{and} \quad \left.\frac{\partial \boldsymbol{\tau}_\alpha}{\partial \boldsymbol{\nu}_\beta}\right|_{\eta,\boldsymbol{\nu}'_\beta} = \left.\frac{\partial \boldsymbol{\tau}_\beta}{\partial \boldsymbol{\nu}_\alpha}\right|_{\eta,\boldsymbol{\nu}'_\alpha}. \tag{5.187}$$

Given a fundamental relation expressed in terms of its $(n+1)$ natural variables, there are $n(n+1)/2$ separate pairs of mixed second derivatives. For example, in the case of a single-component simple system where $n = 2$ and our specific thermostatic volumes correspond to the specific volume, v, and mole number density, n, and the thermostatic tensions correspond to the negative pressure, $-p$, and chemical potential, μ, then we have three Maxwell relations and they are given by

$$\left.\frac{\partial \theta}{\partial \mathsf{v}}\right|_{\eta,\mathsf{n}} = -\left.\frac{\partial p}{\partial \eta}\right|_{\mathsf{v},\mathsf{n}}, \quad \left.\frac{\partial \theta}{\partial \mathsf{n}}\right|_{\eta,\mathsf{v}} = \left.\frac{\partial \mu}{\partial \eta}\right|_{\mathsf{v},\mathsf{n}}, \quad \text{and} \quad -\left.\frac{\partial p}{\partial \mathsf{n}}\right|_{\eta,\mathsf{v}} = \left.\frac{\partial \mu}{\partial \mathsf{v}}\right|_{\eta,\mathsf{n}}.$$

Now it is easy to show that the differentials of the equations of state (5.140), using (5.187) with (5.137) and $(5.165)_1$, can be written as follows:

$$d\theta = \frac{\theta}{c_{\boldsymbol{\nu}_\alpha}} d\eta + \boldsymbol{\varphi}_\alpha \cdot d\boldsymbol{\nu}_\alpha \quad \text{and} \quad d\boldsymbol{\tau}_\alpha = \boldsymbol{\varphi}_\alpha \, d\eta + \boldsymbol{\phi}_{\alpha\beta} \cdot d\boldsymbol{\nu}_\beta. \tag{5.188}$$

All thermodynamic properties that we have introduced are summarized in Table 5.9. In the table, we also denote their physical description, provide their corresponding values for the special case of a single-component simple system with constant mole number density (in which case $n = 1$, where $\boldsymbol{\nu}_1 \to \mathsf{v}$ and $\boldsymbol{\tau}_1 \to -p$), and indicate their units using the International System of units (SI).

5.10.2 Thermodynamic potentials

In both the internal energy and entropy representations, extensive properties play the roles of independent variables, whereas intensive quantities arise as derived concepts. This situation is in contrast to the situation encountered in an experiment. The experimenter usually finds that the intensive quantities are more easily measured and controlled, and therefore, it is easier to think of the intensive quantities as independent variables and the extensive quantities as derived properties.

The question arises as to the possibility of recasting the fundamental relations in such a way that intensive quantities will replace some or all extensive properties as independent variables. This is possible by the use of *Legendre transformations* (see Appendix F). The Legendre transformed functions of the fundamental relation are also called *thermodynamic potentials* and particular transformations subsequently lead to alternate fundamental representations.

To affect such representations, we use the energy form of the fundamental relation, and to simplify the presentation, we first take $\boldsymbol{\nu}_0$ to signify the entropy η and $\boldsymbol{\tau}_0$ to signify the temperature θ, so that the fundamental relation (5.134) is rewritten in the form

$$e = e(\boldsymbol{\nu}_0, \ldots, \boldsymbol{\nu}_n). \tag{5.189}$$

Subsequently, returning to explicit summation convention, the first differential of the fundamental relation (5.139) is now given by

$$de = \sum_{\alpha=0}^{n} \boldsymbol{\tau}_\alpha \cdot d\boldsymbol{\nu}_\alpha, \tag{5.190}$$

where the equations of state (5.136) are given by

$$\boldsymbol{\tau}_\alpha = \left.\frac{\partial e}{\partial \boldsymbol{\nu}_\alpha}\right|_{\boldsymbol{\nu}'_\alpha}, \qquad \alpha = 0, \ldots, n. \tag{5.191}$$

Note that now the heat increment is given by $đq = \boldsymbol{\tau}_0 \cdot d\boldsymbol{\nu}_0$ and the work increment by $đw = \sum_{\alpha=1}^{n} \boldsymbol{\tau}_\alpha \cdot d\boldsymbol{\nu}_\alpha$. Furthermore, the Euler relation (5.154) is now written as

$$e = \sum_{\alpha=0}^{n} \boldsymbol{\tau}_\alpha \cdot \boldsymbol{\nu}_\alpha \tag{5.192}$$

and the Gibbs–Duhem relation (5.155) as

$$\sum_{\alpha=0}^{n} \boldsymbol{\nu}_\alpha \cdot d\boldsymbol{\tau}_\alpha = 0. \tag{5.193}$$

The Maxwell relations (5.187) are subsequently given by

$$\frac{\partial \boldsymbol{\tau}_\alpha}{\partial \boldsymbol{\nu}_\beta} = \frac{\partial \boldsymbol{\tau}_\beta}{\partial \boldsymbol{\nu}_\alpha} \quad \text{for} \quad 0 \le \alpha, \beta \le n, \tag{5.194}$$

where in each of these partial derivatives, the variables to be held constant are all those of the set $\{\boldsymbol{\nu}_0, \ldots, \boldsymbol{\nu}_n\}$ except the variables with respect to which the derivative is taken.

A partial Legendre transformation can subsequently be made by replacing the thermostatic volumes $\boldsymbol{\nu}_0, \ldots, \boldsymbol{\nu}_m$ by the thermostatic tensions $\boldsymbol{\tau}_0, \ldots, \boldsymbol{\tau}_m$ with $m \leq n$. The Legendre transformed function is

$$F \equiv e[\boldsymbol{\tau}_0, \ldots, \boldsymbol{\tau}_m] = F(\boldsymbol{\tau}_0, \ldots, \boldsymbol{\tau}_m, \boldsymbol{\nu}_{m+1}, \ldots, \boldsymbol{\nu}_n) = e - \sum_{\alpha=0}^{m} \boldsymbol{\tau}_\alpha \cdot \boldsymbol{\nu}_\alpha, \tag{5.195}$$

where the quantities inside the brackets denote the new independent intensive quantities in the transformed energy fundamental relation that replace corresponding conjugate variables. The corresponding equations of state are

$$-\boldsymbol{\nu}_\alpha = \left.\frac{\partial F}{\partial \boldsymbol{\tau}_\alpha}\right|_{\boldsymbol{\tau}'_\alpha}, \quad \alpha = 0, \ldots, m, \qquad \text{and} \qquad \boldsymbol{\tau}_\alpha = \left.\frac{\partial F}{\partial \boldsymbol{\nu}_\alpha}\right|_{\boldsymbol{\nu}'_\alpha}, \quad \alpha = m+1, \ldots, n, \tag{5.196}$$

with the first differential of F given by

$$dF = \sum_{\alpha=0}^{m} (-\boldsymbol{\nu}_\alpha) \cdot d\boldsymbol{\tau}_\alpha + \sum_{\alpha=m+1}^{n} \boldsymbol{\tau}_\alpha \cdot d\boldsymbol{\nu}_\alpha. \tag{5.197}$$

Furthermore, the equilibrium values of any unconstrained extensive parameters in a system in contact with reservoirs of constant $\boldsymbol{\tau}_0, \ldots, \boldsymbol{\tau}_m$ minimize F at constant $\boldsymbol{\tau}_0, \ldots, \boldsymbol{\tau}_m$.

Now, as before, given a thermodynamic potential expressed in terms on its $n+1$ natural variables, there are $n(n+1)/2$ separate pairs of mixed partial derivatives that yield $n(n+1)/2$ Maxwell relations. The corresponding equality provided by the Maxwell relations of mixed second derivatives of the potential F become

$$\frac{\partial \boldsymbol{\nu}_\alpha}{\partial \boldsymbol{\tau}_\beta} = \frac{\partial \boldsymbol{\nu}_\beta}{\partial \boldsymbol{\tau}_\alpha} \quad \text{for} \quad 0 \leq \alpha, \beta \leq m, \tag{5.198}$$

$$\frac{\partial \boldsymbol{\nu}_\alpha}{\partial \boldsymbol{\nu}_\beta} = -\frac{\partial \boldsymbol{\tau}_\beta}{\partial \boldsymbol{\tau}_\alpha} \quad \text{for} \quad 0 \leq \alpha \leq m \text{ and } m < \beta \leq n, \quad \text{and} \tag{5.199}$$

$$\frac{\partial \boldsymbol{\tau}_\alpha}{\partial \boldsymbol{\nu}_\beta} = \frac{\partial \boldsymbol{\tau}_\beta}{\partial \boldsymbol{\nu}_\alpha} \quad \text{for} \quad m < \alpha, \beta \leq n. \tag{5.200}$$

In each of these partial derivatives, the variables to be held constant are all those of the set $\{\boldsymbol{\tau}_0, \ldots, \boldsymbol{\tau}_m, \boldsymbol{\nu}_{m+1}, \ldots, \boldsymbol{\nu}_n\}$ except the variables with respect to which the derivative is taken.

Now, using the fundamental relation (5.189) and reverting from $\boldsymbol{\nu}_0$ to η and $\boldsymbol{\tau}_0$ to θ, the following Legendre transformations have been found to be useful:

$$\psi = e[\theta] = \psi(\theta, \boldsymbol{\nu}_\alpha, X) = e - \theta\eta \qquad \text{for} \quad \alpha = 1, \ldots, n, \tag{5.201}$$

$$h = e[\boldsymbol{\tau}_1] = h(\eta, \boldsymbol{\tau}_1, \boldsymbol{\nu}_\alpha, X) = e - \boldsymbol{\tau}_1 \cdot \boldsymbol{\nu}_1 \qquad \text{for} \quad \alpha = 2, \ldots, n, \tag{5.202}$$

$$g = e[\theta, \boldsymbol{\tau}_1] = g(\theta, \boldsymbol{\tau}_1, \boldsymbol{\nu}_\alpha, X) = e - \theta\eta - \boldsymbol{\tau}_1 \cdot \boldsymbol{\nu}_1 \qquad \text{for} \quad \alpha = 2, \ldots, n, \tag{5.203}$$

where ψ is the specific *Helmholtz potential* or *free energy*, h is the specific *enthalpy potential*, and g the specific *Gibbs potential*, respectively. Note that the potentials are related through the identity

$$e - \psi + g - h = 0. \tag{5.204}$$

Other potentials are used infrequently and are mostly unnamed.

Helmholtz potential

In experiments, entropy is generally not a controllable parameter, but temperature often is controllable. As such, it is a more appropriate choice for an independent variable. To change to a temperature representation, we replace the internal energy with its Legendre transform, the Helmholtz potential. Thus, given the fundamental relation (5.134), define the specific Helmholtz potential or free energy through the Legendre transformation (5.201), whose first differential for $\alpha = 1, \ldots, n$ is given by

$$d\psi = \left.\frac{\partial \psi}{\partial \theta}\right|_{\boldsymbol{\nu}_\gamma} d\theta + \left.\frac{\partial \psi}{\partial \boldsymbol{\nu}_\alpha}\right|_{\theta, \boldsymbol{\nu}'_\alpha} \cdot d\boldsymbol{\nu}_\alpha, \tag{5.205}$$

where we have reverted to the implied summation convention and full contraction is also implied. Then, differentiating (5.201) and using (5.139), we obtain

$$d\psi = de - \theta\, d\eta - \eta\, d\theta = -\eta\, d\theta + \boldsymbol{\tau}_\alpha \cdot d\boldsymbol{\nu}_\alpha. \tag{5.206}$$

Subsequently, comparing the above expressions, we see that

$$-\eta = \left.\frac{\partial \psi}{\partial \theta}\right|_{\boldsymbol{\nu}_\gamma} \qquad \text{and} \qquad \boldsymbol{\tau}_\alpha = \left.\frac{\partial \psi}{\partial \boldsymbol{\nu}_\alpha}\right|_{\theta, \boldsymbol{\nu}'_\alpha}, \tag{5.207}$$

so that, from (5.201) and $(5.207)_1$, we have

$$e = \psi - \theta \left.\frac{\partial \psi}{\partial \theta}\right|_{\boldsymbol{\nu}_\gamma}, \tag{5.208}$$

and then from (5.165) and $(5.207)_1$, we see that

$$c_{\boldsymbol{\nu}_\gamma} = -\theta \left.\frac{\partial^2 \psi}{\partial \theta^2}\right|_{\boldsymbol{\nu}_\gamma}. \tag{5.209}$$

The corresponding Maxwell relations are given by

$$\frac{\partial}{\partial \boldsymbol{\nu}_\alpha}\left(\left.\frac{\partial \psi}{\partial \theta}\right|_{\boldsymbol{\nu}_\gamma}\right)\Bigg|_{\theta, \boldsymbol{\nu}'_\alpha} = \frac{\partial}{\partial \theta}\left(\left.\frac{\partial \psi}{\partial \boldsymbol{\nu}_\alpha}\right|_{\theta, \boldsymbol{\nu}'_\alpha}\right)\Bigg|_{\boldsymbol{\nu}_\gamma} \quad \text{and}$$

$$\frac{\partial}{\partial \boldsymbol{\nu}_\beta}\left(\left.\frac{\partial \psi}{\partial \boldsymbol{\nu}_\alpha}\right|_{\theta, \boldsymbol{\nu}'_\alpha}\right)\Bigg|_{\theta, \boldsymbol{\nu}'_\beta} = \frac{\partial}{\partial \boldsymbol{\nu}_\alpha}\left(\left.\frac{\partial \psi}{\partial \boldsymbol{\nu}_\beta}\right|_{\theta, \boldsymbol{\nu}'_\beta}\right)\Bigg|_{\theta, \boldsymbol{\nu}'_\alpha} \tag{5.210}$$

which, upon using (5.207), become

$$-\left.\frac{\partial \eta}{\partial \boldsymbol{\nu}_\alpha}\right|_{\theta, \boldsymbol{\nu}'_\alpha} = \left.\frac{\partial \boldsymbol{\tau}_\alpha}{\partial \theta}\right|_{\boldsymbol{\nu}_\gamma} \qquad \text{and} \qquad \left.\frac{\partial \boldsymbol{\tau}_\alpha}{\partial \boldsymbol{\nu}_\beta}\right|_{\theta, \boldsymbol{\nu}'_\beta} = \left.\frac{\partial \boldsymbol{\tau}_\beta}{\partial \boldsymbol{\nu}_\alpha}\right|_{\theta, \boldsymbol{\nu}'_\alpha}. \tag{5.211}$$

Now it is easy to show that the differentials of the equations of state (5.207), using (5.150), (5.165) and (5.211), can be written as follows:

$$d\eta = \frac{c_{\boldsymbol{\nu}_\alpha}}{\theta} d\theta - \boldsymbol{\beta}_\alpha \cdot d\boldsymbol{\nu}_\alpha \qquad \text{and} \qquad d\boldsymbol{\tau}_\alpha = \boldsymbol{\beta}_\alpha\, d\theta + \boldsymbol{\xi}_{\alpha\beta} \cdot d\boldsymbol{\nu}_\beta. \tag{5.212}$$

In addition, substituting $(5.212)_1$ into (5.139), we obtain

$$de = c_{\boldsymbol{\nu}_\alpha}\, d\theta + (\boldsymbol{\tau}_\alpha - \theta\, \boldsymbol{\beta}_\alpha) \cdot d\boldsymbol{\nu}_\alpha. \tag{5.213}$$

The name of free energy is appropriate for ψ because, as follows from (5.206), it is the portion of the energy available for doing work at constant temperature.

Using $(5.211)_1$ with (5.151), (5.171), and (5.172), we see that we can write

$$\boldsymbol{\lambda}_{\boldsymbol{\nu}_\alpha} = -\theta\, \boldsymbol{\beta}_\alpha \qquad \text{and} \qquad \boldsymbol{\lambda}_{\boldsymbol{\tau}_\alpha} = \theta\, \boldsymbol{\alpha}_\alpha, \tag{5.214}$$

and subsequently, using (5.151), we can rewrite (5.168) and (5.169) as

$$c_{\boldsymbol{\tau}_\alpha} - c_{\boldsymbol{\nu}_\alpha} = -\theta\, \boldsymbol{\alpha}_\alpha \cdot \boldsymbol{\beta}_\alpha = \theta\, \boldsymbol{\alpha}_\alpha \cdot \boldsymbol{\xi}_{\alpha\beta} \cdot \boldsymbol{\alpha}_\beta = \theta\, \boldsymbol{\beta}_\alpha \cdot \boldsymbol{\upsilon}_{\alpha\beta} \cdot \boldsymbol{\beta}_\beta. \tag{5.215}$$

This result represents the generalization of relation (5.163).

An additional thermodynamic property that is often found useful is the Grüneisen parameter, a generalization of which is provided by the following definition of the *Grüneisen tensor*:

$$\boldsymbol{\Gamma}_{\alpha\beta} = \frac{\boldsymbol{\nu}_\alpha}{c_{\boldsymbol{\nu}_\beta}} \left.\frac{\partial \eta}{\partial \boldsymbol{\nu}_\beta}\right|_{\theta, \boldsymbol{\nu}_{\gamma \neq \beta}} = -\frac{\boldsymbol{\nu}_\alpha}{c_{\boldsymbol{\nu}_\beta}} \left.\frac{\partial \boldsymbol{\tau}_\beta}{\partial \theta}\right|_{\boldsymbol{\nu}_\gamma} = -\frac{\boldsymbol{\nu}_\alpha \boldsymbol{\beta}_\beta}{c_{\boldsymbol{\nu}_\beta}}, \tag{5.216}$$

where we have used $(5.211)_1$ and $(5.212)_2$. Now using $(5.214)_1$ and $(5.151)_1$, it is easy to also show that

$$\boldsymbol{\Gamma}_{\alpha\beta} = \frac{\boldsymbol{\nu}_\alpha \boldsymbol{\lambda}_{\boldsymbol{\nu}_\beta}}{\theta\, c_{\boldsymbol{\nu}_\beta}} = \frac{\boldsymbol{\nu}_\alpha \boldsymbol{\xi}_{\beta\gamma} \cdot \boldsymbol{\alpha}_\gamma}{c_{\boldsymbol{\nu}_\beta}}. \tag{5.217}$$

Enthalpy potential

In experiments, the thermostatic volumes may not be controllable parameters, but the thermostatic tensions may be controllable. As such, they would be more appropriate choices for independent variables. To change to a thermostatic tensions representation, we replace the internal energy with its Legendre transform, the enthalpy potential. Thus, given the fundamental relation (5.134), define the specific enthalpy potential as the Legendre transformation (5.202), whose first differential for $\alpha = 2, \ldots, n$ is given by

$$dh = \left.\frac{\partial h}{\partial \eta}\right|_{\boldsymbol{\tau}_1, \boldsymbol{\nu}_\gamma} d\eta + \left.\frac{\partial h}{\partial \boldsymbol{\tau}_1}\right|_{\eta, \boldsymbol{\nu}_\gamma} \cdot d\boldsymbol{\tau}_1 + \left.\frac{\partial h}{\partial \boldsymbol{\nu}_\alpha}\right|_{\eta, \boldsymbol{\tau}_1, \boldsymbol{\nu}'_\alpha} \cdot d\boldsymbol{\nu}_\alpha, \tag{5.218}$$

where we have reverted to the implied summation convention. Then, differentiating (5.202) and using (5.139), we obtain

$$dh = de - \boldsymbol{\nu}_1 \cdot d\boldsymbol{\tau}_1 - \boldsymbol{\tau}_1 \cdot d\boldsymbol{\nu}_1 = \theta\, d\eta - \boldsymbol{\nu}_1 \cdot d\boldsymbol{\tau}_1 + \boldsymbol{\tau}_\alpha \cdot d\boldsymbol{\nu}_\alpha. \tag{5.219}$$

Subsequently, comparing the above expressions, we see that

$$\theta = \left.\frac{\partial h}{\partial \eta}\right|_{\boldsymbol{\tau}_1, \boldsymbol{\nu}_\gamma}, \qquad -\boldsymbol{\nu}_1 = \left.\frac{\partial h}{\partial \boldsymbol{\tau}_1}\right|_{\eta, \boldsymbol{\nu}_\gamma}, \quad \text{and} \quad \boldsymbol{\tau}_\alpha = \left.\frac{\partial h}{\partial \boldsymbol{\nu}_\alpha}\right|_{\eta, \boldsymbol{\tau}_1, \boldsymbol{\nu}'_\alpha}, \tag{5.220}$$

so that

$$e = h - \boldsymbol{\tau}_1 \cdot \left.\frac{\partial h}{\partial \boldsymbol{\tau}_1}\right|_{\eta, \boldsymbol{\nu}_\gamma}, \tag{5.221}$$

and then from $(5.165)_2$, using $(5.220)_2$ and $(5.150)_2$, we see that

$$c_{\boldsymbol{\tau}_1} = \left.\frac{\partial h}{\partial \theta}\right|_{\boldsymbol{\tau}_1}. \tag{5.222}$$

As can be seen from (5.219), the enthalpy h provides the portion of the energy that can be released as heat when the thermostatic tension $\boldsymbol{\tau}_1$ and the specific thermostatic volumes $\boldsymbol{\nu}_\alpha$ for $\alpha \geq 2$ are kept constant.

The corresponding Maxwell relations are given by

$$\frac{\partial}{\partial \boldsymbol{\tau}_1}\left(\left.\frac{\partial h}{\partial \eta}\right|_{\boldsymbol{\tau}_1,\boldsymbol{\nu}_\gamma}\right)\Bigg|_{\eta,\boldsymbol{\nu}_\gamma} = \frac{\partial}{\partial \eta}\left(\left.\frac{\partial h}{\partial \boldsymbol{\tau}_1}\right|_{\eta,\boldsymbol{\nu}_\gamma}\right)\Bigg|_{\boldsymbol{\tau}_1,\boldsymbol{\nu}_\gamma},$$

$$\frac{\partial}{\partial \boldsymbol{\nu}_\alpha}\left(\left.\frac{\partial h}{\partial \eta}\right|_{\boldsymbol{\tau}_1,\boldsymbol{\nu}_\gamma}\right)\Bigg|_{\eta,\boldsymbol{\tau}_1,\boldsymbol{\nu}'_\alpha} = \frac{\partial}{\partial \eta}\left(\left.\frac{\partial h}{\partial \boldsymbol{\nu}_\alpha}\right|_{\eta,\boldsymbol{\tau}_1,\boldsymbol{\nu}'_\alpha}\right)\Bigg|_{\boldsymbol{\tau}_1,\boldsymbol{\nu}_\gamma},$$

$$\frac{\partial}{\partial \boldsymbol{\tau}_1}\left(\left.\frac{\partial h}{\partial \boldsymbol{\nu}_\alpha}\right|_{\eta,\boldsymbol{\tau}_1,\boldsymbol{\nu}'_\alpha}\right)\Bigg|_{\eta,\boldsymbol{\nu}_\gamma} = \frac{\partial}{\partial \boldsymbol{\nu}_\alpha}\left(\left.\frac{\partial h}{\partial \boldsymbol{\tau}_1}\right|_{\eta,\boldsymbol{\nu}_\gamma}\right)\Bigg|_{\eta,\boldsymbol{\tau}_1,\boldsymbol{\nu}'_\alpha},$$

$$\text{and} \quad \frac{\partial}{\partial \boldsymbol{\nu}_\beta}\left(\left.\frac{\partial h}{\partial \boldsymbol{\nu}_\alpha}\right|_{\eta,\boldsymbol{\tau}_1,\boldsymbol{\nu}'_\alpha}\right)\Bigg|_{\eta,\boldsymbol{\tau}_1,\boldsymbol{\nu}'_\beta} = \frac{\partial}{\partial \boldsymbol{\nu}_\alpha}\left(\left.\frac{\partial h}{\partial \boldsymbol{\nu}_\beta}\right|_{\eta,\boldsymbol{\tau}_1,\boldsymbol{\nu}'_\beta}\right)\Bigg|_{\eta,\boldsymbol{\tau}_1,\boldsymbol{\nu}'_\alpha}, \tag{5.223}$$

which become

$$\left.\frac{\partial \theta}{\partial \boldsymbol{\tau}_1}\right|_{\eta,\boldsymbol{\nu}_\gamma} = -\left.\frac{\partial \boldsymbol{\nu}_1}{\partial \eta}\right|_{\boldsymbol{\tau}_1,\boldsymbol{\nu}_\gamma}, \qquad \left.\frac{\partial \theta}{\partial \boldsymbol{\nu}_\alpha}\right|_{\eta,\boldsymbol{\tau}_1,\boldsymbol{\nu}'_\alpha} = \left.\frac{\partial \boldsymbol{\tau}_\alpha}{\partial \eta}\right|_{\boldsymbol{\tau}_1,\boldsymbol{\nu}_\gamma},$$

$$\left.\frac{\partial \boldsymbol{\tau}_\alpha}{\partial \boldsymbol{\tau}_1}\right|_{\eta,\boldsymbol{\nu}_\gamma} = -\left.\frac{\partial \boldsymbol{\nu}_1}{\partial \boldsymbol{\nu}_\alpha}\right|_{\eta,\boldsymbol{\tau}_1,\boldsymbol{\nu}'_\alpha}, \quad \text{and} \quad \left.\frac{\partial \boldsymbol{\tau}_\alpha}{\partial \boldsymbol{\nu}_\beta}\right|_{\eta,\boldsymbol{\tau}_1,\boldsymbol{\nu}'_\beta} = \left.\frac{\partial \boldsymbol{\tau}_\beta}{\partial \boldsymbol{\nu}_\alpha}\right|_{\eta,\boldsymbol{\tau}_1,\boldsymbol{\nu}'_\alpha}. \tag{5.224}$$

Now it is easy to show that the differentials of the equations of state (5.220), using the Maxwell relations (5.224), can be written as follows:

$$d\theta = \left.\frac{\partial \theta}{\partial \eta}\right|_{\boldsymbol{\tau}_1,\boldsymbol{\nu}_\gamma} d\eta - \left.\frac{\partial \boldsymbol{\nu}_1}{\partial \eta}\right|_{\boldsymbol{\tau}_1,\boldsymbol{\nu}_\gamma} \cdot d\boldsymbol{\tau}_1 + \left.\frac{\partial \boldsymbol{\tau}_\alpha}{\partial \eta}\right|_{\boldsymbol{\tau}_1,\boldsymbol{\nu}_\gamma} \cdot d\boldsymbol{\nu}_\alpha,$$

$$d\boldsymbol{\nu}_1 = -\left.\frac{\partial \theta}{\partial \boldsymbol{\tau}_1}\right|_{\eta,\boldsymbol{\nu}_\gamma} d\eta + \left.\frac{\partial \boldsymbol{\nu}_1}{\partial \boldsymbol{\tau}_1}\right|_{\eta,\boldsymbol{\nu}_\gamma} \cdot d\boldsymbol{\tau}_1 - \left.\frac{\partial \boldsymbol{\tau}_\alpha}{\partial \boldsymbol{\tau}_1}\right|_{\eta,\boldsymbol{\nu}_\gamma} \cdot d\boldsymbol{\nu}_\alpha,$$

$$d\boldsymbol{\tau}_\alpha = \left.\frac{\partial \theta}{\partial \boldsymbol{\nu}_\alpha}\right|_{\eta,\boldsymbol{\tau}_1,\boldsymbol{\nu}'_\alpha} d\eta - \left.\frac{\partial \boldsymbol{\nu}_1}{\partial \boldsymbol{\nu}_\alpha}\right|_{\eta,\boldsymbol{\tau}_1,\boldsymbol{\nu}'_\alpha} \cdot d\boldsymbol{\tau}_1 + \left.\frac{\partial \boldsymbol{\tau}_\alpha}{\partial \boldsymbol{\nu}_\beta}\right|_{\eta,\boldsymbol{\tau}_1,\boldsymbol{\nu}'_\beta} \cdot d\boldsymbol{\nu}_\alpha. \tag{5.225}$$

In addition, using (5.202), (5.158)–(5.160), $(5.137)_2$, and $(5.144)_1$, it can be shown that the differential of the internal energy is given by

$$de = (\theta + \boldsymbol{\tau}_1 \cdot \boldsymbol{\zeta}_1)\, d\eta + \left(\boldsymbol{\tau}_1 \cdot \boldsymbol{\phi}_{11}^{-1}\right) \cdot d\boldsymbol{\tau}_1 + \left(\boldsymbol{\tau}_\alpha - \boldsymbol{\tau}_1 \cdot \left.\frac{\partial \boldsymbol{\tau}_\alpha}{\partial \boldsymbol{\tau}_1}\right|_{\eta,\boldsymbol{\nu}_\gamma}\right) \cdot d\boldsymbol{\nu}_\alpha. \tag{5.226}$$

Gibbs potential

In some experiments, neither the entropy nor the thermostatic volumes are controllable parameters, but temperature and thermostatic stresses may be controllable.

As such, they would be more appropriate choices for independent variables. To change to a temperature and thermostatic stresses representation, we replace the internal energy with its Legendre transform, the Gibbs potential. Thus, given the fundamental relation (5.134), define the specific Gibbs potential as the Legendre transformation (5.203), whose first differential for $\alpha = 2, \dots, n$ is given by

$$dg = \left.\frac{\partial g}{\partial \theta}\right|_{\boldsymbol{\tau}_1, \boldsymbol{\nu}_\gamma} d\theta + \left.\frac{\partial g}{\partial \boldsymbol{\tau}_1}\right|_{\theta, \boldsymbol{\nu}_\gamma} \cdot d\boldsymbol{\tau}_1 + \left.\frac{\partial g}{\partial \boldsymbol{\nu}_\alpha}\right|_{\theta, \boldsymbol{\tau}_1, \boldsymbol{\nu}'_\alpha} \cdot d\boldsymbol{\nu}_\alpha, \tag{5.227}$$

where we have reverted to the implied summation convention. Then, differentiating (5.203) and using (5.139), we obtain

$$dg = de - \eta\, d\theta - \theta\, d\eta - \boldsymbol{\tau}_1 \cdot d\boldsymbol{\nu}_1 - \boldsymbol{\nu}_1 \cdot d\boldsymbol{\tau}_1 = -\eta\, d\theta - \boldsymbol{\nu}_1 \cdot d\boldsymbol{\tau}_1 + \boldsymbol{\tau}_\alpha \cdot d\boldsymbol{\nu}_\alpha. \tag{5.228}$$

Subsequently, comparing the above expressions, we see that

$$-\eta = \left.\frac{\partial g}{\partial \theta}\right|_{\boldsymbol{\tau}_1, \boldsymbol{\nu}_\gamma}, \qquad -\boldsymbol{\nu}_1 = \left.\frac{\partial g}{\partial \boldsymbol{\tau}_1}\right|_{\theta, \boldsymbol{\nu}_\gamma} \quad \text{and} \quad \boldsymbol{\tau}_\alpha = \left.\frac{\partial g}{\partial \boldsymbol{\nu}_\alpha}\right|_{\theta, \boldsymbol{\tau}_1, \boldsymbol{\nu}'_\alpha}, \tag{5.229}$$

so that

$$e = g - \theta \left.\frac{\partial g}{\partial \theta}\right|_{\boldsymbol{\tau}_1, \boldsymbol{\nu}_\gamma} - \boldsymbol{\tau}_1 \cdot \left.\frac{\partial g}{\partial \boldsymbol{\tau}_1}\right|_{\theta, \boldsymbol{\nu}_\gamma}, \tag{5.230}$$

and then from (5.165), using $(5.229)_2$ and $(5.150)_2$, we see that

$$c_{\boldsymbol{\tau}_1} = -\theta \left.\frac{\partial^2 g}{\partial \theta^2}\right|_{\boldsymbol{\tau}_1}. \tag{5.231}$$

The corresponding Maxwell relations are given by

$$\begin{gathered}
\left.\frac{\partial}{\partial \boldsymbol{\tau}_1}\left(\left.\frac{\partial g}{\partial \theta}\right|_{\boldsymbol{\tau}_1, \boldsymbol{\nu}_\gamma}\right)\right|_{\theta, \boldsymbol{\nu}_\gamma} = \left.\frac{\partial}{\partial \theta}\left(\left.\frac{\partial g}{\partial \boldsymbol{\tau}_1}\right|_{\theta, \boldsymbol{\nu}_\gamma}\right)\right|_{\boldsymbol{\tau}_1, \boldsymbol{\nu}_\gamma}, \\
\left.\frac{\partial}{\partial \boldsymbol{\nu}_\alpha}\left(\left.\frac{\partial g}{\partial \theta}\right|_{\boldsymbol{\tau}_1, \boldsymbol{\nu}_\gamma}\right)\right|_{\theta, \boldsymbol{\tau}_1, \boldsymbol{\nu}'_\alpha} = \left.\frac{\partial}{\partial \theta}\left(\left.\frac{\partial g}{\partial \boldsymbol{\nu}_\alpha}\right|_{\theta, \boldsymbol{\tau}_1, \boldsymbol{\nu}'_\alpha}\right)\right|_{\boldsymbol{\tau}_1, \boldsymbol{\nu}_\gamma}, \\
\left.\frac{\partial}{\partial \boldsymbol{\nu}_\alpha}\left(\left.\frac{\partial g}{\partial \boldsymbol{\tau}_1}\right|_{\theta, \boldsymbol{\nu}_\gamma}\right)\right|_{\theta, \boldsymbol{\tau}_1, \boldsymbol{\nu}'_\alpha} = \left.\frac{\partial}{\partial \boldsymbol{\tau}_1}\left(\left.\frac{\partial g}{\partial \boldsymbol{\nu}_\alpha}\right|_{\theta, \boldsymbol{\tau}_1, \boldsymbol{\nu}'_\alpha}\right)\right|_{\theta, \boldsymbol{\nu}_\gamma}, \\
\text{and} \quad \left.\frac{\partial}{\partial \boldsymbol{\nu}_\beta}\left(\left.\frac{\partial g}{\partial \boldsymbol{\nu}_\alpha}\right|_{\theta, \boldsymbol{\tau}_1, \boldsymbol{\nu}'_\alpha}\right)\right|_{\theta, \boldsymbol{\tau}_1, \boldsymbol{\nu}'_\beta} = \left.\frac{\partial}{\partial \boldsymbol{\nu}_\alpha}\left(\left.\frac{\partial g}{\partial \boldsymbol{\nu}_\beta}\right|_{\theta, \boldsymbol{\tau}_1, \boldsymbol{\nu}'_\beta}\right)\right|_{\theta, \boldsymbol{\tau}_1, \boldsymbol{\nu}'_\alpha},
\end{gathered} \tag{5.232}$$

which become

$$\begin{gathered}
\left.\frac{\partial \eta}{\partial \boldsymbol{\tau}_1}\right|_{\theta, \boldsymbol{\nu}_\gamma} = \left.\frac{\partial \boldsymbol{\nu}_1}{\partial \theta}\right|_{\boldsymbol{\tau}_1, \boldsymbol{\nu}_\gamma}, \qquad -\left.\frac{\partial \eta}{\partial \boldsymbol{\nu}_\alpha}\right|_{\theta, \boldsymbol{\tau}_1, \boldsymbol{\nu}'_\alpha} = \left.\frac{\partial \boldsymbol{\tau}_\alpha}{\partial \theta}\right|_{\boldsymbol{\tau}_1, \boldsymbol{\nu}_\gamma}, \\
-\left.\frac{\partial \boldsymbol{\nu}_1}{\partial \boldsymbol{\nu}_\alpha}\right|_{\theta, \boldsymbol{\tau}_1, \boldsymbol{\nu}'_\alpha} = \left.\frac{\partial \boldsymbol{\tau}_\alpha}{\partial \boldsymbol{\tau}_1}\right|_{\theta, \boldsymbol{\nu}_\gamma}, \quad \text{and} \quad \left.\frac{\partial \boldsymbol{\tau}_\alpha}{\partial \boldsymbol{\nu}_\beta}\right|_{\theta, \boldsymbol{\tau}_1, \boldsymbol{\nu}'_\beta} = \left.\frac{\partial \boldsymbol{\tau}_\beta}{\partial \boldsymbol{\nu}_\alpha}\right|_{\theta, \boldsymbol{\tau}_1, \boldsymbol{\nu}'_\alpha}.
\end{gathered} \tag{5.233}$$

Now it is easy to show that the differentials of the equations of state (5.229), using the Maxwell relations (5.233), can be written as follows:

$$
\begin{aligned}
d\eta &= \left.\frac{\partial \eta}{\partial \theta}\right|_{\boldsymbol{\tau}_1,\boldsymbol{\nu}_\gamma} d\theta + \left.\frac{\partial \boldsymbol{\nu}_1}{\partial \theta}\right|_{\boldsymbol{\tau}_1,\boldsymbol{\nu}_\gamma} \cdot d\boldsymbol{\tau}_1 - \left.\frac{\partial \boldsymbol{\tau}_\alpha}{\partial \theta}\right|_{\boldsymbol{\tau}_1,\boldsymbol{\nu}_\gamma} \cdot d\boldsymbol{\nu}_\alpha, \\
d\boldsymbol{\nu}_1 &= \left.\frac{\partial \eta}{\partial \boldsymbol{\tau}_1}\right|_{\theta,\boldsymbol{\nu}_\gamma} d\theta + \left.\frac{\partial \boldsymbol{\nu}_1}{\partial \boldsymbol{\tau}_1}\right|_{\theta,\boldsymbol{\nu}_\gamma} \cdot d\boldsymbol{\tau}_1 - \left.\frac{\partial \boldsymbol{\tau}_\alpha}{\partial \boldsymbol{\tau}_1}\right|_{\theta,\boldsymbol{\nu}_\gamma} \cdot d\boldsymbol{\nu}_\alpha, \\
d\boldsymbol{\tau}_\alpha &= -\left.\frac{\partial \eta}{\partial \boldsymbol{\nu}_\alpha}\right|_{\theta,\boldsymbol{\tau}_1,\boldsymbol{\nu}'_\alpha} d\theta - \left.\frac{\partial \boldsymbol{\nu}_1}{\partial \boldsymbol{\nu}_\alpha}\right|_{\theta,\boldsymbol{\tau}_1,\boldsymbol{\nu}'_\alpha} \cdot d\boldsymbol{\tau}_1 + \left.\frac{\partial \boldsymbol{\tau}_\alpha}{\partial \boldsymbol{\nu}_\beta}\right|_{\theta,\boldsymbol{\tau}_1,\boldsymbol{\nu}'_\beta} \cdot d\boldsymbol{\nu}_\beta.
\end{aligned} \tag{5.234}
$$

In addition, using (5.203), (5.158)–(5.160), and (5.150), it can be shown that the differential of the internal energy is given by

$$
de = (c_{\boldsymbol{\tau}_1} + \boldsymbol{\tau}_1 \cdot \boldsymbol{\alpha}_1)\, d\theta + (\boldsymbol{\lambda}_{\boldsymbol{\tau}_1} + \boldsymbol{\tau}_1 \cdot \boldsymbol{v}_{11}) \cdot d\boldsymbol{\tau}_1 + \left(\boldsymbol{\lambda}_{\boldsymbol{\nu}_\alpha} + \boldsymbol{\tau}_\alpha - \boldsymbol{\tau}_1 \cdot \left.\frac{\partial \boldsymbol{\tau}_\alpha}{\partial \boldsymbol{\tau}_1}\right|_{\theta,\boldsymbol{\nu}_\gamma} \right) \cdot d\boldsymbol{\nu}_\alpha. \tag{5.235}
$$

5.10.3 Thermodynamic processes

When any property of a system changes in value, there is a change in state, and the system is said to undergo a process.

Definition: For a given material point X, the thermodynamic state $(\eta(t), \boldsymbol{\nu}_\alpha(t))$, for variable t, defines a *thermodynamic path.* The path on which η = const. is called *isentropic* and that with θ = const. is called *isothermal.*

Now, for fixed X, from the Gibbs equation (5.139), we may also write

$$
\dot{e} = \theta\, \dot{\eta} + \boldsymbol{\tau}_\alpha \cdot \dot{\boldsymbol{\nu}}_\alpha. \tag{5.236}
$$

For thermal changes, diffusion, and chemical phenomena, different types of effects make up the entropy flux $\mathbf{h}$ and the entropy source b in (4.219) and (4.220). It is always possible to express the entropy flux and entropy source as

$$
\mathbf{h} = \frac{\mathbf{q}}{\theta} + \mathbf{h}_1 \qquad \text{and} \qquad b = \frac{r}{\theta} + b_1, \tag{5.237}
$$

where $\mathbf{q}/\theta$ is the entropy flux due to heat input, r/θ is the entropy source supplied by the energy source, and the remaining terms $\mathbf{h}_1$ and b_1 are respectively the entropy flux and source due to all other effects. Above, we have assumed the entropy flux and source due to heat to be of a specific form. Later we shall prove that these forms are indeed correct for a simple material.

Using (5.237) to substitute for r in the energy equation (4.207) and solving the resulting equation for ρb, we obtain

$$
\rho b = \rho b_1 + \rho \frac{\dot{e}}{\theta} - \frac{\Phi}{\theta} + \frac{1}{\theta} \operatorname{div} \mathbf{q}. \tag{5.238}
$$

Now substituting this result as well as the expression for $\mathbf{h}$ into the local entropy inequalities (4.219) and (4.220), we have

$$
\gamma_v \equiv \rho \left(\dot{\eta} - \frac{\dot{e}}{\theta} \right) + \frac{\Phi}{\theta} - \frac{1}{\theta^2} \mathbf{g} \cdot \mathbf{q} + \operatorname{div} \mathbf{h}_1 - \rho b_1 \ge 0, \tag{5.239}
$$

where we recall that

$$\mathbf{g} \equiv \operatorname{grad}\theta, \tag{5.240}$$

and

$$\gamma_s \equiv \left[\!\left[\rho\eta(\mathbf{v}-\mathbf{c}) + \frac{\mathbf{q}}{\theta} + \mathbf{h}_1\right]\!\right]\cdot\mathbf{n} \geq 0. \tag{5.241}$$

Definition: A thermodynamic process in which $\mathbf{h}_1 = \mathbf{0}$ and $b_1 = 0$ is called a *simple thermomechanical process.*

A consequence of the above definition is an expression for temperature as the common ratio of

$$\theta = \frac{|\mathbf{q}|}{|\mathbf{h}|} = \frac{r}{b}. \tag{5.242}$$

For a simple thermomechanical process, the global form of the entropy inequality, using (4.28), (4.29), (4.189)–(4.191), (4.204), and (5.242), is given by

$$\Gamma \equiv \frac{d\mathcal{S}}{dt} + \int_{\mathscr{S}} \frac{\mathbf{q}}{\theta}\cdot d\mathbf{s} - \int_{\mathscr{V}} \frac{\rho r}{\theta}\, dv \geq 0. \tag{5.243}$$

For an *adiabatic process* with no external energy sources $\mathbf{q} = \mathbf{0}$ and $r = 0$, we have that $\dot{\mathcal{S}} \geq 0$, i.e., entropy cannot decrease. The local forms of the entropy inequality for a simple thermomechanical process become

$$\gamma_v \equiv \rho\left(\dot{\eta} - \frac{\dot{e}}{\theta}\right) + \frac{1}{\theta}\mathbf{L}:\boldsymbol{\sigma} - \frac{1}{\theta^2}\mathbf{g}\cdot\mathbf{q} \geq 0$$
$$\text{or}\quad \gamma_v \equiv \rho\left(\dot{\eta} - \frac{\dot{e}}{\theta}\right) + \frac{1}{\theta}L_{lk}\sigma_{lk} - \frac{1}{\theta^2}\theta_{,k}q_k \geq 0, \tag{5.244}$$

and

$$\gamma_s \equiv \left\{\rho\,[\![\eta]\!]\,(\mathbf{v}-\mathbf{c}) + \left[\!\left[\frac{\mathbf{q}}{\theta}\right]\!\right]\right\}\cdot\mathbf{n} \geq 0 \;\text{ or }\; \gamma_s \equiv \left\{\rho\,[\![\eta]\!]\,(v_k - c_k) + \left[\!\left[\frac{q_k}{\theta}\right]\!\right]\right\} n_k \geq 0. \tag{5.245}$$

Definition: A thermodynamic process is said to be *mechanically admissible* if it obeys the conservation of mass, the balance of momenta, and the balance of energy. It is called *constitutively admissible* if it satisfies constitutive restrictions, such as material frame indifference, symmetries, etc.

Definition: A process will be called *thermodynamically admissible* if and only if it obeys the local entropy inequalities and possesses a positive-definite temperature, i.e.,

$$\gamma_v \geq 0, \quad \gamma_s \geq 0 \qquad \text{and} \qquad 0 < \theta < \infty. \tag{5.246}$$

Definition: A process will be called a *reversible process* if and only if $\gamma_v = \gamma_s = 0$. Note that, for an isolated system, a reversible adiabatic process is an isentropic process.

In a simple mechanically admissible process, ρ, $\dot{\mathbf{v}}$, $\boldsymbol{\sigma}$, e, and $\mathbf{q}$ must satisfy the equations of mass, balance of momenta, and balance of energy. Note that our equations do not account for interactions between matter in our system and the external body and energy sources. For a constitutively admissible process, various restrictions also have to be satisfied. One of these is the principle of equipresence.

Thus, e.g., when e is given by (5.134), the constitutive equations for the stress tensor and the heat flux must also use the same independent variables, i.e.,

$$\boldsymbol{\sigma} = \boldsymbol{\sigma}(\eta, \boldsymbol{\nu}_\alpha, X) \qquad \text{and} \qquad \mathbf{q} = \mathbf{q}(\eta, \boldsymbol{\nu}_\alpha, X). \tag{5.247}$$

The above functions, along with that for e, are subject to additional constitutive requirements, so that some of the variables from the chosen list may be shown not to be present in the arguments of some of the constitutive functions in view of other restrictions.

To illustrate this, using the constitutive equation provided by the fundamental relation $e = e(\eta, \boldsymbol{\nu}_\alpha, X)$, we can rewrite the entropy inequality (5.244) in the form

$$\gamma_v \theta \equiv \rho \dot{\eta} \left(\theta - \left. \frac{\partial e}{\partial \eta} \right|_{\boldsymbol{\nu}_\alpha} \right) - \rho \dot{\boldsymbol{\nu}}_\alpha \cdot \left. \frac{\partial e}{\partial \boldsymbol{\nu}_\alpha} \right|_{\eta, \boldsymbol{\nu}'_\alpha} + \mathbf{L} : \boldsymbol{\sigma} - \operatorname{grad}(\log \theta) \cdot \mathbf{q} \geq 0. \tag{5.248}$$

The quantity $\gamma_v \theta$ is called the *dissipation*. Now at material point X, e, θ, $\boldsymbol{\sigma}$, and $\mathbf{q}$ are functions of η and $\boldsymbol{\nu}_\alpha$. This inequality, which is linear in $\dot{\eta}$, must be valid for all values of $\dot{\eta}$. The independent variable $\dot{\eta}$ appears only in the first term. For the equation to be valid for all values of $\dot{\eta}$, we must set its coefficient to zero:

$$\theta = \left. \frac{\partial e}{\partial \eta} \right|_{\boldsymbol{\nu}_\alpha}. \tag{5.249}$$

The above is the same expression as that for the thermostatic temperature (5.136). Thus, we have shown that for a thermodynamically admissible simple thermomechanical process characterized by the set of state variables $(\eta, \boldsymbol{\nu}_\alpha)$ with $\boldsymbol{\nu}_\alpha$ being functionally independent of η and not containing time rates or integrals of η, the temperature and the thermostatic temperature are the same. It subsequently follows that the tensions

$$\boldsymbol{\tau}_\alpha = \left. \frac{\partial e}{\partial \boldsymbol{\nu}_\alpha} \right|_{\eta, \boldsymbol{\nu}'_\alpha} \tag{5.250}$$

and the thermostatic tensions (5.136) are also the same, and the entropy inequality becomes

$$\gamma_v \theta \equiv -\rho \dot{\boldsymbol{\nu}}_\alpha \cdot \boldsymbol{\tau}_\alpha + \mathbf{L} : \boldsymbol{\sigma} - \operatorname{grad}(\log \theta) \cdot \mathbf{q} \geq 0. \tag{5.251}$$

In the expression of the production of entropy, inner products of vectorial and tensorial quantities occur in pairs. In the thermodynamics of irreversible processes, it has become customary to refer to one set of the pairs as *thermodynamic forces* or *affinities* and the conjugate set multiplying the forces as the *thermodynamic fluxes*. For example, from the above entropy inequality, one may select the pairs

Force	*Flux*
$-\rho \dot{\boldsymbol{\nu}}_\alpha$	$\boldsymbol{\tau}_\alpha$
$\mathbf{L}$	$\boldsymbol{\sigma}$
$-\operatorname{grad}(\log \theta)$	$\mathbf{q}$

Linear constitutive equations are then set up between any one of the fluxes and all of the forces with symmetric constitutive coefficients. This is known as *Onsager's principle*. We will not follow this thermodynamic approach to constitutive theory since it is fraught with potential problems, among them the fact that the choice of forces and fluxes is not unique.

We note that for an isolated system, it is more convenient to take the constitutive equation corresponding to the entropy form of the fundamental relation (5.133): $\eta = \eta(e, \boldsymbol{\nu}_\alpha, X)$, where now e, $\boldsymbol{\nu}_\alpha$, and X are the independent variables. For a system in contact with a heat bath at a given temperature, θ replaces e to become an independent variable, or control parameter. The energy e and entropy η will then vary, so that e and η become dependent variables given by equations of state. In this case, it is more appropriate to consider the Helmholtz free energy form of the fundamental relation. A system with fixed external parameters in thermal contact with a heat reservoir at equilibrium has a minimum Helmholtz free energy form of the fundamental relation. Subsequently, a different form of the entropy inequality valid at a regular point is obtained by introducing the specific Helmholtz free energy. Using (5.201) and noting that $\dot{\psi} = \dot{e} - \dot{\theta}\eta - \theta\dot{\eta}$, the entropy inequality (5.239) for a general process, called the *Clausius–Duhem inequality*, becomes

$$-\gamma_v \theta \equiv \rho\left(\dot{\psi} + \eta\dot{\theta}\right) - \mathbf{L} : \boldsymbol{\sigma} + \mathbf{g} \cdot \frac{\mathbf{q}}{\theta} - \theta \operatorname{div}\left(\mathbf{h} - \frac{\mathbf{q}}{\theta}\right) + \rho\theta\left(b - \frac{r}{\theta}\right) \leq 0. \tag{5.252}$$

Note that the above equation is more general than the classical Clausius–Duhem inequality. Classically, the last two terms on the left-hand side of the inequality are set to zero. This implies the fundamental assumptions that for a simple thermomechanical process, $\mathbf{h} = \mathbf{q}/\theta$ and $b = r/\theta$. One can show that generally these assumptions are not valid and a more careful analysis is necessary. However, performing the analysis more carefully, we will prove that they are valid for simple thermoelastic solids and fluids.

5.10.4 Thermodynamic equilibrium and stability

A *nonequilibrium state* is a state where spatial and/or temporal variations of velocity or temperature exist. Alternately, an *equilibrium state* is a persistent state in which no spatial or temporal variations of velocity or temperature exist. A system at equilibrium has no tendency to change when it is isolated from the surroundings. If the complete body is at an equilibrium state, then the body is in *thermodynamic equilibrium*. More specifically, in considering the equilibrium state, we consider the body as being isolated, i.e., free of external supplies, $\mathbf{f} = \mathbf{0}$, $r = b = 0$, and with no mass, momentum, or heat exchange with the surroundings, i.e.,

$$\mathbf{v} \cdot \mathbf{n} = 0, \quad \boldsymbol{\sigma} \cdot \mathbf{n} = \mathbf{0}, \quad \mathbf{q} \cdot \mathbf{n} = 0, \quad \mathbf{h} \cdot \mathbf{n} = 0 \quad \text{on} \quad \mathscr{S}. \tag{5.253}$$

In such case, using (4.187) for a nonpolar material and (4.204) (and using the definitions (4.18), (4.26), and (4.28)), we have that

$$\dot{\mathcal{E}} + \dot{\mathcal{K}} = 0 \qquad \text{and} \qquad \dot{\mathcal{S}} \geq 0. \tag{5.254}$$

In other words, for an isolated body with constant total energy, the total entropy must not decrease in time.

Definition: A body is said to be in *thermodynamic equilibrium* if, when free of external supplies, it is in mechanical and thermal equilibrium and the total entropy remains constant.

Clearly, the above implies that at thermodynamic equilibrium we have zero entropy production, i.e., $\gamma_v = \gamma_s = 0$. In contradistinction, it should be noted that for a reversible process, $\gamma_v = \gamma_s = 0$ but $\dot{\eta} \neq 0$, and in such case, we have a time-dependent nonequilibrium process.

Frequently, it is convenient to study the conditions for equilibrium using a somewhat different perspective. Instead of considering the approach to equilibrium, we focus on the equilibrium state itself. To establish the conditions that are satisfied subject to internal or external constraints, we ask about the characteristics of neighboring imaginary states that can conceivably be reached by displacing the system from equilibrium. Such displacements from equilibrium are called *virtual*. A virtual displacement is a reversible process whereby conditions are created which permit the imaginary insertion of a supplementary internal constraint into the system. The subsequent removal of this internal constraint would induce the system to reach the equilibrium state, assuming that the external constraints have remained unchanged. We employ the symbol δ to denote a small, virtual change of a given quantity associated with the internal constraints, and make use of the equations of thermodynamics in a form appropriate to the reversible process.

We mention five criteria of equilibrium in terms of virtual displacements. Each criterion applies to a different set of constraints. In each case, we use the appropriate thermodynamic potential. To begin with, a fundamental postulate of thermodynamics is that the values assumed by the independent quantities, in the absence of internal constraints, are such as to maximize the entropy. This basic extremum principle implies that $\delta\eta = 0$ and $\delta^2\eta < 0$, i.e., entropy is a maximum at the thermodynamic equilibrium.

Entropy maximum principle: For a given value of the total internal energy, the equilibrium value of any unconstrained internal quantity is such as to maximize the entropy.

An equivalent statement of thermodynamic stability is that the internal energy is a minimum with respect to all virtual displacements at equilibrium: $\delta e = 0$ and $\delta^2 e > 0$.

Internal energy minimum principle: For a given value of the total entropy, the equilibrium value of any unconstrained internal quantity is such as to minimize the internal energy.

The first condition,

$$\delta e = \theta\, \delta\eta + \boldsymbol{\tau}_\alpha \cdot \delta\boldsymbol{\nu}_\alpha = 0, \tag{5.255}$$

for any $\delta\eta$ and $\delta\boldsymbol{\nu}_\alpha$, except the trivial case $\delta\eta = 0$ and $\delta\boldsymbol{\nu}_\alpha = \mathbf{0}$, leads to the equality of temperatures and tensions of unconstrained subsystems and the environment and confirms our original assumption of equilibrium corresponding to a state where no temperature and velocity variations exist. To examine the consequences of the second condition, using the fundamental relation (5.134), for any infinitesimal variation, we have

$$\begin{aligned}
\delta^2 e &= \frac{\partial}{\partial \eta}\left(\left.\frac{\partial e}{\partial \eta}\right|_{\boldsymbol{\nu}_\gamma}\right)\Bigg|_{\boldsymbol{\nu}_\gamma} (\delta\eta)^2 + 2\,\frac{\partial}{\partial \boldsymbol{\nu}_\alpha}\left(\left.\frac{\partial e}{\partial \eta}\right|_{\boldsymbol{\nu}_\gamma}\right)\Bigg|_{\eta,\boldsymbol{\nu}'_\alpha} \cdot \delta\eta\delta\boldsymbol{\nu}_\alpha + \\
&\qquad \frac{\partial}{\partial \boldsymbol{\nu}_\beta}\left(\left.\frac{\partial e}{\partial \boldsymbol{\nu}_\alpha}\right|_{\eta,\boldsymbol{\nu}'_\alpha}\right)\Bigg|_{\eta,\boldsymbol{\nu}'_\beta} : \delta\boldsymbol{\nu}_\alpha\delta\boldsymbol{\nu}_\beta \\
&= \left.\frac{\partial \theta}{\partial \eta}\right|_{\boldsymbol{\nu}_\gamma} (\delta\eta)^2 + 2\left.\frac{\partial \theta}{\partial \boldsymbol{\nu}_\alpha}\right|_{\eta,\boldsymbol{\nu}'_\alpha} \cdot \delta\eta\delta\boldsymbol{\nu}_\alpha + \left.\frac{\partial \boldsymbol{\tau}_\alpha}{\partial \boldsymbol{\nu}_\beta}\right|_{\eta,\boldsymbol{\nu}'_\beta} : \delta\boldsymbol{\nu}_\alpha\delta\boldsymbol{\nu}_\beta \\
&= \frac{\theta}{c_{\boldsymbol{\nu}_\alpha}}(\delta\eta)^2 + 2\,\boldsymbol{\varphi}_\alpha \cdot \delta\eta\delta\boldsymbol{\nu}_\alpha + \boldsymbol{\phi}_{\alpha\beta} : \delta\boldsymbol{\nu}_\alpha\delta\boldsymbol{\nu}_\beta > 0,
\end{aligned} \tag{5.256}$$

where we have used (5.136), (5.137), and (5.165). The stability condition is that the right-hand side must be positive definite for any $\delta\eta$ and $\delta\boldsymbol{\nu}_\alpha$, except the trivial case $\delta\eta = 0$ and $\delta\boldsymbol{\nu}_\alpha = \mathbf{0}$. The right-hand side is a homogeneous quadratic form in the variables $\delta\eta$ and $\delta\boldsymbol{\nu}_\alpha$. The coefficients of the quadratic form and the stability requirement can be rewritten in the symmetric matrix form

$$H_e = \begin{pmatrix} \theta/c_{\boldsymbol{\nu}_\alpha} & \boldsymbol{\varphi}_\alpha \\ \boldsymbol{\varphi}_\alpha & \boldsymbol{\phi}_{\alpha\beta} \end{pmatrix} > 0. \tag{5.257}$$

From previous results, we know that a symmetric matrix has real eigenvalues and it is positive definite if all its eigenvalues are positive. Alternately, we use Sylvester's criterion, which states that a matrix $H_e = [h_{ij}]$ is positive definite if and only if the determinants of all of its principal minors are positive, i.e.,

$$h_{11} > 0, \qquad \begin{vmatrix} h_{11} & h_{12} \\ h_{21} & h_{22} \end{vmatrix} > 0, \qquad \begin{vmatrix} h_{11} & h_{12} & h_{13} \\ h_{21} & h_{22} & h_{23} \\ h_{31} & h_{32} & h_{33} \end{vmatrix} > 0, \qquad \text{etc.} \tag{5.258}$$

To illustrate the application of the stability condition, we take the case where $n = 1$ with $\boldsymbol{\nu}_1$ being the specific volume v and thus write

$$H_e = \begin{pmatrix} \theta/c_{\mathsf{v}} & \varphi \\ \varphi & \phi \end{pmatrix} > 0. \tag{5.259}$$

The trace and determinant of the matrix H_e correspond to the sum and products of the eigenvalues; thus we can write

$$\frac{\theta}{c_{\mathsf{v}}} + \phi = \lambda_1 + \lambda_2 > 0 \qquad \text{and} \qquad \frac{\theta}{c_{\mathsf{v}}}\phi - \varphi^2 = \lambda_1\lambda_2 > 0. \tag{5.260}$$

Now, solving for λ_1 and λ_2, and requiring that $\lambda_1 > 0$ and $\lambda_2 > 0$, we obtain the thermodynamic restrictions. Alternately, using Sylvester's criterion, we easily see that these conditions are satisfied if and only if

$$\frac{\theta}{c_{\mathsf{v}}} > 0, \qquad \phi > 0, \quad \text{and} \quad \frac{\theta}{c_{\mathsf{v}}}\phi > \varphi^2. \tag{5.261}$$

Since $\theta > 0$, the first condition results in

$$c_v > 0. \tag{5.262}$$

The second condition, using (5.136), (5.137), and (5.162), results in

$$\phi = \left.\frac{\partial^2 e}{\partial v^2}\right|_\eta = -\left.\frac{\partial p}{\partial v}\right|_\eta = \frac{1}{v\,\kappa_\eta} > 0, \tag{5.263}$$

or, since $v > 0$,

$$\kappa_\eta > 0. \tag{5.264}$$

To examine the third and last condition, using (5.150), (5.164), $(5.176)_3$, and (5.177), we easily see that

$$\varphi = \frac{\theta\,\alpha}{\kappa_\theta\,c_v}. \tag{5.265}$$

Subsequently, the condition becomes

$$\frac{\theta}{v\,\kappa_\eta c_v} > \left(\frac{\theta\,\alpha}{\kappa_\theta c_v}\right)^2. \tag{5.266}$$

Now, using relations (5.163) and (5.164) along with the previous result that $\kappa_\eta > 0$, it is easy to show that the third condition yields

$$\kappa_\theta > 0. \tag{5.267}$$

Finally, using the relations (5.163) and (5.164) once more, it is easy to see that

$$c_p > c_v > 0 \qquad \text{and} \qquad \kappa_\theta > \kappa_\eta > 0. \tag{5.268}$$

These results, with the use of (5.163), also imply that the magnitude of the coefficient of volume expansion is bounded by

$$\alpha^2 < \frac{c_p\,\kappa_\theta}{\theta\,v}, \tag{5.269}$$

and, with the use of (5.164), that the ratio of specific heats or compressibility moduli is bounded by

$$\gamma > 1. \tag{5.270}$$

We now formally state the additional equivalent principles pertaining to the stability of thermodynamic equilibrium.

Helmholtz free energy minimum principle: In a system in diathermal contact with a heat reservoir, the equilibrium value of any unconstrained internal quantity is such as to minimize the Helmholtz free energy at constant temperature (equal to that of the heat reservoir).

Enthalpy minimum principle: In a system in contact with reservoirs of thermostatic tensions, the equilibrium value of any unconstrained internal quantity is such as to minimize the enthalpy at constant thermostatic tensions (equal to those of the thermostatic tensions reservoirs).

Gibbs free energy minimum principle: In a system in contact with a temperature and thermostatic tensions reservoirs, the equilibrium value of any unconstrained internal quantity is such as to minimize the Gibbs free energy at constant temperature and thermostatic tensions (equal to those of the respective reservoirs).

A process in which a system that is in a thermodynamic equilibrium state proceeds by an infinitely slow (in imagination) evolution to another neighboring state of thermodynamic equilibrium is called a *quasi-static* or *quasi-equilibrium process.*

Definition: A process will be called an *equilibrium process* if and only if the system is at thermodynamic equilibrium.

Definition: A process will be called a *thermostatic process* if the process consists of an infinite sequence of quasi-static processes.

5.10.5 Potential energy and strain energy

When the body forces are *steady and derivable from a potential* $U(\mathbf{x})$, i.e.,

$$\mathbf{f} = -\text{grad } U \qquad \text{or} \qquad f_k = -U_{,k}, \tag{5.271}$$

then the work done by body forces, in using Reynolds' transport theorem (3.466), can be rewritten as

$$\int_{\mathscr{V}} \rho v_k f_k dv = -\int_{\mathscr{V}} \rho v_k U_{,k} dv = -\dot{\mathcal{U}}, \tag{5.272}$$

where

$$\mathcal{U} \equiv \int_{\mathscr{V}} \rho U dv \tag{5.273}$$

is called the *potential energy.* Upon substituting this into the energy balance (4.187) for a nonpolar material, we have

$$\dot{\mathcal{E}} + \dot{\mathcal{K}} + \dot{\mathcal{U}} = \int_{\mathscr{S}} (\mathbf{v} \cdot \boldsymbol{\sigma} - \mathbf{q}) \cdot d\mathbf{s} + \int_{\mathscr{V}} \rho r \; dv. \tag{5.274}$$

When the work of the surface tractions is zero, the body is *insulated*, and there is no external energy supply, we have the *balance of mechanical energy*, which states that the sum of internal, kinetic, and potential energies is constant:

$$\mathcal{E} + \mathcal{K} + \mathcal{U} = \text{const.} \tag{5.275}$$

If $\mathcal{E} = 0$, we obtain the local principle of conservation of energy of classical mechanics, which is obtained by integrating Newton's second law for a particle when the force is derived from a potential.

It is natural to inquire whether a similar situation regarding the existence of a potential which leads to the appearance of a recoverable work term in the form of potential energy can exist in the presence of surface tractions. To this end, we assume that the stress tensor can be rewritten as the sum of two symmetric stress tensors

$$\boldsymbol{\sigma} = \boldsymbol{\sigma}^e + \boldsymbol{\sigma}^d, \tag{5.276}$$

where $\boldsymbol{\sigma}^e$ is the elastic, recoverable or reversible, part, and $\boldsymbol{\sigma}^d$ is the dissipative, or irreversible, part. We assume that the recoverable stress is derivable from a potential $\tau = \tau(\mathbf{F})$ called the *strain energy function* or *elastic potential function*, i.e.,

$$\boldsymbol{\sigma}^e = \rho \frac{\partial \tau(\mathbf{F})}{\partial \mathbf{F}} \cdot \mathbf{F}^T \qquad \text{or} \qquad \sigma^e_{kl} = \rho \frac{\partial \tau(\mathbf{F})}{\partial F_{kL}} F_{lL}. \tag{5.277}$$

We now note that

$$\rho \dot{\tau} = \rho \frac{\partial \tau(\mathbf{F})}{\partial F_{kL}} \dot{F}_{kL} = \rho \frac{\partial \tau(\mathbf{F})}{\partial F_{kL}} L_{kl} F_{lL} = L_{kl} \sigma^e_{kl} = (D_{kl} + W_{kl})\, \sigma^e_{kl} \tag{5.278}$$

or, since $\boldsymbol{\sigma}^e$ is symmetric,

$$\rho \dot{\tau} = \mathbf{D} : \boldsymbol{\sigma}^e. \tag{5.279}$$

Subsequently, again using the fact that the stress tensor is symmetric, we have

$$\begin{aligned}
\int_{\mathscr{S}} (\mathbf{v} \cdot \boldsymbol{\sigma}) \cdot d\mathbf{s} &= \int_{\mathscr{V}} \operatorname{div}(\mathbf{v} \cdot \boldsymbol{\sigma})\, dv \\
&= \int_{\mathscr{V}} [(\operatorname{grad} \mathbf{v}) : \boldsymbol{\sigma} + \mathbf{v} \cdot (\operatorname{div} \boldsymbol{\sigma})]\, dv \\
&= \int_{\mathscr{V}} [\mathbf{L}^T : \boldsymbol{\sigma} + \mathbf{v} \cdot (\operatorname{div} \boldsymbol{\sigma})]\, dv \\
&= \int_{\mathscr{V}} [\mathbf{D} : (\boldsymbol{\sigma}^e + \boldsymbol{\sigma}^d) + \mathbf{v} \cdot (\operatorname{div} \boldsymbol{\sigma})]\, dv
\end{aligned}$$

or

$$\int_{\mathscr{S}} (\mathbf{v} \cdot \boldsymbol{\sigma}) \cdot d\mathbf{s} = \int_{\mathscr{V}} [\rho \dot{\tau} + \mathbf{D} : \boldsymbol{\sigma}^d + \mathbf{v} \cdot (\operatorname{div} \boldsymbol{\sigma})]\, dv. \tag{5.280}$$

Upon substituting the above into the energy balance (4.187) for a nonpolar material (and using (4.173) and (4.186)), we have

$$\dot{\mathcal{E}} + \dot{\mathcal{K}} = \int_{\mathscr{V}} [\rho \dot{\tau} + \mathbf{D} : \boldsymbol{\sigma}^d + \mathbf{v} \cdot (\operatorname{div} \boldsymbol{\sigma})]\, dv + \int_{\mathscr{V}} \rho \mathbf{v} \cdot \mathbf{f}\, dv + \dot{\mathcal{Q}}. \tag{5.281}$$

Now assuming that mass is conserved, we have

$$\dot{\mathcal{K}} = \frac{1}{2} \frac{d}{dt} \int_{\mathscr{V}} \rho \mathbf{v} \cdot \mathbf{v}\, dv = \int_{\mathscr{V}} \rho \mathbf{v} \cdot \mathbf{a}\, dv, \tag{5.282}$$

and define the *total strain energy*, the *total dissipative power*, and the *total thermal energy* as

$$\mathcal{T} \equiv \int_{\mathscr{V}} \rho \tau\, dv, \qquad \mathcal{D} \equiv \int_{\mathscr{V}} \mathbf{D} : \boldsymbol{\sigma}^d\, dv, \quad \text{and} \quad \dot{\mathcal{Q}} \equiv \int_{\mathscr{V}} (-\operatorname{div} \mathbf{q} + \rho r)\, dv. \tag{5.283}$$

Subsequently, rearranging the energy balance, we have

$$\dot{\mathcal{E}} = \dot{\mathcal{T}} + \mathcal{D} + \dot{\mathcal{Q}} - \int_{\mathscr{V}} \mathbf{v} \cdot [\rho(\mathbf{a} - \mathbf{f}) - \operatorname{div} \boldsymbol{\sigma}]\, dv. \tag{5.284}$$

or, making use of the balance of linear momentum (4.109), we obtain

$$\dot{\mathcal{E}} = \dot{\mathcal{T}} + \mathcal{D} + \dot{\mathcal{Q}}. \tag{5.285}$$

In the special case where $\tau = e$, the above equation reduces to $\mathcal{D} + \dot{\mathcal{Q}} = 0$, stating that energy dissipation is totally converted to heat.

Another special case is obtained when the strain energy function depends on $J = \det F$ only. Then, using (3.60), we have

$$\sigma^e_{kl} = \rho \frac{\partial \tau(J)}{\partial J} \frac{\partial J}{\partial F_{lL}} F_{kL} = \rho \frac{\partial \tau(J)}{\partial J} J F^{-1}_{Ll} F_{kL}, \tag{5.286}$$

or since $\rho = \rho_R J^{-1}$, we can write more simply

$$\boldsymbol{\sigma}^e = -p\,\mathbf{1} \qquad \text{or} \qquad \sigma^e_{kl} = -p\,\delta_{kl}, \tag{5.287}$$

where we have defined the *elastic hydrostatic pressure* by

$$p \equiv -\rho_R \frac{\partial \tau(J)}{\partial J}. \tag{5.288}$$

Subsequently, from (5.279), we see that

$$\rho \dot{\tau} = -p\, D_{kk} = -p\, v_{k,k}. \tag{5.289}$$

We now show how the first principle of thermostatics is deduced as a special case of the above result. Classical thermostatics deals with homogeneous systems and the stress consists of a purely hydrostatic pressure that is constant ($\boldsymbol{\sigma}^d = \mathbf{0}$). Under these conditions, since from (3.318) we have $\dot{\overline{dv}} = D_{kk}\, dv$, we have

$$\dot{\mathcal{T}} = \int_{\mathcal{V}} \rho \dot{\tau}\, dv = -\int_{\mathcal{V}} p\, D_{kk}\, dv = -p \int_{\mathcal{V}} \dot{\overline{dv}} = -p \frac{d}{dt} \int_{\mathcal{V}} dv = -p\, \dot{\mathcal{V}}, \tag{5.290}$$

where $\mathcal{V}$ is the volume of the body. Since $\mathcal{D} = 0$, upon integration of (5.285), we have

$$d\mathcal{E} = đ\mathcal{Q} + đ\mathcal{W} \tag{5.291}$$

with the interpretation that the above are considered to be changes between neighboring states, and we wrote $\mathcal{T} = \mathcal{W}$,

$$đ\mathcal{Q} = \dot{\mathcal{Q}}\, dt \qquad \text{and} \qquad đ\mathcal{W} = \dot{\mathcal{W}}\, dt = -p\, d\mathcal{V}. \tag{5.292}$$

Equation (5.291) is the expression for *the first law of thermostatics*, which is valid when the system is uniform and explicitly independent of time, the dissipative part of the stress vanishes, and the elastic part of the stress consists of the hydrostatic pressure only. It may be rewritten into a differential form if an integrating factor $1/\theta$ can be found so that we may express $đ\mathcal{Q}$ as a total differential

$$\frac{đ\mathcal{Q}}{\theta} = d\mathcal{S}, \tag{5.293}$$

where $\mathcal{S}$ is the total entropy. The variable θ is the *absolute temperature*, which is defined to be positive definite, i.e., $\theta > 0$. With this substitution, the equation takes the form

$$d\mathcal{E} = \theta\, d\mathcal{S} - p\, d\mathcal{V}. \tag{5.294}$$

5.11 Entropy and nonequilibrium thermodynamics

5.11.1 Coleman–Noll procedure

Most of the classical work involving the derivation of reduced constitutive functions in nonequilibrium thermodynamics is based on the use of the more restricted Clausius–Duhem inequality (5.252), which makes use of the assumptions that $\mathbf{h} = \mathbf{q}/\theta$ and $b = r/\theta$. In our description and application of the Coleman–Noll procedure, we will not make these assumptions.

The analysis proceeds as follows. First, one writes the constitutive equations for $\mathcal{C} = \{\boldsymbol{\sigma}, \mathbf{q}, \mathbf{h}, \psi, \eta\}$ (see (5.2)) as functions of the appropriate reduced independent variables $\mathcal{I}$ (see (5.1)). For example, for a simple homogeneous thermoelastic solid, these would be $\mathcal{I} = \{\mathbf{F}, {}_{(t)}\mathbf{C}^{(t)}, \theta^{(t)}, {}_{(t)}\mathbf{g}^{(t)}\}$ (see (5.99)), while for a simple fluid, they would be $\mathcal{I} = \{\rho, {}_{(t)}\mathbf{C}^{(t)}, \theta^{(t)}, {}_{(t)}\mathbf{g}^{(t)}\}$ (see (5.114)). These constitutive equations are then introduced into the Clausius–Duhem inequality (5.252) and all differentiations performed so that, with the exception of two sets of terms, the resulting inequality involves a number of terms each multiplied linearly by an independent variable or a derivative of an independent variable. Now since the inequality must be satisfied for arbitrary external sources and arbitrary variations of the independent variables, the terms that involve the sources or that are multiplied linearly by an independent variable or one of its derivatives must be set to zero. Subsequently, one arrives at what is called the *residual inequality.* Analysis of the terms that are set to zero, along with the use of integrability conditions, the residual inequality, and its analysis in the equilibrium limit, provides specific constitutive forms for $\mathcal{C} = \{\boldsymbol{\sigma}, \mathbf{q}, \mathbf{h}, \psi, \eta\}$. Details of the procedure are given in Chapter 7 for thermoelastic solids and Chapter 8 for fluids.

5.11.2 Müller–Liu procedure and Lagrange multipliers

The classical Coleman–Noll procedure based on the standard Clausius–Duhem inequality (5.244) makes use of fundamental assumptions. The first is the necessary presence of energy and entropy external sources. Their presence is essential in being able to manipulate the energy and entropy balance equations (4.207) and (4.219) to arrive at the Clausius–Duhem inequality given in (5.244) (see the discussion surrounding (5.239)). This is perplexing from the standpoint that since the constitutive equations describe material properties, these properties should be independent of external sources, let alone their presence. Second, as noted previously, in the standard Clausius–Duhem inequality, the entropy flux and external source are assumed to be related to the energy flux and external source by $\mathbf{h} = \mathbf{q}/\theta$ and $b = r/\theta$. The above assumptions are not necessary in the Müller–Liu procedure, which we outline next. The procedure is based on the application of four basic principles and applied through the use of Lagrange multipliers.

Müller's entropy principles:

1. *The specific entropy* η *and the entropy flux* $\mathbf{h}$ *are constitutive quantities.*

2. *The entropy production must be nonnegative, i.e.,* $\gamma_v \geq 0$*, for all thermodynamic processes corresponding to solutions of the field equations. Solutions*

of the field equations are assumed to exist and to satisfy all the balance equations, the constitutive equations, and initial and boundary conditions.

3. *The external supply terms appearing in the balance equations cannot influence the material behavior.*

4. *An* empirical temperature, *or* coldness function, *exists and such temperature, along with the tangential velocity components, is continuous across material singular surfaces called ideal boundaries,* i.e., $[\![\mathbf{v}]\!] = \mathbf{0}$ *and* $[\![\theta]\!] = 0$. *At ideal boundaries no entropy is produced:* $\gamma_s = 0$.

Note that continuity of the normal velocity component at a material singular surface is required for mass conservation (see below).

Implementation of the above principles is facilitated through the use of a lemma proved by Liu and which we now state without proof.

Liu's Lemma: Let β be a scalar, $\boldsymbol{\alpha}$, $\mathbf{a}$, and $\mathbf{b}$ vectors, and $\mathbf{A}$ a second-order tensor, all of which are given and whose dimensions are consistent with each other. Then the following three statements are equivalent:

(a) The inequality

$$\boldsymbol{\alpha} \cdot \mathbf{a} + \beta \geq 0 \tag{5.295}$$

holds for all vectors $\mathbf{a}$ that satisfy the equation

$$\mathbf{A} \cdot \mathbf{a} + \mathbf{b} = \mathbf{0}. \tag{5.296}$$

(b) There exists a vector quantity $\boldsymbol{\lambda}$ such that for all $\mathbf{a}$, the inequality

$$\boldsymbol{\alpha} \cdot \mathbf{a} + \beta - \boldsymbol{\lambda} \cdot (\mathbf{A} \cdot \mathbf{a} + \mathbf{b}) \geq 0 \tag{5.297}$$

holds.

(c) There exists a vector quantity $\boldsymbol{\lambda}$ such that

$$\boldsymbol{\alpha} = \boldsymbol{\lambda} \cdot \mathbf{A} \tag{5.298}$$

and

$$\beta - \boldsymbol{\lambda} \cdot \mathbf{b} \geq 0 \tag{5.299}$$

hold.

In the above lemma, $\boldsymbol{\lambda}$ is just a Lagrange multiplier and the contractions are understood to be full contractions.

The application of the principles begins with the statement that the local entropy inequality (4.219) must be satisfied for any fields satisfying the conservation of mass, and balances of momentum and energy, (4.98), (4.109), and (4.207), using the appropriate reduced constitutive equations, and for given initial and boundary conditions. This is nothing but a restatement of the second principle of Müller. It should be noted that such statement also corresponds to the first statement of Liu's lemma where, after introduction of the appropriate reduced constitutive equations into (4.219), one associates the solution fields and their derivatives with $\mathbf{a}$, the local entropy inequality (4.219) with (5.295), and the local conservation of mass, and momentum and energy balance equations, (4.98), (4.109), and (4.207) with (5.296). Equivalently, using the lemma's second statement, (5.297), the entropy inequality can be rewritten as a new inequality that explicitly accounts for the constraints which the solution must satisfy; i.e.,

$$\gamma_v \equiv [\rho(\dot{\eta} - b) + \operatorname{div}\mathbf{h}] - \lambda^\rho [\dot{\rho} + \rho \operatorname{div}\mathbf{v}] - \boldsymbol{\lambda}^\mathrm{v} \cdot [\rho(\dot{\mathbf{v}} - \mathbf{f}) - \operatorname{div}\boldsymbol{\sigma}] - \lambda^e [\rho(\dot{e} - r) - \Phi + \operatorname{div}\mathbf{q}] \geq 0 \quad (5.300)$$

must hold for all fields. The λ's are Lagrange multipliers, which also depend on the appropriate reduced independent variables. Now, as in the Coleman–Noll procedure, and consistent with Müller's first and fourth principles, one writes the constitutive equations for $\mathcal{C} = \{e, \eta, \mathbf{q}, \mathbf{h}, \boldsymbol{\sigma}\}$ as functions of the appropriate reduced independent variables $\mathcal{I}$ in which θ is taken to be the empirical temperature. Subsequently, these constitutive equations are introduced into the above modified entropy inequality and all differentiations performed so that, with the exception of two sets of terms, the resulting inequality involves a number of terms each multiplied linearly by an independent variable or a derivative of an independent variable. The definition of the independent vector **a** in the lemma will depend on the specific reduced independent variables. Now, application of the lemma's third statement (5.298) leads to a number of simplifying relations, while (5.299) leads to the residual inequality. It is noted that in the simplification process, Müller's third and fourth principles are used to help solve for the Lagrange multipliers and to define the empirical temperature. Further analysis in the equilibrium limit provides specific constitutive forms for $\mathcal{C} = \{e, \eta, \mathbf{q}, \mathbf{h}, \boldsymbol{\sigma}\}$. Details of the procedure are given in Chapter 8 for fluids.

5.12 Jump conditions

A surface within a material body across which we have discontinuous fields is one of two types depending on the mass flux m (see (4.101)):

Material singular surface – a surface which is formed by the same material particles at all times; on such a surface, we have that $m = 0$, so that the normal velocity of the medium moves with the normal velocity of the surface. Note that on a material singular surface, the tangential velocity component is generally allowed to slip.

Nonmaterial singular surface – a surface which is not formed by the same material particles, but for which a field variable experiences a jump across it; on such a surface, $m \neq 0$, so that the medium and the surface generally move with different velocities.

The discontinuities of some field variables can have different degrees. For example, a variable can experience a finite jump across such surface. When the motion $\mathbf{x} = \boldsymbol{\chi}(\mathbf{X}, t)$ is discontinuous across the surface, *dislocations* are formed. Also, higher derivatives of the motion (e.g., $\mathbf{v}$, $\mathbf{F}$) can be discontinuous. Examples of such singular surfaces are:

1. A surface separating two immiscible fluids, or two pure solids, or a fluid and a solid without phase change, is a material surface. On such surface, the

normal velocity is continuous. Within this context, we note that a physical boundary on which conditions have to be prescribed can be viewed as a singular surface separating two immiscible materials. Below, we shall take advantage of this fact to elaborate on boundary conditions for general problems.

2. A perfectly sliding surface between two materials is a material singular surface. Such a surface is referred to as a *vortex sheet* and the following conditions are satisfied: $[\![\mathbf{v}]\!] \cdot \mathbf{n} = [\![v_{(n)}]\!] = 0$ and $[\![\mathbf{v}]\!] \cdot \mathbf{s} = [\![v_{(s)}]\!] \neq 0$ (see (4.104) and (4.105)), where $[\![v_{(s)}]\!]$ is the amount of slip.

3. A singular surface in a pure material across which the phase of the material changes between gas, liquid, and solid is generally a nonmaterial surface since the surface does not consist of the same material particles for all times.

4. A *shock* is a surface across which the normal component of velocity experiences a jump and is a nonmaterial singular surface. Such a surface is also referred to as a *shock wave* across which we have $[\![v_{(n)}]\!] \neq 0$ and $[\![\mathsf{v}]\!] \neq 0$ (note that v is the specific volume).

5. An *acceleration wave* is a singular surface across which $\mathbf{v}$ and $\mathbf{F}$ are continuous, but $\dot{\mathbf{v}}$, grad $\mathbf{F}$, and $\dot{\mathbf{F}}$ are not.

6. It is possible to have a nonmaterial vortex surface that is also an acceleration wave. In such case, from the mass balance jump condition, it is easy to see that the specific volume is continuous across such a surface (see (4.103)). On such a surface, $m \neq 0$, $[\![\mathsf{v}]\!] = 0$, $[\![v_{(n)}]\!] = 0$, and $[\![v_{(s)}]\!] \neq 0$.

5.12.1 Characterization of jump conditions

As noted earlier, when the field variables do not satisfy continuity conditions on arbitrary surfaces within a body, then the global balance laws yield jump conditions that must hold across such surfaces. Below, we summarize the general jump conditions for nonpolar materials (see Chapter 4 for their derivations).

Mass: Across a singular surface moving with velocity $\mathbf{c}$, we have that

$$[\![m]\!] = 0, \tag{5.301}$$

where

$$m = \rho(\mathbf{v} - \mathbf{c}) \cdot \mathbf{n} \tag{5.302}$$

is the mass flux (per unit surface area). Note that m is continuous at the jump, so $m^+ = m^- = m$, where we now take m to be the value at the singular surface. Furthermore, we have shown that

$$[\![\mathbf{v}]\!] = m\,[\![\mathsf{v}]\!]\,\mathbf{n} + [\![v_{(s)}]\!]\,\mathbf{s}, \tag{5.303}$$

so that

$$[\![\mathbf{v}]\!] \cdot \mathbf{n} = m\,[\![\mathsf{v}]\!] \qquad \text{and} \qquad [\![\mathbf{v}]\!] \cdot \mathbf{s} = [\![v_{(s)}]\!]\,, \tag{5.304}$$

where $\mathsf{v} = 1/\rho$ is the specific volume, $[\![v_{(s)}]\!]$ is the slip, and $\mathbf{n}$ and $\mathbf{s}$ are the surface unit normal and tangential vectors, respectively. Note that $[\![\mathbf{v}]\!] \cdot \mathbf{n} = 0$ if and only if $[\![\mathsf{v}]\!] = 0$ at nonmaterial surfaces ($m \neq 0$).

Linear momentum: The jump condition across the singular moving surface is

$$m\,[\![\mathbf{v}]\!] - [\![\boldsymbol{\sigma}]\!]\cdot\mathbf{n} = \mathbf{0}. \tag{5.305}$$

Taking the inner product of it with $\mathbf{n}$ and $\mathbf{s}$ respectively, and using (5.304), (4.105), and (4.129), we obtain

$$m^2[\![\mathrm{v}]\!] - [\![\sigma_n]\!] = 0 \qquad \text{and} \qquad m\,[\![v_{(s)}]\!] - [\![\tau]\!] = 0. \tag{5.306}$$

Note that for nonpolar materials, the jump condition for the angular momentum, (4.117), is trivially satisfied.

Energy: We recall that the jump condition for energy, (4.217), can be rewritten as

$$m\left([\![e]\!] - [\![\mathrm{v}]\!]\,\langle\!\langle\sigma_n\rangle\!\rangle\right) - [\![v_{(s)}]\!]\,\langle\!\langle\tau\rangle\!\rangle + [\![\mathbf{q}]\!]\cdot\mathbf{n} = 0, \tag{5.307}$$

or, using (4.212), (5.304), and (5.306), and the fact that $e = \psi + \theta\,\eta$ (see (5.201)), we have

$$m\left(\varepsilon + [\![\theta\,\eta]\!]\right) - [\![v_{(s)}]\!]\,\langle\!\langle\tau\rangle\!\rangle + [\![\mathbf{q}]\!]\cdot\mathbf{n} = 0, \tag{5.308}$$

where

$$\varepsilon = [\![\psi]\!] - [\![\mathrm{v}]\!]\,\langle\!\langle\sigma_n\rangle\!\rangle = [\![\psi]\!] - [\![\mathrm{v}\,\sigma_n]\!] + \frac{1}{2}m^2\,[\![\mathrm{v}^2]\!] = \mathbf{n}\cdot[\![\boldsymbol{\mu}]\!]\cdot\mathbf{n} \tag{5.309}$$

is the *specific energy release rate*, and $\langle\!\langle\tau\rangle\!\rangle$ is the *mean shear stress* or *friction*, over the jump. Note that

$$\boldsymbol{\mu} = \boldsymbol{\Pi} + \frac{1}{2}m^2\mathrm{v}^2\,\mathbf{1} \tag{5.310}$$

is a tensorial *dynamic nonequilibrium chemical potential*, and

$$\boldsymbol{\Pi} = \psi\,\mathbf{1} - \mathrm{v}\,\boldsymbol{\sigma} \tag{5.311}$$

is the Eulerian *Eshelby energy-momentum tensor.*

Entropy: The jump condition across the singular moving surface is

$$\gamma_s \equiv m\,[\![\eta]\!] + \left[\!\!\left[\frac{\mathbf{q}}{\theta}\right]\!\!\right]\cdot\mathbf{n} \geq 0, \tag{5.312}$$

where γ_s represents the entropy production (per unit area) on the singular surface. In writing the above condition, we have assumed a simple thermodynamic process so that $\mathbf{h} = \mathbf{q}/\theta$ (see (5.242)). We will show in Chapters 7 and 8 that both simple thermoelastic solids and fluids satisfy such condition.

Shock surface in an elastic medium

In an elastic medium (where momentum diffusion is neglected), we have that $\boldsymbol{\sigma} = -p\,\mathbf{1}$ (so that $\sigma_n = -p$ and $\tau = 0$), from which it follows from (5.306) that

$$m^2[\![\mathrm{v}]\!] + [\![p]\!] = 0 \qquad \text{and} \qquad [\![v_{(s)}]\!] = 0, \tag{5.313}$$

so that there is no slip at the interface. Note that at a shock surface, $m \neq 0$. In addition, from this fact and (5.303), we now have that

$$[\![\mathbf{v}]\!] = m [\![\mathrm{v}]\!] \mathbf{n}. \tag{5.314}$$

In this case, using (5.313), (5.307) simplifies to

$$m\left([\![e]\!] + [\![\mathrm{v}]\!] \langle\!\langle p \rangle\!\rangle\right) + [\![\mathbf{q}]\!] \cdot \mathbf{n} = m\left([\![h]\!] - \langle\!\langle \mathrm{v} \rangle\!\rangle [\![p]\!]\right) + [\![\mathbf{q}]\!] \cdot \mathbf{n} = m \left[\!\!\left[h + \frac{1}{2} m^2 \mathrm{v}^2 \right]\!\!\right] + [\![\mathbf{q}]\!] \cdot \mathbf{n} = 0, \tag{5.315}$$

where the enthalpy is given by $h = e + \mathrm{v} p$ (see (5.202)). Lastly, by substituting this expression into (5.312), we see that the entropy production across a shock surface is given by

$$\gamma_s \equiv m \left[\!\!\left[\eta - \frac{1}{\theta}\left(h + \frac{1}{2} m^2 \mathrm{v}^2 \right) \right]\!\!\right] \geq 0. \tag{5.316}$$

It should be pointed out that it is often the case that for an elastic medium, one also neglects thermal diffusion (so that $\mathbf{q} = 0$). In this case, from (5.315) and (5.312), we have the following jump conditions for energy and entropy production at the surface:

$$\left[\!\!\left[h + \frac{1}{2} m^2 \mathrm{v}^2 \right]\!\!\right] = 0 \qquad \text{and} \qquad \gamma_s \equiv m [\![\eta]\!] \geq 0. \tag{5.317}$$

Note that at an ideal surface, the entropy jump is zero.

Phase change surface

Here we discuss the jump conditions corresponding to a phase change surface separating the same material, i.e., the two phases correspond to the same material in two different forms of molecular state. For example, such conditions would apply when liquid water's temperature locally falls below 0°C in which case water would change to the solid phase of ice, while if locally the temperature rises above 100°C, the liquid would change into the gaseous phase of water vapor.

In general, there are many types of phase transitions. In addition to melting-solidification and evaporation-condensation, there are also solid-solid as well as other types of transitions. Phase transitions are generally classified according to the *Ehrenfest classification*. The order of a phase transition is defined to be the order of the lowest derivative that varies discontinuously at the phase boundary. The first three orders are given in Table 5.10. Below we shall discuss first-order phase transitions in a simple isotropic material, i.e., those transitions that are characterized by jumps in entropy or specific volume. Other types of second-order transitions are solid-solid (structural) transitions in crystals.

At a phase change surface, the two phases exchange mass. Thus, such a surface is a nonmaterial moving surface with $m \neq 0$. Subsequently, the mass balance and the linear momentum jump conditions require that (5.304) and (5.306) be satisfied. Furthermore, across a phase change surface, the temperature is known to be continuous; thus we take $[\![\theta]\!] = 0$. Lastly, the energy jump condition (5.308) and the entropy inequality (5.312) can be rewritten in the following forms:

$$m\left(\varepsilon + \theta [\![\eta]\!]\right) - [\![v_{(s)}]\!] \langle\!\langle \tau \rangle\!\rangle + [\![\mathbf{q}]\!] \cdot \mathbf{n} = 0 \tag{5.318}$$

and

$$\gamma_s \theta \equiv m \theta [\![\eta]\!] + [\![\mathbf{q}]\!] \cdot \mathbf{n} \geq 0. \tag{5.319}$$

Upon substituting the energy balance jump condition into the entropy inequality, we obtain

$$-\gamma_s \theta \equiv m \varepsilon - \langle\!\langle \tau \rangle\!\rangle [\![v_{(s)}]\!] \leq 0, \tag{5.320}$$

or

$$-\gamma_s \theta \equiv m \left([\![e]\!] - \theta [\![\eta]\!] - \langle\!\langle \sigma_n \rangle\!\rangle [\![\mathrm{v}]\!] \right) - \langle\!\langle \tau \rangle\!\rangle [\![v_{(s)}]\!] \leq 0. \tag{5.321}$$

At equilibrium, $\gamma_s = 0$ and $\boldsymbol{\sigma} = -p\,\mathbf{1}$ (i.e., $\sigma_n = -p$ and $\tau = 0$), so that the linear momentum jump condition (5.306) requires that the phase boundary does not slip, $[\![v_{(s)}]\!] = 0$, and so from (5.321), we have that

$$\langle\!\langle p \rangle\!\rangle = \frac{\theta [\![\eta]\!] - [\![e]\!]}{[\![\mathrm{v}]\!]}. \tag{5.322}$$

If we differentiate the above with respect to the temperature, and use the fundamental relation $de = \theta\, d\eta - p\, d\mathrm{v}$ (see (5.139)), we obtain

$$\frac{d\langle\!\langle p \rangle\!\rangle}{d\theta} = \frac{[\![\eta]\!]}{[\![\mathrm{v}]\!]} + \frac{\theta}{[\![\mathrm{v}]\!]} \frac{d[\![\eta]\!]}{d\theta} - \frac{1}{[\![\mathrm{v}]\!]} \frac{d[\![e]\!]}{d\theta} - \frac{(\theta [\![\eta]\!] - [\![e]\!])}{[\![\mathrm{v}]\!]^2} \frac{d[\![\mathrm{v}]\!]}{d\theta} = \frac{[\![\eta]\!]}{[\![\mathrm{v}]\!]}. \tag{5.323}$$

Now if we define the *latent heat* by

$$L \equiv \theta [\![\eta]\!], \tag{5.324}$$

from (5.325), we obtain the following Clausius–Clapeyron equation

$$\frac{d\langle\!\langle p \rangle\!\rangle}{d\theta} = \frac{L}{\theta [\![\mathrm{v}]\!]}. \tag{5.325}$$

We note that in the limit where the phase change boundary is a material boundary, $m = 0$, the mass jump condition (5.304) further requires that $[\![\mathbf{v}]\!] \cdot \mathbf{n} = 0$ (note that generally $[\![\mathrm{v}]\!] \neq 0$), the linear momentum jump condition becomes $[\![p]\!] = 0$ (and thus $\langle\!\langle p \rangle\!\rangle = p$), and subsequently we obtain the standard relations of thermostatics corresponding to (5.322) and (5.325):

$$p = \frac{\theta [\![\eta]\!] - [\![e]\!]}{[\![\mathrm{v}]\!]} \quad \text{and} \quad \frac{dp}{d\theta} = \frac{L}{\theta [\![\mathrm{v}]\!]}. \tag{5.326}$$

5.12.2 Material singular surface

The mass balance jump condition (5.301) is trivially satisfied at a material singular surface since at such surface, we have that $m = 0$. The condition just states that the normal velocity of a material point at the surface is the same as that imposed by the normal velocity of the surface, $c_{(n)} = \mathbf{c} \cdot \mathbf{n}$, i.e.,

$$\mathbf{v} \cdot \mathbf{n} = c_{(n)}, \tag{5.327}$$

and, subsequently, the normal component of velocity is continuous there (see (5.304)):

$$[\![\mathbf{v}]\!] \cdot \mathbf{n} = 0. \tag{5.328}$$

Note that no condition is imposed on the tangential component; for this reason, it is also called a *free-slip* condition and $[\![v_{(s)}]\!]$ is the amount of slip. If $[\![\mathbf{v}]\!] \neq \mathbf{0}$, then in addition the surface in this case is also a vortex sheet. At the singular surface, using (5.302), the mass balance jump condition requires that

$$[\![\rho\mathbf{v}]\!] \cdot \mathbf{n} = [\![\rho]\!] c_{(n)}. \tag{5.329}$$

If $\mathbf{c} = \mathbf{0}$, the singular surface is a *stationary* surface. If $[\![\rho]\!] = 0$, then the density as well as the linear momentum normal to the surface are continuous. On the other hand, at a *contact discontinuity*, $[\![\rho]\!] \neq 0$ and the linear momentum normal to the surface is discontinuous.

Subsequently, the linear momentum balance tells us that at a material singular surface, we must have (see (5.306))

$$[\![\sigma_n]\!] = 0 \qquad \text{and} \qquad [\![\tau]\!] = 0, \tag{5.330}$$

(note that generally $[\![v]\!] \neq 0$ and $[\![v_{(s)}]\!] \neq 0$) so that the surface normal and tangential components of the stress are continuous. Hence, on a material interface, the surface traction $\mathbf{t}(\mathbf{n}) = \boldsymbol{\sigma} \cdot \mathbf{n}$ is continuous.

The energy jump condition at the material singular surface requires that (see (5.307) and note that now $\langle\!\langle \tau \rangle\!\rangle = \tau$)

$$[\![\mathbf{q}]\!] \cdot \mathbf{n} = [\![v_{(s)}]\!]\,\tau, \tag{5.331}$$

so that the jump in the normal heat flux equals the jump in power of the shear stress. Note that at a contact discontinuity, $[\![e]\!] \neq 0$ and $[\![v]\!] \neq 0$. Clearly if the velocity is continuous on such surface, $[\![\mathbf{v}]\!] = \mathbf{0}$, then the normal component of the heat flux must be continuous as well, i.e.,

$$[\![\mathbf{q}]\!] \cdot \mathbf{n} = 0. \tag{5.332}$$

Lastly, from (5.312) and (5.331), we see that at the material singular surface, the entropy production is due to the jumps in temperature and slip velocity,

$$\gamma_s \equiv \left[\!\!\left[\frac{1}{\theta}\right]\!\!\right] \langle\!\langle \mathbf{q} \rangle\!\rangle \cdot \mathbf{n} + \left\langle\!\!\left\langle \frac{1}{\theta} \right\rangle\!\!\right\rangle [\![v_{(s)}]\!]\,\tau \geq 0. \tag{5.333}$$

Furthermore, if $[\![\eta]\!] = 0$, the entropy is continuous there. On the other hand, at a contact discontinuity, we have that $[\![\eta]\!] \neq 0$.

Contact surface in an inviscid fluid

In this case, we have that $\boldsymbol{\sigma} = -p\mathbf{1}$ (so that $\sigma_n = -p$ and $\tau = 0$), from which it follows from (5.330) that p is continuous, and from (5.331) that $[\![\mathbf{q}]\!] \cdot \mathbf{n} = 0$, which implies that $\langle\!\langle \mathbf{q} \rangle\!\rangle \cdot \mathbf{n} = \mathbf{q} \cdot \mathbf{n}$ is continuous, so that the surface entropy production (5.333) subsequently becomes $\gamma_s \equiv [\![1/\theta]\!]\,\mathbf{q} \cdot \mathbf{n} \geq 0$. Thus, we see that a nonzero entropy production is only possible if we have a temperature jump at the surface, and the temperature must jump from a low to a high value when $\mathbf{q} \cdot \mathbf{n} \geq 0$ and vice versa when $\mathbf{q} \cdot \mathbf{n} \leq 0$. Note that, in general, it is possible to have $[\![v]\!] \neq 0$ and $[\![v_{(s)}]\!] \neq 0$ on such a surface.

Boundary surface

To fully describe a mathematical problem involving a continuous medium, boundary conditions are necessary. In general, physical boundaries are nothing but material singular surfaces. Thus, appropriate boundary conditions are obtained by examining jump conditions at such interfaces. In this subsection, these material interfaces will be called boundaries and we take the region on the positive side of boundaries to correspond to the material being studied, while that on the negative side to the material associated with the boundary. Subsequently, we will then drop the plus superscript on relevant quantities and replace those quantities with the minus superscript by the appropriate prescribed values.

Suppose that we have a boundary that is moving at a prescribed velocity $\mathbf{c}$. Then, since $m = 0$, from (5.302), we must have that at the boundary

$$\mathbf{v} \cdot \mathbf{n} = c_{(n)}, \tag{5.334}$$

where $c_{(n)}$ is now the normal boundary speed. Equation (5.334) is referred to as the *no-penetration* condition. The tangential velocity condition is given by (see (5.304))

$$[\![\mathbf{v}]\!] \cdot \mathbf{s} = [\![v_{(s)}]\!]. \tag{5.335}$$

If $[\![v_{(s)}]\!] = 0$, then we have what is referred to as the *no-slip* condition. If $[\![v_{(s)}]\!] \neq 0$, then we have a *slip* condition. The combined no-penetration and no-slip conditions require that the material velocity be continuous at the boundary (see (5.303)), $[\![\mathbf{v}]\!] = \mathbf{0}$, and it is equal to the boundary velocity, $\mathbf{v} = \mathbf{c}$. This condition is a generally accepted boundary condition derived from experimental observations. In fluid problems with solid boundaries, the no-penetration and no-slip conditions are typically applied.

Now suppose that a force F per unit area acts upon the boundary. Then the jump condition (5.305) requires the stress of the medium at the boundary to satisfy

$$\boldsymbol{\sigma} \cdot \mathbf{n} = \mathsf{F}. \tag{5.336}$$

A boundary is called *free* if $\mathsf{F} = \mathbf{0}$.

Furthermore, the energy jump condition (5.331), in lieu of (5.336) and the definition of shear (4.129), tells us that the heat flux must satisfy the following condition:

$$[\![\mathbf{q}]\!] \cdot \mathbf{n} = [\![v_{(s)}]\!] \mathbf{s} \cdot \mathsf{F} = [\![v_{(s)}]\!] \mathsf{F}_{(s)}. \tag{5.337}$$

Clearly if the boundary is free, $\mathsf{F} = \mathbf{0}$, or the force is normal to the boundary, $\mathsf{F}_{(s)} = 0$ (note that generally $[\![v_{(s)}]\!] \neq 0$ in such cases), or the boundary does not slip, $[\![v_{(s)}]\!] = 0$ (in which case $\mathsf{F} \neq \mathbf{0}$), then the normal component of the heat flux must be continuous, i.e.,

$$\mathbf{q} \cdot \mathbf{n} = \mathsf{G}, \tag{5.338}$$

where G is the normal component of the heat flux imposed at the boundary. A boundary is called *adiabatic* or *thermally isolated* if $\mathsf{G} = 0$, in which case the normal component of the heat flux at the boundary must vanish.

Lastly, from (5.333), we see that the entropy production is given by

$$\gamma_s \equiv \left[\!\!\left[\frac{1}{\theta}\right]\!\!\right] \langle\!\langle \mathbf{q} \rangle\!\rangle \cdot \mathbf{n} + \left\langle\!\!\left\langle \frac{1}{\theta} \right\rangle\!\!\right\rangle [\![v_{(s)}]\!] \mathsf{F}_{(s)} \geq 0. \tag{5.339}$$

If we assume that the temperature is continuous, $[\![\theta]\!] = 0$ (in which case $\langle\!\langle\theta\rangle\!\rangle = \theta$), then

$$\gamma_s \equiv \frac{1}{\theta}[\![v_{(s)}]\!]\,\mathsf{F}_{(s)} \geq 0. \tag{5.340}$$

On the other hand, if we assume that the boundary is free, or the force is normal to the boundary, or the boundary does not slip, upon using (5.338), we see that

$$\gamma_s \equiv \left[\!\!\left[\frac{1}{\theta}\right]\!\!\right] \mathsf{G} \geq 0, \tag{5.341}$$

which tells us that the direction heat must flow depends on the relative magnitudes of temperatures at the boundary and just inside the boundary. Furthermore, if the temperature is continuous and the boundary is free, or the force is normal to the boundary, or the boundary does not slip, then the boundary does not produce any entropy, $\gamma_s = 0$, and the boundary is an ideal boundary.

5.12.3 Equilibrium jump conditions

At equilibrium, we have that $\gamma_s = 0$, $\boldsymbol{\sigma}^e = -p\mathbf{1}$ (so that $\sigma_n = -p$ and $\tau = 0$), and $\mathbf{q}^e = \mathbf{0}$. Thus the entropy jump condition (5.312) reduces to

$$m\,[\![\eta]\!] = 0. \tag{5.342}$$

At a material surface, $m = 0$, and in general $[\![\eta]\!] \neq 0$. In this case, the mass jump condition (5.304) reduces to $[\![\mathbf{v}]\!]\cdot\mathbf{n} = 0$ (note that generally $[\![\mathsf{v}]\!] \neq 0$) and, in conjunction with the linear momentum jump conditions (5.306), we have that $[\![p]\!] = 0$ and in general $[\![v_{(s)}]\!] \neq 0$. Subsequently, from (5.303), we have that $[\![\mathbf{v}]\!] = [\![v_{(s)}]\!]\,\mathbf{s}$. We note that the energy jump condition (5.307) is satisfied identically.

At a nonmaterial surface, $m \neq 0$, and so we must have that $[\![\eta]\!] = 0$. In this case, the mass jump condition (5.304) remains $[\![\mathbf{v}]\!]\cdot\mathbf{n} = m\,[\![\mathsf{v}]\!]$ and, in conjunction with the linear momentum jump conditions (5.306), we have that $m^2[\![\mathsf{v}]\!] + [\![p]\!] = 0$ and $[\![v_{(s)}]\!] = 0$. Using this result and the relation $g = e - \theta\eta + p\mathsf{v}$ (see (5.203)), the energy jump condition (5.307) then requires that

$$\left[\!\!\left[e + p\mathsf{v} + \frac{1}{2}m^2\mathsf{v}^2\right]\!\!\right] = \left[\!\!\left[g + \theta\eta + \frac{1}{2}m^2\mathsf{v}^2\right]\!\!\right] = 0. \tag{5.343}$$

Now, if at the surface $[\![\theta]\!] \to 0$ and $m \to 0$, we then have that $[\![g]\!] = [\![\mu]\!] = 0$, where μ is the *equilibrium chemical potential* or the specific Gibbs free energy.

Table 5.1: Complete and irreducible function basis of isotropic scalar invariants of vectors $\mathbf{v}_\beta$, symmetric tensors $\mathbf{A}_\gamma$, and skew-symmetric tensors $\mathbf{W}_\delta$.

A. One Variable	
Variable	*Invariants*
$\mathbf{v}$	$\mathbf{v}\cdot\mathbf{v}$
A	tr A, tr A^2, tr A^3
W	tr W^2
B. Two Variables, A Assumed	
Variables	*Invariants*
$\mathbf{v}_1$, $\mathbf{v}_2$	$\mathbf{v}_1\cdot\mathbf{v}_2$
$\mathbf{v}$, A	$\mathbf{v}\cdot A\mathbf{v}$, $\mathbf{v}\cdot A^2\mathbf{v}$
$\mathbf{v}$, W	$\mathbf{v}\cdot W^2\mathbf{v}$
A_1, A_2	tr A_1A_2, tr $A_1A_2^2$, tr $A_1^2A_2$, tr $A_1^2A_2^2$
W_1, W_2	tr W_1W_2
A, W	tr AW^2, tr A^2W^2, tr A^2W^2AW
C. Three Variables, B Assumed	
Variables	*Invariants*
$\mathbf{v}_1$, $\mathbf{v}_2$, $\mathbf{v}_3$	0
$\mathbf{v}_1$, $\mathbf{v}_2$, A	$\mathbf{v}_1\cdot A\mathbf{v}_2$, $\mathbf{v}_1\cdot A^2\mathbf{v}_2$
$\mathbf{v}_1$, $\mathbf{v}_2$, W	$\mathbf{v}_1\cdot W\mathbf{v}_2$, $\mathbf{v}_1\cdot W^2\mathbf{v}_2$
$\mathbf{v}$, A_1, A_2	$\mathbf{v}\cdot A_1A_2\mathbf{v}$
$\mathbf{v}$, W_1, W_2	$\mathbf{v}\cdot W_1W_2\mathbf{v}$, $\mathbf{v}\cdot W_1W_2^2\mathbf{v}$, $\mathbf{v}\cdot W_1^2W_2\mathbf{v}$
$\mathbf{v}$, A, W	$\mathbf{v}\cdot AW\mathbf{v}$, $\mathbf{v}\cdot A^2W\mathbf{v}$, $\mathbf{v}\cdot WAW^2\mathbf{v}$
A_1, A_2, A_3	tr $A_1A_2A_3$
W_1, W_2, W_3	tr $W_1W_2W_3$
A_1, A_2, W	tr A_1A_2W, tr $A_1A_2^2W$, tr $A_1^2A_2W$, tr $A_1W^2A_2W$
A, W_1, W_2	tr AW_1W_2, tr $AW_1W_2^2$, tr $AW_1^2W_2$
D. Four Variables, C Assumed	
Variables	*Invariants*
$\mathbf{v}_1$, $\mathbf{v}_2$, A_1, A_2	$\mathbf{v}_1\cdot(A_1A_2-A_2A_1)\mathbf{v}_2$
$\mathbf{v}_1$, $\mathbf{v}_2$, W_1, W_2	$\mathbf{v}_1\cdot(W_1W_2-W_2W_1)\mathbf{v}_2$
$\mathbf{v}_1$, $\mathbf{v}_2$, A, W	$\mathbf{v}_1\cdot(AW+WA)\mathbf{v}_2$

Table 5.2: Generators for $\mathbf{h}$, a vector-valued isotropic function of vector $\mathbf{v}_\beta$, symmetric tensors $\mathbf{A}_\gamma$, and skew-symmetric tensors $\mathbf{W}_\delta$.

A. One Variable	
Variable	*Generator*
$\mathbf{v}$	$\mathbf{v}$
A or W	–
B. Two Variables, A Assumed	
Variables	*Generators*
$\mathbf{v}_1$, $\mathbf{v}_2$	–
$\mathbf{v}$, A	$A\mathbf{v}$, $A^2\mathbf{v}$
$\mathbf{v}$, W	$W\mathbf{v}$, $W^2\mathbf{v}$
A_1, A_2	–
W_1, W_2	–
A, W	–
C. Three Variables, B Assumed	
Variables	*Generators*
$\mathbf{v}_1$, $\mathbf{v}_2$, A	–
$\mathbf{v}_1$, $\mathbf{v}_2$, W	–
$\mathbf{v}$, A_1, A_2	$(A_1A_2 - A_2A_1)\mathbf{v}$
$\mathbf{v}$, W_1, W_2	$(W_1W_2 - W_2W_1)\mathbf{v}$
$\mathbf{v}$, A, W	$(AW + WA)\mathbf{v}$
A_1, A_2, A_3	–

Table 5.3: Generators for $\mathbf{T}$, a symmetric tensor-valued isotropic function of vectors $\mathbf{v}_\beta$, symmetric tensors $\mathbf{A}_\gamma$, and skew-symmetric tensors $\mathbf{W}_\delta$.

A. No Variable	
Variable	*Generator*
0	$\mathbf{1}$
B. One Variable, A Assumed	
Variable	*Generators*
$\mathbf{v}$	$\mathbf{v}\ \mathbf{v}$
A	A, A^2
W	W^2
C. Two Variables, B Assumed	
Variables	*Generators*
$\mathbf{v}_1$, $\mathbf{v}_2$	$\mathbf{v}_1\mathbf{v}_2 + \mathbf{v}_2\mathbf{v}_1$
$\mathbf{v}$, A	$\mathbf{v}A\mathbf{v} + A\mathbf{v}\mathbf{v}$, $\mathbf{v}A^2\mathbf{v} + A^2\mathbf{v}\mathbf{v}$
$\mathbf{v}$, W	$\mathbf{v}W\mathbf{v} + W\mathbf{v}\mathbf{v}$, $W\mathbf{v}W\mathbf{v}$, $W\mathbf{v}W^2\mathbf{v} + W^2\mathbf{v}W\mathbf{v}$
A_1, A_2	$A_1A_2 + A_2A_1$, $A_1^2A_2 + A_2A_1^2$, $A_1A_2^2 + A_2^2A_1$
W_1, W_2	$W_1W_2 + W_2W_1$, $W_1W_2^2 - W_2^2W_1$, $W_1^2W_2 - W_2W_1^2$
A, W	$AW - WA$, WAW, $A^2W - WA^2$, $WAW^2 - W^2AW$
D. Three Variables, C Assumed	
Variables	*Generators*
$\mathbf{v}_1$, $\mathbf{v}_2$, A	$A(\mathbf{v}_1\mathbf{v}_2 - \mathbf{v}_2\mathbf{v}_1) - (\mathbf{v}_1\mathbf{v}_2 - \mathbf{v}_2\mathbf{v}_1)A$
$\mathbf{v}_1$, $\mathbf{v}_2$, W	$W(\mathbf{v}_1\mathbf{v}_2 - \mathbf{v}_2\mathbf{v}_1) + (\mathbf{v}_1\mathbf{v}_2 - \mathbf{v}_2\mathbf{v}_1)W$

Table 5.4: Generators for $\mathbf{T}$, a skew-symmetric tensor-valued isotropic function of vectors $\mathbf{v}_\beta$, symmetric tensors $\mathbf{A}_\gamma$, and skew-symmetric tensors $\mathbf{W}_\delta$.

A. One Variable	
Variable	*Generator*
$\mathbf{v}$ or A	–
W	W
B. Two Variables, A Assumed	
Variables	*Generators*
$\mathbf{v}_1, \mathbf{v}_2$	$\mathbf{v}_1\mathbf{v}_2 - \mathbf{v}_2\mathbf{v}_1$
$\mathbf{v}, A$	$\mathbf{v}A\mathbf{v} - A\mathbf{v}\mathbf{v},\ \mathbf{v}A^2\mathbf{v} - A^2\mathbf{v}\mathbf{v},\ A\mathbf{v}A^2\mathbf{v} - A^2\mathbf{v}A\mathbf{v}$
$\mathbf{v}, W$	$\mathbf{v}W\mathbf{v} - W\mathbf{v}\mathbf{v},\ \mathbf{v}W^2\mathbf{v} - W^2\mathbf{v}\mathbf{v}$
A_1, A_2	$A_1A_2 - A_2A_1,\ A_1A_2^2 - A_2^2A_1,\ A_1^2A_2 - A_2A_1^2,$ $A_1A_2A_1^2 - A_1^2A_2A_1,\ A_2A_1A_2^2 - A_2^2A_1A_2$
W_1, W_2	$W_1W_2 - W_2W_1$
A, W	$AW + WA,\ AW^2 - W^2A$
C. Three Variables, B Assumed	
Variables	*Generators*
$\mathbf{v}_1, \mathbf{v}_2, A$	$A(\mathbf{v}_1\mathbf{v}_2 - \mathbf{v}_2\mathbf{v}_1) + (\mathbf{v}_1\mathbf{v}_2 - \mathbf{v}_2\mathbf{v}_1)A$
$\mathbf{v}_1, \mathbf{v}_2, W$	$W(\mathbf{v}_1\mathbf{v}_2 - \mathbf{v}_2\mathbf{v}_1) - (\mathbf{v}_1\mathbf{v}_2 - \mathbf{v}_2\mathbf{v}_2)W$
$\mathbf{v}, A_1, A_2$	$A_1\mathbf{v}A_2\mathbf{v} - A_2\mathbf{v}A_1\mathbf{v} + \mathbf{v}(A_1A_2 - A_2A_1)\mathbf{v} -$ $(A_1A_2 - A_2A_1)\mathbf{v}\mathbf{v}$
A_1, A_2, A_3	$A_1A_2A_3 + A_2A_3A_1 + A_3A_1A_2 - A_2A_1A_3 -$ $A_1A_3A_2 - A_3A_2A_1$

Table 5.5: Complete and irreducible function basis of hemitropic scalar invariants of vectors $\mathbf{v}_\beta$, symmetric tensors $\mathbf{A}_\gamma$, and skew-symmetric tensors $\mathbf{W}_\delta$.

A. One Variable	
Variable	*Invariants*
$\mathbf{v}$	$\mathbf{v}\cdot\mathbf{v}$
A	tr A, tr A^2, tr A^3
W	tr W^2
B. Two Variables, A Assumed	
Variables	*Invariants*
$\mathbf{v}_1$, $\mathbf{v}_2$	$\mathbf{v}_1\cdot\mathbf{v}_2$
$\mathbf{v}$, A	$\mathbf{v}\cdot A\mathbf{v}$, $\mathbf{v}\cdot A^2\mathbf{v}$, $[\mathbf{v}, A\mathbf{v}, A^2\mathbf{v}]$
$\mathbf{v}$, W	$\mathbf{v}\cdot\langle W\rangle$
A_1, A_2	tr A_1A_2, tr $A_1A_2^2$, tr $A_1^2A_2$, tr $A_1^2A_2^2$
W_1, W_2	tr W_1W_2
A, W	tr AW^2, tr A^2W^2, tr A^2W^2AW
C. Three Variables, B Assumed	
Variables	*Invariants*
$\mathbf{v}_1$, $\mathbf{v}_2$, $\mathbf{v}_3$	$[\mathbf{v}_1, \mathbf{v}_2, \mathbf{v}_3]$
$\mathbf{v}_1$, $\mathbf{v}_2$, A	$\mathbf{v}_1\cdot A\mathbf{v}_2$, $[\mathbf{v}_1, \mathbf{v}_2, A\mathbf{v}_1]$, $[\mathbf{v}_1, \mathbf{v}_2, A\mathbf{v}_2]$
$\mathbf{v}_1$, $\mathbf{v}_2$, W	$\mathbf{v}_1\cdot W\mathbf{v}_2$
$\mathbf{v}$, A_1, A_2	$\mathbf{v}\cdot\langle A_1A_2\rangle$, $\mathbf{v}\cdot\langle A_1A_2^2\rangle$, $\mathbf{v}\cdot\langle A_1^2A_2\rangle$, $[\mathbf{v}, A_1\mathbf{v}, A_2\mathbf{v}]$
$\mathbf{v}$, W_1, W_2	$\mathbf{v}\cdot\langle W_1W_2\rangle$
$\mathbf{v}$, A, W	$\mathbf{v}\cdot AW\mathbf{v}$, $\mathbf{v}\cdot\langle AW\rangle$, $\mathbf{v}\cdot\langle AW^2\rangle$
A_1, A_2, A_3	tr $A_1A_2A_3$
W_1, W_2, W_3	tr $W_1W_2W_3$
A_1, A_2, W	tr A_1A_2W, tr $A_1^2A_2W$, tr $A_1A_2^2W$, tr $A_1W^2A_2W$
A, W_1, W_2	tr AW_1W_2, tr $AW_1^2W_2$, tr $AW_1W_2^2$
D. Four Variables, C Assumed	
Variables	*Invariants*
$\mathbf{v}_1$, $\mathbf{v}_2$, A_1 A_2	–
$\mathbf{v}_1$, $\mathbf{v}_2$, W_1 W_2	–
$\mathbf{v}_1$, $\mathbf{v}_2$, A W	–

Table 5.6: Generators for $\mathbf{h}$, a vector-valued hemitropic function of vector $\mathbf{v}_\beta$, symmetric tensors $\mathbf{A}_\gamma$, and skew-symmetric tensors $\mathbf{W}_\delta$.

A. One Variable	
Variable	*Generator*
$\mathbf{v}$	$\mathbf{v}$
A	–
W	$\langle W \rangle$
B. Two Variables, A Assumed	
Variables	*Generators*
$\mathbf{v}_1$, $\mathbf{v}_2$	$\mathbf{v}_1 \times \mathbf{v}_2$
$\mathbf{v}$, A	$A\mathbf{v}$, $\mathbf{v} \times A\mathbf{v}$
$\mathbf{v}$, W	$W\mathbf{v}$
A_1, A_2	$\langle A_1 A_2 \rangle$, $\langle A_1^2 A_2 \rangle$, $\langle A_1 A_2^2 \rangle$, $\langle A_1 A_2 A_1^2 \rangle$, $\langle A_2 A_1 A_2^2 \rangle$
W_1, W_2	$\langle W_1 W_2 \rangle$
A, W	$\langle AW \rangle$, $\langle AW^2 \rangle$
C. Three Variables, B Assumed	
Variables	*Generator*
$\mathbf{v}_1$, $\mathbf{v}_2$, A	–
$\mathbf{v}_1$, $\mathbf{v}_2$, W	–
$\mathbf{v}$, A_1, A_2	–
$\mathbf{v}$, W_1, W_2	–
$\mathbf{v}$, A, W	–
A_1, A_2, A_3	$\langle A_1 A_2 A_3 + A_2 A_3 A_1 + A_3 A_1 A_2 \rangle$

Table 5.7: Generators for $\mathbf{T}$, a symmetric tensor-valued hemitropic function of vectors $\mathbf{v}_\beta$, symmetric tensors $\mathbf{A}_\gamma$, and skew-symmetric tensors $\mathbf{W}_\delta$.

A. No Variable	
Variable	*Generator*
0	$\mathbf{1}$
B. One Variable, A Assumed	
Variable	*Generators*
$\mathbf{v}$	$\mathbf{v}\,\mathbf{v}$
A	$A,\ A^2$
W	W^2
C. Two Variables, B Assumed	
Variables	*Generators*
$\mathbf{v}_1,\ \mathbf{v}_2$	$\mathbf{v}_1\mathbf{v}_2 + \mathbf{v}_2\mathbf{v}_1,\ \mathbf{v}_1(\mathbf{v}_1 \times \mathbf{v}_2) + (\mathbf{v}_1 \times \mathbf{v}_2)\mathbf{v}_1,$ $\mathbf{v}_2(\mathbf{v}_1 \times \mathbf{v}_2) + (\mathbf{v}_1 \times \mathbf{v}_2)\mathbf{v}_2$
$\mathbf{v},\ A \quad (\mathbf{W} = \mathbf{v} \cdot \boldsymbol{\epsilon})$	$\mathbf{v}A\mathbf{v} + A\mathbf{v}\mathbf{v},\ AW + WA,\ A^2W + WA^2,$ $\mathbf{v}(\mathbf{v} \times A\mathbf{v}) + (\mathbf{v} \times A\mathbf{v})\mathbf{v}$
$\mathbf{v},\ W_1 \quad (\mathbf{W}_2 = \mathbf{v} \cdot \boldsymbol{\epsilon})$	$\mathbf{v}W_1\mathbf{v} + W_1\mathbf{v}\mathbf{v},\ W_1W_2 + W_2W_1,\ W_1^2W_2 + W_2W_1^2$
$A_1,\ A_2$	$A_1A_2 + A_2A_1,\ A_1^2A_2 + A_2A_1^2,\ A_1A_2^2 + A_2^2A_1$
$W_1,\ W_2$	$W_1W_2 + W_2W_1,\ W_1^2W_2 - W_2W_1^2,\ W_1W_2^2 - W_2^2W_1$
$A,\ W$	$AW - WA,\ AW^2 + W^2A,\ WAW^2 - W^2AW,$ $A^2W - WA^2$
D. Three Variables, C Assumed	
Variables	*Generators*
$\mathbf{v}_1,\ \mathbf{v}_2,\ A$	–
$\mathbf{v}_1,\ \mathbf{v}_2,\ W$	–

Table 5.8: Generators for $\mathbf{T}$, a skew-symmetric tensor-valued hemitropic function of vectors $\mathbf{v}_\beta$, symmetric tensors $\mathbf{A}_\gamma$, and skew-symmetric tensors $\mathbf{W}_\delta$.

A. One Variable	
Variable	*Generator*
$\mathbf{v}$ $(\mathbf{W} = \mathbf{v}\cdot\epsilon)$	W
A	–
W	W
B. Two Variables, A Assumed	
Variables	*Generators*
$\mathbf{v}_1$, $\mathbf{v}_2$	$\mathbf{v}_1\mathbf{v}_2 - \mathbf{v}_2\mathbf{v}_1$
$\mathbf{v}$, A $(\mathbf{W} = \mathbf{v}\cdot\epsilon)$	$AW - WA$, $\mathbf{v}A\mathbf{v} - A\mathbf{v}\mathbf{v}$
$\mathbf{v}$, W_1 $(\mathbf{W}_2 = \mathbf{v}\cdot\epsilon)$	$W_1W_2 - W_2W_1$
A_1, A_2	$A_1A_2 - A_2A_1$, $A_1A_2^2 - A_2^2A_1$, $A_1^2A_2 - A_2A_1^2$, $A_1A_2A_1^2 - A_1^2A_2A_1$, $A_2A_1A_2^2 - A_2^2A_1A_2$
W_1, W_2	$W_1W_2 - W_2W_1$
A, W	$AW + WA$, $AW^2 - W^2A$
C. Three Variables, B Assumed	
Variables	*Generators*
$\mathbf{v}_1$, $\mathbf{v}_2$, A	–
$\mathbf{v}_1$, $\mathbf{v}_2$, W	–
$\mathbf{v}$, A_1, A_2	–
A_1, A_2, A_3	$A_1A_2A_3 + A_2A_3A_1 + A_3A_1A_2 - A_2A_1A_3 - A_1A_3A_2 - A_3A_2A_1$

Table 5.9: Thermodynamic property tensors.

General	Description	Special Case ($n=1$)	Units (SI)[1]
η	Entropy	η	J/kg·K
e	Internal energy	e	J/kg
ψ	Helmholtz potential	ψ	J/kg
h	Enthalpy potential	h	J/kg
g	Gibbs potential	g	J/kg
θ	Thermostatic temperature	θ	K
$\boldsymbol{\nu}_\alpha$	Specific thermostatic volume	$\mathsf{v} = 1/\rho$	m^3/kg
$\boldsymbol{\tau}_\alpha$	Thermostatic tension	$-p$	N/m^2, J/m^3
$\boldsymbol{\beta}_\alpha$	Isochoric thermal tension	$-\alpha/\kappa_\theta$	J/m^3·K
$\boldsymbol{\alpha}_\alpha$	Thermal strain	α/ρ	m^3/kg·K
$\boldsymbol{\lambda}_{\boldsymbol{\nu}_\alpha}$	Caloric stiffness	$\theta\,\alpha/\kappa_\theta$	J/m^3
$\boldsymbol{\lambda}_{\boldsymbol{\tau}_\alpha}$	Caloric compliance	$\theta\,\alpha/\rho$	m^3/kg
$c_{\boldsymbol{\nu}_\alpha}$	Specific heat at const. thermostatic volume	c_v	J/kg·K
$c_{\boldsymbol{\tau}_\alpha}$	Specific heat at const. thermostatic tension	c_p	J/kg·K
γ_α	Ratio of specific heats	$\gamma \equiv c_p/c_\mathsf{v}$	—
$\boldsymbol{\zeta}_\alpha$	Isopiestic thermal expansion	$(\gamma-1)\,\kappa_\theta/\gamma\,\alpha$	m^3·K/J
$\boldsymbol{\varphi}_\alpha$	Isentropic thermal stiffness	$-(\gamma-1)\,\rho/\alpha$	kg·K/m^3
$\boldsymbol{\xi}_{\alpha\beta}$	Isothermal elastic stiffness	ρ/κ_θ	J·kg/m^6
$\boldsymbol{\upsilon}_{\alpha\beta}$	Isothermal elastic compliance	κ_θ/ρ	m^6/J·kg
$\boldsymbol{\phi}_{\alpha\beta}$	Isentropic elastic stiffness	$\gamma\,\rho/\kappa_\theta$	J·kg/m^6
$\boldsymbol{\chi}_{\alpha\beta}$	Isentropic elastic compliance	$\kappa_\theta/\gamma\,\rho$	m^6/J·kg
$\boldsymbol{\Gamma}_{\alpha\beta}$	Grüneisen parameter	$\Gamma \equiv \alpha/\rho\, c_\mathsf{v}\,\kappa_\theta$	—

[1] 1 J = 1 N·m, 1 N = 1 kg·m/s^2.

Table 5.10: Ehrenfest classification of phase transitions in a simple isotropic material.

	Discontinuous Quantities	
Order	**Differentials**	**Experimental Quantities**
First	η, v	η, v
Second	$\left.\frac{\partial\eta}{\partial\theta}\right\|_p, \left.\frac{\partial\mathsf{v}}{\partial\theta}\right\|_p, \left.\frac{\partial\eta}{\partial p}\right\|_\theta, \left.\frac{\partial\mathsf{v}}{\partial p}\right\|_\theta$	c_p, α, κ_θ
Third	$\left.\frac{\partial^2\eta}{\partial\theta^2}\right\|_p, \left.\frac{\partial^2\mathsf{v}}{\partial\theta^2}\right\|_p, \frac{\partial^2\eta}{\partial p\partial\theta}, \frac{\partial^2\mathsf{v}}{\partial p\partial\theta}, \left.\frac{\partial^2\eta}{\partial p^2}\right\|_\theta, \left.\frac{\partial^2\mathsf{v}}{\partial p^2}\right\|_\theta$	$\left.\frac{\partial c_p}{\partial\theta}\right\|_p, \left.\frac{\partial\alpha}{\partial\theta}\right\|_p, \left.\frac{\partial\kappa_\theta}{\partial\theta}\right\|_p, \left.\frac{\partial c_p}{\partial p}\right\|_\theta, \left.\frac{\partial\alpha}{\partial p}\right\|_\theta, \left.\frac{\partial\kappa_\theta}{\partial p}\right\|_\theta$

Problems

1. If we have $F(x, y) = 0$, under what conditions, can we write $y = f(x)$?

2. What are the necessary and sufficient conditions for the maximum of $f(x, y, z)$?

3. Find the generators of the hemitropic symmetry group as represented by laminated wood with the geometry shown in Fig. 5.5.

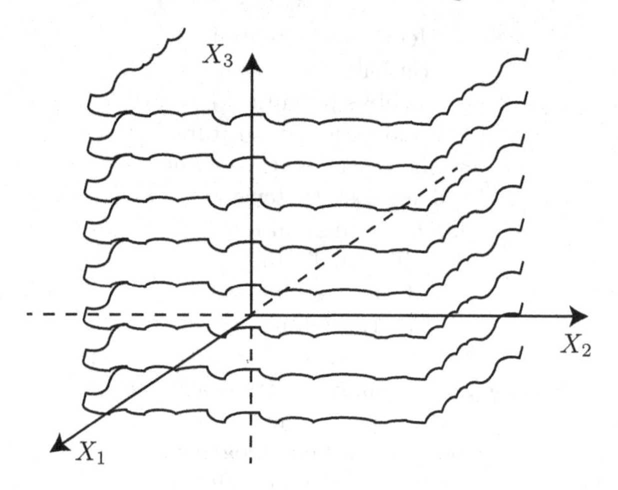

Figure 5.5: Laminated wood geometry.

4. Suppose that a given material is characterized by the constitutive equation for the stress tensor in the following form:

$$\boldsymbol{\sigma} = \boldsymbol{\sigma}(\rho, \mathbf{v}, \mathbf{L}),$$

where $\mathbf{v}$ is the velocity and $\mathbf{L} = (\operatorname{grad} \mathbf{v})^T$. Use the principle of material frame indifference to write the constitutive equation in reduced form.

5. For $\boldsymbol{\sigma} = \boldsymbol{\sigma}(\rho, \theta, \mathbf{g})$, use objectivity to show that

$$\boldsymbol{\sigma} = \alpha \mathbf{1} + \beta \mathbf{g}\mathbf{g},$$

where α and β are functions of ρ, θ, and $\mathbf{g} \cdot \mathbf{g}$ (provide details).

6. Let $\mathbf{T} = \mathbf{T}(\mathbf{A})$ be a continuous isotropic symmetric tensor of rank 2 which is a function of the symmetric tensor $\mathbf{A}$ of rank 2. Clearly, from (5.126) and Tables 5.1 and 5.3, we obtained the representation for $\mathbf{T}$ given in (5.128). Obtain this representation by using the Cayley–Hamilton theorem.

7. Consider a material with the stress tensor

$$\sigma_{ij} = -p\delta_{ij} + K_{ijkl} D_{kl}.$$

Show that, because of the symmetries of the stress and rate of deformation tensors, the tensor of rank 4 K_{ijkl} has at most 36 independent components. Display the components in a 6×6 array.

8. Given the stress tensor in Problem 7, show that objectivity requires that

$$K_{ijkl} = Q_{ip}Q_{jq}Q_{rs}Q_{ls}K_{pqrs},$$

where $\mathbf{Q}$ is any orthogonal tensor.

9. i) If a material is assumed isotropic, show that the rank-4 tensor $\mathbf{K}$ given in Problem 7 only has two independent components. Subsequently, provide the reduced representation of the stress tensor $\boldsymbol{\sigma}$.

   ii) Write the resulting representation into isotropic, symmetric, and deviatoric parts (see Section 2.10).

10. If a material has the constitutive equation for the stress tensor

$$\sigma_{ij} = -p\,\delta_{ij} + \alpha\,D_{ij} + \beta\,D_{ik}D_{kj},$$

show that if the material is incompressible, then

$$\sigma_{ii} = -3\left(p + \frac{2}{3}\,\beta\,D_{(2)}\right),$$

where $D_{(2)}$ is the second invariant of $\mathbf{D}$.

11. Using the definitions of $\boldsymbol{\zeta}_\alpha$ in (5.144), $\boldsymbol{\alpha}_\alpha$ in (5.150), and $c_{\mathcal{T}_\alpha}$ in (5.165), derive (5.177).

12. Prove (5.169).

13. Using the relations (5.175) and (5.177), and the definitions (5.137), (5.150), and (5.158), prove the identities (5.176).

14. Use (5.183) and (5.217) to show that

$$\boldsymbol{\phi}_{\alpha\beta} - \boldsymbol{\xi}_{\alpha\beta} = \frac{\theta}{c_{\mathcal{V}_\alpha}}\,\boldsymbol{\beta}_\alpha\,\boldsymbol{\beta}_\beta = \theta\,c_{\mathcal{V}_\alpha}\,\boldsymbol{\Gamma}_\alpha\,\boldsymbol{\Gamma}_\beta.$$

15. Verify (5.212) by using (5.207).

16. Verify (5.225) by using (5.220).

17. Verify (5.234) by using (5.229).

18. Prove (5.226).

19. Prove (5.235).

20. A *hyperelastic material* is described by the consitutive equation (see (5.277))

$$\boldsymbol{\sigma}^e = \boldsymbol{\sigma}^e(\mathbf{F}) = \rho\frac{\partial\tau}{\partial\mathbf{F}}\cdot\mathbf{F}^T,$$

where $\tau = \tau(\mathbf{F})$ is the strain (or stored) energy function. Show that if τ satisfies objectivity, then so does $\boldsymbol{\sigma}^e$.

21. Consider a general macroscale description of flow through porous media described by the proposed model

$$\nabla p = \mathbf{f}(\gamma, \mathbf{u}, \mathbf{v}, \mathbf{L}),$$

where ∇p is the pressure gradient and $\mathbf{f}$ is a vector-valued function that depends on γ, a scalar property function (more generally it depends on a number of scalar functions representing geometric properties such as porosity and specific surface density, and thermophysical properties of the medium and fluid such as density, viscosity, and compressibility), $\mathbf{u} \equiv \mathbf{v} - \mathbf{v}_s$ is the fluid velocity $\mathbf{v}$ relative to the solid velocity $\mathbf{v}_s$, and $\mathbf{L} \equiv (\operatorname{grad} \mathbf{v})^T$, which is the fluid velocity gradient.

i) Use the principle of frame indifference to show that

$$\nabla p = \mathbf{f}(\gamma, \mathbf{u}, \mathbf{D}).$$

ii) Subsequently, show that the reduced form of this model is given by

$$\nabla p = -\mathbf{H} \cdot \mathbf{u},$$

where

$$\mathbf{H} = \alpha_0 \mathbf{1} + \alpha_1 \mathbf{D} + \alpha_2 \mathbf{D}^2$$

and

$$\alpha_i = \alpha_i(\gamma, |\mathbf{u}|, D_{(1)}, D_{(2)}, D_{(3)}, \mathbf{u} \cdot D\mathbf{u}, \mathbf{u} \cdot D^2\mathbf{u}), \qquad i = 1, 2, 3.$$

We call this the *generalized Darcy's law.*

iii) The second law of thermodynamics requires that the viscous dissipation be such that $\mathbf{f} \cdot \mathbf{u} \leq 0$. Show that the constraints on the α_i's imposed by this restriction are

$$\alpha_0 > 0, \quad \alpha_2 > 0, \quad \text{and} \quad \alpha_1^2 - 4\alpha_0\alpha_2 < 0.$$

iv) If the relative velocity is small but the fluid velocity is not small, show that the generalized Darcy's law linear in $\mathbf{u}$ is given by

$$\nabla p = -\mathbf{H} \cdot \mathbf{u},$$

where

$$\mathbf{H} = \alpha_0 \mathbf{1} + \alpha_1 \mathbf{D} + \alpha_2 \mathbf{D}^2$$

and $\alpha_i = \alpha_i(\gamma, D_{(1)}, D_{(2)}, D_{(3)})$.

v) If the relative velocity and the fluid velocity are both small, show that the generalized Darcy's law linear in $\mathbf{u}$ and $\mathbf{v}$ becomes

$$\nabla p = -\alpha_0 \left(\gamma, D_{(1)}\right) \mathbf{u}.$$

Furthermore, show that if the fluid is incompressible, this reduces to the classical *Darcy's law*

$$\nabla p = -\alpha_0 (\gamma) \mathbf{u},$$

where $\alpha_0(\gamma)$ is the reciprocal of the permeability "coefficient."

22. Show that

$$\mathbf{q} = \mathbf{q}_1 \left(\mathbf{F}, \theta, \mathbf{g} \right)$$

implies

$$\mathbf{q} = \mathbf{F} \cdot \mathbf{q}_2 \left(\mathbf{U}, \theta, \mathbf{R}^T \cdot \mathbf{g} \right)$$

and vice-versa.

23. Show that for a hemitropic material, it is necessary and sufficient that

$$\mathbf{q} = \mathbf{q} \left(\mathbf{B}, \theta, \mathbf{g} \right). \tag{5.344}$$

24. Show that for an isotropic material

$$\mathbf{q} = \left(\alpha_0 \, \mathbf{1} + \alpha_1 \, \mathbf{B} + \alpha_2 \, \mathbf{B}^2 \right) \cdot \mathbf{g}.$$

25. For an isotropic solid, $\psi = \psi(B_{(\alpha)}, \theta)$. Show that for a fluid, $\psi = \psi(B_{(3)}, \theta)$.

26. In the process of obtaining the fundamental relation, it is very convenient in experiments where one can control temperature and thermostatic volumes to be able to decompose the Helmholtz free energy into components that are independent of temperature changes and those that are dependent on temperature changes. With this aim in mind

    i) use the first equation in (5.165) to show that

$$\int_{\theta_0}^{\theta} c_{\boldsymbol{\nu}_\alpha}(\theta', \boldsymbol{\nu}_\alpha) \, d\theta' = e(\theta, \boldsymbol{\nu}_\alpha) - e(\theta_0, \boldsymbol{\nu}_\alpha);$$

    ii) integrate the first equation in (5.209) and use (5.201) and (5.208) to show that

$$\int_{\theta_0}^{\theta} \frac{\theta}{\theta'} c_{\boldsymbol{\nu}_\alpha}(\theta', \boldsymbol{\nu}_\alpha) \, d\theta' = e(\theta, \boldsymbol{\nu}_\alpha) - \psi(\theta, \boldsymbol{\nu}_\alpha) - \theta \, \eta(\theta_0, \boldsymbol{\nu}_\alpha);$$

    iii) subsequently, show that the Hemholtz free energy can be rewritten in the following convenient form:

$$\psi(\theta, \boldsymbol{\nu}_\alpha) = e(\theta_0, \boldsymbol{\nu}_\alpha) - \theta \, \eta(\theta_0, \boldsymbol{\nu}_\alpha) - \int_{\theta_0}^{\theta} \left(\frac{\theta}{\theta'} - 1 \right) c_{\boldsymbol{\nu}_\alpha}(\theta', \boldsymbol{\nu}_\alpha) \, d\theta'. \tag{5.345}$$

27. In the process of obtaining the fundamental relation, it is very convenient in experiments where one can control temperature and thermostatic tensions to be able to decompose the Gibbs free energy into components that are independent of temperature changes and those that are dependent on temperature changes. With this aim in mind

    i) use (5.222) to show that

$$\int_{\theta_0}^{\theta} c_{\boldsymbol{\tau}_\alpha}(\theta', \boldsymbol{\tau}_\alpha) \, d\theta' = h(\theta, \boldsymbol{\tau}_\alpha) - h(\theta_0, \boldsymbol{\tau}_\alpha);$$

ii) integrate (5.231) and use (5.229), (5.230), (5.202), and (5.203) to show that

$$\int_{\theta_0}^{\theta} \frac{\theta}{\theta'} c_{\mathcal{T}_\alpha}(\theta', \boldsymbol{\tau}_\alpha)\, d\theta' = h(\theta, \boldsymbol{\tau}_\alpha) - g(\theta, \boldsymbol{\tau}_\alpha) - \theta\, \eta(\theta_0, \boldsymbol{\tau}_\alpha);$$

iii) subsequently, show that the Gibbs free energy can be rewritten in the following convenient form:

$$g(\theta, \boldsymbol{\tau}_\alpha) = h(\theta_0, \boldsymbol{\tau}_\alpha) - \theta\, \eta(\theta_0, \boldsymbol{\tau}_\alpha) - \int_{\theta_0}^{\theta} \left(\frac{\theta}{\theta'} - 1 \right) c_{\mathcal{T}_\alpha}(\theta', \boldsymbol{\tau}_\alpha)\, d\theta'. \quad (5.346)$$

28. From a set of experiments on a compressible gas at equilibrium, one sees that the specific heats at constant volume and at constant pressure are effectively constant. If we assume that they are truly constant and that the gas is non-dissipative, determine the Helmholtz free energy of the gas, $\psi = \psi(\rho, \theta)$, in terms of unknown constants.

29. Show that

$$[\![\mathrm{v}]\!] \langle\!\langle \sigma_n \rangle\!\rangle = [\![\mathrm{v}\, \sigma_n]\!] - \frac{1}{2} m^2 [\![\mathrm{v}^2]\!] .$$

30. Starting from (4.217), derive (5.308).

Bibliography

M. Bailyn. *A Survey of Thermodynamics.* AIP Press, New York, NY, 1994.

R.C. Batra. *Elements of Continuum Mechanics.* AIAA, Reston, VA, 2006.

J.P. Boehler. On irreducible representations for isotropic scalar functions. *Zeitschrift fur Angewandte Mathematik und Mechanik*, 57(6):323–327, 1977.

R.M. Bowen. *Introduction to Continuum Mechanics for Engineers.* Plenum Press, New York, NY, 1989.

G. Buratti, Y. Huo, and I. Müller. Eshelby tensor as a tensor of free enthalpy. *Journal of Elasticity*, 72:31–42, 2003.

H.B. Callen. *Thermodynamics.* John Wiley & Sons, Inc., New York, NY, 1962.

P. Chadwick. Aspects of dynamics of a rubberlike material. *The Quarterly Journal of Mechanics and Applied Mathematics*, 27(3):263–285, 1974.

P. Chadwick. Thermo-mechanics of rubberlike materials. *Philosophical Transactions of the Royal Society of London. Series A, Mathematical and Physical Sciences*, 276(1260):371–403, 1974.

P. Chadwick. *Continuum Mechanics – Concise Theory and Problems.* Dover Publications, Inc., Mineola, NY, 2nd edition, 1999.

T.J. Chung. *General Continuum Mechanics.* Cambridge University Press, New York, NY, 2007.

B.D. Coleman and W. Noll. The thermodynamics of elastic materials with heat conduction and viscosity. *Archive for Rational Mechanics and Analysis*, 13(1):167–178, 1963.

S.R. de Groot and P. Mazur. *Non-Equilibrium Thermodynamics.* Dover Publications, Inc., Mineola, NY, 1984.

K. Denbigh. *The Principles of Chemical Equilibrium.* Cambridge University Press, Cambridge, England, 1981.

M.R. El-Saden. A thermodynamic formalism based on the fundamental relation and the Legendre transformation. *International Journal of Mechanical Sciences*, 8(1):13–24, 1966.

A.C. Eringen. Irreversible thermodynamics and continuum mechanics. *Physical Review*, 117(5):1174–1183, 1960.

A.C. Eringen. *Nonlinear Theory of Continuous Media.* McGraw-Hill Book Company, Inc., New York, NY, 1962.

A.C. Eringen. Constitutive equations for simple materials: General theory. In A.C. Eringen, editor, *Continuum Physics*, volume II, pages 131–172. Academic Press, Inc., New York, NY, 1975.

A.C. Eringen. *Mechanics of Continua.* R.E. Krieger Publishing Company, Inc., Melbourne, FL, 1980.

E.C. Eringen. A unified theory of thermomechanical materials. *International Journal of Engineering Science*, 4(2):179–202, 1966.

E. Fried. Energy release, friction, and supplemental relations at phase interfaces. *Continuum Mechanics and Thermodynamics*, 7:111–121, 1995.

Y.C. Fung. *A First Course in Continuum Mechanics.* Prentice Hall, Inc., Englewood Cliffs, NJ, 3rd edition, 1994.

O. Gonzalez and A.M. Stuart. *A First Course in Continuum Mechanics.* Cambridge University Press, Cambridge, England, 2008.

M. E. Gurtin. *Introduction to Continuum Mechanics.* Academic Press, New York, 1981.

M.E. Gurtin. *An Introduction to Continuum Mechanics.* Academic Press, San Diego, CA, 2003.

M.E. Gurtin, E. Fried, and L. Anand. *The Mechanics and Thermodynamics of Continua.* Cambridge University Press, Cambridge, UK, 2010.

M.E. Gurtin and W.O. Williams. On the Clausius-Duhem inequality. *Journal of Applied Mathematics and Physics (ZAMP)*, 17:626–633, 1966.

I. Gyarmati. *Non-Equlibrium Thermodynamics.* Springer-Verlag, Berlin, 1970.

M. Hamermesh. *Group Theory and Its Application to Physical Problems.* Addison-Wesley Publishing Company, Reading, Massachusetts, 1962.

R.A. Hauser and N.P. Kirchner. A hystorical note on the entropy principle of Müller and Liu. *Continuum Mechanics and Thermodynamics*, 14(2):223–226, 2002.

G.A. Holzapfel. *Nonlinear Solid Mechanics.* John Wiley & Sons, Ltd., Chichester, England, 2005.

K. Hutter and K. Jöhnk. *Continuum Methods of Physical Modeling.* Springer-Verlag, Berlin, 1981.

W. Jaunzemis. *Continuum Mechanics.* The Macmillan Company, New York, NY, 1967.

J. Jerphagnon, D. Chemla, and R. Bonneville. The description of the physical properties of condensed matter using irreducible tensors. *Advances in Physics*, 27(4):609–650, 1978.

J. Kestin. *A Course in Thermodynamics*, volume 1. McGraw-Hill Book Company, New York, NY, 1979.

J. Kestin. *A Course in Thermodynamics*, volume 2. McGraw-Hill Book Company, New York, NY, 1979.

G.G. Kleinstein. On the derivation of boundary conditions from the global principles of continuum mechanics. *Quarterly of Applied Mathematics*, 63(3):469–478, 2005.

B.H. Lavenda. *Thermodynamics of Irreversible Processes.* Dover Publications, Inc., Mineola, NY, 1978.

I.-S. Liu. Method of Lagrange multipliers for exploitation of the entropy principle. *Archive for Rational Mechanics and Analysis*, 46(2):131–148, 1972.

I.-S. Liu. On the requirement that material symmetries shall preserve density. *Archive for Rational Mechanics and Analysis*, 68(1):19–26, 1978.

I.-S. Liu. On interface equilibrium and inclusion problems. *Continuum Mechanics and Thermodynamics*, 4:177–186, 1992.

I.-S. Liu. On entropy flux-heat flux relation in thermodynamics with Lagrange multipliers. *Continuum Mechanics and Thermodynamics*, 8(4):247–256, 1996.

I.-S. Liu. *Continuum Mechanics.* Springer-Verlag, Berlin, 2002.

L.E. Malvern. *Introduction to the Mechanics of a Continuous Medium.* Prentice-Hall, Inc., Upper Saddle River, NJ, 1969.

A.G. McLellan. *The Classical Thermodynamics of Deformable Materials.* Cambridge University Press, New York, 1980.

I. Müller. On the entropy inequality. *Archive for Rational Mechanics and Analysis*, 26:118–141, 1967.

I. Müller. The coldness, a universal function in thermoelastic bodies. *Archive for Rational Mechanics and Analysis*, 41(5):319–332, 1971.

I. Müller. *Thermodynamics.* Pitman Publishing, Inc., Boston, MA, 1985.

A.I. Murdoch. *Physical Foundations of Continuum Mechanics.* Cambridge University Press, New York, NY, 2012.

W. Muschik. Fundamental remarks on evaluating dissipation inequalities. In J. Casas-Vázquez, D. Jou, and G. Lebon, editors, *Recent Developments in Nonequilibrium Thermodynamics*, Lecture Notes in Physics, pages 388–397. Springer-Verlag, 1984.

H.C. Van Ness and M.M. Abbott. *Classical Thermodynamics of Nonelectrolyte Solutions.* McGraw-Hill Book Company, New York, NY, 2nd edition, 1981.

W. Noll. On the continuity of the solid and fluid states. *Journal of Rational Mechanics and Analysis*, 4(1):3–81, 1955.

W. Noll. A mathematical theory of the mechanical behavior of continuous media. *Archive for Rational Mechanics and Analysis*, 2(1):197–226, 1958.

W. Noll. Lectures on the foundations of continuum mechanics and thermodynamics. *Archive for Rational Mechanics and Analysis*, 52(1):62–92, 1973.

W. Noll. *The Foundations of Mechanics and Thermodynamics – Selected Papers.* Springer-Verlag, New York, 1974.

R.W. Ogden. *Non-Linear Elastic Deformations.* John Wiley & Sons, New York, 1984.

P. Pennisi and M. Trovato. On the irreducibility of Professor G.F. Smith's representation for isotropic functions. *International Journal of Engineering Science*, 25(8):1059–1065, 1987.

W. Prager. *Introduction to Continuum Mechanics.* Dover Publications, Inc., New York, NY, 1961.

I. Prigogine. *Termodinamica dei Processi Irreversibili.* Leonardo Edizione Scientifiche, Roma, 1971.

R.S. Rivlin. Further remarks on the stress-deformation relations for isotropic materials. *Journal of Rational Mechanics and Analysis*, 4(5):681–702, 1955.

R.S. Rivlin. The fundamental equations of nonlinear continuum mechanics. In S.I. Pai, A.J. Faller, T.L. Lincoln, D.A. Tidman, G.N. Trytten, and T.D. Wilkerson, editors, *Dynamics of Fluids in Porous Media*, pages 83–126, Academic Press, New York, 1966.

R.S. Rivlin. An introduction to non-linear continuum mechanics. In R.S. Rivlin, editor, *Non-linear Continuum Theories in Mechanics and Physics and Their Applications*, pages 151–310. Springer-Verlag, Berlin, 1969.

R.S. Rivlin and J.L. Ericksen. Stress-deformation relations for isotropic materials. *Journal of Rational Mechanics and Analysis*, 4(3):323–425, 1955.

M. Silhavy. *The Mechanics and Thermodynamics of Continuous Media.* Springer-Verlag, Berlin, 1997.

D.R. Smith. *An Introduction to Continuum Mechanics.* Kluwer Academic Publishers, Dordrecht, The Netherlands, 1993.

G.F. Smith. On a fundamental error in two papers of C.-C. Wang "On representations for isotropic functions, parts I and II". *Archive for Rational Mechanics and Analysis*, 36:161–165, 1970.

G.F. Smith. On isotropic functions of symmetric tensors, skew-symmetric tensors and vectors. *International Journal of Engineering Science*, 9(10):899–916, 1971.

G.F. Smith. Constitutive equations for anisotropic and isotropic materials. In G.C. Sih, editor, *Mechanics and Physics of Discrete Systems*, volume 3. Elsevier Science B.V., Amsterdam, The Netherlands, 1994.

A.J.M. Spencer. Theory of invariants. In A.C. Eringen, editor, *Continuum Physics*, volume I. Academic Press, New York, 1971.

A.J.M. Spencer. *Continuum Mechanics.* Dover Publications, Inc., Mineola, NY, 1980.

A.J.M. Spencer and R.S. Rivlin. The theory of matrix polynomials and its application to the mechanics of isotropic continua. *Archive for Rational Mechanics and Analysis*, 2(4):309–336, 1959.

A.J.M. Spencer and R.S. Rivlin. Further results in the theory of matrix polynomials. *Archive for Rational Mechanics and Analysis*, 4(3):214–230, 1960.

G. Stephenson. *An Introduction to Matrices, Sets and Groups for Science Students.* Dover Publications, Inc., Mineola, NY, 1965.

E.B. Tadmor, R.E. Miller, and R.S. Elliott. *Continuum Mechanics and Thermodynamics.* Cambridge University Press, New York, NY, 2012.

L. Tisza. *Generalized Thermodynamics.* The M.I.T. Press, Cambridge, MA, 1966.

C. Truesdell. Thermodynamics for beginners. In M. Parkus and L.I. Sedov, editors, *Irreversible Aspects of Continuum Mechanics and Transfer of Physical Characteristics of Moving Fluids*, pages 373–389. Springer, Wien, 1968.

C. Truesdell. *A First Course in Rational Continuum Mechanics*, volume 1. Academic Press, New York, NY, 1977.

C. Truesdell. *Rational Thermodynamics.* Springer-Verlag, New York, NY, 2nd edition, 1984.

C. Truesdell and W. Noll. The non-linear field theories of mechanics. In S. Flügge, editor, *Handbuch der Physik*, volume III/3. Springer, Berlin-Heidelberg-New York, 1965.

D.C. Wallace. *Thermodynamics of Crystals.* Dover Publications, Inc., Mineola, NY, 1972.

C.-C. Wang. A new representation theorem for isotropic functions: An answer to professor G.F. Smith's criticism of my papers on representations for isotropic functions. Part 1. Scalar-valued isotropic functions. *Archive for Rational Mechanics and Analysis*, 36(3):166–197, 1970.

C.-C. Wang. A new representation theorem for isotropic functions: An answer to professor G.F. Smith's criticism of my papers on representations for isotropic functions. Part 2. Vector-valued isotropic functions, symmetric tensor-valued isotropic functions, and skew-symmetric tensor-valued isotropic functions. *Archive for Rational Mechanics and Analysis*, 36(3):198–223, 1970.

C.-C. Wang. Corrigendum to my recent papers on "Representations for isotropic functions". *Archive for Rational Mechanics and Analysis*, 43(3):392–395, 1971.

K. Wilmański. *Thermo-mechanics of Continua*. Springer-Verlag, Berlin, 1998.

Q.-S. Zheng. On the representations for isotropic vector-valued, symmetric tensor-valued and skew-symmetric tensor-valued functions. *International Journal of Engineering Science*, 31(7):1013–1024, 1993.

Q.-S. Zheng. Theory of representations for tensor functions – A unified invariant approach to constitutive equations. *Applied Mechanics Reviews*, 47(11):545–587, 1994.

6

Spatially uniform systems

As a means of exercising the use of our balance equations, initial and boundary conditions, constitutive principles, and thermodynamics relations, we first examine systems that are uniform in space. Note that in order for a system to be spatially uniform, it is necessary but not sufficient for the material to be homogeneous and isotropic. Let us first summarize the mass, momenta, and energy balance laws for nonpolar materials obtained earlier:

$$\dot{\rho} = -\rho \operatorname{div} \mathbf{v}, \tag{6.1}$$

$$\rho \mathbf{a} = \operatorname{div} \boldsymbol{\sigma} + \rho \mathbf{f}, \tag{6.2}$$

$$\boldsymbol{\sigma} = \boldsymbol{\sigma}^T, \tag{6.3}$$

$$\rho \dot{e} = -\operatorname{div} \mathbf{q} + \mathbf{L} : \boldsymbol{\sigma} + \rho r. \tag{6.4}$$

In addition, we have the Clausius–Duhem inequality

$$-\gamma_v \theta \equiv \rho\left(\dot{\psi} + \eta\dot{\theta}\right) - \mathbf{L} : \boldsymbol{\sigma} + \mathbf{g} \cdot \frac{\mathbf{q}}{\theta} - \theta \operatorname{div}\left(\mathbf{h} - \frac{\mathbf{q}}{\theta}\right) + \rho\theta\left(b - \frac{r}{\theta}\right) \leq 0. \tag{6.5}$$

Now since a boundary is a material surface, the conditions there must be

$$\mathbf{v} \cdot \mathbf{n} = c_n, \qquad \boldsymbol{\sigma} \cdot \mathbf{n} = \mathsf{F}, \qquad \mathbf{q} \cdot \mathbf{n} = \mathsf{G}, \qquad \mathbf{h} \cdot \mathbf{n} = \mathsf{R}. \tag{6.6}$$

We consider a system that is closed, and having a boundary that is stationary ($c_n = 0$), free ($\mathsf{F} = \mathbf{0}$), adiabatic ($\mathsf{G} = 0$, thermally isolated), and isentropic ($\mathsf{R} = 0$, no entropy exchange with surroundings). We are thinking of a body having constant mass, a composition that is uniform in space, and state functions that only vary with time. Since the mass is constant, it is convenient to introduce the specific volume $\mathsf{v} = 1/\rho$; subsequently, since $\operatorname{div} \mathbf{v} = \operatorname{tr} \mathbf{D}$, the mass balance (6.1) provides the relation

$$\operatorname{tr} \mathbf{D} = \frac{\dot{\mathsf{v}}}{\mathsf{v}}. \tag{6.7}$$

Furthermore, since we are considering a thermally isolated uniform system undergoing a simple thermomechanical process, then, in conjunction with the boundary conditions and the Clausius–Duhem inequality, we must have

$$\boldsymbol{\sigma} = -p\mathbf{1}, \qquad \mathbf{h} = \frac{\mathbf{q}}{\theta} = \mathbf{0}, \qquad b = \frac{r}{\theta} = 0. \tag{6.8}$$

Now our equations and the Clausius–Duhem inequality (6.2)–(6.6) reduce to

$$\mathbf{a} = \mathbf{f}, \tag{6.9}$$

$$\dot{e} = -p\dot{\mathsf{v}}, \tag{6.10}$$

$$-\gamma_v \theta \mathsf{v} \equiv \dot{\psi} + \eta\dot{\theta} + p\dot{\mathsf{v}} \leq 0. \tag{6.11}$$

The external momentum body source can be an arbitrarily prescribed function of time:

$$\mathbf{f} = \mathbf{f}(t). \tag{6.12}$$

Note that (6.9) and (6.12) require that the acceleration of the body be uniform, $\mathbf{a} = \mathbf{a}(t)$, and from the linear momentum equation (6.9) (this is nothing more than the conventional Newton's second law of motion), the acceleration can be integrated to obtain the velocity, $\mathbf{v} = \mathbf{v}(t)$, as well as the motion, $\mathbf{x} = \boldsymbol{\chi}(t)$, for given initial conditions. We further note that in this case, where the system is spatially uniform, the mechanics and thermodynamics of the system are completely decoupled. In addition, we note that equation (6.10) is nothing more than the first law of thermostatics for a thermally isolated system, and (6.11) is the Clausius–Duhem inequality.

Subsequently, enforcement of the Clausius–Duhem inequality, in conjunction with the balance laws, should provide relations between the dependent constitutive variables

$$\psi = \underset{0\leq s<\infty}{\mathfrak{F}}(\mathsf{v},\theta), \qquad \eta = \underset{0\leq s<\infty}{\mathfrak{G}}(\mathsf{v},\theta), \qquad p = \underset{0\leq s<\infty}{\mathfrak{H}}(\mathsf{v},\theta), \tag{6.13}$$

and the independent variables describing the thermodynamic process

$$\mathsf{v} = \mathsf{v}(t), \qquad \theta = \theta(t). \tag{6.14}$$

In (6.13) our constitutive variables are taken to be functionals of the independent variables (6.14) (see (5.31)), and we note that $\mathsf{v} > 0$ and $\theta > 0$. Also, the dependence on the deformation gradient, $\mathbf{F}$, here reduces to the dependence on $\operatorname{tr}\mathbf{D}$, and thus to the dependence on v through the mass balance equation. Furthermore, we have no dependence on the heat flux, $\mathbf{g}$, in thermally isolated uniform systems.

Now the same material may have different constitutive equations (6.13), which depend on the level of description of the thermodynamic process (6.14). Since we are dealing with a uniform system, what we mean here is the choice of different rate-type materials, i.e., how much memory does the material have (see Section 5.3).

6.1 Material with no memory

Here we take $p = q = 0$ in (5.38) so that

$$\psi = \psi(\mathsf{v},\theta), \qquad \eta = \eta(\mathsf{v},\theta), \qquad p = p(\mathsf{v},\theta). \tag{6.15}$$

Subsequently, (6.11) becomes

$$-\gamma_v \theta \mathsf{v} \equiv \left[\frac{\partial \psi}{\partial \mathsf{v}}(\mathsf{v},\theta) + p(\mathsf{v},\theta)\right]\dot{\mathsf{v}} + \left[\frac{\partial \psi}{\partial \theta}(\mathsf{v},\theta) + \eta(\mathsf{v},\theta)\right]\dot{\theta} \leq 0. \tag{6.16}$$

Now, since $\mathsf{v} > 0$ and $\theta > 0$, for arbitrary processes in which $\dot{\mathsf{v}}$ and $\dot{\theta}$ can be prescribed to be of either sign, and since the terms in square brackets are independent of $\dot{\mathsf{v}}$ and $\dot{\theta}$, (6.16) requires that

$$\frac{\partial \psi}{\partial \mathsf{v}} = -p \qquad \text{and} \qquad \frac{\partial \psi}{\partial \theta} = -\eta, \tag{6.17}$$

in which case we have from (5.206) that the Helmholtz free energy is a thermodynamic potential function of specific volume and temperature,

$$\dot{\psi} = -p\dot{\mathsf{v}} - \eta\,\dot{\theta}, \tag{6.18}$$

and in conjunction with the definition of the Helmholtz potential (5.201), we obtain the Gibbs equation

$$\dot{e} = -p\dot{\mathsf{v}} + \theta\,\dot{\eta}, \tag{6.19}$$

where clearly

$$e = e(\mathsf{v}, \theta) = \psi - \theta\,\frac{\partial \psi}{\partial \theta}. \tag{6.20}$$

Thus, since (6.16) is identically satisfied, processes without memory have zero entropy production, i.e., they are reversible. In addition, since the constitutive relations are independent of $\dot{\mathsf{v}}$ and $\dot{\theta}$, such processes are also equilibrium processes. In fact, classical thermodynamics deals exclusively with reversible equilibrium processes, which, as we have seen, result from spatially uniform systems in which materials have no memory.

Now the internal energy minimum principle tells us that for a given value of the total entropy, the equilibrium value of any unconstrained internal quantity is such as to minimize the internal energy. Thus, it is convenient to substitute the temperature dependence by the entropy dependence:

$$e = e(\eta, \mathsf{v}). \tag{6.21}$$

We now represent the total variation in internal energy as a Taylor series expansion around the equilibrium state. The first-order terms cancel, and for the quadratic term, we have the stability condition:

$$\delta^2 e = \left(\frac{\partial^2 e}{\partial \eta^2}\right)_{\mathsf{v}} (\delta\eta)^2 + 2\left[\frac{\partial}{\partial \mathsf{v}}\left(\frac{\partial e}{\partial \eta}\right)_{\mathsf{v}}\right]_{\eta} \delta\eta\,\delta\mathsf{v} + \left(\frac{\partial^2 e}{\partial \mathsf{v}^2}\right)_{\eta} (\delta\mathsf{v})^2 > 0. \tag{6.22}$$

The stability condition is that the right-hand side must be positive definite for any $\delta\eta$ and $\delta\mathsf{v}$, except of course the trivial case $\delta\eta = \delta\mathsf{v} = 0$, in which case third-order terms must be taken into account. This is a quadratic form which is positive definite if and only if the following two conditions are satisfied:

$$\left(\frac{\partial^2 e}{\partial \eta^2}\right)_{\mathsf{v}} > 0 \qquad \text{and} \qquad \left(\frac{\partial^2 e}{\partial \eta^2}\right)_{\mathsf{v}} \left(\frac{\partial^2 e}{\partial \mathsf{v}^2}\right)_{\eta} - \left[\frac{\partial}{\partial \mathsf{v}}\left(\frac{\partial e}{\partial \eta}\right)_{\mathsf{v}}\right]_{\eta}^2 > 0. \tag{6.23}$$

Now, using standard thermodynamic relations, the first condition becomes

$$\left(\frac{\partial^2 e}{\partial \eta^2}\right)_{\mathsf{v}} = \left(\frac{\partial \theta}{\partial \eta}\right)_{\mathsf{v}} = \frac{\theta}{c_{\mathsf{v}}} > 0, \tag{6.24}$$

which, since $\theta > 0$, implies that the specific heat at constant volume

$$c_{\mathsf{v}} \equiv \theta \left(\frac{\partial \eta}{\partial \theta} \right)_{\mathsf{v}} \tag{6.25}$$

is positive definite: $c_{\mathsf{v}} > 0$. It is easy to see that the two conditions (6.23) also imply that

$$\left(\frac{\partial^2 e}{\partial \mathsf{v}^2} \right)_{\eta} > 0 \tag{6.26}$$

and thus

$$\left(\frac{\partial^2 e}{\partial \mathsf{v}^2} \right)_{\eta} = - \left(\frac{\partial p}{\partial \mathsf{v}} \right)_{\eta} = \frac{1}{\mathsf{v}\, \kappa_{\eta}} > 0. \tag{6.27}$$

Subsequently, since $\mathsf{v} > 0$, the condition implies that the adiabatic compressibility

$$\kappa_{\eta} \equiv -\frac{1}{\mathsf{v}} \left(\frac{\partial \mathsf{v}}{\partial p} \right)_{\eta} \tag{6.28}$$

is positive definite: $\kappa_{\eta} > 0$. Lastly, the second condition in (6.23) can be rewritten as

$$\left(\frac{\partial^2 e}{\partial \eta^2} \right)_{\mathsf{v}} \left(\frac{\partial^2 e}{\partial \mathsf{v}^2} \right)_{\eta} - \left[\frac{\partial}{\partial \mathsf{v}} \left(\frac{\partial e}{\partial \eta} \right)_{\mathsf{v}} \right]_{\eta}^2 = - \left(\frac{\partial \theta}{\partial \eta} \right)_{\mathsf{v}} \left(\frac{\partial p}{\partial \mathsf{v}} \right)_{\eta} - \left(\frac{\partial \theta}{\partial \mathsf{v}} \right)_{\eta}^2 \tag{6.29}$$

and since

$$\left(\frac{\partial \theta}{\partial \mathsf{v}} \right)_{\eta} = - \left(\frac{\partial \theta}{\partial \eta} \right)_{\mathsf{v}} \left(\frac{\partial \eta}{\partial \mathsf{v}} \right)_{\theta}, \tag{6.30}$$

this condition becomes

$$\frac{\theta}{\mathsf{v}\, \kappa_{\eta}\, c_{\mathsf{v}}} - \left(\frac{\theta\, \alpha}{\kappa_{\theta}\, c_{\mathsf{v}}} \right)^2 > 0, \tag{6.31}$$

where the isothermal compressibility and the thermal expansion coefficient are defined by

$$\kappa_{\theta} \equiv -\frac{1}{\mathsf{v}} \left(\frac{\partial \mathsf{v}}{\partial p} \right)_{\theta}, \tag{6.32}$$

and

$$\alpha \equiv \frac{1}{\mathsf{v}} \left(\frac{\partial \mathsf{v}}{\partial \theta} \right)_{p}. \tag{6.33}$$

This condition, in conjunction with the previous conditions, can be easily shown to lead to

$$c_p > c_{\mathsf{v}} > 0 \qquad \text{and} \qquad \kappa_{\theta} > \kappa_{\eta} > 0, \tag{6.34}$$

where the specific heat at constant pressure is defined by

$$c_p \equiv \theta \left(\frac{\partial \eta}{\partial \theta} \right)_{p}, \tag{6.35}$$

and note that the sign of α is arbitrary.

6.2 Material with short memory of volume

Here we take $p = 1$ and $q = 0$ in (5.38) so that

$$\psi = \psi(\mathsf{v}, \dot{\mathsf{v}}, \theta), \qquad \eta = \eta(\mathsf{v}, \dot{\mathsf{v}}, \theta), \qquad p = p(\mathsf{v}, \dot{\mathsf{v}}, \theta). \tag{6.36}$$

Subsequently, (6.11) becomes

$$-\gamma_v \,\theta\, \mathsf{v} \equiv \left[\frac{\partial\psi}{\partial\mathsf{v}}(\mathsf{v}, \dot{\mathsf{v}}, \theta) + p(\mathsf{v}, \dot{\mathsf{v}}, \theta)\right]\dot{\mathsf{v}} + \frac{\partial\psi}{\partial\dot{\mathsf{v}}}(\mathsf{v}, \dot{\mathsf{v}}, \theta)\,\ddot{\mathsf{v}} + \left[\frac{\partial\psi}{\partial\theta}(\mathsf{v}, \dot{\mathsf{v}}, \theta) + \eta(\mathsf{v}, \dot{\mathsf{v}}, \theta)\right]\dot{\theta} \le 0. \tag{6.37}$$

Now, since $\mathsf{v} > 0$ and $\theta > 0$, for arbitrary processes in which $\dot{\mathsf{v}}$, $\ddot{\mathsf{v}}$, and $\dot{\theta}$ can be of either sign, and since ψ is not a function of $\ddot{\mathsf{v}}$, (6.37) requires that

$$\frac{\partial\psi}{\partial\dot{\mathsf{v}}} = 0 \qquad \text{or} \qquad \psi = \psi(\mathsf{v}, \theta), \tag{6.38}$$

so that we can rewrite (6.37) as

$$-\gamma_v \,\theta\, \mathsf{v} \equiv \left[\frac{\partial\psi}{\partial\mathsf{v}}(\mathsf{v}, \theta) + p(\mathsf{v}, \dot{\mathsf{v}}, \theta)\right]\dot{\mathsf{v}} + \left[\frac{\partial\psi}{\partial\theta}(\mathsf{v}, \theta) + \eta(\mathsf{v}, \dot{\mathsf{v}}, \theta)\right]\dot{\theta} \le 0, \tag{6.39}$$

from which, for arbitrary $\dot{\theta}$, since the terms in square brackets are independent of $\dot{\theta}$, we must require that

$$\frac{\partial\psi}{\partial\theta} = -\eta. \tag{6.40}$$

Thus we have that

$$\eta = \eta(\mathsf{v}, \theta) \qquad \text{and} \qquad e = e(\mathsf{v}, \theta), \tag{6.41}$$

and the reduced Clausius–Duhem inequality becomes

$$-\gamma_v \,\theta\, \mathsf{v} \equiv \left[\frac{\partial\psi}{\partial\mathsf{v}}(\mathsf{v}, \theta) + p(\mathsf{v}, \dot{\mathsf{v}}, \theta)\right]\dot{\mathsf{v}} \le 0. \tag{6.42}$$

Now at thermodynamic equilibrium ($\dot{\mathsf{v}} = 0$), the entropy production must be zero, and clearly it is trivially so. But, in addition, from (6.42) we also note that the entropy production must be a minimum at equilibrium. Thus, at equilibrium we must have that

$$\gamma_v(\mathsf{v}, \dot{\mathsf{v}}, \theta)\big|_{\dot{\mathsf{v}}=0} = 0, \quad \frac{\partial\gamma_v}{\partial\dot{\mathsf{v}}}(\mathsf{v}, \dot{\mathsf{v}}, \theta)\bigg|_{\dot{\mathsf{v}}=0} = 0, \quad \text{and} \quad \frac{\partial^2\gamma_v}{\partial\dot{\mathsf{v}}^2}(\mathsf{v}, \dot{\mathsf{v}}, \theta)\bigg|_{\dot{\mathsf{v}}=0} > 0. \tag{6.43}$$

Now, from the first derivative condition, we obtain

$$\frac{\partial\psi}{\partial\mathsf{v}} = -p^e, \tag{6.44}$$

where we have defined the equilibrium pressure

$$p^e = p^e(\mathsf{v}, \theta) \equiv p(\mathsf{v}, 0, \theta). \tag{6.45}$$

Defining the nonequilibrium residual pressure as

$$p^r = p^r(\mathsf{v},\dot{\mathsf{v}},\theta) \equiv p(\mathsf{v},\dot{\mathsf{v}},\theta) - p^e(\mathsf{v},\theta), \tag{6.46}$$

so that $p^r(\mathsf{v},0,\theta) = 0$, the second derivative condition requires that

$$\frac{\partial p^r}{\partial \dot{\mathsf{v}}}(\mathsf{v},0,\theta) < 0. \tag{6.47}$$

Obviously, outside of equilibrium, p^r can have an arbitrary dependence on $\dot{\mathsf{v}}$. To a leading order approximation, it depends linearly on $\dot{\mathsf{v}}$; thus we write it as

$$p^r = -\lambda \frac{\dot{\mathsf{v}}}{\mathsf{v}}, \tag{6.48}$$

where we have defined the *dilatational viscosity*

$$\lambda = \lambda(\mathsf{v},\theta) > 0. \tag{6.49}$$

Note that the positive-definite requirement on the dilatational viscosity is dictated by the second derivative condition.

Summarizing the results of this system, we see that the constitutive quantities are given by

$$\psi = \psi(\mathsf{v},\theta), \qquad \eta = -\frac{\partial \psi}{\partial \theta}, \quad \text{and} \quad p = -\frac{\partial \psi}{\partial \mathsf{v}} - \lambda(\mathsf{v},\theta)\frac{\dot{\mathsf{v}}}{\mathsf{v}}. \tag{6.50}$$

In addition, we find that the Helmholtz free energy is a potential for entropy but only for the equilibrium pressure,

$$\dot{\psi} = -p^e\,\dot{\mathsf{v}} - \eta\,\dot{\theta}, \tag{6.51}$$

and, in conjunction with (5.201), the Gibbs equation is given by

$$\dot{e} = -p^e\,\dot{\mathsf{v}} + \theta\,\dot{\eta}, \tag{6.52}$$

where

$$e = e(\mathsf{v},\theta) = \psi - \theta\frac{\partial \psi}{\partial \theta}. \tag{6.53}$$

For this material, none of the Helmholtz free energy, entropy, or internal energy constitutes fundamental relations since they cannot fully describe the material. On the other hand, the Gibbs free energy, given by

$$g = g(\mathsf{v},\dot{\mathsf{v}},\theta) = e + p\mathsf{v} - \theta\eta = \psi - \frac{\partial \psi}{\partial \mathsf{v}}\mathsf{v} - \lambda(\mathsf{v},\theta)\,\dot{\mathsf{v}}, \tag{6.54}$$

does provide the fundamental relation for this material.

6.3 Material with longer memory of volume

Here we take $p = 2$ and $q = 0$ in (5.38) so that

$$\psi = \psi(\mathsf{v},\dot{\mathsf{v}},\ddot{\mathsf{v}},\theta), \qquad \eta = \eta(\mathsf{v},\dot{\mathsf{v}},\ddot{\mathsf{v}},\theta), \quad \text{and} \quad p = p(\mathsf{v},\dot{\mathsf{v}},\ddot{\mathsf{v}},\theta). \tag{6.55}$$

Subsequently, (6.11) becomes

$$-\gamma_v\,\theta\,\mathsf{v} \equiv \left[\frac{\partial\psi}{\partial\mathsf{v}}(\mathsf{v},\dot{\mathsf{v}},\ddot{\mathsf{v}},\theta) + p(\mathsf{v},\dot{\mathsf{v}},\ddot{\mathsf{v}},\theta)\right]\dot{\mathsf{v}} + \frac{\partial\psi}{\partial\dot{\mathsf{v}}}(\mathsf{v},\dot{\mathsf{v}},\ddot{\mathsf{v}},\theta)\,\ddot{\mathsf{v}} + \frac{\partial\psi}{\partial\ddot{\mathsf{v}}}(\mathsf{v},\dot{\mathsf{v}},\ddot{\mathsf{v}},\theta)\,\dddot{\mathsf{v}} + \left[\frac{\partial\psi}{\partial\theta}(\mathsf{v},\dot{\mathsf{v}},\ddot{\mathsf{v}},\theta) + \eta(\mathsf{v},\dot{\mathsf{v}},\ddot{\mathsf{v}},\theta)\right]\dot{\theta} \le 0. \tag{6.56}$$

Now, since $\mathsf{v} > 0$ and $\theta > 0$, for arbitrary processes in which $\dot{\mathsf{v}}$, $\ddot{\mathsf{v}}$, $\dddot{\mathsf{v}}$, and $\dot{\theta}$ can be of either sign, (6.56) requires that (because of the linearity in $\dddot{\mathsf{v}}$ and $\dot{\theta}$)

$$\frac{\partial\psi}{\partial\ddot{\mathsf{v}}} = 0 \qquad \text{and} \qquad \frac{\partial\psi}{\partial\theta} = -\eta. \tag{6.57}$$

Thus, from the above, as well as (5.201), we have that

$$\psi = \psi(\mathsf{v},\dot{\mathsf{v}},\theta), \qquad \eta = \eta(\mathsf{v},\dot{\mathsf{v}},\theta), \quad \text{and} \quad e = e(\mathsf{v},\dot{\mathsf{v}},\theta), \tag{6.58}$$

and the reduced Clausius–Duhem inequality becomes

$$-\gamma_v\,\theta\,\mathsf{v} \equiv \left[\frac{\partial\psi}{\partial\mathsf{v}}(\mathsf{v},\dot{\mathsf{v}},\theta) + p(\mathsf{v},\dot{\mathsf{v}},\ddot{\mathsf{v}},\theta)\right]\dot{\mathsf{v}} + \frac{\partial\psi}{\partial\dot{\mathsf{v}}}(\mathsf{v},\dot{\mathsf{v}},\theta)\,\ddot{\mathsf{v}} \le 0. \tag{6.59}$$

Note that the above inequality is nonlinear in $\ddot{\mathsf{v}}$; thus to make further progress, we examine the equilibrium behavior of this material. At equilibrium, we must have $\dot{\mathsf{v}} = 0$ and $\ddot{\mathsf{v}} = 0$, and $\gamma_v = 0$ is a minimum:

$$\gamma_v(\mathsf{v},\dot{\mathsf{v}},\ddot{\mathsf{v}},\theta)\big|_{\dot{\mathsf{v}}=\ddot{\mathsf{v}}=0} = 0, \tag{6.60}$$

$$\frac{\partial\gamma_v}{\partial\dot{\mathsf{v}}}(\mathsf{v},\dot{\mathsf{v}},\ddot{\mathsf{v}},\theta)\bigg|_{\dot{\mathsf{v}}=\ddot{\mathsf{v}}=0} = 0, \quad \frac{\partial\gamma_v}{\partial\ddot{\mathsf{v}}}(\mathsf{v},\dot{\mathsf{v}},\ddot{\mathsf{v}},\theta)\bigg|_{\dot{\mathsf{v}}=\ddot{\mathsf{v}}=0} = 0, \tag{6.61}$$

$$\frac{\partial^2\gamma_v}{\partial\dot{\mathsf{v}}^2}(\mathsf{v},\dot{\mathsf{v}},\ddot{\mathsf{v}},\theta)\bigg|_{\dot{\mathsf{v}}=\ddot{\mathsf{v}}=0} > 0, \quad \frac{\partial^2\gamma_v}{\partial\dot{\mathsf{v}}^2}\frac{\partial^2\gamma_v}{\partial\ddot{\mathsf{v}}^2}(\mathsf{v},\dot{\mathsf{v}},\ddot{\mathsf{v}},\theta)\bigg|_{\dot{\mathsf{v}}=\ddot{\mathsf{v}}=0} - \left[\frac{\partial^2\gamma_v}{\partial\dot{\mathsf{v}}\partial\ddot{\mathsf{v}}}(\mathsf{v},\dot{\mathsf{v}},\ddot{\mathsf{v}},\theta)\right]^2_{\dot{\mathsf{v}}=\ddot{\mathsf{v}}=0} > 0. \tag{6.62}$$

Clearly $\gamma_v = 0$ is trivially satisfied at equilibrium. As in the previous section, the first and second derivative conditions sequentially give us that

$$\frac{\partial\psi^r}{\partial\dot{\mathsf{v}}}(\mathsf{v},0,\theta) = 0 \qquad \text{and} \qquad \frac{\partial\psi^e}{\partial\mathsf{v}} = -p^e, \tag{6.63}$$

where we have defined the equilibrium and residual free energy and pressure as follows:

$$\psi^e = \psi^e(\mathsf{v},\theta) \equiv \psi(\mathsf{v},0,\theta), \tag{6.64}$$

$$p^e = p^e(\mathsf{v},\theta) \equiv p(\mathsf{v},0,0,\theta), \tag{6.65}$$

$$\psi^r = \psi^r(\mathsf{v},\dot{\mathsf{v}},\theta) \equiv \psi(\mathsf{v},\dot{\mathsf{v}},\theta) - \psi^e(\mathsf{v},\theta), \tag{6.66}$$

$$p^r = p^r(\mathsf{v},\dot{\mathsf{v}},\ddot{\mathsf{v}},\theta) \equiv p(\mathsf{v},\dot{\mathsf{v}},\ddot{\mathsf{v}},\theta) - p^e(\mathsf{v},\theta), \tag{6.67}$$

and we must have that $\psi^r(\mathsf{v},0,\theta) = 0$ and $p^r(\mathsf{v},0,0,\theta) = 0$. Subsequently, the Clausius–Duhem inequality can be rewritten as

$$-\gamma_v\,\theta\,\mathsf{v} \equiv \left[\frac{\partial\psi^r}{\partial\mathsf{v}}(\mathsf{v},\dot{\mathsf{v}},\theta) + p^r(\mathsf{v},\dot{\mathsf{v}},\ddot{\mathsf{v}},\theta)\right]\dot{\mathsf{v}} + \frac{\partial\psi^r}{\partial\dot{\mathsf{v}}}(\mathsf{v},\dot{\mathsf{v}},\theta)\,\ddot{\mathsf{v}} \le 0. \tag{6.68}$$

The resulting second derivative condition

$$\frac{\partial^2 \psi^r}{\partial \mathsf{v} \partial \dot{\mathsf{v}}}(\mathsf{v},0,\theta) + \frac{\partial p^r}{\partial \dot{\mathsf{v}}}(\mathsf{v},0,0,\theta) < 0 \tag{6.69}$$

does not yield additional simplifications or reduction of (6.68) unless we approximate ψ^r up to some order in $\dot{\mathsf{v}}$, and p^r up to some order in $\dot{\mathsf{v}}$ and $\ddot{\mathsf{v}}$.

In summary, the constitutive equations of this material are given by

$$\psi = \psi^e(\mathsf{v},\theta) + \psi^r(\mathsf{v},\dot{\mathsf{v}},\theta), \qquad \eta = -\frac{\partial \psi}{\partial \theta}, \quad \text{and} \quad p = -\frac{\partial \psi^e}{\partial \mathsf{v}} + p^r(\mathsf{v},\dot{\mathsf{v}},\ddot{\mathsf{v}},\theta). \tag{6.70}$$

In addition, we find that the Helmholtz free energy is a potential for entropy but only the equilibrium part of the free energy,

$$\dot{\psi} = -p^e\,\dot{\mathsf{v}} + \frac{\partial \psi^r}{\partial \mathsf{v}}\,\dot{\mathsf{v}} + \frac{\partial \psi^r}{\partial \dot{\mathsf{v}}}\,\ddot{\mathsf{v}} - \eta\,\dot{\theta}, \tag{6.71}$$

and, in conjunction with (5.201), the Gibbs equation is now given by

$$\dot{e} = -p^e\,\dot{\mathsf{v}} + \frac{\partial \psi^r}{\partial \mathsf{v}}\,\dot{\mathsf{v}} + \frac{\partial \psi^r}{\partial \dot{\mathsf{v}}}\,\ddot{\mathsf{v}} + \theta\,\dot{\eta}, \tag{6.72}$$

where

$$e = e(\mathsf{v},\dot{\mathsf{v}},\theta) = \psi - \frac{\partial \psi}{\partial \theta}. \tag{6.73}$$

Note that the terms involving derivatives of ψ^r contribute to entropy production when the system is not in equilibrium.

Again, for this material, none of the Helmholtz free energy, entropy, or internal energy constitutes fundamental relations since they cannot fully describe this material. On the other hand, the Gibbs free energy

$$g = g(\mathsf{v},\dot{\mathsf{v}},\ddot{\mathsf{v}},\theta) = \psi - \left[\frac{\partial \psi^e}{\partial \mathsf{v}} - p^r(\mathsf{v},\dot{\mathsf{v}},\ddot{\mathsf{v}},\theta)\right]\mathsf{v} \tag{6.74}$$

does provide the fundamental relation for the material.

6.4 Material with short memory

Here we take $p = q = 1$ in (5.38) so that

$$\psi = \psi(\mathsf{v},\dot{\mathsf{v}},\theta,\dot{\theta}), \qquad \eta = \eta(\mathsf{v},\dot{\mathsf{v}},\theta,\dot{\theta}), \quad \text{and} \quad p = p(\mathsf{v},\dot{\mathsf{v}},\theta,\dot{\theta}). \tag{6.75}$$

Subsequently, (6.11) becomes

$$\begin{aligned} -\gamma_v\,\theta\,\mathsf{v} \equiv & \left[\frac{\partial \psi}{\partial \mathsf{v}}(\mathsf{v},\dot{\mathsf{v}},\theta,\dot{\theta}) + p(\mathsf{v},\dot{\mathsf{v}},\theta,\dot{\theta})\right]\dot{\mathsf{v}} + \frac{\partial \psi}{\partial \dot{\mathsf{v}}}(\mathsf{v},\dot{\mathsf{v}},\theta,\dot{\theta})\,\ddot{\mathsf{v}} + \\ & \left[\frac{\partial \psi}{\partial \theta}(\mathsf{v},\dot{\mathsf{v}},\theta,\dot{\theta}) + \eta(\mathsf{v},\dot{\mathsf{v}},\theta,\dot{\theta})\right]\dot{\theta} + \frac{\partial \psi}{\partial \dot{\theta}}(\mathsf{v},\dot{\mathsf{v}},\theta,\dot{\theta})\,\ddot{\theta} \le 0. \end{aligned} \tag{6.76}$$

Now, since $\mathsf{v} > 0$ and $\theta > 0$, for arbitrary processes in which $\dot{\mathsf{v}}$, $\ddot{\mathsf{v}}$, $\dot{\theta}$, and $\ddot{\theta}$ can be of either sign, (6.76) requires that (because of the linearity in $\ddot{\mathsf{v}}$ and $\ddot{\theta}$)

$$\frac{\partial \psi}{\partial \dot{\mathsf{v}}} = 0 \quad \text{and} \quad \frac{\partial \psi}{\partial \dot{\theta}} = 0, \tag{6.77}$$

so that

$$\psi = \psi(\mathsf{v}, \theta) \tag{6.78}$$

and we can rewrite (6.76) as

$$-\gamma_v \theta \mathsf{v} \equiv \left[\frac{\partial \psi}{\partial \mathsf{v}} (\mathsf{v}, \theta) + p(\mathsf{v}, \dot{\mathsf{v}}, \theta, \dot{\theta}) \right] \dot{\mathsf{v}} + \left[\frac{\partial \psi}{\partial \theta} (\mathsf{v}, \theta) + \eta(\mathsf{v}, \dot{\mathsf{v}}, \theta, \dot{\theta}) \right] \dot{\theta} \leq 0. \tag{6.79}$$

Now, at thermodynamic equilibrium, $\dot{\mathsf{v}} = 0$ and $\dot{\theta} = 0$, and the entropy production must be zero, and clearly it is trivially so. But, in addition, from (6.79) we also note that the entropy production must be a minimum at equilibrium. Thus, at equilibrium, we must have that

$$\gamma_v(\mathsf{v}, \dot{\mathsf{v}}, \theta, \dot{\theta})\big|_{\dot{\mathsf{v}}=\dot{\theta}=0} = 0, \tag{6.80}$$

$$\frac{\partial \gamma_v}{\partial \dot{\mathsf{v}}} (\mathsf{v}, \dot{\mathsf{v}}, \theta, \dot{\theta})\bigg|_{\dot{\mathsf{v}}=\dot{\theta}=0} = 0, \quad \frac{\partial \gamma_v}{\partial \dot{\theta}} (\mathsf{v}, \dot{\mathsf{v}}, \theta, \dot{\theta})\bigg|_{\dot{\mathsf{v}}=\dot{\theta}=0} = 0, \tag{6.81}$$

$$\frac{\partial^2 \gamma_v}{\partial \dot{\mathsf{v}}^2} (\mathsf{v}, \dot{\mathsf{v}}, \theta, \dot{\theta})\bigg|_{\dot{\mathsf{v}}=\dot{\theta}=0} > 0, \quad \frac{\partial^2 \gamma_v}{\partial \dot{\mathsf{v}}^2} \frac{\partial^2 \gamma_v}{\partial \dot{\theta}^2} (\mathsf{v}, \dot{\mathsf{v}}, \theta, \dot{\theta})\bigg|_{\dot{\mathsf{v}}=\dot{\theta}=0} - \left[\frac{\partial^2 \gamma_v}{\partial \dot{\mathsf{v}} \partial \dot{\theta}} (\mathsf{v}, \dot{\mathsf{v}}, \theta, \dot{\theta}) \right]^2_{\dot{\mathsf{v}}=\dot{\theta}=0} > 0. \tag{6.82}$$

Satisfaction of the first derivative conditions requires that

$$\frac{\partial \psi}{\partial \mathsf{v}} = -p^e \quad \text{and} \quad \frac{\partial \psi}{\partial \theta} = -\eta^e, \tag{6.83}$$

and allows the inequality to be rewritten as

$$-\gamma_v \theta \mathsf{v} \equiv p^r(\mathsf{v}, \dot{\mathsf{v}}, \theta, \dot{\theta})\, \dot{\mathsf{v}} + \eta^r(\mathsf{v}, \dot{\mathsf{v}}, \theta, \dot{\theta})\, \dot{\theta} \leq 0, \tag{6.84}$$

where we have defined the equilibrium and residual pressure and entropy as follows:

$$p^e = p^e(\mathsf{v}, \theta) \equiv \psi(\mathsf{v}, 0, \theta, 0), \tag{6.85}$$

$$\eta^e = \eta^e(\mathsf{v}, \theta) \equiv \psi(\mathsf{v}, 0, \theta, 0), \tag{6.86}$$

$$p^r = p^r(\mathsf{v}, \dot{\mathsf{v}}, \theta, \dot{\theta}) \equiv p(\mathsf{v}, \dot{\mathsf{v}}, \theta, \dot{\theta}) - p^e(\mathsf{v}, \theta), \tag{6.87}$$

$$\eta^r = \eta^r(\mathsf{v}, \dot{\mathsf{v}}, \theta, \dot{\theta}) \equiv \psi(\mathsf{v}, \dot{\mathsf{v}}, \theta, \dot{\theta}) - \psi^e(\mathsf{v}, \theta), \tag{6.88}$$

and we have that $p^r(\mathsf{v}, 0, \theta, 0) = 0$ and $\eta^r(\mathsf{v}, 0, \theta, 0) = 0$. The second derivative conditions now require that

$$\frac{\partial p^r}{\partial \dot{\mathsf{v}}}(\mathsf{v}, 0, \theta, 0) < 0. \tag{6.89}$$

Additional simplifications and reduction of (6.84) result when p^r and η^r are approximated up to some order in $\dot{\mathsf{v}}$ and $\dot{\theta}$.

To summarize, the constitutive equations of this material are given by

$$\psi = \psi(\mathsf{v}, \theta), \qquad \eta = -\frac{\partial \psi}{\partial \theta} + \eta^r(\mathsf{v}, \dot{\mathsf{v}}, \theta, \dot{\theta}), \quad \text{and} \quad p = -\frac{\partial \psi}{\partial \mathsf{v}} + p^r(\mathsf{v}, \dot{\mathsf{v}}, \theta, \dot{\theta}). \tag{6.90}$$

In addition, we find that the Helmholtz free energy is a potential only for the equilibrium parts of the pressure and entropy,

$$\dot{\psi} = -p^e\, \dot{\mathsf{v}} - \eta^e\, \dot{\theta}, \tag{6.91}$$

and, in conjunction with (5.201), the Gibbs equation is now given by

$$\dot{e} = -p^e \dot{\mathsf{v}} + \eta^r \dot{\theta} + \theta \dot{\eta}, \tag{6.92}$$

where

$$e = e(\mathsf{v}, \dot{\mathsf{v}}, \theta, \dot{\theta}) = \psi - \theta\left(\frac{\partial \psi}{\partial \theta} - \eta^r\right). \tag{6.93}$$

Note that the term involving η^r contributes to entropy production when the system is not in equilibrium.

While the Helmholtz free energy cannot fully describe this material, the entropy, the internal energy, and the Gibbs free energy

$$g = g(\mathsf{v}, \dot{\mathsf{v}}, \theta, \dot{\theta}) = \psi - \left[\frac{\partial \psi}{\partial \mathsf{v}} - p^r(\mathsf{v}, \dot{\mathsf{v}}, \theta, \dot{\theta}),\right] \mathsf{v} \tag{6.94}$$

provide fundamental relations since they can fully describe the material.

Problems

1. The Helmholtz free energy for an ideal gas is given by

$$\psi(\mathsf{v}, \theta) = \psi_0 + \left(\alpha \overline{R} - \eta_0\right)(\theta - \theta_0) - \overline{R}\theta \ln\left[\left(\frac{\theta}{\theta_0}\right)^\alpha \frac{\mathsf{v}}{\mathsf{v}_0}\right],$$

where all quantities with zero subscripts are reference constant quantities, $\overline{R} = R/M$ is the specific gas constant with R the universal gas constant and M the molecular weight of the gas, and α is a parameter related to the internal degrees of freedom of the gas molecule ($\alpha = 3/2$ for monatomic gases, $\alpha = 5/2$ for diatomic gases, and $\alpha = 3$ for polyatomic gases).

Assuming that an ideal gas is a material with no memory, obtain the corresponding constitutive expressions for entropy, internal energy, and pressure from the above fundamental equation.

2. The following fundamental equation for the Helmholtz free energy has been used to determine the properties of water:

$$\psi(\mathsf{v}, \theta) = \psi_0(\theta) - \overline{R}\theta\left[\ln \mathsf{v} - \frac{1}{\mathsf{v}} Q(\mathsf{v}, \theta)\right], \tag{6.95}$$

where $\psi_0(\theta)$ and $Q(\mathsf{v}, \theta)$ are usually given as expansions in their respective arguments that contain adjustable constants.

Assuming that water has no memory, obtain expressions for the entropy, internal energy, and pressure from the above relation.

3. Obtain the constitutive equations for the material with longer memory of volume assuming the linear variations $\psi^r = a(\mathsf{v}, \theta)\,\dot{\mathsf{v}}$ and $p^r = b(\mathsf{v}, \theta)\,\dot{\mathsf{v}} + c(\mathsf{v}, \theta)\,\ddot{\mathsf{v}}$. What are the constraints on $a(\mathsf{v}, \theta)$, $b(\mathsf{v}, \theta)$, and $c(\mathsf{v}, \theta)$?

4. Obtain the constitutive equations for the material with short memory of volume and temperature assuming the linear variations $\eta^r = a(\mathsf{v}, \theta)\,\dot{\mathsf{v}} + b(\mathsf{v}, \theta)\,\dot{\theta}$ and $p^r = c(\mathsf{v}, \theta)\,\dot{\mathsf{v}} + d(\mathsf{v}, \theta)\,\dot{\theta}$. What are the constraints on $a(\mathsf{v}, \theta)$, $b(\mathsf{v}, \theta)$, $c(\mathsf{v}, \theta)$, and $d(\mathsf{v}, \theta)$?

Bibliography

M. Bailyn. *A Survey of Thermodynamics.* AIP Press, New York, NY, 1994.

H.B. Callen. *Thermodynamics.* John Wiley & Sons, Inc., New York, NY, 1962.

M.R. El-Saden. A thermodynamic formalism based on the fundamental relation and the Legendre transformation. *International Journal of Mechanical Sciences*, 8(1):13–24, 1966.

J. Kestin. *A Course in Thermodynamics*, volume 1. McGraw-Hill Book Company, New York, NY, 1979.

I. Samohýl. *Thermodynamics of Irreversible Processes in Fluid Mixtures.* B.G. Teubner Verlagsgesellschaft, Leipzig, 1987.

7

Thermoelastic solids

We have shown that the constitutive functional of an objective simple homogeneous solid in the current configuration is given by (5.99):

$$T(\mathbf{x},t) = \underset{0\le s<\infty}{\mathfrak{F}} \{\mathbf{F}(\mathbf{x},t), {}_{(t)}\mathbf{C}^{(t)}(\mathbf{x},s), \theta^{(t)}(\mathbf{x},s), {}_{(t)}\mathbf{g}^{(t)}(\mathbf{x},s)\}. \tag{7.1}$$

From (5.96) and (5.97), and suppressing spatial dependencies, we note that

$$\begin{aligned} {}_{(t)}\mathbf{C}^{(t)}(s) &= \left(\mathbf{F}^T(t)\right)^{-1} \cdot \mathbf{C}^{(t)}(s) \cdot \mathbf{F}^{-1}(t), && (7.2)\\ {}_{(t)}\mathbf{g}^{(t)}(s) &= \left(\mathbf{F}^T(t)\right)^{-1} \cdot \mathbf{G}^{(t)}(s). && (7.3) \end{aligned}$$

Now, assuming continuous material derivatives with respect to s at $s = 0$, we can write

$$\begin{aligned} \mathbf{C}^{(t)}(s) &= \mathbf{C}(t) - \dot{\mathbf{C}}(t)\ s + \cdots, && (7.4)\\ \theta^{(t)}(s) &= \theta(t) - \dot{\theta}(t)\ s + \cdots, && (7.5)\\ \mathbf{G}^{(t)}(s) &= \mathbf{G}(t) - \dot{\mathbf{G}}(t)\ s + \cdots, && (7.6) \end{aligned}$$

where we recall that $\mathbf{C}(t) = \mathbf{F}^T(t) \cdot \mathbf{F}(t)$ and $\mathbf{G}(t) = \mathbf{F}^T(t) \cdot \mathbf{g}(t)$, so that

$$\begin{aligned} {}_{(t)}\mathbf{C}^{(t)}(s) &= \left(\mathbf{F}^T(t)\right)^{-1} \cdot \left[\mathbf{C}(t) - \dot{\mathbf{C}}(t)\, s + \cdots\right] \cdot \mathbf{F}^{-1}(t)\\ &= \mathbf{1} - \left(\mathbf{F}^T(t)\right)^{-1} \cdot \left[\dot{\mathbf{F}}^T(t) \cdot \mathbf{F}(t) + \mathbf{F}^T(t) \cdot \dot{\mathbf{F}}(t)\right] \cdot \mathbf{F}^{-1}(t)\, s + \cdots\\ &= \mathbf{1} - \left[\left(\dot{\mathbf{F}}(t) \cdot \mathbf{F}^{-1}(t)\right)^T + \left(\dot{\mathbf{F}}(t) \cdot \mathbf{F}^{-1}(t)\right)\right] s + \cdots\\ &= \mathbf{1} - \left[\mathbf{L}^T(t) + \mathbf{L}(t)\right] s + \cdots\\ &= \mathbf{1} - 2\, \mathbf{D}(t)\, s + \cdots, && (7.7) \end{aligned}$$

and

$$\begin{aligned} {}_{(t)}\mathbf{g}^{(t)}(s) &= \left(\mathbf{F}^T(t)\right)^{-1} \cdot \left[\mathbf{G}(t) - \dot{\mathbf{G}}(t)\, s + \cdots\right]\\ &= \mathbf{g}(t) - \left(\mathbf{F}^T(t)\right)^{-1} \cdot \left[\dot{\mathbf{F}}^T(t) \cdot \mathbf{g}(t) + \mathbf{F}^T(t) \cdot \dot{\mathbf{g}}(t)\right] s + \cdots\\ &= \mathbf{g}(t) - \left[\left(\dot{\mathbf{F}}(t) \cdot \mathbf{F}(t)^{-1}\right)^T \cdot \mathbf{g}(t) + \dot{\mathbf{g}}(t)\right] s + \cdots\\ &= \mathbf{g}(t) - \left[\mathbf{L}^T(t) \cdot \mathbf{g}(t) + \dot{\mathbf{g}}(t)\right] s + \cdots. && (7.8) \end{aligned}$$

Subsequently, using (7.5), (7.7), and (7.8) in (7.1), and considering a material with no memory ($p = q = r = 0$), we obtain the following reduced constitutive equation for a simple homogeneous solid (see (5.39)):

$$T = T(\mathbf{F}, \theta, \mathbf{g}). \tag{7.9}$$

Note that there is no loss of generality in expressing this relation as a function of $\mathbf{g}$ or $\mathbf{G}$ since they are related by $\mathbf{G} = \mathbf{F}^T \cdot \mathbf{g}$ and since the constitutive function already depends on $\mathbf{F}$.

A homogeneous solid material which only remembers its natural state is called a *thermoelastic solid* and is subsequently defined by

$$\psi = \psi(\mathbf{F}, \theta, \mathbf{g}), \tag{7.10}$$
$$\eta = \eta(\mathbf{F}, \theta, \mathbf{g}), \tag{7.11}$$
$$\mathbf{q} = \mathbf{q}(\mathbf{F}, \theta, \mathbf{g}), \tag{7.12}$$
$$\mathbf{h} = \mathbf{h}(\mathbf{F}, \theta, \mathbf{g}), \tag{7.13}$$
$$\boldsymbol{\sigma} = \boldsymbol{\sigma}(\mathbf{F}, \theta, \mathbf{g}). \tag{7.14}$$

The above equations are required to satisfy the following frame-invariance conditions:

$$\psi(\mathbf{Q} \cdot \mathbf{F}, \theta, \mathbf{Q} \cdot \mathbf{g}) = \psi(\mathbf{F}, \theta, \mathbf{g}), \tag{7.15}$$
$$\eta(\mathbf{Q} \cdot \mathbf{F}, \theta, \mathbf{Q} \cdot \mathbf{g}) = \eta(\mathbf{F}, \theta, \mathbf{g}), \tag{7.16}$$
$$\mathbf{q}(\mathbf{Q} \cdot \mathbf{F}, \theta, \mathbf{Q} \cdot \mathbf{g}) = \mathbf{Q} \cdot \mathbf{q}(\mathbf{F}, \theta, \mathbf{g}), \tag{7.17}$$
$$\mathbf{h}(\mathbf{Q} \cdot \mathbf{F}, \theta, \mathbf{Q} \cdot \mathbf{g}) = \mathbf{Q} \cdot \mathbf{h}(\mathbf{F}, \theta, \mathbf{g}), \tag{7.18}$$
$$\boldsymbol{\sigma}(\mathbf{Q} \cdot \mathbf{F}, \theta, \mathbf{Q} \cdot \mathbf{g}) = \mathbf{Q} \cdot \boldsymbol{\sigma}(\mathbf{F}, \theta, \mathbf{g}) \cdot \mathbf{Q}^T, \tag{7.19}$$

for all $\mathbf{Q} \in \mathscr{O}(\mathscr{V})$. Material symmetry further requires that ψ, η, $\mathbf{q}$, $\mathbf{h}$, and $\boldsymbol{\sigma}$ be such that $\mathbf{H} \in \mathscr{G}_{\boldsymbol{\kappa}} \subseteq \mathscr{O}(\mathscr{V})$.

A thermoelastic material is a highly idealized material that has perfect memory of only its *natural* or *preferred state.* This material remembers precisely that state and when the forces maintaining a different state are removed, it always returns to its configuration in its natural state. The deformation history of all intermediate states leaves no trace on its memory. Thus, as evident from the above, the constitutive equations of thermoelastic materials depend only on the present deformation gradients relative to the natural state.

7.1 Clausius–Duhem inequality

We restrict the material response functions to only those that satisfy the more general Clausius–Duhem inequality (5.252). For the moment, we define the vector quantity

$$\mathbf{K} \equiv \mathbf{h} - \frac{\mathbf{q}}{\theta} = \mathbf{K}(\mathbf{F}, \theta, \mathbf{g}), \tag{7.20}$$

which satisfies the frame-invariance condition

$$\mathbf{K}(\mathbf{Q} \cdot \mathbf{F}, \theta, \mathbf{Q} \cdot \mathbf{g}) = \mathbf{Q} \cdot \mathbf{K}(\mathbf{F}, \theta, \mathbf{g}) \tag{7.21}$$

since such condition is satisfied by both $\mathbf{q}$ and $\mathbf{h}$. It should be noted that the constitutive quantities in this case are equivalently given by

$$\mathcal{C} = \{\psi, \eta, \mathbf{q}, \mathbf{K}, \boldsymbol{\sigma}\} \tag{7.22}$$

and are functions of the independent basic fields

$$\mathcal{I} = \{\mathbf{F}, \theta, \mathbf{g}\}. \tag{7.23}$$

Now we can write

$$\dot{\psi} = \frac{\partial \psi}{\partial F_{kK}} \overline{F_{kK}}^{\cdot} + \frac{\partial \psi}{\partial \theta}\dot{\theta} + \frac{\partial \psi}{\partial \theta_{,k}} \overline{\theta_{,k}}^{\cdot}, \tag{7.24}$$

and

$$K_{k,k} = \frac{\partial K_k}{\partial F_{lL}} F_{lL,k} + \frac{\partial K_k}{\partial \theta}\theta_{,k} + \frac{\partial K_k}{\partial \theta_{,l}}\theta_{,lk}, \tag{7.25}$$

and since $\overline{F_{kK}}^{\cdot} = v_{k,l}F_{lK}$ and $F_{lL,k} = F_{lL,K}F^{-1}_{Kk}$, (5.252) becomes

$$\begin{aligned} -\gamma_v\,\theta \equiv \rho\left(\eta + \frac{\partial \psi}{\partial \theta}\right)\dot{\theta} - \left(\sigma_{kl} - \rho\frac{\partial \psi}{\partial F_{kK}}F_{lK}\right)v_{k,l} + \rho\frac{\partial \psi}{\partial \theta_{,k}}\overline{\theta_{,k}}^{\cdot} - \\ \frac{1}{2}\theta\left(F^{-1}_{Kk}\frac{\partial K_k}{\partial F_{lL}} + F^{-1}_{Lk}\frac{\partial K_k}{\partial F_{lK}}\right)F_{lL,K} - \frac{1}{2}\theta\left(\frac{\partial K_k}{\partial \theta_{,l}} + \frac{\partial K_l}{\partial \theta_{,k}}\right)\theta_{,lk} + \\ \left(\frac{q_k}{\theta} - \theta\frac{\partial K_k}{\partial \theta}\right)\theta_{,k} + \rho\theta\left(b - \frac{r}{\theta}\right) \le 0. \end{aligned} \tag{7.26}$$

In writing the above, we have accounted for the symmetry of second derivatives $F_{lL,K}$ and $\theta_{,lk}$. The above inequality must hold for every thermodynamic process, which means that special cases may be chosen which might result in further restrictions on the constitutive functions $\mathcal{C}$. Pursuing this fact, it is clear that $\mathcal{I}$ can be independently chosen (after all, that is why they are called independent fields). (Note that for given $\rho_0(\mathbf{X})$, by choosing $\mathbf{F}$ we are also choosing $\rho > 0$, since $\mathbf{F}$ is non-singular.) After a choice, it should be clear that $\mathcal{C}$ is fixed. Furthermore, the time and space derivatives of the independent fields, i.e.,

$$\mathbf{a} \equiv \left\{\dot{\theta}, \operatorname{grad}\mathbf{v}, \dot{\mathbf{g}}, \operatorname{Grad}\mathbf{F}, \operatorname{grad}\mathbf{g}\right\}, \tag{7.27}$$

may then be arbitrarily chosen to be of any magnitude or sign. This is the case because we are considering only simple homogenous materials ($P = Q = 1$ in (5.30); see also (5.33)) with no memory ($p = q = r = 0$ in (5.38); see also (5.39)). Note that, for fixed $\mathbf{F}$, choosing $\operatorname{grad}\mathbf{v}$ is equivalent to choosing $\dot{\mathbf{F}}$. First consider the equilibrium state where $\{\mathbf{F} = \mathbf{1}, \theta = \theta_0, \mathbf{g} = \mathbf{0}\}$, where $\theta_0 > 0$ is an arbitrary constant temperature. In this case, from (7.27), we see that $\mathbf{a} = \mathbf{0}$, and thus from (7.26), we must have

$$\left(b - \frac{r}{\theta_0}\right) \le 0. \tag{7.28}$$

Since r can be chosen in general to be of any magnitude and sign, in order for the inequality to be always satisfied for any θ_0, we must have that

$$b = \frac{r}{\theta}. \tag{7.29}$$

If we now define the vector

$$\boldsymbol{\alpha} \equiv \left\{ \rho\left(\eta + \frac{\partial\psi}{\partial\theta}\right), -\left(\sigma_{kl} - \rho\frac{\partial\psi}{\partial F_{kK}}F_{lK}\right), \rho\frac{\partial\psi}{\partial\theta_{,k}}, \right.$$
$$\left. -\frac{1}{2}\theta\left(F^{-1}_{Kk}\frac{\partial K_k}{\partial F_{lL}} + F^{-1}_{Lk}\frac{\partial K_k}{\partial F_{lK}}\right), -\frac{1}{2}\theta\left(\frac{\partial K_k}{\partial\theta_{,l}} + \frac{\partial K_l}{\partial\theta_{,k}}\right)\right\} \tag{7.30}$$

and the scalar

$$\beta \equiv \left(\frac{q_k}{\theta} - \theta\frac{\partial K_k}{\partial\theta}\right)\theta_{,k}, \tag{7.31}$$

and use the result (7.29), then the inequality (7.26) can be rewritten as

$$\boldsymbol{\alpha}\cdot\mathbf{a} + \beta \le 0. \tag{7.32}$$

Since the inequality is linear in **a**, and since the variables in **a** can take on values of any magnitude and sign, one would be able to violate (7.32) unless

$$\boldsymbol{\alpha} = \mathbf{0} \quad \text{and} \quad \beta \le 0. \tag{7.33}$$

More explicitly, this provides

$$\eta = -\frac{\partial\psi}{\partial\theta}, \tag{7.34}$$

$$\sigma_{kl} = \rho\frac{\partial\psi}{\partial F_{kK}}F_{lK}, \tag{7.35}$$

$$\frac{\partial\psi}{\partial\theta_{,k}} = 0, \tag{7.36}$$

$$\left(F^{-1}_{Kk}\frac{\partial K_k}{\partial F_{lL}} + F^{-1}_{Lk}\frac{\partial K_k}{\partial F_{lK}}\right) = 0, \tag{7.37}$$

$$\left(\frac{\partial K_k}{\partial\theta_{,l}} + \frac{\partial K_l}{\partial\theta_{,k}}\right) = 0, \tag{7.38}$$

and

$$\left(\frac{q_k}{\theta} - \theta\frac{\partial K_k}{\partial\theta}\right)\theta_{,k} \le 0. \tag{7.39}$$

This last inequality is known as the *residual entropy inequality.* It is straightforward to show that the constraint (7.37) can be rewritten in the following form if it is multiplied by J, and (3.60) and (3.64) are used:

$$\left[\frac{\partial(JF^{-1}_{Kk}K_k)}{\partial F_{lL}} + \frac{\partial(JF^{-1}_{Lk}K_k)}{\partial F_{lK}}\right] = 0. \tag{7.40}$$

Now the constraint differential equations (7.38) and (7.40) are of the form

$$\left(\frac{\partial f_i}{\partial y_j} + \frac{\partial f_j}{\partial y_i}\right) = 0. \tag{7.41}$$

It can be readily verified that the general solution of this equation is

$$f_i = \Omega_{ij} y_j + \omega_i, \tag{7.42}$$

with $\Omega_{ij} = -\Omega_{ji}$ and ω_i an axial vector independent of y_k. Using this result, the solution of (7.38) and (7.40) is subsequently given by

$$\mathbf{K} = \mathbf{h} - \frac{\mathbf{q}}{\theta} = J^{-1}\mathbf{F} \cdot \left[\boldsymbol{\Omega}(\theta) \cdot \mathbf{F}^T \cdot \mathbf{g} + \boldsymbol{\omega}(\theta)\right], \tag{7.43}$$

where $\boldsymbol{\Omega}(\theta) = -\boldsymbol{\Omega}^T(\theta)$ is an arbitrary skew-symmetric rank 2 tensor and $\boldsymbol{\omega}(\theta)$ is an arbitrary axial vector. Note that at this point, $\mathbf{K}$ is not necessarily zero and, subsequently, $\mathbf{h}$ does not necessarily equal $\mathbf{q}/\theta$! However, most importantly, we note that there are no objective vectors or skew-symmetric rank 2 tensors that are only functions of θ, thus

$$\boldsymbol{\omega}(\theta) = \mathbf{0} \qquad \text{and} \qquad \boldsymbol{\Omega}(\theta) = \mathbf{0} \tag{7.44}$$

in (7.43). Subsequently, we have that

$$\mathbf{h} = \frac{\mathbf{q}}{\theta}. \tag{7.45}$$

With the above results, we can now rewrite

$$\psi = \psi(\mathbf{F}, \theta), \tag{7.46}$$

$$\eta = -\frac{\partial \psi}{\partial \theta}, \tag{7.47}$$

$$\mathbf{q} = \mathbf{q}(\mathbf{F}, \theta, \mathbf{g}), \tag{7.48}$$

$$\mathbf{h} = \mathbf{h}(\mathbf{F}, \theta, \mathbf{g}) = \frac{\mathbf{q}}{\theta}, \tag{7.49}$$

$$\boldsymbol{\sigma} = \rho \frac{\partial \psi}{\partial \mathbf{F}} \cdot \mathbf{F}^T = \rho \mathbf{F} \cdot \frac{\partial \psi}{\partial \mathbf{F}}, \tag{7.50}$$

where (7.50) follows from the symmetry of the stress tensor, and using (7.39), (7.26) results in the reduced entropy inequality

$$-\gamma_v \theta^2 \equiv \mathbf{q} \cdot \mathbf{g} \leq 0. \tag{7.51}$$

By comparing (7.50) with (5.277), we note that the strain energy function τ associated with the elastic part of the stress tensor $\boldsymbol{\sigma}^e$ is nothing more than the recoverable part of the Helmoltz free energy density function. Here, the differential of the free energy density function (7.46), using (7.47) and (7.50), is now given by

$$d\psi = \frac{1}{\rho} \boldsymbol{\sigma} \cdot \left(\mathbf{F}^T\right)^{-1} \cdot d\mathbf{F} - \eta \, d\theta, \tag{7.52}$$

and subsequently, using the definition of the Helmholtz free energy density (5.201), we have the *Gibbs equation* for thermoelastic materials

$$de = \frac{1}{\rho}\boldsymbol{\sigma}\cdot\left(\mathbf{F}^T\right)^{-1}\cdot d\mathbf{F} + \theta\, d\eta. \tag{7.53}$$

Equations (7.52) and (7.53) should be compared to (5.206) and (5.139).

Now the frame-invariance conditions (7.15)–(7.19) require the reduced constitutive relations to be (see (5.94)–(5.95))

$$\psi = \psi(\mathbf{C},\theta), \tag{7.54}$$

$$\eta = -\frac{\partial\psi}{\partial\theta}, \tag{7.55}$$

$$\overline{\mathbf{q}} = \overline{\mathbf{q}}(\mathbf{C},\theta,\mathbf{G}), \tag{7.56}$$

$$\overline{\mathbf{h}} = \overline{\mathbf{h}}(\mathbf{C},\theta,\mathbf{G}) = \frac{\overline{\mathbf{q}}}{\theta}, \tag{7.57}$$

$$\overline{\boldsymbol{\sigma}} = 2\rho\mathbf{C}\cdot\frac{\partial\psi}{\partial\mathbf{C}}\cdot\mathbf{C}, \tag{7.58}$$

where

$$\overline{\mathbf{q}}(\mathbf{C},\theta,\mathbf{G}) = \mathbf{F}^T\cdot\mathbf{q}(\mathbf{F},\theta,\mathbf{g}), \tag{7.59}$$

$$\overline{\mathbf{h}}(\mathbf{C},\theta,\mathbf{G}) = \mathbf{F}^T\cdot\mathbf{h}(\mathbf{F},\theta,\mathbf{g}), \tag{7.60}$$

$$\overline{\boldsymbol{\sigma}}(\mathbf{C},\theta,\mathbf{G}) = \mathbf{F}^T\cdot\boldsymbol{\sigma}(\mathbf{F},\theta,\mathbf{g})\cdot\mathbf{F}, \tag{7.61}$$

are the convected heat flux, entropy flux, and stress tensor. The reduced entropy inequality (7.51) is now given by

$$-\gamma_v\,\theta^2 \equiv \mathbf{q}(\mathbf{F},\theta,\mathbf{g})\cdot\mathbf{g} = \overline{\mathbf{q}}(\mathbf{C},\theta,\mathbf{G})\cdot\mathbf{G} \equiv A(\mathbf{C},\theta,\mathbf{G}) \le 0 \tag{7.62}$$

and is called *Fourier's inequality.* It states that the angle between a nonzero temperature gradient and a nonzero heat flux is greater than or equal to 90°, or that heat flows from a high temperature to a lower temperature.

Recall that a state with no entropy production is a *thermodynamic equilibrium state.* We now see that, for fixed independent variables $\mathcal{I}$, A is maximum (or γ_v is a minimum) when $\mathbf{G} = \mathbf{0}$, i.e., at $A(\mathbf{C},\theta,\mathbf{0})$. Subsequently, we must have

$$\left.\frac{\partial A}{\partial\theta_{,J}}\right|_{\mathbf{G}=\mathbf{0}} = \left[\overline{q}_J + \frac{\partial\overline{q}_I}{\partial\theta_{,J}}\theta_{,I}\right]_{\mathbf{G}=\mathbf{0}} = 0 \tag{7.63}$$

or

$$\overline{q}_J(\mathbf{C},\theta,\mathbf{0}) = 0, \tag{7.64}$$

and

$$\left.\frac{\partial^2 A}{\partial\theta_{,K}\partial\theta_{,J}}\right|_{\mathbf{G}=\mathbf{0}} = \left.\left[\frac{\partial\overline{q}_J}{\partial\theta_{,K}} + \frac{\partial\overline{q}_K}{\partial\theta_{,J}} + \frac{\partial^2\overline{q}_I}{\partial\theta_{,K}\partial\theta_{,J}}\theta_{,I}\right]\right|_{\mathbf{G}=\mathbf{0}} = \left[\frac{\partial\overline{q}_J}{\partial\theta_{,K}} + \frac{\partial\overline{q}_K}{\partial\theta_{,J}}\right]_{\mathbf{G}=\mathbf{0}} \le 0. \tag{7.65}$$

When we have no deformation, $\mathbf{F} = \mathbf{R}$, $J = 1$, and $\mathbf{C} = \mathbf{1}$. Then

$$\psi = \psi(\theta), \quad e = e(\theta), \quad \mathbf{q} = \mathbf{q}(\theta,\mathbf{g}), \quad \mathbf{h} = \frac{\mathbf{q}}{\theta}(\theta,\mathbf{g}), \quad \boldsymbol{\sigma} = \mathbf{0}, \tag{7.66}$$

and the balance equations of mass, momentum, and energy become

$$\rho = \rho_R, \quad \mathbf{a} = \mathbf{f}, \quad \rho_R c_v \dot{\theta} = -\text{div } \mathbf{q} + \rho_R r, \tag{7.67}$$

where $c_v = \partial e / \partial \theta|_v$ is the specific heat at constant volume (see (5.165)). Note that the linear momentum is just Newton's equation of motion, while the energy equation is just the equation of heat conduction with an external heat source.

In passing, we note that some materials are defined as elastic in a special sense. In this regard, a material is said to be a *hyperelastic material* if the stress tensor can be represented through an energy function, such as in (7.50), where ψ is the Helmholtz free energy function. In the special case where ψ is only a function of the deformation gradient or some strain tensor, the Helmholtz free energy function is referred to as the stored energy function or strain energy function, such as in (7.46) or (5.277). In contrast, a *hypo-elastic material* is distinct from a hyperelastic material in that, except under special circumstances, it cannot be derived from a (scalar-valued) energy function.

7.2 Material symmetries

To further reduce the constitutive equations, it is first necessary to consider symmetries of materials. Material symmetry requires that the constitutive functions (7.54)–(7.61) for a homogeneous solid must be symmetric with respect to $\mathbf{H} \in \mathscr{G}_\kappa \subseteq \mathscr{O}(\mathscr{V})$ for all $(\mathbf{C}, \theta, \mathbf{G})$. Such transformations are applied to the undistorted reference state and makes the resulting configuration indistinguishable from the original configuration. Different transformation subgroups distinguish the different material classes.

Here we will discuss *anisotropic solids* that satisfy *crystal symmetries* (32 crystal classes forming the *crystallographic group*) and *transverse isotropy* in which any rotation of material coordinate axes about a preferred direction does not alter the material properties. Most laminated materials exhibit this type of anisotropy.

All possible crystal classes are described by six parameters: unit cell translation vectors $(\mathbf{a}, \mathbf{b}, \mathbf{c})$ and angles (α, β, γ), as illustrated in Fig. 7.1. The specific choices of these parameters are indicated in Table 7.1. The possible *lattice variations* that are obtained are called *primitive* (P), *face-centered* (F), *body-centered* (I), and *base-centered* (C). These give rise to 14 space lattices called *Bravais lattices* into which all crystal structures fall. The space lattices are generated by the translation vector

$$\mathbf{t} = m\,\mathbf{a} + n\,\mathbf{b} + p\,\mathbf{c}, \tag{7.68}$$

where m, n, and p are integers. The magnitudes of the translation vectors $(\mathbf{a}, \mathbf{b}, \mathbf{c})$ are called *lattice parameters*, which, together with the *lattice angles* (α, β, γ), define the unit cell of a crystal.

The symmetry of a crystal is related to the symmetry of its physical properties. A fundamental postulate of crystal physics is known as *Neumann's principle* or *postulate*:

Neumann's Principle: *The symmetry elements of any physical property tensor of a crystal must include all the symmetry elements of the point group of the crystal.*

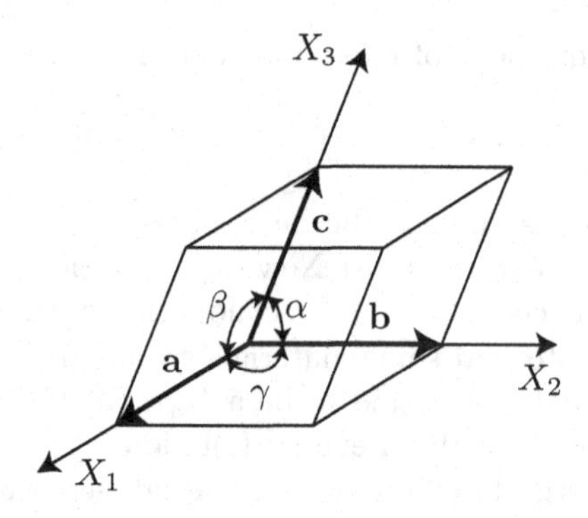

Figure 7.1: General space lattice showing translation vectors and angles.

The Neumann principle does not state that the symmetry elements of a physical property are the same as those of the point group. It just says that the symmetry elements of a physical property must include those of the point group. The physical properties may, and often do, possess more symmetry than the point group. For example, the elasticity property tensor relates the field tensors of stress and strain. The symmetries of both stress and strain tensors require that the elasticity tensor possess higher symmetries than that possessed by the crystal.

It is well known from group theory that for various crystal classes, every symmetry operation may be deduced from a few basic symmetry operations. The application of matrices corresponding to these basic operations (the *generating transformations*) are sufficient to obtain the effect due to the symmetry of a crystal class on a given property tensor. Table 7.2 lists the generating matrices and Table 7.3 summarizes the symmetries for the crystal classes. All n symmetry transformations for each of the 32 crystal classes are given in the table.

In Table 7.2, I is the identity transformation, C is the central inversion, R_n ($n = 1, 2, 3$) is the reflection in a plane normal to X_n, D_n rotates the coordinate system through 180° about the X_n-axis, T_n is a reflection in the plane through the X_n-axis bisecting the angle between the other two axes, M_1 and M_2 are rotations of the axes through 120° and 240°, respectively, about a line passing through the origin and the point $(1, 1, 1)$, and the transformations S_1 and S_2 are rotations of the axes through 120° and 240°, respectively, about the X_3-axis.

Now, for every property tensor of any rank, the tensor components must satisfy all symmetries associated with a particular crystal class. The procedure that one uses is as follows. Let's assume that the rank-2 symmetric property tensor $\mathbf{A}$ relates the two vector fields $\mathbf{q}$ and $\mathbf{g}$:

$$\mathbf{q} = \mathbf{A} \cdot \mathbf{g}. \tag{7.69}$$

Now if the material is symmetric with respect to an orthogonal transformation $\mathbf{H}$, then we must have that the property tensor does not change when both $\mathbf{q}$ and $\mathbf{g}$ are subjected to the orthogonal symmetry transformation $\mathbf{H}$:

$$\left(\mathbf{H}^T \cdot \mathbf{q}\right) = \mathbf{A} \cdot \left(\mathbf{H}^T \cdot \mathbf{g}\right),$$

or

$$\mathbf{q} = \left(\mathbf{H} \cdot \mathbf{A} \cdot \mathbf{H}^T\right) \cdot \mathbf{g}. \tag{7.70}$$

Subsequently, the property tensor must satisfy

$$\mathbf{A} = \left(\mathbf{H}\cdot\mathbf{A}\cdot\mathbf{H}^T\right). \tag{7.71}$$

Example

To illustrate the method, a 90° rotation about the X_3-axis ($H = R_2T_3$) is a symmetry element of a tetragonal-pyramidal (Class # 9) material (see Table 7.3):

$$\begin{aligned} A = [a_{ij}] &= \begin{bmatrix} a_{11} & a_{12} & a_{13} \\ a_{21} & a_{22} & a_{23} \\ a_{31} & a_{32} & a_{33} \end{bmatrix} \\ &= \begin{bmatrix} 0 & 1 & 0 \\ -1 & 0 & 0 \\ 0 & 0 & 1 \end{bmatrix} \begin{bmatrix} a_{11} & a_{12} & a_{13} \\ a_{21} & a_{22} & a_{23} \\ a_{31} & a_{32} & a_{33} \end{bmatrix} \begin{bmatrix} 0 & -1 & 0 \\ 1 & 0 & 0 \\ 0 & 0 & 1 \end{bmatrix} \\ &= \begin{bmatrix} a_{22} & -a_{21} & a_{23} \\ -a_{12} & a_{11} & -a_{13} \\ a_{32} & -a_{31} & a_{33} \end{bmatrix}. \end{aligned}$$

The above equality implies that $a_{11} = a_{22}$, $a_{21} = -a_{12}$, $a_{13} = a_{23} = 0$, $a_{31} = a_{32} = 0$, and $a_{33} = a_{33}$. But since $\mathbf{A}$ is assumed symmetric, we also have that $a_{12} = a_{21} = 0$. Subsequently, we have that

$$A = [a_{ij}] = \begin{bmatrix} a_{11} & 0 & 0 \\ 0 & a_{11} & 0 \\ 0 & 0 & a_{33} \end{bmatrix}.$$

The other symmetry transformations for the tetragonal-pyramidal class (D_3 and R_1T_3) do not reduce the components of $\mathbf{A}$ any further. This result provides the most general symmetric property tensor for this material class (see Table 7.4).

To arrive at the symmetries of the components, considerable time can be saved by using the *method of direct inspection* of Fumi, which is a modified procedure that one can use to arrive at the same result obtained as described above. In this method, one can deduce the value of a transformed property tensor component by inspection. However, we should note that this method is not directly applicable to crystal systems whose symmetries include generating matrices S_1 and S_2 (see Tables 7.2 and 7.3).

Consider the components of a rank-2 tensor $[a_{ij}]$. It is clear that each subscript in the tensor component is associated with a coordinate direction. Specifically, using the bases, the tensor is fully represented by $\mathbf{A} = a_{ij}\mathbf{e}_i\mathbf{e}_j$, and associated with each basis, we have the corresponding coordinates X_i and X_j. To examine material symmetries, we apply all coordinate transformations corresponding to symmetry transformations of a specific crystal class. Subsequently, let's examine the product of the coordinate pair X_iX_j. If we apply a coordinate transformation, the new coordinate pair would become $X_i'X_j'$. Now assuming that H_{ij} corresponds to one of the crystal class symmetries, then we have

$$X_i' X_j' = (H_{ik} X_k)(H_{jl} X_l) = H_{ik} H_{jl} X_k X_l. \tag{7.72}$$

Now the components of the objective property tensor **A** must transform as

$$a_{ij} = Q_{ik} Q_{jl} a_{kl}, \tag{7.73}$$

which is *analogous* to the way in which the product of coordinates transforms. Note that they are not identical since interchanging H_{ik} and H_{jl} does not change the value of the product of the coordinates, but doing so for a_{ij} has implications associated with bases in different directions. Therefore, the elements of a tensor transform, upon a change of coordinate axes, in exactly the same way as the product of corresponding coordinates *provided* we maintain the correct order of terms.

Example

Here we repeat the previous example, but use the modified procedure to arrive at the same result. That is, to illustrate the method, a 90° rotation about the X_3-axis ($H = R_2 T_3$) is a symmetry element of a tetragonal-pyramidal (Class # 9) material (see Table 7.3). Now the material symmetry for this crystal class requires that

$$\begin{aligned} X_1' &= X_2 \\ X_2' &= -X_1 \\ X_3' &= X_3 \end{aligned}$$

or in a more concise way

$$X_1 \to -X_2, \qquad X_2 \to X_1, \qquad X_3 \to X_3.$$

Now to determine the new value for, say, a_{12}, we evaluate

$$X_1' X_2' = X_2 (-X_1).$$

As the tensor elements transform like the product of coordinates, we see by inspection that

$$a_{12}' = -a_{21};$$

that is, upon this change of axes, the number that appears in the subscripts of the new tensor a_{ij}' is the negative of the number that appeared in the subscripts of a_{ij}. Using the above procedure, it easily follows that

$$\begin{array}{lll} 11 \to 22 & 12 \to -21 & 13 \to -23 \\ 21 \to -12 & 22 \to 11 & 23 \to 13 \\ 31 \to -32 & 32 \to 31 & 33 \to 33 \end{array}$$

that is

$$\begin{array}{lll} a_{11} \to a_{22} & a_{12} \to -a_{21} & a_{13} \to -a_{23} \\ a_{21} \to -a_{12} & a_{22} \to a_{11} & a_{23} \to a_{13} \\ a_{31} \to -a_{32} & a_{32} \to a_{31} & a_{33} \to a_{33} \end{array}$$

Then, as a consequence of Newmann's principle that every component should transform into itself, we must have that $a_{11} = a_{22}$, $a_{21} = -a_{12}$, and $a_{13} = a_{31} = a_{23} = a_{32} = 0$. But, since A is assumed to be symmetric, we must also have that $a_{12} = a_{21} = 0$ and subsequently, we obtain

$$A = [a_{ij}] = \begin{bmatrix} a_{11} & 0 & 0 \\ 0 & a_{11} & 0 \\ 0 & 0 & a_{33} \end{bmatrix}.$$

As before, the other symmetry transformations for the tetragonal-pyramidal class (D_3 and $R_1 T_3$) do not reduce a_{ij} further.
The method is directly applicable to tensors of all ranks. For example, if we wish to determine the new value of, say, the elastic stiffness element c_{1213} for this same material, we would examine the transformation of the product $X_1 X_2 X_1 X_3$:

$$X_1' X_2' X_1' X_3' = X_2 (-X_1) X_2 X_3;$$

then the value of c'_{1213} is thus $-c_{2123}$.

As noted, we can use either of the above procedures to write the general forms of property tensors of varying ranks for different crystal classes. For example, consider the property tensor $\mathbf{A}$ of rank 2 that relates two field vectors $\boldsymbol{q}$ and $\boldsymbol{g}$:

$$q_i = a_{ij}\, g_j, \qquad i,j = 1,2,3. \tag{7.74}$$

If $\mathbf{q}$ is the heat flux and $\mathbf{g}$ is the temperature gradient, then if we take $a_{ij} \to -k_{ij}$, the symmetric property tensor $\mathbf{K}$ corresponds to the thermal conductivity. It is also possible to have a rank-2 property tensor $\mathbf{A}$, which relates a scalar field θ to a rank-2 tensor field $\mathbf{e}$:

$$e_{ij} = a_{ij}\, \Delta\theta, \qquad i,j = 1,2,3. \tag{7.75}$$

For example, if $\mathbf{e}$ is the linear strain tensor and $\Delta\theta$ is the temperature difference, then the symmetric property tensor $\mathbf{A}$ corresponds to the thermal expansion tensor. Now, requiring that the material tensor component matrix satisfies all symmetries corresponding to each crystal class, in addition to the symmetry dictated by physical restrictions on the field tensors, results in the reduced forms of the component matrices shown in Table 7.4.

In providing a comprehensive example for a rank-4 tensor, below we introduce a convenient notation due to Voigt that will allow us to write the generally sparse property tensor components into convenient smaller matrices. The sparsity arises from the fact that quite often, one or both of the field tensors that the property tensor relates will satisfy symmetry properties arising from physical restrictions.

Consider the property tensor of rank 4, $\mathbf{C}$, that relates two symmetric rank-2 field tensors $\boldsymbol{\sigma}$ and $\mathbf{e}$:

$$\sigma_{ij} = c_{ijkl}\, e_{kl} \qquad i,j,k,l = 1,2,3. \tag{7.76}$$

Since $\sigma_{ij} = \sigma_{ji}$ and $e_{kl} = e_{lk}$, this requires that $c_{ijkl} = c_{mn}$, which permits us to write in condensed form as

$$\sigma_m = c_{mn}\, e_n \qquad m,n = 1,\ldots,6, \tag{7.77}$$

thereby taking direct advantage of the reduction in the number of components from 81 to 21. To convert between the tensor notation (7.76) and the matrix notation (7.77), we adopt the following conventions:

$$\begin{bmatrix} \sigma_{11} & \sigma_{12} & \sigma_{13} \\ & \sigma_{22} & \sigma_{23} \\ \text{Sym} & & \sigma_{33} \end{bmatrix} \Longleftrightarrow \begin{bmatrix} \sigma_1 & \sigma_6 & \sigma_5 \\ & \sigma_2 & \sigma_4 \\ \text{Sym} & & \sigma_3 \end{bmatrix}, \quad \text{i.e.,} \quad \sigma_{ij} \Leftrightarrow \sigma_m, \tag{7.78}$$

$$\begin{bmatrix} e_{11} & e_{12} & e_{13} \\ & e_{22} & e_{23} \\ \text{Sym} & & e_{33} \end{bmatrix} \Longleftrightarrow \begin{bmatrix} e_1 & \frac{1}{2}e_6 & \frac{1}{2}e_5 \\ & e_2 & \frac{1}{2}e_4 \\ \text{Sym} & & e_3 \end{bmatrix}, \quad \text{i.e.,} \quad e_{ij} \Leftrightarrow \begin{cases} e_m & \text{if } i = j, \\ \frac{1}{2}e_m & \text{if } i \neq j, \end{cases} \tag{7.79}$$

and

$$c_{ijkl} \Leftrightarrow c_{mn}. \tag{7.80}$$

Example

To write the tensor components σ_{21} in the matrix representation, we proceed as follows:

$$\begin{aligned} \sigma_{21} &= c_{2111}e_{11} + c_{2122}e_{22} + \cdots + c_{2123}e_{23} + c_{2132}e_{32} + \cdots, & (7.81) \\ \sigma_6 &= c_{2111}e_1 + c_{2122}e_2 + \cdots + c_{2123}(\tfrac{1}{2}e_4) + c_{2132}(\tfrac{1}{2}e_4) + \cdots, & (7.82) \\ \sigma_6 &= c_{61}e_1 + c_{62}e_2 + \cdots + \tfrac{1}{2}c_{64}e_4 + \tfrac{1}{2}c_{64}e_4 + \cdots, & (7.83) \\ \sigma_6 &= c_{61}e_1 + c_{62}e_2 + \cdots + c_{64}e_4 + \cdots. & (7.84) \end{aligned}$$

The reverse representation of σ_6 from matrix form to tensor component form σ_{21} is given by

$$\begin{aligned} \sigma_6 &= c_{61}e_1 + c_{62}e_2 + \cdots + c_{64}e_4 + \cdots, & (7.85) \\ \sigma_6 &= c_{61}e_1 + c_{62}e_2 + \cdots + \tfrac{1}{2}c_{64}e_4 + \tfrac{1}{2}c_{64}e_4 + \cdots, & (7.86) \\ \sigma_{21} &= c_{61}e_{11} + c_{62}e_{22} + \cdots + c_{64}e_{23} + c_{64}e_{32} + \cdots, & (7.87) \\ \sigma_{21} &= c_{2111}e_{11} + c_{2122}e_{22} + \cdots + c_{2123}e_{23} + c_{2132}e_{32} + \cdots. & (7.88) \end{aligned}$$

Now, requiring that the material tensor component matrix C satisfy all symmetries corresponding to each crystal class results in the reduced forms of the component matrices using Voigt's notation shown in Table 7.5.

The other class of materials that we would like to discuss are those that have *transverse isotropy.* A material with a single preferred direction which is the same at every point is said to be transversely isotropic. For such material, constitutive relations are invariant under rotations about the preferred direction. If the X_3 direction is chosen as the preferred direction as illustrated in Fig. 7.2, then such rotations are represented by the matrix

$$M_\Theta = \begin{bmatrix} \cos\Theta & \sin\Theta & 0 \\ -\sin\Theta & \cos\Theta & 0 \\ 0 & 0 & 1 \end{bmatrix}, \qquad 0 \leq \Theta \leq 2\pi. \tag{7.89}$$

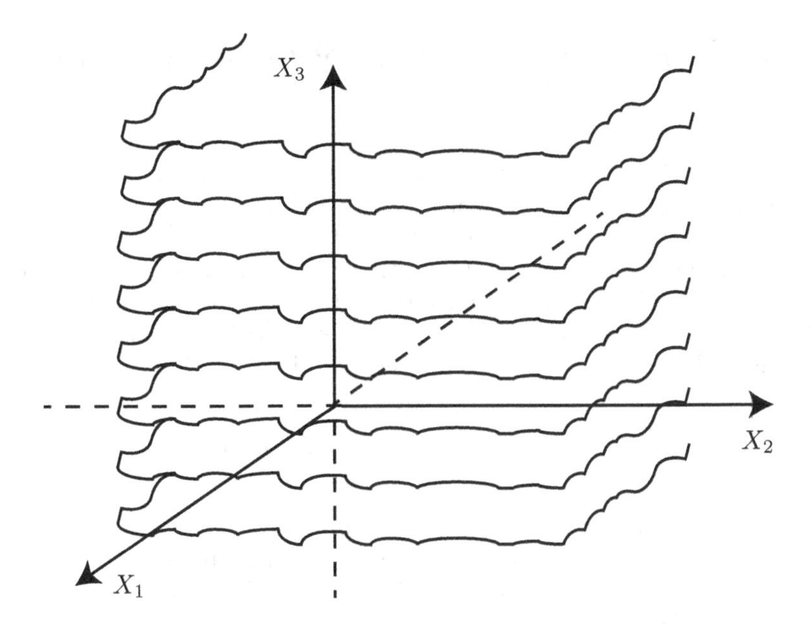

Figure 7.2: Laminated material geometry.

Now, five cases can arise depending on whether or not certain other transformations are permitted. These cases are characterized by the transformation groups generated by the following matrices: (i) M_Θ, (ii) M_Θ, R_1, (iii) M_Θ, R_3, (iv) M_Θ, D_2, and (v) M_Θ, R_1, R_3, D_2, where the transformations R_1, R_3, D_2 are given in Table 7.2. Case (i), in which only rotations about the preferred direction are allowed, is sometimes said to characterize rotational symmetry. Now, in view of the requirement to satisfy the symmetries $(M_\Theta, R_1, R_3, D_2)$, it is easy to show that the corresponding components of the rank-2 and rank-4 property tensors are given as shown in Tables 7.4 and 7.5. Note that we respectively have 2 and 5 independent parameters in the corresponding component matrices of the rank-2 and rank-4 property tensors of transversely isotropic materials.

An important point that needs to be made is that the symmetry of a crystal depends on the state of the crystal. We note that our assessment of the material symmetries have been based on the state of the material without considering external influences (symmetries of property tensors at equilibrium). If, due to some external influence (represented by field tensors), there is a change in the state of the crystal, there may also be a change in the crystal symmetry. The symmetry of a given state of the crystal may be determined using the Curie principle from the symmetry of the crystal free of any external influence (symmetry of property tensors) and from the symmetry of the external influences (symmetry of field tensors).

Curie's Principle: *A crystal under an external influence will exhibit only those symmetry elements that are common to the crystal without the influence and those of the influence without the crystal.*

In the above example, and in Tables 7.4 and 7.5, the symmetries associated with external influences have been used through the symmetry assumptions of the corresponding field tensors.

7.3 Linear deformations of anisotropic materials

For the subsequent discussion, we consider the constitutive equations (7.46)–(7.48), (7.50), and (7.45). Due to the requirement of frame indifference, the dependence of the Helmholtz free energy density and heat flux on $\mathbf{F}$ must reduce to the dependence on the right Cauchy–Green strain tensor $\mathbf{C}$ (see (7.54)–(7.45)). However, in the linear limit, since $\mathbf{C} = (\mathbf{1} - 2\,\mathbf{E})$ and the left Cauchy–Green strain tensor is $\mathbf{B} = (\mathbf{1} - 2\,\mathbf{e})^{-1} = \mathbf{1} + 2\,\mathbf{e} + O(\mathbf{e}^2)$ (see (3.178)), for convenience in arriving at corresponding results for isotropic materials, we write the linear approximation in terms of the Almansi–Hamel strain tensor $\mathbf{e}$, i.e.,

$$\psi = \psi(\theta, \mathbf{e}) \qquad \text{and} \qquad \mathbf{q} = \mathbf{q}(\theta, \mathbf{e}, \mathbf{g}), \tag{7.90}$$

where, in relating (7.90) to (5.201), (5.206), and (5.207), we note that the specific thermostatic volume is related to the linear strain tensor by $\boldsymbol{\nu}_1 \to \mathbf{e}/\rho_0$, where ρ_0 is a constant reference density, which we take here to be that at the reference state, and the thermostatic tension is related to the stress tensor, $\boldsymbol{\tau}_1 \to \boldsymbol{\sigma}$. We now assume that the strain $\mathbf{e}$, the temperature difference $\tilde{\theta} = \theta - \theta_0$, and the temperature gradient $\mathbf{g}$ are small, where θ_0 is the temperature of the body at the reference state. Since we address small deformations from the reference state, we write the following expansions of the Helmholtz free energy density and heat flux about the reference state (see (3.189), (3.190), and (3.192)):

$$\psi(\theta, \mathbf{e}) = \psi_0 - \eta_0\,\tilde{\theta} + \frac{1}{\rho_0}\boldsymbol{\sigma}_0 : \tilde{\mathbf{e}} - \frac{c_{e_0}}{2\,\theta_0}\tilde{\theta}^2 + \frac{1}{\rho_0}\tilde{\mathbf{e}} : \boldsymbol{\beta}_0\,\tilde{\theta} + \frac{1}{2\rho_0^2}\tilde{\mathbf{e}} : \boldsymbol{\xi}_0 : \tilde{\mathbf{e}} + \cdots \tag{7.91}$$

and

$$\mathbf{q}(\theta, \mathbf{e}, \mathbf{g}) = \mathbf{q}_0 - \mathbf{i}_0\,\tilde{\theta} + \frac{1}{\rho_0}\mathbf{j}_0 : \tilde{\mathbf{e}} - \mathbf{k}_0 \cdot \tilde{\mathbf{g}} + \cdots, \tag{7.92}$$

where we have used the following quantities evaluated at the reference state ($\theta = \theta_0, \mathbf{e} = \mathbf{0}, \mathbf{g} = \mathbf{0}$), which is denoted by the zero subscript:

$$\psi_0 = \psi\,(\theta = \theta_0, \mathbf{e} = \mathbf{0})\,, \tag{7.93}$$

$$\eta_0 = -\left.\frac{\partial \psi}{\partial \theta}\right|_{\theta=\theta_0, \mathbf{e}=\mathbf{0}}, \qquad \text{(see (5.207))} \tag{7.94}$$

$$\boldsymbol{\sigma}_0 = \rho_0 \left.\frac{\partial \psi}{\partial \mathbf{e}}\right|_{\theta=\theta_0, \mathbf{e}=\mathbf{0}}, \qquad \text{(see (5.207))} \tag{7.95}$$

$$c_{e_0} = -\,\theta_0 \left.\frac{\partial^2 \psi}{\partial \theta^2}\right|_{\theta=\theta_0, \mathbf{e}=\mathbf{0}} = \theta_0 \left.\frac{\partial \eta}{\partial \theta}\right|_{\theta=\theta_0, \mathbf{e}=\mathbf{0}}, \qquad \text{(see (5.209))} \tag{7.96}$$

$$\boldsymbol{\beta}_0 = \rho_0 \left.\frac{\partial^2 \psi}{\partial \mathbf{e} \partial \theta}\right|_{\theta=\theta_0, \mathbf{e}=\mathbf{0}} = -\rho_0 \left.\frac{\partial \eta}{\partial \mathbf{e}}\right|_{\theta=\theta_0, \mathbf{e}=\mathbf{0}} = \left.\frac{\partial \boldsymbol{\sigma}}{\partial \theta}\right|_{\theta=\theta_0, \mathbf{e}=\mathbf{0}}, \quad \text{(see (5.211))} \tag{7.97}$$

$$\begin{aligned}
\boldsymbol{\xi}_0 &= \rho_0^2 \left.\frac{\partial^2 \psi}{\partial \mathbf{e}^2}\right|_{\theta=\theta_0,\mathbf{e}=\mathbf{0}} = \rho_0 \left.\frac{\partial \boldsymbol{\sigma}}{\partial \mathbf{e}}\right|_{\theta=\theta_0,\mathbf{e}=\mathbf{0}}, \qquad \text{(see (5.150))} && (7.98)\\
\mathbf{q}_0 &= \mathbf{q}(\theta=\theta_0, \mathbf{e}=\mathbf{0}, \mathbf{g}=\mathbf{0}), && (7.99)\\
\mathbf{i}_0 &= -\left.\frac{\partial \mathbf{q}}{\partial \theta}\right|_{\theta=\theta_0,\mathbf{e}=\mathbf{0},\mathbf{g}=\mathbf{0}}, && (7.100)\\
\mathbf{j}_0 &= \rho_0 \left.\frac{\partial \mathbf{q}}{\partial \mathbf{e}}\right|_{\theta=\theta_0,\mathbf{e}=\mathbf{0},\mathbf{g}=\mathbf{0}}, && (7.101)\\
\mathbf{k}_0 &= -\left.\frac{\partial \mathbf{q}}{\partial \mathbf{g}}\right|_{\theta=\theta_0,\mathbf{e}=\mathbf{0},\mathbf{g}=\mathbf{0}}. && (7.102)
\end{aligned}$$

We now note that the reference state is an equilibrium state and thus $\psi(\theta,\mathbf{e}) \geq \psi_0 \geq 0$, $\boldsymbol{\sigma}_0 = \mathbf{0}$, $\mathbf{q}_0 = \mathbf{0}$, $\mathbf{i}_0 = \mathbf{0}$, and $\mathbf{j}_0 = \mathbf{0}$. Furthermore, we recognize that

$$\begin{aligned}
\boldsymbol{\beta}_0 &= -\boldsymbol{\xi}_0 : \boldsymbol{\alpha}_0, \qquad \text{(see (5.151))} && (7.103)\\
\boldsymbol{\alpha}_0 &= \frac{1}{\rho_0} \left.\frac{\partial \mathbf{e}}{\partial \theta}\right|_{\theta=\theta_0,\mathbf{e}=\mathbf{0}}. \qquad \text{(see (5.150))} && (7.104)
\end{aligned}$$

Subsequently, if we truncate the expansion for the Helmholtz free energy density to second order and the heat flux to first order, the constitutive equations become

$$\begin{aligned}
\psi &= \psi_0 - \eta_0\,\tilde{\theta} - \frac{c_{\mathbf{e}_0}}{2\,\theta_0}\,\tilde{\theta}^2 - \frac{1}{\rho_0}\widetilde{\mathbf{e}} : \boldsymbol{\xi}_0 : \boldsymbol{\alpha}_0\,\tilde{\theta} + \frac{1}{2\rho_0^2}\widetilde{\mathbf{e}} : \boldsymbol{\xi}_0 : \widetilde{\mathbf{e}}, && (7.105)\\
\eta &= -\frac{\partial \psi}{\partial \tilde{\theta}} = \eta_0 + \frac{c_{\mathbf{e}_0}}{\theta_0}\tilde{\theta} + \frac{1}{\rho_0}\widetilde{\mathbf{e}} : \boldsymbol{\xi}_0 : \boldsymbol{\alpha}_0, && (7.106)\\
\mathbf{q} &= -\mathbf{k}_0 \cdot \widetilde{\mathbf{g}}, && (7.107)\\
\mathbf{h} &= -\frac{\mathbf{k}_0}{\theta_0} \cdot \widetilde{\mathbf{g}}, && (7.108)\\
\boldsymbol{\sigma} &= \rho_0 \frac{\partial \psi}{\partial \widetilde{\mathbf{e}}} = \frac{1}{\rho_0}\boldsymbol{\xi}_0 : \left(\widetilde{\mathbf{e}} - \rho_0\,\boldsymbol{\alpha}_0\,\tilde{\theta}\right). && (7.109)
\end{aligned}$$

Now, defining the linear *thermal* and *mechanical strain tensors*

$$\widetilde{\mathbf{e}}_\theta = \rho_0\,\boldsymbol{\alpha}_0\,\tilde{\theta} = \mathbf{a}_0\,\tilde{\theta} \qquad \text{and} \qquad \widetilde{\mathbf{e}}_M = \widetilde{\mathbf{e}} - \widetilde{\mathbf{e}}_\theta, \tag{7.110}$$

we can rewrite the above constitutive equations in their final form:

$$\begin{aligned}
\psi &= \psi_0 - \eta_0\,\tilde{\theta} - \frac{c_{\mathbf{e}_0}}{2\,\theta_0}\,\tilde{\theta}^2 - \frac{1}{\rho_0}\widetilde{\mathbf{e}} : \mathbf{C}_0 : \mathbf{a}_0\,\tilde{\theta} + \frac{1}{2\,\rho_0}\widetilde{\mathbf{e}} : \mathbf{C}_0 : \widetilde{\mathbf{e}}\\
&= \psi_0 - \eta_0\,\tilde{\theta} - \frac{c_{\boldsymbol{\sigma}_0}}{2\,\theta_0}\,\tilde{\theta}^2 + \frac{1}{2\rho_0}\widetilde{\mathbf{e}}_M : \mathbf{C}_0 : \widetilde{\mathbf{e}}_M, && (7.111)\\
\eta &= \eta_0 + \frac{c_{\mathbf{e}_0}}{\theta_0}\,\tilde{\theta} + \frac{1}{\rho_0}\widetilde{\mathbf{e}} : \mathbf{C}_0 : \mathbf{a}_0\\
&= \eta_0 + \frac{c_{\boldsymbol{\sigma}_0}}{\theta_0}\,\tilde{\theta} + \frac{1}{\tilde{\theta}}\widetilde{\mathbf{e}}_M : \mathbf{C}_0 : \widetilde{\mathbf{e}}_\theta, && (7.112)\\
\mathbf{q} &= -\mathbf{k}_0 \cdot \widetilde{\mathbf{g}}, && (7.113)\\
\mathbf{h} &= -\frac{\mathbf{k}_0}{\theta_0} \cdot \widetilde{\mathbf{g}}, && (7.114)\\
\boldsymbol{\sigma} &= \mathbf{C}_0 : \left(\widetilde{\mathbf{e}} - \mathbf{a}_0\,\tilde{\theta}\right) = \mathbf{C}_0 : \widetilde{\mathbf{e}}_M. && (7.115)
\end{aligned}$$

In the above equations, $c_{\mathbf{e}_0}$ is the specific heat capacity at constant strain, $c_{\boldsymbol{\sigma}_0}$ is the specific heat capacity at constant stress (see (5.215)) given by

$$c_{\boldsymbol{\sigma}_0} - c_{\mathbf{e}_0} = \theta_0\, \boldsymbol{\alpha}_0 : \boldsymbol{\xi}_0 : \boldsymbol{\alpha}_0 = \frac{\theta_0}{\rho_0}\, \mathbf{a}_0 : \mathbf{C}_0 : \mathbf{a}_0, \tag{7.116}$$

$\mathbf{a}_0$ and $\boldsymbol{\beta}_0$ are the symmetric second rank *thermal expansion* and *thermal stress tensors* related by

$$\boldsymbol{\beta}_0 = -\mathbf{C}_0 : \mathbf{a}_0 \tag{7.117}$$

and satisfying the symmetry conditions

$$(a_0)_{ij} = (a_0)_{ji} \qquad \text{and} \qquad (\beta_0)_{ij} = (\beta_0)_{ji}, \tag{7.118}$$

$$\mathbf{C}_0 = \frac{1}{\rho_0}\, \boldsymbol{\xi}_0 \tag{7.119}$$

is the fully symmetric rank-4 *elastic stiffness tensor* satisfying the following requirements due to the symmetries of the stress and mechanical strain tensors,

$$(c_0)_{ijkl} = (c_0)_{jikl} = (c_0)_{jilk} = (c_0)_{ijlk} = (c_0)_{klij}, \tag{7.120}$$

and $\mathbf{k}_0$ is the rank-2 *thermal conductivity tensor* for which we note that, using (7.113) in the residual inequality (7.62), the symmetric part of $\mathbf{k}_0$ must be positive semi-definite:

$$\operatorname{sym} \mathbf{k}_0 \geq \mathbf{0}. \tag{7.121}$$

Now, the thermal conductivity can be decomposed into symmetric and skew-symmetric parts, and it can be shown that the divergence of the heat flux component associated with the skew-symmetric part does not contribute to heat flow and thus is unmeasurable or not observable. Based on this fact, it is generally assumed that the thermal conductivity tensor is symmetric:

$$\mathbf{k}_0 = \mathbf{k}_0^T \qquad \text{and} \qquad \mathbf{k}_0 \geq \mathbf{0}. \tag{7.122}$$

It should be pointed out that the symmetries in the material tensors reduce the number of independent components from 9 to 6 for $\mathbf{a}_0$, $\boldsymbol{\beta}_0$, and $\mathbf{k}_0$, and from 81 to 21 for $\mathbf{C}_0$.

Lastly, the stability condition that the internal energy must be a minimum at equilibrium requires that the matrix (5.257) be positive. The first condition resulting from this requires that

$$c_{\mathbf{e}_0} > 0. \tag{7.123}$$

The second condition of $\phi_0 > 0$, using (5.137) and (7.115), requires that the isothermal elastic stiffness tensor be positive definite

$$\mathbf{C}_0 > \mathbf{0}. \tag{7.124}$$

Again, since C_0 is a real symmetric matrix, this condition requires that all the eigenvalues of C_0 be real and positive definite, or alternately, all the determinants of the principal minors of C_0 be positive definite (see (5.258)).

The specific requirements on the components of $\mathbf{C}_0$ depend on the material crystal class. In general, such restrictions are fairly complex, but, first translating

$\mathbf{C}_0$ to Voigt's notation, we see from Table 7.5 that progress can be made for the material classes of orthorhombic, tetragonal (Classes # 12–15), hexagonal, cubic, hemitropic and isotropic since in those cases, the component matrix of $\mathbf{C}_0$ has nonzero elements only along the main diagonal and in the principal 3×3 minor. For example, consider a material having cubic crystal symmetry. From Table 7.5, we see that the eigenvalues of $[C_0]$ are easily found:

$$\lambda_1 = \lambda_2 = (c_0)_{11} - (c_0)_{12}, \; \lambda_3 = (c_0)_{11} + 2\,(c_0)_{12}, \text{ and } \lambda_4 = \lambda_5 = \lambda_6 = (c_0)_{44}. \quad (7.125)$$

Equilibrium stability then requires that

$$(c_0)_{11} - (c_0)_{12} > 0, \quad (c_0)_{11} + 2\,(c_0)_{12} > 0, \quad \text{and} \quad (c_0)_{44} > 0. \quad (7.126)$$

Alternately, the determinants of the principal minors of $[C_0]$ lead to

$$(c_0)_{11} > 0, \; (c_0)_{11}^2 - (c_0)_{12}^2 > 0, \; [(c_0)_{11} - (c_0)_{12}]^2 \, [(c_0)_{11} + 2\,(c_0)_{12}] > 0, \; (c_0)_{44} > 0. \quad (7.127)$$

These are equivalent to (7.126) since the first three inequalities in (7.127) can be replaced by

$$(c_0)_{11} + (c_0)_{12} > 0, \qquad (c_0)_{11} - (c_0)_{12} > 0, \qquad (c_0)_{11} + 2\,(c_0)_{12} > 0, \quad (7.128)$$

which in turn can be replaced by the first two inequalities in (7.126). For these crystal classes, the positive semi-definite condition on the thermal conductivity also requires that

$$(k_0)_{11} \geq 0, \qquad (k_0)_{22} \geq 0, \quad \text{and} \quad (k_0)_{33} \geq 0. \quad (7.129)$$

For a cubic crystal, this reduces to the restriction that $(k_0)_{11} \geq 0$.

7.3.1 Propagation of elastic waves in crystals

We consider only the case of isothermal conditions ($\tilde{\theta} = 0$). The procedure is readily extended to non-isothermal conditions.

Since the rank-2 stress and strain field tensors $\boldsymbol{\sigma}$ and $\widetilde{\mathbf{e}}$ are symmetric, using Voigt's notation, we observe from Table 7.5 that the elasticity or stiffness property tensor $\mathbf{C}$ relations are given by

$$\boldsymbol{\sigma} = \mathbf{C} \cdot \widetilde{\mathbf{e}}, \quad (7.130)$$

where $\boldsymbol{\sigma} = (\sigma_1, \sigma_2, \sigma_3, \sigma_4, \sigma_5, \sigma_6)^T$, $\widetilde{\mathbf{e}} = (\widetilde{e}_1, \widetilde{e}_2, \widetilde{e}_3, \widetilde{e}_4, \widetilde{e}_5, \widetilde{e}_6)^T$, and, to simplify notations, we have suppressed the zero subscripts indicating the reference density and the linearity of the isothermal elastic stiffness tensor (not to be confused with the right Cauchy–Green strain tensor). As we want an equation which involves displacements, we substitute the definition of $\widetilde{e}_m$ in terms of displacements gradients $h_{ij} = u_{i,j}$ (see (3.192) and (7.79)):

$$\widetilde{\mathbf{e}} = \blacktriangledown^T \cdot \mathbf{u}, \quad (7.131)$$

where $\mathbf{u} = (u_1, u_2, u_3)^T$ is the material displacement vector, and, using the abbreviated notation $\partial_i = \partial/\partial x_i$ $(i = 1, 2, 3)$,

$$\blacktriangledown \equiv \begin{pmatrix} \partial_1 & 0 & 0 & 0 & \partial_3 & \partial_2 \\ 0 & \partial_2 & 0 & \partial_3 & 0 & \partial_1 \\ 0 & 0 & \partial_3 & \partial_2 & \partial_1 & 0 \end{pmatrix}. \quad (7.132)$$

Substituting (7.131) into (7.130), we obtain

$$\boldsymbol{\sigma} = \mathbf{C} \cdot \blacktriangledown^T \cdot \mathbf{u}. \tag{7.133}$$

Now using the linear momentum balance (4.109), assuming that we have no body forces, using Voigt's notation, and considering only linear terms, upon using the abbreviation $\partial_{tt} = \partial^2/\partial t^2$, we have

$$\rho\, \partial_{tt} \mathbf{u} = \blacktriangledown \cdot \boldsymbol{\sigma}, \tag{7.134}$$

or, using (7.133),

$$\rho\, \partial_{tt} \mathbf{u} = \boldsymbol{\mathcal{C}} \cdot \mathbf{u}, \tag{7.135}$$

where we have defined the symmetric operator

$$\boldsymbol{\mathcal{C}} \equiv \blacktriangledown \cdot \mathbf{C} \cdot \blacktriangledown^T. \tag{7.136}$$

The system of partial differential equations (7.135) represents the equations governing the displacement of a differential volume element. While the equations look formidable, they are nothing more than wave equations for each component of the displacement vector. These equations involve components of the property tensor. Now we assume a solution in the form of a wave propagating in three dimensions, i.e.,

$$\mathbf{u}(\mathbf{x}, t) = e^{i(\omega t - \mathbf{k}\cdot\mathbf{x})} \mathbf{U}, \tag{7.137}$$

where ω is the wave frequency, $\mathbf{k} = (k_1, k_2, k_3)$ is the wave vector corresponding to the direction of wave propagation, and $\mathbf{U} = (U_1, U_2, U_3)^T$ is the displacement wave vector. Substituting this assumed solution form into the wave equation (7.135), we obtain

$$\left(k^2 \mathbf{A} - \rho\omega^2 \mathbf{1}\right) \cdot \widehat{\mathbf{U}} = \mathbf{0}, \tag{7.138}$$

where

$$\mathbf{A} = \mathbf{L} \cdot \mathbf{C} \cdot \mathbf{L}^T, \tag{7.139}$$

$$\mathbf{L} \equiv \begin{pmatrix} l_1 & 0 & 0 & 0 & l_3 & l_2 \\ 0 & l_2 & 0 & l_3 & 0 & l_1 \\ 0 & 0 & l_3 & l_2 & l_1 & 0 \end{pmatrix}, \tag{7.140}$$

$\mathbf{l} = (l_1, l_2, l_3)^T = \mathbf{k}/k$ is the wave direction cosine vector with $k = |\mathbf{k}|$, and $\widehat{\mathbf{U}} = (\widehat{U}_1, \widehat{U}_2, \widehat{U}_3)^T = \mathbf{U}/|\mathbf{U}|$ is the displacement direction cosine vector, as illustrated in Fig. 7.3. The wave (or phase) velocity is given by

$$\mathbf{c}_w = \frac{\omega}{\mathbf{k}} = \frac{\omega}{k} \mathbf{l}. \tag{7.141}$$

Note that the crystal gives us $\mathbf{C}$ in one of the forms given in Table 7.5. For a

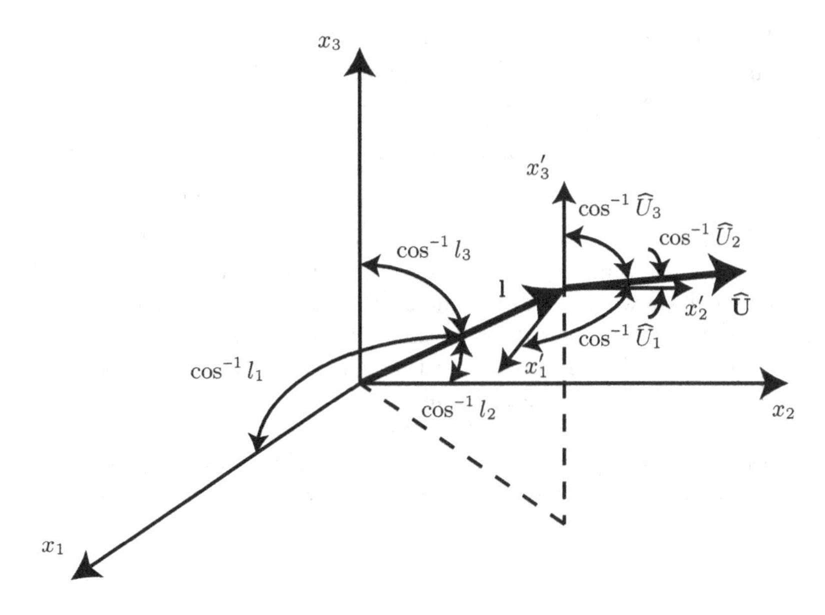

Figure 7.3: The relationship between the wave and displacement direction cosine vectors.

triclinic crystal, the symmetric matrix $\mathbf{A}$ has components

$$a_{11} = c_{11}l_1^2 + c_{66}l_2^2 + c_{55}l_3^2 + 2\left(c_{56}l_2l_3 + c_{15}l_3l_1 + c_{16}l_1l_2\right), \tag{7.142}$$

$$a_{22} = c_{66}l_1^2 + c_{22}l_2^2 + c_{44}l_3^2 + 2\left(c_{24}l_2l_3 + c_{46}l_3l_1 + c_{26}l_1l_2\right), \tag{7.143}$$

$$a_{33} = c_{55}l_1^2 + c_{44}l_2^2 + c_{33}l_3^2 + 2\left(c_{34}l_2l_3 + c_{35}l_3l_1 + c_{45}l_1l_2\right), \tag{7.144}$$

$$a_{12} = c_{16}l_1^2 + c_{26}l_2^2 + c_{45}l_3^2 + \left(c_{46} + c_{25}\right)l_2l_3 + \left(c_{14} + c_{56}\right)l_3l_1 + \left(c_{12} + c_{66}\right)l_1l_2, \tag{7.145}$$

$$a_{13} = c_{15}l_1^2 + c_{46}l_2^2 + c_{35}l_3^2 + \left(c_{45} + c_{36}\right)l_2l_3 + \left(c_{13} + c_{55}\right)l_3l_1 + \left(c_{14} + c_{56}\right)l_1l_2, \tag{7.146}$$

$$a_{23} = c_{56}l_1^2 + c_{24}l_2^2 + c_{34}l_3^2 + \left(c_{44} + c_{23}\right)l_2l_3 + \left(c_{36} + c_{45}\right)l_3l_1 + \left(c_{25} + c_{46}\right)l_1l_2. \tag{7.147}$$

Next we pick the direction $\mathbf{l}$ (or $\mathbf{k}$) in which we wish to propagate the wave. Subsequently, the wave frequency ω, and thus the wave speed $c_w = \omega/k$, and the direction of the displacement vector $\widehat{\mathbf{U}}$ are the unknown variables. The above problem corresponds to a standard eigenvalue problem with $\rho\omega^2$ (or ρc_w^2) being the eigenvalue and $\widehat{\mathbf{U}}$ the eigenvector. A nontrivial solution ($\widehat{\mathbf{U}} \neq \mathbf{0}$) exists only if

$$\Omega(\omega, \mathbf{k}) \equiv \det\left[k^2\,\mathbf{A}(l_1, l_2, l_3) - \rho\omega^2\mathbf{1}\right] = \mathbf{0} \tag{7.148}$$

(a trivial solution is impossible since $\widehat{U}_1^2 + \widehat{U}_2^2 + \widehat{U}_3^2 = 1$). This gives rise to the following characteristic cubic polynomial (dispersion relation) for $\rho\omega^2$,

$$\Omega(\omega, \mathbf{k}) \equiv \left(k^2a_{11} - \rho\omega^2\right)\left(k^2a_{22} - \rho\omega^2\right)\left(k^2a_{33} - \rho\omega^2\right) - k^4\left[k^2\left(a_{11}a_{23}^2 + a_{22}a_{13}^2 + a_{33}a_{12}^2 - 2\,a_{12}a_{13}a_{23}\right) - \rho\omega^2\left(a_{12}^2 + a_{13}^2 + a_{23}^2\right)\right] = 0, \tag{7.149}$$

which provides three wave speeds, say $c_w^{(i)} = \omega^{(i)}/k$, $i = 1, 2, 3$. For each wave speed, we can subsequently obtain the corresponding displacement direction cosine vector $\widehat{\mathbf{U}}^{(i)}$ from (7.138) in the form

$$\frac{\widehat{U}_2}{\widehat{U}_1} = \frac{a_{23}\left(a_{11} - \rho c_w^2\right) - a_{12}a_{13}}{a_{13}\left(a_{22} - \rho c_w^2\right) - a_{12}a_{23}} \quad \text{and} \quad \frac{\widehat{U}_3}{\widehat{U}_1} = \frac{\left(a_{11} - \rho c_w^2\right)\left(a_{22} - \rho c_w^2\right) - a_{12}^2}{a_{13}\left(a_{22} - \rho c_w^2\right) - a_{12}a_{23}}. \tag{7.150}$$

Such displacements, which are functions of the phase velocity, propagation direction, and elastic stiffness constants, will, in general, constitute quasi-longitudinal and quasi-shear waves. Numerical computations are required to obtain the wave speeds and the displacement directions. However, general features of the dispersion relation (7.149) can be deduced by considering special propagation directions for which the dispersion relation factors.

Another important aspect of elastic wave propagation in crystals is that the direction of energy propagation may not coincide with the propagation direction $\mathbf{l}$ of the wave velocity. Indeed, it can be shown that energy in an elastic medium propagates with the group velocity

$$\mathbf{c}_g = \frac{\partial \omega}{\partial \mathbf{k}} = -\frac{1}{k}\frac{\partial \Omega/\partial \mathbf{l}}{\partial \Omega/\partial \omega}. \tag{7.151}$$

The above results for a triclinic crystal can be specialized to the other crystal classes by appropriately setting some of the elastic parameters to zero. Say that we are studying wave propagation in aluminum. Aluminum has a cubic crystal symmetry so that $c_{14} = c_{15} = c_{16} = 0$, $c_{24} = c_{25} = c_{26} = 0$, $c_{34} = c_{35} = c_{36} = 0$, $c_{45} = c_{46} = 0$, $c_{56} = 0$, $c_{11} = c_{22} = c_{33}$, $c_{12} = c_{13} = c_{23}$, and $c_{44} = c_{55} = c_{66}$, with specific values of the elastic parameters given by $(c_{11}, c_{12}, c_{44}) = (10.80, 6.13, 2.85) \times 10^{10}$ N/m^2. In this case, it is easy to see that

$$\mathbf{A} = \begin{bmatrix} c_{11}l_1^2 + c_{44}\left(1 - l_1^2\right) & \left(c_{12} + c_{44}\right)l_1 l_2 & \left(c_{12} + c_{44}\right)l_3 l_1 \\ \left(c_{12} + c_{44}\right)l_1 l_2 & c_{11}l_2^2 + c_{44}\left(1 - l_2^2\right) & \left(c_{12} + c_{44}\right)l_2 l_3 \\ \left(c_{12} + c_{44}\right)l_3 l_1 & \left(c_{12} + c_{44}\right)l_2 l_3 & c_{11}l_3^2 + c_{44}\left(1 - l_3^2\right) \end{bmatrix}. \tag{7.152}$$

If the propagation direction in a cubic crystal is along any crystal axis, we obtain

$$\Omega(\omega, \mathbf{k}) \equiv \left(k^2 c_{11} - \rho\omega^2\right)\left(k^2 c_{44} - \rho\omega^2\right)^2 = 0, \tag{7.153}$$

and the wave displacements are pure longitudinal and shear modes. Also for such crystal, it is easy to see that if the propagation direction is in the (x_1, x_3)-plane ($l_2 = 0$), the dispersion relation (7.149) becomes a product of a linear term (in $\rho\omega^2$ or, equivalently, in ρc_w^2) and a quadratic term

$$\Omega(\omega, \mathbf{k}) \equiv \left[k^2 c_{44} - \rho\omega^2\right] \cdot$$
$$\left\{\left[\left(c_{11}k_1^2 + c_{44}k_3^2\right) - \rho\omega^2\right]\left[\left(c_{11}k_3^2 + c_{44}k_1^2\right) - \rho\omega^2\right] - \left(c_{12} + c_{44}\right)^2 k_1^2 k_3^2\right\} = 0. \tag{7.154}$$

Subsequently, we have that

$$c_w^{(1)} = \left(\frac{c_{44}}{\rho}\right)^{1/2}, \tag{7.155}$$

$$c_w^{(2,3)} = \left[\frac{c_{11} + c_{44} \mp \sqrt{\left(c_{11} - c_{44}\right)^2 \cos^2 2\phi + \left(c_{11} + c_{44}\right)^2 \sin^2 2\phi}}{2\rho}\right]^{1/2}, \tag{7.156}$$

where $\cos\phi = l_1$ (note that in this case $l_1^2 + l_3^2 = 1$). Now it is easy to see that $\widehat{\mathbf{U}}^{(1)}$ has a particle velocity that is normal to the (x_1, x_3)-plane and is therefore normal to $\mathbf{k}$. This is a pure shear wave polarized along x_2. On the other hand, $\widehat{\mathbf{U}}^{(2,3)}$ are quasi-shear and quasi-longitudinal waves, respectively, but they reduce to pure modes for special propagation directions: $\phi = 0$ ($l_3 = 0$) or $\phi = \pi/2$ ($l_1 = 0$). Pure mode solutions also occur at $\phi = \pi/4$ ($l_1 = l_3 = 1/\sqrt{2}$) and $\phi = 3\pi/4$ ($l_1 = -l_3 = -1/\sqrt{2}$), in which case we respectively have

$$\frac{\widehat{U}_1}{\widehat{U}_3} = \mp \frac{(c_{12} + c_{44})}{(c_{11} + c_{44}) - 2\rho c_w^2}. \tag{7.157}$$

But, from (7.155) and (7.156) for $\phi = \pi/4$ and $\phi = 3\pi/4$, we have that

$$\begin{aligned}
c_w^{(1)} &= \left(\frac{c_{44}}{\rho}\right)^{1/2}, && \text{pure shear} \\
c_w^{(2)} &= \left(\frac{c_{11} - c_{12}}{2\rho}\right)^{1/2}, && \text{pure shear} \\
c_w^{(3)} &= \left(\frac{c_{11} + c_{12} + 2c_{44}}{2\rho}\right)^{1/2}, && \text{pure longitudinal}
\end{aligned} \tag{7.158}$$

for these propagation directions. Substitution of (7.158) into (7.157) gives

$$\left(\frac{\widehat{U}_1}{\widehat{U}_3}\right)^{(2,3)} = \mp 1 \quad \text{at} \quad \phi = \frac{\pi}{4} \qquad \text{and} \qquad \left(\frac{\widehat{U}_1}{\widehat{U}_3}\right)^{(2,3)} = \pm 1 \quad \text{at} \quad \phi = \frac{3\pi}{4}. \tag{7.159}$$

Also, from (7.154), we have that

$$\frac{\partial\Omega}{\partial k_1} = 2c_{11}k_1\left(c_{11}k_3^2 + c_{44}k_1^2 - \rho\omega^2\right) + 2c_{44}k_1\left(c_{11}k_1^2 + c_{44}k_3^2 - \rho\omega^2\right) - 2\left(c_{12} + c_{44}\right)^2 k_1 k_3^2, \tag{7.160}$$

$$\frac{\partial\Omega}{\partial k_2} = 0, \tag{7.161}$$

$$\frac{\partial\Omega}{\partial k_3} = 2c_{44}k_3\left(c_{11}k_3^2 + c_{44}k_1^2 - \rho\omega^2\right) + 2c_{11}k_3\left(c_{11}k_1^2 + c_{44}k_3^2 - \rho\omega^2\right) - 2\left(c_{12} + c_{44}\right)^2 k_1^2 k_3, \tag{7.162}$$

$$\frac{\partial\Omega}{\partial\omega} = -2\rho\omega\left[(c_{11} + c_{44})k^2 - 2\rho\omega^2\right]. \tag{7.163}$$

Substitution of these derivatives into (7.151) gives

$$(c_g)_1 = \frac{l_1}{\rho c_w} \cdot \frac{\left[c_{11}\left(c_{11}l_3^2 + c_{44}l_1^2 - \rho c_w^2\right) + c_{44}\left(c_{11}l_1^2 + c_{44}l_3^2 - \rho c_w^2\right) - (c_{12} + c_{44})^2 l_3^2\right]}{\left[(c_{11} + c_{44}) - 2\rho c_w^2\right]}, \tag{7.164}$$

$$(c_g)_2 = 0, \tag{7.165}$$

$$(c_g)_3 = \frac{l_3}{\rho c_w} \cdot \frac{\left[c_{44}\left(c_{11}l_3^2 + c_{44}l_1^2 - \rho c_w^2\right) + c_{11}\left(c_{11}l_1^2 + c_{44}l_3^2 - \rho c_w^2\right) - (c_{12} + c_{44})^2 l_1^2\right]}{\left[(c_{11} + c_{44}) - 2\rho c_w^2\right]}. \tag{7.166}$$

In summary, (i) there are three different types of waves that are propagated along any given direction in the crystal; (ii) the speeds of these waves are different and are functions of the wave direction $\mathbf{l}$, the density ρ, and the elastic stiffness $\mathbf{C}$; (iii) the direction $\widehat{\mathbf{U}}$ of the displacement amplitude $|\mathbf{U}|$ of the wave is *not* arbitrary, but is fixed by the direction of propagation $\mathbf{l}$ and the elastic stiffness $\mathbf{C}$, and moreover, it is different for the three types of waves. Lastly, we should note that the eigenvalue problem, and thus the dispersion relation, will take on a much simpler form and be more amenable to explicit solution for special directions $\mathbf{l}$ in the crystal. That is, for certain propagation directions, usually along directions of high crystalline symmetry, the three normal modes are found to be one purely longitudinal ($\widehat{\mathbf{U}}^{(1)} \parallel \mathbf{l}$) and two purely transverse ($\widehat{\mathbf{U}}^{(2,3)} \perp \mathbf{l}$); these are called pure modes. Except for pure mode directions, the eigenvectors $\widehat{\mathbf{U}}^{(i)}$ depend explicitly on the values of the elastic constants. Lastly, the group velocity direction may not coincide with the propagation direction $\mathbf{l}$ of the wave velocity.

7.4 Nonlinear deformations of anisotropic materials

We recall from (7.54) that $\psi = \psi(\mathbf{C}, \theta)$ and the right Cauchy–Green tensor $\mathbf{C}$ is symmetric. We assume that the Helmholtz free energy is a polynomial in the six components of $\mathbf{C}$ without limitation on its degree, i.e., we assume that ψ is an analytic function. The material symmetry determines a polynomial basis $I_1, \ldots, I_n$ in the components of $\mathbf{C}$ which are invariant under the specific symmetry group. Hence, any polynomial in $I_1, \ldots, I_n$ is also invariant under the symmetry group. Conversely, it can be proved that any polynomial in the components of $\mathbf{C}$ that is invariant under the the material symmetry group is expressible as a polynomial in the polynomial basis. Hence, we can write

$$\psi = \psi(\theta, I_1, \ldots, I_n). \tag{7.167}$$

Therefore, the constitutive equation for the stress tensor, given by (7.58) and (7.61), can be expressed as

$$\boldsymbol{\sigma} = 2\,\rho \mathbf{F} \cdot \frac{\partial \psi}{\partial \mathbf{C}} \cdot \mathbf{F}^T = 2\,\rho \sum_{\alpha=1}^{n} \frac{\partial \psi}{\partial I_\alpha} \boldsymbol{\Lambda}_\alpha, \tag{7.168}$$

where

$$\boldsymbol{\Lambda}_\alpha = \mathbf{F} \cdot \frac{\partial I_\alpha}{\partial \mathbf{C}} \cdot \mathbf{F}^T. \tag{7.169}$$

The functions $\boldsymbol{\Lambda}_\alpha$ do not depend on the Helmholtz free enery; they depend only on the symmetry of the material. Consequently, for different materials exhibiting the same type of anisotropy, only the coefficients $\partial \psi / \partial I_\alpha$ take different values.

Using known theorems of invariant theory, polynomial bases for the Helmholtz free energy for various crystal classes can be obtained. The free energy will then be an arbitrary polynomial in the appropriate basis. It has been found that there are only 11 different sets of bases which characterize the elastic properties of crystals and they are associated with the 11 types of crystals. Table 7.6 summarizes the basis for each class of crystals along with those of transverse isotropic and isotropic materials.

Example

The Helmholtz free energy for an orthorhombic material is a polynomial in the invariants $\{c_{11}, c_{22}, c_{33}, c_{23}^2, c_{13}^2, c_{12}^2, c_{12}c_{23}c_{13}\}$. Now, it is straightforward to see that the general form of ψ is

$$\psi = \psi_1 + c_{12}\,c_{23}\,c_{13}\,\psi_2, \tag{7.170}$$

where ψ_1 and ψ_2 are functions of θ and the six invariants $\{c_{11}, c_{22}, c_{33}, c_{23}^2, c_{13}^2, c_{12}^2\}$. We then obtain from (7.168) that

$$\begin{aligned}\sigma_{ij} \;=\; 2\,\rho\Bigg[&\frac{\partial\psi}{\partial c_{11}}x_{i,1}x_{j,1} + \frac{\partial\psi}{\partial c_{22}}x_{i,2}x_{j,2} + \frac{\partial\psi}{\partial c_{33}}x_{i,3}x_{j,3}+\\ &\left(2\,c_{12}\frac{\partial\psi}{\partial c_{12}^2} + c_{23}c_{13}\psi_2\right)(x_{i,1}x_{j,2} + x_{i,2}x_{j,1}) +\\ &\left(2\,c_{23}\frac{\partial\psi}{\partial c_{23}^2} + c_{12}c_{13}\psi_2\right)(x_{i,2}x_{j,3} + x_{i,3}x_{j,2}) +\\ &\left(2\,c_{13}\frac{\partial\psi}{\partial c_{13}^2} + c_{12}c_{23}\psi_2\right)(x_{i,1}x_{j,3} + x_{i,3}x_{j,1})\Bigg].\end{aligned} \tag{7.171}$$

7.5 Linear deformations of isotropic materials

For linear deformations of an isotropic solid, using the symmetry conditions (7.120) and using Voigt's notation with $(c_0)_{11} = 2\,\mu_0 + \lambda_0$ and $(c_0)_{12} = \lambda_0$ (see Table 7.5), we must have (see Appendix B)

$$\begin{aligned}(c_0)_{ijkl} &= \lambda_0\,\delta_{ij}\delta_{kl} + \mu_0\,(\delta_{ik}\delta_{jl} + \delta_{il}\delta_{jk}), &(7.172)\\ (a_0)_{kl} &= a_0\,\delta_{kl}, &(7.173)\\ (\beta_0)_{ij} &= -(c_0)_{ijkl}(a_0)_{kl} &(7.174)\\ &= -\left[\lambda_0\delta_{ij}\delta_{kl} + \mu_0\,(\delta_{ik}\delta_{jl} + \delta_{il}\delta_{jk})\right]a_0\,\delta_{kl} &(7.175)\\ &= -3\left(\lambda_0 + \frac{2}{3}\,\mu_0\right)a_0\,\delta_{ij} &(7.176)\\ &= 3\,\beta_0\,\delta_{ij}, &(7.177)\\ (k_0)_{ij} &= k_0\,\delta_{ij}. &(7.178)\end{aligned}$$

The constants λ_0 and μ_0 are known as *isothermal Lamé's constants*, and they are related to the *isothermal Young modulus* E_0 and the *isothermal Poisson ratio* ν_0 as well as to the other parameters, as indicated in Table 7.7. The parameter μ_0 is the *isothermal shear modulus* and the parameter λ_0 is related to the *isothermal bulk modulus* $b_0 = \lambda_0 + 2/3\,\mu_0$. Note that the isothermal bulk modulus is the reciprocal of the isothermal compressibility, i.e., $b_0 = 1/\kappa_\theta$. In addition, we have defined the *isothermal coefficient of thermal expansion* a_0 and the *thermal stress coefficient*

$\beta_0 = -b_0\, a_0$. Now, our constitutive relations (7.105)–(7.109) become

$$\begin{aligned}
\psi &= \psi_0 - \eta_0\,\tilde{\theta} - \frac{c_{\mathbf{e}_0}}{2\,\theta_0}\,\tilde{\theta}^2 - \frac{3\,b_0\,a_0}{\rho_0}\,(\mathrm{tr}\,\widetilde{\mathbf{e}})\,\tilde{\theta} + \frac{1}{2\,\rho_0}\left[2\,\mu_0\,\widetilde{\mathbf{e}}:\widetilde{\mathbf{e}} + \lambda_0\,(\mathrm{tr}\,\widetilde{\mathbf{e}})^2\right], && (7.179)\\
\eta &= \eta_0 + c_{\mathbf{e}_0}\,\frac{\tilde{\theta}}{\theta_0} + \frac{3\,b_0\,a_0}{\rho_0}\,(\mathrm{tr}\,\widetilde{\mathbf{e}}), && (7.180)\\
\mathbf{q} &= -k_0\,\widetilde{\mathbf{g}}, && (7.181)\\
\mathbf{h} &= -\frac{k_0}{\theta_0}\,\widetilde{\mathbf{g}}, && (7.182)\\
\boldsymbol{\sigma} &= (\lambda_0\,\mathrm{tr}\,\widetilde{\mathbf{e}} - 3\,b_0\,a_0\,\tilde{\theta})\mathbf{1} + 2\,\mu_0\,\widetilde{\mathbf{e}}. && (7.183)
\end{aligned}$$

Equation (7.183) corresponds to the thermoelastic generalization of Hooke's law. In addition, it is easy to show that

$$c_{\boldsymbol{\sigma}_0} - c_{\mathbf{e}_0} = \frac{9\,b_0\,a_0^2\,\theta_0}{\rho_0} = \frac{3\,E_0\,a_0^2\,\theta_0}{\rho_0\,(1 - 2\,\nu_0)} \tag{7.184}$$

and that the Grüneisen tensor (5.217) in this case simplifies to yield the conventional formula

$$\Gamma = \frac{3\,\rho_0\,a_0\,b_0}{c_{\mathbf{e}_0}}. \tag{7.185}$$

Thermodynamic stability at equilibrium in this case requires that

$$c_{\mathbf{e}_0} > 0, \qquad (c_0)_{11} - (c_0)_{12} > 0, \qquad \text{and} \qquad (c_0)_{11} + 2\,(c_0)_{12} > 0. \tag{7.186}$$

or, equivalently,

$$c_{\mathbf{e}_0} > 0, \qquad \mu_0 > 0, \qquad \text{and} \qquad b_0 > 0. \tag{7.187}$$

These in turn restrict the Poisson ratio to be within the range

$$-1 < \nu_0 < \frac{1}{2}. \tag{7.188}$$

Lastly, the reduced inequality provides a restriction for the thermal conductivity:

$$k_0 \geq 0. \tag{7.189}$$

7.6 Nonlinear deformations of isotropic materials

More generally, we have shown that the constitutive functional of an objective simple homogeneous solid that is isotropic or hemitropic is given by (see (5.109))

$$T(\mathbf{x},t) = \underset{0\leq s<\infty}{\mathfrak{F}}\,\{\mathbf{B}, {}_{(t)}\mathbf{C}^{(t)}, \theta^{(t)}, {}_{(t)}\mathbf{g}^{(t)}\}. \tag{7.190}$$

Subsequently, if we consider a material with no memory ($p = q = r = 0$), and use relations (7.46)–(7.48), (7.50), and (7.45), the constitutive functional reduces to the following constitutive functions:

$$\psi = \psi(\mathbf{B}, \theta), \tag{7.191}$$

$$\eta = -\frac{\partial \psi}{\partial \theta}, \tag{7.192}$$

$$\mathbf{q} = \mathbf{q}(\mathbf{B}, \theta, \mathbf{g}), \tag{7.193}$$

$$\mathbf{h} = \frac{\mathbf{q}}{\theta}, \tag{7.194}$$

$$\boldsymbol{\sigma} = 2\rho \frac{\partial \psi}{\partial \mathbf{B}} \cdot \mathbf{B}, \tag{7.195}$$

with $\mathbf{H}\ (= \mathbf{R}^T = \mathbf{Q}) \in \mathscr{G}_{\kappa} = \mathscr{O}(\mathscr{V})$ or $\mathbf{H}\ (= \mathbf{R}^T = \mathbf{Q}) \in \mathscr{G}_{\kappa} = \mathscr{O}^+(\mathscr{V})$ for all $(\mathbf{B}, \theta, \mathbf{g})$. Recall from (7.50) that

$$\sigma_{ij} = \rho \frac{\partial \psi}{\partial F_{iI}} F_{jI} = 2\rho \frac{\partial \psi}{\partial B_{ik}} B_{kj}, \tag{7.196}$$

where we have used the fact that $B_{ik} = F_{iI} F_{kI}$.

From Tables 5.1 and 5.5, and using the relations (3.98)–(3.100), we see that for isotropic and hemitropic solids,

$$\psi = \psi(\mathbf{B}_{(\gamma)}, \theta), \tag{7.197}$$

where $\mathbf{B}_{(\gamma)}$, $\gamma = 1, 2, 3$, correspond to the invariants of $\mathbf{B}$.

From Tables 5.1 and 5.2, we see that the isotropic representation for the heat flux is given by

$$\mathbf{q} = \left(\alpha_0 I + \alpha_1 B + \alpha_2 B^2\right) \mathbf{g}, \tag{7.198}$$

with

$$\alpha_j = \alpha_j \left(B_{(\gamma)}, \theta, \mathbf{g} \cdot \mathbf{g}, \mathbf{g} \cdot B\,\mathbf{g}, \mathbf{g} \cdot B^2 \mathbf{g}\right), \tag{7.199}$$

while for a hemitropic solid, from Tables 5.5 and 5.6, we have

$$\mathbf{q} = (\alpha_0 I + \alpha_1 B)\,\mathbf{g} + \alpha_2\ \mathbf{g} \times B\,\mathbf{g}, \tag{7.200}$$

with

$$\alpha_i = \alpha_i \left(B_{(\gamma)}, \theta, \mathbf{g} \cdot \mathbf{g}, \mathbf{g} \cdot B\,\mathbf{g}, \mathbf{g} \cdot B^2 \mathbf{g}, \left[\mathbf{g}, B\,\mathbf{g}, B^2 \mathbf{g})\right]\right). \tag{7.201}$$

Note that since

$$d\psi = \frac{\partial \psi}{\partial \mathbf{B}} \cdot d\mathbf{B} + \frac{\partial \psi}{\partial \theta}\, d\theta = \frac{1}{2\rho} \boldsymbol{\sigma} \cdot \mathbf{B}^{-1} \cdot d\mathbf{B} - \eta\, d\theta, \tag{7.202}$$

we have the Gibbs equation

$$de = \frac{1}{2\rho} \boldsymbol{\sigma} \cdot \mathbf{B}^{-1} \cdot d\mathbf{B} + \theta\, d\eta. \tag{7.203}$$

Equations (7.202) and (7.203) should be compared to (5.206) and (5.139).

Now, using (7.197), we can write

$$\frac{\partial \psi}{\partial B_{ik}} = \sum_{\gamma=1}^{3} \frac{\partial \psi}{\partial B_{(\gamma)}} \frac{\partial B_{(\gamma)}}{\partial B_{ik}}. \tag{7.204}$$

It is easy to see from (3.88) or (3.89) that

$$\frac{\partial B_{(1)}}{\partial B_{ik}} = \delta_{ik}, \quad \frac{\partial B_{(2)}}{\partial B_{ik}} = B_{(1)}\delta_{ik} - B_{ki}, \quad \text{and} \quad \frac{\partial B_{(3)}}{\partial B_{ik}} = B_{(3)}B_{ki}^{-1}. \tag{7.205}$$

Subsequently, since $\mathbf{B} = \mathbf{F} \cdot \mathbf{F}^T$, we have

$$\boldsymbol{\sigma} = \beta_0 \mathbf{1} + \beta_1 \mathbf{B} + \beta_2 \mathbf{B}^2, \tag{7.206}$$

where

$$\beta_0 = 2\rho B_{(3)} \frac{\partial \psi}{\partial B_{(3)}}, \tag{7.207}$$

$$\beta_1 = 2\rho \left(\frac{\partial \psi}{\partial B_{(1)}} + B_{(1)} \frac{\partial \psi}{\partial B_{(2)}} \right), \tag{7.208}$$

$$\beta_2 = -2\rho \frac{\partial \psi}{\partial B_{(2)}}. \tag{7.209}$$

Alternatively, by taking the inner product of Cayley–Hamilton's relation (3.95),

$$\mathbf{B}^3 - B_{(1)}\mathbf{B}^2 + B_{(2)}\mathbf{B} - B_{(3)}\mathbf{1} = \mathbf{0}, \tag{7.210}$$

with $\mathbf{B}^{-1}$, solving the resulting equation for $\mathbf{B}^2$, and substituting the result into the constitutive equation (7.206), we can write equivalently

$$\boldsymbol{\sigma} = \overline{\beta}_0 \mathbf{1} + \overline{\beta}_1 \mathbf{B} + \overline{\beta}_{-1} \mathbf{B}^{-1}, \tag{7.211}$$

where

$$\overline{\beta}_0 = \beta_0 - \beta_2 B_{(2)} = 2\rho \left(B_{(2)} \frac{\partial \psi}{\partial B_{(2)}} + B_{(3)} \frac{\partial \psi}{\partial B_{(3)}} \right), \tag{7.212}$$

$$\overline{\beta}_1 = \beta_1 + \beta_2 B_{(1)} = 2\rho \frac{\partial \psi}{\partial B_{(1)}}, \tag{7.213}$$

$$\overline{\beta}_{-1} = \beta_2 B_{(3)} = -2\rho B_{(3)} \frac{\partial \psi}{\partial B_{(2)}}. \tag{7.214}$$

From (7.211) we see that

$$\boldsymbol{\sigma} \cdot \mathbf{B} = \mathbf{B} \cdot \boldsymbol{\sigma}. \tag{7.215}$$

Note that for a deformation such that $B_{13} = B_{23} = 0$, we see that $\sigma_{13} = \sigma_{23} = 0$ as well. Then, the only nontrivial relation from (7.215) can be written in the form

$$\frac{\sigma_{11} - \sigma_{22}}{\sigma_{12}} = \frac{B_{11} - B_{22}}{B_{12}}. \tag{7.216}$$

Relations (7.215) and (7.216) between stresses and deformations that are independent of any specific constitutive relation are called *universal relations*. Such relations play important roles in experimental validation of material models since they reflect a direct consequence of material symmetry without having to know the constitutive function itself, as long as the material is classified as a simple isotropic homogeneous thermoelastic one.

7.6.1 Special nonlinear deformations

For isothermal conditions, the steady balance equations are identically satisfied in the absence of external supplies $\mathbf{f}$ and r when $\operatorname{div}\boldsymbol{\sigma} = \mathbf{0}$. Boundary conditions can also be satisfied by appropriate application of forces on the boundary of the body. We would like to find all possible material deformations in such case. Specifically, we would like to find solutions of $\sigma_{ik,k} = 0$ for all $\psi = \psi(B_{(\gamma)}, \theta)$. First we note that for the deformation function $x_i = \chi_i(X_I)$, we can write

$$\boldsymbol{\sigma} = J^{-1}\overline{\boldsymbol{\sigma}}\cdot\mathbf{F}^T \qquad \text{or} \qquad \sigma_{ik} = J^{-1}\overline{\sigma}_{iK}x_{k,K}, \tag{7.217}$$

where the quantity $\overline{\boldsymbol{\sigma}}$, called the first *Piola–Kirchhoff* stress tensor, is given by

$$\overline{\boldsymbol{\sigma}} = J\boldsymbol{\sigma}\cdot(\mathbf{F}^T)^{-1} \qquad \text{or} \qquad \overline{\sigma}_{iK} = J\sigma_{ik}X_{K,k}. \tag{7.218}$$

Unlike the Cauchy stress tensor, the first Piola–Kirchhoff stress tensor is not symmetric in general. If follows that

$$\mathbf{t} = \boldsymbol{\sigma}\cdot\mathbf{n} = J^{-1}\overline{\boldsymbol{\sigma}}\cdot\mathbf{F}^T\cdot\mathbf{n} = \frac{1}{\eta}\overline{\boldsymbol{\sigma}}\cdot\mathbf{N} \qquad \text{or}$$

$$t_i = \sigma_{ik}n_k = J^{-1}\overline{\sigma}_{iK}x_{k,K}n_k = \frac{1}{\eta}\overline{\sigma}_{iK}N_K, \tag{7.219}$$

where

$$\mathbf{N} = \eta\, J^{-1}\mathbf{F}^T\cdot\mathbf{n} \qquad \text{or} \qquad N_K = \eta\, J^{-1}x_{k,K}n_k, \tag{7.220}$$

and η is the area stretch ratio (see (3.45)).

Now, using mass conservation,

$$\sigma_{ik} = \rho\frac{\partial\psi}{\partial x_{i,K}}x_{k,K} = \rho_0 J^{-1}\frac{\partial\psi}{\partial x_{i,K}}x_{k,K}; \tag{7.221}$$

thus

$$\overline{\sigma}_{iK} = \rho_0\frac{\partial\psi}{\partial x_{i,K}} = \frac{\partial\psi_0}{\partial x_{i,K}}, \tag{7.222}$$

where $\psi_0 = \rho_0\psi$ and we have assumed that $\rho_0 = \text{const}$. Hence $\sigma_{ik,k} = 0$ implies that $\overline{\sigma}_{iK,K} = 0$, since

$$\sigma_{ik,k} = \left(J^{-1}\overline{\sigma}_{iK}x_{k,K}\right)_{,k} = \left(J^{-1}x_{k,K}\right)_{,k}\overline{\sigma}_{iK} + J^{-1}\overline{\sigma}_{iK,L}x_{k,K}X_{L,k} = J^{-1}\overline{\sigma}_{iK,K}, \tag{7.223}$$

where above we have made use of the Euler–Piola–Jacobi identity (3.67).

In summary, for the deformation function $x_i = \chi_i(X_K)$, the original problem has been reformulated into

$$\overline{\sigma}_{iK,K} = 0, \qquad \text{where} \qquad \overline{\sigma}_{iK} = \frac{\partial\psi_0}{\partial x_{i,K}}, \qquad \text{for all} \quad \psi_0 = \psi_0(B_{(\gamma)}, \theta). \tag{7.224}$$

In particular, if we choose $\psi_0 = B_{(1)} = \operatorname{tr}B = \operatorname{tr}(FF^T) = x_{i,K}x_{i,K}$, then $\overline{\sigma}_{iK} = 2x_{i,K}$, and subsequently, the requirement that $x_{i,KK} = 0$ implies that x_i is harmonic and infinitely differentiable.

Another particular choice is $\psi_0 = B_{(3)} = \det B = \det(FF^T) = J^2$. Now using (3.60), we have $\overline{\sigma}_{iK} = 2J^2 X_{K,i}$ and subsequently we require that $\overline{\sigma}_{iK,K} = (2J^2 X_{K,i})_{,K} = 0$. Now using the Euler–Piola–Jacobi identity (3.66), we have that $2JX_{K,i}J_{,K} = 0$ or $J = \text{const.}$

Now for any $\psi_0 = J\overline{\psi}$, using (3.60), we get

$$\overline{\sigma}_{iK} = \frac{\partial(J\overline{\psi})}{\partial x_{i,K}} = JX_{K,i}\overline{\psi} + J\frac{\partial \overline{\psi}}{\partial x_{i,K}}, \tag{7.225}$$

and so, using the Euler–Piola–Jacobi identity (3.66) again and our previous results, we have

$$0 = \overline{\sigma}_{iK,K} = JX_{K,i}\frac{\partial \overline{\psi}}{\partial X_K} + J\left(\frac{\partial \overline{\psi}}{\partial x_{i,K}}\right)_{,K} = JX_{K,i}\frac{\partial \overline{\psi}}{\partial X_K} + J\overline{\Sigma}_{iK,K} = JX_{K,i}\frac{\partial \overline{\psi}}{\partial X_K}; \tag{7.226}$$

thus $\overline{\psi} = \text{const.}$, which implies that $B_{(1)} = \text{const.}$, $B_{(2)} = \text{const.}$, and $B_{(3)} = \text{const.}$ Since $B_{(1),KK} = 0$, then

$$0 = (x_{i,IK}x_{i,I} + x_{i,I}x_{i,IK})_{,K} = 2(x_{i,IK}x_{i,I})_{,K} = 2(x_{i,IKK}x_{i,I} + x_{i,IK}x_{i,IK}) = 2x_{i,IK}x_{i,IK};$$

thus

$$x_{i,IK} = 0 \qquad \text{or} \qquad x_{i,I} = \text{const.} \tag{7.227}$$

Hence, the deformations possible in materials subject to surface traction alone are homogeneous.

A. Homogeneous deformations

A homogeneous deformation is one such that

$$\mathbf{x} = \mathbf{x}_0 + \mathbf{F}\cdot(\mathbf{X} - \mathbf{X}_0), \tag{7.228}$$

where the deformation gradient $\mathbf{F}$ is a constant two-point tensor, and $\mathbf{X}_0$ and $\mathbf{x}_0$ correspond to the origins in the reference and deformed configurations. Such deformation is a universal solution of thermoelastic materials since it does not depend on any particular form of the constitutive relation. Therefore, a homogeneous deformation is an important controllable class of deformation which can be exploited in conjunction with experiments to determine specific material parameters.

a) **Uniaxial stretch:** The homogeneous deformation of uniaxial stretch is given by

$$x_1 = \alpha_1 X_1, \tag{7.229}$$

$$x_2 = \alpha_2 X_2, \tag{7.230}$$

$$x_3 = \alpha_3 X_3, \tag{7.231}$$

where the stretches α_i, $i = 1, 2, 3$, are constant. Then

$$F = [\text{Grad}\,\mathbf{x}] = \begin{bmatrix} \alpha_1 & 0 & 0 \\ 0 & \alpha_2 & 0 \\ 0 & 0 & \alpha_3 \end{bmatrix}, \tag{7.232}$$

$$B = FF^T = \begin{bmatrix} \alpha_1^2 & 0 & 0 \\ 0 & \alpha_2^2 & 0 \\ 0 & 0 & \alpha_3^2 \end{bmatrix}, \quad \text{and} \quad B^{-1} = \begin{bmatrix} \alpha_1^{-2} & 0 & 0 \\ 0 & \alpha_2^{-2} & 0 \\ 0 & 0 & \alpha_3^{-2} \end{bmatrix}. \tag{7.233}$$

Furthermore,

$$B_{(1)} = \alpha_1^2 + \alpha_2^2 + \alpha_3^2, \tag{7.234}$$

$$B_{(2)} = \alpha_1^2\alpha_2^2 + \alpha_2^2\alpha_3^2 + \alpha_3^2\alpha_1^2, \tag{7.235}$$

$$B_{(3)} = \alpha_1^2\alpha_2^2\alpha_3^2. \tag{7.236}$$

Subsequently,

$$[\sigma_{ij}] = \overline{\beta}_0 \begin{bmatrix} 1 & 0 & 0 \\ 0 & 1 & 0 \\ 0 & 0 & 1 \end{bmatrix} + \overline{\beta}_1 \begin{bmatrix} \alpha_1^2 & 0 & 0 \\ 0 & \alpha_2^2 & 0 \\ 0 & 0 & \alpha_3^2 \end{bmatrix} + \overline{\beta}_{-1} \begin{bmatrix} \alpha_1^{-2} & 0 & 0 \\ 0 & \alpha_2^{-2} & 0 \\ 0 & 0 & \alpha_3^{-2} \end{bmatrix}. \tag{7.237}$$

where $\overline{\beta}_k = \overline{\beta}_k(\alpha_\gamma^2, \theta)$, with $\gamma = 1, 2, 3$ and $k = 0, 1, -1$.

For uniaxial stretch in the X_1-direction, the lateral stresses must vanish:

$$\sigma_{22} = \overline{\beta}_0 + \overline{\beta}_1\alpha_2^2 + \overline{\beta}_{-1}\alpha_2^{-2} = 0, \tag{7.238}$$

$$\sigma_{33} = \overline{\beta}_0 + \overline{\beta}_1\alpha_3^2 + \overline{\beta}_{-1}\alpha_3^{-2} = 0, \tag{7.239}$$

so that $\sigma_{22} - \sigma_{33}$ results in

$$\left(\alpha_2^2 - \alpha_3^2\right)\left(\overline{\beta}_1 - \overline{\beta}_{-1}\frac{1}{\alpha_2^2\alpha_3^2}\right) = 0, \tag{7.240}$$

which has a symmetric solution, $\alpha_2 = \alpha_3$, and an asymmetric solution, $\alpha_2 \neq \alpha_3$. Since $\sigma_{22} = 0$, we can also write the uniaxial stress in the form

$$\sigma_{11} = \left(\alpha_1^2 - \alpha_2^2\right)\left(\overline{\beta}_1 - \overline{\beta}_{-1}\frac{1}{\alpha_1^2\alpha_2^2}\right). \tag{7.241}$$

Now since the stress σ_{11} and stretches α_1 and α_2 can be measured, then (7.241) can be used to determine $\overline{\beta}_1$ and $\overline{\beta}_{-1}$, while $\overline{\beta}_0$ can subsequently be determined from (7.238) and (7.239).

b) Simple extension: The deformation

$$x_1 = \alpha v X_1, \tag{7.242}$$

$$x_2 = \alpha v X_2, \tag{7.243}$$

$$x_3 = v X_3, \tag{7.244}$$

where $0 < \alpha \leq 1$ and $v \geq 1$ is called a *simple extension* in the X_3-direction. Note that

$$F = v \begin{bmatrix} \alpha & 0 & 0 \\ 0 & \alpha & 0 \\ 0 & 0 & 1 \end{bmatrix}, \tag{7.245}$$

$$B = v^2 \begin{bmatrix} \alpha^2 & 0 & 0 \\ 0 & \alpha^2 & 0 \\ 0 & 0 & 1 \end{bmatrix}, \quad \text{and} \quad B^{-1} = v^{-2} \begin{bmatrix} \alpha^{-2} & 0 & 0 \\ 0 & \alpha^{-2} & 0 \\ 0 & 0 & 1 \end{bmatrix}. \tag{7.246}$$

Furthermore,

$$B_{(1)} = v^2(1 + 2\alpha^2), \qquad B_{(2)} = \alpha^2 v^4 (2 + \alpha^2), \qquad B_{(3)} = \alpha^4 v^6. \tag{7.247}$$

Subsequently,

$$\sigma_{11} = \sigma_{22} = \overline{\beta}_0 + \alpha^2 v^2 \overline{\beta}_1 + \alpha^{-2} v^{-2} \overline{\beta}_{-1}, \tag{7.248}$$

$$\sigma_{33} = \overline{\beta}_0 + v^2 \overline{\beta}_1 + v^{-2} \overline{\beta}_{-1}, \tag{7.249}$$

$$\sigma_{ij} = 0 \quad \text{for} \quad i \neq j, \tag{7.250}$$

where $\overline{\beta}_k = \overline{\beta}_k(B_{(\gamma)}, \theta)$, with $\gamma = 1, 2, 3$ and $k = 0, 1, -1$.

For a simple extension, we must have

$$\sigma_{11} = \sigma_{22} = 0, \tag{7.251}$$

so that

$$\alpha^2 = f(v^2, \theta), \tag{7.252}$$

and thus

$$\sigma_{33} = \overline{\beta}_0 + v^2 \overline{\beta}_1 + v^{-2} \overline{\beta}_{-1} = \sigma_{33}(v^2, \theta). \tag{7.253}$$

c) **Simple dilatation or compression:** The deformation function for a *simple dilatation* or *compression* is given by

$$x_1 = vX_1, \tag{7.254}$$

$$x_2 = vX_2, \tag{7.255}$$

$$x_3 = vX_3, \tag{7.256}$$

depending on whether $v \geq 1$ or $v \leq 1$. Then $\mathbf{F} = v\mathbf{1}$, $\mathbf{B} = v^2\mathbf{1}$, $\mathbf{B}^{-1} = v^{-2}\mathbf{1}$, $B_{(1)} = 3v^2$, $B_{(2)} = 3v^4$, and $B_{(3)} = v^6$, and subsequently, we have

$$\sigma_{11} = \sigma_{22} = \sigma_{33} = \overline{\beta}_0 + v^2 \overline{\beta}_1 + v^{-2} \overline{\beta}_{-1} \equiv -p \quad \text{and} \quad \sigma_{ij} = 0 \quad \text{for} \quad i \neq j, \tag{7.257}$$

where $p = p(v^2, \theta)$.

d) Simple shear: The deformation function is given by

$$x_1 = X_1 + \kappa X_2, \tag{7.258}$$
$$x_2 = X_2, \tag{7.259}$$
$$x_3 = X_3, \tag{7.260}$$

where $\kappa \geq 0$ is the amount of shear. Then

$$F = \begin{bmatrix} 1 & \kappa & 0 \\ 0 & 1 & 0 \\ 0 & 0 & 1 \end{bmatrix}, \tag{7.261}$$

$$B = \begin{bmatrix} 1+\kappa^2 & \kappa & 0 \\ \kappa & 1 & 0 \\ 0 & 0 & 1 \end{bmatrix}, \quad \text{and} \quad B^{-1} = \begin{bmatrix} 1 & -\kappa & 0 \\ -\kappa & 1+\kappa^2 & 0 \\ 0 & 0 & 1 \end{bmatrix}. \tag{7.262}$$

Furthermore, $B_{(1)} = B_{(2)} = 3 + \kappa^2$ and $B_{(3)} = 1$ so that $\overline{\beta}_k = \overline{\beta}_k(B_{(\gamma)}, \theta) = \overline{\beta}_k(\kappa^2, \theta)$, for $\gamma = 1, 2, 3$ and $k = 0, 1, -1$. Note that simple shear is an isochoric or incompressible deformation since $B_{(3)} = 1$. Now, as can be easily verified, we have

$$\begin{aligned} [\sigma_{ij}] = {} & \left(\overline{\beta}_0 + \overline{\beta}_1 + \overline{\beta}_{-1}\right) \begin{bmatrix} 1 & 0 & 0 \\ 0 & 1 & 0 \\ 0 & 0 & 1 \end{bmatrix} + \kappa\left(\overline{\beta}_1 - \overline{\beta}_{-1}\right) \begin{bmatrix} 0 & 1 & 0 \\ 1 & 0 & 0 \\ 0 & 0 & 0 \end{bmatrix} + \\ & \kappa^2 \overline{\beta}_1 \begin{bmatrix} 1 & 0 & 0 \\ 0 & 0 & 0 \\ 0 & 0 & 0 \end{bmatrix} + \kappa^2 \overline{\beta}_{-1} \begin{bmatrix} 0 & 0 & 0 \\ 0 & 1 & 0 \\ 0 & 0 & 0 \end{bmatrix}, \end{aligned} \tag{7.263}$$

and thus

$$\frac{\sigma_{21}}{\kappa} = \frac{\sigma_{12}}{\kappa} = \overline{\beta}_1 - \overline{\beta}_{-1} \equiv \mu(\kappa^2), \tag{7.264}$$

where $\mu(\kappa^2)$ is called the *generalized shear modulus*. Note that for small κ,

$$\mu(\kappa^2) = \mu(0) + O(\kappa^2), \tag{7.265}$$

where $\mu(0)$ is called the *ordinary shear modulus* of the material. The principal stretches obtained earlier (see (3.167)) are

$$\left(\lambda^{(1)}\right)^2 = 1 + \frac{1}{2}\kappa^2 + \kappa\sqrt{1 + \frac{1}{4}\kappa^2}, \tag{7.266}$$

$$\left(\lambda^{(2)}\right)^2 = 1 + \frac{1}{2}\kappa^2 - \kappa\sqrt{1 + \frac{1}{4}\kappa^2} = \frac{1}{\left(\lambda^{(1)}\right)^2}, \tag{7.267}$$

$$\left(\lambda^{(3)}\right)^2 = 1, \tag{7.268}$$

and the principal stresses are easily seen to be given by

$$\sigma^{(i)} = \overline{\beta}_0 + \left(\lambda^{(i)}\right)^2 \overline{\beta}_1 + \left(\lambda^{(i)}\right)^{-2} \overline{\beta}_{-1}, \quad i = 1, 2, 3. \tag{7.269}$$

It can now be shown that

$$\frac{\sigma^{(1)} - \sigma^{(2)}}{\left(\lambda^{(1)}\right)^2 - \left(\lambda^{(2)}\right)^2} = \mu(\kappa^2) \tag{7.270}$$

or

$$\sigma^{(1)} - \sigma^{(2)} = 2\kappa\sqrt{1 + \frac{1}{4}\kappa^2}\; \mu(\kappa^2), \tag{7.271}$$

so that $\mu \geq 0$ only if the direction of the greatest principal stress is the same as that of the greatest principal stretches. We also see that the shear stresses alone do not suffice to maintain simple shear since

$$\sigma_{11} = \overline{\beta}_0 + (1 + \kappa^2)\overline{\beta}_1 + \overline{\beta}_{-1}, \tag{7.272}$$

$$\sigma_{22} = \overline{\beta}_0 + \overline{\beta}_1 + (1 + \kappa^2)\overline{\beta}_{-1}, \tag{7.273}$$

$$\sigma_{33} = \overline{\beta}_0 + \overline{\beta}_1 + \overline{\beta}_{-1}, \tag{7.274}$$

and $\sigma_{11} = \sigma_{22} = \sigma_{33} = 0$ if and only if $\overline{\beta}_0 = \overline{\beta}_1 = \overline{\beta}_{-1} = 0$, and thus $\mu \equiv 0$. Furthermore, we have

$$\frac{\sigma_{11} - \sigma_{33}}{\kappa^2} = \overline{\beta}_1, \tag{7.275}$$

$$\frac{\sigma_{22} - \sigma_{33}}{\kappa^2} = \overline{\beta}_{-1}, \tag{7.276}$$

$$\frac{\sigma_{11} - \sigma_{22}}{\kappa^2} = \overline{\beta}_1 - \overline{\beta}_{-1} = \frac{\sigma_{12}}{\kappa}. \tag{7.277}$$

We see that $\sigma_{11} = \sigma_{22}$ implies that $\mu = 0$. The existence of a normal stress difference is usually known as the *Poynting effect.* The mean hydrostatic pressure is given by

$$\overline{p} \equiv -\frac{1}{3}\sigma_{ii} = -\left(\overline{\beta}_0 + \overline{\beta}_1 + \overline{\beta}_{-1}\right) - \frac{1}{3}\left(\overline{\beta}_1 + \overline{\beta}_{-1}\right)\,\kappa^2. \tag{7.278}$$

When there is no shear, $\kappa = 0$, and so from (7.263), we see that

$$\lim_{\kappa \to 0} \overline{p} = -\lim_{\kappa \to 0}\left[\overline{\beta}_0(\kappa^2) + \overline{\beta}_1(\kappa^2) + \overline{\beta}_{-1}(\kappa^2)\right] = 0. \tag{7.279}$$

Thus, a mean hydrostatic pressure, which is of second order in the amount of shear, is necessary to generate a simple shear. This is the so-called *Kelvin effect.* Note that the equation

$$\frac{\sigma_{11} - \sigma_{22}}{\sigma_{12}} = \kappa \tag{7.280}$$

provides a universal relation for this deformation.

B. Incompressible deformations

For an isothermal incompressible (or isochoric) deformation, we have $J = B_{(3)} = 1$, and so from (7.197), we have $\psi = \psi(B_{(1)}, B_{(2)}, \theta)$. We note from (3.101) that $B_{(2)} = \operatorname{tr} B^{-1}$. The constitutive equation is given by

$$\sigma_{ij} = -\overline{p}\delta_{ij} + \overline{\beta}_1 B_{ij} + \overline{\beta}_{-1} B^{-1}_{ij}, \tag{7.281}$$

where

$$\overline{\beta}_1 = 2\rho\frac{\partial\psi}{\partial B_{(1)}} \qquad \text{and} \qquad \overline{\beta}_{-1} = -2\rho\frac{\partial\psi}{\partial B_{(2)}}. \tag{7.282}$$

In this case $\overline{p}$ cannot be determined from the deformation. Subsequently, since the mass density is constant and the hydrostatic pressure undetermined, $\overline{\beta}_0$ becomes superfluous.

B.1 Homogeneous strain

Assuming that no external body forces are present, it is clear that any homogeneous isochoric strain is a steady configuration subject to surface tractions only for any isothermal, homogeneous, incompressible, elastic body, and that the pressure $\overline{p}$ is an arbitrary constant that is to be determined. To illustrate, consider the simple extension deformation

$$x_1 = \alpha v X_1, \tag{7.283}$$
$$x_2 = \alpha v X_2, \tag{7.284}$$
$$x_3 = v X_3, \tag{7.285}$$

and then

$$F = v\begin{bmatrix} \alpha & 0 & 0 \\ 0 & \alpha & 0 \\ 0 & 0 & 1 \end{bmatrix}. \tag{7.286}$$

For the deformation to be isochoric, $J = \det F = 1$; thus $\alpha^2 v^3 = 1$, or $\alpha = v^{-3/2}$. Then

$$B = \begin{bmatrix} v^{-1} & 0 & 0 \\ 0 & v^{-1} & 0 \\ 0 & 0 & v^2 \end{bmatrix} \qquad \text{and} \qquad B^{-1} = \begin{bmatrix} v & 0 & 0 \\ 0 & v & 0 \\ 0 & 0 & v^{-2} \end{bmatrix}, \tag{7.287}$$

and

$$B_{(1)} = 2v^{-1} + v^2, \qquad B_{(2)} = 2v + v^{-2}. \tag{7.288}$$

Subsequently,

$$\sigma_{11} = \sigma_{22} = -\overline{p} + v^{-1}\overline{\beta}_1 + v\overline{\beta}_{-1}, \tag{7.289}$$
$$\sigma_{33} = -\overline{p} + v^2\overline{\beta}_1 + v^{-2}\overline{\beta}_{-1}, \tag{7.290}$$

with $\overline{\beta}_1 = \overline{\beta}_1(B_{(1)}, B_{(2)}, \theta)$ and $\overline{\beta}_{-1} = \overline{\beta}_{-1}(B_{(1)}, B_{(2)}, \theta)$. Clearly this deformation satisfies the field equation

$$\sigma_{ik,k} = 0. \tag{7.291}$$

If we require that $\sigma_{11} = \sigma_{22} = 0$, we then have

$$\overline{p} = v^{-1}\overline{\beta}_1 + v\overline{\beta}_{-1}. \tag{7.292}$$

Thus $\overline{p}$ is determined once we have the solution to the boundary value problem. Then

$$\sigma_{33} = \sigma_{33}(v) = \left(v^2 - v^{-1}\right)\overline{\beta}_1 + \left(v^{-2} - v\right)\overline{\beta}_{-1}. \tag{7.293}$$

The simplest isothermal incompressible deformation is given by a linear variation of the free energy with respect to $B_{(1)}$ and $B_{(2)}$:

$$\psi = \alpha\left(B_{(1)} - 3\right) + \beta\left(B_{(2)} - 3\right), \tag{7.294}$$

where $\alpha = \alpha(\theta)$ and $\beta = \beta(\theta)$. Subsequently, we have that $\overline{\beta}_1 = 2\rho\alpha$ and $\overline{\beta}_{-1} = -2\rho\beta$. This defines a *Mooney–Rivlin* material and is considered to be a good model for rubber-like materials. The case $\beta = 0$ (and thus $\overline{\beta}_{-1} = 0$) defines a *neo-Hookean* material, which provides a reasonable approximation for the behavior of rubber under small strains. Experimental data indicate that both α and β are positive, leading to what are called the *empirical inequalities*

$$\overline{\beta}_1 > 0 \qquad \text{and} \qquad \overline{\beta}_{-1} \leq 0. \tag{7.295}$$

B.2 Nonhomogeneous strain

The steady linear momentum equation is given by

$$\operatorname{div}\boldsymbol{\sigma} + \rho\mathbf{f} = \mathbf{0} \qquad \text{or} \qquad \sigma^k_{l,k} + \rho f_l = 0, \tag{7.296}$$

where

$$\boldsymbol{\sigma} = -\overline{p}\mathbf{1} + \overline{\beta}_1\mathbf{B} + \overline{\beta}_{-1}\mathbf{B}^{-1} \equiv -\overline{p}\mathbf{1} + \widehat{\boldsymbol{\sigma}} \tag{7.297}$$

or

$$\sigma^k_l = -\overline{p}\delta^k_l + \overline{\beta}_1 B^k_l + \overline{\beta}_{-1}(B^{-1})^k_l \equiv -\overline{p}\delta^k_l + \widehat{\sigma}^k_l, \tag{7.298}$$

$B^k_l = g_{ml}B^{km}$, and $B^{km} = g^{IJ}F^k_I F^m_J$. Note that since we consider deformations involving non-Cartesian coordinates, we have written the linear momentum equation for covariant components and have used the mixed-component form of the Cauchy stress tensor. Furthermore, for convenience, we have defined the *extra stress* $\widehat{\boldsymbol{\sigma}}$. If we assume that the external body force is conservative, and thus derivable from a potential, i.e.,

$$\mathbf{f} = -\operatorname{grad} U \qquad \text{or} \qquad f_l = -U_{,l}, \tag{7.299}$$

then for an incompressible material with $\rho_0(\mathbf{X}) = \text{const.}$, we have

$$-\operatorname{grad}\left(\overline{p} + \rho U\right) + \operatorname{div}\widehat{\boldsymbol{\sigma}} = \mathbf{0} \qquad \text{or} \qquad -\left(\overline{p} + \rho U\right)_{,l} + \widehat{\sigma}^k_{l,k} = 0, \tag{7.300}$$

where $(\overline{p} + \rho U)$ is just a modified pressure. To eliminate it, we take the curl of the above equation, so we have

$$\operatorname{curl}\operatorname{div}\widehat{\boldsymbol{\sigma}} = \mathbf{0} \qquad \text{or} \qquad \epsilon^{ijl}\widehat{\sigma}^k_{l,kj} = 0 \tag{7.301}$$

or

$$\widehat{\sigma}^k_{l,kj} = \widehat{\sigma}^k_{j,kl}. \tag{7.302}$$

We see that the presence of a conservative body force does not make the solution of problems for incompressible materials more difficult since all it does is modify the pressure $\overline{p}$. Of course, the surface tractions that must be applied are such that the deformation will depend on U, but the compatibility of any given deformation is independent of whether or not a conservative body force is present. Thus, for simplicity, only steady solutions subjected to no external body forces are usually considered, i.e., $\mathbf{a} = \mathbf{0}$ and $\mathbf{f} = \mathbf{0}$, so that $\operatorname{div} \boldsymbol{\sigma} = \mathbf{0}$, or in terms of physical components, from (2.283) and (2.284),

$$\sigma_{<kl,l>} = \frac{h_{\underline{k}}}{h_1 h_2 h_3} \frac{\partial}{\partial x^l} \left(\frac{h_1 h_2 h_3}{h_{\underline{k}} h_{\underline{l}}} \sigma_{<kl>} \right) + \frac{h_{\underline{k}}}{h_{\underline{m}} h_{\underline{l}}} \Gamma^k_{lm} \sigma_{<ml>} = 0, \tag{7.303}$$

where from (2.285) we have

$$\sigma_{<kl>} = \frac{h_{\underline{k}}}{h_{\underline{l}}} \sigma^k_l = -\overline{p} \delta_{<kl>} + \widehat{\sigma}_{<kl>} \tag{7.304}$$

and

$$\widehat{\sigma}_{<kl>} = \frac{h_{\underline{k}}}{h_{\underline{l}}} \widehat{\sigma}^k_l. \tag{7.305}$$

Given a particular tensor field $\mathbf{B}$, whether or not (7.302) is satisfied will depend, in general, upon $\overline{\beta}_1(B_{(1)}, B_{(2)}, \theta)$ and $\overline{\beta}_{-1}(B_{(1)}, B_{(2)}, \theta)$. However, there are certain forms of $\mathbf{B}$ such that (7.302) is satisfied independent of the specific forms of $\overline{\beta}_1$ and $\overline{\beta}_{-1}$. The deformations that lead to such forms of $\mathbf{B}$ can be effected in every homogeneous, incompressible, isotropic, isothermal, elastic body by application of suitable surface tractions. In such case, we shall drop the dependences of $\overline{\beta}_1$ and $\overline{\beta}_{-1}$ on θ. All solutions are such that appropriate physical components of the stress are constant on each member of a family of parallel planes, coaxial cylinders, or concentric spheres. There are five families of such deformations. They are the following:

a) **Bending, stretching, and shearing of a rectangular block:** Let the coordinate in the initial reference configuration be Cartesian (X, Y, Z), and those in the deformed configuration cylindrical, (r, θ, z). Consider the deformation

$$r = \sqrt{2\mathsf{A}X}, \tag{7.306}$$

$$\theta = \mathsf{B}Y, \tag{7.307}$$

$$z = \frac{Z}{\mathsf{AB}} - \mathsf{BC}Y, \tag{7.308}$$

where A, B, and C are constants, and $\mathsf{AB} \neq 0$. If $\mathsf{C} = 0$, this deformation carries the block bounded by the planes $X = X_1$, $X = X_2$, $Y = \pm Y_0$, $Z = \pm Z_0$ into the annular wedge bounded by the cylinders $r = r_1 = \sqrt{2\mathsf{A}X_1}$ and $r = r_2 = \sqrt{2\mathsf{A}X_2}$, and the planes $\theta = \pm\theta_0 = \pm\mathsf{B}Y_0$, $z = \pm z_0 = \pm Z_0/(\mathsf{A}\ \mathsf{B})$. If we think of B as given, then an arbitrary axial stretch $1/(\mathsf{A}\ \mathsf{B})$ is allowed, and the radial stretch is adjusted to render the deformation isochoric. In the general case, the deformation may be effected in two steps, the first of which is the bending and axial stretch just described, while the second is a homogeneous strain, in fact a simple shear, which carries the body into the solid bounded by the cylindrical surfaces $r = r_1$ and $r = r_2$, the planes $\theta = \pm\theta_0$, and the helicoidal surfaces $z + \mathsf{C}\theta = \pm z_0$.

b) Straightening, stretching, and shearing of a sector of a hollow circular cylinder: Using a cylindrical coordinate system (R, Θ, Z), consider a body that in the reference configuration is bounded by the cylinders $R = R_1$ and $R = R_2$, the planes $\Theta = \pm\Theta_0$, and the planes $Z = \pm Z_0$. We use Cartesian coordinates (x, y, z) for the deformed state. Let the body first be straightened by the deformation $\overline{x} = \frac{1}{2}\mathsf{A}R^2$, $\overline{y} = \Theta/\mathsf{A}$, $\overline{z} = Z$. It then becomes the block bounded by the planes $\overline{x} = \frac{1}{2}\mathsf{A}R_1^2$, $\overline{x} = \frac{1}{2}\mathsf{A}R_2^2$, $\overline{y} = \pm\Theta_0/\mathsf{A}$, $\overline{z} = \pm Z_0$. Now stretch the block along the x-axis with equal transverse contractions: $x' = \mathsf{B}^2\overline{x}$, $y' = \overline{y}/\mathsf{B}$, $z' = \overline{z}/\mathsf{B}$. Finally, effect a shear in the z-y planes: $x = x'$, $y = y'$, $z = z' + \mathsf{C}y'$. The resulting deformation is

$$x = \frac{1}{2}\mathsf{A}\ \mathsf{B}^2 R^2, \tag{7.309}$$

$$y = \frac{\Theta}{\mathsf{AB}}, \tag{7.310}$$

$$z = \frac{Z}{\mathsf{B}} + \frac{\mathsf{C}\Theta}{\mathsf{AB}}, \tag{7.311}$$

where $R_1 \le R \le R_2$, $-\Theta_0 \le \Theta \le \Theta_0$, $-Z_0 \le Z \le Z_0$, and $\mathsf{AB} \ne 0$.

c) Inflation, bending, torsion, extension, and shearing of an annular wedge: The deformation carries the particle with cylindrical coordinates (R, Θ, Z) into the deformed cylindrical coordinates (r, θ, z) as follows:

$$r = \sqrt{\mathsf{A}R^2 + \mathsf{B}}, \tag{7.312}$$

$$\theta = \mathsf{C}\Theta + \mathsf{D}Z, \tag{7.313}$$

$$z = \mathsf{E}\Theta + \mathsf{F}Z, \tag{7.314}$$

where B, C, D, E, F are arbitrary constants, A and B have values such that $\mathsf{A}R^2 + \mathsf{B} > 0$ when R is in some interval $R_1 \le R \le R_2$, and $\mathsf{A}(\mathsf{CF}\text{-}\mathsf{DE}) = 1$. This deformation may be regarded as resulting from a succession of four simpler ones. First, the cylinder defined by $R_1 \le R \le R_2$ is inflated uniformly: $r' = \sqrt{\mathsf{A}R^2 + \mathsf{B}}$, $\theta' = \Theta$, $z' = Z/\mathsf{A}$. Second, the inflated cylinder is subjected to a uniform longitudinal stretch of amount $1/\mathsf{C}$, so that $\overline{r} = r'$, $\overline{\theta} = \mathsf{C}\theta'$, and $\overline{z} = z'/\mathsf{C}$. Third, the inflated and stretched cylinder is twisted with a torsion of amount ACD; thus $\widehat{r} = \overline{r}$, $\widehat{\theta} = \overline{\theta} + \mathsf{ACD}\overline{z}$, and $\widehat{z} = \overline{z}$. Fourth, a kind of shear of the azimuthal planes is effected: $r = \widehat{r}$, $\theta = \widehat{\theta}$, $z = \widehat{z} + \mathsf{K}\widehat{\theta}$. As the result of this last deformation, the planes Z = const. are deformed into the helicoidal surfaces $z - \mathsf{E}\theta/\mathsf{C}$ = const. The constants E and F of the resulting deformation are given by $\mathsf{E} = \mathsf{CK}$, $\mathsf{F} = \mathsf{DK} + 1/(\mathsf{AC})$. Note that if $\mathsf{A} > 0$, r is an increasing function of R, while if $\mathsf{A} < 0$, r is a decresing function of R; thus the case when $\mathsf{A} < 0$ represents a hollow cylinder, or a part of one, which is turned inside out, or everted. If $\mathsf{D} = \mathsf{E} = 0$, the deformation is a plane strain superposed upon a uniform extension perpendicular to the plane.

d) Inflation or eversion of a sector of a spherical shell: Let the point with spherical coordinates (R, Θ, Φ) be deformed into one with spherical coordinates (r, θ, ϕ), where

$$r = \left(\pm R^3 + \mathsf{A}\right)^{1/3}, \tag{7.315}$$

$$\theta = \pm\Theta, \tag{7.316}$$

$$\phi = \Phi. \tag{7.317}$$

This deformation carries the region between two concentric spheres into a region between two other concentric spheres; any angular boundaries are preserved. The ± signs are associated; if both are taken as +, the curvature of the shell is unchanged in sign, while if both are taken as −, the shell is everted.

e) Inflation, bending, extension, and azimuthal shearing of an annular wedge: The deformation carries the material particle with cylindrical coordinates (R, Θ, Z) into the deformed cylindrical coordinates (r, θ, z) as follows:

$$r = \mathsf{A}\,R, \tag{7.318}$$

$$\theta = \mathsf{B}\ln(R) + \mathsf{C}\,\Theta, \tag{7.319}$$

$$z = \mathsf{D}\,Z, \tag{7.320}$$

where A, B, C, and D are arbitrary constants with the restriction that $\mathsf{A}^2\mathsf{C}\,\mathsf{D} = 1$, and $R_1 \le R \le R_2$, $-\Theta_0 \le \Theta \le \Theta_0$, and $-Z_0 \le Z \le Z_0$. This deformation may also be viewed as resulting from a succession of four simpler deformations.

Below we will describe in detail only the bending, stretching, and shearing of a rectangular block as an illustration of the solution of a problem involving non-Cartesian coordinates. The deformation (7.306)–(7.308) looks as shown in Fig. 7.4. The deformation gradient, with $(r, \theta, z) = (x^1, x^2, x^3)$, is given by

$$\begin{aligned}\left[F^k_I\right] = \left[\frac{\partial x^k}{\partial X^I}\right] &= \begin{bmatrix} \mathsf{A}/\sqrt{2\mathsf{A}X} & 0 & 0 \\ 0 & \mathsf{B} & 0 \\ 0 & -\mathsf{BC} & 1/(\mathsf{AB}) \end{bmatrix} \\ &= \begin{bmatrix} \mathsf{A}/r & 0 & 0 \\ 0 & \mathsf{B} & 0 \\ 0 & -\mathsf{BC} & 1/(\mathsf{AB}) \end{bmatrix},\end{aligned} \tag{7.321}$$

and subsequently

$$\begin{aligned}\left[B^{km}\right] = \left[F^k_I\right]\left[F^m_J\right]\left[g^{IJ}\right], &= \left[F^k_I\right]\left[F^m_I\right]^T \\ &= \begin{bmatrix} \mathsf{A}^2/r^2 & 0 & 0 \\ 0 & \mathsf{B}^2 & -\mathsf{B}^2\mathsf{C} \\ 0 & -\mathsf{B}^2\mathsf{C} & \mathsf{B}^2\mathsf{C}^2 + 1/(\mathsf{A}^2\mathsf{B}^2) \end{bmatrix},\end{aligned} \tag{7.322}$$

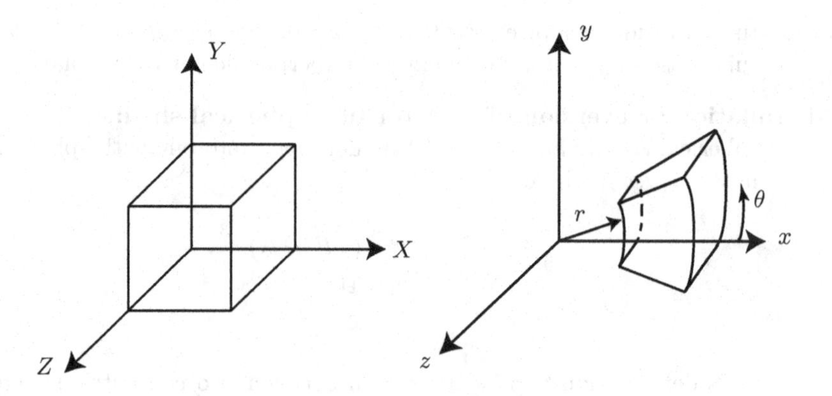

Figure 7.4: Bending and stretching of a rectangular block.

where we have made use of the fact that in the undeformed configuration, $[g^{IJ}] = I$. Now

$$[B_l^k] = [B^{km}][g_{ml}] = \begin{bmatrix} \mathsf{A}^2/r^2 & 0 & 0 \\ 0 & \mathsf{B}^2 r^2 & -\mathsf{B}^2\mathsf{C} \\ 0 & -\mathsf{B}^2\mathsf{C}r^2 & \mathsf{B}^2\mathsf{C}^2 + 1/(\mathsf{A}^2\mathsf{B}^2) \end{bmatrix} \tag{7.323}$$

and

$$[(B^{-1})_l^k] = \begin{bmatrix} r^2/\mathsf{A}^2 & 0 & 0 \\ 0 & (1+\mathsf{A}^2\mathsf{B}^4\mathsf{C}^2)/(\mathsf{B}^2 r^2) & (\mathsf{A}^2\mathsf{B}^2\mathsf{C})/r^2 \\ 0 & \mathsf{A}^2\mathsf{B}^2\mathsf{C} & \mathsf{A}^2\mathsf{B}^2 \end{bmatrix}, \tag{7.324}$$

where we note that $[g_{ml}] = \text{diag}\,(1, r^2, 1)$ and $[g^{ml}] = \text{diag}\,(1, r^{-2}, 1)$. In addition, we have the invariants

$$B_{(1)} = B_k^k = \frac{\mathsf{A}^2}{r^2} + \mathsf{B}^2 r^2 + \mathsf{B}^2\mathsf{C}^2 + \frac{1}{\mathsf{A}^2\mathsf{B}^2}, \tag{7.325}$$

$$B_{(2)} = \frac{1}{2}\left[B_{(1)}^2 - B_m^k B_k^m\right] = \frac{r^2}{\mathsf{A}^2} + \frac{1+\mathsf{A}^2\mathsf{B}^4\mathsf{C}^2}{r^2\mathsf{B}^2} + \mathsf{A}^2\mathsf{B}^2, \tag{7.326}$$

$$B_{(3)} = \det\left[B_l^k\right] = 1, \tag{7.327}$$

so that $\overline{\beta}_1 = \overline{\beta}_1(B_{(1)}, B_{(2)}, B_{(3)}) = \overline{\beta}_1(r)$ and $\overline{\beta}_{-1} = \overline{\beta}_{-1}(B_{(1)}, B_{(2)}, B_{(3)}) = \overline{\beta}_{-1}(r)$.

We now have

$$\sigma_l^k = -\overline{p}\delta_l^k + \overline{\beta}_1 B_l^k + \overline{\beta}_{-1}\left(B^{-1}\right)_l^k = -\overline{p}\delta_l^k + \widehat{\sigma}_l^k(r), \tag{7.328}$$

where

$$[\widehat{\sigma}_l^k](r) = \overline{\beta}_1 \begin{bmatrix} \mathsf{A}^2/r^2 & 0 & 0 \\ 0 & \mathsf{B}^2 r^2 & -\mathsf{B}^2\mathsf{C} \\ 0 & -\mathsf{B}^2\mathsf{C}r^2 & \mathsf{B}^2\mathsf{C}^2 + 1/(\mathsf{A}^2\mathsf{B}^2) \end{bmatrix} + \overline{\beta}_{-1} \begin{bmatrix} r^2/\mathsf{A}^2 & 0 & 0 \\ 0 & (1+\mathsf{A}^2\mathsf{B}^4\mathsf{C}^2)/(\mathsf{B}^2 r^2) & (\mathsf{A}^2\mathsf{B}^2\mathsf{C})/r^2 \\ 0 & \mathsf{A}^2\mathsf{B}^2\mathsf{C} & \mathsf{A}^2\mathsf{B}^2 \end{bmatrix}, \tag{7.329}$$

or in terms of physical components, using (7.304) and (7.305), we have

$$\sigma_{<kl>} = -\overline{p}\delta_{<kl>} + \widehat{\sigma}_{<kl>}(r), \tag{7.330}$$

where

$$\begin{aligned} [\widehat{\sigma}_{<kl>}](r) &= \overline{\beta}_1 \begin{bmatrix} \mathsf{A}^2/r^2 & 0 & 0 \\ 0 & \mathsf{B}^2 r^2 & -\mathsf{B}^2\mathsf{C}r \\ 0 & -\mathsf{B}^2\mathsf{C}r & \mathsf{B}^2\mathsf{C}^2 + 1/(\mathsf{A}^2\mathsf{B}^2) \end{bmatrix} + \\ & \overline{\beta}_{-1} \begin{bmatrix} r^2/\mathsf{A}^2 & 0 & 0 \\ 0 & (1+\mathsf{A}^2\mathsf{B}^4\mathsf{C}^2)/(\mathsf{B}^2 r^2) & (\mathsf{A}^2\mathsf{B}^2\mathsf{C})/r \\ 0 & (\mathsf{A}^2\mathsf{B}^2\mathsf{C})/r & \mathsf{A}^2\mathsf{B}^2 \end{bmatrix}. \end{aligned} \tag{7.331}$$

In this case, from (7.303), the linear momentum balance becomes

$$\frac{1}{r}\frac{\partial}{\partial r}(r\sigma_{<rr>}) - \frac{1}{r}\sigma_{<\theta\theta>} = 0, \tag{7.332}$$

$$\frac{1}{r^2}\frac{\partial}{\partial r}(r^2\cancelto{0}{\sigma_{<r\theta>}}) - \frac{1}{r}\frac{\partial p}{\partial \theta} = 0, \tag{7.333}$$

$$\frac{1}{r}\frac{\partial}{\partial r}(r\cancelto{0}{\sigma_{<rz>}}) - \frac{\partial p}{\partial z} = 0, \tag{7.334}$$

where we have used (2.261), (2.266), and (2.267). It is now evident from (7.330), (7.331), (7.333), and (7.334) that $\overline{p} = \overline{p}(r)$ and subsequently, from (7.330), $\sigma_{<kl>} = \sigma_{<kl>}(r)$. Note that, from (7.330), we can rewrite (7.332) in the form

$$\frac{d\sigma_{<rr>}}{dr} + \frac{1}{r}(\widehat{\sigma}_{<rr>} - \widehat{\sigma}_{<\theta\theta>}) = 0 \tag{7.335}$$

$$\tag{7.336}$$

so that, for $r_1 \neq 0$,

$$\begin{aligned} \sigma_{<rr>}(r) &= \sigma_{<rr>}(r_1) - \int_{r_1}^{r} \frac{1}{\xi}[\widehat{\sigma}_{<rr>}(\xi) - \widehat{\sigma}_{<\theta\theta>}(\xi)]\,d\xi & (7.337) \\ &= \sigma_{<rr>}(r_1) - \int_{r_1}^{r}\left[\left(\frac{\mathsf{A}^2}{\xi^3} - \mathsf{B}^2\xi\right)\overline{\beta}_1(\xi) - \left(\frac{1+\mathsf{A}^2\mathsf{B}^4\mathsf{C}^2}{\mathsf{B}^2\xi^3} - \frac{\xi}{\mathsf{A}^2}\right)\overline{\beta}_{-1}(\xi)\right]d\xi & (7.338) \\ &= \sigma_{<rr>}(r_1) + \frac{1}{2}\int_{r_1}^{r}\left[\frac{dB_{(1)}(\xi)}{d\xi}\overline{\beta}_1(\xi) - \frac{dB_{(2)}(\xi)}{d\xi}\overline{\beta}_{-1}(\xi)\right]d\xi & (7.339) \end{aligned}$$

and, from (7.330),

$$\begin{aligned} \overline{p} &= \widehat{\sigma}_{<rr>}(r) + \int_{r_1}^{r}\left[\left(\frac{\mathsf{A}^2}{\xi^3} - \mathsf{B}^2\xi\right)\overline{\beta}_1(\xi) - \left(\frac{1+\mathsf{A}^2\mathsf{B}^4\mathsf{C}^2}{\mathsf{B}^2\xi^3} - \frac{\xi}{\mathsf{A}^2}\right)\overline{\beta}_{-1}(\xi)\right]d\xi - \\ &\quad \sigma_{<rr>}(r_1). \end{aligned} \tag{7.340}$$

Subsequently, from (7.332) and (7.338), we have

$$\begin{aligned} \sigma_{<\theta\theta>} &= \frac{d}{dr}(r\sigma_{<rr>}) \\ &= \sigma_{<rr>} - \left(\frac{\mathsf{A}^2}{r^2} - \mathsf{B}^2 r^2\right)\overline{\beta}_1(r) + \left(\frac{1+\mathsf{A}^2\mathsf{B}^4\mathsf{C}^2}{\mathsf{B}^2 r^2} - \frac{r^2}{\mathsf{A}^2}\right)\overline{\beta}_{-1}(r), \end{aligned} \tag{7.341}$$

while from (7.330) and (7.331), we have

$$\begin{aligned}\sigma_{<zz>} &= -\overline{p} + \widehat{\sigma}_{<zz>} \\ &= \sigma_{<rr>} + (\widehat{\sigma}_{<zz>} - \widehat{\sigma}_{<rr>}) \\ &= \sigma_{<rr>} + \left(\mathsf{B}^2\mathsf{C}^2 + \frac{1}{\mathsf{A}^2\mathsf{B}^2} - \frac{\mathsf{A}^2}{r^2}\right)\overline{\beta}_1(r) - \left(\frac{r^2}{\mathsf{A}^2} - \mathsf{A}^2\mathsf{B}^2\right)\overline{\beta}_{-1}(r), \quad (7.342) \\ \sigma_{<\theta z>} &= \sigma_{<z\theta>} = \widehat{\sigma}_{<z\theta>} \\ &= -\mathsf{B}^2\mathsf{C}r\overline{\beta}_1(r) + \frac{\mathsf{A}^2\mathsf{B}^2\mathsf{C}}{r}\overline{\beta}_{-1}(r). \quad (7.343)\end{aligned}$$

Now, since $t_{<k>} = \sigma_{<kl>}n_{<l>}$, by the choice of $\sigma_{<rr>}(r_1) = 0$, the surface $r = r_1$ may be rendered free of traction. Subsequently, the resultant normal force R per unit height on the part of the axial plane θ = const. cut off by the cylinders $r = r_1$ and $r = r_2 > r_1$ is given by

$$\mathsf{R} = \int_{r_1}^{r_2} \sigma_{<\theta\theta>}\, dr = r_2\, \sigma_{<rr>}(r_2), \tag{7.344}$$

where we have used (7.332). Thus, the faces θ = const. are free of resultant normal traction if and only if the surface $r = r_2$ is free of normal traction. In order that this surface also be free of traction, so that the block may be bent and sheared by forces applied to its plane and helicoidal faces only, we see from (7.338) that it is necessary that

$$\int_{r_1}^{r_2} \left[\left(\frac{\mathsf{A}^2}{r^3} - \mathsf{B}^2 r\right)\overline{\beta}_1(r) - \left(\frac{1 + \mathsf{A}^2\mathsf{B}^4\mathsf{C}^2}{\mathsf{B}^2 r^3} - \frac{r}{\mathsf{A}^2}\right)\overline{\beta}_{-1}(r)\right] dr = 0. \tag{7.345}$$

For given r_1 and r_2, this is a relation among the constants A, B, and C. Whether or not this equation has any real roots depends on the nature of $\overline{\beta}_1$ and $\overline{\beta}_{-1}$.

It is now clear that the helicoidal faces cannot be free of traction. We readily calculate the normal and tangential tractions N and T that have to be applied on these faces to maintain the deformation:

$$\begin{aligned}\mathsf{N} &= \frac{1}{1 + \mathsf{C}^2/r^2}\left[\sigma_{<zz>} + 2\frac{\mathsf{C}}{r}\sigma_{<\theta z>} + \frac{\mathsf{C}^2}{r^2}\sigma_{<\theta\theta>}\right], \quad (7.346) \\ \mathsf{T} &= \frac{1}{1 + \mathsf{C}^2/r^2}\left[\left(1 - \frac{\mathsf{C}^2}{r^2}\right)\sigma_{<\theta z>} + \frac{\mathsf{C}}{r}\left(\sigma_{<\theta\theta>} - \sigma_{<zz>}\right)\right], \quad (7.347)\end{aligned}$$

where the values of $\sigma_{<kl>}$ are to be taken from (7.341)–(7.343). In the case of pure bending, $\mathsf{C} = 0$, and then $\mathsf{N} = \sigma_{<zz>}$ and $\mathsf{T} = 0$, but in general both normal and tangential tractions must be applied. The presence of these tractions is the Poynting effect for bending.

The tractions that must be applied upon the plane ends $\theta = \pm\theta_0$ may be read off from (7.341) to (7.343), and the resultant normal force on them is given by (7.344). The resultant moment M per unit height, with respect to a point on the axis $r = 0$, exerted by the normal stress acting upon these faces, is given by

$$\begin{aligned}
\mathsf{M} &= \int_{r_1}^{r_2} r\sigma_{<\theta\theta>}(r)\,dr \\
&= \int_{r_1}^{r_2} r\frac{d}{dr}\left[r\sigma_{<rr>}(r)\right]dr \\
&= r^2\sigma_{<rr>}(r)\Big|_{r_1}^{r_2} - \int_{r_1}^{r_2} r\sigma_{<rr>}(r)\,dr \\
&= \frac{1}{2}\,r^2\sigma_{<rr>}(r)\Big|_{r_1}^{r_2} + \frac{1}{2}\int_{r_1}^{r_2} r^2\frac{d\sigma_{<rr>}(r)}{dr}\,dr \\
&= \frac{1}{2}\,r^2\sigma_{<rr>}(r)\Big|_{r_1}^{r_2} - \frac{1}{2}\int_{r_1}^{r_2} r\left[\hat{\sigma}_{<rr>}(r) - \hat{\sigma}_{<\theta\theta>}(r)\right]dr \\
&= \frac{1}{2}\,r^2\sigma_{<rr>}(r)\Big|_{r_1}^{r_2} - \frac{1}{2}\int_{r_1}^{r_2}\Bigg[\left(\frac{\mathsf{A}^2}{r} - \mathsf{B}^2r^3\right)\overline{\beta}_1(r) - \\
&\qquad\left(\frac{1+\mathsf{A}^2\mathsf{B}^4\mathsf{C}^2}{\mathsf{B}^2r} - \frac{r^3}{\mathsf{A}^2}\right)\overline{\beta}_{-1}(r)\Bigg]dr,
\end{aligned} \tag{7.348}$$

where we have used (7.341) in the last step.

All the results are referred to the deformed body but can easily be expressed in terms of the undeformed body by use of (7.306)–(7.308).

The resultant normal traction N acting upon the plane annular wedge $r_1 \le r \le r_2$, z = const., and $-\theta_0 \le \theta \le \theta_0$ may be calculated as follows. We would like to solve a problem with boundary conditions $t_{<k>} = 0$ on $r = r_1, r_2$. Now, since $t_{<k>} = \sigma_{<kl>}n_{<l>}$, this requires that $\sigma_{<rr>} = 0$ on $r = r_1, r_2$. These conditions determine $\sigma_{<rr>}(r_1)$ and provide a relationship between A, B, and C. In the z-direction, we have $t_{<k>} = \sigma_{<kz>}n_{<z>}$, so that the conditions $t_{<r>} = t_{<\theta>} = 0$ and $t_{<z>} = \mathsf{N}$ become $\sigma_{<rz>} = \sigma_{<\theta z>} = 0$ and $\sigma_{<zz>}(r) = \mathsf{N}$. Lastly, at $\theta = \theta_0$ we require that $t_{<r>} = t_{<z>} = 0$ and $t_{<\theta>} = \mathsf{L}$; since $t_{<k>} = \sigma_{<k\theta>}n_{<\theta>}$, we have $\sigma_{<r\theta>} = \sigma_{<z\theta>} = 0$ and $\sigma_{<\theta\theta>}(r) = \mathsf{L}$. But the resultant normal force is then

$$\mathsf{R} = \int_{r_1}^{r_2} \sigma_{<\theta\theta>}dr = r\sigma_{<rr>}\Big|_{r_1}^{r_2} = 0, \tag{7.349}$$

since

$$\frac{1}{r}\frac{\partial}{\partial r}(r\sigma_{<rr>}) - \frac{\sigma_{<\theta\theta>}}{r} = 0, \tag{7.350}$$

and we have used the boundary conditions. Thus, in the case of pure bending ($\mathsf{C} = 0$), we subsequently have from (7.348) that

$$\mathsf{M}_0 = -\frac{1}{2}\int_{r_1}^{r_2}\left[\left(\frac{\mathsf{A}^2}{r^2} - \mathsf{B}^2r^3\right)\overline{\beta}_1(r) + \left(\frac{r^3}{\mathsf{A}^2} - \frac{1}{r\mathsf{B}^2}\right)\overline{\beta}_{-1}(r)\right]dr, \tag{7.351}$$

where we have integrated by parts in the second step.

Table 7.1: Unit cell axes and angles associated with crystal systems comprising the 14 Bravais lattices.

Crystal System	Unit Cell Axes	Unit Cell Angles	Lattice Variations
Triclinic	$a \neq b \neq c$	$\alpha \neq \beta \neq \gamma$	P
Monoclinic	$a \neq b \neq c$	$\alpha = \gamma = 90° \neq \beta$	P, C
Orthorhombic	$a \neq b \neq c$	$\alpha = \beta = \gamma = 90°$	P, F, I, C
Tetragonal	$a = b \neq c$	$\alpha = \beta = \gamma = 90°$	P, I
Trigonal	$a = b \neq c$	$\alpha = \beta = \gamma \neq 90°$	P
Hexagonal	$a = b \neq c$	$\alpha = \beta = 90°$, $\gamma = 120°$	P
Cubic	$a = b = c$	$\alpha = \beta = \gamma = 90°$	P, F, I

Table 7.2: Generating transformations.

$$I = \begin{bmatrix} 1 & 0 & 0 \\ 0 & 1 & 0 \\ 0 & 0 & 1 \end{bmatrix}, \qquad C = \begin{bmatrix} -1 & 0 & 0 \\ 0 & -1 & 0 \\ 0 & 0 & -1 \end{bmatrix},$$

$$R_1 = \begin{bmatrix} -1 & 0 & 0 \\ 0 & 1 & 0 \\ 0 & 0 & 1 \end{bmatrix}, \qquad R_2 = \begin{bmatrix} 1 & 0 & 0 \\ 0 & -1 & 0 \\ 0 & 0 & 1 \end{bmatrix}, \qquad R_3 = \begin{bmatrix} 1 & 0 & 0 \\ 0 & 1 & 0 \\ 0 & 0 & -1 \end{bmatrix},$$

$$D_1 = \begin{bmatrix} 1 & 0 & 0 \\ 0 & -1 & 0 \\ 0 & 0 & -1 \end{bmatrix}, \qquad D_2 = \begin{bmatrix} -1 & 0 & 0 \\ 0 & 1 & 0 \\ 0 & 0 & -1 \end{bmatrix}, \qquad D_3 = \begin{bmatrix} -1 & 0 & 0 \\ 0 & -1 & 0 \\ 0 & 0 & 1 \end{bmatrix},$$

$$T_1 = \begin{bmatrix} 1 & 0 & 0 \\ 0 & 0 & 1 \\ 0 & 1 & 0 \end{bmatrix}, \qquad T_2 = \begin{bmatrix} 0 & 0 & 1 \\ 0 & 1 & 0 \\ 1 & 0 & 0 \end{bmatrix}, \qquad T_3 = \begin{bmatrix} 0 & 1 & 0 \\ 1 & 0 & 0 \\ 0 & 0 & 1 \end{bmatrix},$$

$$M_1 = \begin{bmatrix} 0 & 1 & 0 \\ 0 & 0 & 1 \\ 1 & 0 & 0 \end{bmatrix}, \qquad M_2 = \begin{bmatrix} 0 & 0 & 1 \\ 1 & 0 & 0 \\ 0 & 1 & 0 \end{bmatrix},$$

$$S_1 = \begin{bmatrix} -1/2 & \sqrt{3}/2 & 0 \\ -\sqrt{3}/2 & -1/2 & 0 \\ 0 & 0 & 1 \end{bmatrix}, \quad S_2 = \begin{bmatrix} -1/2 & -\sqrt{3}/2 & 0 \\ \sqrt{3}/2 & -1/2 & 0 \\ 0 & 0 & 1 \end{bmatrix}.$$

Table 7.3: Transformations that characterize the 32 crystal classes.

Crystal System	Type	Class	Class #	Symmetry Transformations
Triclinic:	1	Pedial	1	I
		Pinacoidal	2	I, C
Monoclinic:	2	Sphenoidal	3	I, D_1
		Domatic	4	I, R_1
		Prismatic	5	I, C, R_1, D_1
Orthorhombic:	3	Rhombic-disphenoidal	6	I, D_1, D_2, D_3
		Rhombic-pyramidal	7	I, R_2, R_3, D_1
		Rhombic-dipyramidal	8	$I, C, R_1, R_2, R_3, D_1, D_2, D_3$
Tetragonal:	4	Tetragonal-pyramidal	9	$I, D_3, (R_1, R_2)\,T_3$
		Tetragonal-disphenoidal	10	$I, D_3, (D_1, D_2)\,T_3$
		Tetragonal-dipyramidal	11	$I, C, R_3, D_3, (R_1, R_2, D_1, D_2$
	5	Tetragonal-trapezohedral	12	$I, D_1, D_2, D_3, (C, R_1, R_2, R_3$
		Ditetragonal-pyramidal	13	$(I, R_1, R_2, D_3)\,(I, T_3)$
		Tetragonal-scalenohedral	14	$(I, D_1, D_2, D_3)\,(I, T_3)$
		Ditetragonal-dipyramidal	15	$(I, C, R_1, R_2, R_3, D_1, D_2, D_3$ (I, T_3)
Trigonal:	6	Trigonal-pyramidal	16	I, S_1, S_2
		Rhombohedral	17	$(I, C)\,(I, S_1, S_2)$
	7	Trigonal-trapezohedral	18	$(I, D_1)\,(I, S_1, S_2)$
		Ditrigonal-pyramidal	19	$(I, R_1)\,(I, S_1, S_2)$
		Trigonal-dipyramidal	20	$(I, C, R_1, D_1)\,(I, S_1, S_2)$
Hexagonal:	8	Hexagonal-pyramidal	21	$(I, D_3)\,(I, S_1, S_2)$
		Trigonal-scalenohedral	22	$(I, R_3)\,(I, S_1, S_2)$
		Hexagonal-dipyramidal	23	$(I, C, R_3, D_3)\,(I, S_1, S_2)$
	9	Hexagonal-trapezohedral	24	$(I, D_1, D_2, D_3)\,(I, S_1, S_2)$
		Dihexagonal-pyramidal	25	$(I, R_1, R_2, D_3)\,(I, S_1, S_2)$
		Ditragonal-dipyramidal	26	$(I, R_1, R_3, D_3)\,(I, S_1, S_2)$
		Dihexagonal-dipyramidal	27	$(I, C, R_1, R_2, R_3, D_1, D_2, D_3$ (I, S_1, S_2)
Cubic:	10	Tetartoidal	28	$(I, D_1, D_2, D_3)\,(I, M_1, M_2)$
		Diploidal	29	$(I, C, R_1, R_2, R_3, D_1, D_2, D_3$ (I, M_1, M_2)
	11	Gyroidal	30	$(I, D_1, D_2, D_3)\,(I, M_1, M_2),$ $(C, R_1, R_2, R_3)\,(T_1, T_2, T_3)$
		Hextetrahedral	31	$(I, D_1, D_2, D_3)\cdot$ $(I, T_1, T_2, T_3, M_1, M_2)$
		Hexoctahedral	32	$(I, C, R_1, R_2, R_3, D_1, D_2, D_3)$ $(I, T_1, T_2, T_3, M_1, M_2)$

Table 7.4: Crystal symmetry on the component matrix A of a symmetric property tensor relating two vector fields or a scalar field to a rank-2 symmetric tensor field.

Triclinic (6 parameters)

$$\begin{bmatrix} a_{11} & a_{12} & a_{13} \\ & a_{22} & a_{23} \\ \text{Sym} & & a_{33} \end{bmatrix}$$

Monoclinic (4 parameters)

$$\begin{bmatrix} a_{11} & 0 & a_{13} \\ & a_{22} & 0 \\ \text{Sym} & & a_{33} \end{bmatrix}$$

Orthorhombic (3 parameters)

$$\begin{bmatrix} a_{11} & 0 & 0 \\ & a_{22} & 0 \\ \text{Sym} & & a_{33} \end{bmatrix}$$

Tetragonal, Trigonal, Hexagonal, and Transverse isotropic (2 parameters)

$$\begin{bmatrix} a_{11} & 0 & 0 \\ & a_{11} & 0 \\ \text{Sym} & & a_{33} \end{bmatrix}$$

Cubic and Isotropic (1 parameter)

$$\begin{bmatrix} a_{11} & 0 & 0 \\ & a_{11} & 0 \\ \text{Sym} & & a_{11} \end{bmatrix}$$

Table 7.5: Crystal symmetry on the symmetric component matrix C of a rank-4 property tensor relating two symmetric rank-2 tensor fields using Voigt's representation.

Triclinic (21 parameters)

$$\begin{bmatrix} c_{11} & c_{12} & c_{13} & c_{14} & c_{15} & c_{16} \\ & c_{22} & c_{23} & c_{24} & c_{25} & c_{26} \\ & & c_{33} & c_{34} & c_{35} & c_{36} \\ & & & c_{44} & c_{45} & c_{46} \\ & \text{Sym} & & & c_{55} & c_{56} \\ & & & & & c_{66} \end{bmatrix}$$

Monoclinic (13 parameters)

$$\begin{bmatrix} c_{11} & c_{12} & c_{13} & 0 & c_{15} & 0 \\ & c_{22} & c_{23} & 0 & c_{25} & 0 \\ & & c_{33} & 0 & c_{35} & 0 \\ & & & c_{44} & 0 & c_{46} \\ & \text{Sym} & & & c_{55} & 0 \\ & & & & & c_{66} \end{bmatrix}$$

Orthorhombic (9 parameters)

$$\begin{bmatrix} c_{11} & c_{12} & c_{13} & 0 & 0 & 0 \\ & c_{22} & c_{23} & 0 & 0 & 0 \\ & & c_{33} & 0 & 0 & 0 \\ & & & c_{44} & 0 & 0 \\ & \text{Sym} & & & c_{55} & 0 \\ & & & & & c_{66} \end{bmatrix}$$

Tetragonal (7 or 6 parameters)

Classes 9–11, Classes 12–15

$$\begin{bmatrix} c_{11} & c_{12} & c_{13} & 0 & 0 & c_{16} \\ & c_{11} & c_{13} & 0 & 0 & -c_{16} \\ & & c_{33} & 0 & 0 & 0 \\ & & & c_{44} & 0 & 0 \\ & \text{Sym} & & & c_{44} & 0 \\ & & & & & c_{66} \end{bmatrix}, \begin{bmatrix} c_{11} & c_{12} & c_{13} & 0 & 0 & 0 \\ & c_{11} & c_{13} & 0 & 0 & 0 \\ & & c_{33} & 0 & 0 & 0 \\ & & & c_{44} & 0 & 0 \\ & \text{Sym} & & & c_{44} & 0 \\ & & & & & c_{66} \end{bmatrix}$$

Table 7.5: (continued) Crystal symmetry on the component matrix C of a rank-4 property tensor relating two symmetric rank-2 tensor fields using Voigt's representation. In the matrices below $c_{66} = \frac{1}{2}(c_{11} - c_{12})$.

Trigonal (7 or 6 parameters)

Classes 16–17, Classes 18–20

$$\begin{bmatrix} c_{11} & c_{12} & c_{13} & c_{14} & c_{15} & 0 \\ & c_{11} & c_{13} & -c_{14} & -c_{15} & 0 \\ & & c_{33} & 0 & 0 & 0 \\ & & & c_{44} & 0 & -c_{15} \\ & \text{Sym} & & & c_{44} & c_{14} \\ & & & & & c_{66} \end{bmatrix}, \begin{bmatrix} c_{11} & c_{12} & c_{13} & c_{14} & 0 & 0 \\ & c_{11} & c_{13} & -c_{14} & 0 & 0 \\ & & c_{33} & 0 & 0 & 0 \\ & & & c_{44} & 0 & 0 \\ & \text{Sym} & & & c_{44} & c_{14} \\ & & & & & c_{66} \end{bmatrix}$$

Hexagonal and Transverse isotropic (5 parameters)

$$\begin{bmatrix} c_{11} & c_{12} & c_{13} & 0 & 0 & 0 \\ & c_{11} & c_{13} & 0 & 0 & 0 \\ & & c_{33} & 0 & 0 & 0 \\ & & & c_{44} & 0 & 0 \\ & \text{Sym} & & & c_{44} & 0 \\ & & & & & c_{66} \end{bmatrix}$$

Cubic (3 parameters)

$$\begin{bmatrix} c_{11} & c_{12} & c_{12} & 0 & 0 & 0 \\ & c_{11} & c_{12} & 0 & 0 & 0 \\ & & c_{11} & 0 & 0 & 0 \\ & & & c_{44} & 0 & 0 \\ & \text{Sym} & & & c_{44} & 0 \\ & & & & & c_{44} \end{bmatrix}$$

Isotropic (2 parameters)

$$\begin{bmatrix} c_{11} & c_{12} & c_{12} & 0 & 0 & 0 \\ & c_{11} & c_{12} & 0 & 0 & 0 \\ & & c_{11} & 0 & 0 & 0 \\ & & & c_{66} & 0 & 0 \\ & \text{Sym} & & & c_{66} & 0 \\ & & & & & c_{66} \end{bmatrix}$$

Table 7.6: Complete irreducible polynomial basis of scalar invariants of the elastic stiffness tensor for different crystal systems.

Crystal System	Type	Invariants
Triclinic	1	$c_{11},\ c_{22},\ c_{33},\ c_{12},\ c_{13},\ c_{23}$
Monoclinic	2	$c_{11},\ c_{22},\ c_{33},\ c_{12}^2,\ c_{13}^2,\ c_{23},\ c_{13}c_{12}$
Orthorhombic	3	$c_{11},\ c_{22},\ c_{33},\ c_{23}^2,\ c_{13}^2,\ c_{12}^2,\ c_{12}c_{23}c_{13}$
Tetragonal	4	$c_{11}+c_{22},\ c_{33},\ c_{13}^2+c_{23}^2,\ c_{12}^2,\ c_{11}c_{22},\ c_{12}(c_{11}-c_{22}),$ $c_{13}c_{23}(c_{11}-c_{22}),\ c_{12}c_{23}c_{13},\ c_{12}(c_{13}^2-c_{23}^2),$ $c_{11}c_{23}^2+c_{22}c_{13}^2,\ c_{23}c_{13}(c_{13}^2-c_{23}^2),\ c_{13}^2c_{23}^2$
	5	$c_{11}+c_{22},\ c_{33},\ c_{12}^2,\ c_{13}^2+c_{23}^2,\ c_{11}c_{22},\ c_{12}c_{23}c_{13},$ $c_{11}c_{23}^2+c_{22}c_{13}^2,\ c_{13}^2c_{23}^2$
Trigonal	6	$c_{33},\ c_{11}+c_{22},\ c_{11}c_{22}-c_{12}^2,\ c_{11}[(c_{11}+3c_{22})^2-12c_{12}^2],$ $c_{13}^2+c_{23}^2,\ c_{13}(c_{13}^2-3c_{23}^2),\ (c_{11}-c_{22})c_{13}-2c_{12}c_{23},$ $(c_{22}-c_{11})c_{23}-2c_{12}c_{13},\ 3c_{12}(c_{11}-c_{22})^2-4c_{12}^3,$ $c_{23}(c_{23}^2-3c_{13}^2),\ c_{22}c_{13}^2+c_{11}c_{23}^2-2c_{23}c_{13}c_{12},$ $c_{13}[(c_{11}+c_{22})^2+4(c_{12}^2-c_{22}^2)]-8c_{11}c_{12}c_{23},$ $c_{23}[(c_{11}+c_{22})^2+4(c_{12}^2-c_{22}^2)]+8c_{11}c_{12}c_{13},$ $(c_{11}-c_{22})c_{23}c_{13}+c_{12}(c_{23}^2-c_{13}^2)$
	7	$c_{33},\ c_{11}+c_{22},\ c_{11}c_{22}-c_{12}^2,\ c_{11}[(c_{11}+3c_{22})^2-12c_{12}^2],$ $c_{13}^2+c_{23}^2,\ c_{23}(c_{23}^2-3c_{13}^2),\ (c_{11}-c_{22})c_{23}+2c_{12}c_{13},$ $c_{11}c_{13}^2+c_{22}c_{23}^2+2c_{23}c_{13}c_{12},$ $c_{23}[(c_{11}+c_{22})^2+4(c_{12}^2-c_{22}^2)]+8c_{11}c_{12}c_{13}$
Hexagonal	8	$c_{33},\ c_{11}+c_{22},\ c_{11}c_{22}-c_{12}^2,\ c_{11}[(c_{11}+3c_{22})^2-12c_{12}^2],$ $c_{13}^2+c_{23}^2,\ c_{13}^2(c_{13}^2-3c_{23}^2)^2,\ c_{11}c_{23}^2+c_{22}c_{13}^2-2c_{23}c_{13}c_{12},$ $c_{12}(c_{13}^2-c_{23}^2)+(c_{22}-c_{11})c_{13}c_{23},$ $3c_{12}(c_{11}-c_{22})^2-4c_{12}^3,$ $c_{13}c_{23}[3(c_{13}^2-c_{23}^2)^2-4c_{13}^2c_{23}^2],$ $c_{11}(c_{13}^4+3c_{23}^4)+2c_{22}c_{13}^2(c_{13}^2+3c_{23}^2)-8c_{12}c_{23}c_{13}^3,$ $c_{13}^2[(c_{11}+c_{22})^2+4(c_{12}^2-c_{22}^2)]-$ $2c_{11}[(c_{11}+3c_{22})(c_{13}^2+c_{23}^2)-4c_{23}c_{13}c_{12}],$ $c_{23}c_{13}[(c_{11}+c_{22})^2+4(c_{12}^2-c_{22}^2)]+4c_{11}c_{12}(c_{23}^2-c_{13}^2),$ $c_{12}[(c_{13}^2+c_{23}^2)^2+4c_{23}^2(c_{13}^2-c_{23})]-4c_{13}^3c_{23}(c_{11}-c_{22})$
	9	$c_{33},\ c_{11}+c_{22},\ c_{11}c_{22}-c_{12}^2,\ c_{11}[(c_{11}+3c_{22})^2-12c_{12}^2],$ $c_{13}^2+c_{23}^2,\ c_{13}^2(c_{13}^2-3c_{23}^2)^2,$ $c_{11}c_{23}^2+c_{22}c_{13}^2-2c_{23}c_{13}c_{12},$ $c_{11}(c_{13}^4+3c_{23}^4)+2c_{22}c_{13}^2(c_{13}^2+3c_{23}^2)-8c_{12}c_{23}c_{13}^3,$ $c_{13}^2[(c_{11}+3c_{22})^2+4(c_{12}^2-c_{22}^2)]-$ $2c_{11}[(c_{11}+3c_{22})(c_{13}^2+c_{23}^2)-4c_{23}c_{13}c_{12}]$

Table 7.6: (continued) Complete irreducible polynomial basis of scalar invariants of the elastic stiffness tensor for different crystal systems.

Crystal System	Type	Invariants
Cubic	10	$c_{11}+c_{22}+c_{33}$, $c_{22}c_{33}+c_{33}c_{11}+c_{11}c_{22}$, $c_{11}c_{22}c_{33}$, $c_{23}^2+c_{13}^2+c_{12}^2$, $c_{13}^2c_{12}^2+c_{12}^2c_{23}^2+c_{23}^2c_{13}^2$, $c_{23}c_{13}c_{12}$, $c_{22}c_{12}^2+c_{33}c_{23}^2+c_{11}c_{13}^2$, $c_{13}^2c_{33}+c_{12}^2c_{11}+c_{23}^2c_{22}$, $c_{33}c_{22}^2+c_{11}c_{33}^2+c_{22}c_{11}^2$, $c_{12}^2c_{13}^4+c_{23}^2c_{12}^4+c_{13}^2c_{23}^4$, $c_{11}c_{13}^2c_{12}^2+c_{22}c_{12}^2c_{23}^2+c_{33}c_{23}^2c_{13}^2$, $c_{23}^2c_{22}c_{33}+$ $c_{13}^2c_{33}c_{11}+c_{12}^2c_{11}c_{22}$, $c_{11}c_{22}c_{13}^2+c_{22}c_{33}c_{12}^2+$ $c_{33}c_{11}c_{23}^2$, $c_{23}^2c_{13}^2c_{22}+c_{13}^2c_{12}^2c_{33}+c_{12}^2c_{23}^2c_{11}$
	11	$c_{11}+c_{22}+c_{33}$, $c_{22}c_{33}+c_{33}c_{11}+c_{11}c_{22}$, $c_{11}c_{22}c_{33}$, $c_{23}^2+c_{13}^2+c_{12}^2$, $c_{13}^2c_{12}^2+c_{12}^2c_{23}^2+c_{23}^2c_{13}^2$, $c_{23}c_{13}c_{12}$, $c_{22}(c_{12}^2+c_{23}^2)+c_{33}(c_{13}^2+c_{23}^2)+c_{11}(c_{12}^2+c_{13}^2)$, $c_{11}c_{13}^2c_{12}^2+c_{22}c_{12}^2c_{23}^2+c_{33}c_{23}^2c_{13}^2$, $c_{23}^2c_{22}c_{33}+c_{13}^2c_{33}c_{11}+c_{12}^2c_{11}c_{22}$
Transverse isotropic		tr C, tr C^2, tr C^3, c_{33}, $c_{13}^2+c_{23}^2$
Isotropic		tr C, tr C^2, tr C^3

Table 7.7: Relations for the elastic properties in terms of different independent parameters.

	λ_0, μ_0	$(c_0)_{11}$, $(c_0)_{12}$	E_0, ν_0	E_0, μ_0
λ_0	λ_0	$(c_0)_{12}$	$\frac{E_0\,\nu_0}{(1+\nu_0)(1-2\,\nu_0)}$	$\frac{\mu_0\,(E_0-2\,\mu_0}{3\,\mu_0-E_0}$
μ_0	μ_0	$\frac{1}{2}[(c_0)_{11}-(c_0)_{12}]$	$\frac{E_0}{2\,(1+\nu_0)}$	μ_0
E_0	$\frac{\mu_0\,(3\,\lambda_0+2\,\mu_0)}{\lambda_0+\mu_0}$	$\frac{[(c_0)_{11}-(c_0)_{12}][(c_0)_{11}+2\,(c_0)_{12}]}{[(c_0)_{11}+(c_0)_{12}]}$	E_0	E_0
b_0	$\lambda_0+\frac{2}{3}\,\mu_0$	$\frac{1}{3}\,[(c_0)_{11}+2\,(c_0)_{12}]$	$\frac{E_0}{3\,(1-2\,\nu_0)}$	$\frac{E_0\,\mu_0}{3\,(3\,\mu_0-E_0)}$
ν_0	$\frac{\lambda_0}{2\,(\lambda_0+\mu_0)}$	$\frac{[(c_0)_{12}]}{[(c_0)_{11}+(c_0)_{12}]}$	ν_0	$\frac{E_0}{2\,\mu_0}-1$

Problems

1. Show that (7.40) and subsequently (7.43) are true.

2. First show that (7.50) is true and subsequently provide the details to obtain (7.205) and (7.206).

3. The Cartesian coordinates of a point, X_i, transforms upon a change of coordinate system according to

$$X'_i = a_{ij} X_j,$$

where a_{ij} is the direction cosine which relates the axes of the two coordinate systems. In the method of direct inspection, we have stated that if one can write down a relation between the axes of the two coordinate systems X_i and X_j (e.g., $X'_1 = -X_1$, $X'_2 = -X_2$), then this provides a convenient method for extracting the direction cosine associated with material tensors (e.g., $c_{11} = -1, c_{12} = 0, c_{13} = 0$, etc.).

   The above statement is rather cavalier, however, and not strictly correct: the first equation involves coordinates of a point, while the latter equations, written by inspection, involve a vector relationship between reference axes in the two coordinate systems.

   Demonstrate (by means of a sketch or otherwise) that the procedure is correct – i.e., that the coordinates of a point transform in exactly the same way as the axes of the references system.

4. Show that for transversely isotropic materials, the requirement of satisfying the symmetries $(M_\Theta, R_1, R_3, D_2)$, the rank-2 and rank-4 property tensors have components respectively given in Tables 7.4 and 7.5.

5. The trace of a rank-2 tensor is defined as the sum of its diagonal elements, $\operatorname{tr} \mathbf{C} = c_{11} + c_{22} + c_{33}$. Show that the trace of a tensor remains invariant upon a change of reference axes which is specified by a general transformation with direction cosines $[a_{ij}]$.

6. The engineering moduli of crystals may all be expressed in terms of the elements of the stiffness matrix constants c_{ij} (in Voigt's notation). Show, for example, that for a cubic crystal the isothermal bulk modulus is given by

$$b_0 = \frac{1}{\kappa_\theta} = \frac{1}{3}\left(c_{11} + 2\, c_{12}\right).$$

7. Repeat the analysis in Section 7.3.1 of the propagation of elastic waves in a cubic crystals, but without making the isothermal assumption.

8. Solve for $\widetilde{\mathbf{e}}$ in (7.183) to obtain the following result

$$\operatorname{tr} \widetilde{\mathbf{e}} = 3\, a_0\, \tilde{\theta} - \frac{\overline{p}}{b_0} = 3\, a_0\, \tilde{\theta} - \frac{3\,(1 - 2\,\nu_0)}{E_0}\, \overline{p},$$

and subsequently show that

$$\widetilde{\mathbf{e}} = \left(a_0\,\tilde{\theta} + \frac{\lambda_0}{2\,\mu_0\,b_0}\,\overline{p}\right)\mathbf{1} + \frac{1}{2\,\mu_0}\,\boldsymbol{\sigma} = \left(a_0\,\tilde{\theta} + \frac{3\,\nu_0}{E_0}\,\overline{p}\right)\mathbf{1} + \frac{1+\nu_0}{E_0}\,\boldsymbol{\sigma},$$

where we have defined the mechanical pressure

$$\overline{p} = -\frac{1}{3}\,\mathrm{tr}\,\boldsymbol{\sigma}.$$

9. Eliminate $\tilde{\theta}$ between (7.180) and (7.183) to show that

$$\boldsymbol{\sigma} = \left[\left(b_0' - \frac{2}{3}\,\mu_0\right)\mathrm{tr}\,\widetilde{\mathbf{e}} - \frac{3\,\theta_0\,b_0\,a_0}{c_{\mathbf{e}_0}}\,(\eta - \eta_0)\right]\mathbf{1} + 2\,\mu_0\,\widetilde{\mathbf{e}}, \tag{7.352}$$

where

$$b_0' = b_0 + \frac{9\,\theta_0\,b_0^2\,a_0^2}{\rho_0\,c_{\mathbf{e}_0}} \tag{7.353}$$

is the *isentropic bulk modulus.*

10. Solve (7.352) for $\widetilde{\mathbf{e}}$ to show that

$$\widetilde{\mathbf{e}} = \left(\frac{\theta_0\,a_0'}{c_{\mathbf{e}_0}}\,\tilde{\theta} - \frac{\nu_0'}{E_0'}\,\mathrm{tr}\,\boldsymbol{\sigma}\right)\mathbf{1} + \frac{1+\nu_0'}{E_0'}\,\boldsymbol{\sigma},$$

where, using (7.353),

$$\begin{aligned} a_0' &= \frac{b_0\,a_0}{b_0'}, \\ \nu_0' &= \frac{3\,b_0' - 2\,\mu_0}{2\,(3\,b_0' + \mu_0)}, \\ E_0' &= \frac{9\,b_0'\,\mu_0}{3\,b_0' + \mu_0}, \end{aligned}$$

are the *isentropic coefficient of thermal expansion, isentropic Poisson ratio,* and *isentropic Young modulus,* respectively.

11. Show that in an elastic isotropic medium, the three acoustic wave speeds consist of the longitudinal wave

$$v_w^{(1)} = \left(\frac{c_{11}}{\rho}\right)^{1/2} = \left(\frac{\lambda + 2\,\mu}{\rho}\right)^{1/2}$$

and the two shear waves

$$v_w^{(2,3)} = \left(\frac{c_{11} - c_{12}}{2\,\rho}\right)^{1/2} = \left(\frac{\mu}{\rho}\right)^{1/2}.$$

12. Use (7.215) to show (7.216) with $B_{13} = B_{23} = 0$.

13. Prove (7.270).

14. The representation for the stress tensor for a homogeneous isothermal isotropic elastic solid is given by

$$\boldsymbol{\sigma} = \beta_0\left(B_{(1)}, B_{(2)}, B_{(3)}\right)\mathbf{1} + \beta_1\left(B_{(1)}, B_{(2)}, B_{(3)}\right)\mathbf{B} + \beta_2\left(B_{(1)}, B_{(2)}, B_{(3)}\right)\mathbf{B}^2. \tag{7.354}$$

i) Show that when $\mathbf{B} = \mathbf{1}$, the above equation yields the *residual stress*

$$\boldsymbol{\sigma} = -\overline{p}\,\mathbf{1},$$

where $\overline{p}$ is the residual pressure given by

$$\overline{p} = -\beta_0(3,3,1) - \beta_1(3,3,1) - \beta_2(3,3,1).$$

ii) Expand (7.354) about the state $\mathbf{B} = \mathbf{1}$ and show that the linear approximation is given by

$$\boldsymbol{\sigma} = -\overline{p}\,\mathbf{1} + \frac{1}{2}(\lambda + \overline{p})(\operatorname{tr}\mathbf{B}')\mathbf{1} + (\mu - \overline{p})\,\mathbf{B}',$$

where $\mathbf{B}' = \mathbf{B} - \mathbf{1}$,

$$\lambda + \overline{p} = 2\left(\frac{\partial}{\partial B_{(1)}} + 2\,\frac{\partial}{\partial B_{(2)}} + \frac{\partial}{\partial B_{(3)}}\right)(\beta_0 + \beta_1 + \beta_2)\Bigg|_{B_{(1)}=B_{(2)}=3,\,B_{(3)}=1},$$

and

$$\mu - \overline{p} = \beta_1(3,3,1) + 2\,\beta_2(3,3,1).$$

15. Using the Mooney–Rivlin expression for the free energy

$$\rho\psi = \frac{1}{2}\mu\left[\left(\frac{1}{2} + \beta\right)\left(B_{(1)} - 3\right) + \left(\frac{1}{2} - \beta\right)\left(B_{(2)} - 3\right)\right], \quad \mu > 0, \quad -\frac{1}{2} \le \beta \le \frac{1}{2},$$

look at (i) simple extension, (ii) dilatation or compression, and (iii) simple shear, to see what is predicted. As a special case, what does the neo-Hookean material ($\beta = 1/2$) say.

16. Use the residual inequality to find the restrictions on the coefficients of

$$\mathbf{q} = \alpha_{00}\,\mathbf{g} + \alpha_{01}\,\mathbf{1}\cdot\mathbf{g} + \alpha_{10}\,\mathbf{B}\cdot\mathbf{g}.$$

17. Determine C and provide an explicit relationship between A and B emerging from the conditions $\sigma_{<rr>} = 0$ at $r = r_1, r_2$, for a Mooney–Rivlin material under the deformation of bending, stretching, and shearing of a rectangular block given by (7.306)–(7.308). What about if the material is neo-Hookean?

18. Analyze fully the deformation

$$r = \sqrt{\mathsf{A}R^2 + \mathsf{B}},$$
$$\theta = \Theta + \mathsf{D}Z,$$
$$z = \mathsf{F}Z.$$

What are the consequences for a Mooney–Rivlin material? What about for a neo-Hookean?

19. Obtain the stress tensor for the deformation of straightening, stretching, and shearing of a sector of a hollow circular cylinder given by (7.309)–(7.311).

20. Obtain the stress tensor for the deformation of inflation, bending, torsion, extension, and shearing of an annular wedge given by (7.312)–(7.314).

21. Obtain the stress tensor for the deformation of inflation or eversion of a sector of a spherical shell given by (7.315)–(7.317).

22. Obtain the stress tensor for the deformation of inflation, bending, extension, and azimuthal shearing of an annular wedge given by (7.318)–(7.320).

23. In an effort to produce a theory of heat conduction for which thermal disturbances propagate with finite speed, it is reasonable to investigate a homogeneous material for which $p = q = 1$ and $r = 0$ in (5.38), so that (5.40) becomes

$$T(\mathbf{X},t) = T\{\mathbf{F}(\mathbf{X},t), \dot{\mathbf{F}}(\mathbf{X},t), \theta(\mathbf{X},t), \dot{\theta}(\mathbf{X},t), \mathbf{G}(\mathbf{X},t)\}.$$

    i) Derive the reduced constitutive equations in this case.
    ii) Derive the thermodynamic restrictions.

24. In order to include additional memory effects, it is reasonable to investigate a homogeneous material for which $p = 2$ and $q = r = 0$ in (5.38), so that (5.40) becomes

$$T(\mathbf{X},t) = T\{\mathbf{F}(\mathbf{X},t), \dot{\mathbf{F}}(\mathbf{X},t), \ddot{\mathbf{F}}(\mathbf{X},t), \theta(\mathbf{X},t), \mathbf{G}(\mathbf{X},t)\}.$$

    i) Derive the reduced constitutive equations in this case.
    ii) Derive the thermodynamic restrictions.

25. In order to include additional spatial effects, it is reasonable to investigate a homogeneous material for which $P = 2$ and $Q = 1$ in (5.30) and $p = 1$, $q = r = 0$ in (5.38), so that (5.33) becomes

$$T(\mathbf{X},t) = T\{\mathbf{F}(\mathbf{X},t), \mathrm{Grad}\,\mathbf{F}(\mathbf{X},t), \theta(\mathbf{X},t), \mathbf{G}(\mathbf{X},t)\}.$$

    i) Derive the reduced constitutive equations in this case.
    ii) Show that the thermodynamic restrictions are such as to reduce it to the case of the thermoelastic material with heat conduction considered.

Bibliography

N.W. Ashcroft and N.D. Mermin. *Solid State Physics.* Brooks/Cole Cengage Lerning, 1976.

M. Bailyn. *A Survey of Thermodynamics.* AIP Press, New York, NY, 1994.

M.A. Biot. Thermoelasticity and irreversible thermodynamics. *Journal of Applied Physics*, 27(3):240–253, 1956.

P.J. Blatz and W.L. Ko. Application of finite elastic theory to the deformation of rubbery materials. *Transactions of the Society of Rheology*, 6(1):223–251, 1962.

M.J. Buerger. *Elementary Crystallography.* The M.I.T. Press, Cambridge, MA, 1978.

H.B. Callen. *Thermodynamics.* John Wiley & Sons, Inc., New York, NY, 1962.

A. Campanella and M.L. Tonon. A note on the Cauchy relations. *Meccanica*, 29(1):105–108, 1994.

M.M. Carroll. Controllable deformations of incompressible simple materials. *International Journal of Engineering Science*, 5(6):515–525, 1967.

P. Chadwick and I.N. Sneddon. Plane waves in an elastic solid conducting heat. *Journal of the Mechanics and Physics of Solids*, 6:223–230, 1958.

T.J. Chung. *General Continuum Mechanics.* Cambridge University Press, New York, NY, 2007.

S.C. Cowin and M.M. Mehrabadi. On the identification of material symmetry for anisotropic elastic materials. *Quarterly of Applied Mathematics*, 40(4):451–476, 1987.

R.S. Dhaliwal and H.H. Sherief. Generalized thermoelasticity for anisotropic media. *Quarterly of Applied Mathematics*, 38(1):1–8, 1980.

J.L. Ericksen. Deformations possible in every isotropic, incompressible, perfectly elastic body. *Journal of Applied Mathematics and Physics (ZAMP)*, 5(6):466–489, 1954.

A.C. Eringen. *Nonlinear Theory of Continuous Media.* McGraw-Hill Book Company, Inc., New York, NY, 1962.

A.C. Eringen. *Mechanics of Continua.* R.E. Krieger Publishing Company, Inc., Melbourne, FL, 1980.

F.G. Fumi. Physical properties of crystals: The direct-inspection method. *Acta Crystallographica*, 5(1):44–48, 1952.

M.E. Gurtin, E. Fried, and L. Anand. *The Mechanics and Thermodynamics of Continua.* Cambridge University Press, Cambridge, UK, 2010.

E. Hartmann. *An Introduction to Crystal Physics.* International Union of Crystallography, 2001.

P. Haupt. *Continuum Mechanics and Theory of Materials.* Springer-Verlag, Berlin, 2000.

G.A. Holzapfel. *Nonlinear Solid Mechanics.* John Wiley & Sons, Ltd., Chichester, England, 2005.

C.B. Kadafar. Methods of solution: Exact solutions in fluids and solids. In E.H. Dill, editor, *Continuum Physics*, volume II, pages 407–448. Academic Press, Inc., New York, NY, 1975.

A.D. Kovalenko. *Thermoelasticity.* Wolters-Noordhoff Publishing, Groningen, 1969.

W.M. Lai, D. Rubin, and E. Krempl. *Introduction to Continuum Mechanics.* Butterworth-Heinemann, Burlington, MA, 2010.

L.D. Landau and E.M. Lifshitz. *Statistical Physics*, volume 5 of *Course of Theoretical Physics.* Pergamon Press, Burlington, MA, 1980.

I.-S. Liu. On representations of anisotropic invariants. *International Journal of Engineering Science*, 20(10):1099–1109, 1982.

V.A. Lubarda. On thermodynamic potentials in linear thermoelasticity. *International Journal of Solids and Structures*, 41(26):7377–7398, 2004.

I. Müller. The coldness, a universal function in thermoelastic bodies. *Archive for Rational Mechanics and Analysis*, 41(5):319–332, 1971.

A.S. Nowick. *Crystal Properties via Group Theory.* Cambridge University Press, Cambridge, England, 1995.

A. Nussbaum. *Applied Group Theory for Chemists, Physicists and Engineers.* Prentice-Hall, Inc., Englewood Cliffs, NJ, 1971.

J.F. Nye. *Physical Properties of Crystals.* Oxford University Press, Oxford, England, 2008.

R.W. Ogden. *Non-Linear Elastic Deformations.* John Wiley & Sons, New York, 1984.

P. Podio-Guidugli. A primer in elasticity. *Journal of Elasticity*, 58(1):1–104, 2000.

R.S. Rivlin. The fundamental equations of nonlinear continuum mechanics. In S.I. Pai, A.J. Faller, T.L. Lincoln, D.A. Tidman, G.N. Trytten, and T.D. Wilkerson, editors, *Dynamics of Fluids in Porous Media*, pages 83–126, Academic Press, New York, 1966.

D.E. Sands. *Introduction to Crystallography.* Dover Publications, Inc., Mineola, NY, 1975.

J.N. Sharma and H. Singh. Generalized thermoelastic waves in anisotropic media. *Journal of the Acoustical Society of America*, 85(4):1407–1413, 1989.

R.T. Shield. Deformations possible in every compressible, isotropic, perfectly elastic material. *Journal of Elasticity*, 1(1):91–92, 1971.

G.F. Smith. Constitutive equations for anisotropic and isotropic materials. In G.C. Sih, editor, *Mechanics and Physics of Discrete Systems*, volume 3. Elsevier Science B.V., Amsterdam, The Netherlands, 1994.

G.F. Smith and R.S. Rivlin. The anisotropic tensors. *Quarterly of Applied Mathematics*, 15(3):308–314, 1957.

G.F. Smith and R.S. Rivlin. The strain-energy function for anisotropic elastic materials. *Transactions of the American Mathematical Society*, 88(1):175–193, 1958.

A.J.M. Spencer. Theory of invariants. In A.C. Eringen, editor, *Continuum Physics*, volume I. Academic Press, New York, 1971.

E.S. Suhubi. Constitutive equations for simple materials: Thermoelastic solids. In A.C. Eringen, editor, *Continuum Physics*, volume II, chapter 2, pages 173–265. Academic Press, Inc., New York, NY, 1975.

R.F. Tinder. *Tensor Properties of Solids.* Morgan & Claypool Publishers, San Rafael, CA, 2008.

T.W. Ting and J.C.M. Li. Thermodynamics for elastic solids. General formulation. *Physical Review*, 106(6):1165–1167, 1957.

C. Truesdell and W. Noll. The non-linear field theories of mechanics. In S. Flügge, editor, *Handbuch der Physik*, volume III/3. Springer, Berlin-Heidelberg-New York, 1965.

D.C. Wallace. Thermoelastic theory of stressed crystals and higher-order elastic constants. *Solid State Physics*, 25(1):301–404, 1970.

D.C. Wallace. *Thermodynamics of Crystals.* Dover Publications, Inc., Mineola, NY, 1972.

C.-C. Wang. On the symmetry of the heat conduction tensor. In C. Truesdell, editor, *Rational Thermodynamics*, pages 396–401. Springer-Verlag, New York, NY, 1984.

8

Fluids

We have shown earlier (see (5.114)) that the response functional for a simple fluid is given by

$$T(\mathbf{x},t) = \underset{0\le s<\infty}{\mathfrak{F}} \{\rho(\mathbf{x},t), {}_{(t)}\mathbf{C}^{(t)}(\mathbf{x},s), \theta^{(t)}(\mathbf{x},s), {}_{(t)}\mathbf{g}^{(t)}(\mathbf{x},s)\}. \tag{8.1}$$

We now assume that continuous material derivatives exist, and consider the simplest constitutive function of a thermoviscous fluid. From (7.5), (7.7), and (7.8), we have shown that, assuming continuous material derivatives with respect to s at $s = 0$, we have

$$\theta^{(t)}(s) = \theta(t) - \dot{\theta}(t)\ s + \cdots, \tag{8.2}$$

$${}_{(t)}\mathbf{g}^{(t)}(s) = \mathbf{g}(t) - \left[\mathbf{L}^T(t)\cdot\mathbf{g}(t) + \dot{\mathbf{g}}(t)\right] s + \cdots, \tag{8.3}$$

$${}_{(t)}\mathbf{C}^{(t)}(s) = \mathbf{1} - 2\,\mathbf{D}(t)\,s + \cdots. \tag{8.4}$$

Subsequently, if we consider rates of deformation gradients up to first order ($p = 1$), and temperature and its gradients up to zero order ($q = r = 0$), we have the following constitutive equations for a thermoviscous fluid:

$$\psi = \psi\left(\rho(\mathbf{x},t), \mathbf{D}(\mathbf{x},t), \theta(\mathbf{x},t), \mathbf{g}(\mathbf{x},t)\right), \tag{8.5}$$

$$\eta = \eta\left(\rho(\mathbf{x},t), \mathbf{D}(\mathbf{x},t), \theta(\mathbf{x},t), \mathbf{g}(\mathbf{x},t)\right), \tag{8.6}$$

$$\mathbf{q} = \mathbf{q}\left(\rho(\mathbf{x},t), \mathbf{D}(\mathbf{x},t), \theta(\mathbf{x},t), \mathbf{g}(\mathbf{x},t)\right), \tag{8.7}$$

$$\mathbf{h} = \mathbf{h}\left(\rho(\mathbf{x},t), \mathbf{D}(\mathbf{x},t), \theta(\mathbf{x},t), \mathbf{g}(\mathbf{x},t)\right), \tag{8.8}$$

$$\boldsymbol{\sigma} = \boldsymbol{\sigma}\left(\rho(\mathbf{x},t), \mathbf{D}(\mathbf{x},t), \theta(\mathbf{x},t), \mathbf{g}(\mathbf{x},t)\right), \tag{8.9}$$

which are required to satisfy the following frame-invariance conditions:

$$\psi\left(\rho, \mathbf{Q}\cdot\mathbf{D}\cdot\mathbf{Q}^T, \theta, \mathbf{Q}\cdot\mathbf{g}\right) = \psi\left(\rho, \mathbf{D}, \theta, \mathbf{g}\right), \tag{8.10}$$

$$\eta\left(\rho, \mathbf{Q}\cdot\mathbf{D}\cdot\mathbf{Q}^T, \theta, \mathbf{Q}\cdot\mathbf{g}\right) = \eta\left(\rho, \mathbf{D}, \theta, \mathbf{g}\right), \tag{8.11}$$

$$\mathbf{q}\left(\rho, \mathbf{Q}\cdot\mathbf{D}\cdot\mathbf{Q}^T, \theta, \mathbf{Q}\cdot\mathbf{g}\right) = \mathbf{Q}\cdot\mathbf{q}\left(\rho, \mathbf{D}, \theta, \mathbf{g}\right), \tag{8.12}$$

$$\mathbf{h}\left(\rho, \mathbf{Q}\cdot\mathbf{D}\cdot\mathbf{Q}^T, \theta, \mathbf{Q}\cdot\mathbf{g}\right) = \mathbf{Q}\cdot\mathbf{h}\left(\rho, \mathbf{D}, \theta, \mathbf{g}\right), \tag{8.13}$$

$$\boldsymbol{\sigma}\left(\rho, \mathbf{Q}\cdot\mathbf{D}\cdot\mathbf{Q}^T, \theta, \mathbf{Q}\cdot\mathbf{g}\right) = \mathbf{Q}\cdot\boldsymbol{\sigma}\left(\rho, \mathbf{D}, \theta, \mathbf{g}\right)\cdot\mathbf{Q}^T, \tag{8.14}$$

for all $\mathbf{Q} \in \mathscr{O}(\mathscr{V})$. As noted earlier, material symmetry requires that ψ, η, $\mathbf{q}$, $\mathbf{h}$, and $\boldsymbol{\sigma}$ be isotropic scalar, vector, and tensor functions that satisfy the above frame-indifference equations.

8.1 Coleman–Noll procedure

From (5.252), and since when $\boldsymbol{\sigma} = \boldsymbol{\sigma}^T$, it follows that $\mathbf{L} : \boldsymbol{\sigma} = \mathbf{D} : \boldsymbol{\sigma}$, we have

$$-\gamma_v \theta \equiv \rho\left(\dot{\psi} + \eta \dot{\theta}\right) - D_{ik}\, \sigma_{ik} + \frac{q_i}{\theta}\, g_i - \theta\left(h_i - \frac{q_i}{\theta}\right)_{,i} + \rho\theta\left(b - \frac{r}{\theta}\right) \le 0. \qquad (8.15)$$

For simplicity, we have suppressed the functional dependencies in the Clausius–Duhem inequality so that only the constitutive dependent and independent variables appear explicitly. Using the chain rule, we can write

$$\dot{\psi} = \frac{\partial \psi}{\partial \rho}\, \dot{\rho} + \frac{\partial \psi}{\partial D_{ik}}\, \dot{D}_{ik} + \frac{\partial \psi}{\partial \theta}\, \dot{\theta} + \frac{\partial \psi}{\partial g_i}\, \dot{g}_i, \qquad (8.16)$$

so that substituting this into the inequality, and considering that from the local mass balance (4.98) we have $\dot{\rho} = -\rho \operatorname{div} \mathbf{v} = -\rho \operatorname{tr} \mathbf{D}$, we arrive at

$$-\gamma_v \theta \equiv \rho\left(\eta + \frac{\partial \psi}{\partial \theta}\right)\dot{\theta} - \left(\sigma_{ik} + \rho^2 \frac{\partial \psi}{\partial \rho}\delta_{ik}\right) D_{ik} + \rho\, \frac{\partial \psi}{\partial D_{ik}}\, \dot{D}_{ik} + \rho\, \frac{\partial \psi}{\partial g_i}\, \dot{g}_i +$$
$$\frac{q_i}{\theta}\, g_i - \theta\left(h_i - \frac{q_i}{\theta}\right)_{,i} + \rho\theta\left(b - \frac{r}{\theta}\right) \le 0. \qquad (8.17)$$

Now the response functions $\{\psi, \eta, q_i, h_i, \sigma_{ik}\}$ are fixed if we fix the independent variables $\{\rho, D_{ik}, \theta, g_i\}$. Subsequently, the above equation will then depend linearly on $\dot{\theta}$, $\dot{g}_i$, and $\dot{D}_{ik}$, which can still be made to vary.

By requiring that the inequality hold for arbitrary choices of $\{\rho, D_{ik}, \theta, g_i\}$ and assuming that

$$h_i = \frac{q_i}{\theta} \qquad \text{and} \qquad b = \frac{r}{\theta} \qquad (8.18)$$

(these can in fact be proved as in the previous chapter or as in the next section), we get that

$$\rho\left(\eta + \frac{\partial \psi}{\partial \theta}\right) = 0, \quad \rho\, \frac{\partial \psi}{\partial g_i} = 0, \quad \rho\, \frac{\partial \psi}{\partial D_{ik}} = 0, \quad -\left(\sigma_{ik} + \rho^2 \frac{\partial \psi}{\partial \rho}\delta_{ik}\right) D_{ik} + \frac{q_i}{\theta}\, g_i \le 0, (8.19)$$

or

$$\psi = \psi(\rho, \theta), \qquad (8.20)$$

$$\eta = -\frac{\partial \psi}{\partial \theta}, \qquad (8.21)$$

$$-\gamma_v \theta \equiv -\left(\sigma_{ik} + \rho^2 \frac{\partial \psi}{\partial \rho}\delta_{ik}\right) D_{ik} + \frac{q_i}{\theta}\, g_i \equiv A(\rho, D_{ik}, \theta, g_i) \le 0. \qquad (8.22)$$

Equation (8.22) is called the *residual entropy inequality*. If we have no deformation, then $D_{ik} = 0$ and obtain the *heat conduction inequality*

$$-\frac{q_i}{\theta}\, g_i \ge 0, \qquad (8.23)$$

which states that the angle between a nonzero temperature gradient and a nonzero heat flux is greater than or equal to 90°. If the material is isothermal, then $g_i = 0$ and the residual entropy inequality becomes the *mechanical dissipation inequality*

$$\left(\sigma_{ik} + \rho^2 \frac{\partial \psi}{\partial \rho} \delta_{ik}\right) D_{ik} \geq 0. \tag{8.24}$$

We recall that a state with no entropy production is a thermodynamic equilibrium state. At the equilibrium state, $D_{ik} = g_i = 0$ and we denote the equilibrium stress and heat flux by $\sigma^e_{ik} = \sigma^e_{ik}(\rho, \mathbf{0}, \theta, \mathbf{0})$ and $q^e_i = q^e_i(\rho, \mathbf{0}, \theta, \mathbf{0})$, so that $A(\rho, \mathbf{0}, \theta, \mathbf{0}) = 0$. Then, assuming that A is continuous with D_{ik} and g_i at the equilibrium state, the necessary conditions for A to be a maximum there are given by

$$\left.\frac{\partial A}{\partial D_{ik}}\right|_{\mathbf{D}=\mathbf{0},\ \mathbf{g}=\mathbf{0}} = \sigma^e_{ik} + \rho^2 \frac{\partial \psi}{\partial \rho} \delta_{ik} = 0 \qquad \text{and} \qquad \left.\frac{\partial A}{\partial g_i}\right|_{\mathbf{D}=\mathbf{0},\ \mathbf{g}=\mathbf{0}} = \frac{q^e_i}{\theta} = 0. \tag{8.25}$$

Thus, at equilibrium, we must have

$$\sigma^e_{ik} = -\rho^2 \frac{\partial \psi}{\partial \rho} \delta_{ik} \qquad \text{and} \qquad q^e_i = 0. \tag{8.26}$$

We note that this agrees with equilibrium thermodynamics, since if we take v to be the specific volume, i.e., $\mathsf{v} = 1/\rho$, then

$$d\mathsf{v} = -\frac{1}{\rho^2} d\rho, \tag{8.27}$$

and since

$$\sigma^e_{ik} = -p\, \delta_{ik} \qquad \text{and} \qquad p = -\frac{\partial \psi}{\partial \mathsf{v}}, \tag{8.28}$$

where p is the thermodynamic pressure, we get

$$\sigma^e_{ik} = -\rho^2 \frac{\partial \psi}{\partial \rho} \delta_{ik}. \tag{8.29}$$

The requirement for the matrix of second derivatives of A with respect to $\mathbf{D}$ and $\mathbf{g}$ for A to be a maximum at a thermodynamic equilibrium state leads to additional restrictions. These reduce exactly to the requirement (5.259), whose analysis we have previously discussed and whose resulting restrictions are given by (5.268).

Subsequently, we take

$$\sigma_{ik} = \sigma^e_{ik} + \sigma^d_{ik} \qquad \text{and} \qquad q_i = q^e_i + q^d_i, \tag{8.30}$$

where σ^d_{ik} and q^d_i represent dissipative, or nonequilibrium, parts of the stress tensor and heat flux, thus allowing us to rewrite the residual entropy inequality as

$$-\gamma_v\, \theta \equiv -\sigma^d_{ik} D_{ik} + \frac{q^d_i}{\theta} g_i \leq 0. \tag{8.31}$$

This inequality clearly states that heat, in the presence of deformation, does not, in general, flow from higher to lower temperatures. From the above definition of a thermodynamic equilibrium state, we note that σ^d_{ik} and q^d_i must vanish when D_{ik} and g_i vanish, respectively.

8.2 Müller–Liu procedure

In the Müller–Liu procedure, we require that the entropy inequality (4.219) be satisfied for all tensor fields satisfying the conservation of mass (4.98), and linear momentum (4.109), angular momentum (for a nonpolar material) (4.120), and energy (4.207) balances. Satisfaction of these equations is enforced through the use of Lagrange multipliers (see (5.300), and note that $\operatorname{div}\mathbf{v} = \operatorname{tr}\mathbf{D}$ and $\Phi = \mathbf{L} : \boldsymbol{\sigma} = \mathbf{D} : \boldsymbol{\sigma}$):

$$\gamma_v \equiv [\rho(\dot{\eta} - b) + \operatorname{div}\mathbf{h}] - \lambda^\rho [\dot{\rho} + \rho \operatorname{tr}\mathbf{D}] - \boldsymbol{\lambda}^{\mathrm{v}} \cdot [\rho(\dot{\mathbf{v}} - \mathbf{f}) - \operatorname{div}\boldsymbol{\sigma}] - \lambda^e [\rho(\dot{e} - r) - \mathbf{D} : \boldsymbol{\sigma} + \operatorname{div}\mathbf{q}] \geq 0. \quad (8.32)$$

Here, we take the constitutive quantities to be given by

$$\mathcal{C} = \{e, \eta, \mathbf{q}, \mathbf{h}, \boldsymbol{\sigma}\} \quad (8.33)$$

and which are functions of the independent basic fields

$$\mathcal{I} = \{\rho, \mathbf{D}, \theta, \mathbf{g}\}. \quad (8.34)$$

By the principle of equipresence, the Lagrange multipliers $\{\lambda^\rho, \boldsymbol{\lambda}^{\mathrm{v}}, \lambda^e\}$ are required to be functions of the same independent basic fields.

As in the Coleman–Noll procedure, we rewrite the derivatives of the constitutive dependent variables appearing in (8.32), $\{\dot{e}, \dot{\eta}, \operatorname{grad}\mathbf{q}, \operatorname{grad}\mathbf{h}, \operatorname{grad}\boldsymbol{\sigma}\}$, in terms of derivatives of the independent basic fields, $\{\dot{\rho}, \dot{\mathbf{D}}, \dot{\theta}, \dot{\mathbf{g}}, \operatorname{grad}\rho, \operatorname{grad}\mathbf{D}, \operatorname{grad}\theta, \operatorname{grad}\mathbf{g}\}$:

$$\dot{e} = \frac{\partial e}{\partial \rho}\dot{\rho} + \frac{\partial e}{\partial \mathbf{D}} : \dot{\mathbf{D}} + \frac{\partial e}{\partial \theta}\dot{\theta} + \frac{\partial e}{\partial \mathbf{g}} \cdot \dot{\mathbf{g}}, \quad (8.35)$$

$$\dot{\eta} = \frac{\partial \eta}{\partial \rho}\dot{\rho} + \frac{\partial \eta}{\partial \mathbf{D}} : \dot{\mathbf{D}} + \frac{\partial \eta}{\partial \theta}\dot{\theta} + \frac{\partial \eta}{\partial \mathbf{g}} \cdot \dot{\mathbf{g}}, \quad (8.36)$$

$$\operatorname{grad}\mathbf{q} = \frac{\partial \mathbf{q}}{\partial \rho}\operatorname{grad}\rho + \frac{\partial \mathbf{q}}{\partial \mathbf{D}} : \operatorname{grad}\mathbf{D} + \frac{\partial \mathbf{q}}{\partial \theta}\operatorname{grad}\theta + \frac{\partial \mathbf{q}}{\partial \mathbf{g}} \cdot \operatorname{grad}\mathbf{g}, \quad (8.37)$$

$$\operatorname{grad}\mathbf{h} = \frac{\partial \mathbf{h}}{\partial \rho}\operatorname{grad}\rho + \frac{\partial \mathbf{h}}{\partial \mathbf{D}} : \operatorname{grad}\mathbf{D} + \frac{\partial \mathbf{h}}{\partial \theta}\operatorname{grad}\theta + \frac{\partial \mathbf{h}}{\partial \mathbf{g}} \cdot \operatorname{grad}\mathbf{g}, \quad (8.38)$$

$$\operatorname{grad}\boldsymbol{\sigma} = \frac{\partial \boldsymbol{\sigma}}{\partial \rho}\operatorname{grad}\rho + \frac{\partial \boldsymbol{\sigma}}{\partial \mathbf{D}} : \operatorname{grad}\mathbf{D} + \frac{\partial \boldsymbol{\sigma}}{\partial \theta}\operatorname{grad}\theta + \frac{\partial \boldsymbol{\sigma}}{\partial \mathbf{g}} \cdot \operatorname{grad}\mathbf{g}. \quad (8.39)$$

Subsequently, identifying (8.32) with (5.297), it is easy to verify that

$$\beta = \frac{\partial \mathbf{h}}{\partial \theta} \cdot \operatorname{grad}\theta - \rho b, \quad (8.40)$$

$$\boldsymbol{\lambda} = (\lambda^\rho, \boldsymbol{\lambda}^{\mathrm{v}}, \lambda^e), \quad (8.41)$$

$$\boldsymbol{\alpha} = \left(\rho\frac{\partial \eta}{\partial \rho}, \mathbf{0}, \rho\frac{\partial \eta}{\partial \mathbf{D}}, \rho\frac{\partial \eta}{\partial \theta}, \rho\frac{\partial \eta}{\partial \mathbf{g}}, \frac{\partial \mathbf{h}}{\partial \rho}, \frac{\partial \mathbf{h}}{\partial \mathbf{D}}, \frac{\partial \mathbf{h}}{\partial \mathbf{g}}\right), \quad (8.42)$$

$$\mathbf{a} = \left(\dot{\rho}, \dot{\mathbf{v}}, \dot{\mathbf{D}}, \dot{\theta}, \dot{\mathbf{g}}, \operatorname{grad}\rho, \operatorname{grad}\mathbf{D}, \operatorname{grad}\mathbf{g}\right)^T, \quad (8.43)$$

$$\mathbf{b} = \left(\rho \operatorname{tr}\mathbf{D}, -\frac{\partial \boldsymbol{\sigma}}{\partial \theta} \cdot \operatorname{grad}\theta - \rho\mathbf{f}, \frac{\partial \mathbf{q}}{\partial \theta} \cdot \operatorname{grad}\theta - \mathbf{D} : \boldsymbol{\sigma} - \rho r\right)^T \quad (8.44)$$

and

$$\mathbf{A} = \begin{pmatrix} 1 & 0 & 0 & 0 & 0 & 0 & 0 & 0 \\ \mathbf{0} & \rho\mathbf{1} & \mathbf{0} & \mathbf{0} & \mathbf{0} & -\frac{\partial\boldsymbol{\sigma}}{\partial\rho} & -\frac{\partial\boldsymbol{\sigma}}{\partial\mathbf{D}} & -\frac{\partial\boldsymbol{\sigma}}{\partial\mathbf{g}} \\ \rho\frac{\partial e}{\partial\rho} & \mathbf{0} & \rho\frac{\partial e}{\partial\mathbf{D}} & \rho\frac{\partial e}{\partial\theta} & \rho\frac{\partial e}{\partial\mathbf{g}} & \frac{\partial\mathbf{q}}{\partial\rho} & \frac{\partial\mathbf{q}}{\partial\mathbf{D}} & \frac{\partial\mathbf{q}}{\partial\mathbf{g}} \end{pmatrix}. \tag{8.45}$$

It should be noted from the above definitions that the entropy inequality (4.219) corresponds to (5.295), and the conservation of mass (4.98), and linear momentum (4.109), angular momentum (for a nonpolar material) (4.120), and energy (4.207) balances correspond to the system (5.296). Subsequently, from (5.298), and using the fact that $\rho > 0$, we obtain the result that $\boldsymbol{\lambda}^{\mathbf{v}} = \mathbf{0}$, and the following additional relations:

$$\frac{\partial\eta}{\partial\rho} - \lambda^{\rho}\frac{1}{\rho} - \lambda^{e}\frac{\partial e}{\partial\rho} = 0, \tag{8.46}$$

$$\frac{\partial\eta}{\partial D_{kl}} - \lambda^{e}\frac{\partial e}{\partial D_{kl}} = 0, \tag{8.47}$$

$$\frac{\partial\eta}{\partial\theta} - \lambda^{e}\frac{\partial e}{\partial\theta} = 0, \tag{8.48}$$

$$\frac{\partial\eta}{\partial g_k} - \lambda^{e}\frac{\partial e}{\partial g_k} = 0, \tag{8.49}$$

$$\frac{\partial h_j}{\partial\rho} - \lambda^{e}\frac{\partial q_j}{\partial\rho} = 0, \tag{8.50}$$

$$K''_{jkl} + K''_{lkj} = 0, \tag{8.51}$$

$$K'_{jk} + K'_{kj} = 0, \tag{8.52}$$

where, in writing (8.51) and (8.52), we have defined

$$K''_{jkl} \equiv \frac{\partial h_j}{\partial D_{kl}} - \lambda^{e}\frac{\partial q_j}{\partial D_{kl}} \quad \text{and} \quad K'_{jk} \equiv \frac{\partial h_j}{\partial g_k} - \lambda^{e}\frac{\partial q_j}{\partial g_k}, \tag{8.53}$$

and used the symmetries $D_{kl} = D_{lk}$ and

$$\frac{\partial D_{kl}}{\partial x_j} = \frac{1}{2}\left(\frac{\partial v_{k,l}}{\partial x_j} + \frac{\partial v_{l,k}}{\partial x_j}\right) = \frac{1}{2}\left(\frac{\partial^2 v_k}{\partial x_j\partial x_l} + \frac{\partial^2 v_l}{\partial x_j\partial x_k}\right) \quad \text{and} \quad \frac{\partial\theta_{,k}}{\partial x_j} = \frac{\partial\theta_{,j}}{\partial x_k}. \tag{8.54}$$

Furthermore, we note that $K''_{jkl} = K''_{jlk}$ and from (8.51), we have

$$K''_{jkl} = -K''_{lkj} = -K''_{ljk} = K''_{kjl} = K''_{klj} = -K''_{jlk} = -K''_{jkl}. \tag{8.55}$$

Thus, (8.51) further reduces to

$$K''_{jkl} = 0. \tag{8.56}$$

In addition, from (5.299), we obtain the following reduced entropy inequality:

$$\gamma_v \equiv \left(\frac{\partial h_j}{\partial \theta} - \lambda^e \frac{\partial q_j}{\partial \theta}\right)\frac{\partial \theta}{\partial x_j} - \rho\left(b - \lambda^e r\right) - \left(\lambda^\rho \rho\, \delta_{ij} - \lambda^e \sigma_{ij}\right) D_{ij} \geq 0. \tag{8.57}$$

Note that for the vector fields $\mathbf{q}$ and $\mathbf{h}$ to be frame indifferent, they must satisfy (8.12) and (8.13), or, suppressing for the moment dependencies on ρ and θ, we write, e.g.,

$$f_i\left(Q_{jk}\right) \equiv h_i\left(Q_{km}Q_{ln}D_{mn}, Q_{mn}g_n\right) - Q_{ij}h_j\left(D_{kl}, g_m\right) = 0. \tag{8.58}$$

Now, differentiating with respect to Q_{pq} and subsequently taking $Q_{ij} = \delta_{ij}$, we obtain

$$\left.\frac{\partial f_i}{\partial Q_{pq}}\right|_{\mathbf{Q}=\mathbf{1}} = \frac{\partial h_i}{\partial D_{pl}} D_{ql} + \frac{\partial h_i}{\partial D_{kp}} D_{kq} + \frac{\partial h_i}{\partial g_p} g_q - \delta_{ip} h_q = 0. \tag{8.59}$$

It is evident that

$$\left.\frac{\partial f_i}{\partial Q_{pq}}\right|_{\mathbf{Q}=\mathbf{1}} = \left.\frac{\partial f_i}{\partial Q_{qp}}\right|_{\mathbf{Q}=\mathbf{1}}, \tag{8.60}$$

so that we can write

$$\left(\delta_{ip}h_q - \delta_{iq}h_p\right) = \left(\frac{\partial h_i}{\partial D_{pl}} D_{ql} + \frac{\partial h_i}{\partial D_{kp}} D_{kq} - \frac{\partial h_i}{\partial D_{ql}} D_{pl} - \frac{\partial h_i}{\partial D_{kq}} D_{kp}\right) + \left(\frac{\partial h_i}{\partial g_p} g_q - \frac{\partial h_i}{\partial g_q} g_p\right), \tag{8.61}$$

and similarly

$$\left(\delta_{ip}q_q - \delta_{iq}q_p\right) = \left(\frac{\partial q_i}{\partial D_{pl}} D_{ql} + \frac{\partial q_i}{\partial D_{kp}} D_{kq} - \frac{\partial q_i}{\partial D_{ql}} D_{pl} - \frac{\partial q_i}{\partial D_{kq}} D_{kp}\right) + \left(\frac{\partial q_i}{\partial g_p} g_q - \frac{\partial q_i}{\partial g_q} g_p\right). \tag{8.62}$$

Now multiplying the latter equation by λ^e, subtracting it from the first equation, using the definitions (8.53) and equations (8.52) and (8.56), we obtain

$$\left(\delta_{ip}K_q - \delta_{iq}K_p\right) = \left(K'_{ip}g_q - K'_{iq}g_p\right), \tag{8.63}$$

where we have defined

$$K_j \equiv h_j - \lambda^e q_j. \tag{8.64}$$

Using the fact that K'_{jk} is skew-symmetric (see (8.52)), system (8.63) yields the following equations:

$$K_1 = K'_{12}\, g_2, \qquad K_2 = K'_{23}\, g_3, \qquad K_3 = K'_{31}\, g_1, \tag{8.65}$$

$$0 = K'_{12}\, g_2 + K'_{31}\, g_3, \quad 0 = K'_{12}\, g_1 + K'_{23}\, g_3, \quad 0 = K'_{31}\, g_1 + K'_{23}\, g_2, \tag{8.66}$$

$$0 = K'_{23}\, g_1 + K'_{12}\, g_3, \quad 0 = K'_{31}\, g_2 + K'_{12}\, g_3, \quad 0 = K'_{23}\, g_1 + K'_{31}\, g_2. \tag{8.67}$$

Now, by appropriately linearly combining the three equations (8.67), we obtain

$$K'_{12}\,g_3 = 0, \qquad K'_{23}\,g_1 = 0, \qquad K'_{31}\,g_2 = 0, \tag{8.68}$$

and since $g_k \neq 0$ in general, then we must have that

$$K'_{jk} = 0, \tag{8.69}$$

and subsequently from (8.65), we have that

$$K_j = 0, \tag{8.70}$$

and from (8.64), we have that

$$h_j = \lambda^e(\rho, D_{kl}, \theta, g_k)\, q_j. \tag{8.71}$$

Now, using (8.56), (8.69), and (8.70), equations (8.53) become

$$K''_{jkl} = \frac{\partial K_j}{\partial D_{kl}} + \frac{\partial \lambda^e}{\partial D_{kl}}\, q_j = \frac{\partial \lambda^e}{\partial D_{kl}}\, q_j = 0 \text{ and } K'_{jk} = \frac{\partial K_j}{\partial g_k} + \frac{\partial \lambda^e}{\partial g_k}\, q_j = \frac{\partial \lambda^e}{\partial g_k}\, q_j = 0, \tag{8.72}$$

and since $q_j \neq 0$ in general, we must have that λ^e is independent of D_{kl} and g_k, i.e., $\lambda^e = \lambda^e(\rho, \theta)$, and thus

$$h_j = \lambda^e(\rho, \theta)\, q_j. \tag{8.73}$$

Using this result in (8.50), we obtain

$$\frac{\partial \lambda^e}{\partial \rho}\, q_j = 0. \tag{8.74}$$

Now, since $q_j \neq 0$ in general, we must come to the conclusion that

$$\lambda^e = \lambda^e(\theta), \tag{8.75}$$

and subsequently, we can rewrite (8.73) as

$$h_j = \lambda^e(\theta)\, q_j. \tag{8.76}$$

Now since $\lambda^e = \lambda^e(\theta)$, from (8.47) and (8.49), we see that the quantity $(\eta - \lambda^e\, e)$ is independent of $\mathbf{D}$ and $\mathbf{g}$. Furthermore, from (8.48), we have that

$$\frac{\partial}{\partial \theta}(\eta - \lambda^e\, e) = -e\,\frac{\lambda^e}{\theta}, \tag{8.77}$$

so that it follows that e is independent of $\mathbf{D}$ and $\mathbf{g}$ (if we exclude the possibility of λ^e being a constant, in general). Subsequently, we have that

$$e = e(\rho, \theta) \qquad \text{and} \qquad \eta = \eta(\rho, \theta), \tag{8.78}$$

and the relation (8.46) can be rewritten as

$$de = \frac{1}{\lambda^e}\left(d\eta - \frac{\lambda^\rho}{\rho}\, d\rho\right), \tag{8.79}$$

which also shows that

$$\lambda^\rho = \lambda^\rho(\rho, \theta). \tag{8.80}$$

Note that since $\psi = e - \theta\,\eta$, we also have that

$$\psi = \psi(\rho, \theta). \tag{8.81}$$

Using (8.76), the reduced inequality (8.57) becomes

$$\gamma_v \equiv \frac{d\lambda^e}{d\theta}\, q_j\, g_j - \rho\,(b - \lambda^e\, r) - (\lambda^\rho\, \rho\, \delta_{ij} - \lambda^e\, \sigma_{ij})\, D_{ij} \geq 0. \tag{8.82}$$

We recall that a state with no entropy production is a thermodynamic equilibrium state. At the equilibrium state, $D_{ij} = g_j = 0$, and we denote the equilibrium stress and heat flux by $\sigma^e_{ij} = \sigma^e_{ij}(\rho, \mathbf{0}, \theta, \mathbf{0}) = -p(\rho, \theta)\, \delta_{ij}$ and $q^e_i = q^e_i(\rho, \mathbf{0}, \theta, \mathbf{0})$, so that $\gamma_v(\rho, \mathbf{0}, \theta, \mathbf{0}) = 0$. Noting that the independent fields are assumed to not depend on the external energy supply r, and since r can be of arbitrary sign, we see from (8.82) that we must have that

$$b - \lambda^e\, r = 0. \tag{8.83}$$

Subsequently, the reduced entropy inequality becomes

$$\gamma_v \equiv \left(\frac{\partial h_j}{\partial \theta} - \lambda^e\, \frac{\partial q_j}{\partial \theta}\right) \frac{\partial \theta}{\partial x_j} - (\lambda^\rho\, \rho\, \delta_{ij} - \lambda^e\, \sigma_{ij})\, D_{ij} \geq 0. \tag{8.84}$$

Then, assuming that γ_v is continuous with D_{ij} and g_j at the equilibrium state, the necessary conditions for γ_v to be a minimum there are given by

$$\left.\frac{\partial \gamma_v}{\partial D_{ij}}\right|_{\mathbf{D}=\mathbf{0},\ \mathbf{g}=\mathbf{0}} = (\lambda^e\, p + \lambda^\rho\, \rho)\, \delta_{ij} = 0 \quad \text{and} \quad \left.\frac{\partial \gamma_v}{\partial g_j}\right|_{\mathbf{D}=\mathbf{0},\ \mathbf{g}=\mathbf{0}} = \frac{d\lambda^e}{d\theta}\, q^e_j = 0. \tag{8.85}$$

Subsequently, we have that

$$\lambda^\rho = -\lambda^e\, \frac{p(\rho, \theta)}{\rho} \qquad \text{and} \qquad q^e_j = 0. \tag{8.86}$$

Substituting the first equation into (8.79), we have

$$de = \frac{1}{\lambda^e}\, d\eta + \frac{p}{\rho^2}\, d\rho. \tag{8.87}$$

By comparison with the Gibbs equation (5.138) (and noting that in our case $n = 1$, $\boldsymbol{\nu}_1 \to 1/\rho$ and $\boldsymbol{\tau}_1 \to -p$), we see that

$$\lambda^e = \lambda^e(\theta) = \frac{1}{\theta}, \tag{8.88}$$

and subsequently,

$$h_j = \frac{1}{\theta}\, q_j \tag{8.89}$$

and from (8.83) we also have that

$$b = \frac{1}{\theta}\, r. \tag{8.90}$$

In addition, we now have from $(8.86)_1$ that

$$\lambda^\rho = -\frac{p(\rho,\theta)}{\rho\,\theta}, \tag{8.91}$$

which for an ideal gas, where the thermal equation of state is $p(\rho,\theta) = \rho\,R\,\theta$, reduces to $\lambda^\rho = -R$, where R is the ideal gas constant. More generally, λ^ρ will be a function of ρ and θ as given by (8.91). We can also rewrite (8.87) in terms of the Helmholtz free energy,

$$d\psi = -\eta\,d\theta + \frac{p}{\rho^2}\,d\rho, \tag{8.92}$$

from which we easily see that

$$\eta = -\left.\frac{\partial\psi}{\partial\theta}\right|_\rho \qquad \text{and} \qquad p = \rho^2\left.\frac{\partial\psi}{\partial\rho}\right|_\theta. \tag{8.93}$$

The requirement for the matrix of second derivatives of γ_v (or A in the Coleman–Noll procedure) with respect to $\mathbf{D}$ and $\mathbf{g}$ be positive definite at equilibrium for the entropy production to be a minimum provides additional restrictions. These reduce exactly to the requirement (5.259), whose analysis we have previously discussed and whose resulting restrictions are given by (5.268).

Finally, taking

$$\sigma_{ij} = \sigma^e_{ij} + \sigma^d_{ij} \qquad \text{and} \qquad q_j = q^e_j + q^d_j, \tag{8.94}$$

where σ^d_{ij} and q^d_j represent dissipative, or nonequilibrium, parts of the stress tensor and heat flux, and using (8.89), the residual entropy inequality (8.84) becomes

$$\gamma_v\,\theta \equiv \sigma^d_{ij}\,D_{ij} - \frac{1}{\theta}\,q^d_j\,g_j \geq 0. \tag{8.95}$$

This inequality clearly states that heat, in the presence of deformation, does not, in general, flow from higher to lower temperatures. From the above definition of a thermodynamic equilibrium state, we note that σ^d_{ij} and q^d_j must vanish when D_{ij} and g_j vanish, respectively.

8.3 Representations of $\mathbf{q}^d$ and $\boldsymbol{\sigma}^d$

From Tables 5.1–5.3, we can immediately write down the representations for q^d_i and σ^d_{ik} corresponding to isotropic vector-valued and symmetric tensor-valued functions of vector $\theta_{,i}$ and symmetric second-order tensor D_{ik}. These are

$$q^d_i = -\left(\kappa_0\,\delta_{il} + \kappa_1\,D_{il} + \kappa_2\,D_{ik}D_{kl}\right)\theta_{,l}, \tag{8.96}$$

$$\begin{aligned}\sigma^d_{ik} = {} & \alpha_0\,\delta_{ik} + \alpha_1\,D_{ik} + \alpha_2\,D_{il}D_{lk} + \alpha_3\,\theta_{,i}\theta_{,k} + \frac{1}{2}\alpha_4\left(\theta_{,i}D_{kl} + \theta_{,k}D_{il}\right)\theta_{,l} + {}\\ & \frac{1}{2}\alpha_5\left(\theta_{,i}D_{kl} + \theta_{,k}D_{il}\right)D_{lm}\theta_{,m},\end{aligned} \tag{8.97}$$

where κ_p $(p = 0, 1, 2)$ and α_q $(q = 0, \ldots, 5)$ are functions of ρ and θ, and the isotropic scalar invariants $\{I_1, \ldots, I_6\}$, where

$$I_1 = D_{ii} = D_{(1)} = d^{(1)} + d^{(2)} + d^{(3)}, \tag{8.98}$$

$$I_2 = D_{ik}D_{ki} = D_{(1)}^2 - 2D_{(2)} = {d^{(1)}}^2 + {d^{(2)}}^2 + {d^{(3)}}^2, \tag{8.99}$$

$$I_3 = D_{ik}D_{kl}D_{li} = D_{(1)}^3 - 3D_{(1)}D_{(2)} + 3D_{(3)} = {d^{(1)}}^3 + {d^{(2)}}^3 + {d^{(3)}}^3, \tag{8.100}$$

$$I_4 = \theta_{,i}\theta_{,i}, \quad I_5 = \theta_{,i}D_{ik}\theta_{,k}, \quad I_6 = \theta_{,i}D_{ik}D_{kl}\theta_{,l}, \tag{8.101}$$

and, using (3.135)–(3.137),

$$D_{(1)} = d^{(1)} + d^{(2)} + d^{(3)}, \quad D_{(2)} = d^{(1)}d^{(2)} + d^{(2)}d^{(3)} + d^{(3)}d^{(1)}, \quad D_{(3)} = d^{(1)}d^{(2)}d^{(3)}, \tag{8.102}$$

where $d^{(i)}$, $i = 1, 2, 3$, are the principal deformation rates. We note that we have used (3.98)–(3.100) and (8.102) in relating the isotropic scalar invariants I_1, I_2, and I_3 to $D_{(1)}$, $D_{(2)}$, and $D_{(3)}$, and $d^{(1)}$, $d^{(2)}$, and $d^{(3)}$. In the representation (8.97), we require that $\alpha_0 \to 0$ as $\theta_{,i} \to 0$ and $D_{ik} \to 0$, since as we approach equilibrium, we must have $\sigma_{ik}^d \to 0$.

Now, substituting the representations (8.96) and (8.97) into the reduced entropy inequality (8.95), and using the Cayley–Hamilton theorem (3.95) and the definitions of the invariants (8.98)–(8.101), we obtain

$$\begin{aligned} 0 \leq \gamma_v\,\theta \quad \equiv \quad & \alpha_0 I_1 + \alpha_1 I_2 + \alpha_2 I_3 + \left[\frac{1}{6}\alpha_5\left(I_1^3 - 3I_1I_2 + 2I_3\right) + \frac{\kappa_0}{\theta}\right] I_4 + \\ & \left[\alpha_3 - \frac{1}{2}\alpha_5\left(I_1^2 - I_2\right) + \frac{\kappa_1}{\theta}\right] I_5 + \left(\alpha_4 + \alpha_5 I_1 + \frac{\kappa_2}{\theta}\right) I_6, \end{aligned} \tag{8.103}$$

$$\begin{aligned} = \quad & \alpha_0 D_{(1)} + \alpha_1\left(D_{(1)}^2 - 2D_{(2)}\right) + \alpha_2\left(D_{(1)}^3 - 3D_{(1)}D_{(2)} + 3D_{(3)}\right) + \\ & \left(\alpha_5 D_{(3)} + \frac{\kappa_0}{\theta}\right) I_4 + \left(\alpha_3 - \alpha_5 D_{(2)} + \frac{\kappa_1}{\theta}\right) I_5 + \\ & \left(\alpha_4 + \alpha_5 D_{(1)} + \frac{\kappa_2}{\theta}\right) I_6. \end{aligned} \tag{8.104}$$

The constitutive equations (8.96) and (8.97) are too complicated for solutions of fluid problems. Generally, polynomial approximations of various degrees in $\mathbf{D}$ and $\mathbf{g}$ are used. We note that $\deg I_1 = \deg D_{(1)} = 1$, $\deg I_2 = \deg D_{(2)} = 2$, $\deg I_3 = \deg D_{(3)} = 3$, $\deg I_4 = 2$, $\deg I_5 = 3$, and $\deg I_6 = 4$. Subsequently, for q_i^d to be of order N, κ_0, κ_1, and κ_2 can at most be of degrees $N-1$, $N-2$, and $N-3$, respectively, while for σ_{ik}^d to be of order N, α_0, α_1, α_2, α_3, α_4, and α_5 can at most be of degrees N, $N-1$, $N-2$, $N-2$, $N-3$, and $N-4$, respectively.

Below, as alternative to using directly the above general representations, we will construct the following specific representations in detail to illustrate the use of the frame-indifference restrictions:

a) Rigid motion, thus $D_{ik} = 0$

b) Isothermal conditions, thus $\theta_{,i} = 0$

c) Zeroth-order representations in D_{ik} and $\theta_{,i}$

d) First-order representations in D_{ik} and $\theta_{,i}$

e) Second-order representations in D_{ik} and $\theta_{,i}$

The results that we obtain will be seen to correspond to specific simplifications of (8.96), (8.97), and (8.103) or (8.104).

a) Rigid motion

In this case, since $D_{ik} = 0$, we have $q_i^d = q_i^d(\rho, \theta, \theta_{,k})$ and $\sigma_{ik}^d = \sigma_{ik}^d(\rho, \theta, \theta_{,k})$. From Tables 5.1–5.3, and as can be readily verified from (8.96) and (8.97) when $D_{ik} = 0$, the most general representations consistent with frame indifference are

$$q_i^d = -\kappa_0\,\theta_{,i}, \tag{8.105}$$

$$\sigma_{ik}^d = \alpha_0\,\delta_{ik} + \alpha_3\,\theta_{,i}\theta_{,k}, \tag{8.106}$$

where

$$\kappa_0 = \kappa_0(\rho, \theta, I_4), \qquad \alpha_0 = \alpha_0(\rho, \theta, I_4), \quad \alpha_3 = \alpha_3(\rho, \theta, I_4). \tag{8.107}$$

Now, substituting the representations (8.105) and (8.106) into the reduced entropy inequality (8.95), or from (8.103), we see that the reduced entropy inequality in this case is

$$\gamma_v\,\theta \equiv \frac{\kappa_0}{\theta} I_4 \geq 0, \tag{8.108}$$

or, since $\theta > 0$ and $I_4 > 0$,

$$0 \leq \kappa_0 < \infty. \tag{8.109}$$

Note that the reduced entropy inequality imposes no restrictions on α_0 and α_3 since $D_{ik} = 0$. Subsequently, we have

$$q_i = -\kappa_0\,\theta_{,i}, \tag{8.110}$$

$$\sigma_{ik} = (-p + \alpha_0)\,\delta_{ik} + \alpha_3\,\theta_{,i}\theta_{,k}, \tag{8.111}$$

which represent the constitutive equations that account for heat conduction and thermoelastic stress.

If we linearize q_i in the limit $\theta_{,i} \to 0$, we obtain *Fourier's law of heat conduction*

$$q_i = -k\;\theta_{,i}, \tag{8.112}$$

where $k = k(\rho, \theta) \equiv \kappa_0(\rho, \theta, 0)$. In addition, if we linearize σ_{ik} in the limit $\theta_{,i} \to 0$ and require that $\sigma_{ik}^d \to 0$ as $\theta_{,i} \to 0$, we have that $\alpha_0(\rho, \theta, 0) = 0$, and thus we obtain the stress tensor for a perfect fluid

$$\sigma_{ik} = -p\,\delta_{ik}, \tag{8.113}$$

where $p = p(\rho, \theta)$. Balance equations using (8.112) and (8.113) are referred to as *Euler–Fourier equations.*

b) Isothermal conditions

In this case, since $\theta_{,i} = 0$, it is easy to see that

$$q_i = q_i^d = q_i^d(\rho, D_{ik}, \theta) = 0 \qquad \text{and} \qquad \sigma_{ik}^d = \sigma_{ik}^d(\rho, D_{ik}, \theta). \tag{8.114}$$

From Tables 5.1–5.3, and as can be readily verified from (8.97) when $\theta_{,i} = 0$, the most general representation consistent with frame indifference is

$$\sigma_{ik}^d = \alpha_0 \delta_{ik} + \alpha_1 D_{ik} + \alpha_2 D_{il} D_{lk}, \tag{8.115}$$

where $\alpha_i = \alpha_i(\rho, \theta, D_{(1)}, D_{(2)}, D_{(3)})$, $i = 0, 1, 2$, and $\alpha_0(\rho, \theta, 0, 0, 0) = 0$ when $D_{ik} = 0$. In order that α_0, α_1, and α_2 be continuous at the coalescence of the principal values of D_{ik}, it is required that σ_{ik}^d be three times continuously differentiable in the components of D_{ik}.

Subsequently, suppressing all dependencies of α_j's on ρ, θ, and $D_{(j)}$'s, from (8.26) and (8.30), we can write

$$\sigma_{ik} = \left(-\rho^2 \frac{\partial \psi}{\partial \rho} + \alpha_0\right) \delta_{ik} + \alpha_1 D_{ik} + \alpha_2 D_{il} D_{lk}. \tag{8.116}$$

This is the constitutive relation for what is referred to as a *Reiner–Rivlin fluid.* If we let $\overline{p}$ be the *mechanical pressure* defined by

$$\overline{p} = -\frac{1}{3}\sigma_{kk}, \tag{8.117}$$

and since from (8.26) and (8.28), the *thermodynamic pressure* is

$$p = \rho^2 \frac{\partial \psi}{\partial \rho}, \tag{8.118}$$

we have, using (3.98) and (3.99), that

$$-\overline{p} = -p + \alpha_0 + \frac{1}{3}\alpha_1 D_{(1)} + \frac{1}{3}\alpha_2 \left(D_{(1)}^2 - 2D_{(2)}\right). \tag{8.119}$$

Note that the mechanical and thermodynamic pressures are generally different!

From (8.104), we see that the reduced entropy inequality in this case is

$$\gamma_v \theta \equiv \alpha_0 D_{(1)} + \alpha_1 (D_{(1)}^2 - 2D_{(2)}) + \alpha_2 (D_{(1)}^3 - 3D_{(1)} D_{(2)} + 3D_{(3)}) \geq 0. \tag{8.120}$$

If we expand the coefficients in power series of the invariants

$$\alpha_i = \sum_{p=0}^{\infty} \sum_{q=0}^{\infty} \sum_{r=0}^{\infty} a_{ipqr} D_{(1)}^p D_{(2)}^q D_{(3)}^r, \tag{8.121}$$

where $a_{ipqr} = a_{ipqr}(\rho, \theta)$, then the above inequality will impose certain restrictions on the coefficients. Clearly, we must have that $a_{0000} = 0$. However, it is not obvious how one finds a general inequality that a_{ipqr} must satisfy for arbitrary orders.

c) Zeroth-order representations

From (8.96) and (8.97), with $D_{ik} = 0$ and $\theta_{,i} = 0$, we see that the zeroth-order representations are given by

$$q_i^d = 0 \qquad \text{and} \qquad \sigma_{ik}^d = 0; \tag{8.122}$$

thus

$$q_i = 0 \qquad \text{and} \qquad \sigma_{ik} = -p\,\delta_{ik}, \tag{8.123}$$

where $p = p(\rho, \theta)$. This corresponds to *inviscid fluids* and p is a thermodynamic property. The corresponding linear momentum equation is called *Euler's equation of motion.* Fluids in which $p = p(\rho)$ are called *barotropic fluids.*

d) First-order representations

If we linearize (8.96) and (8.97) in the limits $\theta_{,i} \to 0$ and $D_{ik} \to 0$, and noting that for first-order representations $\kappa_i = \kappa_i(\rho, \theta, D_{(1)})$ and $\alpha_i = \alpha_i(\rho, \theta, D_{(1)})$, we then have that

$$q_i^d = -\kappa_{00}\,\theta_{,i} \tag{8.124}$$

and

$$\sigma_{ik}^d = \big(\alpha_{00} + \alpha_{01}\,D_{(1)}\big)\,\delta_{ik} + \alpha_{10}\,D_{ik}, \tag{8.125}$$

where now κ_{00}, α_{00}, α_{01}, and α_{10} are only functions of ρ and θ. Since the q_i^d and σ_{ik}^d must vanish when $\theta_{,i}$ and D_{ik} vanish, we have that $\alpha_{00} = 0$. From the reduced entropy inequality (8.104), we now have

$$\gamma_v\,\theta \equiv (\alpha_{01} + \alpha_{10})\,D_{(1)}^2 - 2\,\alpha_{10}D_{(2)} + \frac{\kappa_{00}}{\theta}\,I_4 \geq 0. \tag{8.126}$$

Rewriting $D_{(2)}$ in terms of the invariant of the deviatoric component using (3.108), we have

$$\gamma_v\,\theta \equiv \left(\alpha_{01} + \frac{1}{3}\,\alpha_{10}\right) D_{(1)}^2 - 2\,\alpha_{10}D'_{(2)} + \frac{\kappa_{00}}{\theta}\,I_4 \geq 0, \tag{8.127}$$

or, rewriting the above in terms of principal invariants using (3.135) and (3.114), we have

$$\gamma_v\,\theta \equiv \left(\alpha_{01} + \frac{1}{3}\,\alpha_{10}\right)\left(\lambda^{(1)} + \lambda^{(2)} + \lambda^{(3)}\right)^2 + \\ \frac{1}{3}\,\alpha_{10}\left[\left(\lambda^{(1)} - \lambda^{(2)}\right)^2 + \left(\lambda^{(2)} - \lambda^{(3)}\right)^2 + \left(\lambda^{(3)} - \lambda^{(1)}\right)^2\right] + \frac{\kappa_{00}}{\theta}\,I_4 \geq 0. \tag{8.128}$$

Since the inequality must hold for any deformation, then we require that

$$\alpha_{01} + \frac{1}{3}\,\alpha_{10} \geq 0, \qquad \alpha_{10} \geq 0, \quad \text{and} \quad \kappa_{00} \geq 0. \tag{8.129}$$

Now calling $\alpha_{01} = \lambda$, $\alpha_{10} = 2\mu$, and $\kappa_{00} = k$, where $\lambda = \lambda(\rho, \theta)$ and $\mu = \mu(\rho, \theta)$ are the *dilatational* and *shear viscosities,* and $k = k(\rho, \theta)$ is the *thermal conductivity,* the restrictions (8.129) become

$$0 \leq \zeta(\rho, \theta) < \infty, \qquad 0 \leq \mu(\rho, \theta) < \infty, \quad \text{and} \quad 0 \leq k < \infty, \tag{8.130}$$

where $\zeta \equiv (\lambda + \frac{2}{3}\mu)$ is called the *bulk viscosity.*

Subsequently, the constitutive equations (8.94) and the reduced entropy inequality (8.127) become

$$q_i = -k\,\theta_{,i}, \tag{8.131}$$

$$\sigma_{ik} = \left(-p + \lambda D_{(1)}\right)\delta_{ik} + 2\mu D_{ik}, \tag{8.132}$$

and

$$\gamma_v\,\theta \equiv \zeta\, D_{(1)}^2 - 4\,\mu\, D'_{(2)} + \frac{k}{\theta}\, I_4 \geq 0. \tag{8.133}$$

These are the constitutive equations for what is referred to as a thermoviscous *Newtonian fluid.* The balance equations for linear momentum and energy using these constitutive equations are referred as the *Newton–Fourier equations.* Note from (8.119) that for a thermoviscous Newtonian fluid, the mechanical and thermodynamic pressures are related by

$$-\overline{p} = -p + \zeta\, D_{(1)}. \tag{8.134}$$

In addition, note that the bulk viscosity measures the excess between the mechanical and thermodynamic pressures. These pressures are equal under any of the following conditions:

i) $D_{(1)} \neq 0$ but the bulk viscosity vanishes, i.e.,

$$\zeta = 0. \tag{8.135}$$

This represents *Stokes' hypothesis.* It is not valid in general (it vanishes for a monatomic gas), and indeed, there is indication that Stokes himself questioned it. Then, the constitutive equation is given by (8.132) with $\lambda = -\frac{2}{3}\mu$ and the pressure is, as usual, given by $p = p(\rho, \theta)$. The linear momentum equations with this stress tensor is referred to as the *Navier–Stokes equations.*

ii) $D_{(1)} \neq 0$ but the dilatational and shear viscosities vanish, i.e., $\lambda = \mu = 0$. In this case, the constitutive equations reduces to

$$\sigma_{ik} = -p\,\delta_{ik}, \tag{8.136}$$

and the pressure is again given by $p = p(\rho, \theta)$. The linear momentum equations with this stress tensor is referred to as *Euler's equations of motion.*

iii) $D_{(1)} = 0$, in which case the fluid is isochoric (or incompressible), the constitutive equation (8.132) becomes

$$\sigma_{ik} = -p\,\delta_{ik} + 2\mu\, D_{ik}, \tag{8.137}$$

the dilatational viscosity λ drops out from the equations, and p corresponds to the mean or hydrostatic pressure.

iv) $D_{ik} = 0$, in which case we have rigid motion with

$$\sigma_{ik} = -p\,\delta_{ik}, \tag{8.138}$$

both terms involving viscosities drop out, and p also corresponds to the mean or "hydrostatic" pressure.

e) Second-order representations

If we write second-order expansions of (8.96) and (8.97) in the limits $\theta_{,i} \to 0$ and $D_{ik} \to 0$, we require that $\kappa_0 = \kappa_{00} + \kappa_{01} D_{(1)}$, $\kappa_1 = \kappa_{10}$, $\kappa_2 = 0$, $\alpha_0 = \alpha_{0000} + \alpha_{0100} D_{(1)} + \alpha_{0200} D_{(1)}^2 + \alpha_{0010} D_{(2)} + \alpha_{0001} I_4$, $\alpha_1 = \alpha_{10} + \alpha_{11} D_{(1)}$, $\alpha_2 = \alpha_{20}$, $\alpha_3 = \alpha_{30}$, and $\alpha_4 = \alpha_5 = 0$. Note that now κ_{00}, κ_{01}, κ_{10}, α_{0000}, α_{0100}, α_{0200}, α_{0010}, α_{10}, α_{11}, α_{20}, and α_{30} are just functions of ρ and θ. Subsequently, since q_i^d and D_{ik}^d must vanish as $\theta_{,i} \to 0$ and $D_{ik} \to 0$, we have that $\alpha_{0000} = 0$ and thus

$$q_i^d = -\left[\left(\kappa_{00} + \kappa_{01} D_{(1)}\right) \delta_{il} + \kappa_{10} D_{il}\right] \theta_{,l}, \tag{8.139}$$

$$\begin{aligned} \sigma_{ik}^d = & \left(\alpha_{0100} D_{(1)} + \alpha_{0200} D_{(1)}^2 + \alpha_{0010} D_{(2)} + \alpha_{0001} I_4\right) \delta_{ik} + \\ & \left(\alpha_{10} + \alpha_{11} D_{(1)}\right) D_{ik} + \alpha_{20} D_{il} D_{lk} + \alpha_{30} \theta_{,i} \theta_{,k}. \end{aligned} \tag{8.140}$$

Remark

Before addressing constraints on the coefficients from the above inequality, it is useful to derive the representations (8.139) and (8.140) using simple arguments. We start by noting that since $D_{ik} \neq 0$ and $\theta_{,l} \neq 0$, we have the general representations

$$q_i^d = q_i^d(\rho, D_{lm}, \theta, \theta_{,n}), \tag{8.141}$$

$$\sigma_{ik}^d = \sigma_{ik}^d(\rho, D_{lm}, \theta, \theta_{,n}). \tag{8.142}$$

If we limit ourselves to second-order expansions in D_{lm} and $\theta_{,n}$, we have

$$\begin{aligned} q_i^d = & -(G_{ik}\theta_{,k} + H_{ikl} D_{kl} + J_{iklm}\theta_{,k} D_{lm} + K_{iklmn} D_{kl} D_{mn} + \\ & M_{ikl}\theta_{,k}\theta_{,l}), \end{aligned} \tag{8.143}$$

$$\begin{aligned} \sigma_{ik}^d = & A_{ikl}\theta_{,l} + B_{iklm} D_{lm} + C_{iklmn}\theta_{,l} D_{mn} + E_{iklmst} D_{lm} D_{st} + \\ & F_{iklm}\theta_{,l}\theta_{,m}, \end{aligned} \tag{8.144}$$

where all the coefficients are functions of ρ and θ, and we have used the equilibrium conditions that $\sigma_{ik}^d \to 0$ and $q_i^d \to 0$ when $D_{ik} \to 0$ and $\theta_{,i} \to 0$. Since σ_{ik}^d and D_{lm} are symmetric, we must have

$$A_{ikl} = A_{kil}, \tag{8.145}$$

$$B_{iklm} = B_{kilm} = B_{kiml} = B_{ikml}, \tag{8.146}$$

$$C_{iklmn} = C_{kilmn} = C_{kilnm} = C_{iklnm}, \tag{8.147}$$

$$\begin{aligned} E_{iklmst} = & E_{kilmst} = E_{kimlst} = E_{kimlts} = E_{ikmlts} = E_{iklmts} = E_{ikstml} = \\ & E_{kistml} = E_{kitsml} = E_{kitslm} = E_{iktslm} = E_{ikstlm}, \end{aligned} \tag{8.148}$$

$$F_{iklm} = F_{kilm} = F_{kiml} = F_{ikml}, \tag{8.149}$$

$$H_{ikl} = H_{ilk}, \tag{8.150}$$

$$J_{iklm} = J_{ikml}, \tag{8.151}$$

$$K_{iklmn} = K_{ilkmn} = K_{iklnm} = K_{ilknm}, \tag{8.152}$$

$$M_{ikl} = M_{ilk}. \tag{8.153}$$

Frame indifference requires that $\mathbf{q}^d$ and $\boldsymbol{\sigma}^d$ satisfy

$$\mathbf{q}^d(\rho, \mathbf{Q}\cdot\mathbf{D}\cdot\mathbf{Q}^T, \theta, \mathbf{Q}\cdot\text{grad }\theta) = \mathbf{Q}\cdot\mathbf{q}^d(\rho, \mathbf{D}, \theta, \text{grad }\theta), \tag{8.154}$$

$$\boldsymbol{\sigma}^d(\rho, \mathbf{Q}\cdot\mathbf{D}\cdot\mathbf{Q}^T, \theta, \mathbf{Q}\cdot\text{grad }\theta) = \mathbf{Q}\cdot\boldsymbol{\sigma}^d(\rho, \mathbf{D}, \theta, \text{grad }\theta)\cdot\mathbf{Q}^T, \tag{8.155}$$

for all $\mathbf{Q} \in \mathscr{O}(\mathscr{V})$. From objectivity, we have seen that in general

$$A_{ik\cdots st} = Q_{ip}Q_{kq}\cdots Q_{sl}Q_{tm}A_{pq\cdots lm}. \tag{8.156}$$

In addition, material symmetry requires that $\boldsymbol{\sigma}^d$ and $\mathbf{q}^d$ be isotropic tensors. Below, we shall make use of the representations of isotropic tensors given in Appendix B.

We examine first the constitutive equation for the heat flux. Now

$$G_{ik} = Q_{ip}Q_{kq}G_{pq}, \tag{8.157}$$

and since δ_{ik} is the only isotropic second-order tensor as indicated in (B.15), we must have

$$G_{ik} = g\delta_{ik}. \tag{8.158}$$

Similarly, from (B.18), we have

$$H_{ikl} = h\epsilon_{ikl} \tag{8.159}$$

since ϵ_{ikl} is the only isotropic third-order tensor. Now, from the symmetries (8.150) and (8.153),

$$H_{ikl} = h\epsilon_{ikl} = H_{ilk} = h\epsilon_{ilk} = 0, \tag{8.160}$$

and

$$M_{ikl} = m\epsilon_{ikl} = M_{ilk} = m\epsilon_{ilk} = 0. \tag{8.161}$$

In addition, from (B.20) to (B.22), we have

$$J_{iklm} = j_1\delta_{ik}\delta_{lm} + j_2\delta_{il}\delta_{km} + j_3\delta_{im}\delta_{kl}, \tag{8.162}$$

so that from the symmetry (8.151), it follows that $j_2 = j_3$ and thus

$$J_{iklm} = j_1\delta_{ik}\delta_{lm} + j_2\left(\delta_{il}\delta_{km} + \delta_{im}\delta_{kl}\right). \tag{8.163}$$

Furthermore, using the symmetries (8.152), after some algebra we find that

$$K_{iklmn} = k\left(\epsilon_{iln}\delta_{km} + \epsilon_{ilm}\delta_{kn}\right) = K_{ilknm} = k\left(\epsilon_{ikm}\delta_{ln} + \epsilon_{ikn}\delta_{lm}\right) = 0. \tag{8.164}$$

Thus, substituting the above results into (8.143), we obtain

$$q_i^d = -g\theta_{,k}\delta_{ik} - \left[j_1\delta_{ik}\delta_{lm} + j_2\left(\delta_{il}\delta_{km} + \delta_{im}\delta_{kl}\right)\right]\theta_{,k}D_{lm}, \tag{8.165}$$

or finally

$$q_i^d = -\left(g + j_1D_{(1)}\delta_{ik} + 2j_2D_{ik}\right)\theta_{,k}, \tag{8.166}$$

where $g = g(\rho,\theta)$, $j_1 = j_1(\rho,\theta)$, and $j_2 = j_2(\rho,\theta)$.

We now proceed in a similar fashion to obtain the coefficients of σ_{ik}^d. Applying the symmetries (8.145)–(8.149), after some lengthy algebra, we obtain

$$
\begin{aligned}
A_{ikl} &= a\epsilon_{ikl} = 0, && (8.167)\\
B_{iklm} &= b_1\delta_{ik}\delta_{lm} + b_2\left(\delta_{il}\delta_{km} + \delta_{im}\delta_{kl}\right), && (8.168)\\
C_{iklmn} &= c\left(\epsilon_{kln}\delta_{im} + \epsilon_{iln}\delta_{km} + \epsilon_{klm}\delta_{in} + \epsilon_{ilm}\delta_{kn}\right) = 0, && (8.169)\\
E_{iklmst} &= e_1\delta_{ik}\delta_{lm}\delta_{st} + e_2\left(\delta_{ls}\delta_{mt} + \delta_{ms}\delta_{lt}\right)\delta_{ik} + \\
&\quad e_3\left[\left(\delta_{il}\delta_{km} + \delta_{im}\delta_{kl}\right)\delta_{st} + \left(\delta_{is}\delta_{kt} + \delta_{it}\delta_{ks}\right)\delta_{lm}\right] + \\
&\quad e_4\left[\left(\delta_{is}\delta_{mt} + \delta_{it}\delta_{ms}\right)\delta_{kl} + \left(\delta_{ks}\delta_{mt} + \delta_{kt}\delta_{ms}\right)\delta_{il} + \right.\\
&\quad \left.\left(\delta_{it}\delta_{ls} + \delta_{is}\delta_{lt}\right)\delta_{km} + \left(\delta_{ks}\delta_{lt} + \delta_{kt}\delta_{ls}\right)\delta_{im}\right], && (8.170)\\
F_{iklm} &= f_1\delta_{ik}\delta_{lm} + f_2\left(\delta_{il}\delta_{km} + \delta_{im}\delta_{kl}\right). && (8.171)
\end{aligned}
$$

Substituting the above coefficients in the constitutive equation (8.144), we obtain

$$
\begin{aligned}
\sigma_{ik}^d = {} & \left[b_1 D_{(1)} + (e_1 + 2e_2)D_{(1)}^2 - 4e_2 D_{(2)} + f_1\theta_{,j}\theta_{,j}\right]\delta_{ik} + \\
& 2\left(b_2 + 2e_3 D_{(1)}\right)D_{ik} + 8e_4 D_{il}D_{lk} + f_2\theta_{,i}\theta_{,k}, \qquad (8.172)
\end{aligned}
$$

where the coefficients are functions of ρ and θ. It is clear that equations (8.166) and (8.172) that we have just derived are the same as (8.139) and (8.140), respectively.

Now, using (3.108), the residual entropy inequality (8.104) for this order of approximation becomes

$$
\begin{aligned}
\gamma_v\,\theta \equiv {} & \left(\alpha_{0100} + \frac{1}{3}\alpha_{10}\right)D_{(1)}^2 - 2\,\alpha_{10}D'_{(2)} + \left(\alpha_{0200} + \frac{1}{3}\alpha_{11} + \frac{1}{3}\alpha_{0010}\right)D_{(1)}^3 + \\
& \left(\alpha_{0010} - 2\,\alpha_{11} - 3\,\alpha_{20}\right)D_{(1)}D'_{(2)} + 3\,\alpha_{20}D_{(3)} + \\
& \left[\frac{\kappa_{00}}{\theta} + \left(\alpha_{0001} + \frac{\kappa_{01}}{\theta}\right)D_{(1)}\right]I_4 + \left(\alpha_{30} + \frac{\kappa_{10}}{\theta}\right)I_5 \geq 0. \qquad (8.173)
\end{aligned}
$$

We now examine the restrictions imposed by the reduced entropy inequality (8.173) on the coefficients in (8.139) and (8.140). The inequality must be satisfied for any D_{ik} and any $\theta_{,i}$. Specifically, it must be satisfied for incompressible constant density isothermal deformations for which $D_{(1)} = 0$ and $\theta_{,i} = 0$. In this case, (8.173) becomes

$$
\gamma_v\,\theta \equiv -2\,\alpha_{10}D'_{(2)} + 3\,\alpha_{20}D_{(3)} \geq 0. \qquad (8.174)
$$

Now, since $D'_{(2)} \leq 0$ (see (3.108)) and $D_{(3)}$ can be of any magnitude and sign depending on the deformation, in order for the above inequality to be always satisfied, we must have that

$$
\alpha_{10} \geq 0 \qquad \text{and} \qquad \alpha_{20} = 0. \qquad (8.175)
$$

If we now consider an incompressible constant density non-isothermal motion and use the above results, then we must satisfy

$$
\gamma_v\,\theta \equiv -2\,\alpha_{10}D'_{(2)} + \frac{\kappa_{00}}{\theta}I_4 + \left(\alpha_{30} + \frac{\kappa_{10}}{\theta}\right)I_5 \geq 0. \qquad (8.176)
$$

Recalling that $-2\,\alpha_{10}D'_{(2)} \geq 0$, noting that $I_4 \geq 0$, and since I_5 can be of any magnitude and sign depending on D_{ik} and $\theta_{,i}$, then we must have that

$$\kappa_{00} \geq 0 \qquad \text{and} \qquad \alpha_{30} = -\frac{\kappa_{10}}{\theta}. \tag{8.177}$$

If we next consider a compressible isothermal motion and again use previous results, the reduced entropy inequality (8.173) reduces to

$$\gamma_v\,\theta \;\equiv\; \left(\alpha_{0100} + \frac{1}{3}\alpha_{10}\right) D^2_{(1)} - 2\,\alpha_{10}D'_{(2)} + \left[\left(\alpha_{0200} + \frac{1}{3}\alpha_{11} + \frac{1}{3}\alpha_{0010}\right) D^2_{(1)} + \left(\alpha_{0010} - 2\,\alpha_{11}\right) D'_{(2)}\right] D_{(1)} \geq 0. \tag{8.178}$$

Now, since $-2\,\alpha_{10}D'_{(2)} \geq 0$ and since $D_{(1)}$ can be chosen to be of any magnitude and sign, for arbitrary motions, we must have that

$$\alpha_{0100} + \frac{1}{3}\alpha_{10} \geq 0, \qquad \alpha_{0200} + \frac{1}{3}\alpha_{11} + \frac{1}{3}\alpha_{0010} = 0, \quad \text{and} \quad \alpha_{0010} - 2\,\alpha_{11} = 0. \tag{8.179}$$

Lastly, considering compressible non-isothermal motion and upon using the previous results, the reduced entropy inequality becomes

$$\gamma_v\,\theta \equiv \left(\alpha_{0100} + \frac{1}{3}\alpha_{10}\right) D^2_{(1)} - 2\,\alpha_{10}D'_{(2)} + \left[\frac{\kappa_{00}}{\theta} + \left(\alpha_{0001} + \frac{\kappa_{01}}{\theta}\right) D_{(1)}\right] I_4 \geq 0. \tag{8.180}$$

It is now easy to see that for the inequality to be satisfied for arbitrary motions, we must have that

$$\alpha_{0001} = -\frac{\kappa_{01}}{\theta}. \tag{8.181}$$

In summary, we have that

$$\kappa_{00} \geq 0, \quad \alpha_{0100} + \frac{1}{3}\alpha_{10} \geq 0, \quad \alpha_{10} \geq 0, \quad \alpha_{0010} = 2\,\alpha_{11}, \quad \alpha_{20} = 0,$$
$$\alpha_{0200} = -\alpha_{11}, \quad \alpha_{0001} = -\frac{\kappa_{01}}{\theta}, \quad \alpha_{30} = -\frac{\kappa_{10}}{\theta}, \tag{8.182}$$

and κ_{01}, κ_{10}, and α_{11} can be of any sign. If we rewrite the above by taking $\alpha_{0100} = \lambda$, $\alpha_{10} = 2\,\mu$, $\alpha_{11} = \mu'$, $\kappa_{00} = k$, $\kappa_{01} = k'$, and $\kappa_{10} = k''$, then we write the quadratic constitutive representations (8.94), (8.99), (8.139), and (8.140) as

$$\sigma_{ik} = \left(-p + \lambda D_{(1)} - \mu' D_{jl}D_{lj} - \frac{k'}{\theta}\theta_{,j}\theta_{,j}\right)\delta_{ik} + \left(2\,\mu + \mu' D_{(1)}\right) D_{ik} - \frac{k''}{\theta}\theta_{,i}\theta_{,k}, \tag{8.183}$$

and

$$q_i = -\left[\left(k + k' D_{(1)}\right)\delta_{il} + k'' D_{il}\right]\theta_{,l}, \tag{8.184}$$

and the reduced entropy inequality remains (8.133), where

$$\zeta = \lambda + \frac{2}{3}\mu \geq 0, \qquad \mu \geq 0, \qquad k \geq 0, \tag{8.185}$$

are the same restrictions obtained from the linear theory, and we have one additional viscous property, μ', and two additional thermal properties, k' and k'';

these last three properties can be of any sign and do not contribute to entropy production. However, note that they do contribute to the mechanical pressure since

$$-\overline{p} = -p + \zeta\, D_{(1)} - \frac{2}{3}\,\mu'\, D_{(1)}^2 + 2\,\mu'\, D_{(2)} - \frac{\left(k' + \frac{1}{3}\,k''\right)}{\theta}\,\theta_{,i}\theta_{,i}, \tag{8.186}$$

and in this case, the mechanical pressure is not equal to the thermodynamic pressure even when the fluid is incompressible (i.e., when $D_{(1)} = 0$). Furthermore, while in the classical linear theory, we have that $p \geq \overline{p}$ in an expansion and $p \leq \overline{p}$ in a contraction motion (see (8.134)), the above equation shows that the sign of $(p - \overline{p})$ does not necessarily follow the intuitive notion of being positive on expansion and negative on contraction motion.

8.4 Propagation of sound

The fundamental local laws of conservation of mass and balances of linear momentum and energy with linear constitutive equations that satisfy the entropy inequality for an isotropic fluid, ignoring external sources of momentum and energy, are given by

$$\dot{\rho} + \rho\,\mathrm{div}\,\mathbf{v} = 0, \tag{8.187}$$

$$\rho\,\dot{\mathbf{v}} - \mathrm{div}\,\boldsymbol{\sigma} = \mathbf{0}, \tag{8.188}$$

$$\rho\,\dot{e} - \Phi + \mathrm{div}\,\mathbf{q} = 0, \tag{8.189}$$

where

$$\boldsymbol{\sigma} = -p\,\mathbf{1} + \boldsymbol{\sigma}^d = -p\,\mathbf{1} + \left[2\,\mu\,\mathbf{D} + \left(\zeta - \frac{2}{3}\,\mu\right) D_{(1)}\mathbf{1}\right], \tag{8.190}$$

$$\mathbf{q} = -k\,\mathrm{grad}\,\theta, \tag{8.191}$$

$$\Phi = \mathbf{D} : \boldsymbol{\sigma} = -p\,D_{(1)} + \mathbf{D} : \boldsymbol{\sigma}^d = -p\,D_{(1)} + \left[2\,\mu\,\mathbf{D} : \mathbf{D} + \left(\zeta - \frac{2}{3}\,\mu\right) D_{(1)}^2\right], \tag{8.192}$$

and

$$\mu(\rho,\theta) \geq 0, \qquad \zeta(\rho,\theta) \geq 0, \quad \text{and} \quad k(\rho,\theta) \geq 0 \tag{8.193}$$

are the dynamic shear viscosity, dynamic bulk viscosity, and thermal conductivity, respectively.

There are too many unknowns in the above equations. To eliminate these additional unknowns, we need thermodynamic constitutive relations. To this end, from a given fundamental relation $\psi = \psi(\rho, \theta)$, it would be useful to introduce thermal and caloric equations of state for the pressure and internal energy by writing (see $(5.207)_2$, (5.208), and Table 8.1 for $n = 1$)

$$p = p\,(\rho,\theta) = \rho^2\,\left.\frac{\partial\psi}{\partial\rho}\right|_\theta \qquad \text{and} \qquad e = e\,(\rho,\theta) = \psi - \theta\,\left.\frac{\partial\psi}{\partial\theta}\right|_\rho. \tag{8.194}$$

In general, we can easily show that (see $(5.212)_2$, (5.213), and Table 8.1 for $n = 1$)

$$dp = \frac{1}{\kappa_\theta}\left(\frac{1}{\rho}\,d\rho + \alpha\,d\theta\right) \qquad \text{and} \qquad \rho\,de = \frac{1}{\rho}\left(p - \frac{\alpha}{\kappa_\theta}\,\theta\right) d\rho + \rho\,c_v\,d\theta, \tag{8.195}$$

where (see (5.161))

$$c_v \equiv \left.\frac{\partial e}{\partial \theta}\right|_\rho, \qquad \kappa_\theta \equiv \frac{1}{\rho}\left.\frac{\partial \rho}{\partial p}\right|_\theta, \quad \text{and} \quad \alpha \equiv -\frac{1}{\rho}\left.\frac{\partial \rho}{\partial \theta}\right|_p \tag{8.196}$$

are the coefficients of specific heat at constant volume, isothermal compressibility, and volume expansion, respectively. From equilibrium thermodynamics considerations, we also have the restrictions (see (5.268)–(5.270))

$$c_v > 0, \qquad \kappa_\theta > 0, \quad \gamma \equiv \frac{c_p}{c_v} > 1, \quad \text{and} \quad \alpha^2 < \frac{\rho\, c_p\, \kappa_\theta}{\theta}, \tag{8.197}$$

where

$$c_p \equiv \left.\frac{\partial h}{\partial \theta}\right|_p \tag{8.198}$$

is the specific heat at constant pressure.

Upon substituting the thermodynamic relations (8.195) into (8.187)–(8.189), it is straightforward to see that we obtain the following closed set of equations for the unknowns ρ, $\mathbf{v}$, and θ once all properties, which are functions of ρ and θ, are specified:

$$\dot{\rho} + \rho \operatorname{div} \mathbf{v} = 0, \tag{8.199}$$

$$\rho \dot{\mathbf{v}} + \frac{1}{\kappa_\theta}\left(\frac{1}{\rho}\operatorname{grad}\rho + \alpha \operatorname{grad}\theta\right) - \operatorname{div}\boldsymbol{\sigma}^d = \mathbf{0}, \tag{8.200}$$

$$\rho c_v \dot{\theta} + \frac{\alpha}{\kappa_\theta}\theta D_{(1)} - \mathbf{D} : \boldsymbol{\sigma}^d + \operatorname{div}\mathbf{q} = 0. \tag{8.201}$$

These equations are fairly complicated. To investigate the propagation of acoustics in a fluid, we make the following simplifying assumption. The density and temperature are assumed to deviate only slightly from constant equilibrium values of ρ_0 and θ_0, while the equilibrium value of velocity is taken to be zero:

$$\rho = \rho_0 + \rho', \qquad \mathbf{v} = \mathbf{0} + \mathbf{v}', \quad \text{and} \quad \theta = \theta_0 + \theta', \tag{8.202}$$

where $\rho'/\rho_0 \ll 1$, $\mathbf{v}' \ll 1$, and $\theta'/\theta_0 \ll 1$. Note that due to Galilean frame indifference, the results will also apply to the case where the equilibrium value of velocity is a constant. Substituting (8.202) into (8.190)–(8.191) and (8.199)–(8.201), and linearizing the resulting equations, we have

$$\frac{\partial \rho'}{\partial t} + \rho_0 \operatorname{div}\mathbf{v}' = 0, \tag{8.203}$$

$$\rho_0 \frac{\partial \mathbf{v}'}{\partial t} + \frac{1}{\kappa_{\theta 0}}\left(\frac{1}{\rho_0}\operatorname{grad}\rho' + \alpha_0 \operatorname{grad}\theta'\right) - \mu_0 \operatorname{div}(\operatorname{grad}\mathbf{v}') - \left(\zeta_0 + \frac{1}{3}\mu_0\right)\operatorname{grad}(\operatorname{div}\mathbf{v}') = \mathbf{0}, \tag{8.204}$$

$$\rho_0 c_{v0} \frac{\partial \theta'}{\partial t} + \frac{\alpha_0 \theta_0}{\kappa_{\theta 0}} \operatorname{div}\mathbf{v}' - k_0 \operatorname{div}(\operatorname{grad}\theta') = 0, \tag{8.205}$$

where all properties with zero subscripts are evaluated at ρ_0 and θ_0.

These linearized Newton–Fourier equations are the basic tool to understanding acoustic propagation in fluids. Their structure is rather remarkable. We can clearly recognize two types of terms. In a first class, the coefficients of ρ', $\mathbf{v}'$, and

θ' depend only on the equilibrium thermodynamic properties of ρ_0, θ_0, α_0, $\kappa_{\theta 0}$, and c_{v0}. In the second class, the terms depend on the nonequilibrium transport properties of μ_0, ζ_0, and k_0. If the latter are set equal to zero, we obtain the ideal or non-dissipative equations. The additional terms are responsible for the dissipation through the viscosities and heat conduction. They clearly are also the only terms giving rise to nonzero entropy production.

Of special interest to us are those motions with very long characteristic spatial scales. This is certainly consistent with the use of constitutive equations that are linear in the deformation and temperature gradients. The study of such motions will lead to an eigenvalue problem whose solution is given by a superposition of elementary long-range motions. With such study in mind, we assume that our linear equations (8.203)–(8.205) describe the motion of a fluid in an infinite domain. We investigate their solution in the forms

$$\rho'(\mathbf{x},t) = e^{i\mathbf{k}\cdot\mathbf{x}+\lambda t}\,\rho_{\mathbf{k}}, \tag{8.206}$$

$$\mathbf{v}'(\mathbf{x},t) = e^{i\mathbf{k}\cdot\mathbf{x}+\lambda t}\,\mathbf{v}_{\mathbf{k}}, \tag{8.207}$$

$$\theta'(\mathbf{x},t) = e^{i\mathbf{k}\cdot\mathbf{x}+\lambda t}\,\theta_{\mathbf{k}}, \tag{8.208}$$

where $\mathbf{k}$ is the (real) wavenumber vector and λ is the (complex) frequency. Substituting these forms in (8.203)–(8.205), and using (5.163) and (5.164), we obtain the following set of linear algebraic equations:

$$\lambda\,\rho_{\mathbf{k}} + \rho_0\,i\,\mathbf{k}\cdot\mathbf{v}_{\mathbf{k}} = 0, \tag{8.209}$$

$$\lambda\,\mathbf{v}_{\mathbf{k}} + \frac{1}{\rho_0^2\,\kappa_{\theta 0}}\,i\,\mathbf{k}\,\rho_{\mathbf{k}} + \frac{\alpha_0}{\rho_0\,\kappa_{\theta 0}}\,i\,\mathbf{k}\,\theta_{\mathbf{k}} + \nu_0\,k^2\,\mathbf{v}_{\mathbf{k}} + \left(\nu_0' + \frac{1}{3}\,\nu_0\right)\mathbf{k}\,(\mathbf{k}\cdot\mathbf{v}_{\mathbf{k}}) = \mathbf{0}, \tag{8.210}$$

$$\lambda\,\theta_{\mathbf{k}} + \frac{\gamma_0 - 1}{\alpha_0}\,i\,\mathbf{k}\cdot\mathbf{v}_{\mathbf{k}} + \gamma_0\,\chi_0\,k^2\,\theta_{\mathbf{k}} = 0, \tag{8.211}$$

where we have used the following abbreviations:

$$\gamma_0 = \frac{c_{p0}}{c_{v0}}, \qquad \nu_0 \equiv \frac{\mu_0}{\rho_0}, \qquad \nu_0' \equiv \frac{\zeta_0}{\rho_0}, \qquad \chi_0 \equiv \frac{k_0}{\rho_0\,c_{p0}}, \tag{8.212}$$

with γ_0 the ratio of specific heats, ν_0 the kinematic shear viscosity, ν_0' the kinematic bulk viscosity, and χ_0 the thermal diffusivity.

Now, without loss of generality, we can choose the $\mathbf{x}$-axis to be oriented along the wavenumber vector $\mathbf{k}$, thus taking

$$\mathbf{k} = (k, 0, 0) \qquad \text{and} \qquad \mathbf{v}_{\mathbf{k}} = (u_{\mathbf{k}}, v_{\mathbf{k}}, w_{\mathbf{k}}), \tag{8.213}$$

we can rewrite the system (8.209)–(8.211) in the following form:

$$\begin{pmatrix} \lambda & \rho_0\,i\,k & 0 & 0 & 0 \\ \dfrac{1}{\rho_0^2\,\kappa_{\theta 0}}\,i\,k & \lambda + \left(\frac{4}{3}\,\nu_0 + \nu_0'\right)k^2 & 0 & 0 & \dfrac{\alpha_0}{\rho_0\,\kappa_{\theta 0}}\,i\,k \\ 0 & 0 & \lambda + \nu_0\,k^2 & 0 & 0 \\ 0 & 0 & 0 & \lambda + \nu_0\,k^2 & 0 \\ 0 & \dfrac{\gamma_0 - 1}{\alpha_0}\,i\,k & 0 & 0 & \lambda + \gamma_0\,\chi_0\,k^2 \end{pmatrix} \begin{pmatrix} \rho_{\mathbf{k}} \\ u_{\mathbf{k}} \\ v_{\mathbf{k}} \\ w_{\mathbf{k}} \\ \theta_{\mathbf{k}} \end{pmatrix} = 0. \tag{8.214}$$

For a nontrivial solution to exist, the determinant of the square matrix must be zero, giving rise to a characteristic polynomial for λ, which represents the *dispersion relation*,

$$(\lambda + \nu_0 k^2)^2 \left\{ \lambda^3 + \left(\frac{4}{3}\nu_0 + \nu_0' + \gamma_0 \chi_0 \right) k^2 \lambda^2 + \left[c_0^2 + \gamma_0 \chi_0 \left(\frac{4}{3}\nu_0 + \nu_0' \right) k^2 \right] k^2 \lambda + c_0^2 \chi_0 k^4 \right\} = 0, \quad (8.215)$$

where

$$c_0 \equiv \left(\frac{\gamma_0}{\rho_0 \kappa_{\theta 0}} \right)^{1/2} \quad (8.216)$$

is the fluid's sound speed.

Two of the roots of (8.215) are the same and immediately obvious: $\lambda_{3,4} = -\nu_0 k^2$. These two roots are real and correspond to purely dissipative shear modes since they depend only on the shear viscosity. The other three roots satisfy the cubic equation within the braces. Two of them are a complex conjugate pair and correspond to dissipative acoustic modes since they involve the sound speed to leading order, but also involve dissipative terms at higher orders due to shear viscosity, bulk viscosity, and thermal diffusivity. These two acoustic modes travel in opposite directions. The third root is real and corresponds to a purely dissipative thermal mode since it only involves the thermal diffusivity. These roots can be written explicitly, but they are fairly complicated algebraically. Since the constitutive equations are linear in the deformation and temperature gradients, it makes sense to look at these three roots in the small wavenumber limit. In such limit, these roots are given in Table 8.1 up to second order in k. In the table, we also see that when the fluid is ideal, the eigenvalues of the shear and thermal modes are zero, and the acoustic modes travel at the sound speed without any damping.

8.5 Classifications of fluid motions

8.5.1 Restrictions on the type of motion

Restrictions on geometry

Various special motions are defined by geometric restrictions on the velocity field.

1. *Homogeneous* motion:

$$\mathbf{v} = \dot{\mathbf{x}} = \mathbf{A}(t) \cdot \mathbf{x} + \mathbf{b}(t). \quad (8.217)$$

Since any continuous motion may be approximated in the neighborhood of any point by an appropriate homogeneous motion, this motion is very important. In this case, using (8.217), the stretch and spin tensors are given by

$$\mathbf{D} = \operatorname{sym} \mathbf{A} \quad \text{and} \quad \mathbf{W} = \operatorname{skw} \mathbf{A}. \quad (8.218)$$

It now follows that any homogenous motion may be regarded at any instant as a rigid translation, a rigid rotation, and an irrotational homogeneous stretchings along the three principal axes of $\mathbf{D}$.

2. *Lineal* motion:

$$v_x = v_x(x,t), \quad v_y = 0, \quad v_z = 0, \quad \rho = \rho(x,t). \tag{8.219}$$

In this case, the velocity is constant over each member of a family of parallel planes, and normal to those planes, the motion is lineal, where in the above characterization, these planes are the surfaces x = const.

3. *Pseudo-lineal* motion of the *first kind:*

$$v_x = v_x(x,y,z,t), \quad v_y = 0, \quad v_z = 0, \quad \rho = \rho(x,y,z,t). \tag{8.220}$$

4. *Pseudo-lineal* motion of the *second kind:*

$$v_x = v_x(x,t), \quad v_y = v_y(x,t), \quad v_z = v_z(x,t), \quad \rho = \rho(x,t). \tag{8.221}$$

We note that for lineal and pseudo-lineal motions of the first and second kind, the local mass balance equation becomes

$$\frac{\partial \rho}{\partial t} + \frac{\partial}{\partial x}(\rho v_x) = 0. \tag{8.222}$$

Thus, there always exists a function $Q(x,t)$ (or $Q(x,y,z,t)$) such that

$$\rho = -\frac{\partial Q}{\partial x}, \qquad \rho v_x = \frac{\partial Q}{\partial t}. \tag{8.223}$$

5. *Plane* motion:

$$v_x = v_x(x,y,t), \quad v_y = v_y(x,y,t), \quad v_z = 0, \quad \rho = \rho(x,y,t). \tag{8.224}$$

A motion is said to be plane if its velocity field is a plane field and thus the vorticity $\boldsymbol{\omega}$ satisfies

$$\mathbf{v} \cdot \boldsymbol{\omega} = 0, \tag{8.225}$$

with the surfaces z = const. being a family of parallel planes. The third dimension is obtained by rotating the plane of motions about the z-axis.

The idea of plane motion is easily generalized by replacing z = const. with any curved surface $x^3 = 0$, upon which x^1 and x^2 may be any curvilinear coordinates. The equations

$$\dot{x}^k = \dot{x}^k(x^1, x^2, t), \qquad k = 1,2, \tag{8.226}$$

now describe a strictly two-dimensional motion. To generate a three-dimensional motion in a region of space near $x^3 = 0$, choose a coordinate system in which the surface x^3 = const. are surfaces parallel to $x^3 = 0$, while the surfaces x^1 = const. and x^2 = const. are those swept out by the normals to $x^3 = 0$ along the x^1 and x^2 coordinate curves upon it.

6. *Pseudo-plane* motion of the *first kind*:

$$v_x = v_x(x, y, z, t), \quad v_y = v_y(x, y, z, t), \quad v_z = 0, \quad \rho = \rho(x, y, z, t). \tag{8.227}$$

Such motions result if the first and second conditions $(8.224)_{1,2}$ are removed and the third condition $(8.224)_3$ is retained.

7. *Pseudo-plane* motion of the *second kind*:

$$v_x = v_x(x, y, t), \quad v_y = v_y(x, y, t), \quad v_z = v_z(x, y, t), \quad \rho = \rho(x, y, t). \tag{8.228}$$

Such motions result if the first and second conditions $(8.224)_{1,2}$ are retained while the third condition $(8.224)_3$ is removed.

8. *Rotationally symmetric* motion:

$$v_r = v_r(r, z, t), \quad v_\theta = 0, \quad v_z = v_z(r, z, t), \quad \rho = \rho(r, z, t). \tag{8.229}$$

A motion is said to be rotationally symmetric if its velocity field is a symmetrically rotational field with the surfaces z = const. being a family of coaxial planes.

Restrictions on steadiness

1. *Steady* motion:

$$\mathbf{v} = \mathbf{v}(\mathbf{x}). \tag{8.230}$$

The ultimate steady flow is when $\mathbf{v} = \mathbf{0}$, in which case the problem reduces to one of *hydrostatics*.

Here, we just note that the definition of steady motion refers to a particular frame and thus is not frame invariant.

2. *Steady* motion with *steady density*:

The requirement of steady motion does not require that ρ be steady, but $\dot{\overline{\log \rho}}$ must be steady. Furthermore, if g is any steady quantity that is constant on each stream line, then $\dot{g} = 0$ and

$$\text{div}\,(\rho g \mathbf{v}) = \rho \mathbf{v} \cdot \text{grad}\, g + g\, \text{div}\,(\rho \mathbf{v}) = 0. \tag{8.231}$$

Thus, by assigning a steady quantity g a constant value on each stream line, we obtain a steady density ρg, and thus

$$\mathbf{v} = \mathbf{v}(\mathbf{x}), \quad \rho g = f(\mathbf{x}), \quad \dot{g} = 0, \tag{8.232}$$

where g is an arbitrary function.

3. Motion with *steady stream lines*:

   We first note that in order that the flux of a vector field $\mathbf{u}(\mathbf{x},t)$ across every material surface remains constant in time, it is necessary and sufficient that Zorawski's criterion (3.458) be satisfied. It follows from the *Helmholtz–Zorawski criterion* that the necessary and sufficient condition that the vector lines of $\mathbf{u}$ be material lines is that they satisfy

$$\mathbf{u}\times\left[\frac{\partial\mathbf{u}}{\partial t}+\mathbf{v}\operatorname{div}\mathbf{u}+\operatorname{curl}(\mathbf{u}\times\mathbf{v})\right]=\mathbf{0}. \tag{8.233}$$

   By taking $\mathbf{u}=\mathbf{v}$, we have our result

$$\mathbf{v}\times\frac{\partial\mathbf{v}}{\partial t}=\mathbf{0}, \tag{8.234}$$

   or, equivalently,

$$\frac{\partial\mathbf{v}}{\partial t}=C(\mathbf{x},t)\,\mathbf{v}. \tag{8.235}$$

4. *D'Alembert* motion:

$$\mathbf{v}(\mathbf{x},t)=f(t)\,\mathbf{u}(\mathbf{x}). \tag{8.236}$$

   These motions have steady stream lines, they are isochoric if and only if $\mathbf{u}$ is solenoidal, and their *kinematic vorticity number* A (see (3.502)) is steady. Now, since

$$\dot{\mathbf{v}}=\dot{f}\,\mathbf{u}+f^2\,\mathbf{u}\cdot\operatorname{grad}\mathbf{u}, \tag{8.237}$$

   we have

$$\boldsymbol{\omega}^\star\equiv\operatorname{curl}\dot{\mathbf{v}}=\dot{f}\operatorname{curl}\mathbf{u}+f^2\operatorname{curl}\left[(\operatorname{curl}\mathbf{u})\times\mathbf{u}\right]. \tag{8.238}$$

   Now, if $\dot{f}/f^2\neq$ const., then $\boldsymbol{\omega}^\star=\mathbf{0}$ cannot be satisfied unless $\boldsymbol{\omega}=\mathbf{0}$. Thus, besides the case when

$$f=\frac{1}{a\,t+b},\qquad a,b=\text{const.},\qquad a\,b\neq 0, \tag{8.239}$$

   a D'Alembert motion is circulation-preserving if and only if it is irrotational. In the case where (8.239) holds, the condition for $\boldsymbol{\omega}^\star=\mathbf{0}$ is that $a\operatorname{curl}\mathbf{u}=\operatorname{curl}\left[(\operatorname{curl}\mathbf{u})\times\mathbf{u}\right]$. If $a=0$, this is just a condition that $\mathbf{u}$ be the velocity field of a circulation-preserving motion, and this is the case if and only if $f=$ const.

5. Motion *without acceleration*:

$$\dot{\mathbf{v}}=\mathbf{0}. \tag{8.240}$$

   In this case, every particle initially at $\mathbf{X}$ travels in a straight line at a uniform velocity $\mathbf{v}(\mathbf{X})$. Steady motion without acceleration is typified by rectilinear shear motions. But when the stream lines are not steady, they are not straight, and the straight path lines may cross each other in a very large number of ways. A functional form of (8.240) is given by

$$\mathbf{x}=\mathbf{f}(\mathbf{X})\,t+\mathbf{X}, \tag{8.241}$$

and since $\mathbf{v} = \mathbf{f}(\mathbf{X})$, this functional equation may be rewritten in the form

$$\mathbf{v} = \mathbf{f}(\mathbf{x} - \mathbf{v}t). \tag{8.242}$$

Homogeneous motion without acceleration is a special case. In this case, substituting (8.217) into

$$\dot{\mathbf{v}} = \frac{\partial \mathbf{v}}{\partial t} + \mathbf{v} \cdot (\operatorname{grad} \mathbf{v}), \tag{8.243}$$

we obtain

$$\dot{\mathbf{A}} + \mathbf{A}^2 = \mathbf{0} \qquad \text{and} \qquad \dot{\mathbf{b}} = \mathbf{A} \cdot \mathbf{b}. \tag{8.244}$$

In the steady case, we have that

$$\mathbf{A}^2 = \mathbf{0} \qquad \text{and} \qquad \mathbf{A} \cdot \mathbf{b} = \mathbf{0}. \tag{8.245}$$

In the unsteady case, the general solution is given by the algebraic system

$$\mathbf{A} = (\mathbf{1} + \mathbf{A}_0 t)^{-1} \cdot \mathbf{A}_0 \qquad \text{and} \qquad \mathbf{b} = (\mathbf{1} + \mathbf{A}_0 t)^{-1} \cdot \mathbf{b}_0, \tag{8.246}$$

where $\mathbf{A}_0$ and $\mathbf{b}_0$ are the initial values of $\mathbf{A}$ and $\mathbf{b}$. In general, these motions develop singularities in a finite time, determined by the initial velocity and velocity gradient.

Using (8.217) and (8.241), we also have that

$$\mathbf{x} = (\mathbf{A} \cdot \mathbf{x} + \mathbf{b})\, t + \mathbf{X}. \tag{8.247}$$

Now, multiplying both sides by $(\mathbf{1} + \mathbf{A}_0 t)$ and using (8.246),

$$\mathbf{x} = (\mathbf{1} + \mathbf{A}_0 t) \cdot \mathbf{X} + \mathbf{b}_0 t, \tag{8.248}$$

which results in

$$\mathbf{v} = \mathbf{A}_0 \cdot \mathbf{X} + \mathbf{b}_0, \tag{8.249}$$

so that $\mathbf{v}$ is linear in $\mathbf{X}$ as well as in $\mathbf{x}$.

Example

Take

$$\mathbf{A}(t) = k(t) \begin{pmatrix} -\sigma(t) & 0 & 0 \\ 0 & -\sigma(t) & 0 \\ 0 & & -1 \end{pmatrix}, \qquad \mathbf{b}(t) = \mathbf{0}, \tag{8.250}$$

where, we see that to satisfy (8.244), we must have

$$k(t) = \frac{k_0}{1 + k_0\, t} \qquad \text{and} \qquad \sigma(t) = \sigma_0 \frac{1 + k_o\, t}{1 - k_0\, \sigma_0\, t}. \tag{8.251}$$

In this motion, a rectangular region of fluid with edges parallel to the coordinate planes is extended along the x_3-direction at the rate $k(t)$ and is contracted transversely in the ratio $\sigma(t)$. The stream lines in the $x_2 = 0$ plane are the curves $x_3\, x_1^{1/\sigma} = \text{const.}$, and the ratio of the current volume v to the initial volume V is given by $v/V = (1 + k_0 t)(1 - k_0 \sigma_0 t)^2$. We see that the motion develops a singularity when the volume is reduced to zero.

Restrictions on stretch

1. *Rigid* motion:

   A motion is rigid if all material lengths remain unchanged by the motion. The necessary and sufficient condition for this to be the case is given by

$$\mathbf{D} = \mathbf{0}. \tag{8.252}$$

   Its general solution is

$$\mathbf{v} = \dot{\mathbf{c}} + \mathbf{W}(t) \cdot (\mathbf{x} - \mathbf{c}(t)), \tag{8.253}$$

   where $\mathbf{W}$ is skew-symmetric and the corresponding axial vector is the angular velocity of the motion, and $\dot{\mathbf{c}}$ is the linear velocity of the origin of the frame relative to an arbitrary frame. If $\mathbf{W} = \mathbf{0}$, the rigid motion is just a translation.

2. *Isochoric* motion:

   A motion such that the volume occupied by any material region is unaltered, however, that region may change its shape in the course of time, is isochoric and subsequently

$$\mathbf{D}_{(1)} = \operatorname{div} \mathbf{v} = 0. \tag{8.254}$$

3. *Dilatational* motion:

   Since $\mathbf{D}$ is a real symmetric tensor, its real and orthogonal principal axes are the *axes of stretching*, and its real eigenvalues $d^{(i)}$ are the *principal stretchings*. When $\mathbf{D}$ is spherical, then the three principle stretchings are equal and the motion is a *dilatation*:

$$d^{(1)} = d^{(2)} = d^{(3)} = d. \tag{8.255}$$

   Equivalently, for a dilatation, the conditions on the invariants of $\mathbf{D}$ are (see (3.135)–(3.137))

$$\left(\frac{1}{3}\mathbf{D}_{(1)}\right)^3 = \left(\frac{1}{3}\mathbf{D}_{(2)}\right)^{3/2} = \mathbf{D}_{(3)}. \tag{8.256}$$

   For a general motion, the six components of $\mathbf{D}$ cannot be assigned arbitrarily. In order that there exists a field $\mathbf{v}$ such that $\mathbf{D}$ is symmetric, the integrability condition for the components is given by

$$\delta^{km}_{pq}\, \delta^{ln}_{rs}\, D_{kl,mn} = 0. \tag{8.257}$$

   Now, substituting (8.255) into (8.257), it becomes evident that $\mathbf{D}$ is a linear function, and therefore, the velocity is given, to within a rigid motion, by

$$\mathbf{v} = [\mathbf{a}(t) \cdot \mathbf{x} + b(t)]\,\mathbf{x} - \frac{1}{2}(\mathbf{x} \cdot \mathbf{x})\,\mathbf{a}(t). \tag{8.258}$$

4. *Shearing* motion:

A *rectilinear shearing* is a steady motion in which the paths of the material particles are parallel straight lines and the speed of each particle is constant in time. Thus, in the rectangular Cartesian system, we have

$$v_x = v_x(y,z), \qquad v_y = 0, \qquad v_z = 0. \tag{8.259}$$

We see that a pseudo-lineal motion of the first kind is isochoric if and only if it is a rectilinear shearing motion. Elements parallel or normal to the direction of motion are not stretched. However, elements normal to the direction of motion are rotated, and thus at a later instant are no longer normal. Very typical of this motion is simple shearing (see (3.162)–(3.164)):

$$v_x = \kappa\, y, \qquad v_y = 0, \qquad v_z = 0. \tag{8.260}$$

For this special case, we recall that the principal axes of stretching are the z-axis and the bisectors of the x- and y-axes, the principal stretchings are $\pm\frac{1}{2}\,\kappa$, the maximum orthogonal shears are experienced by elements in the x and y coordinate directions, the axis of spin is the z-axis, and both the vorticity and the maximum shearing have the value κ. Note that simple shearing is a special case of homogeneous motion.

For shearing, we must have that

$$d^{(1)} = -d^{(3)} \quad \text{and} \quad d^{(2)} = 0. \tag{8.261}$$

But this is not enough, since in simple shearing motion the spin is parallel to the second principal axis of stretching, and its magnitude satisfies $|\boldsymbol{\omega}| = 2\,d^{(1)}$. Therefore, the necessary and sufficient conditions for a shearing motion are that:

i) $\mathbf{D}_{(1)} = \mathbf{D}_{(3)} = 0$,

ii) $\boldsymbol{\omega}$ is parallel to the principal axis of the stretch along which the stretching is zero, and

iii) $|\boldsymbol{\omega}| = \sqrt{-4\,\mathbf{D}_{(2)}} = 2\,d^{(1)} =$ *amount of shearing*,

or, equivalently, note that any shearing motion may be regarded locally as a simple shearing superposed upon a rigid rotation.

Restrictions on spin

1. *Irrotational* motion:

A motion is irrotational if and only if its velocity field is lamellar, i.e.,

$$\mathbf{v} = -\operatorname{grad}\phi, \tag{8.262}$$

where ϕ is called the *velocity potential.* Thus, it follows that a motion for which the spin vanishes is irrotational:

$$\boldsymbol{\omega} = \operatorname{curl}\mathbf{v} = -\operatorname{curl}(\operatorname{grad}\phi) = \mathbf{0} \tag{8.263}$$

or, equivalently,

$$\mathbf{W} = \mathbf{0}. \tag{8.264}$$

Motions for which $\mathbf{W} \neq \mathbf{0}$ are called rotational.

Since $\mathbf{L} \neq \mathbf{0}$ is symmetric if and only if $\mathbf{W} = \mathbf{0}$, we conclude that a necessary and sufficient condition for a motion to be instantaneously irrotational at a point is that there exist three mutually orthogonal directions undergoing no instantaneous rotation. This follows from the fact that the symmetric tensor has three distinct eigenvalues. From this it also follows that a motion is locally a translation if and only if it is irrotational and rigid.

We note that the derivative of the velocity potential in the direction of the velocity is given by

$$\frac{d\phi}{ds} = \frac{\mathbf{v}}{|\mathbf{v}|} \cdot \operatorname{grad}\phi = -|\mathbf{v}| \leq 0. \tag{8.265}$$

Hence, the velocity potential can never increase in the direction of motion along a stream line, and in steady irrotational motion, a particle always moves toward a region of lower velocity potential. A stagnation point is a stationary point for the velocity potential along the stream line on which it occurs. In addition, using the definition of the material derivative in the Euler description, we also have that

$$\frac{\partial\phi}{\partial t} - \dot{\phi} = |\mathbf{v}|^2 \geq 0. \tag{8.266}$$

Hence, the squared speed is the excess of the local over the material time derivative of the velocity potential.

In an irrotational motion, since

$$\mathbf{D} = -\operatorname{grad}(\operatorname{grad}\phi), \tag{8.267}$$

we easily see upon contraction that

$$\mathbf{D}_{(1)} = -\nabla^2\phi. \tag{8.268}$$

Now, ϕ is sub-harmonic, harmonic, or super-harmonic according to $\mathbf{D}_{(1)} < 0$, $\mathbf{D}_{(1)} = 0$, or $\mathbf{D}_{(1)} > 0$. Therefore, in the interior of a region where the material volumes are not decreasing, the velocity potential cannot have a minimum, and in a region where the material volumes are not increasing, the velocity potential cannot have a maximum. Lastly, we see that an irrotational motion is isochoric if and only if the velocity potential is harmonic. Thus, all properties of *isochoric irrotational motions* follow from *potential theory.*

2. *Complex-lamellar* motion:

A motion is complex-lamellar if the velocity field is complex-lamellar:

$$\mathbf{v} = -f \operatorname{grad}\phi \tag{8.269}$$

or, analogously,

$$\boldsymbol{\omega} \cdot \mathbf{v} = 0. \tag{8.270}$$

Hence, a rotational motion is complex-lamellar if and only if its vortex lines are orthogonal to its stream lines. Both plane motions and rotationally symmetric motions are complex-lamellar. Complex-lamellar motions have some of the same properties as irrotational motions. The function ϕ is somewhat analogous to the velocity potential. Equations

$$f\,\frac{\partial \phi}{\partial s} = -|\mathbf{v}| \le 0 \qquad \text{and} \qquad f\left(\frac{\partial \phi}{\partial t} - \dot{\phi}\right) = |\mathbf{v}|^2 \ge 0 \tag{8.271}$$

are generalizations of (8.265) and (8.266). Thus, the conclusions regarding the stream lines of an irrotational motion derived from these equations carry over to a region of complex-lamellar motion where f is of one sign. Now, by continuity of the velocity field, f can be of opposite sign at two points on a stream line only if there is an intermediate point where $f = 0$, and subsequently, such point is a stagnation point. Consequently, in a region of complex-lamellar motion without stagnation points, the conclusions regarding the speed and the stream line pattern of an irrotational motion continue to hold, provided that $-\phi$ is substituted for ϕ if $f < 0$.

3. *Screw* motion:

To understand screw motions, we first note that we can write the acceleration field in the following form known as *Lagrange's* formula:

$$\begin{aligned} \dot{\mathbf{v}} = \frac{\partial \mathbf{v}}{\partial t} + \mathbf{v}\cdot \operatorname{grad}\mathbf{v} &= \frac{\partial \mathbf{v}}{\partial t} + 2\,\mathbf{W}\cdot\mathbf{v} + \operatorname{grad}\left(\frac{1}{2}\mathbf{v}\cdot\mathbf{v}\right) \\ &= \frac{\partial \mathbf{v}}{\partial t} + \boldsymbol{\omega}\times\mathbf{v} + \operatorname{grad}\left(\frac{1}{2}\mathbf{v}\cdot\mathbf{v}\right). \end{aligned} \tag{8.272}$$

This equation shows that in three-dimensional space, the acceleration is expressed in terms of four vectors: the local acceleration $\partial\mathbf{v}/\partial t$, the vorticity vector $\boldsymbol{\omega}$, the velocity vector $\mathbf{v}$, and the gradient of the specific kinetic energy $\operatorname{grad}\left(\frac{1}{2}\mathbf{v}\cdot\mathbf{v}\right)$. The vector $\boldsymbol{\omega}\times\mathbf{v}$ is called the *Lamb vector.* In a rotational motion with vanishing Lamb vector,

$$\boldsymbol{\omega}\times\mathbf{v} = \mathbf{0} \qquad \text{and} \qquad \boldsymbol{\omega} \neq \mathbf{0}, \tag{8.273}$$

the velocity field is a screw field and the motion is called a screw motion. As readily apparent, screw motions and complex-lamellar motions are mutually exclusive motions, but each shares some of the properties of irrotational motions. In a screw motion, just as in an irrotational motion, Lagrange's equation reduces to

$$\dot{\mathbf{v}} = \frac{\partial \mathbf{v}}{\partial t} + \operatorname{grad}\left(\frac{1}{2}\mathbf{v}\cdot\mathbf{v}\right). \tag{8.274}$$

Subsequently, we see that the vortex lines in this case coincide with the stream lines.

4. *Complex-screw* motion:

To understand complex-screw motions, it is useful to introduce a *generalized convection vector* field that is proportional to the velocity field

$$\mathbf{v} = v_0\,\boldsymbol{v}, \tag{8.275}$$

where v_0 is any non-vanishing scalar field whose substantial derivative is zero:

$$\dot{v}_0 = 0, \qquad v_0 \neq 0. \tag{8.276}$$

It is now easy to see that

$$\boldsymbol{\omega} = \operatorname{curl}(v_0 \boldsymbol{v}) = v_0 \boldsymbol{\varpi} + \operatorname{grad} v_0 \times \boldsymbol{v}, \tag{8.277}$$

where we have defined

$$\boldsymbol{\varpi} \equiv \operatorname{curl} \boldsymbol{v}. \tag{8.278}$$

We now see that

$$\mathbf{v} \cdot \boldsymbol{\omega} = v_0^2 \boldsymbol{v} \cdot \boldsymbol{\varpi}. \tag{8.279}$$

It follows that the generalized convection vector is complex-lamellar if and only if the motion is complex-lamellar.

Now motions in which

$$\boldsymbol{v} \cdot \boldsymbol{\varpi} = \mathbf{0} \tag{8.280}$$

are called complex-screw motions. To clarify such motions, first note that

$$\mathbf{v} \times \boldsymbol{\omega} = v_0^2 \boldsymbol{v} \times \boldsymbol{\varpi} + \mathbf{v} \cdot \mathbf{v} \operatorname{grad}(\log v_0) + \left(\frac{\partial \log v_0}{\partial t}\right) \mathbf{v}. \tag{8.281}$$

From this identity, we note that in a complex-screw motion, for which

$$\mathbf{v} \times \boldsymbol{\omega} = \mathbf{v} \cdot \mathbf{v} \operatorname{grad}(\log v_0) + \left(\frac{\partial \log v_0}{\partial t}\right) \mathbf{v}, \tag{8.282}$$

if v_0 is either uniform or steady, we have an irrotational or screw motion if and only if v_0 is both uniform and steady. This result implies that the class of complex-screw motions is more extensive than that of irrotational and screw motions. If we substitute this equation into Lagrange's formula (8.272), we obtain

$$\begin{aligned} \dot{\mathbf{v}} &= \frac{\partial \mathbf{v}}{\partial t} - \mathbf{v} \cdot \mathbf{v} \operatorname{grad}(\log v_0) - \left(\frac{\partial \log v_0}{\partial t}\right) \mathbf{v} + \operatorname{grad}\left(\frac{1}{2} \mathbf{v} \cdot \mathbf{v}\right) \\ &= v_0 \frac{\partial \boldsymbol{v}}{\partial t} + v_0^2 \operatorname{grad}\left(\frac{1}{2} \boldsymbol{v} \cdot \boldsymbol{v}\right). \end{aligned} \tag{8.283}$$

Hence, in a complex-screw motion whose convection vector is steady, the acceleration is complex-lamellar.

Taking the curl of the above equation, we have

$$\operatorname{curl} \dot{\mathbf{v}} = \operatorname{grad} v_0 \times \frac{\partial \boldsymbol{v}}{\partial t} + v_0 \frac{\partial \boldsymbol{\varpi}}{\partial t} + \frac{1}{2} \operatorname{grad} v_0^2 \times \operatorname{grad}(\boldsymbol{v} \cdot \boldsymbol{v}). \tag{8.284}$$

If we assume that the local time derivatives are zero: (i) if v_0 is uniform, then from $\boldsymbol{\varpi} \times \boldsymbol{v} = \mathbf{0}$ it follows that $\boldsymbol{\omega} \times \mathbf{v} = \mathbf{0}$; (ii) we may have $\boldsymbol{v} \cdot \boldsymbol{v} =$ const.; (iii) if the surfaces $v_0 =$ const. coincide with the surfaces $|\boldsymbol{v}| =$ const., these in turn are surfaces of constant speed. In summary, a complex-screw motion whose convection vector is steady is a circulation-preserving motion if and only if (a) it is an irrotational motion, or (b) its convection vector is of uniform magnitude, or (c) at each fixed time, v_0 is a function of the speed alone. When the motion itself is steady, then using a previous result, (c) is replaced by (c′) the surfaces of constant speed are Lamb surfaces, and the acceleration is normal to them.

5. Motion with *steady vorticity*:

The vortex lines are steady if and only if

$$\frac{\partial \boldsymbol{\omega}}{\partial t} \times \boldsymbol{\omega} = \mathbf{0}. \tag{8.285}$$

For the vortex lines to be steady, it is sufficient, but not necessary, that the vorticity itself be steady:

$$\frac{\partial \boldsymbol{\omega}}{\partial t} = \operatorname{curl}\left(\frac{\partial \mathbf{v}}{\partial t}\right) = \mathbf{0}. \tag{8.286}$$

From (8.285), it follows that

$$\frac{\partial \boldsymbol{\omega}}{\partial t} = C\,\boldsymbol{\omega}, \tag{8.287}$$

where C is a scalar quantity. Thus, we have

$$\operatorname{curl}\left(\frac{\partial \mathbf{v}}{\partial t}\right) = \operatorname{curl}(C\,\mathbf{v}) - \operatorname{grad} C \times \mathbf{v}. \tag{8.288}$$

Taking the divergence of this equation, we obtain

$$0 = \operatorname{div}(\operatorname{grad} C \times \mathbf{v}) = -\operatorname{grad} C \cdot \boldsymbol{\omega}. \tag{8.289}$$

Hence, it follows that in a motion where the vortex lines are steady but the vorticity is not steady, the surfaces upon which $\partial\boldsymbol{\omega}/\partial t$ has a constant ratio to $\boldsymbol{\omega}$ are vortex surfaces.

Now it is clear that (8.285) is satisfied if and only if there exists a scalar field $U(\mathbf{x}, t)$ such that

$$\frac{\partial \mathbf{v}}{\partial t} = \operatorname{grad} U. \tag{8.290}$$

This relation may be substituted into Lagrange's equation (8.272) to obtain

$$\dot{\mathbf{v}} = \boldsymbol{\omega} \times \mathbf{v} + \operatorname{grad}\left(\frac{1}{2}\mathbf{v}\cdot\mathbf{v} + U\right). \tag{8.291}$$

6. *Circulation-preserving* motion:

In order to investigate the possibility that the vortex lines be material lines, we substitute $\mathbf{u} = \boldsymbol{\omega}$ into the Helmholtz–Zorawski criterion (8.233) to obtain

$$\boldsymbol{\omega} \times \left[\frac{\partial \boldsymbol{\omega}}{\partial t} + \operatorname{curl}(\boldsymbol{\omega} \times \mathbf{v})\right] = \mathbf{0}. \tag{8.292}$$

Also, by taking the curl of Lagrange's formula (8.272), we obtain

$$\boldsymbol{\omega}^{\star} \equiv \operatorname{curl}\dot{\mathbf{v}} = \frac{\partial \boldsymbol{\omega}}{\partial t} + \operatorname{curl}(\boldsymbol{\omega} \times \mathbf{v}) \tag{8.293}$$

for any motion. Hence, (8.292) becomes

$$\boldsymbol{\omega} \times \boldsymbol{\omega}^{\star} = \mathbf{0}. \tag{8.294}$$

Thus, a necessary and sufficient condition that the vortex lines be material lines is that the acceleration be lamellar or that its curl be parallel to the vorticity. By putting $\mathbf{u} = \boldsymbol{\omega}$ into the Zorawski criterion (3.458), we similarly obtain

$$\boldsymbol{\omega}^{\star} = \mathbf{0}. \tag{8.295}$$

Subsequently, a necessary and sufficient condition that the strengths of all vortex tubes remain constant in time is that the acceleration be lamellar; this condition is also sufficient that the vortex tubes be material tubes.

8.5.2 Specializations of the equations of motion

Creeping flow

It is sometimes justifiable to assume that the velocity is small so that terms involving velocity of second and higher order are negligible in comparison to linear terms of velocity. Invoking this approximation leads to a linear equation for the velocity field, which can be solved more readily, particularly in light of the fact that such flows are often isochoric. For the Navier–Stokes equations, this leads to *Stokes' flow* about various bodies.

Perturbed flow

Another similar specialization arises in stability theory when a basic flow is known and is perturbed by a small amount. Here, the squares and higher products of the small perturbations are regarded as negligible. Again, a linear problem results, which for the Navier–Stokes equations leads to problems such as in acoustics, as we have seen earlier, and *Orr–Sommerfeld equations* in hydrodynamic stability.

Boundary layer flow

When large gradients of dependent variables are confined to the neighborhood of a boundary, we refer to the region close to the boundary as a boundary layer. The boundary layer is characterized by a sharp change in the velocity in the direction normal to the boundary (*shear layer*) or a sharp change in the temperature (*thermal boundary layer*). It usually happens that the thickness of this layer is small in comparison with the longitudinal dimensions. Advantage is taken of this fact to reduce the equations to simpler forms by neglecting certain terms in the balance equations by comparison with others. The simplifications lead to approximate equations called the *boundary layer equations.* More generally, the boundary layer phenomenon arises in narrow zones near a boundary as well as in the interior of a fluid where such surfaces of "discontinuity" are referred to as shocks.

The concept of a boundary layer was introduced by L. Prandtl in connection with the hydrodynamics of viscous liquids. Although the concept has its origin in fluid dynamics, the mathematical developments of this concept has led to the field of *perturbation theory.*

8.5.3 Specializations of the constitutive equations

Newtonian fluid

Here, the assumption of a linear relation between stress and strain rate, as we have seen, leads to the constitutive equation

$$\boldsymbol{\sigma} = (-p + \lambda \operatorname{div} \mathbf{v})\,\mathbf{1} + 2\,\mu\,\mathbf{D}, \tag{8.296}$$

where $\mu = \mu(\rho, \theta)$ and $\lambda = \lambda(\rho, \theta)$. The equations of motion then become the *Navier equations.*

Stokesian fluid

A Newtonian fluid that incorporates the Stokes hypothesis of a vanishing bulk viscosity (it vanishes for a monatomic gas), i.e.,

$$\zeta = \lambda + \frac{2}{3}\,\mu = 0, \tag{8.297}$$

is referred to as a Stokesian fluid and the linear momentum equations with this simplification of the stress tensor is referred to as the *Navier–Stokes equations.*

Ideal fluid

An ideal fluid has no viscosity, i.e., $\lambda = \mu = 0$, so that

$$\boldsymbol{\sigma} = -p\,\mathbf{1} \tag{8.298}$$

and

$$\rho\,(\mathbf{a} - \mathbf{f}) = -\operatorname{grad} p. \tag{8.299}$$

This is known as *Euler's equations of motion.* In addition, the ideal fluid has zero conductivity, k, and if there is no external heat supply, the energy equation becomes

$$\dot{\eta} = 0 \tag{8.300}$$

so that the flow is isentropic.

Ideal gas

The free energy of an *ideal gas* is given by

$$\psi(\rho, \theta) = e_0 - (\theta - \theta_0)\,\eta_0 + \theta\left[\int_{\theta_0}^{\theta}\left(\frac{1}{\theta} - \frac{1}{\theta'}\right) c_v(\theta')\,d\theta' + R\ln\rho\right], \tag{8.301}$$

where θ_0, e_0, and η_0 are constants, and $R = c_p - c_v$. In this case, we have that

$$\eta(\rho, \theta) = \eta_0 + \int_{\theta_0}^{\theta} \frac{c_v(\theta')}{\theta'}\,d\theta' - R\ln\rho, \tag{8.302}$$

and it is easy to see that the caloric and thermal equations of state are

$$e(\rho, \theta) = e(\theta) = e_0 + \int_{\theta_0}^{\theta} c_v(\theta')\,d\theta' \quad \text{and} \quad p(\rho, \theta) = R\,\rho\,\theta. \tag{8.303}$$

Ideal gases do not exist in nature, but the properties of all known gases approach those of the respective ideal gas as the pressure is extrapolated to zero. We say that these equations of state constitute the asymptotic forms for all real gases. In spite of this, the ideal gas equations of state can be used as approximations to represent the properties of real gases even at comparatively high pressures. When used judiciously, the equations provide a high degree of accuracy, and many practical calculations are based on them. Values of the gas constant and several other properties of ideal gases are usually collected in tables to facilitate calculations.

Piezotropic fluid and barotropic flow

A fluid is said to be piezotropic if

$$e = e(\eta, \rho) = e_1(\eta) + e_2(\rho), \tag{8.304}$$

in which case we have that

$$\theta = e_1'(\eta) \qquad \text{and} \qquad p = -e_2'(\rho), \tag{8.305}$$

where the primes indicate derivatives with respect to the corresponding independent quantities. Many liquids can be represented in a first approximation as piezotropic. A fluid in which $p \equiv 0$, i.e., $e_2 =$ const., is incompressible; what is called "pressure" in an incompressible fluid is not given by the thermodynamic definition and is not a thermodynamic variable at all. It is easy to see also that a fluid is piezotropic if and only if $\gamma = 1$, or, equivalently, $\alpha = 0$ (see (5.161)–(5.164)).

If the motion is such that the density and pressure are directly related (e.g., as in $(8.305)_2$) the motion is called barotropic. Thus, all piezotropic fluids (which includes incompressible fluids as a special case) flow barotropically, but other fluids may also do so. The terms piezotropic and barotropic thus stand in the same relationship as incompressible and isochoric. The relation between p and ρ allows us to write

$$\frac{1}{\rho}\operatorname{grad} p = \operatorname{grad} P(p) = \operatorname{grad} \int \frac{1}{\rho}\, dp, \tag{8.306}$$

and such relation becomes useful in irrotational motion. The quantity P is called the pressure function.

Incompressible fluid

All real fluids subjected to a sufficiently large uniform pressure change their specific volume, and hence their density. Depending on the properties of the fluid and the range of variation of pressure in a particular process, the specific volume may undergo small changes. Sometimes this change is so small that disregarding it does not lead to serious errors in the solution of a particular problem. In the case of liquids, it is necessary to apply very large changes in pressure to effect a significant change in specific volume. In the case of gases, negligible changes in specific volume result only from sufficiently small changes in pressure. Similarly, in many cases, the temperature changes during a process are so small that their effect upon the specific volume can also be disregarded. In all such cases, we then say that the fluid is incompressible. In terms of thermodynamics, this is equivalent to replacing the thermal equation of state of the system by

$$\mathsf{v} = \text{const.} \qquad \text{or} \qquad \rho = \text{const.} \tag{8.307}$$

The constant is selected as a suitable average in a particular problem with reference to the true thermal equation of state

$$F(\rho,\theta,p)=0. \tag{8.308}$$

Geometrically, the assumption (8.307) is equivalent to saying that the surface of states is a plane perpendicular to the ρ-axis. Thus, the real surface $F(\rho,\theta,p)=0$ is replaced by the plane $\rho=$ const.; in general, the latter is not tangent to the former and does not constitute a mathematical approximation in the same sense that a tangent surface would, and it is necessary to realize that we are just dealing with a physical approximation. Consequently, the relations between the thermodynamic properties of the approximation need not be the same as those between the thermodynamic properties of the real fluid.

Writing the total differential

$$d\rho=\left.\frac{\partial\rho}{\partial\theta}\right|_p d\theta+\left.\frac{\partial\rho}{\partial p}\right|_\theta dp, \tag{8.309}$$

we notice that (8.307) implies $d\rho=0$ for any values of the increments of $d\theta$ and dp, so that

$$\left.\frac{\partial\rho}{\partial\theta}\right|_p=\left.\frac{\partial\rho}{\partial p}\right|_\theta=0, \tag{8.310}$$

which says that ρ remains constant regardless of the variation in pressure and temperature. It follows that for an incompressible fluid, the coefficient of thermal expansion, the isothermal compressibility, and the isochoric pressure coefficient (or coefficient of tension) acquire the singular values

$$\alpha=-\frac{1}{\rho}\left.\frac{\partial\rho}{\partial\theta}\right|_p=0,\quad \kappa_\theta=-\frac{1}{\rho}\left.\frac{\partial\rho}{\partial p}\right|_\theta=0,\quad \pi=\frac{1}{p}\left.\frac{\partial p}{\partial\theta}\right|_\rho=\text{indeterminate}. \tag{8.311}$$

In solving a problem, the above values must be used for consistency.

The internal energy of an incompressible fluid is a function of temperature alone,

$$e=e(\theta), \tag{8.312}$$

and the specific heat at constant volume also becomes a function of temperature alone,

$$c_v=\left.\frac{\partial e}{\partial\theta}\right|_\rho=c_v(\theta). \tag{8.313}$$

The form of the function must be obtained from experiments. Since the enthalpy is related to the internal energy by the relation $h=e+p/\rho$, and since the quantity p/ρ is independent of temperature, the specific heat at constant pressure can be calculated by direct differentiation to obtain

$$c_p=\left.\frac{\partial h}{\partial\theta}\right|_p=\left.\frac{\partial e}{\partial\theta}\right|_p. \tag{8.314}$$

Since, however, the internal energy is a function of temperature alone, we have

$$\left.\frac{\partial e}{\partial\theta}\right|_p=\left.\frac{\partial e}{\partial\theta}\right|_\rho=c_v, \tag{8.315}$$

and thus

$$c_p = c_v = c. \tag{8.316}$$

Hence, there is no distinction between the two specific heats and the symbol c is usually used to replace both c_p and c_v.

In some problems, it is found necessary to take into account the effect of a change in temperature on the specific volume, while the effect of change in pressure may still be disregarded:

$$\left.\frac{\partial \rho}{\partial \theta}\right|_p \neq 0 \qquad \text{and} \qquad \left.\frac{\partial \rho}{\partial p}\right|_\theta = 0. \tag{8.317}$$

Consequently,

$$\alpha = -\frac{1}{\rho}\left.\frac{\partial \rho}{\partial \theta}\right|_p \neq 0, \tag{8.318}$$

and since in this case the density is a function of temperature only, we must also assume that

$$\alpha = \alpha(\theta). \tag{8.319}$$

In other words, the coefficient of thermal expansion, as well as the density, must be averaged with respect to the variation of pressure in the problem, but their dependence on temperature is accounted for. This fluid model is also described as incompressible. The difference between this and the previous case is that now the effect of thermal expansion is allowed for.

Geometrically, the equation $F(\rho, \theta, p) = 0$ of an incompressible but thermally expanding fluid is described by a cylindrical surface, since $\rho = \text{const.}$ at $\theta = \text{const.}$ This surface averages, but does not approximate, the real ρ, θ, p surface.

In addition, we note that in the first model of incompressible flow, the work per unit mass in any reversible process vanishes since from (5.156) with $\boldsymbol{\tau}_1 = -p$ and $\boldsymbol{\nu}_1 = \mathrm{v}$, we have

$$đw = -p\, d\mathrm{v} = \frac{p}{\rho^2}\, d\rho = 0. \tag{8.320}$$

In the second model, where we allow thermal expansion, the work per unit mass is nonzero:

$$đw = -p\, d\mathrm{v} = -\frac{\alpha p}{\rho}\, d\theta. \tag{8.321}$$

The motion of an incompressible fluid for which $\rho = \text{const.}$ is always isochoric and previous considerations of isochoric motion apply. Subsequently, the linear momentum equations with this simplification of the stress tensor is referred to as the *Navier–Stokes equations*. It should be remembered that for an incompressible medium, the pressure is not defined thermodynamically, but is an independent variable of the motion.

Linear liquid

For some substances, including many liquids (and solids), since their volume change is small, it is useful to use a thermal equation of state which is linearized about the fixed state (θ_0, p_0):

$$\rho = \rho(\theta, p) = \rho_0\left[1 - \alpha_0\left(\theta - \theta_0\right) + \kappa_{\theta 0}\left(p - p_0\right)\right], \tag{8.322}$$

where the reference density, coefficient of volume expansion, and isothermal compressibility are given by

$$\rho_0 = \rho_0(\theta_0, p_0), \qquad \alpha_0 = -\frac{1}{\rho}\frac{\partial \rho}{\partial \theta}\bigg|_{\theta_0, p_0}, \qquad \text{and} \qquad \kappa_{\theta 0} = \frac{1}{\rho}\frac{\partial \rho}{\partial p}\bigg|_{\theta_0, p_0}. \tag{8.323}$$

The corresponding caloric equation of state is given by

$$\begin{aligned} e = e(\theta, p) = e_0 + \int_{\theta_0}^{\theta} c_v(\theta')\, d\theta' - \frac{1}{\rho_0\, \kappa_{\theta 0}} \{ \alpha_0\, \theta - \kappa_{\theta 0}\, p + \\ (1 + \alpha_0\, \theta_0 - \kappa_{\theta 0}\, p_0) \ln [1 - \alpha_0\, (\theta - \theta_0) + \kappa_{\theta 0}\, (p - p_0)] \}. \end{aligned} \tag{8.324}$$

To be able to obtain all properties, it is now just necessary to specify $c_v(\theta)$ or $c_p(\theta)$. For liquids, values are usually obtained experimentally and subsequently tabled (for solids, approximate formulas from statistical mechanics exist). Here we only note that for many liquids, $c_p(\theta)$ is almost independent of temperature (for liquid water $c_p \approx 4.1855$ J/g·K). Then, the temperature dependence of $c_v(\theta)$ is obtained from

$$c_p - c_v = \frac{\alpha_0^2\, \theta}{\rho_0\, \kappa_{\theta 0}\, [1 - \alpha_0\, (\theta - \theta_0) + \kappa_{\theta 0}\, (p - p_0)]}. \tag{8.325}$$

Note that even for those liquids with constant c_p, there consequently is a temperature dependence of $c_v(\theta)$.

Table 8.1: Normal modes of acoustic propagation.

Ideal Fluid	Dissipative Fluid	Normal Mode
$\lambda_1 = i\,c_0\,k$	$\lambda_1 = i\,c_0\,k - \frac{1}{2}\left[\frac{4}{3}\,\nu_0 + \nu_0' + (\gamma_0 - 1)\,\chi_0\right]k^2 + O(k^3)$	Acoustic
$\lambda_2 = -i\,c_0\,k$	$\lambda_2 = -i\,c_0\,k - \frac{1}{2}\left[\frac{4}{3}\,\nu_0 + \nu_0' + (\gamma_0 - 1)\,\chi_0\right]k^2 + O(k^3)$	Acoustic
$\lambda_3 = 0$	$\lambda_3 = -\nu_0\,k^2$	Shear
$\lambda_4 = 0$	$\lambda_4 = -\nu_0\,k^2$	Shear
$\lambda_5 = 0$	$\lambda_5 = -\chi_0\,k^2 + O(k^4)$	Thermal

Problems

1. For a scalar quantity such as entropy, frame invariance requires that
$$\eta\left(\rho, \mathbf{Q}\cdot\mathbf{D}\cdot\mathbf{Q}^T, \theta, \mathbf{Q}\cdot\mathbf{g}\right) = \eta\left(\rho, \mathbf{D}, \theta, \mathbf{g}\right).$$
Use the same procedure used for the frame invariance of a vector quantity (see (8.58)) to obtain a result analogous to (8.61).

2. Derive (8.167)–(8.171).

3. Determine $\sigma_{11} - \sigma_{22}$ and $\sigma_{11} - \sigma_{33}$ for a simple shear flow, where
$$\boldsymbol{\sigma} = -p\,\mathbf{1} + 2\,\mu\,\mathbf{D} + \nu\,\mathbf{D}^2.$$

4. i) Show that for a linear compressible viscous fluid described by the stress tensor
$$\boldsymbol{\sigma} = -p(\rho,\theta)\,\mathbf{1} + \lambda(\rho,\theta)\,(\mathrm{tr}\,\mathbf{D})\,\mathbf{1} + 2\,\mu\,(\rho,\theta)\,\mathbf{D},$$
the viscous dissipation is given by
$$\Phi = \lambda\,(\mathrm{tr}\,\mathbf{D})^2 + 2\,\mu\,\mathrm{tr}\,(\mathbf{D}^2). \tag{8.326}$$
ii) Show that (8.326) can be rewritten as
$$\Phi = \left(\lambda + \frac{2}{3}\,\mu\right)(\mathrm{tr}\,\mathbf{D})^2 + 2\,\mu\,\mathrm{tr}\left[\mathbf{D} - \frac{1}{3}(\mathrm{tr}\,\mathbf{D})\mathbf{1}\right]^2. \tag{8.327}$$
iii) Show that (8.326) can be rewritten as
$$\Phi = (\lambda + 2\,\mu)\,D_{(1)}^2 - 4\,\mu\,D_{(2)}, \tag{8.328}$$
where $D_{(1)}$ and $D_{(2)}$ are the first two invariants of $\mathbf{D}$.

5. For a Newtonian fluid, show that the mechanical energy can be written as
$$\Phi = \zeta\,D_{(1)}^2 + 2\mu\,D'_{ik}D'_{ik}.$$
Furthermore, writing the Cauchy stress tensor as $\sigma_{ij} = -p\,\delta_{ij} + \sigma_{ij}^d$, show that $\partial\Phi/\partial D_{ij} = \sigma_{ij}^d/2$.

6. Show that for an incompressible, irrotational flow with velocity potential ϕ given by $\mathbf{v} = -\nabla\phi$, the mechanical energy for a Newtonian fluid is
$$\Phi = \nabla^2\,(\nabla\phi)^2 = \frac{1}{2}\nabla^4\phi^2.$$

7. i) Verify (8.199)–(8.201) starting from (8.187)–(8.189).
ii) Verify (8.203)–(8.205) starting from (8.199)–(8.201).
iii) Verify (8.209)–(8.211) starting from (8.203)–(8.205).
iv) Rewrite (8.209)–(8.211) in the from (8.214) and subsequently verify the dispersion relation (8.215).

v) Plot λ_i $(i = 1, \ldots, 5)$ vs. k for air at a temperature of 300 K and a pressure of 1 atm, treating air as an ideal gas, and compare with the approximate expressions for λ_i.

8. If a fluid motion is very slow so that higher order terms in the velocity are negligible, the limiting case of creeping flow results. For this case, show that in a steady incompressible flow with zero body forces, the pressure is a harmonic function, i.e., $\nabla^2 p = 0$.

9. Assuming constant viscosity, show that the Navier–Stokes equations for isochoric irrotational flow reduce to the Euler equations of motion.

10. Express the mass conservation and Navier–Stokes equations in terms of the velocity potential ϕ for an irrotational motion assuming constant viscosities.

11. Show that if the body force is conservative so that $\mathbf{f} = -\text{grad}\ \Omega$, the compressible Navier–Stokes equations for the irrotational motion of an inviscid barotropic fluid may be integrated to yield

$$-\frac{\partial \phi}{\partial t} + \frac{1}{2}(\nabla \phi)^2 + P + \Omega = f(t),$$

where ϕ is the velocity potential given by $\mathbf{v} = -\nabla \phi$, P is the pressure function, and f is an arbitrary function of time. This equation is known as *Bernoulli's equation*. Subsequently, show that for steady flow of an ideal gas, it takes the form

a) for isothermal flow (p/ρ = const.)

$$\frac{1}{2}v^2 + \Omega + \frac{p}{\rho}\ln p = \text{const.}$$

and

b) for isentropic flow (p/ρ^γ = const.)

$$\frac{1}{2}v^2 + \Omega + \left(\frac{\gamma}{\gamma - 1}\right)\left(\frac{p}{\rho}\right) = \text{const.}$$

12. Starting from the Euler equations of motion for a fluid of constant density with a potential force per unit mass $\mathbf{f} = -\text{grad}\ \Omega$, show that for a fixed volume $\mathscr{V}$ enclosed by a surface $\mathscr{S}$, we have

$$\frac{d}{dt}\int_{\mathscr{V}} \frac{1}{2}\rho v^2 dv + \int_{\mathscr{S}} \rho H \mathbf{v} \cdot d\mathbf{s},$$

where

$$H = \frac{1}{2}v^2 + \Omega + \frac{p}{\rho}$$

is the Bernoulli quantity.

13. A two-dimensional irrotational flow given by the velocity potential $\phi = e^{kx_2}\sin kx_1$, with $k > 0$, occupies the half-space $x_2 < 0$. Show that the flow is incompressible. Calculate the velocity field $\mathbf{v}$ and the stream function $\psi(x_1, x_2)$. Sketch the stream lines.

14. Calculate the vorticity $\boldsymbol{\omega} = \operatorname{curl} \mathbf{v}$ of the velocity field

$$v_1 = -\alpha\, x_1 - r\, x_2\, f(t), \qquad v_2 = -\alpha\, x_2 + r\, x_1\, f(t), \qquad v_3 = 2\,\alpha\, x_3,$$

where $x_1^2 + x_2^2 = r^2$. Show that $\operatorname{div} \mathbf{v} = 0$ for any function $f(t)$, and that the vorticity equation for a Newtonian fluid is satisfied if and only if $f(t) \propto e^{2\,\alpha\, t}$.

15. Show that for an inviscid flow with a conservative body force and constant density, the following relation between vorticity and velocity is satisfied:

$$\dot{\omega}_i - \omega_j v_{i,j} = 0.$$

For steady flow, show that

$$v_j \omega_{i,j} = \omega_i v_{i,j}.$$

16. Determine the pressure function $P(p)$ for a piezotropic fluid having the equation of state $p = k\,\rho^\gamma$, where k and γ are constants.

17. A non-dissipative van der Waal compressible gas is characterized by the Helmholtz free energy

$$\psi(\mathsf{v}, \theta) = -R\,\theta\, \log(\mathsf{v} - b) - \frac{a}{\mathsf{v}} - c_\mathsf{v}\,\theta\, \log\left(\frac{\theta}{\theta_0}\right),$$

where v is the specific volume, and a and b are constants. Determine explicit expressions for $e(\mathsf{v}, \theta)$, $p(\mathsf{v}, \theta)$, and $\eta(\mathsf{v}, \theta)$.

18. Show that for an ideal gas under isothermal conditions and with body force $\mathbf{f} = -g\,\mathbf{i}_3$, we have

$$\frac{\rho}{\rho_0} = \frac{p}{p_0} = e^{-(g/R\,\theta\, x_3)},$$

where ρ_0 and p_0 are the density and pressure at $x_3 = 0$.

Bibliography

R. Aris. *Vectors, Tensors and the Basic Equations of Fluid Mechanics.* Dover Publications, Inc., Mineola, NY, 1962.

T.S. Chang. Constitutive equations for simple materials: Thermoelastic fluids. In A.C. Eringen, editor, *Continuum Physics*, volume II, chapter 3, pages 267–281. Academic Press, Inc., New York, NY, 1975.

B.D. Coleman and W. Noll. The thermodynamics of elastic materials with heat conduction and viscosity. *Archive for Rational Mechanics and Analysis*, 13(1):167–178, 1963.

A.C. Eringen. *Mechanics of Continua.* R.E. Krieger Publishing Company, Inc., Melbourne, FL, 1980.

C.B. Kadafar. Methods of solution: Exact solutions in fluids and solids. In E.H. Dill, editor, *Continuum Physics*, volume II, pages 407–448. Academic Press, Inc., New York, NY, 1975.

I.-S. Liu. On entropy flux-heat flux relation in thermodynamics with Lagrange multipliers. *Continuum Mechanics and Thermodynamics*, 8(4):247–256, 1996.

I.-S. Liu. *Continuum Mechanics.* Springer-Verlag, Berlin, 2002.

C. Truesdell and W. Noll. The non-linear field theories of mechanics. In S. Flügge, editor, *Handbuch der Physik*, volume III/3. Springer, Berlin-Heidelberg-New York, 1965.

C. Truesdell and K.R. Rajagopal. *An Introduction to the Mechanics of Fluids.* Birkhauser, Boston, MA, 2000.

9

Viscoelasticity

9.1 Introduction

The stress tensor for an incompressible Newtonian fluid is given by

$$\sigma_{ik} = -p\,\delta_{ik} + 2\,\mu\,D_{ik}, \tag{9.1}$$

where

$$D_{ik} = \frac{1}{2}\left(\frac{\partial v_i}{\partial x_k} + \frac{\partial v_k}{\partial x_i}\right) \tag{9.2}$$

and $v_i = v_i(x_j, t)$. For the simple shear (Couette) flow illustrated in Fig. 9.1, we have

$$v_1 = 0, \qquad v_2 = \kappa\,x_1, \quad v_3 = 0, \tag{9.3}$$

where $\kappa = V/d$ is the amount of shear, so that $\sigma_{12} = \sigma_{21} = \mu\,\kappa$ with all other dissipative stress components being zero.

In fact, when we do an experiment, we find that

$$\sigma_{12} = \sigma_{21} = \mu\,(\kappa)\;\kappa, \tag{9.4}$$

where $\mu = \mu\,(\kappa)$ is called the *apparent shear viscosity.* Different variations of viscosity have been observed as shown in Fig. 9.2. *Shear thinning* and *shear thickening*

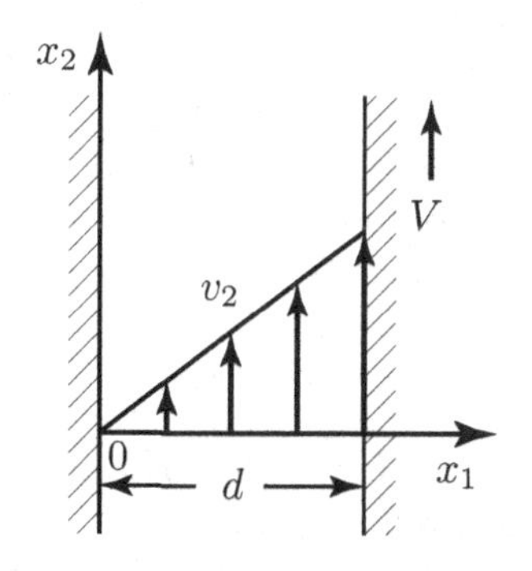

Figure 9.1: Simple shear flow.

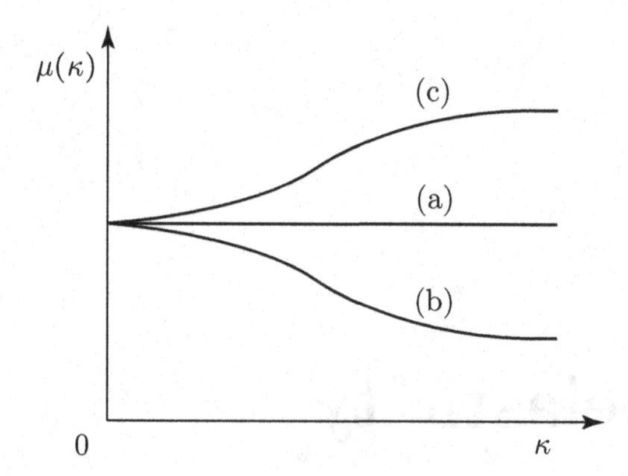

Figure 9.2: Variations of shear viscosity: (a) Newtonian fluid; (b) shear thinning fluid; and (c) shear thickening fluid.

fluids are also referred to as *pseudoplastic* and *dilatant* fluids, respectively, and such fluids are called *non-Newtonian* fluids.

We note that Newtonian fluids, such as air and water, tend to be low-molecular-weight substances with little internal structure. On the other hand, non-Newtonian fluids, including plastics, food materials, paints, inks, biological fluids, and lubricants just to name a few, tend to have some internal structure that causes non-Newtonian behavior. Other anomalous effects of non-Newtonian fluids are the climbing on a rotating rod, called the *Weissenberg effect*, and the *die swell effect* of a fluid exiting from a tube. In contrast to simple shear, both of these phenomena are normal stress effects.

To reconcile with experiments, one might consider the Reiner–Rivlin fluid whose stress tensor for an incompressible fluid is given by (see (8.116))

$$\sigma_{ik} = \sigma_{ik}\left(D_{lm}\right) = -p\,\delta_{ik} + \alpha_1\,D_{ik} + \alpha_2\,D_{il}\,D_{lk}, \tag{9.5}$$

where $\sigma_{ik} = \sigma_{ki}$, and α_1 and α_2 are functions of the invariants $D_{(2)}$ and $D_{(3)}$ (note that $D_{(1)} = 0$ for an incompressible fluid). Now, for the simple shear flow given by (9.3), we find that

$$[\sigma_{ik}] = -p\begin{pmatrix} 1 & 0 & 0 \\ 0 & 1 & 0 \\ 0 & 0 & 1 \end{pmatrix} + \alpha_1\begin{pmatrix} 0 & \frac{1}{2}\kappa & 0 \\ \frac{1}{2}\kappa & 0 & 0 \\ 0 & 0 & 0 \end{pmatrix} + \alpha_2\begin{pmatrix} \frac{1}{4}\kappa^2 & 0 & 0 \\ 0 & \frac{1}{4}\kappa^2 & 0 \\ 0 & 0 & 0 \end{pmatrix}, \tag{9.6}$$

the invariants are $D_{(1)} = D_{(3)} = 0$, $D_{(2)} = -\frac{1}{4}\kappa^2$, and thus $\alpha_1 = \alpha_1(\kappa^2)$ and $\alpha_2 = \alpha_2(\kappa^2)$. For this flow, assuming that no body forces are present, the mass and linear momentum balances are automatically satisfied. To maintain this motion, we need to apply the following stresses:

$$\sigma_{12} = \sigma_{21} = \frac{1}{2}\,\alpha_1(\kappa^2)\kappa, \qquad \sigma_{11} = \sigma_{22} = -p + \frac{1}{4}\,\alpha_2(\kappa^2)\kappa^2, \qquad \sigma_{33} = -p. \tag{9.7}$$

However, experimentally it is found that σ_{11} and σ_{22} are not equal. Consequently, this extension of Newtonian flow fails.

In passing, we note that special flows that enable the measurement of viscosities are referred to as *viscometric flows*. In addition to the simple shearing flows, there are other incompressible steady flows that can be solved exactly, and the resulting stress and velocity profiles are related to the same material functions. Such problems include the flow through a circular pipe (Poiseuille flow), the flow between two concentric pipes, and the flow between two such pipes but in which the pipes rotate with different constant velocities.

To gain some insight on why and when the Reiner–Rivlin fluid fails, let us examine the incompressible *elongational* or *stagnation flow*

$$v_1 = -a\,x_1, \qquad v_2 = a\,x_2, \qquad v_3 = 0. \tag{9.8}$$

Now we find that

$$[\sigma_{ik}] = -p\begin{pmatrix} 1 & 0 & 0 \\ 0 & 1 & 0 \\ 0 & 0 & 1 \end{pmatrix} + \alpha_1\begin{pmatrix} -a & 0 & 0 \\ 0 & a & 0 \\ 0 & 0 & 0 \end{pmatrix} + \alpha_2\begin{pmatrix} a^2 & 0 & 0 \\ 0 & a^2 & 0 \\ 0 & 0 & 0 \end{pmatrix}, \tag{9.9}$$

where $D_{(1)} = D_{(3)} = 0$, $D_{(2)} = -a^2$, and thus $\alpha_1 = \alpha_1(a^2)$ and $\alpha_2 = \alpha_2(a^2)$. Subsequently, we have the following normal stress differences:

$$\sigma_{22} - \sigma_{11} = 2\,\alpha_1 a, \qquad \sigma_{33} - \sigma_{11} = \alpha_1 a - \alpha_2\,a^2, \qquad \sigma_{33} - \sigma_{22} = -\alpha_1 a - \alpha_2\,a^2. \tag{9.10}$$

In this case, we see normal stress differences that are observed in experiments. We note that the above flow is equivalent to the shearing flow

$$v_1 = \kappa\,x_2, \qquad v_2 = \kappa\,x_1, \qquad v_3 = 0 \tag{9.11}$$

with $\kappa = -a$. This is seen by rotating the x_1-x_2 axes by $\pi/2$.

At this stage, to gain additional insight, we could superpose the shearing and elongational flows as in

$$v_1 = -a\,x_1, \qquad v_2 = \kappa\,x_1 + a\,x_2, \qquad v_3 = 0. \tag{9.12}$$

However, although the above generalization is interesting, in practice it is found that markedly non-Newtonian fluids have a more complex behavior than is permitted by the Reiner–Rivlin constitutive equation. Subsequently, we wish to introduce the effect of history on such flow. Thus, we examine the flow whose material points were at $\xi_i(\tau)$ at time τ, which at time t are at $x_i(t)$, i.e., we examine the motion $\xi_i = \xi_i(\tau)$ satisfying the following flow:

$$\frac{d\xi_1}{d\tau} = -a\,\xi_1, \qquad \frac{d\xi_2}{d\tau} = \kappa\,\xi_1 + a\,\xi_2, \qquad \frac{d\xi_3}{d\tau} = 0, \tag{9.13}$$

with end conditions

$$\xi_i(t) = x_i. \tag{9.14}$$

The solution is easily obtained by solving the system and applying the end conditions:

$$\xi_1 = x_1\,e^{a(t-\tau)}, \tag{9.15}$$

$$\begin{aligned} \xi_2 &= \left(x_2 + \frac{\kappa\,x_1}{2\,a}\right)e^{-a(t-\tau)} - \frac{\kappa\,x_1}{2\,a}\,e^{a(t-\tau)} \\ &= x_2\,e^{-a(t-\tau)} - \frac{\kappa\,x_1}{a}\,\sinh\left[a(t-\tau)\right], \end{aligned} \tag{9.16}$$

$$\xi_3 = x_3. \tag{9.17}$$

Before examining the effects of history on the constitutive equations, we pause to examine such effects on kinematic quantities.

9.2 Kinematics

Recall that a material point X located at $\mathbf{x}$ at time t, which was at $\boldsymbol{\xi}$ at time τ, is described by the relative motion (see (3.266))

$$\boldsymbol{\xi} = {}_{(t)}\boldsymbol{\chi}(\mathbf{x}, \tau). \tag{9.18}$$

Then, the velocity field given as a function of the position $\boldsymbol{\xi}$ and time τ,

$$\mathbf{v} = \mathbf{v}(\boldsymbol{\xi}(\tau), \tau), \tag{9.19}$$

can be found if its position $\mathbf{x}$ at time t is known. In such case, it is given by

$$\frac{d\boldsymbol{\xi}(\tau)}{d\tau} = \mathbf{v}(\boldsymbol{\xi}(\tau), \tau) \tag{9.20}$$

satisfying the end condition

$$\boldsymbol{\xi}(\tau)|_{\tau=t} = {}_{(t)}\boldsymbol{\chi}(\mathbf{x}, t) = \mathbf{x}. \tag{9.21}$$

Thus, knowledge of the position $\mathbf{x}$ of each material point at time t and the velocity field (9.19) for all τ is equivalent to knowledge of the relative motion (9.18).

The gradient of the relative motion (9.18) yields the relations (3.268) and (3.303) for the history of the relative deformation gradient:

$${}_{(t)}\mathbf{F}(\mathbf{x}, \tau) = (\operatorname{grad} \boldsymbol{\xi})^T \qquad \text{and} \qquad d\boldsymbol{\xi} = {}_{(t)}\mathbf{F}(\mathbf{x}, \tau) \cdot d\mathbf{x}. \tag{9.22}$$

We also recall that the history of the relative deformation gradient ${}_{(t)}\mathbf{F}(\tau)$ is related to the deformation gradient in the reference configuration by (see (3.274); here we take $t_0 = 0$)

$${}_{(t)}\mathbf{F}(\mathbf{x}, \tau) = {}_{(0)}\mathbf{F}(\mathbf{X}, \tau) \cdot {}_{(0)}\mathbf{F}^{-1}(\mathbf{X}, t) = \mathbf{F}(\mathbf{X}, \tau) \cdot \mathbf{F}^{-1}(\mathbf{X}, t), \tag{9.23}$$

where ${}_{(t)}\mathbf{F}(\mathbf{x}, t) = \mathbf{1}$ (see (3.269)).

In what follows, to simplify notations, we will not display the explicit dependence of quantities on the motion $\mathbf{x}$; this dependence will be made explicit later when it becomes relevant. Furthermore, as in Section 5.1, it is often convenient to write the history of quantities in terms of the time shift or history variable $s = t - \tau$ instead of the current variable τ, in which case $0 \le s < \infty$. That is, we can write

$${}_{(t)}\mathbf{F}(\tau) = {}_{(t)}\mathbf{F}(t - s) = {}_{(t)}\mathbf{F}^{(t)}(s), \tag{9.24}$$

where we recall from Sections 3.2.3 and 5.1 that the preceding subscript "(t)" denotes the relative description of the motion (relative to t) while the superscript "(t)" denotes the history up to time t. Note that we are and will be using the same symbols for the functions of τ and s. This should lead to no confusion since we make the function dependencies explicit.

Now, from $(9.22)_2$, we have that the length of a material element is provided by

$$d\boldsymbol{\xi}^T \cdot d\boldsymbol{\xi} = d\mathbf{x}^T \cdot {}_{(t)}\mathbf{F}^T(\tau) \cdot {}_{(t)}\mathbf{F}(\tau) \cdot d\mathbf{x} = d\mathbf{x}^T \cdot {}_{(t)}\mathbf{C}(\tau) \cdot d\mathbf{x} = d\mathbf{x}^T \cdot {}_{(t)}\mathbf{C}^{(t)}(s) \cdot d\mathbf{x}, \tag{9.25}$$

where we have used (3.283). If we assume that the deformation gradient is infinitely differentiable in time, we can write the history of the relative right Cauchy–Green tensor as

$$
{}_{(t)}\mathbf{C}(\tau) = \sum_{n=0}^{\infty} \frac{1}{n!} \left. \frac{\partial^n {}_{(t)}\mathbf{C}(\tau)}{\partial \tau^n} \right|_{\tau=t} (\tau - t)^n = \sum_{n=0}^{\infty} \frac{1}{n!} \mathbf{A}^{(n)}(t)\,(\tau - t)^n = \sum_{n=0}^{\infty} \frac{(-s)^n}{n!} \mathbf{A}^{(n)}(t), \quad (9.26)
$$

where we have defined the nth *Rivlin–Ericksen tensor*

$$
\mathbf{A}^{(n)}(t) \equiv \left. \frac{\partial^n {}_{(t)}\mathbf{C}(\tau)}{\partial \tau^n} \right|_{\tau=t} = (-1)^n \left. \frac{\partial^n {}_{(t)}\mathbf{C}^{(t)}(s)}{\partial s^n} \right|_{s=0}, \qquad n = 0, 1, 2, \ldots \quad (9.27)
$$

Subsequently, we can write that

$$
d\boldsymbol{\xi}^T \cdot d\boldsymbol{\xi} = d\mathbf{x}^T \cdot \sum_{n=0}^{\infty} \frac{(-s)^n}{n!} \mathbf{A}^{(n)}(t) \cdot d\mathbf{x}. \quad (9.28)
$$

Note that if $\mathbf{A}^{(n)}(t) = \mathbf{0}$ for all t, then ${}_{(t)}\mathbf{C}^{(t)}(s)$ is a polynomial in s of degree less than n.

From the definition of the relative right Cauchy–Green tensor (3.283), it is easy to show that

$$
\frac{\partial^n {}_{(t)}\mathbf{C}(\tau)}{\partial \tau^n} = \sum_{k=0}^{n} \binom{n}{k} \frac{\partial^k {}_{(t)}\mathbf{F}^T(\tau)}{\partial \tau^k} \cdot \frac{\partial^{n-k} {}_{(t)}\mathbf{F}(\tau)}{\partial \tau^{n-k}}, \qquad n = 0, 1, 2, \ldots, \quad (9.29)
$$

where the binomial coefficients are given by

$$
\binom{n}{k} = \frac{n!}{(n-k)!\,k!}. \quad (9.30)
$$

Evaluating (9.29) at $\tau = t$, and using (9.27), we subsequently have that

$$
\mathbf{A}^{(n)}(t) = \sum_{k=0}^{n} \binom{n}{k} {}_{(t)}\overset{(k)}{\mathbf{F}}{}^T(t) \cdot {}_{(t)}\overset{(n-k)}{\mathbf{F}}(t), \qquad n = 0, 1, 2, \ldots \quad (9.31)
$$

We also define the nth *acceleration gradient* by

$$
\mathbf{L}^{(n)}(t) \equiv \left(\operatorname{grad} \overset{(n)}{\mathbf{x}}(t)\right)^T, \qquad n = 0, 1, 2, \ldots \quad (9.32)
$$

Note that $\overset{(0)}{\mathbf{x}} = \mathbf{x}$, $\overset{(1)}{\mathbf{x}} = \dot{\mathbf{x}} = \mathbf{v}$ and $\overset{(2)}{\mathbf{x}} = \ddot{\mathbf{x}} = \mathbf{a}$, and that $\mathbf{L}^{(0)} = \mathbf{1}$, $\mathbf{L}^{(1)} = \mathbf{L}$, and the spatial gradient of the acceleration is given by

$$
\mathbf{L}^{(2)}(t) \equiv \left(\operatorname{grad} \ddot{\mathbf{x}}(t)\right)^T. \quad (9.33)
$$

Also note that since (see (3.290))

$$
\mathbf{L}(t) = {}_{(t)}\dot{\mathbf{F}}(t) = \left. \frac{\partial}{\partial \tau} {}_{(t)}\mathbf{F}(\tau) \right|_{\tau=t} = - \left. \frac{\partial}{\partial s} {}_{(t)}\mathbf{F}^{(t)}(s) \right|_{s=0}, \quad (9.34)
$$

the spatial gradient of the acceleration is related to ${}_{(t)}\mathbf{F}(t)$ by

$$\mathbf{L}^{(2)}(t) = {}_{(t)}\ddot{\mathbf{F}}(t). \tag{9.35}$$

Subsequently, we see that the nth *acceleration gradient* is related to ${}_{(t)}\mathbf{F}(t)$ by

$$\mathbf{L}^{(n)}(t) = {}_{(t)}\overset{(n)}{\mathbf{F}}(t) = \frac{\partial^n}{\partial \tau^n}{}_{(t)}\mathbf{F}(\tau)\bigg|_{\tau=t} = (-1)^n \frac{\partial^n}{\partial s^n}{}_{(t)}\mathbf{F}^{(t)}(s)\bigg|_{s=0}, \qquad n = 0, 1, \ldots. \tag{9.36}$$

Now using (9.27) and (9.29), we easily see that at $\tau = t$ the nth Rivlin–Ericksen tensor is also given by

$$\mathbf{A}^{(n)}(t) = \sum_{k=0}^{n} \binom{n}{k} \left[\mathbf{L}^{(k)}(t)\right]^T \cdot \mathbf{L}^{(n-k)}(t), \qquad n = 0, 1, 2, \ldots. \tag{9.37}$$

Note that $\mathbf{A}^{(0)} = \mathbf{1}$ and $\mathbf{A}^{(1)} = \mathbf{L} + \mathbf{L}^T = 2\mathbf{D}$.

It follows from (9.25) that

$$\frac{d^n}{d\tau^n}\left(d\boldsymbol{\xi}^T \cdot d\boldsymbol{\xi}\right) = d\mathbf{x}^T \cdot \frac{\partial^n {}_{(t)}\mathbf{C}(\tau)}{\partial \tau^n} \cdot d\mathbf{x} = (-1)^n d\mathbf{x}^T \cdot \frac{\partial^n {}_{(t)}\mathbf{C}^{(t)}(s)}{\partial s^n} \cdot d\mathbf{x}. \tag{9.38}$$

Subsequently, evaluating this result at $\tau = t$ or $s = 0$, we have that

$$\overset{(n)}{\overline{d\mathbf{x}^T \cdot d\mathbf{x}}} = \frac{d^n}{dt^n}\left(d\mathbf{x}^T \cdot d\mathbf{x}\right) = d\mathbf{x}^T \cdot \mathbf{A}^{(n)}(t) \cdot d\mathbf{x}. \tag{9.39}$$

Differentiating (9.39) with respect to t, we obtain

$$\overset{(n+1)}{\overline{d\mathbf{x}^T \cdot d\mathbf{x}}} = \dot{\overline{d\mathbf{x}^T}} \cdot \mathbf{A}^{(n)}(t) \cdot d\mathbf{x} + d\mathbf{x}^T \cdot \dot{\mathbf{A}}^{(n)}(t) \cdot d\mathbf{x} + d\mathbf{x}^T \cdot \mathbf{A}^{(n)}(t) \cdot \dot{\overline{d\mathbf{x}}}. \tag{9.40}$$

Now, using (9.39) and since $\dot{\overline{d\mathbf{x}}} = \mathbf{L} \cdot d\mathbf{x}$ (see (3.305)), we have that

$$\mathbf{A}^{(n+1)}(t) = \dot{\mathbf{A}}^{(n)}(t) + \mathbf{L}^T(t) \cdot \mathbf{A}^{(n)}(t) + \mathbf{A}^{(n)}(t) \cdot \mathbf{L}(t), \qquad n = 0, 1, 2, \ldots, \tag{9.41}$$

where

$$\dot{\mathbf{A}}^{(n)}(t) = \frac{\partial \mathbf{A}^{(n)}(t)}{\partial t} + (\mathbf{v}(t) \cdot \mathrm{grad})\, \mathbf{A}^{(n)}(t). \tag{9.42}$$

Recalling the Cotter–Rivlin tensor (3.425), we see that

$$\mathbf{A}^{(n+1)}(t) = \overset{\triangle}{\mathbf{A}}^{(n)}(t). \tag{9.43}$$

Furthermore, using the polar decomposition expressed through $(3.282)_1$, and generalizing (3.293) and (3.294) by defining the nth *stretch* (or rate of strain) and *spin tensors* by

$$\mathbf{D}^{(n)}(t) \equiv {}_{(t)}\overset{(n)}{\mathbf{U}}(t) = \frac{\partial^n}{\partial \tau^n}{}_{(t)}\mathbf{U}(\tau)\bigg|_{\tau=t} = (-1)^n \frac{\partial^n}{\partial s^n}{}_{(t)}\mathbf{U}^{(t)}(s)\bigg|_{s=0}, \qquad n = 0, 1, \ldots, \tag{9.44}$$

and

$$\mathbf{W}^{(n)}(t) \equiv {}_{(t)}\overset{(n)}{\mathbf{R}}(t) = \left.\frac{\partial^n}{\partial \tau^n}{}_{(t)}\mathbf{R}(\tau)\right|_{\tau=t} = (-1)^n \left.\frac{\partial^n}{\partial s^n}{}_{(t)}\mathbf{R}^{(t)}(s)\right|_{s=0}, \qquad n = 0,1,\ldots, \tag{9.45}$$

it is easy to show that the nth acceleration gradient and stretch and spin tensors are respectively given by

$$\mathbf{L}^{(n)}(t) = \sum_{k=0}^{n} \binom{n}{k} \mathbf{W}^{(k)}(t) \cdot \mathbf{D}^{(n-k)}(t), \qquad n = 0,1,2,\ldots, \tag{9.46}$$

$$\mathbf{D}^{(n)}(t) = \frac{1}{2}\left[\mathbf{A}^{(n)}(t) - \sum_{k=1}^{n-1} \binom{n}{k} \mathbf{D}^{(k)}(t) \cdot \mathbf{D}^{(n-k)}(t)\right], \qquad n = 1,2,\ldots, \tag{9.47}$$

and

$$\mathbf{W}^{(n)}(t) = \mathbf{L}^{(n)}(t) - \mathbf{D}^{(n)}(t) - \sum_{k=1}^{n-1} \binom{n}{k} \mathbf{W}^{(k)}(t) \cdot \mathbf{D}^{(n-k)}(t), \qquad n = 1,2,\ldots. \tag{9.48}$$

Note that $\mathbf{D}^{(0)} = \mathbf{1}$, $\mathbf{W}^{(0)} = \mathbf{1}$, $\mathbf{D}^{(1)} = \mathbf{D}$, and $\mathbf{W}^{(1)} = \mathbf{W}$. Equation (9.47) is a recursion formula which can be used to find explicit expressions for $\mathbf{D}^{(n)}$ as polynomials in $\mathbf{A}^{(k)}$, $k = 1,2,\ldots,n$. Hence, after substitution of (9.37), one would also obtain explicit expressions for $\mathbf{D}^{(n)}$ as polynomials in $\mathbf{L}^{(k)}$ and $\mathbf{L}^{(k)^T}$, $k = 1,2,\ldots,n$. Equation (9.48) is also a recursion formula which permits us to express $\mathbf{W}^{(n)}$ as a polynomial in $\mathbf{L}^{(k)}$ and $\mathbf{D}^{(k)}$. Since $\mathbf{D}^{(k)}$ are polynomials in $\mathbf{L}^{(k)}$ and $\mathbf{L}^{(k)^T}$, we can also find expressions for $\mathbf{W}^{(n)}$ as polynomials in $\mathbf{L}^{(k)}$ and $\mathbf{L}^{(k)^T}$, $k = 1,2,\ldots,n$. In the special case $n = 1$, we find that $\mathbf{D}^{(1)} = \frac{1}{2}\mathbf{A}^{(1)} = \frac{1}{2}(\mathbf{L} + \mathbf{L}^T)$ and $\mathbf{W}^{(1)} = \mathbf{L}^{(1)} - \mathbf{D}^{(1)} = \mathbf{L} - \frac{1}{2}(\mathbf{L} + \mathbf{L}^T) = \frac{1}{2}(\mathbf{L} - \mathbf{L}^T)$.

Note that while the spin $\mathbf{W}$ is skew-symmetric, the higher spins $\mathbf{W}^{(n)}$, $n > 1$, are not necessarily skew-symmetric. Also note that the symmetric Rivlin–Ericksen tensors $\mathbf{A}^{(n)}$ and stretching tensors $\mathbf{D}^{(n)}$ are frame indifferent, while the velocity gradients $\mathbf{L}^{(n)}$ and spin tensors $\mathbf{W}^{(n)}$ are not.

In the above definitions of $\mathbf{A}^{(n)}$, $\mathbf{D}^{(n)}$, and $\mathbf{W}^{(n)}$, the present configuration has been chosen as the reference configuration. If a fixed reference configuration is used, more complicated formulae result (e.g., see (3.491) and (3.492)).

Example

Here we obtain $\mathbf{A}^{(2)}$, $\mathbf{D}^{(2)}$, and $\mathbf{W}^{(2)}$ in terms of acceleration gradients. We readily see from (9.37), (9.47), and (9.48) that

$$\begin{aligned}
\mathbf{A}^{(2)} &= \mathbf{L}^{(2)} + \mathbf{L}^{(2)^T} + 2\mathbf{L}^T \cdot \mathbf{L}, && (9.49)\\
\mathbf{D}^{(2)} &= \frac{1}{2}\left(\mathbf{A}^{(2)} - 2\mathbf{D}^{(1)^2}\right)\\
&= \frac{1}{2}\left(\mathbf{A}^{(2)} - \frac{1}{2}\mathbf{A}^{(1)^2}\right)\\
&= \frac{1}{2}\left(\mathbf{L}^{(2)} + \mathbf{L}^{(2)^T}\right) - \frac{1}{4}\left(\mathbf{L} + \mathbf{L}^T\right)^2 + \mathbf{L}^T \cdot \mathbf{L}, && (9.50)\\
\mathbf{W}^{(2)} &= \mathbf{L}^{(2)} - \mathbf{D}^{(2)} - 2\,\mathbf{W}^{(1)}\,\mathbf{D}^{(1)}\\
&= \frac{1}{2}\left(\mathbf{L}^{(2)} - \mathbf{L}^{(2)^T}\right) + \frac{1}{4}\left(\mathbf{L} + \mathbf{L}^T\right)^2 - \mathbf{L}^T \cdot \mathbf{L}. && (9.51)
\end{aligned}$$

Lastly, we define the *right Cauchy–Green tensor difference history*, or *history tensor* for short, as follows:

$$\boldsymbol{G}^{(t)}(s) \equiv {}_{(t)}\mathbf{C}^{(t)}(s) - \mathbf{1} \quad \text{or} \quad G^{(t)}(s) = {}_{(t)}C^{(t)}(s) - I \quad \text{or} \quad G^{(t)}_{ik}(s) \equiv {}_{(t)}C^{(t)}_{ik}(s) - \delta_{ik}, \tag{9.52}$$

where we see that $\boldsymbol{G}^{(t)}(0) = \mathbf{0}$. Then, from (9.26) and (9.52), we have

$$\boldsymbol{G}^{(t)}(s) = \sum_{n=1}^{\infty} \frac{(-s)^n}{n!}\mathbf{A}^{(n)}(t), \tag{9.53}$$

where

$$\mathbf{A}^{(n)}(t) \equiv (-1)^n \left.\frac{\partial^n \boldsymbol{G}^{(t)}(s)}{\partial s^n}\right|_{s=0}, \qquad n = 1, 2, \ldots. \tag{9.54}$$

9.2.1 Motion with constant stretch history

In many motions, that are important theoretically and experimentally, the histories frequently take particular forms enabling considerable simplifications. A *motion with constant stretch history* (MWCSH) results when the present deformation is obtained through a sequence of past deformations that remain constant. That is, a motion, along the path of a material point, has constant stretch history if there exists an orthogonal tensor function $\mathbf{Q}(t)$ such that the histories of the right Cauchy–Green tensors ${}_{(t)}\mathbf{C}^{(t)}(s)$ and ${}_{(0)}\mathbf{C}^{(0)}(s)$ are related as

$${}_{(t)}\mathbf{C}^{(t)}(s) = \mathbf{Q}(t) \cdot {}_{(0)}\mathbf{C}^{(0)}(s) \cdot \mathbf{Q}^T(t), \qquad 0 \le s < \infty, \tag{9.55}$$

with $\mathbf{Q}(0) = \mathbf{1}$.

It can be proved that a motion is an MWCSH if and only if there exist an orthogonal tensor $\mathbf{Q}(t)$, a scalar κ (called the *shearing*), and a constant second rank tensor $\mathbf{N}_0$ such that

$${}_{(0)}\mathbf{F}(\tau) = \mathbf{F}(\tau) = \mathbf{Q}(\tau) \cdot e^{\tau\kappa\mathbf{N}_0}, \qquad \mathbf{Q}(0) = \mathbf{1}, \quad |\mathbf{N}_0| = 1, \tag{9.56}$$

where $|\mathbf{N}_0|^2 = \mathrm{tr}\,(\mathbf{N}_0^T \cdot \mathbf{N}_0)$. From the polar decomposition of the deformation gradient, it is seen that $\mathbf{Q}(\tau)$ corresponds to the rotation tensor $\mathbf{R}(\tau)$ and $e^{\tau\kappa\mathbf{N}_0}$ corresponds to the right stretch tensor $\mathbf{U}(\tau)$. It is easy to note that in the MWCSH deformation, the principal stretch histories remain constant but the principal axes rotate.

We remark that since (see (3.239))

$$\dot{\mathbf{F}}(t) = \mathbf{L}(t) \cdot \mathbf{F}(t), \tag{9.57}$$

if we have no deformation initially, i.e., $\mathbf{F}(0) = \mathbf{1}$, and if the velocity gradient $\mathbf{L}$ is constant, then we have

$$\mathbf{F}(t) = e^{t\,\mathbf{L}} \tag{9.58}$$

and

$$_{(t)}\mathbf{F}(\tau) = e^{(\tau - t)\,\mathbf{L}}. \tag{9.59}$$

Now it is straightforward to show that

$$_{(t)}\mathbf{C}^{(t)}(s) = {}_{(0)}\mathbf{C}^{(0)}(s) \tag{9.60}$$

so, when compared to the more general definition, we see that in this case $\mathbf{Q}(t) = \mathbf{1}$ and $\kappa\,\mathbf{N}_0 = \mathbf{L}$. When $\mathbf{Q}(t) = \mathbf{1}$, the rotation associated with the relative deformation gradient reduces to the rotation of the eigenvectors associated with the rate of strain.

Before proceeding with our discussion, it is important to recall the following relations involving the exponential of a tensor:

$$e^{\mathbf{M}} = \sum_{n=0}^{\infty} \frac{1}{n!} \mathbf{M}^n, \tag{9.61}$$

$$\left(e^{\mathbf{M}}\right)^T = e^{\mathbf{M}^T}, \qquad \left(e^{\mathbf{M}}\right)^{-1} = e^{-\mathbf{M}}, \qquad e^{\mathbf{0}} = \mathbf{1}, \tag{9.62}$$

$$e^{\mathbf{M}+\mathbf{N}} = e^{\mathbf{M}} e^{\mathbf{N}} \quad \text{if} \quad \mathbf{M} \cdot \mathbf{N} = \mathbf{N} \cdot \mathbf{M}, \tag{9.63}$$

$$\left(\mathbf{Q} \cdot \mathbf{M} \cdot \mathbf{Q}^T\right)^n = \mathbf{Q} \cdot \mathbf{M}^n \cdot \mathbf{Q}^T \quad \text{if} \quad \mathbf{Q} \in \mathscr{O}, \tag{9.64}$$

$$e^{\mathbf{Q}\cdot\mathbf{M}\cdot\mathbf{Q}^T} = \mathbf{Q} \cdot e^{\mathbf{M}} \cdot \mathbf{Q}^T \quad \text{if} \quad \mathbf{Q} \in \mathscr{O}, \tag{9.65}$$

$$\frac{d}{d\tau} e^{\tau\,\mathbf{M}} = \mathbf{M} \cdot e^{\tau\,\mathbf{M}} = e^{\tau\,\mathbf{M}} \cdot \mathbf{M}. \tag{9.66}$$

Furthermore, a tensor $\mathbf{M}$ is nilpotent if $\mathbf{M}^n = \mathbf{0}$ for some integer $n \geq 0$. The exponential $e^{\tau\,\mathbf{M}}$ is a finite polynomial in τ if and only if $\mathbf{M}$ is nilpotent. It can be shown that in three dimensions, if $\mathbf{M}$ is nilpotent, then $n \leq 3$, and in this case,

$$e^{\tau\,\mathbf{M}} = \mathbf{1} + \tau\,\mathbf{M} + \frac{1}{2}\tau^2\,\mathbf{M}^2. \tag{9.67}$$

Moreover, it can be shown that the components of $\mathbf{M}$ relative to an appropriate orthonormal basis have the form

$$M = \begin{pmatrix} 0 & 0 & 0 \\ \chi & 0 & 0 \\ \lambda & \nu & 0 \end{pmatrix}. \tag{9.68}$$

If $\mathbf{M}^2 = \mathbf{0}$, then the basis may be chosen such that

$$M = \chi \begin{pmatrix} 0 & 0 & 0 \\ 1 & 0 & 0 \\ 0 & 0 & 0 \end{pmatrix}. \tag{9.69}$$

Subsequently, from (9.23), (9.56), and (9.65), we have

$$ {}_{(t)}\mathbf{F}(\tau) = \mathbf{Q}(\tau)\cdot e^{(\tau-t)\,\kappa\,\mathbf{N}_0}\cdot\mathbf{Q}^T(t) = \mathbf{Q}(\tau)\cdot\mathbf{Q}^T(t)\cdot e^{(\tau-t)\,\kappa\,\mathbf{N}} = \mathbf{Q}(\tau)\cdot\mathbf{Q}^T(t)\cdot e^{(\tau-t)\,\mathbf{M}}, \tag{9.70} $$

and recalling that $\tau = t - s$, we can also write

$$ {}_{(t)}\mathbf{F}^{(t)}(s) = \mathbf{Q}(t-s)\cdot\mathbf{Q}^T(t)\cdot e^{-s\,\mathbf{M}}, \tag{9.71} $$

where the tensors $\mathbf{N}$ and $\mathbf{M}$ are defined by

$$ \mathbf{M} \equiv \kappa\,\mathbf{N} \equiv \kappa\,\mathbf{Q}(t)\cdot\mathbf{N}_0\cdot\mathbf{Q}^T(t), \qquad |\mathbf{N}| = 1. \tag{9.72} $$

We now observe that

$$ {}_{(0)}\mathbf{C}(\tau) = {}_{(0)}\mathbf{F}^T(\tau)\cdot{}_{(0)}\mathbf{F}(\tau) = e^{\tau\,\kappa\,\mathbf{N}_0{}^T}\cdot e^{\tau\,\kappa\,\mathbf{N}_0}, \tag{9.73} $$

$$ {}_{(0)}\mathbf{C}^{(0)}(s) = \left({}_{(0)}\mathbf{F}^{(0)}(s)\right)^T\cdot{}_{(0)}\mathbf{F}^{(0)}(s) = e^{-s\,\kappa\,\mathbf{N}_0{}^T}\cdot e^{-s\,\kappa\,\mathbf{N}_0}, \tag{9.74} $$

and, using (9.55),

$$ {}_{(t)}\mathbf{C}(\tau) = {}_{(t)}\mathbf{C}^{(t)}(s) = \mathbf{Q}(t)\cdot e^{-s\,\kappa\,\mathbf{N}_0^T}\cdot e^{-s\,\kappa\,\mathbf{N}_0}\cdot\mathbf{Q}^T(t) = e^{-s\,\mathbf{M}^T}\cdot e^{-s\,\mathbf{M}}. \tag{9.75} $$

Furthermore, associated with the relative deformation gradient, we have that the relative right stretch tensor is given by

$$ {}_{(t)}\mathbf{U}(\tau) = {}_{(t)}\mathbf{U}^{(t)}(s) = \left({}_{(t)}\mathbf{C}^{(t)}(s)\right)^{1/2} = e^{-\frac{1}{2}\,s\,\mathbf{M}^T}\cdot e^{-\frac{1}{2}\,s\,\mathbf{M}} = \mathbf{Q}(t)\cdot{}_{(0)}\mathbf{U}^{(0)}(s)\cdot\mathbf{Q}^T(t), \tag{9.76} $$

the relative rate of rotation tensor by

$$ {}_{(t)}\mathbf{R}(\tau) = {}_{(t)}\mathbf{R}^{(t)}(s) = {}_{(t)}\mathbf{F}^{(t)}(s)\cdot\left({}_{(t)}\mathbf{U}^{(t)}(s)\right)^{-1} = \mathbf{Q}(\tau)\cdot\mathbf{Q}^T(t)\cdot e^{-\frac{1}{2}\,s\,\mathbf{M}}\cdot e^{\frac{1}{2}\,s\,\mathbf{M}^T} = \mathbf{Q}(\tau)\cdot{}_{(0)}\mathbf{R}^{(0)}(s)\cdot\mathbf{Q}^T(t), \tag{9.77} $$

the velocity gradient by (see (3.289))

$$ \mathbf{L}(t) = {}_{(t)}\dot{\mathbf{F}}(\tau)\big|_{\tau=t} = \mathbf{M} + \boldsymbol{\Omega}, \quad \text{with} \quad \boldsymbol{\Omega} = \dot{\mathbf{Q}}(t)\cdot\mathbf{Q}^T(t), \tag{9.78} $$

the stretch and spin tensors by (see (3.293) and (3.294))

$$ \mathbf{D}(t) = {}_{(t)}\dot{\mathbf{U}}(\tau)\big|_{\tau=t} = \frac{1}{2}\left(\mathbf{M}+\mathbf{M}^T\right) = \frac{\kappa}{2}\,\mathbf{Q}(t)\cdot\left(\mathbf{N}_0+\mathbf{N}_0^T\right)\cdot\mathbf{Q}^T(t) \tag{9.79} $$

and

$$ \mathbf{W}(t) = {}_{(t)}\dot{\mathbf{R}}(\tau)\big|_{\tau=t} = \frac{1}{2}\left(\mathbf{M}-\mathbf{M}^T\right) + \boldsymbol{\Omega} = \frac{\kappa}{2}\,\mathbf{Q}(t)\cdot\left(\mathbf{N}_0-\mathbf{N}_0^T\right)\cdot\mathbf{Q}^T(t) + \boldsymbol{\Omega}, \tag{9.80} $$

and, since for an MWCSH $\dot{\mathbf{A}}^{(n)}(t) = \mathbf{0}$, the Rivlin–Ericksen tensors by

$$ \mathbf{A}^{(n+1)}(t) = \mathbf{L}^T(t)\cdot\mathbf{A}^{(n)}(t) + \mathbf{A}^{(n)}(t)\cdot\mathbf{L}(t). \tag{9.81} $$

Note that by using (9.37), we also have that

$$\mathbf{A}^{(1)}(t) = \mathbf{L}^T + \mathbf{L} = \mathbf{M}^T + \mathbf{M}, \tag{9.82}$$

$$\mathbf{A}^{(2)}(t) = \mathbf{M}^T \cdot \mathbf{A}^{(1)} + \mathbf{A}^{(1)} \cdot \mathbf{M} = \left(\mathbf{M}^T\right)^2 + 2\,\mathbf{M}^T \cdot \mathbf{M} + \mathbf{M}^2, \tag{9.83}$$

$$\vdots$$

$$\begin{aligned} \mathbf{A}^{(n)}(t) &= \mathbf{M}^T \cdot \mathbf{A}^{(n-1)} + \mathbf{A}^{(n-1)} \cdot \mathbf{M} = \sum_{k=0}^{n} \binom{n}{k} \left(\mathbf{M}^{n-k}\right)^T \cdot \mathbf{M}^k, \\ &= \mathbf{Q}(t) \cdot \mathbf{A}^{(n)}(0) \cdot \mathbf{Q}^T(t). \end{aligned} \tag{9.84}$$

Since, as remarked earlier, if $\mathbf{M}$ is nilpotent, then necessarily $\mathbf{M}^3 = \mathbf{0}$ in a three-dimensional Euclidean space, then it can be shown that all MWCSH flows can be divided in the following three classes depending on the properties of $\mathbf{M}$ (or $\mathbf{N}$ or $\mathbf{N}_0$):

I. $\mathbf{M}^2 = \mathbf{0}$;

II. $\mathbf{M}^2 \neq \mathbf{0}$ but $\mathbf{M}^3 = \mathbf{0}$;

III. $\mathbf{M}^n \neq \mathbf{0}$ for all $n = 1, 2, \ldots$.

The consequences of this result are as follows. If $\mathbf{M}^3 = \mathbf{0}$, then $\mathbf{A}^{(n)} = \mathbf{0}$ for $n \geq 5$, and $e^{-s\mathbf{M}}$ is at most a quadratic polynomial in $\mathbf{M}$. If $\mathbf{M}^2 = \mathbf{0}$, then $\mathbf{A}^{(n)} = \mathbf{0}$ for $n \geq 3$, and $e^{-s\mathbf{M}}$ is at most a linear polynomial in $\mathbf{M}$. If $\mathbf{M}$ is not nilpotent, then $e^{-s\mathbf{M}}$ is not a finite polynomial in $\mathbf{M}$.

Class I includes all steady *viscometric flows* (e.g., simple shear) and in particular flows defined by a steady velocity field. From (9.75), in this case we have

$${}_{(t)}\mathbf{C}^{(t)}(s) = \left[\mathbf{1} - s\,\mathbf{M}^T\right] \cdot \left[\mathbf{1} - s\,\mathbf{M}\right]. \tag{9.85}$$

For Class II, we have

$${}_{(t)}\mathbf{C}^{(t)}(s) = \left[\mathbf{1} - s\,\mathbf{M}^T + \frac{1}{2}\,s^2\left(\mathbf{M}^T\right)^2\right] \cdot \left[\mathbf{1} - s\,\mathbf{M} + \frac{1}{2}\,s^2\,\mathbf{M}^2\right]. \tag{9.86}$$

This class contains the subclass of *doubly superposed viscometric flows*. These flows are generated by a superposition of two steady viscometric flows; however, we note that not all such superpositions belong to Class II.

Class III, for which

$${}_{(t)}\mathbf{C}^{(t)}(s) = \left[\sum_{n=0}^{\infty} \frac{(-1)^n}{n!}\,s^n\left(\mathbf{M}^T\right)^n\right] \cdot \left[\sum_{n=0}^{\infty} \frac{(-1)^n}{n!}\,s^n\,\mathbf{M}^n\right], \tag{9.87}$$

contains the subclass of *triply superposed viscometric flows* (e.g., steady pure shear, steady simple extension).

We note that a motion is irrotational if the velocity gradient is symmetric. All MWCSH described by a constant symmetric velocity gradient are customarily called *extensional* or *elongational* flows. Since every symmetric tensor $\mathbf{L}$ can always be represented in a diagonal form, then one finds that all irrotational MWCSH are extensional or elongational flows and belong to Class III. For such flows, in the appropriate reference frame for which $\mathbf{L}$ is diagonal, we have

$$\mathbf{L} = \mathbf{M} = \kappa\,\mathbf{N} = \kappa\,\mathbf{N}_0, \qquad \mathbf{Q}(t) = \mathbf{1}, \tag{9.88}$$

with $\mathbf{N}_0 = \mathbf{N}_0^T$ (thus $\mathbf{N} = \mathbf{N}^T$, $\mathbf{M} = \mathbf{M}^T$, and $\mathbf{L} = \mathbf{L}^T$), and subsequently, from (9.75), we have that

$$_{(t)}\mathbf{C}^{(t)}(s) = {}_{(0)}\mathbf{C}^{(0)}(s) = e^{-2\,s\,\mathbf{M}}, \tag{9.89}$$

and, from (9.54) and (9.82), we see that

$$\mathbf{A}^{(n)} = \left(\mathbf{A}^{(1)}\right)^n = 2^n\,\mathbf{M}, \qquad n = 1, 2, \ldots. \tag{9.90}$$

Furthermore, an MWCSH is isochoric if $\operatorname{tr}\mathbf{L} = 0$, and thus if and only if $\operatorname{tr}\mathbf{M} = \operatorname{tr}\mathbf{N} = \operatorname{tr}\mathbf{N}_0 = 0$. Now, if $\mathbf{M}^3 = \mathbf{0}$, from the Cayley–Hamilton theorem (see (3.95)), we have that

$$(\operatorname{tr}\mathbf{M})\,\mathbf{M}^2 = \frac{1}{2}\left[(\operatorname{tr}\mathbf{M})^2 - \operatorname{tr}\mathbf{M}^2\right]\mathbf{M}. \tag{9.91}$$

If $\mathbf{M} \neq \mathbf{0}$ and $\mathbf{M}^2 = \mathbf{0}$, it implies that $\operatorname{tr}\mathbf{M}^2 = \mathbf{0}$ and $\operatorname{tr}\mathbf{M} = 0$, while if $\mathbf{M}^2 \neq \mathbf{0}$, then

$$(\operatorname{tr}\mathbf{M})\,\mathbf{M}^3 = \mathbf{0} = \frac{1}{2}\left[(\operatorname{tr}\mathbf{M})^2 - \operatorname{tr}\mathbf{M}^2\right]\mathbf{M}^2, \tag{9.92}$$

which implies that $(\operatorname{tr}\mathbf{M})^2 = \operatorname{tr}\mathbf{M}^2$, which in turn implies that $\operatorname{tr}\mathbf{M} = 0$ and thus $\operatorname{tr}\mathbf{M}^2 = 0$. Subsequently, we deduce that all flows corresponding to Classes I and II are isochoric.

Before presenting an import theorem, the following lemma is necessary.

Lemma: Suppose that S is a 3×3 diagonal matrix, and W is a 3×3 skew-symmetric matrix, i.e.,

$$S = \begin{pmatrix} a & 0 & 0 \\ 0 & b & 0 \\ 0 & 0 & c \end{pmatrix} \quad \text{and} \quad W = \begin{pmatrix} 0 & x & y \\ -x & 0 & z \\ -y & -z & 0 \end{pmatrix}. \tag{9.93}$$

Now if

i) $a \neq b \neq c$, then $S\,W = W\,S$ if and only if $x = y = z = 0$ (i.e., $W = 0$);

ii) $a = b \neq c$, then $S\,W = W\,S$ if and only if $y = z = 0$ (where x is arbitrary);

iii) $a = b = c$, then $S\,W = W\,S$ for all x, y, and z (i.e., any W).

Proof: By multiplication

$$S\,W = \begin{pmatrix} 0 & a\,x & a\,y \\ -b\,x & 0 & b\,z \\ -c\,y & -c\,z & 0 \end{pmatrix} \quad \text{and} \quad W\,S = \begin{pmatrix} 0 & b\,x & c\,y \\ -a\,x & 0 & c\,z \\ -a\,y & -b\,z & 0 \end{pmatrix}. \tag{9.94}$$

Therefore, $S\,W = W\,S$ if and only if

$$(a - b)\,x = 0, \quad (a - c)\,y = 0, \quad (b - c)\,z = 0. \tag{9.95}$$

Hence, the lemma follows.

Theorem: For an MWCSH, the relative history of $_{(t)}\mathbf{C}^{(t)}(s)$ is determined uniquely by $\mathbf{A}^{(1)}$, $\mathbf{A}^{(2)}$, and $\mathbf{A}^{(3)}$.

Proof:

i) If $\mathbf{A}^{(1)}$ has three distinct eigenvalues, then it is claimed that $\mathbf{M}$ is determined uniquely by $\mathbf{A}^{(1)}$ and $\mathbf{A}^{(2)}$. To see this, suppose it is false. Then, from (9.82) and (9.83), we have that

$$\mathbf{A}^{(1)} = \mathbf{M}^T + \mathbf{M} = \overline{\mathbf{M}}^T + \overline{\mathbf{M}} \quad \text{and}$$
$$\mathbf{A}^{(2)} = \mathbf{M}^T \cdot \mathbf{A}^{(1)} + \mathbf{A}^{(1)} \cdot \mathbf{M} = \overline{\mathbf{M}}^T \cdot \mathbf{A}^{(1)} + \mathbf{A}^{(1)} \cdot \overline{\mathbf{M}}. \tag{9.96}$$

But, from the first relation, we see that

$$\mathbf{M} - \overline{\mathbf{M}} = -\left(\mathbf{M} - \overline{\mathbf{M}}\right)^T \tag{9.97}$$

is skew-symmetric, and from the second relation, we have

$$\left(\mathbf{M} - \overline{\mathbf{M}}\right) \cdot \mathbf{A}^{(1)} = \mathbf{A}^{(1)} \cdot \left(\mathbf{M} - \overline{\mathbf{M}}\right), \tag{9.98}$$

i.e., the symmetric tensor $\mathbf{A}^{(1)}$ commutes with $\mathbf{M} - \overline{\mathbf{M}}$; hence $\mathbf{M} - \overline{\mathbf{M}} = \mathbf{0}$ since the eigenvalues are distinct. That is, $\mathbf{M}$ is uniquely determined by $\mathbf{A}^{(1)}$ and $\mathbf{A}^{(2)}$.

ii) If $\mathbf{A}^{(1)}$ has only two distinct eigenvalues, then, relative to a principal orthonormal basis, the component matrix of $\mathbf{A}^{(1)}$ can be written in the following form

$$A^{(1)} = \begin{pmatrix} a & 0 & 0 \\ 0 & a & 0 \\ 0 & 0 & b \end{pmatrix}, \qquad a \neq b. \tag{9.99}$$

We consider two subcases.

a) Relative to the principal orthonormal basis of $\mathbf{A}^{(1)}$, the component matrix of $\mathbf{A}^{(2)}$ is of the form

$$A^{(2)} = \begin{pmatrix} u & 0 & 0 \\ 0 & u & 0 \\ 0 & 0 & v \end{pmatrix}, \tag{9.100}$$

where u may or may not be equal to v. In this case, we deduce that $u = a^2$, $v = b^2$, and thus ${}_{(t)}\mathbf{C}^{(t)}(s)$ is determined by $\mathbf{A}^{(1)}$ ($\mathbf{M}$ is not unique, but ${}_{(t)}\mathbf{C}^{(t)}(s)$ is unique), and then it can be easily shown that

$${}_{(t)}\mathbf{C}^{(t)}(s) = e^{-s\,\mathbf{A}^{(1)}}. \tag{9.101}$$

b) If $\mathbf{A}^{(2)}$ is not of the above form, then ${}_{(t)}\mathbf{C}^{(t)}(s)$ is a function of $\mathbf{A}^{(1)}$, $\mathbf{A}^{(2)}$, and $\mathbf{A}^{(3)}$, and $\mathbf{M}$ is unique, i.e., the solution of the system

$$\mathbf{A}^{(1)} = \mathbf{M}^T + \mathbf{M}, \tag{9.102}$$
$$\mathbf{A}^{(2)} = \mathbf{M}^T \cdot \mathbf{A}^{(1)} + \mathbf{A}^{(1)} \cdot \mathbf{M}, \tag{9.103}$$
$$\mathbf{A}^{(3)} = \mathbf{M}^T \cdot \mathbf{A}^{(2)} + \mathbf{A}^{(2)} \cdot \mathbf{M}, \tag{9.104}$$

is unique. To see this, suppose that $\overline{\mathbf{M}}$ is another solution of the system. Then, from (9.102), the difference between $\mathbf{M}$ and $\overline{\mathbf{M}}$ is a skew-symmetric tensor. From (9.103), $(\mathbf{M}-\overline{\mathbf{M}})$ commutes with $\mathbf{A}^{(1)}$. Thus, from the previous lemma, we have

$$M-\overline{M}=\begin{pmatrix} 0 & x & 0 \\ -x & 0 & 0 \\ 0 & 0 & 0 \end{pmatrix}. \tag{9.105}$$

From (9.104), $(\mathbf{M}-\overline{\mathbf{M}})$ also commutes with $\mathbf{A}^{(2)}$. Since by assumption $\mathbf{A}^{(2)}$ is not of the form (9.100), by multiplication it is easily seen that $x=0$. Thus $\mathbf{M}=\overline{\mathbf{M}}$.

iii) Here we claim that if all eigenvalues of $\mathbf{A}^{(1)}$ are equal, then $\mathbf{A}^{(1)}=a\,\mathbf{1}$. It is easily seen that in this case

$$_{(t)}\mathbf{C}^{(t)}(s)=e^{-a\,s\,\mathbf{1}}. \tag{9.106}$$

Again, the tensor $\mathbf{M}$ is not unique, but any solution $\mathbf{M}$ of (9.102) yields the same relative right Cauchy–Green tensor (9.106).

Thus, the theorem is proved.

9.3 Constitutive equations

The most general constitutive equation that we have considered is of the form (see (5.14))

$$T(\mathbf{X},t)=\underset{\substack{\mathbf{Y}\in V\\0\le s<\infty}}{\mathfrak{F}}\{[\mathbf{x}^{(t)}(\mathbf{Y},s)-\mathbf{x}^{(t)}(\mathbf{X},s)],\theta^{(t)}(\mathbf{Y},s),\mathbf{X}\}, \tag{9.107}$$

subject to restrictions arising from the rigid rotation of the frame, such as (5.13) for a second order tensor. Reasonable assumptions can be made that values of constitutive variables at distant materials points from $\mathbf{X}$ do not affect appreciably the values of the constitutive variables at $\mathbf{X}$. Under such assumption, two possible avenues for developing more constructive forms of the constitutive equation (9.107) have been pursued. The first is based on smoothness of neighborhood, where derivatives up to some order in space are assumed to exist (see Section 5.4). In such case, considering homogeneous materials and applying the frame-invariance requirement as done in Section 5.7, one obtains the following reduced constitutive equation

$$\overline{T}(\mathbf{X},t)=\underset{0\le s<\infty}{\mathfrak{G}}\{{}^{i}\mathbf{C}^{(t)}(\mathbf{X},s),\theta^{(t)}(\mathbf{X},s),{}^{j}\mathbf{G}^{(t)}(\mathbf{X},s)\}, \tag{9.108}$$

where

$$\overline{T}\equiv\mathbf{F}^{T}\cdot T\cdot\mathbf{F} \tag{9.109}$$

is the convected tensor, and $i=1,2,\dots,P$, $j=1,2,\dots,Q$. Using (5.23), we note that

$${}^{i}\mathbf{C}^{(t)}\equiv\mathbf{F}^{T}\cdot{}^{i}\mathbf{F}^{(t)} \qquad \text{or} \qquad {}^{i}C^{(t)}_{KK_1K_2\cdots K_i}\equiv F_{kK}\,{}^{i}F^{(t)}_{kK_1K_2\cdots K_i}. \tag{9.110}$$

The above constitutive functional generalizes the expression given in (5.94), which applies to a simple material. Subsequently, using the same procedure as in Section 5.7, one finds that the corresponding constitutive functional for a homogeneous non-simple isotropic solid in the current configuration is given by

$$T(\mathbf{x},t) = \underset{0\le s<\infty}{\mathfrak{F}} \{\mathbf{B}, {}_{(t)}^{\;i}\mathbf{C}^{(t)}(\mathbf{x},s), \theta^{(t)}(\mathbf{x},s), {}_{(t)}^{\;j}\mathbf{g}^{(t)}(\mathbf{x},s)\} \tag{9.111}$$

and that for a non-simple (isotropic) fluid by

$$T(\mathbf{x},t) = \underset{0\le s<\infty}{\mathfrak{F}} \{\rho, {}_{(t)}^{\;i}\mathbf{C}^{(t)}(\mathbf{x},s), \theta^{(t)}(\mathbf{x},s), {}_{(t)}^{\;j}\mathbf{g}^{(t)}(\mathbf{x},s)\}, \tag{9.112}$$

where

$${}_{(t)}^{\;i}\mathbf{C}^{(t)}(\mathbf{x},s) = {}_{(t)}^{\;i}\mathbf{C}(\mathbf{x},\tau) \qquad \text{and} \qquad {}_{(t)}^{\;j}\mathbf{g}^{(t)}(\mathbf{x},s) = {}_{(t)}^{\;j}\mathbf{g}(\mathbf{x},\tau). \tag{9.113}$$

The constitutive functionals (9.111) and (9.112) are required to satisfy appropriate frame invariance conditions depending on the order of tensor T analogous to (5.110) and (5.115) for a simple solid and fluid, respectively.

For the rest of the chapter, we will limit our discussion to constitutive equations of isotropic simple materials undergoing isothermal deformations, and more specifically, we focus on the constitutive equation of the stress tensor and suppress the dependency on θ. Furthermore, continuing in suppressing the dependencies of independent quantities on $\mathbf{x}$, we recall that the constitutive equation of an isotropic simple solid (see (5.109) or (9.111) for $P = 1$) is given by

$$\boldsymbol{\sigma}(t) = \underset{0\le s<\infty}{\mathfrak{F}} \{\mathbf{B}(t), {}_{(t)}\mathbf{C}^{(t)}(s)\}, \tag{9.114}$$

while for a simple fluid the dependence on $\mathbf{B}(t)$ reduces to a dependence on the density, $\rho(t)\,\mathbf{1}$ (see (5.114)). Here, $\mathfrak{F}$ is a tensor-valued functional (i.e., an operator, not necessarily linear, mapping tensor-valued functions into tensors). The above constitutive functional is required to satisfy the invariance condition

$$\mathbf{Q}(t)\cdot \underset{0\le s<\infty}{\mathfrak{F}} \{\mathbf{B}(t), {}_{(t)}\mathbf{C}^{(t)}(s)\}\cdot\mathbf{Q}^T(t) = \underset{0\le s<\infty}{\mathfrak{F}} \{\mathbf{Q}(t)\cdot\mathbf{B}(t)\cdot\mathbf{Q}^T(t), \mathbf{Q}(t)\cdot{}_{(t)}\mathbf{C}^{(t)}(s)\cdot\mathbf{Q}^T(t)\}. \tag{9.115}$$

It is useful to rewrite (9.114) in a slightly different form by decomposing the right-hand side into an "equilibrium" function and a functional that vanishes when the material has always been at rest:

$$\boldsymbol{\sigma}(t) = \mathfrak{h}\{\mathbf{B}(t)\} + \underset{0\le s<\infty}{\mathfrak{I}} \{\mathbf{B}(t), \boldsymbol{G}^{(t)}(s)\}, \tag{9.116}$$

where, having used the definition (9.52), we require that

$$\underset{0\le s<\infty}{\mathfrak{I}} \{\mathbf{B}(t), \mathbf{0}(s)\} = \mathbf{0}, \tag{9.117}$$

and it is understood that if the material has always been in equilibrium, then

$$\boldsymbol{G}^{(t)}(s) = \mathbf{0} \qquad \text{for } 0 \le s < \infty. \tag{9.118}$$

The function

$$\mathbf{0}(s) = \mathbf{0}, \qquad 0 \le s < \infty, \tag{9.119}$$

is called the *zero function.*

Now the frame invariance condition for the equilibrium state requires that

$$\mathbf{Q}(t) \cdot \mathfrak{h}\{\mathbf{B}(t)\} \cdot \mathbf{Q}^T(t) = \mathfrak{h}\{\mathbf{Q}(t) \cdot \mathbf{B}(t) \cdot \mathbf{Q}^T(t)\}. \tag{9.120}$$

Subsequently, satisfaction of this condition yields the representation

$$\mathfrak{h}\{\mathbf{B}(t)\} = h_0\,\mathbf{1} + h_1\,\mathbf{B} + h_2\,\mathbf{B}^2, \tag{9.121}$$

where $h_i = h_i(\operatorname{tr}\mathbf{B}, \operatorname{tr}\mathbf{B}^2, \operatorname{tr}\mathbf{B}^3)$, $i = 0, 1, 2$, (see Tables 5.1 and 5.3).

Correspondingly, for a fluid the frame invariance condition becomes

$$\mathbf{Q}(t) \cdot \mathfrak{h}\{\rho(t)\} \cdot \mathbf{Q}^T(t) = \mathfrak{h}\{\rho(t)\}, \tag{9.122}$$

whose representation is given by

$$\mathfrak{h}\{\rho(t)\} = -p\,(\rho(t))\,\mathbf{1}, \tag{9.123}$$

and (9.116) becomes

$$\boldsymbol{\sigma}(t) = -p\,(\rho(t))\,\mathbf{1} + \underset{0\le s<\infty}{\mathfrak{J}} \{\rho(t), \boldsymbol{G}^{(t)}(s)\} \tag{9.124}$$

or

$$\boldsymbol{\sigma}^d(t) = \boldsymbol{\sigma}(t) + p\,(\rho(t))\,\mathbf{1} = \underset{0\le s<\infty}{\mathfrak{J}} \{\rho(t), \boldsymbol{G}^{(t)}(s)\}, \tag{9.125}$$

where $\boldsymbol{\sigma}^d(t)$ is the dissipative or extra stress.

If the fluid is incompressible, then $\rho(t) = \rho_R$, where we assume that $\rho_R = \text{const.}$, and the stress tensor is determined by the history of the motion only up to a scalar pressure p. Thus, for incompressible fluids, we can replace (9.125) by

$$\boldsymbol{\sigma}^d(t) = \boldsymbol{\sigma}(t) + p\,\mathbf{1} = \underset{0\le s<\infty}{\mathfrak{J}} \{\boldsymbol{G}^{(t)}(s)\}, \tag{9.126}$$

and the tensor-valued functional is now determined only up to an arbitrary scalar-valued functional of $\boldsymbol{G}^{(t)}(s)$. The indeterminacy is removed by the normalization

$$\operatorname{tr}\boldsymbol{\sigma}^d(t) = \operatorname{tr} \underset{0\le s<\infty}{\mathfrak{J}} \{\boldsymbol{G}^{(t)}(s)\} = 0. \tag{9.127}$$

If follows from (9.126) and (9.127) that p is given by the mean pressure

$$p = -\frac{1}{3}\operatorname{tr}\boldsymbol{\sigma}, \tag{9.128}$$

and the extra stress is equal to the deviatoric part of $\boldsymbol{\sigma}$, i.e., $\boldsymbol{\sigma}^d = \boldsymbol{\sigma}'$.

Note the frame invariance condition (9.115) and the equilibrium condition (9.117) reduce to

$$\mathbf{Q}(t) \cdot \underset{0\le s<\infty}{\mathfrak{F}} \{\rho(t), \boldsymbol{G}^{(t)}(s)\} \cdot \mathbf{Q}^T(t) = \underset{0\le s<\infty}{\mathfrak{F}} \{\rho(t), \mathbf{Q}(t) \cdot \boldsymbol{G}^{(t)}(s) \cdot \mathbf{Q}^T(t)\} \tag{9.129}$$

and

$$\underset{0\le s<\infty}{\mathfrak{J}} \{\rho(t), \mathbf{0}(s)\} = \mathbf{0}. \tag{9.130}$$

9.3.1 Constitutive equations for motion with constant stretch history

For an MWCSH, the constitutive equation (9.116), upon using (9.52) and (9.75), becomes

$$\boldsymbol{\sigma}(t) = \mathfrak{h}\{\mathbf{B}\} + \mathfrak{f}(\mathbf{B}, \mathbf{M}), \tag{9.131}$$

where

$$\mathfrak{f}(\mathbf{B}, \mathbf{M}) = \underset{0 \leq s < \infty}{\mathfrak{I}} \{\mathbf{B}(t), e^{-s\mathbf{M}^T} \cdot e^{-s\mathbf{M}} - \mathbf{1}\}, \tag{9.132}$$

and, from (9.117),

$$\mathfrak{f}(\mathbf{B}, \mathbf{0}) = \mathbf{0}. \tag{9.133}$$

The implication of the theorem in Section 9.2.1 is that we can write (9.131) in the most general form as

$$\boldsymbol{\sigma}(t) = \mathfrak{h}\{\mathbf{B}\} + \mathfrak{f}\left(\mathbf{B}, \mathbf{A}^{(1)}, \mathbf{A}^{(2)}, \mathbf{A}^{(3)}\right). \tag{9.134}$$

For a fluid, this becomes

$$\boldsymbol{\sigma}(t) = -p(\rho)\,\mathbf{1} + \mathfrak{f}\left(\rho, \mathbf{A}^{(1)}, \mathbf{A}^{(2)}, \mathbf{A}^{(3)}\right), \tag{9.135}$$

and if the fluid is incompressible (see (9.126)),

$$\boldsymbol{\sigma}^d(t) = \mathfrak{f}\left(\mathbf{A}^{(1)}, \mathbf{A}^{(2)}, \mathbf{A}^{(3)}\right). \tag{9.136}$$

Note that for simplicity we have used the same tensor function symbol $\mathfrak{f}$ in all of the above cases; clearly they would correspond to a different function in each case. Now objectivity requires that

$$\begin{aligned}\mathbf{Q} \cdot \mathfrak{f}\left(\mathbf{B}, \mathbf{A}^{(1)}, \mathbf{A}^{(2)}, \mathbf{A}^{(3)}\right) \cdot \mathbf{Q}^T &= \\ \mathfrak{f}\left(\mathbf{Q} \cdot \mathbf{B} \cdot \mathbf{Q}^T, \mathbf{Q} \cdot \mathbf{A}^{(1)} \cdot \mathbf{Q}^T, \mathbf{Q} \cdot \mathbf{A}^{(2)} \cdot \mathbf{Q}^T, \mathbf{Q} \cdot \mathbf{A}^{(3)} \cdot \mathbf{Q}^T\right)\end{aligned} \tag{9.137}$$

or for a fluid

$$\mathbf{Q} \cdot \mathfrak{f}\left(\rho, \mathbf{A}^{(1)}, \mathbf{A}^{(2)}, \mathbf{A}^{(3)}\right) \cdot \mathbf{Q}^T = \mathfrak{f}\left(\rho, \mathbf{Q} \cdot \mathbf{A}^{(1)} \cdot \mathbf{Q}^T, \mathbf{Q} \cdot \mathbf{A}^{(2)} \cdot \mathbf{Q}^T, \mathbf{Q} \cdot \mathbf{A}^{(3)} \cdot \mathbf{Q}^T\right). \tag{9.138}$$

The representation of (9.135) satisfying (9.138) is fairly lengthy. However, from the theorem, we see that for a large majority of MWCSH flows, we have that the relative history is uniquely determined by $\mathbf{A}^{(1)}$ and $\mathbf{A}^{(2)}$. In such cases, the representation for a fluid is given by (see Table 5.3)

$$\begin{aligned}\mathfrak{f} &= \mathfrak{f}\left(\rho, \mathbf{A}^{(1)}, \mathbf{A}^{(2)}\right) \\ &= \alpha_0\,\mathbf{1} + \alpha_1\,\mathbf{A}^{(1)} + \alpha_2\,\mathbf{A}^{(2)} + \alpha_3\left(\mathbf{A}^{(1)}\right)^2 + \alpha_4\left(\mathbf{A}^{(2)}\right)^2 + \\ &\quad \alpha_5\left(\mathbf{A}^{(1)} \cdot \mathbf{A}^{(2)} + \mathbf{A}^{(2)} \cdot \mathbf{A}^{(1)}\right) + \alpha_6\left[\left(\mathbf{A}^{(1)}\right)^2 \cdot \mathbf{A}^{(2)} + \mathbf{A}^{(2)} \cdot \left(\mathbf{A}^{(1)}\right)^2\right] + \\ &\quad \alpha_7\left[\mathbf{A}^{(1)} \cdot \left(\mathbf{A}^{(2)}\right)^2 + \left(\mathbf{A}^{(2)}\right)^2 \cdot \mathbf{A}^{(1)}\right],\end{aligned} \tag{9.139}$$

where $\alpha_0, \ldots, \alpha_7$ are functions of ρ and the invariants (see Table 5.1)

$$\operatorname{tr}\mathbf{A}^{(1)}, \operatorname{tr}(\mathbf{A}^{(1)})^2, \operatorname{tr}(\mathbf{A}^{(1)})^3, \operatorname{tr}\mathbf{A}^{(2)}, \operatorname{tr}(\mathbf{A}^{(2)})^2, \operatorname{tr}(\mathbf{A}^{(2)})^3,$$
$$\operatorname{tr}(\mathbf{A}^{(1)}\cdot\mathbf{A}^{(2)}), \operatorname{tr}[\mathbf{A}^{(1)}\cdot(\mathbf{A}^{(2)})^2], \operatorname{tr}[(\mathbf{A}^{(1)})^2\cdot\mathbf{A}^{(2)}], \operatorname{tr}[(\mathbf{A}^{(1)})^2\cdot(\mathbf{A}^{(2)})^2]. \tag{9.140}$$

From (9.127) and (9.130), we must require that

$$\operatorname{tr}\mathfrak{f}\left(\rho, \mathbf{A}^{(1)}, \mathbf{A}^{(2)}\right) = 0 \qquad \text{and} \qquad \mathfrak{f}(\rho, \mathbf{0}, \mathbf{0}) = \mathbf{0}. \tag{9.141}$$

Example

To illustrate the concepts introduced in previous sections, we reconsider the isochoric simple shear flow discussed in Section 9.1:

$$v_1 = 0, \quad v_2 = \kappa x_1, \quad v_3 = 0. \tag{9.142}$$

Subsequently, we write

$$\frac{d\xi_1}{d\tau} = 0, \qquad \frac{d\xi_2}{d\tau} = \kappa\,\xi_1, \qquad \frac{d\xi_3}{d\tau} = 0, \tag{9.143}$$

with final conditions

$$\xi_i(t) = x_i. \tag{9.144}$$

The solutions are

$$\xi_1 = x_1, \quad \xi_2 = x_2 + (\tau - t)\,\kappa x_1 = x_2 - s\,\kappa x_1, \quad \xi_3 = x_3. \tag{9.145}$$

Note from (9.142) that since

$$L = \kappa \begin{pmatrix} 0 & 0 & 0 \\ 1 & 0 & 0 \\ 0 & 0 & 0 \end{pmatrix} \tag{9.146}$$

is constant, then from (9.59), (9.71), (9.72), and (9.78), we have an MWCSH with $Q(t) = I$,

$$L = M = \kappa N = \kappa N_0 = \kappa \begin{pmatrix} 0 & 0 & 0 \\ 1 & 0 & 0 \\ 0 & 0 & 0 \end{pmatrix}, \qquad M^2 = N^2 = N_0^2 = 0, \tag{9.147}$$

$$N_0^T N_0 = \begin{pmatrix} 1 & 0 & 0 \\ 0 & 0 & 0 \\ 0 & 0 & 0 \end{pmatrix}, \quad \text{and} \quad |N_0|^2 = \operatorname{tr}\left(N_0^T N_0\right) = 1. \tag{9.148}$$

From (9.82)–(9.84), we have

$$A^{(1)} = \kappa \begin{pmatrix} 0 & 1 & 0 \\ 1 & 0 & 0 \\ 0 & 0 & 0 \end{pmatrix}, \quad A^{(2)} = 2\kappa^2 \begin{pmatrix} 1 & 0 & 0 \\ 0 & 0 & 0 \\ 0 & 0 & 0 \end{pmatrix}, \text{ and } A^{(n)} = 0 \text{ for } n \geq 3. \tag{9.149}$$

Note that since $M^2 = 0$, then e^{-sM} is at most a linear polynomial in M. Furthermore, since $A^{(1)}$ has three distinct eigenvalues, (0, $\pm\kappa$), then M is uniquely determined by $A^{(1)}$ and $A^{(2)}$. Subsequently, from (9.71), we have that

$$ {}_{(t)}F^{(t)}(s) = \left[\frac{\partial \xi_i}{\partial x_k}\right] = e^{-s\mathbf{M}} = I - s\,M = \begin{pmatrix} 1 & 0 & 0 \\ -\kappa\,s & 1 & 0 \\ 0 & 0 & 1 \end{pmatrix}, \tag{9.150} $$

from (9.85) we have that

$$ {}_{(t)}C^{(t)}(s) = \left(I - s\,M^T\right)\left(I + s\,M\right) = \begin{pmatrix} 1 & -\kappa\,s & 0 \\ -\kappa\,s & 1 + (\kappa\,s)^2 & 0 \\ 0 & 0 & 1 \end{pmatrix}, \tag{9.151} $$

and from (9.52) we have that

$$ G^{(t)}(s) = \begin{pmatrix} 0 & -\kappa\,s & 0 \\ -\kappa\,s & (\kappa\,s)^2 & 0 \\ 0 & 0 & 0 \end{pmatrix}. \tag{9.152} $$

Hence, we see that simple shearing flow is completely determined by

$$ \boldsymbol{G}^{(t)}(\mathbf{x}, s) = -s\,\mathbf{A}^{(1)}(x_1) + \frac{1}{2}\,s^2\mathbf{A}^{(2)}(x_1). \tag{9.153} $$

Since simple shear flow is isochoric, from (9.82) and (9.83), we have that

$$ \operatorname{tr} A^{(1)} = \operatorname{tr} A^{(2)} - \operatorname{tr}\left(A^{(1)}\right)^2 = 0. \tag{9.154} $$

Hence, from (9.139) and (9.140), absorbing α_0 in the arbitrary pressure p, the constitutive equation is

$$ \boldsymbol{\sigma} = -p\,\mathbf{1} + \mathfrak{f}\left(\mathbf{A}^{(1)}, \mathbf{A}^{(2)}\right) = -p\,\mathbf{1} + \beta_1\mathbf{A}^{(1)} + \beta_2\mathbf{A}^{(2)} + \beta_3\left(\mathbf{A}^{(1)}\right)^2, \tag{9.155} $$

where β_1, β_2, and β_3 are functions of κ, given by

$$ \beta_1 = \alpha_1 + 2\,\alpha_5\,\kappa^2 + 4\,\alpha_7\,\kappa^4, \quad \beta_2 = \alpha_2 + 2\,(\alpha_4 + \alpha_6)\,\kappa^2, \quad \beta_3 = \alpha_3 \tag{9.156} $$

with (using (9.140))

$$ \alpha_i = \alpha_i\left(0, 2\,\kappa^2, 0, 2\,\kappa^2, 4\,\kappa^4, 8\,\kappa^6, 0, 0, 2\,\kappa^4, 4\,\kappa^6\right), \quad i = 1, \ldots, 7. \tag{9.157} $$

It follows from (9.155) that since $\mathbf{A}^{(1)}$ and $\mathbf{A}^{(2)}$ have the form (9.149), then the components of $\boldsymbol{\sigma}^d(t)$ must have the form

$$ \left[\sigma^d_{jk}(t)\right] = \begin{pmatrix} \sigma^d_{11} & \sigma^d_{12} & 0 \\ \sigma^d_{21} & \sigma^d_{22} & 0 \\ 0 & 0 & \sigma^d_{33} \end{pmatrix}, \tag{9.158} $$

where $\sigma^d_{21} = \sigma^d_{12}$ and σ^d_{jk} are only functions of κ. Furthermore, from $(9.141)_1$ we must have that

$$ \sigma^d_{11} + \sigma^d_{22} + \sigma^d_{33} = 0. \tag{9.159} $$

We have taken the liberty to write (9.158) and (9.159) in a slightly more general form that applies to isochoric flows (more general than for (9.142), for which $\sigma^d_{33} = 0$). Now (9.155), (9.158), and (9.159) allow us to write the components of the extra stress in terms of three independent functions of κ:

$$\sigma^d_{21} = \sigma^d_{12} = \kappa\,\beta_1(\kappa) \equiv \mu_1(\kappa), \tag{9.160}$$
$$\sigma^d_{11} - \sigma^d_{33} = \kappa^2\,(2\,\beta_2(\kappa) + \beta_3(\kappa)) \equiv \mu_2(\kappa), \tag{9.161}$$
$$\sigma^d_{22} - \sigma^d_{33} = \kappa^2\,\beta_3(\kappa) \equiv \mu_3(\kappa). \tag{9.162}$$

Furthermore, we see that the components of the stress tensor itself are given by

$$\sigma_{21} = \sigma_{12} = \mu_1(\kappa), \qquad \sigma_{11} - \sigma_{33} = \mu_2(\kappa), \qquad \sigma_{22} - \sigma_{33} = \mu_3(\kappa). \tag{9.163}$$

We note that because of the isotropy condition, μ_1, μ_2, and μ_3 cannot depend on any Cartesian coordinate or which direction the material is sheared. Thus, these three functions are material functions. These functions are not completely arbitrary since it can be seen from (9.156), (9.157), and (9.161)–(9.162) that the representation restricts the shear and normal component functions to be odd and even functions, respectively:

$$\mu_1(-\kappa) = -\mu_1(\kappa) \qquad \text{and} \qquad \mu_{2,3}(-\kappa) = \mu_{2,3}(\kappa). \tag{9.164}$$

Furthermore, $(9.141)_2$ implies that these functions must vanish for $\kappa = 0$:

$$\mu_{1,2,3}(0) = 0. \tag{9.165}$$

Lastly, it can be easily shown that the reduced entropy inequality (8.31) requires that the mechanical work must be positive semi-definite:

$$\kappa\,\mu_1(\kappa) \geq 0, \tag{9.166}$$

i.e., $\mu_1(\kappa)$ must have the same sign as κ. If also $\mu_1'(0) > 0$, it is often convenient to introduce the inverse function ζ:

$$\kappa = \zeta(\mu_1) \tag{9.167}$$

called the *shear-rate function.*
A shear-dependent viscosity can be defined as

$$\mu(\kappa) = \frac{\mu_1(\kappa)}{\kappa} \geq 0. \tag{9.168}$$

It follows that μ must be an even function of κ. If we assume that μ is twice differentiable at $\kappa = 0$, and since $\mu_1(0) = 0$, then μ is differentiable at $\kappa = 0$ and

$$\mu'(0) = 0. \tag{9.169}$$

If we also assume that $\mu_{2,3}$ are differentiable, then we also have that

$$\mu_{2,3}'(0) = 0. \tag{9.170}$$

If approximations based on Taylor series expansions of μ and $\mu_{2,3}$ about $\kappa = 0$ are used to fit experimental data, it can be noted that only even powers of κ can occur. This is consistent with our results (9.156) and (9.157).
Let us compare the forms taken by the general material functions $\mu_{1,2,3}$ with those obtained from other approximations of an incompressible fluid. Specifically, for a perfect fluid, we have

$$\mu_{1,2,3}(\kappa) = 0$$

for all κ. For a Newtonian fluid, we have

$$\mu_1(\kappa) = \mu\,\kappa \qquad \text{and} \qquad \mu_{2,3} = 0,$$

where $\mu > 0$ is constant. Note than in both of the above cases, the normal stress differences in (9.163) are zero. Lastly, the incompressible Reiner–Rivlin theory (see (8.116)) places no restriction on $\mu(\kappa)$, but yields the result that $\sigma_{11} = \sigma_{22}$, which in turn gives

$$\mu_2(\kappa) = \mu_3(\kappa).$$

Example

Consider the following elongational flow:

$$v_1 = a\,x_1, \quad v_2 = a\,x_2, \quad v_3 = -2\,a\,x_3, \tag{9.171}$$

so that

$$\frac{d\xi_1}{d\tau} = a\,\xi_1, \quad \frac{d\xi_2}{d\tau} = a\,\xi_2, \quad \frac{d\xi_3}{d\tau} = -2\,a\,\xi_3, \tag{9.172}$$

with final conditions $\xi_i(t) = x_i$. The solutions are

$$\xi_1 = e^{a(\tau-t)}x_1 = e^{-a\,s}x_1, \quad \xi_2 = e^{-a\,s}x_2, \quad \xi_3 = e^{2\,a\,s}x_3. \tag{9.173}$$

In this case,

$$L = a\begin{pmatrix} 1 & 0 & 0 \\ 0 & 1 & 0 \\ 0 & 0 & -2 \end{pmatrix} = L^T, \tag{9.174}$$

the flow is isochoric since $\operatorname{tr} L = 0$, L is constant and thus we have an MWCSH with $Q = I$, and

$$L = M = a\,N = \kappa\,N_0 = a\begin{pmatrix} 1 & 0 & 0 \\ 0 & 1 & 0 \\ 0 & 0 & -2 \end{pmatrix}, \tag{9.175}$$

where

$$\kappa = \sqrt{6}\,a \qquad \text{and} \qquad N_0 = N_0^T = \frac{1}{\sqrt{6}}\begin{pmatrix} 1 & 0 & 0 \\ 0 & 1 & 0 \\ 0 & 0 & -2 \end{pmatrix}. \tag{9.176}$$

We see that in this case M is not nilpotent. It then follows from (9.71) that

$$ {}_{(t)}F^{(t)}(s) = \left[\frac{\partial \xi_i}{\partial x_k}\right] = e^{-s\,\mathbf{M}} = \begin{pmatrix} e^{-a\,s} & 0 & 0 \\ 0 & e^{-a\,s} & 0 \\ 0 & 0 & e^{2\,a\,s} \end{pmatrix}, \tag{9.177} $$

from (9.87) that

$$ {}_{(t)}C^{(t)}(s) = \begin{pmatrix} e^{-2\,a\,s} & 0 & 0 \\ 0 & e^{-2\,a\,s} & 0 \\ 0 & 0 & e^{4\,a\,s} \end{pmatrix}, \tag{9.178} $$

and from (9.82) to (9.84) that

$$ A^{(1)} = a \begin{pmatrix} 2 & 0 & 0 \\ 0 & 2 & 0 \\ 0 & 0 & -4 \end{pmatrix} \quad \text{and} \quad A^{(n)} = \left(A^{(1)}\right)^n \quad \text{for} \quad n \geq 2, \tag{9.179} $$

since $A^{(1)}$ has two distinct eigenvalues, ($2\,a$ and $-4\,a$). Subsequently, since the flow is isochoric, from (9.82) we have that

$$ \operatorname{tr} A^{(1)} = 0 \tag{9.180} $$

and the constitutive equation is uniquely determined by $A^{(1)}$:

$$ \boldsymbol{\sigma} = -p\,\mathbf{1} + \alpha_1 \mathbf{A}^{(1)} + \alpha_2 \left(\mathbf{A}^{(1)}\right)^2, \tag{9.181} $$

where we have absorbed α_0 in the arbitrary pressure p, and where α_1 and α_2 are functions of of the invariants $\operatorname{tr} A^{(1)}$, $\operatorname{tr}(A^{(1)})^2$, and $\operatorname{tr}(A^{(1)})^3$:

$$ \alpha_i = \alpha_i\left(0, 24\,a^2, -48\,a^3\right). \tag{9.182} $$

Furthermore, we see that the components of the stress tensor itself are given by

$$ \sigma_{21} = \sigma_{12} = 0, \qquad \sigma_{11} - \sigma_{33} = \sigma_{22} - \sigma_{33} = 4\,a\,(\alpha_1 - 3\,a\,\alpha_2) \equiv \mu(a). \tag{9.183} $$

In this case, we have only one material function.
Note that, since $A^{(1)} = 2\,D$, we can represent the constitutive equation in the form

$$ \boldsymbol{\sigma} = -p\,\mathbf{1} + \mathfrak{f}(\mathbf{D}) = -p\,\mathbf{1} + \beta_1 \mathbf{D} + \beta_2 \mathbf{D}^2, \tag{9.184} $$

where β_1 and β_2 are functions of $\operatorname{tr} D^2$ and $\det D$, i.e., functions of a.

Example

Here we combine the elongation and shear flows:

$$ v_1 = a\,x_1, \quad v_2 = \kappa\,x_1 + a\,x_2, \quad v_3 = -2\,a\,x_3 \tag{9.185} $$

so that

$$\frac{d\xi_1}{d\tau} = a\,\xi_1, \quad \frac{d\xi_2}{d\tau} = \kappa\,\xi_1 + a\,\xi_2, \quad \frac{d\xi_3}{d\tau} = -2\,a\,\xi_3, \tag{9.186}$$

with final conditions $\xi_i(t) = x_i$. Then, if we let $s = t - \tau$, the solutions can be written as

$$\xi_1 = e^{-a\,s}x_1, \quad \xi_2 = -s\,\kappa\,e^{-a\,s}x_1 + e^{-a\,s}x_2, \quad \xi_3 = e^{2\,a\,s}x_3. \tag{9.187}$$

Note from (9.185) that since

$$L = \begin{pmatrix} a & 0 & 0 \\ \kappa & a & 0 \\ 0 & 0 & -2\,a \end{pmatrix} \tag{9.188}$$

is constant, then from (9.59), (9.71), (9.72), and (9.78), we have that $Q(t) = I$,

$$L = M = \begin{pmatrix} a & 0 & 0 \\ \kappa & a & 0 \\ 0 & 0 & -2\,a \end{pmatrix}, \tag{9.189}$$

and M is not nilpotent. From (9.82) and (9.83), we have

$$A^{(1)} = \begin{pmatrix} 2\,a & \kappa & 0 \\ \kappa & 2\,a & 0 \\ 0 & 0 & -4\,a \end{pmatrix} \quad \text{and} \quad A^{(2)} = \begin{pmatrix} 2\,\kappa^2 + 4\,a^2 & 4\,\kappa\,a & 0 \\ 4\,\kappa\,a & 4\,a^2 & 0 \\ 0 & 0 & 16\,a^2 \end{pmatrix}. \tag{9.190}$$

Now since $A^{(1)}$ has three distinct eigenvalues, $(-4\,a,\ 2\,a \pm \kappa)$, then M is uniquely determined by $A^{(1)}$ and $A^{(2)}$. Subsequently, from (9.71) we have that

$${}_{(t)}F^{(t)}(s) = \left[\frac{\partial \xi_i}{\partial x_k}\right] = e^{-s\,L} = \begin{pmatrix} e^{-a\,s} & 0 & 0 \\ -s\,\kappa\,e^{-a\,s} & e^{-a\,s} & 0 \\ 0 & 0 & e^{2\,a\,s} \end{pmatrix}, \tag{9.191}$$

so that

$${}_{(t)}C^{(t)}(s) = \begin{pmatrix} e^{-2\,a\,s} & -s\,\kappa\,e^{-2\,a\,s} & 0 \\ -s\,\kappa\,e^{-2\,a\,s} & \left[1 + (s\,\kappa)^2\right] e^{-2\,a\,s} & 0 \\ 0 & 0 & e^{4\,a\,s} \end{pmatrix}. \tag{9.192}$$

For the simple shear-elongational flow, we have (see (9.139))

$$\begin{aligned} \boldsymbol{\sigma}^d(t) = \boldsymbol{\sigma}(t) + p\,\mathbf{1} = \alpha_1\,\mathbf{A}^{(1)} + \alpha_2\,\mathbf{A}^{(2)} + \alpha_3\left(\mathbf{A}^{(1)}\right)^2 + \alpha_4\left(\mathbf{A}^{(2)}\right)^2 + \\ \alpha_5\left(\mathbf{A}^{(1)}\cdot\mathbf{A}^{(2)} + \mathbf{A}^{(2)}\cdot\mathbf{A}^{(1)}\right) + \alpha_6\left[\left(\mathbf{A}^{(1)}\right)^2\cdot\mathbf{A}^{(2)} + \mathbf{A}^{(2)}\cdot\left(\mathbf{A}^{(1)}\right)^2\right] + \\ \alpha_7\left[\mathbf{A}^{(1)}\cdot\left(\mathbf{A}^{(2)}\right)^2 + \left(\mathbf{A}^{(2)}\right)^2\cdot\mathbf{A}^{(1)}\right], \end{aligned} \tag{9.193}$$

where we have absorbed α_0 in the arbitrary pressure p, and α_i are functions of the invariants (see (9.140))

$$\begin{gathered}\operatorname{tr}\mathbf{A}^{(1)}, \operatorname{tr}(\mathbf{A}^{(1)})^2, \operatorname{tr}(\mathbf{A}^{(1)})^3, \operatorname{tr}\mathbf{A}^{(2)}, \operatorname{tr}(\mathbf{A}^{(2)})^2, \operatorname{tr}(\mathbf{A}^{(2)})^3,\\ \operatorname{tr}(\mathbf{A}^{(1)}\cdot\mathbf{A}^{(2)}), \operatorname{tr}[\mathbf{A}^{(1)}\cdot(\mathbf{A}^{(2)})^2], \operatorname{tr}[(\mathbf{A}^{(1)})^2\cdot\mathbf{A}^{(2)}],\\ \operatorname{tr}[(\mathbf{A}^{(1)})^2\cdot(\mathbf{A}^{(2)})^2], \end{gathered} \tag{9.194}$$

which in turn are functions of κ and a. In this case, we see that

$$\sigma_{12} = \sigma_{21} = \mu_1(\kappa, a), \quad \sigma_{11} - \sigma_{33} = \mu_2(\kappa, a), \quad \sigma_{22} - \sigma_{33} = \mu_3(\kappa, a). \tag{9.195}$$

From (9.141) and (9.190), we also have the requirements that

$$\operatorname{tr}\boldsymbol{\sigma}^d(\kappa, a) = 0 \qquad \text{and} \qquad \boldsymbol{\sigma}^d(0,0) = \mathbf{0}. \tag{9.196}$$

9.3.2 Fading memory

We assume that the memory of a material fades in time. This assumption is appropriate for most simple materials (a notable exception are materials that are hypo-elastic). To characterize how the memory fades, we introduce the *memory influence function* $h(s)$ of order $r > 0$ satisfying the following conditions:

a) $h(s)$ is defined for $0 \le s < \infty$ and is real and positive definite, $h(s) > 0$;

b) $h(s)$ is normalized by the condition $h(0) = 1$;

c) $h(s)$ decays to zero monotonically for large s according to

$$\lim_{s\to\infty} s^r h(s) = 0.$$

For example,

$$h(s) = (1+s)^{-p}, \qquad p > 1,$$

is a memory influence function of order r if $r < p$. The exponential

$$h(s) = e^{-\alpha s}, \qquad \alpha > 0,$$

is a memory influence function of any order.

Now, let a memory influence function be given. We define the magnitude of the history tensor $\boldsymbol{G}^{(t)}(s)$ by

$$\left|\boldsymbol{G}^{(t)}(s)\right|^2 \equiv \operatorname{tr}\left[(\boldsymbol{G}^{(t)}(s))^T\cdot\boldsymbol{G}^{(t)}(s)\right] \tag{9.197}$$

and its norm by

$$\left\|\boldsymbol{G}^{(t)}(s)\right\|_h^2 = \int_0^\infty \left|\boldsymbol{G}^{(t)}(s)\right|^2 h^2(s)\,ds. \tag{9.198}$$

The influence function $h(s)$ determines the influence of the values of $\boldsymbol{G}^{(t)}(s)$ in computing the norm $\left\|\boldsymbol{G}^{(t)}(s)\right\|_h$. Since $h(s)\to 0$ as $s\to\infty$, the values of $\boldsymbol{G}^{(t)}(s)$ for small s (recent past) have a greater weight than the values for large s (distant past). The collection of all histories in the constitutive functional with finite norm (9.198) forms a Hilbert space $\mathcal{H}_h$ with weight $h(s)$ and inner product defined by

$$\left\langle \boldsymbol{G}_1^{(t)}(s) \cdot \boldsymbol{G}_2^{(t)}(s) \right\rangle_h = \int_0^\infty \operatorname{tr}\left[\left(\boldsymbol{G}_1^{(t)}(s) \right)^T \cdot \boldsymbol{G}_2^{(t)}(s) \right] h^2(s)\, ds. \tag{9.199}$$

The history tensor $\boldsymbol{G}^{(t)}(s)$ belongs to the space $\mathcal{H}_h$ if it does not grow too fast as $s \to \infty$.

Now consider an influence function $h(s)$ and the functional in (9.116), i.e.,

$$\underset{0 \le s < \infty}{\mathfrak{J}} \{\mathbf{B}(t), \boldsymbol{G}^{(t)}(s)\},$$

which is defined in a neighborhood of the zero history in the Hilbert space $\mathcal{H}_h$ and whose value is a rank 2 tensor quantity that at zero history is zero (see (9.117)), i.e.,

$$\underset{0 \le s < \infty}{\mathfrak{J}} \{\mathbf{B}(t), \mathbf{0}(s)\} = \mathbf{0}.$$

If the history tensor $\boldsymbol{G}^{(t)}(s)$ is sufficiently smooth, then we say that $\mathfrak{J}$ is *Fréchet differentiable* at the zero history (see (2.190)) so that

$$\underset{0 \le s < \infty}{\mathfrak{J}} \{\mathbf{B}(t), \boldsymbol{G}^{(t)}(s)\} = \sum_{n=1}^{m} \frac{1}{n!} \delta^n \underset{0 \le s < \infty}{\mathfrak{J}} \left\{\mathbf{B}(t), \boldsymbol{G}^{(t)}(s)\right\} + \mathbf{o}\left(\left\| \boldsymbol{G}^{(t)}(s) \right\|_h^m \right) \tag{9.200}$$

$$= \sum_{n=1}^{m} \boldsymbol{\mathcal{F}}^{(n)} \left\{ \mathbf{B}(t), \mathbf{0} \left| \underset{0 \le s_1 < \infty}{\mathfrak{J}} \boldsymbol{G}^{(t)}(s_1) \right| \cdots \left| \underset{0 \le s_n < \infty}{\mathfrak{J}} \boldsymbol{G}^{(t)}(s_n) \right. \right\} + \mathbf{o}\left(\left\| \boldsymbol{G}^{(t)}(s) \right\|_h^m \right), \tag{9.201}$$

where

$$\lim_{\left\|\boldsymbol{G}^{(t)}(s)\right\|_h \to 0} \frac{\mathbf{o}\left(\left\| \boldsymbol{G}^{(t)}(s) \right\|_h^m \right)}{\left\| \boldsymbol{G}^{(t)}(s) \right\|_h} = \mathbf{0}. \tag{9.202}$$

The linear functional $\delta^n \mathfrak{J}$ is called the nth *Fréchet derivative* of $\mathfrak{J}$ at the zero history (it is essentially a generalization of the derivative of a real-valued function to the derivative of a functional), and it can be shown that the functional is Fréchet differentiable at the zero history for any value of the tensor $\mathbf{B}$ (or $\rho \mathbf{1}$ for a simple fluid) for an influence function $h(s)$ of order $r > m + 1/2$. The Fréchet functional derivative of order n evaluated on $\boldsymbol{G}^{(t)}(s) = \mathbf{0}$ (the second variable) enables us to define $\boldsymbol{\mathcal{F}}^{(n)}$, a tensor-valued function of $\mathbf{B}(t)$ (or $\rho \mathbf{1}$ for a simple fluid) and a multilinear continuous tensor-valued form in $\boldsymbol{G}^{(t)}(\bullet)$, whose argument functions are histories. A material obeying a constitutive equation of the form (9.201) is called a simple material of order m.

To clarify the notation, suppose $\{\boldsymbol{G}^{(t)}(s), \boldsymbol{J}^{(t)}(s), \boldsymbol{H}^{(t)}(s)\} \in \mathcal{H}_h$. Then

$$\delta \underset{0 \le s < \infty}{\mathfrak{J}} \left\{\mathbf{B}(t), \boldsymbol{G}^{(t)}(s)\right\} = \frac{\partial}{\partial \lambda_1} \underset{0 \le s < \infty}{\mathfrak{J}} \left\{\mathbf{B}(t), \boldsymbol{G}^{(t)}(s) + \lambda_1 \boldsymbol{J}^{(t)}(s)\right\}\Big|_{\lambda_1 = 0} \equiv$$

$$\boldsymbol{\mathcal{F}}^{(1)} \left\{ \mathbf{B}(t), \boldsymbol{G}^{(t)}(s) \left| \underset{0 \le s_1 < \infty}{\mathfrak{J}} \boldsymbol{J}^{(t)}(s_1) \right. \right\} \tag{9.203}$$

is linear and continuous in $\boldsymbol{J}^{(t)}(s)$. The second derivative is defined by

$$\delta^2 \underset{0\le s<\infty}{\mathfrak{J}} \left\{ \mathbf{B}(t), \boldsymbol{G}^{(t)}(s) \right\} =$$

$$\frac{\partial^2}{\partial\lambda_1\,\partial\lambda_2} \underset{0\le s<\infty}{\mathfrak{J}} \left\{ \mathbf{B}(t), \boldsymbol{G}^{(t)}(s) + \lambda_1 \boldsymbol{J}^{(t)}(s) + \lambda_2 \boldsymbol{K}^{(t)}(s) \right\}\bigg|_{\lambda_1=\lambda_2=0} \equiv$$

$$\boldsymbol{\mathcal{F}}^{(2)} \left\{ \mathbf{B}(t), \boldsymbol{G}^{(t)}(s) \middle| \underset{0\le s_1<\infty}{\mathfrak{J}} \boldsymbol{J}^{(t)}(s_1) \middle| \underset{0\le s_1<\infty}{\mathfrak{J}} \boldsymbol{K}^{(t)}(s_2) \right\}, \quad (9.204)$$

where $\boldsymbol{\mathcal{F}}^{(2)}$ is linear and continuous in $\boldsymbol{J}^{(t)}(s)$ and $\boldsymbol{K}^{(t)}(s)$. Higher derivatives are computed in the same way. We obtain derivatives on the zero history by letting $\boldsymbol{G}^{(t)}(s) \to \mathbf{0}$, $\boldsymbol{J}^{(t)}(s) \to \boldsymbol{G}^{(t)}(s)$, and $\boldsymbol{K}^{(t)}(s) \to \boldsymbol{G}^{(t)}(s)$ in the above equations and subsequently obtain (9.201). It is obvious that the functional derivatives are symmetric to all transpositions of their linear arguments in (9.201).

9.3.3 Constitutive equations of differential type

For slow motions which have derivatives of $\boldsymbol{G}^{(t)}(s)$ at $s = 0$, we can use the Taylor series (9.53) to expand (9.201):

$$\boldsymbol{\mathcal{F}}^{(n)} \left\{ \mathbf{B}(t), \mathbf{0}(s) \middle| \underset{0\le s_1<\infty}{\mathfrak{J}} \boldsymbol{G}^{(t)}(s_1) \middle| \cdots \middle| \underset{0\le s_n<\infty}{\mathfrak{J}} \boldsymbol{G}^{(t)}(s_n) \right\} =$$

$$\boldsymbol{\mathcal{F}}^{(n)} \left\{ \mathbf{B}(t), \mathbf{0}(s) \middle| \underset{0\le s_1<\infty}{\mathfrak{J}} \sum_{j_1=1}^{\infty} \frac{(-1)^{j_1}}{j_1!} \mathbf{A}^{(j_1)}(t)\, s_1^{j_1} \middle| \cdots \right.$$

$$\left. \middle| \underset{0\le s_n<\infty}{\mathfrak{J}} \sum_{j_n=1}^{\infty} \frac{(-1)^{j_n}}{j_n!} \mathbf{A}^{(j_n)}(t)\, s_n^{j_n} \right\} =$$

$$\sum_{(j_1,\ldots,j_n)} \mathfrak{f}^{(j_1,\ldots,j_n)} \left\{ \mathbf{B}(t), \left[\mathbf{A}^{(j_1)}(t), \ldots, \mathbf{A}^{(j_n)}(t) \right] \right\}, \quad (9.205)$$

where each $\mathfrak{f}^{(j_1,\ldots,j_n)} \left\{ \mathbf{B}(t), \left[\mathbf{A}^{(j_1)}(t), \ldots, \mathbf{A}^{(j_n)}(t) \right] \right\}$, for each choice of $\mathbf{B}$ (or $\rho(t)\,\mathbf{1}$ for a fluid), is a multilinear isotropic tensor-valued function of n tensor variables. The summation is to be extended over all sets of n indices $(j_1, \ldots, j_n)$ satisfying the inequalities

$$1 \le j_1 \le \cdots \le j_n \le m, \qquad j_1 + \cdots + j_n \le m. \quad (9.206)$$

Subsequently, for an isotropic and homogeneous simple solid, we have

$$\boldsymbol{\sigma} = \mathfrak{h}\{\mathbf{B}(t)\} + \sum_{n=1}^{m} \sum_{(j_1,\ldots,j_n)} \mathfrak{f}^{(j_1,\ldots,j_n)} \left\{ \mathbf{B}(t), \left[\mathbf{A}^{(j_1)}(t), \ldots, \mathbf{A}^{(j_n)}(t) \right] \right\}, \quad (9.207)$$

while for a fluid we have

$$\boldsymbol{\sigma} = -p(\rho(t))\,\mathbf{1} + \sum_{n=1}^{m} \sum_{(j_1,\ldots,j_n)} \mathfrak{f}^{(j_1,\ldots,j_n)} \left\{ \rho(t), \left[\mathbf{A}^{(j_1)}(t), \ldots, \mathbf{A}^{(j_n)}(t) \right] \right\}. \quad (9.208)$$

Materials satisfying (9.207) or (9.208) are called materials of differential type of order m.

Example

We would like to write the explicit form of the constitutive equation (9.207) for an isotropic and homogeneous simple viscoelastic solid of order $m = 1$. In this case, we have

$$\boldsymbol{\sigma}(t) = \mathfrak{h}\{\mathbf{B}(t)\} + \mathfrak{f}^{(1)}\{\mathbf{B}(t), [\mathbf{A}^{(1)}]\} = \mathfrak{h}\{\mathbf{B}(t)\} + \mathfrak{l}\{\mathbf{B}(t), \mathbf{D}\}, \tag{9.209}$$

where we have exploited the fact that $\mathbf{A}^{(1)} = 2\,\mathbf{D}^{(1)} = 2\,\mathbf{D}$. Now, using (9.121) and Tables 5.1 and 5.3, we can write

$$\begin{aligned}\boldsymbol{\sigma}(t) = \left(h_0\,\mathbf{1} + h_1\,\mathbf{B} + h_2\,\mathbf{B}^2\right) + \left[l_0\,\mathbf{1} + l_1\,\mathbf{B} + l_2\,\mathbf{B}^2 + l_3\,\mathbf{D} + l_4\,\mathbf{D}^2\right] + \\ l_5\left(\mathbf{B}\cdot\mathbf{D} + \mathbf{D}\cdot\mathbf{B}\right) + l_6\left(\mathbf{B}^2\cdot\mathbf{D} + \mathbf{D}\cdot\mathbf{B}^2\right) + l_7\left(\mathbf{B}\cdot\mathbf{D}^2 + \mathbf{D}^2\cdot\mathbf{B}\right),\end{aligned} \tag{9.210}$$

where the h_i's are functions of the invariants of $\mathbf{B}$, i.e.,

$$\{\mathrm{tr}\,\mathbf{B}, \mathrm{tr}\,\mathbf{B}^2, \mathrm{tr}\,\mathbf{B}^3\}, \tag{9.211}$$

and the l_j's are functions of the invariants of $\mathbf{B}$ and $\mathbf{D}$, i.e.,

$$\begin{aligned}\{\mathrm{tr}\,\mathbf{B}, \mathrm{tr}\,\mathbf{B}^2, \mathrm{tr}\,\mathbf{B}^3, \mathrm{tr}\,\mathbf{D}, \mathrm{tr}\,\mathbf{D}^2, \mathrm{tr}\,\mathbf{D}^3, \mathrm{tr}\,(\mathbf{B}\cdot\mathbf{D}), \mathrm{tr}\,\left(\mathbf{B}\cdot\mathbf{D}^2\right), \\ \mathrm{tr}\,\left(\mathbf{B}^2\cdot\mathbf{D}\right), \mathrm{tr}\,\left(\mathbf{B}^2\cdot\mathbf{D}^2\right)\}.\end{aligned} \tag{9.212}$$

Example

Correspondingly, the explicit form of the constitutive equation (9.208) for a simple fluid of order $m = 1$ is given by

$$\boldsymbol{\sigma}(t) = -p(\rho(t))\,\mathbf{1} + \mathfrak{f}^{(1)}\{\rho(t), [\mathbf{A}^{(1)}]\} = -p(\rho(t))\,\mathbf{1} + \mathfrak{l}\{\rho(t), \mathbf{D}\}, \tag{9.213}$$

where we have exploited the fact that $\mathbf{A}^{(1)} = 2\,\mathbf{D}^{(1)} = 2\,\mathbf{D}$. Now, using Tables 5.1 and 5.3, we can write

$$\boldsymbol{\sigma}(t) = [-p(\rho(t)) + l_0]\,\mathbf{1} + l_1\,\mathbf{D} + l_2\,\mathbf{D}^2, \tag{9.214}$$

where the l_i's are functions of ρ and the invariants of $\mathbf{D}$, i.e., $\{\mathrm{tr}\,\mathbf{D}, \mathrm{tr}\,\mathbf{D}^2, \mathrm{tr}\,\mathbf{D}^3\}$. As we see from (8.116) and (8.118), this is just the representation of a Reiner–Rivlin fluid.

In general, (9.207) and (9.208) are not good representations, because (9.53) is not a convenient way to represent the history when higher order terms are not negligible. The series representation (9.53) is made useful for *retarded motion* in which the first few terms dominate. The retarded history of $\boldsymbol{G}^{(t)}(s)$ is obtained by replacing s with ϵs:

$$\boldsymbol{G}^{(t)}_{\epsilon}(s) = \boldsymbol{G}^{(t)}(\epsilon s) = \sum_{n=1}^{\infty} \frac{(-\epsilon s)^n}{n!}\mathbf{A}^{(n)}(t) \tag{9.215}$$

and thus diminish the importance of the higher $\mathbf{A}^{(n)}$ when ϵ is small. Then, following the above procedure, constitutive equations for retarded motion (nearly steady slow motion) of order m are obtained by identifying increasing powers of ϵ.

Example

The retarded motion of order $m = 1$ of a simple isotropic solid is obtained by retaining terms in (9.210)–(9.212) that are of $O(\epsilon)$ in $\mathbf{A}^{(1)}$ or $\mathbf{D}$, i.e.,

$$\begin{aligned}\boldsymbol{\sigma}(t) = &\left(h_0\,\mathbf{1} + h_1\,\mathbf{B} + h_2\,\mathbf{B}^2\right) + \left[k_1 \operatorname{tr}\mathbf{D} + k_2 \operatorname{tr}(\mathbf{B}\cdot\mathbf{D}) + \right.\\ &\left. k_3 \operatorname{tr}\left(\mathbf{B}^2\cdot\mathbf{D}\right)\right]\mathbf{1} + \left[k_4 \operatorname{tr}\mathbf{D} + k_5 \operatorname{tr}(\mathbf{B}\cdot\mathbf{D}) + k_6 \operatorname{tr}\left(\mathbf{B}^2\cdot\mathbf{D}\right)\right]\mathbf{B} + \\ &\left[k_7 \operatorname{tr}\mathbf{D} + k_8 \operatorname{tr}(\mathbf{B}\cdot\mathbf{D}) + k_9 \operatorname{tr}\left(\mathbf{B}^2\cdot\mathbf{D}\right)\right]\mathbf{B}^2 + \\ &k_{10}\,\mathbf{D} + k_{11}\left(\mathbf{B}\cdot\mathbf{D} + \mathbf{D}\cdot\mathbf{B}\right) + k_{12}\left(\mathbf{B}^2\cdot\mathbf{D} + \mathbf{D}\cdot\mathbf{B}^2\right), \qquad (9.216)\end{aligned}$$

where the h_i's and k_j's are functions of the invariants of $\mathbf{B}$, i.e., $\{\operatorname{tr}\mathbf{B}, \operatorname{tr}\mathbf{B}^2, \operatorname{tr}\mathbf{B}^3\}$.

Example

The corresponding explicit form of the constitutive equation for retarded motion of a simple fluid of order $m = 1$ is obtained by retaining the terms that are liner in $\mathbf{D}$ in (9.214). In this case, we obtain the Newtonian stress tensor

$$\boldsymbol{\sigma}(t) = \left[-p(\rho(t)) + \lambda \operatorname{tr}\mathbf{D}\right]\mathbf{1} + 2\,\mu\,\mathbf{D}, \qquad (9.217)$$

where λ and μ are functions of ρ.

9.3.4 Constitutive equations of integral type

More generally, if we assume that the motion is not slow or differentiable, we can make use of *Riesz' representation theorem*, which states that every continuous functional may be written as an inner product. This allows us to write (9.200) and (9.201) in the form

$$\begin{aligned}\frac{1}{n!}\,\delta^n \underset{0\le s<\infty}{\mathfrak{J}} \{\mathbf{B}(t), \boldsymbol{G}^{(t)}(s)\} = \boldsymbol{\mathcal{F}}^{(n)}\left\{\mathbf{B}(t), \mathbf{0}\,\middle|\, \underset{0\le s_1<\infty}{\mathfrak{J}}\, \boldsymbol{G}^{(t)}(s_1)\middle|\cdots\middle|\, \underset{0\le s_n<\infty}{\mathfrak{J}}\, \boldsymbol{G}^{(t)}(s_n)\right\} = \\ \int_0^\infty\!\!\cdots\!\int_0^\infty \mathbf{K}^{(n)}\left(s_1,\ldots,s_n;\mathbf{B}(t)\right)\cdot\left[\boldsymbol{G}^{(t)}(s_1)\cdots\boldsymbol{G}^{(t)}(s_n)\right] ds_1\cdots ds_n. \qquad (9.218)\end{aligned}$$

Substitution of (9.201) and (9.218) into (9.116) provides an approximation to the constitutive equation of a simple material with fading memory in the form

$$\begin{aligned}&\boldsymbol{\sigma}(t) = \mathfrak{h}\{\mathbf{B}(t)\} + \\ &\sum_{n=1}^{m}\int_0^\infty\!\!\cdots\!\int_0^\infty \mathbf{K}^{(n)}\left(s_1,\ldots,s_n;\mathbf{B}(t)\right)\cdot\left[\boldsymbol{G}^{(t)}(s_1)\cdots\boldsymbol{G}^{(t)}(s_n)\right] ds_1\cdots ds_n. \ (9.219)\end{aligned}$$

The tensor functions $\mathbf{K}^{(n)}\left(s_1,\ldots,s_n;\mathbf{B}(t)\right)\cdot\left[\boldsymbol{G}^{(t)}(s_1)\cdots\boldsymbol{G}^{(t)}(s_n)\right]$ are bounded isotropic second-order tensor polynomials that are multilinear in the n tensor variables $\boldsymbol{G}^{(t)}(s_1)\cdots\boldsymbol{G}^{(t)}(s_n)$. Furthermore, they are completely symmetric under any permutations of $1, \ldots, n$. The values of $\mathbf{K}^{(n)}(s_1, \ldots, s_n; \mathbf{B}(t))$ are isotropic tensors of order $2(n+1)$ that are continuous with the tensor parameter $\mathbf{B}(t)$. The error of this approximation approaches zero faster than the mth power of the history tensor norm (9.198). A material obeying a constitutive function of the form (9.219) is called a simple material of integral type of order m.

For example, when $m = 1$ we have that

$$\boldsymbol{\sigma}(t) = \mathfrak{h}\{\mathbf{B}(t)\} + \int_0^\infty \mathbf{K}^{(1)}(s_1; \mathbf{B}(t)) \cdot \left[\boldsymbol{G}^{(t)}(s_1)\right] ds_1, \tag{9.220}$$

where $\mathbf{K}^{(1)}(s_1; \mathbf{B}(t)) \cdot \left[\boldsymbol{G}^{(t)}(s_1)\right]$, for each choice of $s_1 \geq 0$ and $\mathbf{B}(t)$, is a linear function of the tensor variable $\boldsymbol{G}^{(t)}(s_1)$. The quantity $\mathbf{K}^{(1)}(s_1; \mathbf{B}(t))$ is a fourth-order tensor whose magnitude satisfies

$$\int_0^\infty \left|\mathbf{K}^{(1)}(s_1; \mathbf{B}(t))\right|^2 h^{-2}(s_1)\, ds_1 < \infty. \tag{9.221}$$

Simple materials satisfying the constitutive equation (9.220) are called *finite linear viscoelastic materials.*

For $m = 2$ we have

$$\begin{aligned}\boldsymbol{\sigma}(t) = \mathfrak{h}\{\mathbf{B}(t)\} &+ \int_0^\infty \mathbf{K}^{(1)}(s_1; \mathbf{B}(t)) \cdot \left[\boldsymbol{G}^{(t)}(s_1)\right] ds_1 + \\ &\int_0^\infty \int_0^\infty \mathbf{K}^{(2)}(s_1, s_2; \mathbf{B}(t)) \cdot \left[\boldsymbol{G}^{(t)}(s_1)\, \boldsymbol{G}^{(t)}(s_2)\right] ds_1 ds_2,\end{aligned} \tag{9.222}$$

where, in addition to the conditions on $\mathbf{K}^{(1)}$ noted above, we have that $\mathbf{K}^{(2)}(s_1, s_2; \mathbf{B}(t)) \cdot \left[\boldsymbol{G}^{(t)}(s_1) \boldsymbol{G}^{(t)}(s_2)\right]$, for each choice of $s_1 \geq 0$, $s_2 \geq 0$, and $\mathbf{B}(t)$, is a function of the tensor variables $\boldsymbol{G}^{(t)}(s_1)$ and $\boldsymbol{G}^{(t)}(s_2)$. The quantity $\mathbf{K}^{(2)}(s_1, s_2; \mathbf{B}(t))$ is a sixth-order tensor whose magnitude satisfies

$$\int_0^\infty \int_0^\infty \left|\mathbf{K}^{(2)}(s_1, s_2; \mathbf{B}(t))\right|^2 h^{-2}(s)\, ds_1\, ds_2 < \infty. \tag{9.223}$$

Correspondingly, the general constitutive equation of integral type for a compressible fluid is given by

$$\begin{aligned}&\boldsymbol{\sigma}(t) = -p(\rho(t))\, \mathbf{1} + \\ &\sum_{n=1}^{m} \int_0^\infty \cdots \int_0^\infty \mathbf{K}^{(n)}(s_1, \ldots, s_n; \rho(t)) \cdot \left[\boldsymbol{G}^{(t)}(s_1) \cdots \boldsymbol{G}^{(t)}(s_n)\right] ds_1 \cdots ds_n\end{aligned} \tag{9.224}$$

and for an incompressible fluid by

$$\begin{aligned}&\boldsymbol{\sigma}(t) = -p\, \mathbf{1} + \\ &\sum_{n=1}^{m} \int_0^\infty \cdots \int_0^\infty \mathbf{K}^{(n)}(s_1, \ldots, s_n) \cdot \left[\boldsymbol{G}^{(t)}(s_1) \cdots \boldsymbol{G}^{(t)}(s_n)\right] ds_1 \cdots ds_n.\end{aligned} \tag{9.225}$$

Example

For a linear viscoelastic incompressible fluid, we have

$$\sigma_{ik}(t) = -p\,\delta_{ik} + \int_0^\infty K^{(1)}_{iklm}(s)\; G^{(t)}_{lm}(s)\,ds. \tag{9.226}$$

From objectivity, $K^{(1)}_{iklm}$ is isotropic, and since the stress tensor and $G^{(t)}_{lm}$ are symmetric, we must have

$$K^{(1)}_{iklm} = K^{(1)}_{kilm} = K^{(1)}_{ikml} = K^{(1)}_{kiml}. \tag{9.227}$$

Furthermore, $K^{(1)}_{iklm}$ is a symmetric linear transformation if and only if the components obey the relations

$$K^{(1)}_{iklm} = K^{(1)}_{lmik}. \tag{9.228}$$

The most general fourth rank tensor satisfying the above symmetries can be written as (see (B.20)–(B.22))

$$K^{(1)}_{iklm}(s) = K^{(1)}_1(s)\,\delta_{ik}\delta_{lm} + K^{(1)}_2(s)\,(\delta_{il}\delta_{km} + \delta_{kl}\delta_{im}). \tag{9.229}$$

For an incompressible fluid, $K^{(1)}_1$ can be combined with the term involving p, so setting $K^{(1)}_2 = \frac{1}{2}\beta^{(1)}$, we can write

$$\sigma_{ik}(t) = -p\,\delta_{ik} + \int_0^\infty \beta^{(1)}(s)\; G^{(t)}_{ik}(s)\,ds, \tag{9.230}$$

for $\beta^{(1)}(s)$ decaying sufficiently fast with s. Now substituting (9.53), we have

$$\sigma_{ik}(t) = -p\,\delta_{ik} + \sum_{n=1}^{\infty} \frac{(-1)^n}{n!} A^{(n)}_{ik}(t) \int_0^\infty \beta^{(1)}(s)\, s^n ds. \tag{9.231}$$

Possible choices of $\beta^{(1)}(s)$ are

$$\beta^{(1)}(s) = \begin{cases} K e^{-\lambda s}, \\ \sum_{l=1}^{r} K_l\, e^{-\lambda_l s}, \\ \int_0^\infty K(r)\, e^{-\lambda(r) s} dr, \end{cases} \tag{9.232}$$

where the λ's represent relaxation times of the material. If we assume that $\beta^{(1)}(s) = K e^{-\lambda s}$, then

$$\sigma_{ik}(t) = -p\,\delta_{ik} + \sum_{n=1}^{\infty} \frac{(-1)^n}{\lambda^{n+1}} K A^{(n)}_{ik}(t). \tag{9.233}$$

Using the isotropy and symmetry conditions, there exist certain relations among the kernels $\mathbf{K}^{(n)}(s_1,\ldots,s_n;\rho(t))\cdot\left[\boldsymbol{G}^{(t)}(s_1)\cdots\boldsymbol{G}^{(t)}(s_n)\right]$ that reduce substantially the number of independent stress-relaxation moduli that can appear in a given order of approximation. It can be shown that such product can be expressed as a sum of multilinear products of tensors, trace of tensors, and trace of products of tensors:

$$\alpha_i^{(n)}(s_1,\ldots,s_n;\rho(t))\ \mathrm{tr}\left[\boldsymbol{G}^{(t)}(s_1)\cdots\boldsymbol{G}^{(t)}(s_{n_1})\right]\mathrm{tr}\left[\boldsymbol{G}^{(t)}(s_{n_1+1})\cdots\boldsymbol{G}^{(t)}(s_{n_2})\right]\cdots$$
$$\mathrm{tr}\left[\boldsymbol{G}^{(t)}(s_{n_{l-1}+1})\cdots\boldsymbol{G}^{(t)}(s_{n_l})\right]\boldsymbol{G}^{(t)}(s_{n_l+1})\cdots\boldsymbol{G}^{(t)}(s_n),\quad(9.234)$$

where $1\le n_1<n_2<\cdots<n_l\le n$, and where $n_1\le 6$, $n_k-n_{k-1}\le 6$, $n-n_l\le 5$. If $n_l=n$, then the product at the end must be replaced by the unit tensor. Thus, using (9.234) in (9.224), we have

$$\boldsymbol{\sigma}(t)=-p(\rho(t))+\sum_{n=1}^{m}\int_0^\infty\!\!\cdots\!\int_0^\infty\sum_i\alpha_i^{(n)}(s_1,\ldots,s_n;\rho(t))\,\mathrm{tr}\left[\boldsymbol{G}^{(t)}(s_1)\cdots\right.$$
$$\left.\boldsymbol{G}^{(t)}(s_{n_1})\right]\mathrm{tr}\left[\boldsymbol{G}^{(t)}(s_{n_1+1})\cdots\boldsymbol{G}^{(t)}(s_{n_2})\right]\cdots\mathrm{tr}\left[\boldsymbol{G}^{(t)}(s_{n_{l-1}+1})\cdots\boldsymbol{G}^{(t)}(s_{n_l})\right]$$
$$\boldsymbol{G}^{(t)}(s_{n_l+1})\cdots\boldsymbol{G}^{(t)}(s_n)\,ds_1\cdots ds_n.\quad(9.235)$$

For fluids of first order ($m=1$), the constitutive equation is explicitly given by

$$\boldsymbol{\sigma}(t)=-p(\rho(t))\,\mathbf{1}+\int_0^\infty\left[\alpha_1^{(1)}(s_1;\rho(t))\,\mathrm{tr}\,\boldsymbol{G}^{(t)}(s_1)\,\mathbf{1}+\alpha_2^{(1)}(s_1;\rho(t))\,\boldsymbol{G}^{(t)}(s_1)\right]ds_1.\quad(9.236)$$

For a second-order fluid ($m=2$), we have

$$\boldsymbol{\sigma}(t)=-p(\rho(t))\,\mathbf{1}+\int_0^\infty\left[\alpha_1^{(1)}(s_1;\rho(t))\,\mathrm{tr}\,\boldsymbol{G}^{(t)}(s_1)\,\mathbf{1}+\alpha_2^{(1)}(s_1;\rho(t))\,\boldsymbol{G}^{(t)}(s_1)\right]ds_1+$$
$$\int_0^\infty\!\!\int_0^\infty\Big\{\Big[\alpha_1^{(2)}(s_1,s_2;\rho(t))\ \mathrm{tr}\,\boldsymbol{G}^{(t)}(s_1)\,\mathrm{tr}\,\boldsymbol{G}^{(t)}(s_2)+$$
$$\alpha_2^{(2)}(s_1,s_2;\rho(t))\,\mathrm{tr}\left(\boldsymbol{G}^{(t)}(s_1)\cdot\boldsymbol{G}^{(t)}(s_2)\right)\Big]\mathbf{1}+$$
$$\alpha_3^{(2)}(s_1,s_2;\rho(t))\,\mathrm{tr}\,\boldsymbol{G}^{(t)}(s_1)\,\boldsymbol{G}^{(t)}(s_2)+$$
$$\alpha_4^{(2)}(s_1,s_2;\rho(t))\,\boldsymbol{G}^{(t)}(s_1)\cdot\boldsymbol{G}^{(t)}(s_2)\Big\}\,ds_1ds_2.\quad(9.237)$$

For an incompressible fluid, all terms in (9.236) and (9.237) that are scalar multiples of the unit tensor can be absorbed in the pressure term resulting in the following corresponding constitutive equations for orders $m=1$ and $m=2$, respectively:

$$\boldsymbol{\sigma}(t)=-p\,\mathbf{1}+\int_0^\infty\beta_1^{(1)}(s_1)\,\boldsymbol{G}^{(t)}(s_1)\,ds_1\quad(9.238)$$

and

$$\boldsymbol{\sigma}(t)=-p\,\mathbf{1}+\int_0^\infty\beta_1^{(1)}(s_1)\,\boldsymbol{G}^{(t)}(s_1)\,ds_1+$$
$$\int_0^\infty\!\!\int_0^\infty\Big\{\beta_1^{(2)}(s_1,s_2)\,\mathrm{tr}\,\boldsymbol{G}^{(t)}(s_1)\,\boldsymbol{G}^{(t)}(s_2)+$$
$$\beta_2^{(2)}(s_1,s_2)\,\boldsymbol{G}^{(t)}(s_1)\cdot\boldsymbol{G}^{(t)}(s_2)\Big\}\,ds_1ds_2.\quad(9.239)$$

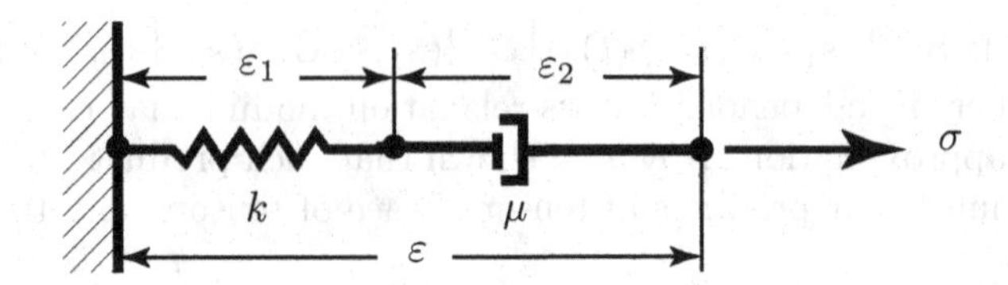

Figure 9.3: Schematic representation of the Maxwell model.

Note that (9.238) is the same equation as (9.230), which was obtained by following an alternate procedure.

Explicit representations may also be obtained for general isotropic materials of the integral type provided that the tensor functions $\mathbf{K}^{(n)}(s_1,\ldots,s_n;\mathbf{B}(t))\cdot\left[\boldsymbol{G}^{(t)}(s_1)\cdots\boldsymbol{G}^{(t)}(s_n)\right]$ appearing in (9.219) are polynomial functions of $\mathbf{B}$. For example, the constitutive function for an isotropic material of first order ($m = 1$) is given by

$$\begin{aligned}\boldsymbol{\sigma}(t) = \mathfrak{h}\{\mathbf{B}(t)\} + \int_0^\infty \Big\{&\boldsymbol{\alpha}_1^{(1)}(s_1;\mathbf{B}(t))\cdot\boldsymbol{G}^{(t)}(s_1) + \boldsymbol{G}^{(t)}(s_1)\cdot\boldsymbol{\alpha}_1^{(1)}(s_1;\mathbf{B}(t)) + \\ &\operatorname{tr}\left[\boldsymbol{G}^{(t)}(s_1)\cdot\boldsymbol{\alpha}_2^{(1)}(s_1;\mathbf{B}(t))\right]\mathbf{1} + \operatorname{tr}\left[\boldsymbol{G}^{(t)}(s_1)\cdot\boldsymbol{\alpha}_3^{(1)}(s_1;\mathbf{B}(t))\right]\mathbf{B} + \\ &\operatorname{tr}\left[\boldsymbol{G}^{(t)}(s_1)\cdot\boldsymbol{\alpha}_4^{(1)}(s_1;\mathbf{B}(t))\right]\mathbf{B}^2\Big\}\,ds_1, \end{aligned} \tag{9.240}$$

where $\mathfrak{h}\{\mathbf{B}(t)\}$ is given by (9.121), and the isotropic tensor functions $\boldsymbol{\alpha}_k^{(1)}$ have the representations

$$\boldsymbol{\alpha}_k^{(1)}\{s_1;\mathbf{B}(t)\} = \alpha_{1k}^{(1)}\,\mathbf{1} + \alpha_{2k}^{(1)}\,\mathbf{B} + \alpha_{3k}^{(1)}\,\mathbf{B}^2, \quad k = 1,2,3,4, \tag{9.241}$$

with $\alpha_{ik}^{(1)} = \alpha_{ik}^{(1)}(s_1;\mathbf{B}_{(1)},\mathbf{B}_{(2)},\mathbf{B}_{(3)})$, $i = 1,2,3$. This is readily verified by using Tables 5.1 and 5.3.

Example

A stress tensor approximation for an incompressible fluid, referred as the Maxwell model, is given by

$$\boldsymbol{\sigma} = -p\mathbf{1} - \int_0^\infty k\,e^{-s/\lambda}\,\frac{d}{ds}\boldsymbol{G}^{(t)}(s)\,ds. \tag{9.242}$$

It is noted, by integrating by parts, that

$$\begin{aligned}\boldsymbol{\sigma} &= -p\mathbf{1} - \cancel{k\,e^{-s/\lambda}\boldsymbol{G}^{(t)}(s)}\Big|_0^{\infty\;0} - \int_0^\infty \frac{k}{\lambda}\,e^{-s/\lambda}\,\boldsymbol{G}^{(t)}(s)\,ds \\ &= -p\mathbf{1} - \int_0^\infty \frac{k}{\lambda}\,e^{-s/\lambda}\,\boldsymbol{G}^{(t)}(s)\,ds, \end{aligned} \tag{9.243}$$

which is just (9.230) or (9.238) with

$$\beta^{(1)}(s) = -\frac{k}{\lambda}\,e^{-s/\lambda}. \tag{9.244}$$

The model in one dimension is illustrated in Fig. 9.3 where σ is the tension force, k the spring constant, μ the dashpot damping coefficient, and $\varepsilon = \varepsilon_1 + \varepsilon_2$ the total displacement, with ε_1 and ε_2 being the displacements due to the spring and dashpot, respectively. Now we can write

$$\sigma = k\,\varepsilon_1 = \mu\,\dot{\varepsilon}_2 \tag{9.245}$$

so

$$\sigma = \mu\,(\dot{\varepsilon} - \dot{\varepsilon}_1) = \mu\dot{\varepsilon} - \frac{\mu}{k}\,\dot{\sigma}. \tag{9.246}$$

If we let $\lambda = \mu/k$, then we can write

$$\dot{\sigma} + \frac{1}{\lambda}\,\sigma = k\,\dot{\varepsilon}(t) \tag{9.247}$$

and we take $\sigma(-\infty) = 0$. So assuming we know $\dot{\varepsilon}(t)$, we have

$$\sigma(t) = \int_{-\infty}^{t} k\,e^{-(t-\tau)/\lambda}\dot{\varepsilon}(\tau)\,d\tau, \tag{9.248}$$

or, upon taking $s = t - \tau$,

$$\sigma(t) = -\int_{0}^{\infty} k\,e^{-s/\lambda}\,\frac{d}{ds}\,\varepsilon(t-s)\,ds, \tag{9.249}$$

which is a one-dimensional model of (9.242) with $\boldsymbol{G}^{(t)}(s) \to \varepsilon(t-s)$.

Example

We would like to examine the application of the Maxwell model to simple shear. For simple shear, we have

$$v_1 = 0, \quad v_2 = \kappa\,x_1, \quad v_3 = 0. \tag{9.250}$$

Thus,

$$\frac{d\xi_1}{d\tau} = 0, \quad \frac{d\xi_2}{d\tau} = \kappa\,\xi_1, \quad \frac{d\xi_3}{d\tau} = 0, \tag{9.251}$$

and so

$$\xi_1 = x_1, \quad \xi_2 = x_2 + \kappa\,(\tau - t)\,x_1 = x_2 - \kappa\,s\,x_1, \quad \xi_3 = x_3. \tag{9.252}$$

Now

$${}_{(t)}F_{ik}^{(t)}(s) = \frac{\partial \xi_i}{\partial x_k} = \begin{pmatrix} 1 & 0 & 0 \\ -\kappa\,s & 1 & 0 \\ 0 & 0 & 1 \end{pmatrix}, \tag{9.253}$$

$${}_{(t)}C^{(t)}(s) = ({}_{(t)}F^{(t)}(s))^T\,{}_{(t)}F^{(t)}(s) = \begin{pmatrix} 1 & -\kappa\,s & 0 \\ -\kappa\,s & 1 + (\kappa\,s)^2 & 0 \\ 0 & 0 & 1 \end{pmatrix}, \tag{9.254}$$

and from (9.52) we have that

$$G^{(t)}(s) = \begin{pmatrix} 0 & -\kappa s & 0 \\ -\kappa s & (\kappa s)^2 & 0 \\ 0 & 0 & 0 \end{pmatrix}. \tag{9.255}$$

Now, using Maxwell's model (9.243), for this incompressible fluid, we have

$$\boldsymbol{\sigma} = -p\mathbf{1} - \int_0^\infty \frac{k}{\lambda} e^{-s/\lambda} \boldsymbol{G}^{(t)}(s)\, ds, \tag{9.256}$$

where λ is the material relaxation time. Thus,

$$\sigma_{12} = \sigma_{21} = \frac{k\kappa}{\lambda} \int_0^\infty e^{-s/\lambda} s\, ds = k\,\kappa\,\lambda = \mu\,\kappa \tag{9.257}$$

and

$$\sigma_{22} + p = -\frac{k\kappa^2}{\lambda} \int_0^\infty e^{-s/\lambda} s^2\, ds = -2\,k\,\kappa^2\,\lambda^2 = -2\,\mu\,\lambda\,\kappa^2. \tag{9.258}$$

Example

Here we would like to apply the Maxwell model to the elongational flow

$$v_1 = 2\,a\,x_1, \quad v_2 = -a\,x_2, \quad v_3 = -a\,x_3. \tag{9.259}$$

In this case, we have

$$\frac{d\xi_1}{d\tau} = 2\,a\,\xi_1, \quad \frac{d\xi_2}{d\tau} = -a\,\xi_2, \quad \frac{d\xi_3}{d\tau} = -a\,\xi_3, \tag{9.260}$$

and

$$\xi_1 = e^{2\,a(\tau - t)} x_1 = e^{-2\,a\,s} x_1, \quad \xi_2 = e^{a\,s} x_2, \quad \xi_3 = e^{a\,s} x_3. \tag{9.261}$$

Thus

$$_{(t)}C^{(t)}(s) = \begin{pmatrix} e^{-4\,a\,s} & 0 & 0 \\ 0 & e^{2\,a\,s} & 0 \\ 0 & 0 & e^{2\,a\,s} \end{pmatrix} \text{ and }$$

$$G^{(t)}(s) = \begin{pmatrix} e^{-4\,a\,s} - 1 & 0 & 0 \\ 0 & e^{2\,a\,s} - 1 & 0 \\ 0 & 0 & e^{2\,a\,s} - 1 \end{pmatrix}, \tag{9.262}$$

and so on. We finally get

$$\sigma_{11}(t) = -p + \frac{4\,a\,k}{\lambda^{-1} + 4\,a} \quad \text{and} \quad \sigma_{22}(t) = \sigma_{33}(t) = -p + \frac{2\,a\,k}{\lambda^{-1} - 2\,a}. \tag{9.263}$$

If $a = 1/(2\,\lambda)$, then $\sigma_{22}, \sigma_{33} \to \infty$, so we expect this model to fail. However, the model can be fixed up.

Many other first-order models, more complex than the Maxwell model, can also be shown to correspond to specific cases of (9.236) or (9.240) for different choices of $\alpha_i^{(1)}(s_1;\rho)$ or $\boldsymbol{\alpha}_i^{(1)}(s_1;\mathbf{B})$. Their functional dependence on ρ for a compressible fluid and on $\mathbf{B}$ and its invariants for an isotropic solid (see (9.241)) is exploited in many of the more complex models where specific choices are made (e.g., the Bird–Carreau model).

9.3.5 Constitutive equations of rate type

We recall that the response functional (5.3) was sufficiently general so that, e.g., the stress could depend on the histories of stress, heat flux, entropy flux, free energy, entropy, as well as the motion and temperature at all other points in the body and their rates. Here, we continue to limit the discussion to constitutive equations of homogeneous simple materials that are independent of temperature gradient, and suppress the functional dependence on temperature. Subsequently, we focus on the constitutive equation for the stress tensor. If we now consider materials of rate type, and more specifically, the constitutive equation of rate n and mechanical rate p, the response functional (5.3) takes the form

$$\underset{0\le s<\infty}{\mathfrak{F}}\{\boldsymbol{\sigma}(\mathbf{X},s),\dot{\boldsymbol{\sigma}}(\mathbf{X},s),\ldots,\overset{(n)}{\boldsymbol{\sigma}}(\mathbf{X},s);\mathbf{F}(\mathbf{X},s),\dot{\mathbf{F}}(\mathbf{X},s),\ldots,\overset{(p)}{\mathbf{F}}(\mathbf{X},s)\}=\mathbf{0}. \tag{9.264}$$

Assuming that the functional for $\overset{(n)}{\boldsymbol{\sigma}}$ is non-singular and single valued, we have the following equivalent formulation for the stress tensor of a homogeneous simple material at material point $\mathbf{X}$ (see (5.33)):

$$\boldsymbol{\sigma}(t)=\underset{0\le s<\infty}{\mathfrak{F}}\{\mathbf{F}^{(t)}(s)\}, \tag{9.265}$$

where it is now assumed that $\boldsymbol{\sigma}=\boldsymbol{\sigma}(t)$ satisfies a differential equation of the form

$$\overset{(n)}{\boldsymbol{\sigma}}(t)=\mathfrak{g}\left(\boldsymbol{\sigma}(t),\dot{\boldsymbol{\sigma}}(t),\ldots,\overset{(n-1)}{\boldsymbol{\sigma}}(t);\mathbf{F}(t),\dot{\mathbf{F}}(t),\ldots,\overset{(p)}{\mathbf{F}}(t)\right) \tag{9.266}$$

with initial conditions

$$\boldsymbol{\sigma}(t_0)=\boldsymbol{\sigma}_0,\qquad \dot{\boldsymbol{\sigma}}(t_0)=\dot{\boldsymbol{\sigma}}_0,\qquad \ldots,\qquad \overset{(n-1)}{\boldsymbol{\sigma}}(t_0)=\overset{(n-1)}{\boldsymbol{\sigma}_0}. \tag{9.267}$$

It should be noted that (9.266) is not really a complete constitutive equation. In addition to the initial conditions (9.267) at some initial time t_0, we also require the history of the deformation from (9.265) up to time t_0. Nevertheless, the constitutive differential equation (9.266) is called a *constitutive equations of rate type*. Above, it is assumed that $\mathfrak{g}$ is a tensor-valued function that is sufficiently smooth, and that the differential equation has a unique solution. It should be noted that the stress tensor $\boldsymbol{\sigma}(t)$ depends not only on the initial data but also on values of $\mathbf{F}(\tau)$ for $t_0\le\tau\le t$. It is also noted that the constitutive equation which is not of the rate type corresponds to the case where $n=0$.

Frame-invariant forms of (9.265) under a Euclidean transformation for a homogeneous isotropic simple viscoelastic material have been discussed in Sections 9.3

and 9.3.3. Subsequently, (9.265) takes the form given in (9.207). We also recall from Section 3.3 that rates of second rank tensors are not objective under such transformation. Nevertheless, the corresponding frame-indifferent constitutive equation of rate type (9.266) is obtained by following the same procedure as in Sections 5.7 and 9.3. In doing this, by using the polar decomposition of $\mathbf{F}$ and choosing $\mathbf{Q} = \mathbf{R}^T$, we arrive at the following objective form:

$$\overset{\triangle}{\boldsymbol{\sigma}}_n(t) = \mathfrak{f}\left(\boldsymbol{\sigma}(t), \overset{\triangle}{\boldsymbol{\sigma}}_1(t), \ldots, \overset{\triangle}{\boldsymbol{\sigma}}_{n-1}(t); \mathbf{B}(t); \mathbf{A}^{(1)}(t), \mathbf{A}^{(2)}(t), \ldots, \mathbf{A}^{(p)}(t)\right), \quad (9.268)$$

where $\overset{\triangle}{\boldsymbol{\sigma}}_i$ is the ith convected stress rate defined by

$$\overset{\triangle}{\boldsymbol{\sigma}}_i(t) \equiv \frac{\partial^i}{\partial \tau^i} \left[{}_{(t)}\mathbf{F}^T(\tau) \cdot \boldsymbol{\sigma}(\tau) \cdot {}_{(t)}\mathbf{F}(\tau)\right]\Big|_{\tau=t} = \sum_{\substack{a+b+c=i \\ a,b,c=0,1,\ldots,i}} \frac{i!}{a!\,b!\,c!} \left(\mathbf{L}^{(a)}(t)\right)^T \cdot \overset{(b)}{\boldsymbol{\sigma}}(t) \cdot \mathbf{L}^{(c)}(t). \quad (9.269)$$

Note that for $i = 0$ we have $\overset{\triangle}{\boldsymbol{\sigma}}_0(t) = \boldsymbol{\sigma}(t)$ and for $i = 1$ we obtain the Cotter–Rivlin convected stress rate (see (3.425)):

$$\overset{\triangle}{\boldsymbol{\sigma}}_1 = \overset{\triangle}{\boldsymbol{\sigma}} = \dot{\boldsymbol{\sigma}} + \mathbf{L}^T \cdot \boldsymbol{\sigma} + \boldsymbol{\sigma} \cdot \mathbf{L}. \quad (9.270)$$

The corresponding constitutive equation for a simple *fluid of rate type* is of the following form:

$$\overset{\triangle}{\boldsymbol{\sigma}}_n(t) = \mathfrak{f}\left(\boldsymbol{\sigma}(t), \overset{\triangle}{\boldsymbol{\sigma}}_1(t), \ldots, \overset{\triangle}{\boldsymbol{\sigma}}_{n-1}(t); \rho(t); \mathbf{A}^{(1)}(t), \mathbf{A}^{(2)}(t), \ldots, \mathbf{A}^{(p)}(t)\right). \quad (9.271)$$

The simplest form for a fluid of rate type ($n = 1$) is given by

$$\overset{\triangle}{\boldsymbol{\sigma}}_1(t) = \mathfrak{f}\left(\boldsymbol{\sigma}(t); \rho(t); \mathbf{A}^{(1)}(t)\right), \quad (9.272)$$

which can be recast as

$$\overset{\triangle}{\boldsymbol{\sigma}}{}_1^d(t) = \mathfrak{g}\left(\boldsymbol{\sigma}^d(t); \rho(t); \mathbf{A}^{(1)}(t)\right) \quad (9.273)$$

in terms of the extra, or dissipative part, of the stress by writing

$$\boldsymbol{\sigma}^d(t) = \boldsymbol{\sigma}(t) + p(\rho(t))\,\mathbf{1}. \quad (9.274)$$

Note that under hydrostatic conditions, we must have that

$$\mathfrak{g}\left(\mathbf{0}; \rho(t); \mathbf{0}\right) = \mathbf{0}. \quad (9.275)$$

This class of materials include materials which have properties of both fluids and solids and are called *hygrosteric materials.*

Now, using Tables 5.1 and 5.3, and the fact that $\mathbf{A}^{(1)} = 2\mathbf{D}$, (9.273) has the following representation:

$$\overset{\triangle}{\boldsymbol{\sigma}}{}_1^d(t) = \alpha_0\,\mathbf{1} + \alpha_1\,\boldsymbol{\sigma}^d + \alpha_2\,(\boldsymbol{\sigma}^d)^2 + \alpha_3\,\mathbf{D} + \alpha_4\,\mathbf{D}^2 + \alpha_5\left(\boldsymbol{\sigma}^d \cdot \mathbf{D} + \mathbf{D} \cdot \boldsymbol{\sigma}^d\right) + \alpha_6\left((\boldsymbol{\sigma}^d)^2 \cdot \mathbf{D} + \mathbf{D} \cdot (\boldsymbol{\sigma}^d)^2\right) + \alpha_7\left(\boldsymbol{\sigma}^d \cdot \mathbf{D}^2 + \mathbf{D}^2 \cdot \boldsymbol{\sigma}^d\right), \quad (9.276)$$

where α_i, $i = 0, \ldots, 7$, are functions of ρ and the invariants $\operatorname{tr}\mathbf{D}$, $\operatorname{tr}\mathbf{D}^2$, $\operatorname{tr}\mathbf{D}^3$, $\operatorname{tr}\boldsymbol{\sigma}^d$, $\operatorname{tr}(\boldsymbol{\sigma}^d)^2$, $\operatorname{tr}(\boldsymbol{\sigma}^d)^3$, $\operatorname{tr}(\boldsymbol{\sigma}^d \cdot \mathbf{D})$, $\operatorname{tr}(\boldsymbol{\sigma}^d \cdot \mathbf{D}^2)$, $\operatorname{tr}((\boldsymbol{\sigma}^d)^2 \cdot \mathbf{D})$, and $\operatorname{tr}((\boldsymbol{\sigma}^d)^2 \cdot \mathbf{D}^2)$. From (9.275), it follows that $\alpha_0 = 0$ for $\boldsymbol{\sigma}^d = \mathbf{D} = \mathbf{0}$ and ρ arbitrary.

Example

We develop the linear constitutive equation of rate $n = 1$. In this case, we assume that $\mathfrak{g}\left(\boldsymbol{\sigma}^d; \rho(t); \mathbf{D}\right)$ is linear in $\boldsymbol{\sigma}^d$ and $\mathbf{D}$. Subsequently, from (9.276), we see that $\overset{\triangle}{\boldsymbol{\sigma}}{}_1^d$ must have the form

$$\overset{\triangle}{\boldsymbol{\sigma}}{}_1^d(t) = \overset{\triangle}{\boldsymbol{\sigma}}{}^d = \dot{\boldsymbol{\sigma}}^d + \mathbf{L}^T \cdot \boldsymbol{\sigma}^d + \boldsymbol{\sigma}^d \cdot \mathbf{L} = \alpha_0\, \mathbf{1} + \alpha_1\, \boldsymbol{\sigma}^d + \alpha_3\, \mathbf{D} + \alpha_5\left(\boldsymbol{\sigma}^d \cdot \mathbf{D} + \mathbf{D} \cdot \boldsymbol{\sigma}^d\right), \qquad (9.277)$$

where

$$\alpha_0 = \lambda_2 \operatorname{tr} \boldsymbol{\sigma}^d + \mu_2 \operatorname{tr} \mathbf{D} + \beta_1 \operatorname{tr}\left(\mathbf{D} \cdot \boldsymbol{\sigma}^d\right) + \beta_4 \left(\operatorname{tr} \boldsymbol{\sigma}^d\right)(\operatorname{tr} \mathbf{D}),$$
$$\alpha_1 = \lambda_1 + \beta_5 \operatorname{tr} \mathbf{D}, \quad \alpha_3 = \mu_1 + \beta_2 \operatorname{tr} \boldsymbol{\sigma}^d, \quad \alpha_5 = \beta_3, \qquad (9.278)$$

and μ_i, λ_i, and β_i are functions of ρ.
It should be noted that, since $\mathbf{L} = \mathbf{D} + \mathbf{W}$, then it follows that $\overset{\triangle}{\boldsymbol{\sigma}}{}^d = \overset{\circ}{\boldsymbol{\sigma}}{}^d + \boldsymbol{\sigma}^d \cdot \mathbf{D} + \mathbf{D} \cdot \boldsymbol{\sigma}^d$, where $\overset{\circ}{\boldsymbol{\sigma}}{}^d$ is the Jaumann corotational stress rate (see (3.422)), and thus we can also write, without loss of generality (by subsequently taking $(\alpha_5 - 1) \to \alpha_5$),

$$\overset{\circ}{\boldsymbol{\sigma}}{}^d = \dot{\boldsymbol{\sigma}}^d - \mathbf{W} \cdot \boldsymbol{\sigma}^d + \boldsymbol{\sigma}^d \cdot \mathbf{W} = \alpha_0\, \mathbf{1} + \alpha_1\, \boldsymbol{\sigma}^d + \alpha_3\, \mathbf{D} + \alpha_5\left(\boldsymbol{\sigma}^d \cdot \mathbf{D} + \mathbf{D} \cdot \boldsymbol{\sigma}^d\right). \qquad (9.279)$$

Such constitutive equation models linear fluent materials, where by requiring that $\lambda_2(\operatorname{tr} \boldsymbol{\sigma}^d) + \lambda_1\, \boldsymbol{\sigma}^d \neq 0$ for $\boldsymbol{\sigma}^d \neq 0$, we must have that

$$\lambda_1 \neq 0, \qquad \lambda_1 + 3\,\lambda_2 \neq 0, \qquad \text{and} \quad p = p(\rho). \qquad (9.280)$$

It also models linear hypo-elastic materials, where the material is not a function of ρ, in which case we have

$$\lambda_1 = \lambda_2 = 0 \quad \text{and} \quad p = 0. \qquad (9.281)$$

Note that the requirement of $p = 0$ for linear hypo-elastic materials is equivalent to having $\boldsymbol{\sigma}^d = \boldsymbol{\sigma}$.

Example

Consider the steady linear flow in the Cartesian coordinates (x_1, x_2, x_3) of the form

$$\mathbf{v} = (0, v_2(x_1), 0)\,. \qquad (9.282)$$

Note that for this flow the mass conservation equation becomes

$$\operatorname{div} \mathbf{v} = 0, \qquad (9.283)$$

i.e., the fluid is incompressible where we take $\rho = \rho_R = \text{const.}$, and thus the pressure is an unknown function unrelated to thermodynamic equations of state. Furthermore, since $\dot{\mathbf{v}} = \mathbf{0}$, the linear momentum equation becomes

$$\operatorname{div} \boldsymbol{\sigma}^d - \operatorname{grad} \varphi = \mathbf{0}, \qquad (9.284)$$

where

$$\varphi = p + \rho_R U, \qquad (9.285)$$

and we have assumed that the body force derives from a potential, i.e.,

$$\mathbf{g} = -\operatorname{grad} U. \qquad (9.286)$$

Now, it is clear that the extra stress in this case is of the form

$$\boldsymbol{\sigma}^d = \begin{pmatrix} t_1(x_1) & s(x_1) & 0 \\ s(x_1) & t_2(x_1) & 0 \\ 0 & 0 & t_3(x_1) \end{pmatrix}, \qquad (9.287)$$

where t_i's are normal stresses and s is the shear stress. Subsequently,

$$\operatorname{div} \boldsymbol{\sigma}^d = (t_1', s', 0), \qquad (9.288)$$

where the primes denote derivatives with respect to x_1, and the linear momentum equation (9.284) reduces to

$$t_1'(x_1) - \frac{\partial \varphi}{\partial x_1} = 0, \qquad s'(x_1) - \frac{\partial \varphi}{\partial x_2} = 0, \qquad \frac{\partial \varphi}{\partial x_3} = 0. \qquad (9.289)$$

From these equations, we infer that

$$s = -a\, x_1 + b \qquad \text{and} \qquad \varphi = -a\, x_2 + t_1(x_1) + c, \qquad (9.290)$$

where a, b, and c are constants that depend on constitutive parameters. Now, the velocity gradient is given by

$$\mathbf{L} = v_2' \begin{pmatrix} 0 & 0 & 0 \\ 1 & 0 & 0 \\ 0 & 0 & 0 \end{pmatrix}, \qquad (9.291)$$

and it then follows from (9.277) and (9.278) that

$$v_2' \begin{pmatrix} 2\,s & t_2 & 0 \\ t_2 & 0 & 0 \\ 0 & 0 & 0 \end{pmatrix} = (\lambda_2\, t_0 + \beta_1\, s\, v_2') \begin{pmatrix} 1 & 0 & 0 \\ 0 & 1 & 0 \\ 0 & 0 & 1 \end{pmatrix} + \lambda_1 \begin{pmatrix} t_1 & s & 0 \\ s & t_2 & 0 \\ 0 & 0 & t_3 \end{pmatrix} +$$

$$\frac{1}{2} v_2' \,(\mu_1 + \beta_2\, t_0) \begin{pmatrix} 0 & 1 & 0 \\ 1 & 0 & 0 \\ 0 & 0 & 0 \end{pmatrix} + \frac{1}{2} \beta_3\, v_2' \begin{pmatrix} 2\,s & t_1 + t_2 & 0 \\ t_1 + t_2 & 2\,s & 0 \\ 0 & 0 & 0 \end{pmatrix}, \qquad (9.292)$$

where

$$t_0 = t_1 + t_2 + t_3 = \operatorname{tr} \boldsymbol{\sigma}^d. \qquad (9.293)$$

Note that since the material is incompressible, λ_i, β_j, and μ_k are purely constant. The system (9.292) provides four nontrivial coupled equations.

To simplify the solution of such equations, and without loss of generality, we take $\beta_3 \to 1 + \beta_3$. Subsequently, the four independent equations are

$$\lambda_1 t_1 + \lambda_2 t_0 + (\beta_1 + \beta_3 - 1)\, s\, v_2' = 0, \tag{9.294}$$
$$\lambda_1 t_2 + \lambda_2 t_0 + (\beta_1 + \beta_3 + 1)\, s\, v_2' = 0, \tag{9.295}$$
$$\lambda_1 t_3 + \lambda_2 t_0 + \beta_1\, s\, v_2' = 0, \tag{9.296}$$
$$\lambda_1 s + \frac{1}{2}\left[\mu_1 + \beta_2 t_0 + \beta_3 (t_1 + t_2) + (t_1 - t_2)\right] v_2' = 0. \tag{9.297}$$

Adding (9.294)–(9.296) results in

$$(\lambda_1 + 3\lambda_2)\, t_0 + (3\beta_1 + 2\beta_3)\, s\, v_2' = 0. \tag{9.298}$$

Now, since from (9.280) $\lambda_1 \neq 0$, we can solve for $t_1 + t_2$ and $t_1 - t_2$ from (9.294) and (9.295) and substitute these into (9.297), to obtain

$$\lambda_1 s + \frac{1}{2}\left\{\mu_1 + \frac{1}{\lambda_1}(\lambda_1 \beta_2 - 2\lambda_2 \beta_3)\, t_0 - \frac{2}{\lambda_1}\left[\beta_3 (\beta_1 + \beta_3) - 1\right] s\, v_2'\right\} v_2' = 0. \tag{9.299}$$

Again, since from (9.280) $\lambda_1 + 3\lambda_2 \neq 0$, we can solve (9.298) for t_0 and substitute this into (9.299), to obtain the following expressions for v_2' and the shear stress

$$v_2' = \frac{4s}{\eta}\,\frac{1}{1 \pm \sqrt{1 - 16\alpha\tau_d^2\eta^{-2}s^2}} \tag{9.300}$$

and

$$s = \frac{1}{2}\eta\, v_2'\,\frac{1}{1 + \beta\,\tau_d^2 (v_2')^2}, \tag{9.301}$$

where

$$\eta = -\frac{\mu_1}{\lambda_1}, \quad \beta = 1 - \frac{\lambda_1 \beta_2 - 2\lambda_2 \beta_3}{2(\lambda_1 + 3\lambda_2)}(3\beta_1 + 2\beta_3) - \beta_3(\beta_1 + \beta_3), \quad \tau_d = -\frac{1}{\lambda_1}. \tag{9.302}$$

In addition, from (9.294) to (9.296) and (9.298), we obtain the normal stresses

$$t_1 = \tau_d(-\beta_0 + \beta_1 + \beta_3 - 1)\, s\, v_2', \tag{9.303}$$
$$t_2 = \tau_d(-\beta_0 + \beta_1 + \beta_3 + 1)\, s\, v_2', \tag{9.304}$$
$$t_3 = \tau_d(-\beta_0 + \beta_1)\, s\, v_2', \tag{9.305}$$

where

$$\beta_0 = \frac{\lambda_2}{\lambda_1 + 3\lambda_2}(3\beta_1 + 2\beta_3). \tag{9.306}$$

For simple shearing flow, where $v_2(0) = 0$ and $v_2(d) = V$ and with φ having no gradient in the flow direction, we have that $v_2(x_1) = \kappa x_1$, with shear rate $\kappa = V/d = \text{const.}$ and $v_2' = \kappa$, so that from (9.301) and (9.290) we must have that $a = 0$ and from (9.303) to (9.305) that the normal stresses are constant. Subsequently,

$$s = b = \text{const.} \qquad \text{and} \qquad \varphi = p + \rho_R U = t_1 + c = \text{const.} \tag{9.307}$$

Let us examine the shear stress given by (9.301). We first note that the classical result of $s = b = \frac{1}{2}\eta\kappa$ is obtained when $\beta = 0$. Now, β can be either positive or negative. If $\beta > 0$, and defining a shearing yield rate by $\tau_s = \sqrt{\beta}\,\tau_d$, we can rewrite (9.301) in the form

$$s = b = \frac{1}{2}\eta\kappa\frac{1}{1+\tau_s^2\kappa^2}, \tag{9.308}$$

as illustrated in Fig. 9.4(a). We note that for $\kappa \le \frac{1}{4}\tau_s^{-1}$ the error in the classical formula is less than 6%. For $\frac{1}{4}\tau_s^{-1} \le \kappa \le \tau_s^{-1}$, the rate of shearing is greater than for that provided by the classical result, and indeed at $\kappa_c = \tau_s^{-1}$, a yield-like phenomenon is observed and there we have that the shear stress is given by $s_c = \frac{1}{4}\eta\,\tau_s^{-1}$.

On the other hand, if $\beta < 0$, and defining a shearing limit time by $\overline{\tau}_s = \sqrt{|\beta|}\,\tau_d$, we can rewrite (9.301) in the form

$$s = b = \frac{1}{2}\eta\kappa\frac{1}{1-\overline{\tau}_s^2\kappa^2}, \tag{9.309}$$

as illustrated in Fig. 9.4(b). In this case, for $\kappa \le \frac{1}{4}\overline{\tau}_s^{-1}$ the error in the classical formula is less than 7% and for $\frac{1}{4}\overline{\tau}_s^{-1} \le \kappa \le \overline{\tau}_s^{-1}$ the shearing stress is substantially higher than the classical result. In this case, the material becomes more and more resistant to shearing stress. We note that the region where $\kappa > \kappa_l \equiv \overline{\tau}_s^{-1}$ is presumed to be unphysical since the rate of shear in this branch cannot be reached through a continuous increase of the rate.

Since s and t_i are constant, it follows from (9.303) to (9.305) that the normal extra stresses, t_i are also constant. In the case where we have no external body forces ($U = 0$), from (9.285) and (9.290), the pressure p is an arbitrary constant, and thus the normal stresses are also constant. We can choose

$$p = t_2 \tag{9.310}$$

so that there is no normal stress in the flow direction. Thus, from (9.303) to (9.305), we can write that

$$\sigma_{11} = -2\,\tau_d\,s\,\kappa, \qquad \sigma_{22} = 0, \qquad \sigma_{33} = -\tau_d\,(1+\beta_3)\,s\,\kappa. \tag{9.311}$$

Now in the case of a classical viscous fluid, no normal stresses are necessary for steady flow. Here, if $\tau_d > 0$, the stresses σ_{11} normal to the plates must be negative. This indicates that the plates would tend to spread apart in the absence of such stresses and leads to flow swelling at the exit of such plates. The stress σ_{33} normal to the direction of flow and parallel to the plates is also nonzero – its sign depends on the sign of $1+\beta_3$. For slow flow, where $\kappa \ll \tau_d^{-1}$, the normal stresses will be small compared to the shearing stresses (assuming for σ_{33} that $\beta_3 = O(1)$).

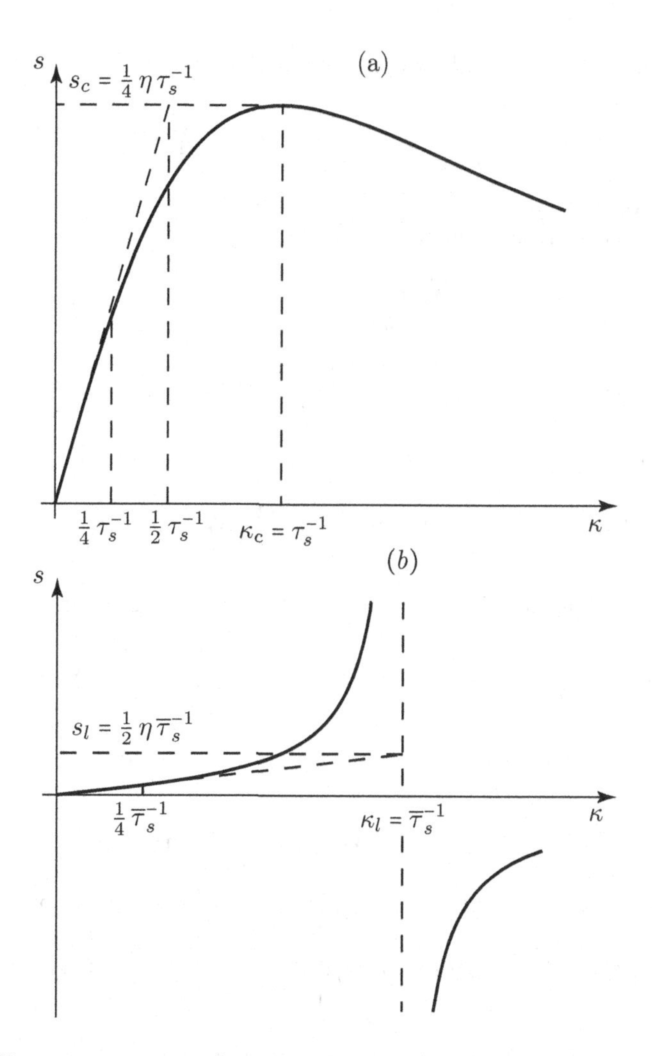

Figure 9.4: Shear stress, s, as a function of shear rate, κ, for the shear flow resulting from a linear constitutive equation of rate type of order one: (a) $\beta > 0$, (b) $\beta < 0$.

Problems

1. From the definition of the relative right Cauchy–Green tensor (3.283), show that

$$\frac{\partial^n {}_{(t)}\mathbf{C}(\tau)}{\partial\tau^n} = \sum_{k=0}^{n} \binom{n}{k} \frac{\partial^k {}_{(t)}\mathbf{F}^T(\tau)}{\partial\tau^k} \cdot \frac{\partial^{n-k} {}_{(t)}\mathbf{F}(\tau)}{\partial\tau^{n-k}}, \qquad n = 0,1,2,\ldots,$$

where the binomial coefficients are given by

$$\binom{n}{k} = \frac{n!}{(n-k)!\,k!}.$$

2. Show that the nth acceleration gradient and stretch and spin tensors can be respectively written as

$$\mathbf{L}^{(n)}(t) = \sum_{k=0}^{n} \binom{n}{k} \mathbf{W}^{(k)}(t) \cdot \mathbf{D}^{(n-k)}(t), \qquad n = 0,1,2,\ldots,$$

$$\mathbf{D}^{(n)}(t) = \frac{1}{2}\left[\mathbf{A}^{(n)}(t) - \sum_{k=1}^{n-1} \binom{n}{k} \mathbf{D}^{(k)}(t) \cdot \mathbf{D}^{(n-k)}(t)\right], \qquad n = 1,2,\ldots,$$

and

$$\mathbf{W}^{(n)}(t) = \mathbf{L}^{(n)}(t) - \mathbf{D}^{(n)}(t) - \sum_{k=1}^{n-1} \binom{n}{k} \mathbf{W}^{(k)}(t) \cdot \mathbf{D}^{(n-k)}(t), \qquad n = 1,2,\ldots.$$

3. Show that the ith corotational rate of vector $\mathbf{u}$, defined by

$$\mathring{\mathbf{u}}_i(t) \equiv \frac{\partial^i}{\partial\tau^i}\left[{}_{(t)}\mathbf{R}^T(\tau) \cdot \mathbf{u}(\tau)\right]\bigg|_{\tau=t} = \sum_{\substack{a+b=i \\ a,b=0,1,\ldots,i}} \frac{i!}{a!\,b!} \left(\mathbf{W}^{(a)}(t)\right)^T \cdot \overset{(b)}{\mathbf{u}}(t), \tag{9.312}$$

is objective. Note that for $i = 1$ it corresponds to the corotational rate of the vector (see (3.419)):

$$\mathring{\mathbf{u}}_1 = \mathring{\mathbf{u}} = \dot{\mathbf{u}} - \mathbf{W} \cdot \mathbf{u}. \tag{9.313}$$

4. Show that the ith convected rate of vector $\mathbf{u}$, defined by

$$\overset{\otimes}{\mathbf{u}}_i(t) \equiv \frac{\partial^i}{\partial\tau^i}\left[{}_{(t)}\mathbf{F}^T(\tau) \cdot \mathbf{u}(\tau)\right]\bigg|_{\tau=t} = \sum_{\substack{a+b=i \\ a,b=0,1,\ldots,i}} \frac{i!}{a!\,b!} \left(\mathbf{L}^{(a)}(t)\right)^T \cdot \overset{(b)}{\mathbf{u}}(t), \tag{9.314}$$

is objective. Note that for $i = 1$ it corresponds to the convected vector rate

$$\overset{\otimes}{\mathbf{u}}_1 = \overset{\otimes}{\mathbf{u}} = \dot{\mathbf{u}} + \mathbf{L}^T \cdot \mathbf{u}. \tag{9.315}$$

5. Show that the ith corotational stress rate, defined by

$$\mathring{\boldsymbol{\sigma}}_i(t) \equiv \frac{\partial^i}{\partial\tau^i}\left[{}_{(t)}\mathbf{R}^T(\tau) \cdot \boldsymbol{\sigma}(\tau) \cdot {}_{(t)}\mathbf{R}(\tau)\right]\bigg|_{\tau=t} =$$

$$\sum_{\substack{a+b+c=i \\ a,b,c=0,1,\ldots,i}} \frac{i!}{a!\,b!\,c!} \left(\mathbf{W}^{(a)}(t)\right)^T \cdot \overset{(b)}{\boldsymbol{\sigma}}(t) \cdot \mathbf{W}^{(c)}(t), \tag{9.316}$$

is objective. Note that for $i = 1$ it corresponds to the Jaumann corotational stress rate (see (3.422)):

$$\overset{\circ}{\boldsymbol{\sigma}}_1 = \overset{\circ}{\boldsymbol{\sigma}} = \dot{\boldsymbol{\sigma}} - \mathbf{W}\cdot\boldsymbol{\sigma} + \boldsymbol{\sigma}\cdot\mathbf{W}. \tag{9.317}$$

6. Show that the ith convected stress rate, defined by

$$\overset{\triangle}{\boldsymbol{\sigma}}_i(t) \equiv \frac{\partial^i}{\partial\tau^i}\left[{}_{(t)}\mathbf{F}^T(\tau)\cdot\boldsymbol{\sigma}(\tau)\cdot{}_{(t)}\mathbf{F}(\tau)\right]\bigg|_{\tau=t} =$$

$$\sum_{\substack{a+b+c=i\\ a,b,c=0,1,\ldots,i}} \frac{i!}{a!\,b!\,c!}\left(\mathbf{L}^{(a)}(t)\right)^T\cdot\overset{(b)}{\boldsymbol{\sigma}}(t)\cdot\mathbf{L}^{(c)}(t), \tag{9.318}$$

is objective. Note that for $i = 1$ it corresponds to the Cotter–Rivlin convected stress rate (see (3.425)):

$$\overset{\triangle}{\boldsymbol{\sigma}}_1 = \overset{\triangle}{\boldsymbol{\sigma}} = \dot{\boldsymbol{\sigma}} + \mathbf{L}^T\cdot\boldsymbol{\sigma} + \boldsymbol{\sigma}\cdot\mathbf{L}. \tag{9.319}$$

7. Prove that a motion is an MWCSH if and only if there exist an orthogonal tensor $\mathbf{Q}(t)$, a scalar κ (called the *shearing*), and a constant second rank tensor $\mathbf{N}_0$ such that

$${}_{(0)}\mathbf{F}(\tau) = \mathbf{F}(\tau) = \mathbf{Q}(\tau)\cdot e^{\tau\kappa\mathbf{N}_0}, \qquad \mathbf{Q}(0) = \mathbf{1}, \quad |\mathbf{N}_0| = 1,$$

where $|\mathbf{N}_0|^2 = \operatorname{tr}(\mathbf{N}_0^T\cdot\mathbf{N}_0)$.

8. Show that in any MWCSH,

$$\mathbf{A}_2 - \mathbf{A}_1^2 = \kappa^2\left(\mathbf{N}^T\cdot\mathbf{N} - \mathbf{N}\cdot\mathbf{N}^T\right), \tag{9.320}$$

and hence

$$\operatorname{tr}\mathbf{A}_1^2 = \operatorname{tr}\mathbf{A}_2 = 2\kappa^2\left(1 + \operatorname{tr}\mathbf{N}^2\right). \tag{9.321}$$

Subsequently, show that in a viscometric flow and in a motion of Class II,

$$\kappa^2 = \frac{1}{2}\operatorname{tr}\mathbf{A}_1^2 = \frac{1}{2}\operatorname{tr}\mathbf{A}_2. \tag{9.322}$$

9. A tensor $\mathbf{M}$ is nilpotent if $\mathbf{M}^n = \mathbf{0}$ for some integer $n \geq 0$.

   i) Show that the exponential $e^{\tau\mathbf{M}}$ is a finite polynomial in τ if and only if $\mathbf{M}$ is nilpotent.

   ii) Show that in three dimensions, if $\mathbf{M}$ is nilpotent, then $n \leq 3$, and in this case,

   $$e^{\tau\mathbf{M}} = \mathbf{1} + \tau\mathbf{M} + \frac{1}{2}\tau^2\mathbf{M}^2.$$

   iii) Show that the components of $\mathbf{M}$ relative to an appropriate orthonormal basis have the form

   $$M = \begin{pmatrix} 0 & 0 & 0 \\ \chi & 0 & 0 \\ \lambda & \nu & 0 \end{pmatrix},$$

   and if $\mathbf{M}^2 = \mathbf{0}$, then the basis may be chosen such that

   $$M = \chi\begin{pmatrix} 0 & 0 & 0 \\ 1 & 0 & 0 \\ 0 & 0 & 0 \end{pmatrix}.$$

10. As shown in Problem 9, if $\mathbf{M}$ is nilpotent, then necessarily $\mathbf{M}^3 = \mathbf{0}$ in a three-dimensional Euclidean space. Subsequently, show that all MWCSH flows can be divided in the following three classes depending on the properties of $\mathbf{M}$:

    I. $\mathbf{M}^2 = \mathbf{0}$;

    II. $\mathbf{M}^2 \neq \mathbf{0}$ but $\mathbf{M}^3 = \mathbf{0}$;

    III. $\mathbf{M}^n \neq \mathbf{0}$ for all $n = 1, 2, \ldots$.

11. The representation for the stress for a third-order linear viscoelastic fluid is of the form

$$\boldsymbol{\sigma} = -p\,\mathbf{1} + \mathbf{S}^{(1)} + \mathbf{S}^{(2)} + \mathbf{S}^{(3)},$$

where it was shown that

$$\begin{aligned}
&\mathbf{S}^{(1)} = \mu\,\mathbf{A}^{(1)},\\
&\mathbf{S}^{(2)} = b_1 \mathrm{tr}\,\mathbf{A}^{(2)}\mathbf{1} + b_2\,\mathbf{A}^{(2)} + c_1\,\mathbf{A}^{(1)} : \mathbf{A}^{(1)}\,\mathbf{1} + c_2\,\mathbf{A}^{(1)} \cdot \mathbf{A}^{(1)}.
\end{aligned}$$

Determine $\mathbf{S}^{(3)}$.

12. For the flow

$$v_1 = a\,x_1, \quad v_2 = -a\,x_2, \quad v_3 = 0,$$

show that the stress tensor is of the form

$$\boldsymbol{\sigma} = -p\,\mathbf{1} + \beta_1\,\mathbf{A}^{(1)} + \beta_2\left(\mathbf{A}^{(1)}\right)^2$$

and calculate $\mathbf{A}^{(1)}$.

13. Show that in an isochoric motion, the first three invariants of $\mathbf{A}^{(n)}$ are given by

$$\mathrm{tr}\,\mathbf{A}^{(1)} = 0, \tag{9.323}$$

$$\mathrm{tr}\,\mathbf{A}^{(2)} = \mathrm{tr}\left(\mathbf{A}^{(1)}\right)^2, \tag{9.324}$$

$$\mathrm{tr}\,\mathbf{A}^{(3)} = -2\,\mathrm{tr}\left(\mathbf{A}^{(1)}\right)^3 + 3\,\mathrm{tr}\left(\mathbf{A}^{(1)}\cdot\mathbf{A}^{(2)}\right), \tag{9.325}$$

and in general $\mathrm{tr}\,\mathbf{A}^{(n)}$ is a linear combination of traces of products formed from $\mathbf{A}^{(1)}, \ldots, \mathbf{A}^{(n-1)}$.

14. Show that the most general constitutive equation of the stress tensor of a simple fluid for the viscometric flow $\mathbf{v} = v(x_1)\,\mathbf{i}_2$ is given by

$$\boldsymbol{\sigma} = -p(\kappa)\,\mathbf{1} + \mu_1(\kappa)\left(\mathbf{N} + \mathbf{N}^T\right) + \mu_2(\kappa)\,\mathbf{N}\cdot\mathbf{N}^T + \mu_3(\kappa)\,\mathbf{N}^T\cdot\mathbf{N}, \tag{9.326}$$

where

$$\kappa \equiv \frac{dv(x_1)}{dx_1}. \tag{9.327}$$

Find $\mathbf{N}$ and show that the three viscometric functions μ_1, μ_2, and μ_3 satisfy the conditions

$$\mu_1(-\kappa) = -\mu_1(\kappa) \qquad \text{and} \qquad \mu_{2,3}(-\kappa) = \mu_{2,3}(\kappa). \tag{9.328}$$

15. Consider the steady Poiseuille flow of a simple fluid in an infinite circular tube of radius R, and use cylindrical coordinates with the axis of the tube coincident with the z-axis. We take the velocity field $\mathbf{v} = (v_r, v_z, v_\theta)$ to have the form

$$v_r = 0, \qquad v_z = v(r), \qquad v_\theta = 0,$$

and the fluid to satisfy the no-penetration and no-slip conditions along the tube wall, $v(R) = 0$. Take the body force to act in the z-direction.

i) Use the linear momentum balance to show that the *driving force density* along the tube (pressure gradient) is given by

$$f = \frac{F}{\pi R^2 (z_2 - z_1)},$$

where the total force is given by

$$F = \int_A \left[(\sigma_{zz} - \rho_R U)|_{z=z_2} - (\sigma_{zz} - \rho_R U)|_{z=z_1} \right] dA,$$

where U is the gravitational potential. Prove that f is constant; i.e., it is independent of the choice of z_1 and z_2.

ii) Show that the velocity profile is given by

$$v(r) = \int_r^R \zeta\left(\frac{1}{2} f r'\right) dr',$$

where ζ is the shear-rate function (see (9.167)), and from this it follows that the volume discharge per unit time through a cross-section of the tube,

$$Q = 2\pi \int_0^R v(r)\, r\, dr,$$

is given by

$$Q = \frac{8\pi}{f^3} \int_0^{fR/2} \zeta(r)\, r^2 dr.$$

iii) Show that the stress tensor has the form

$$[\sigma_{jk}] = \begin{pmatrix} \sigma_{rr} & \sigma_{rz} & 0 \\ \sigma_{zr} & \sigma_{zz} & 0 \\ 0 & 0 & \sigma_{\theta\theta} \end{pmatrix},$$

where

$$\sigma_{zr} = \sigma_{rz} = -\frac{1}{2} f r,$$

$$\sigma_{rr} = f z + \rho_R U + \int_r^R \frac{1}{r'} \mu_2 \left[\zeta\left(\frac{1}{2} f r'\right) \right] dr' + c,$$

$$\sigma_{rr} - \sigma_{\theta\theta} = \mu_2 \left[\zeta\left(\frac{1}{2} f r\right) \right],$$

$$\sigma_{zz} - \sigma_{\theta\theta} = \mu_3 \left[\zeta\left(\frac{1}{2} f r\right) \right],$$

and c is a constant. Note that because of the assumed incompressibility of the fluid, when only f, R, and U are given, the normal pressures are determined only up to a constant hydrostatic pressure of magnitude c. Also note that the normal stress in the axial direction,

$$\sigma_{zz} = \sigma_{rr} + \mu_3 \left[\zeta\left(\frac{1}{2} f\, r \right) \right] - \mu_2 \left[\zeta\left(\frac{1}{2} f\, r \right) \right],$$

generally depends on r.

iv) What is the implication that $\sigma_{zz} \neq \sigma_{rr}$ upon a stream of fluid exiting the tube?

16. Consider the steady flow of an incompressible simple fluid between two fixed coaxial circular cylinders of radii R_1 and R_2 ($R_1 < R_2$) and use cylindrical coordinates with the axis of the tubes coincident with the z-axis. We take the velocity field $\mathbf{v} = (v_r, v_z, v_\theta)$ to have the form

$$v_r = 0, \qquad v_z = v(r), \qquad v_\theta = 0,$$

and the fluid to satisfy the no-penetration and no-slip conditions along the tube walls, $v(R_1) = v(R_2) = 0$. Take the body force to act in the z-direction.

i) Use the linear momentum balance to show that the *driving force density* along and between the tubes (pressure gradient) is given by

$$f = \frac{F}{\pi\,(R_2^2 - R_1^2)\,(z_2 - z_1)},$$

where the total force is given by

$$F = \int_A \left[(\sigma_{zz} - \rho_R\, U)|_{z=z_2} - (\sigma_{zz} - \rho_R\, U)|_{z=z_1} \right] dA,$$

where U is the gravitational potential. Prove that f is constant; i.e., it is independent of the choice of z_1 and z_2.

ii) Show that the velocity profile is given by

$$v(r) = \int_{R_1}^{r} \zeta\,[\alpha\,(r')]\, dr',$$

where ζ is the shear-rate function (see (9.167)),

$$\alpha(r) = \frac{a}{r} - \frac{1}{2} f\, r^2,$$

and the constant a is chosen so that

$$\int_{R_1}^{R_2} \zeta\,[\alpha\,(r)]\, dr = 0.$$

From this, show that it follows that the volume discharge per unit time through a cross-section of the tube,

$$Q = 2\,\pi \int_{R_1}^{R_2} v(r)\, r\, dr,$$

is given by

$$Q = -\pi \int_{R_1}^{R_2} \zeta\,[\alpha(r)]\, r^2 dr.$$

iii) Show that the stress tensor has the form

$$[\sigma_{jk}] = \begin{pmatrix} \sigma_{rr} & \sigma_{rz} & 0 \\ \sigma_{zr} & \sigma_{zz} & 0 \\ 0 & 0 & \sigma_{\theta\theta} \end{pmatrix},$$

where

$$\begin{aligned} \sigma_{zr} &= \sigma_{rz} = \alpha(r), \\ \sigma_{rr} &= f\,z + \rho_R\,U + \int_{R_1}^{r} \frac{1}{r'}\,\mu_2\left\{\zeta\left[\alpha\left(r'\right)\right]\right\} dr' + c, \\ \sigma_{rr} - \sigma_{\theta\theta} &= \mu_2\left\{\zeta\left[\alpha\left(r'\right)\right]\right\}, \\ \sigma_{zz} - \sigma_{\theta\theta} &= \mu_3\left\{\zeta\left[\alpha\left(r'\right)\right]\right\}, \end{aligned}$$

and c is a constant. Note that because of the assumed incompressibility of the fluid, when only f, R, and U are given, the normal pressures are determined only up to a constant hydrostatic pressure of magnitude c. Also note that the stress difference in the radial direction between the two radii is given by

$$\sigma_{rr}(R_2) - \sigma_{rr}(R_1) = -\int_{R_1}^{R_2} \frac{1}{r}\,\mu_2\left\{\zeta\left[\alpha\left(r\right)\right]\right\} dr.$$

iv) What is the implication that $\sigma_{rr}(R_2) \neq \sigma_{rr}(R_1)$ upon a stream of fluid exiting the tubes?

17. Consider the steady Couette flow of an incompressible simple fluid between two coaxial circular cylinders of radii R_1 and R_2 ($R_1 < R_2$) and use cylindrical coordinates with the axis of the tubes coincident with the z-axis. The two tubes rotate with constant angular velocities Ω_1 and Ω_2, respectively. We take the velocity field $\mathbf{v} = (v_r, v_z, v_\theta)$ to have the form

$$v_r = 0, \qquad v_z = 0, \qquad v_\theta = \omega(r),$$

and the fluid to satisfy the no-penetration and no-slip conditions along the tube walls,

$$\omega(R_1) = \Omega_1 \qquad \text{and} \qquad \omega(R_2) = \Omega_2.$$

i) Show that the velocity profile obeys the equation

$$\frac{d\omega(r)}{dr} = \frac{1}{r}\,\zeta\left(\frac{M}{2\,\pi\,r}\right),$$

where ζ is the shear-rate function (see (9.167)) and M is the torque per unit length required to maintain the relative motion between the cylinders.

ii) By integrating the above equation, show that the angular velocity difference and the torque are related by

$$\Omega_2 - \Omega_1 = \frac{1}{2}\int_{M/(2\,\pi\,R_2^2)}^{M/(2\,\pi\,R_1^2)} \frac{1}{s}\,\zeta(s)\,ds.$$

iii) Show that the stress tensor has the form

$$[\sigma_{jk}] = \begin{pmatrix} \sigma_{rr} & 0 & \sigma_{r\theta} \\ 0 & \sigma_{zz} & 0 \\ \sigma_{\theta r} & 0 & \sigma_{\theta\theta} \end{pmatrix},$$

where

$$\sigma_{\theta r} = \sigma_{r\theta} = \frac{M}{2\pi r^2},$$

$$\sigma_{rr} = \rho_R U - \int_{R_1}^{r} \left\{ \rho_R r' \omega^2(r') + \frac{1}{r'} \mu_2 \left[\zeta \left(\frac{M}{2\pi r'^2} \right) \right] - \frac{1}{r'} \mu_3 \left[\zeta \left(\frac{M}{2\pi r'^2} \right) \right] \right\} dr' + c,$$

$$\sigma_{rr} - \sigma_{zz} = \mu_2 \left[\zeta \left(\frac{M}{2\pi r^2} \right) \right],$$

$$\sigma_{\theta\theta} - \sigma_{zz} = \mu_3 \left[\zeta \left(\frac{M}{2\pi r^2} \right) \right],$$

where U is the gravitational potential and c is a constant. Note that because of the assumed incompressibility of the fluid, the normal pressures are determined only up to a constant hydrostatic pressure of magnitude c. Also note that the stress difference in the radial direction between the two radii given by

$$\sigma_{rr}(R_2) - \sigma_{rr}(R_1) = -\int_{R_1}^{R_2} \left\{ \rho_R r \omega^2(r) - \frac{1}{r} \mu_2 \left[\zeta \left(\frac{M}{2\pi r^2} \right) \right] - \frac{1}{r} \mu_3 \left[\zeta \left(\frac{M}{2\pi r^2} \right) \right] \right\} dr$$

generally depends on r.

iv) What is the implication that $\sigma_{rr}(R_2) \neq \sigma_{rr}(R_1)$?

18. Consider the steady Poiseuille flow of a simple fluid in an infinite circular tube of radius R, and use cylindrical coordinates with the axis of the tube coincident with the z-axis. Take the velocity field $\mathbf{v} = (v_r, v_\theta, v_z)$ to have the form

$$v_r = 0, \qquad v_\theta = 0, \qquad v_z = v(r),$$

with the fluid moving in the positive z-direction and satisfying the no-penetration and no-slip conditions along the tube wall, $v(R) = 0$. Take the body force to act in the z-direction. In addition, take the stress tensor to be linear of rate type and of rate 1 and the extra stress given in the form

$$[\sigma^d_{jk}] = \begin{pmatrix} t_r & 0 & s \\ 0 & r^{-2} t_\theta & 0 \\ 0 & 0 & t_z \end{pmatrix},$$

where $t_i = t_i(r)$ $(i = r, \theta, z)$ and $s = s(r)$.

i) Show that $s = -\frac{1}{2} a r$, where a is a positive constant.

ii) Show that $p = -(a - \rho g)z + f(r)$.

iii) Show that the velocity is given by

$$v = \frac{2}{\nu \eta a}\left[(u - u_R) - \log\left(\frac{u}{u_R}\right)\right],$$

where

$$u = 1+\sqrt{1-\nu a^2 r^2},\; u_R = 1+\sqrt{1-\nu a^2 R^2},\; \nu = 4\beta\tau_d^2\eta^{-2} = \frac{1}{4}\kappa,\; a = \rho g+\frac{\Delta\overline{p}_z}{l},$$

η, β, and τ_d are given in (9.302), $\Delta\overline{p}_z = \overline{p}_{z1} - \overline{p}_{z2}$, and $\overline{p}_{z1}$ and $\overline{p}_{z2}$ are the average pressures at two cross-sections of the tube separated by a distance l. Note that for steady flow to be possible, it is necessary that $a^2 \le \nu^{-1}R^{-2}$.

iv) Show that the volume discharge per unit time and the average velocity are given by

$$Q = \frac{\pi R^4 a (2u_R - 1)}{3\eta u_R^2} \qquad \text{and} \qquad \overline{v} = \frac{Q}{\pi R^2}.$$

Note that when $\beta = 0$, $u_R = 2$, and then

$$\overline{v} = \overline{v}_c \equiv \frac{R^2 a}{4\eta},$$

which is the same equation as for classical viscous fluids. For the sake of simplicity, from now on, assume that we have no body forces so that $a = \Delta\overline{p}_z/l$ is the pressure gradient in the flow direction.

v) Show that for $\beta > 0$

$$\frac{\overline{v}}{\overline{v}_c} = 1 + \frac{\xi^2\left(2 - \frac{9}{4}\xi^2\right)}{6\left[\left(1-\xi^2\right)^{3/2} + \left(1 - \frac{3}{2}\xi^2 + \frac{3}{8}\xi^4\right)\right]},$$

where

$$\xi = \frac{2\tau_s}{\eta} a R \qquad \text{and} \qquad \tau_s = \sqrt{\beta}\,\tau_d,$$

with τ_s being the shearing yield time. Note that real values of $\overline{v}$ are obtained only if $|\xi| \le 1$, so that the largest possible values of the pressure gradient and mean velocity are given by

$$a_{\mathrm{cr}} = \frac{\eta}{2\tau_s R} \qquad \text{and} \qquad \overline{v}_{\mathrm{cr}} = \frac{R}{6\tau_s}.$$

When the pressure gradient exceeds a_{cr}, the flow necessarily becomes unsteady. Also, when $a = a_{\mathrm{cr}}$, we have that $\overline{v}/\overline{v}_{cr} = 4/3$. Plot $\overline{v}/\overline{v}_c$ as a function of ξ and comment on the accuracy of the classical solution.

vi) Show that for $\beta < 0$

$$\frac{\overline{v}}{\overline{v}_c} = 1 - \frac{\zeta^2\left(2 + \frac{9}{4}\zeta^2\right)}{6\left[\left(1+\zeta^2\right)^{3/2} + \left(1 + \frac{3}{2}\zeta^2 + \frac{3}{8}\zeta^4\right)\right]},$$

where

$$\zeta = \frac{2\tau_s'}{\eta} a R \qquad \text{and} \qquad \tau_s' = \sqrt{|\beta|}\,\tau_d,$$

with τ_s' being the shearing limit time. Note that no matter how large the pressure gradient is, the mean velocity never exceeds the limiting value

$$\overline{v}_{\text{lim}} = \frac{R}{3\tau_s'}.$$

Plot $\overline{v}/\overline{v}_c$ as a function of ζ and comment on the accuracy of the classical solution.

vii) Show that the normal stresses are given by

$$\sigma_{rr} = -p + t_r = a\,z + \frac{1}{2} a\,\tau_d\,(\beta_3 - 1)\,v,$$

$$\sigma_{\theta\theta} = -p + t_\theta = a\,z + \frac{1}{2} a\,\tau_d\,(\beta_3 - 1)\left(v - \frac{2\,a\,r^2}{\eta\,u}\right),$$

$$\sigma_{zz} = -p + t_z = a\,z + \frac{1}{2} a\,\tau_d\left[(\beta_3 - 1)\,v + \frac{4\,a\,r^2}{\eta\,u}\right],$$

and note that these stresses deviate from those of the classical normal stresses, which are given by

$$\sigma_{rr} = \sigma_{\theta\theta} = \sigma_{zz} = -p = a\,z.$$

Bibliography

R.B. Bird, R.C. Armstrong, and O. Hassager. *Dynamics of Polymeric Liquids*, volume 1. Wiley, New York, 1977.

T.S. Chang. Constitutive equations for simple materials: Simple materials with fading memory. In E.H. Dill, editor, *Continuum Physics*, volume II, pages 283–403. Academic Press, Inc., New York, NY, 1975.

B.D. Coleman. Kinematical concepts with applications in the mechanics and thermodynamics of incompressible viscoelastic fluids. *Archive for Rational Mechanics and Analysis*, 9(1):273–300, 1962.

B.D. Coleman. Substantially stagnant motions. *Transactions of the Society of Rheology*, 6(1):293–300, 1962.

B.D. Coleman. On thermodynamics, strain impulses, and viscoelasticity. *Archive for Rational Mechanics and Analysis*, 17(3):230–254, 1964.

B.D. Coleman. Thermodynamics of materials with memory. *Archive for Rational Mechanics and Analysis*, 17:1–46, 1964.

B.D. Coleman, H. Markovitz, and W. Noll. *Viscometric Flows of Non-Newtonian Fluids.* Springer-Verlag, Berlin, 1966.

B.D. Coleman and W. Noll. On certain steady flows of general fluids. *Archive for Rational Mechanics and Analysis*, 3(1):289–303, 1959.

B.D. Coleman and W. Noll. An approximation theorem for functionals, with applications in continuum mechanics. *Archive for Rational Mechanics and Analysis*, 6(1):355–370, 1960.

B.D. Coleman and W. Noll. Foundations of linear viscoelasticity. *Reviews of Modern Physics*, 33(2):239–249, 1961.

B.D. Coleman and W. Noll. Recent results in the continuum theory of viscoelastic fluids. *Annals of the New York Academy of Sciences*, 89(4):672–714, 1961.

B.D. Coleman and W. Noll. Steady extension of incompressible simple fluids. *Physics of Fluids*, 5(7):840–843, 1962.

B.A. Cotter and R.S. Rivlin. Tensors associated with time-dependent stress. *Quarterly of Applied Mathematics*, 13(2):177–182, 1955.

W.O. Criminale, J.L. Ericksen, and G.L. Filbey. Steady shear flow of non-Newtonian fluids. *Archive for Rational Mechanics and Analysis*, 1(1):410–417, 1957.

W.A. Day. *The Thermodynamics of Simple Materials with Fading Memory.* Springer-Verlag, Berlin, 1972.

M.O. Deville and T.B. Gatski. *Mathematical Modeling for Complex Fluids and Flows.* Springer-Verlag, Berlin, 2012.

A.C. Eringen. Constitutive equations for simple materials: General theory. In A.C. Eringen, editor, *Continuum Physics*, volume II, pages 131–172. Academic Press, Inc., New York, NY, 1975.

E.C. Eringen. A unified theory of thermomechanical materials. *International Journal of Engineering Science*, 4(2):179–202, 1966.

R.L. Fosdick and K.R. Rajagopal. Thermodynamics and stability of fluids of third grade. *Proceedings of the Royal Society of London A: Mathematical, Physical and Engineering Sciences*, 369(1738):351–377, 1980.

A.E. Green and R.S. Rivlin. The mechanics of non-linear materials with memory. part i. *Archive for Rational Mechanics and Analysis*, 1(1):1–21, 1957.

A.E. Green, R.S. Rivlin, and A.J.M. Spencer. The mechanics of non-linear materials with memory. Part II. *Archive for Rational Mechanics and Analysis*, 3(1):82–90, 1959.

R.R. Huilgol. Recent advances in the continuum mechanics of viscoelastic liquids. *International Journal of Engineering Science*, 24(2):161–251, 1986.

D.D. Joseph. Instability of the rest state of fluids of arbitrary grade greater than one. *Archive for Rational Mechanics and Analysis*, 75(3):251–256, 1981.

D.C. Leigh. *Nonlinear Continuum Mechanics.* McGraw-Hill Book Company, New York, NY, 1968.

W. Noll. A mathematical theory of the mechanical behavior of continuous media. *Archive for Rational Mechanics and Analysis*, 2(1):197–226, 1958.

W. Noll. Motions with constant stretch history. *Archive for Rational Mechanics and Analysis*, 11(1):97–105, 1962.

A. Pipkin. Small finite deformation of viscoelastic solids. *Reviews of Modern Physics*, 36(4):1034–1041, 1964.

M. Reiner. Second-order effects. In R.J. Seeger and G. Temple, editors, *Research Frontiers in Fluids Dynamics*, pages 193–211. Interscience Publishers, New York, 1965.

R.S. Rivlin. Viscoelastic fluids. In R.J. Seeger and G. Temple, editors, *Research Frontiers in Fluids Dynamics*, pages 144–192. Interscience Publishers, New York, 1965.

R.S. Rivlin. The fundamental equations of nonlinear continuum mechanics. In S.I. Pai, A.J. Faller, T.L. Lincoln, D.A. Tidman, G.N. Trytten, and T.D. Wilkerson, editors, *Dynamics of Fluids in Porous Media*, pages 83–126, Academic Press, New York, 1966.

R.S. Rivlin. An introduction to non-linear continuum mechanics. In R.S. Rivlin, editor, *Non-linear Continuum Theories in Mechanics and Physics and Their Applications*, pages 151–310. Springer-Verlag, Berlin, 1969.

R.S. Rivlin and K.N. Sawyers. Nonlinear continuum mechanics of viscoelastic fluids. *Annual Review of Fluid Mechanics*, pages 117–146, 1971.

J.C. Saut and D.D. Joseph. Fading memory. *Archive for Rational Mechanics and Analysis*, 81(1):53–95, 1983.

W.R. Schowalter. *Mechanics of Non-Newtonian Fluids.* Pergamon Press, Oxford–Frankfurt, 1978.

C. Truesdell. *A First Course in Rational Continuum Mechanics*, volume 1. Academic Press, New York, NY, 1977.

C. Truesdell and W. Noll. The non-linear field theories of mechanics. In S. Flügge, editor, *Handbuch der Physik*, volume III/3. Springer, Berlin-Heidelberg-New York, 1965.

C. Truesdell and R.A. Toupin. The classical field theories. In S. Flügge, editor, *Handbuch der Physik*, volume III/1. Springer, Berlin-Heidelberg-New York, 1960.

C.-C. Wang. A representation theorem for the constitutive equation of a simple material in motions with constant stretch history. *Archive for Rational Mechanics and Analysis*, 20(5):329–340, 1965.

A. Wineman. Nonlinear viscoelastic solids – A review. *Mathematics and Mechanics of Solids*, 14(3):300–366, 2009.

S. Zahorski. Flows with proportional stretch history. *Archives of Mechanics*, 24:681–699, 1972.

S. Zahorski. *Mechanics of Viscoelastic Fluids.* Martinus Nijhoff Publishers, The Hague, 1981.

Appendices

A Summary of Cartesian tensor notation

The standard Kronecker delta symbol is defined as

$$\delta_{ij} = \begin{cases} 1 & \text{if } i = j, \\ 0 & \text{if } i \neq j. \end{cases} \tag{A.1}$$

Note that

$$\delta_{ii} = 3 \tag{A.2}$$

and

$$\delta_{ij}\delta_{ij} = \delta_{ii} = 3. \tag{A.3}$$

The standard Levi–Civita symbol is given by

$$\epsilon_{ijk} = \begin{cases} 1 & \text{if } (i,j,k) \text{ is an even permutation of } (1,2,3), \\ -1 & \text{if } (i,j,k) \text{ is an odd permutation of } (1,2,3), \\ 0 & \text{if any two labels are the same.} \end{cases} \tag{A.4}$$

Useful relations between the Levi–Civita symbol and the generalized Kronecker delta symbol (see (2.91)) are

$$\begin{aligned} \epsilon_{ijk}\epsilon_{lmn} &= \delta_{iljmkn} = \det\begin{bmatrix} \delta_{il} & \delta_{im} & \delta_{in} \\ \delta_{jl} & \delta_{jm} & \delta_{jn} \\ \delta_{kl} & \delta_{km} & \delta_{kn} \end{bmatrix}, \\ &= \delta_{il}\delta_{jm}\delta_{kn} + \delta_{im}\delta_{jn}\delta_{kl} + \delta_{in}\delta_{jl}\delta_{km} - \delta_{im}\delta_{jl}\delta_{kn} - \delta_{il}\delta_{jn}\delta_{km} - \\ &\qquad \delta_{in}\delta_{jm}\delta_{kl}, \end{aligned} \tag{A.5}$$

$$\epsilon_{ijk}\epsilon_{lmk} = 1!\,\delta_{iljm} = \det\begin{bmatrix} \delta_{il} & \delta_{im} \\ \delta_{jl} & \delta_{jm} \end{bmatrix} = \delta_{il}\delta_{jm} - \delta_{im}\delta_{jl}, \tag{A.6}$$

$$\epsilon_{ijk}\epsilon_{ijl} = 2!\,\delta_{kl} = 2\,\delta_{kl}, \tag{A.7}$$

$$\epsilon_{ijk}\epsilon_{ijk} = 3! = 6. \tag{A.8}$$

The generalized Levi–Civita (or permutation) symbol (see (2.88)) is given by

$$\epsilon_{ijkl\ldots} = \begin{cases} +1 & \text{if } (i,j,k,l,\ldots) \text{ is an even permutation of } (1,2,3,4,\ldots), \\ -1 & \text{if } (i,j,k,l,\ldots) \text{ is an odd permutation of } (1,2,3,4,\ldots), \\ 0 & \text{if any two labels are the same.} \end{cases} \tag{A.9}$$

Furthermore, it can be shown that

$$\epsilon_{i_1 i_2 i_3\ldots}\epsilon_{j_1 j_2 j_3\ldots} = \delta_{i_1 j_1 i_2 j_2 i_3 j_3\ldots} = \det\begin{bmatrix} \delta_{i_1 j_1} & \delta_{i_1 j_2} & \delta_{i_1 j_3} & \cdots \\ \delta_{i_2 j_1} & \delta_{i_2 j_2} & \delta_{i_2 j_3} & \cdots \\ \delta_{i_3 j_1} & \delta_{i_3 j_2} & \delta_{i_3 j_3} & \cdots \\ \vdots & \vdots & \vdots & \ddots \end{bmatrix}, \tag{A.10}$$

$$\epsilon_{i_1,\ldots,i_n}\epsilon_{i_1,\ldots,i_n} = n!, \tag{A.11}$$

and

$$a = \det A = \epsilon_{j_1 \ldots j_n} a_{1j_1} \cdots a_{nj_n} = \epsilon_{i_1 \ldots i_n} a_{i_1 1} \cdots a_{i_n n} = \frac{1}{n!} \epsilon_{i_1 \ldots i_n} \epsilon_{j_1 \ldots j_n} a_{i_1 j_1} \cdots a_{i_n j_n}. \quad (A.12)$$

From the aforementioned, it is easy to show, e.g., that

$$\epsilon_{ij}\epsilon_{lm} = \delta_{iljm} = \det \begin{bmatrix} \delta_{il} & \delta_{im} \\ \delta_{jl} & \delta_{jm} \end{bmatrix} = \delta_{il}\delta_{jm} - \delta_{im}\delta_{jl}, \quad (A.13)$$

$$\epsilon_{ij}\epsilon_{il} = 1!\,\delta_{jl} = \delta_{jl}, \quad (A.14)$$

$$\epsilon_{ij}\epsilon_{ij} = 2! = 2. \quad (A.15)$$

A vector is represented by its typical components relative to three mutually orthogonal unit vectors of a right-handed Cartesian coordinate system with components (x_1, x_2, x_3). Let

$$\mathbf{u} = u_i \mathbf{i}_i \qquad \text{and} \qquad \mathbf{v} = v_j \mathbf{i}_j.$$

Then

$$\mathbf{u} \cdot \mathbf{v} = u_i v_i \qquad \text{and} \qquad \mathbf{u} \times \mathbf{v} = \epsilon_{ijk} u_j v_k,$$

where repeated subscripts are summed from 1 to 3, i.e., $u_i v_i = \sum_{i=1}^{3} u_i v_i$.

Repeated subscripts are dummy subscripts, in the sense that they may be replaced by another letter without affecting the value of the sum:

$$u_i v_i = u_j v_j = u_k v_k.$$

If δ_{ij} appears in an expression summing both i and j, j may be set equal to i and the δ_{ij} removed:

$$u_i v_j \delta_{ij} = u_i v_i.$$

Example

Prove that $\mathbf{u} \times (\mathbf{v} \times \mathbf{w}) = \mathbf{v}(\mathbf{u} \cdot \mathbf{w}) - \mathbf{w}(\mathbf{u} \cdot \mathbf{v})$:

$$\begin{aligned} \mathbf{u} \times (\mathbf{v} \times \mathbf{w}) &= \epsilon_{ijk} u_j (\epsilon_{klm} v_l w_m) \\ &= \epsilon_{ijk}\epsilon_{klm} u_j v_l w_m \\ &= \epsilon_{ijk}\epsilon_{lmk} u_j v_l w_m \\ &= (\delta_{il}\delta_{jm} - \delta_{im}\delta_{jl})\, u_j v_l w_m \\ &= v_i u_j w_j - w_i u_j v_j \\ &= \mathbf{v}(\mathbf{u} \cdot \mathbf{w}) - \mathbf{w}(\mathbf{u} \cdot \mathbf{v}). \end{aligned}$$

Let

$$\nabla = \mathbf{i}_i \frac{\partial}{\partial x_i} = \mathbf{i}_i \partial_i.$$

Then

$$\operatorname{grad} \phi = \nabla \phi = \mathbf{i}_i \partial_i \phi,$$

$$\operatorname{div}\mathbf{u} = \nabla \cdot \mathbf{u} = \partial_i \mathbf{u}_i,$$

$$\operatorname{curl}\mathbf{u} = \nabla \times \mathbf{u} = \epsilon_{ijk}\partial_j \mathbf{u}_k.$$

Example

Prove that $\nabla \times \nabla\phi = \mathbf{0}$:

$$\begin{aligned} \nabla \times \nabla\phi &= \epsilon_{ijk}\partial_j\partial_k\phi \\ &= \epsilon_{ijk}\partial_k\partial_j\phi \quad \text{(order of differentiation may be changed)} \\ &= \epsilon_{ikj}\partial_j\partial_k\phi \quad (j \text{ and } k \text{ are dummy indices}) \\ &= -\epsilon_{ijk}\partial_j\partial_k\phi \quad (\epsilon_{ijk} = -\epsilon_{ikj}) \\ &= -\nabla \times \nabla\phi. \end{aligned}$$

Subsequently, we must have that $\nabla \times \nabla\phi = \mathbf{0}$.

Problems

1. Show that $(\mathbf{a} \times \mathbf{b}) \cdot (\mathbf{c} \times \mathbf{d}) = (\mathbf{a} \cdot \mathbf{c})(\mathbf{b} \cdot \mathbf{d}) - (\mathbf{a} \cdot \mathbf{d})(\mathbf{b} \cdot \mathbf{c})$.
2. Show that $\nabla \cdot (\nabla \times \mathbf{u}) = 0$.
3. Show that $\nabla \times (\mathbf{u} \times \mathbf{v}) = \mathbf{v} \cdot \nabla\mathbf{u} - \mathbf{v}(\nabla \cdot \mathbf{u}) + \mathbf{u}(\nabla \cdot \mathbf{v}) - \mathbf{u} \cdot \nabla\mathbf{v}$.
4. Show that $\nabla \times (\nabla \times \mathbf{u}) = \nabla(\nabla \cdot \mathbf{u}) - \nabla^2\mathbf{u}$.

Bibliography

H. Jeffreys. *Cartesian Tensors.* Cambridge University Press, London, 1969.

G. Temple. *Cartesian Tensors.* Methuuen & Co. Ltd., London, 1960.

B Isotropic tensors

Let's assume that we have a linear relationship between the second rank tensors fields $\boldsymbol{\sigma}$ and $\mathbf{D}$:

$$\sigma_{ik} = A_{iklm} D_{lm}. \tag{B.1}$$

Now if the material is homogeneous and isotropic, its properties will be the same at all points and in all frames of reference. Hence, the above equation must remain invariant under rigid rotations of the frame of reference. Now under such a transformation, the tensors $\boldsymbol{\sigma}$ and $\mathbf{D}$ become $\boldsymbol{\sigma}'$ and $\mathbf{D}'$ where

$$\sigma'_{pq} = Q_{pi} Q_{qk} \sigma_{ik}, \tag{B.2}$$
$$D'_{rs} = Q_{rl} Q_{sm} D_{lm}, \tag{B.3}$$

or

$$D_{lm} = Q_{ms} Q_{lr} D'_{rs}, \tag{B.4}$$

since $\mathbf{Q}^{-1} = \mathbf{Q}^T$. Hence,

$$\sigma'_{pq} = Q_{pi} Q_{qk} \sigma_{ik} = Q_{pi} Q_{qk} A_{iklm} D_{lm} = Q_{pi} Q_{qk} A_{iklm} Q_{ms} Q_{lr} D'_{rs} = A'_{pqrs} D'_{rs}, \tag{B.5}$$

where

$$A'_{pqrs} = Q_{pi} Q_{qk} Q_{rl} Q_{sm} A_{iklm}. \tag{B.6}$$

The coefficients A'_{pqrs} are therefore the components of a tensor of rank 4. But if the linear relations remain invariant, the coefficients must remain unaltered, i.e.,

$$A'_{pqrs} = A_{pqrs}. \tag{B.7}$$

Thus the tensor A'_{pqrs} must have the same set of coefficients in all bases. Such tensors are called *isotropic* tensors.

In general, we define an isotropic tensor of any rank by the criterion that its components form the same set of numbers in all bases, or that it is invariant under any rotation.

It is evident that all rank-0 tensors (scalars) are isotropic. There are no isotropic rank-1 tensors (vectors). To see that this is the case, consider a small rotation, expressed by the skew-symmetric tensor Q_{ik}, of the vector $\mathbf{v}$. Then in the new coordinate system,

$$v'_i = (\delta_{ij} - Q_{ij}) v_j = v_i - Q_{ij} v_j, \tag{B.8}$$

and this can be equal to v_i only if

$$Q_{ij} v_j = 0. \tag{B.9}$$

Now the above corresponds to a system of three linear equations. But $Q_{ii} = 0$ and $Q_{ij} = -Q_{ji}$, $j \neq i$, are linearly independent. Hence, the above can be satisfied only if $v_i = 0$, and therefore, there is no isotropic tensor of first order other than zero.

The number of linearly independent isotropic tensors of rank $n = \{0, 1, 2, 3, 4, 5, 6, \ldots\}$ are $a_n = \{1, 0, 1, 1, 3, 6, 15, \ldots\}$. These numbers are called the Motzkin sum numbers and are given by the recurrence relation

$$a_n = \frac{n-1}{n+1}(2a_{n-1} + 3a_{n-2}) \qquad \text{with} \quad a_0 = 1, \ a_1 = 0. \tag{B.10}$$

Starting at rank 5, *syzygies* play a role in restricting the number of independent isotropic tensors. In particular, syzygies occur at rank 5, 7, 8, and all higher ranks. A syzygy is a mathematical object defined in terms of a polynomial ring of n variables. An example of a rank-5 syzygy is

$$\epsilon_{ijk}\delta_{lm} - \epsilon_{jkl}\delta_{im} + \epsilon_{ikl}\delta_{jm} - \epsilon_{ijl}\delta_{km} = 0, \tag{B.11}$$

which can be easily verified.

One effective approach of generating isotropic tensors of arbitrary rank in three dimensions is based on Weyl's theory of invariant polynomials. This approach reduces to noting that any even rank isotropic tensor must be expressed as a linear combination of products of the unit tensor $\mathbf{1}$, and odd tensors as a linear combination of products of the unit tensor and the alternating tensor $\boldsymbol{\epsilon}$. In practice, this means finding every possible way of writing inner products between pairs of vectors. For example, for a rank-2 isotropic tensor $\mathbf{A}$, we have only one possible inner product between two vectors, so

$$\mathbf{A} : \mathbf{uv} = \alpha(\mathbf{u} \cdot \mathbf{v}) \tag{B.12}$$

or

$$A_{ij}u_iv_j = \alpha u_kv_k = \alpha\delta_{ik}\delta_{jk}u_iv_j = \alpha\delta_{ij}u_iv_j. \tag{B.13}$$

Subsequently, A_{ij} is given by a linear relation with the one isotropic tensor of second rank $\mathbf{I}^{(2,1)} = \mathbf{1}$:

$$A_{ij} = \alpha I_{ij}^{(2,1)}, \tag{B.14}$$

where

$$I_{ij}^{(2,1)} = \delta_{ij}. \tag{B.15}$$

The unit tensor is the only isotropic tensor of second rank.

For a rank-3 isotropic tensor $\mathbf{A}$, again we only have one linear combination of products of three vectors, so

$$\mathbf{A} \vdots \mathbf{uvw} = \alpha\left[\mathbf{u} \cdot (\mathbf{v} \times \mathbf{w})\right] \tag{B.16}$$

or

$$A_{ijk}u_iv_jw_k = \alpha u_k\epsilon_{klm}v_lw_m = \alpha\epsilon_{klm}\delta_{ik}\delta_{jl}\delta_{km}u_iv_jw_k = \alpha\epsilon_{ijk}u_iv_jw_k. \tag{B.17}$$

Subsequently, we have that the Levi–Civita tensor is the only isotropic tensor of rank-3, i.e., $\mathbf{I}^{(3,1)} = \boldsymbol{\epsilon}$ or

$$I_{ijk}^{(3,1)} = \epsilon_{ijk}. \tag{B.18}$$

For a rank-4 isotropic tensor $\mathbf{A}$, we have three linear combinations of inner products of four vectors:

$$\mathbf{A} :: \mathbf{uvwx} = \alpha_1(\mathbf{u}\cdot\mathbf{v})\ (\mathbf{w}\cdot\mathbf{x}) + \alpha_2(\mathbf{u}\cdot\mathbf{w})\ (\mathbf{v}\cdot\mathbf{x}) + \alpha_3(\mathbf{u}\cdot\mathbf{x})\ (\mathbf{v}\cdot\mathbf{w}). \tag{B.19}$$

Following the same procedure as above, we obtain the three isotropic tensors

$$I^{(4,1)}_{ijkl} = \delta_{ij}\delta_{kl}, \tag{B.20}$$

$$I^{(4,2)}_{ijkl} = \delta_{ik}\delta_{jl}, \tag{B.21}$$

$$I^{(4,3)}_{ijkl} = \delta_{il}\delta_{jk}. \tag{B.22}$$

For a rank-5 isotropic tensor $\mathbf{A}$, we have ten linear combinations of inner products of five vectors:

$$\begin{aligned}\mathbf{A} \vdots : \mathbf{uvwxy} \;=\;& \alpha_1(\mathbf{u}\cdot\mathbf{v})\ (\mathbf{w}\cdot\mathbf{x}\times\mathbf{y}) + \alpha_2(\mathbf{u}\cdot\mathbf{w})\ (\mathbf{v}\cdot\mathbf{x}\times\mathbf{y}) + \\ & \alpha_3(\mathbf{u}\cdot\mathbf{x})\ (\mathbf{v}\cdot\mathbf{w}\times\mathbf{y}) + \alpha_4(\mathbf{u}\cdot\mathbf{y})\ (\mathbf{v}\cdot\mathbf{w}\times\mathbf{x}) + \\ & \alpha_5(\mathbf{v}\cdot\mathbf{w})\ (\mathbf{u}\cdot\mathbf{x}\times\mathbf{y}) + \alpha_6(\mathbf{v}\cdot\mathbf{x})\ (\mathbf{u}\cdot\mathbf{w}\times\mathbf{y}) + \\ & \alpha_7(\mathbf{v}\cdot\mathbf{y})\ (\mathbf{u}\cdot\mathbf{w}\times\mathbf{x}) + \alpha_8(\mathbf{w}\cdot\mathbf{x})\ (\mathbf{u}\cdot\mathbf{v}\times\mathbf{y}) + \\ & \alpha_9(\mathbf{w}\cdot\mathbf{y})\ (\mathbf{u}\cdot\mathbf{v}\times\mathbf{x}) + \alpha_{10}(\mathbf{x}\cdot\mathbf{y})\ (\mathbf{u}\cdot\mathbf{v}\times\mathbf{w}).\end{aligned} \tag{B.23}$$

Following the same procedure as above, we can basically read off the ten isotropic tensors $\{\epsilon_{klm}\delta_{ij}, \epsilon_{jlm}\delta_{ik}, \epsilon_{jkm}\delta_{il}, \epsilon_{jkl}\delta_{im}, \epsilon_{ilm}\delta_{jk}, \epsilon_{ikm}\delta_{jl}, \epsilon_{ikl}\delta_{jm}, \epsilon_{ijm}\delta_{kl}, \epsilon_{ijl}\delta_{km}, \epsilon_{ijk}\delta_{lm}\}$ corresponding to the above ten products. However, not all these isotropic tensors are independent. Specifically, using the syzygy that we noted earlier, we see that $\epsilon_{ijl}\delta_{km}$ is related to $\epsilon_{ijk}\delta_{lm}$, $\epsilon_{jkl}\delta_{im}$, and $\epsilon_{ikl}\delta_{jm}$. Subsequently, we can reduce the isotropic tensors to the following nine: $\{\epsilon_{klm}\delta_{ij}, \epsilon_{jlm}\delta_{ik}, \epsilon_{jkm}\delta_{il}, \epsilon_{jkl}\delta_{im}, \epsilon_{ilm}\delta_{jk}, \epsilon_{ikm}\delta_{jl}, \epsilon_{ikl}\delta_{jm}, \epsilon_{ijm}\delta_{kl}, \epsilon_{ijk}\delta_{lm}\}$. We note that by appropriately interchanging subscripts m and l in the previous syzygy, we have the syzygy

$$\epsilon_{ijk}\delta_{ml} - \epsilon_{jkm}\delta_{il} + \epsilon_{ikm}\delta_{jl} - \epsilon_{ijm}\delta_{kl} = 0. \tag{B.24}$$

Now we see that the isotropic tensor $\epsilon_{ijm}\delta_{kl}$ can be written in terms of the isotropic tensors $\epsilon_{ijk}\delta_{ml}$, $\epsilon_{jkm}\delta_{il}$, and $\epsilon_{ikm}\delta_{jl}$. Subsequently, since $\delta_{ml} = \delta_{lm}$, we can reduce the number of isotropic tensors to the following eight: $\{\epsilon_{klm}\delta_{ij}, \epsilon_{jlm}\delta_{ik}, \epsilon_{jkm}\delta_{il}, \epsilon_{jkl}\delta_{im}, \epsilon_{ilm}\delta_{jk}, \epsilon_{ikm}\delta_{jl}, \epsilon_{ikl}\delta_{jm}, \epsilon_{ijk}\delta_{lm}\}$. Now interchanging l and j in the previous syzygy, we have the syzygy

$$\epsilon_{ilk}\delta_{mj} - \epsilon_{lkm}\delta_{ij} + \epsilon_{ikm}\delta_{lj} - \epsilon_{ilm}\delta_{kj} = 0, \tag{B.25}$$

or

$$-\epsilon_{ikl}\delta_{mj} + \epsilon_{klm}\delta_{ij} + \epsilon_{ikm}\delta_{lj} - \epsilon_{ilm}\delta_{kj} = 0. \tag{B.26}$$

Noting that $\epsilon_{ilm}\delta_{kj}$ can be written in terms of $\epsilon_{ikl}\delta_{mj}$, $\epsilon_{klm}\delta_{ij}$, and $\epsilon_{ikm}\delta_{lj}$, and using the symmetry of the Kronecker delta, we can reduce the number of isotropic tensors to the following seven: $\{\epsilon_{klm}\delta_{ij}, \epsilon_{jlm}\delta_{ik}, \epsilon_{jkm}\delta_{il}, \epsilon_{jkl}\delta_{im}, \epsilon_{ikm}\delta_{jl}, \epsilon_{ikl}\delta_{jm}, \epsilon_{ijk}\delta_{lm}\}$. We can reduce the number of isotropic tensor again by using the following syzygy, which is obtained by interchanging the subscripts j and i in the last syzygy:

$$-\epsilon_{jkl}\delta_{mi} + \epsilon_{klm}\delta_{ji} + \epsilon_{jkm}\delta_{li} - \epsilon_{jlm}\delta_{ki} = 0. \tag{B.27}$$

Now noting that $\epsilon_{jlm}\delta_{ki}$ can be written in terms of $\epsilon_{jkl}\delta_{mi}$, $\epsilon_{klm}\delta_{ji}$, and $\epsilon_{jkm}\delta_{li}$, and using the symmetry of the Kronecker delta, allows us to reduce the number of independent isotropic tensors to the following six: $\{\epsilon_{klm}\delta_{ij}, \epsilon_{jkm}\delta_{il}, \epsilon_{jkl}\delta_{im}, \epsilon_{ikm}\delta_{jl}, \epsilon_{ikl}\delta_{jm}, \epsilon_{ijk}\delta_{lm}\}$. No further reductions are possible. Thus, for a rank-5 tensor, we obtain the following six linearly independent isotropic tensors:

$$I^{(5,1)}_{ijklm} = \epsilon_{klm}\delta_{ij}, \tag{B.28}$$

$$I^{(5,2)}_{ijklm} = \epsilon_{jkm}\delta_{il}, \tag{B.29}$$

$$I^{(5,3)}_{ijklm} = \epsilon_{jkl}\delta_{im}, \tag{B.30}$$

$$I^{(5,4)}_{ijklm} = \epsilon_{ikm}\delta_{jl}, \tag{B.31}$$

$$I^{(5,5)}_{ijklm} = \epsilon_{ikl}\delta_{jm}, \tag{B.32}$$

$$I^{(5,6)}_{ijklm} = \epsilon_{ijk}\delta_{lm}. \tag{B.33}$$

For a rank-6 isotropic tensor $\mathbf{A}$, we have fifteen linear combinations of inner products of six vectors:

$$\begin{aligned} \mathbf{A}::\mathbf{uvwxyz} = {} & \alpha_1(\mathbf{u}\cdot\mathbf{v})\ (\mathbf{w}\cdot\mathbf{x})\ (\mathbf{y}\cdot\mathbf{z}) + \alpha_2(\mathbf{u}\cdot\mathbf{w})\ (\mathbf{v}\cdot\mathbf{x})\ (\mathbf{y}\cdot\mathbf{z}) + \\ & \alpha_3(\mathbf{u}\cdot\mathbf{x})\ (\mathbf{v}\cdot\mathbf{w})\ (\mathbf{y}\cdot\mathbf{z}) + \alpha_4(\mathbf{u}\cdot\mathbf{y})\ (\mathbf{v}\cdot\mathbf{w})\ (\mathbf{x}\cdot\mathbf{z}) + \\ & \alpha_5(\mathbf{u}\cdot\mathbf{z})\ (\mathbf{v}\cdot\mathbf{w})\ (\mathbf{x}\cdot\mathbf{y}) + \alpha_6(\mathbf{u}\cdot\mathbf{w})\ (\mathbf{v}\cdot\mathbf{y})\ (\mathbf{x}\cdot\mathbf{z}) + \\ & \alpha_7(\mathbf{u}\cdot\mathbf{w})\ (\mathbf{v}\cdot\mathbf{z})\ (\mathbf{x}\cdot\mathbf{y}) + \alpha_8(\mathbf{u}\cdot\mathbf{v})\ (\mathbf{w}\cdot\mathbf{y})\ (\mathbf{x}\cdot\mathbf{z}) + \\ & \alpha_9(\mathbf{u}\cdot\mathbf{v})\ (\mathbf{x}\cdot\mathbf{y})\ (\mathbf{w}\cdot\mathbf{z}) + \alpha_{10}(\mathbf{u}\cdot\mathbf{y})\ (\mathbf{w}\cdot\mathbf{x})\ (\mathbf{v}\cdot\mathbf{z}) + \\ & \alpha_{11}(\mathbf{u}\cdot\mathbf{z})\ (\mathbf{w}\cdot\mathbf{x})\ (\mathbf{v}\cdot\mathbf{y}) + \alpha_{12}(\mathbf{u}\cdot\mathbf{z})\ (\mathbf{v}\cdot\mathbf{x})\ (\mathbf{w}\cdot\mathbf{y}) + \\ & \alpha_{13}(\mathbf{u}\cdot\mathbf{y})\ (\mathbf{v}\cdot\mathbf{x})\ (\mathbf{w}\cdot\mathbf{z}) + \alpha_{14}(\mathbf{u}\cdot\mathbf{x})\ (\mathbf{v}\cdot\mathbf{y})\ (\mathbf{w}\cdot\mathbf{z}) + \\ & \alpha_{15}(\mathbf{u}\cdot\mathbf{x})\ (\mathbf{v}\cdot\mathbf{z})\ (\mathbf{w}\cdot\mathbf{y}). \end{aligned} \tag{B.34}$$

Following the same procedure as above, since no syzygies exist for rank-6 tensors, we can read off the fifteen linearly independent isotropic tensors

$$I^{(6,1)}_{ijklmn} = \delta_{ij}\delta_{kl}\delta_{mn}, \tag{B.35}$$

$$I^{(6,2)}_{ijklmn} = \delta_{ik}\delta_{jl}\delta_{mn}, \tag{B.36}$$

$$I^{(6,3)}_{ijklmn} = \delta_{il}\delta_{jk}\delta_{mn}, \tag{B.37}$$

$$I^{(6,4)}_{ijklmn} = \delta_{im}\delta_{jk}\delta_{ln}, \tag{B.38}$$

$$I^{(6,5)}_{ijklmn} = \delta_{in}\delta_{jk}\delta_{lm}, \tag{B.39}$$

$$I^{(6,6)}_{ijklmn} = \delta_{ik}\delta_{jm}\delta_{ln}, \tag{B.40}$$

$$I^{(6,7)}_{ijklmn} = \delta_{ik}\delta_{jn}\delta_{lm}, \tag{B.41}$$

$$I^{(6,8)}_{ijklmn} = \delta_{ij}\delta_{km}\delta_{ln}, \tag{B.42}$$

$$I^{(6,9)}_{ijklmn} = \delta_{ij}\delta_{lm}\delta_{kn}, \tag{B.43}$$

$$I^{(6,10)}_{ijklmn} = \delta_{im}\delta_{kl}\delta_{jn}, \tag{B.44}$$

$$I^{(6,11)}_{ijklmn} = \delta_{in}\delta_{kl}\delta_{jm}, \tag{B.45}$$

$$I^{(6,12)}_{ijklmn} = \delta_{in}\delta_{jl}\delta_{km}, \tag{B.46}$$

$$I^{(6,13)}_{ijklmn} = \delta_{im}\delta_{jl}\delta_{kn}, \tag{B.47}$$

$$I^{(6,14)}_{ijklmn} = \delta_{il}\delta_{jm}\delta_{kn}, \tag{B.48}$$

$$I^{(6,15)}_{ijklmn} = \delta_{il}\delta_{jn}\delta_{km}. \tag{B.49}$$

We conclude by noting that while the above representations of isotropic tensors are linearly independent, these representations are not unique. For example, for the isotropic tensor of rank-4, it is typical to also see the following representation:

$$I^{(4,1)}_{ijkl} = \delta_{ij}\delta_{kl}, \tag{B.50}$$

$$I^{(4,2)}_{ijkl} = \delta_{ik}\delta_{jl} + \delta_{il}\delta_{jk}, \tag{B.51}$$

$$I^{(4,3)}_{ijkl} = \delta_{ik}\delta_{jl} - \delta_{il}\delta_{jk}. \tag{B.52}$$

Clearly, while one may find the above representation more convenient, nevertheless it is equivalent to our previous representation.

Problems

1. Verify (B.11).
2. Verify (B.19) and (B.20)–(B.22).
3. Verify (B.23) and (B.28)–(B.33).
4. Verify (B.34) and (B.35)–(B.49).

Bibliography

B.C. Eu. A complete set of irreducible isotropic tensors of rank six. *Canadian Journal of Physics*, 58(7):931–932, 1980.

A.J.M. Spencer. Theory of invariants. In A.C. Eringen, editor, *Continuum Physics*, volume I. Academic Press, New York, 1971.

C Balance laws in material coordinates

Sometimes, for solid bodies, it is more convenient to use the material description of balance laws. The corresponding relations for the general balance equation and jump condition obtained in Section 4.1 can be derived in a similar manner. Specifically, using the relationships between differential surface and volume elements (3.36) and (3.51) between the current and reference configurations, we have

$$\frac{d}{dt}\int_V \psi J\, dV = \int_S \mathbf{t}\cdot J\left(\mathbf{F}^{-1}\right)^T \cdot d\mathbf{S} + \int_V gJ\, dV. \tag{C.1}$$

Now writing

$$\Psi = J\psi, \qquad \mathbf{T} = J\,\mathbf{t}\cdot\left(\mathbf{F}^{-1}\right)^T, \quad \text{and} \quad G = Jg, \tag{C.2}$$

we can rewrite

$$\frac{d}{dt}\int_V \Psi\, dV = \int_S \mathbf{T}\cdot d\mathbf{S} + \int_V G\, dV. \tag{C.3}$$

The transport theorem (3.466) remains valid for $\Psi(\mathbf{X},t)$ in a moving region $V(t)$:

$$\frac{d}{dt}\int_V \Psi\, dV = \int_V \dot{\Psi}\, dV + \int_S \Psi\mathbf{v}\cdot d\mathbf{S}, \tag{C.4}$$

where $\mathbf{v}(\mathbf{X},t)\cdot\mathbf{N}$ is the outward speed of a surface point $\mathbf{X}$ on the surface S with unit normal $\mathbf{N}$. Now, since the material region illustrated in Fig. 4.1 is fixed in the reference configuration, the corresponding generalized transport theorem becomes

$$\frac{d}{dt}\int_{V-\varsigma} \Psi\, dV = \int_{V-\varsigma} \dot{\Psi}\, dV + \int_\varsigma [\![\Psi\mathbf{c}]\!]\cdot d\boldsymbol{\varsigma}, \tag{C.5}$$

where $\mathbf{c}(\mathbf{X},t)$ is the speed of the singular surface $\boldsymbol{\varsigma}(t)$ with unit normal $\mathbf{N}$, and since the divergence theorem (2.299) is independent of the current or reference configuration, we have

$$\int_{V-\varsigma} \operatorname{Div}\mathbf{T}\, dV = \int_{S-\varsigma} \mathbf{T}\cdot d\mathbf{S} - \int_\varsigma [\![\mathbf{T}]\!]\cdot d\boldsymbol{\varsigma}. \tag{C.6}$$

Subsequently, using (C.5) and (C.6), the integral balance law in material coordinates (C.3) for a singular surface which does not possess any properties of its own becomes

$$\int_{V-\varsigma} \left[\dot{\Psi} - \operatorname{Div}\mathbf{T} - G\right] dV + \int_\varsigma [\![\Psi\mathbf{c} - \mathbf{T}]\!]\cdot d\boldsymbol{\varsigma} = 0. \tag{C.7}$$

Using the same arguments used in Section 4.1, we obtain the local balance equation and jump condition in material coordinates:

$$\dot{\Psi} - \operatorname{Div}\mathbf{T} - G = 0, \tag{C.8}$$

$$[\![\Psi\mathbf{c} - \mathbf{T}]\!]\cdot\mathbf{N} = 0. \tag{C.9}$$

Now the corresponding local conservation of mass, and balances of linear momentum, angular momentum, energy, and entropy for a nonpolar material can be easily written down:

$$\dot{\rho}_R = 0, \tag{C.10}$$

$$\rho_R (\ddot{\mathbf{x}} - \mathbf{f}) = \operatorname{Div} \overline{\boldsymbol{\sigma}}, \tag{C.11}$$

$$\overline{\boldsymbol{\sigma}} \cdot \mathbf{F}^T = \mathbf{F} \cdot \overline{\boldsymbol{\sigma}}^T, \tag{C.12}$$

$$\rho_R (\dot{e} - r) = \Phi_R - \operatorname{Div} \mathbf{q}_R, \tag{C.13}$$

$$\gamma_v \equiv \rho_R (\dot{\eta} - b) + \operatorname{Div} \mathbf{h}_R \geq 0, \tag{C.14}$$

where we recall that $\rho_R = J\rho$, and the following definitions have been introduced:

$$\overline{\boldsymbol{\sigma}} \equiv \mathbf{F} \cdot \overline{\overline{\boldsymbol{\sigma}}} \equiv J\boldsymbol{\sigma} \cdot \left(\mathbf{F}^{-1}\right)^T, \quad \Phi_R \equiv \dot{\mathbf{F}} : \overline{\boldsymbol{\sigma}} \equiv \dot{\mathbf{E}} : \overline{\overline{\boldsymbol{\sigma}}}, \quad \mathbf{q}_R \equiv J\mathbf{F}^{-1} \cdot \mathbf{q}, \quad \mathbf{h}_R \equiv J\mathbf{F}^{-1} \cdot \mathbf{h}, \tag{C.15}$$

which correspond to the first and second *Piola–Kirchhoff stress tensors*, the *material stress power*, the *Piola–Kirchhoff heat flux*, and the *Piola–Kirchhoff entropy flux*. Note that, in a nonpolar medium, unlike the Cauchy stress tensor $\boldsymbol{\sigma}$, the first Piola–Kirchhoff stress tensor $\overline{\boldsymbol{\sigma}}$ is not symmetric. We also note that the second Piola–Kirchhoff stress tensor $\overline{\overline{\boldsymbol{\sigma}}}$ is symmetric.

Similarly, we obtain the following jump conditions for a nonpolar material at a singular surface for the conservation of mass, and balances of linear momentum, energy, and entropy in material coordinates :

$$[\![\rho_R \mathbf{c}]\!] \cdot \mathbf{N} = 0, \tag{C.16}$$

$$[\![\rho_R \dot{\mathbf{x}} \mathbf{c} + \overline{\boldsymbol{\sigma}}]\!] \cdot \mathbf{N} = \mathbf{0}, \tag{C.17}$$

$$\left[\!\!\left[\rho_R \left(e + \frac{1}{2} \dot{\mathbf{x}} \cdot \dot{\mathbf{x}}\right) \mathbf{c} + \overline{\boldsymbol{\sigma}}^T \cdot \dot{\mathbf{x}} - \mathbf{q}_R\right]\!\!\right] \cdot \mathbf{N} = 0, \tag{C.18}$$

$$\gamma_s \equiv [\![\rho_R \eta \, \mathbf{c} - \mathbf{h}_R]\!] \cdot \mathbf{N} \geq 0. \tag{C.19}$$

Bibliography

R.M. Bowen. *Introduction to Continuum Mechanics for Engineers.* Plenum Press, New York, NY, 1989.

P. Haupt. *Continuum Mechanics and Theory of Materials.* Springer-Verlag, Berlin, 2000.

G.A. Holzapfel. *Nonlinear Solid Mechanics.* John Wiley & Sons, Ltd., Chichester, England, 2005.

K. Hutter and K. Jöhnk. *Continuum Methods of Physical Modeling.* Springer-Verlag, Berlin, 1981.

D Curves and surfaces in space

Quite often we are required to deal with the continuum mechanics of material curves and surfaces that are embedded within the three-dimensional Euclidean space $\mathcal{E}^3$. In such case, we have to consider fields that are defined on such curves and surfaces as functions of time. With this goal in mind, we first require a description of their geometry.

One's first thought might be that a curve or surface can be thought of as a $\mathcal{E}^1$ or $\mathcal{E}^2$ Euclidean space, which is a subset of $\mathcal{E}^3$. This certainly works well for straight lines or plane surfaces since they are Euclidean spaces. But arbitrary curves and surfaces are non-Euclidean or Riemannian, because, e.g., a vector consisting of the sum of two surface vectors generally does not lie on the surface. Furthermore, the distance between two points measured along a curved line or surface is in general not equal to the distance between these same two points measured in the Euclidean space $\mathcal{E}^3$. Subsequently, we have to consider the Riemannian geometry of spaces $\mathcal{V}^1$ and $\mathcal{V}^2$. In passing, we note that consideration of the Riemannian geometry of space-time $\mathcal{V}^4$ is essential in general relativity.

In the following sections, we will examine some fundamental properties of curves and surfaces. For example, at each point of a space curve, we can construct a moving coordinate system consisting of a tangent vector, a normal vector, and a binormal vector which is perpendicular to both the tangent and normal vectors. How these vectors change as we move along the curve in space brings up the subjects of curvature and torsion associated with the space curve. The curvature is a measure of how the tangent vector to the curve is changing and the torsion is a measure of the twisting of the curve out of a plane. We will find that straight lines have zero curvature and plane curves have zero torsion.

In a similar fashion, associated with every smooth surface there are two coordinate surface curves and a normal surface vector through each point on the surface. The coordinate surface curves have tangent vectors which together with the normal surface vector, create a set of basis vectors and form a coordinate system at each point of the surface. These vectors can be used to define such things as a two-dimensional surface metric and a second-order curvature tensor. How these surface vectors change brings into consideration two different curvatures: a normal curvature and a tangential curvature. How these curvatures are related to the curvature tensor and to the Riemann–Christoffel tensor, as well as other interesting relationships between the various surface vectors and curvatures, is the subject of the differential geometry of curves and surfaces, which we discuss below.

Before embarking on this discussion, we find it convenient to define the *intrinsic* or *absolute derivative* of a vector $\mathbf{u} = u_i(x^j(y^\alpha))\,\mathbf{e}^i = u^i(x^j(y^\alpha))\,\mathbf{e}_i$ taken along the direction y^α:

$$\frac{\delta u_i}{\delta y^\alpha} = u_{i,j} a^j_\alpha = \left[\frac{\partial u_i}{\partial x^j} - \Gamma^k_{ij} u_k\right] a^j_\alpha = \frac{\partial u_i}{\partial y^\alpha} - \Gamma^k_{ij} a^j_\alpha u_k, \tag{D.1}$$

and similarly

$$\frac{\delta u^i}{\delta y^\alpha} = u^i_{,j} a^j_\alpha = \left[\frac{\partial u^i}{\partial x^j} + \Gamma^i_{jk} u^k\right] a^j_\alpha = \frac{\partial u^i}{\partial y^\alpha} + \Gamma^i_{jk} a^j_\alpha u^k, \tag{D.2}$$

where

$$a^j_\alpha \equiv \frac{\partial x^j}{\partial y^\alpha} \tag{D.3}$$

are the contravariant components of the coordinate transformation. The absolute derivative of higher order tensors is similarly defined. For example, to differentiate the mixed components $T^{ij}_{klm} = T^{ij}_{klm}(x^p(y^\alpha))$ in the direction tangent to the curve $x^p = x^p(y^\alpha)$, we have

$$\frac{\delta T^{ij}_{klm}}{\delta y^\alpha} = T^{ij}_{klm,p} a^p_\alpha. \tag{D.4}$$

In addition, the rule for taking the absolute derivatives of sums and products of tensors is the same as for ordinary derivatives. For example, the second absolute derivative is given by the absolute derivative of the absolute derivative. To illustrate, if we have the scalar field $f = f(x^j(y^\alpha))$, then

$$\frac{\delta f}{\delta y^\alpha} = f_{,j} a^j_\alpha, \tag{D.5}$$

and since

$$a^j_\gamma a^\gamma_l = \delta^j_l, \quad \text{and thus} \quad \frac{\partial a^j_\alpha}{\partial x^k} = -\frac{\partial a^\gamma_l}{\partial x^k} a^l_\alpha a^j_\gamma, \tag{D.6}$$

it is easy to show that

$$\frac{\delta}{\delta y^\beta}\left(\frac{\delta f}{\delta y^\alpha}\right) = f_{,jk} a^j_\alpha a^k_\beta - \frac{\partial a^\gamma_l}{\partial x^k} a^l_\alpha a^k_\beta \frac{\delta f}{\delta y^\gamma}. \tag{D.7}$$

Note that, since j and k are dummy indices, we have

$$\frac{\delta}{\delta y^\beta}\left(\frac{\delta f}{\delta y^\alpha}\right) - \frac{\delta}{\delta y^\alpha}\left(\frac{\delta f}{\delta y^\beta}\right) = \left(\frac{\partial a^\gamma_k}{\partial x^j} - \frac{\partial a^\gamma_j}{\partial x^k}\right) a^j_\alpha a^k_\beta \frac{\delta f}{\delta y^\gamma} \equiv \mathscr{F}^\gamma_{\alpha\beta} \frac{\delta f}{\delta y^\gamma}, \tag{D.8}$$

where the geometrical tensor of rank 3 whose components are $\mathscr{F}^\gamma_{\alpha\beta}$ is called the *object of anholonominity* of y^γ. The components are skew-symmetric with respect to the indices α and β. Subsequently, the components of the associated axial tensor of rank 2 are given by $\overline{\mathscr{F}}^{\delta\gamma} = \frac{1}{2}\varepsilon^{\delta\alpha\beta}\mathscr{F}^\gamma_{\alpha\beta}$. This object arises since, in general, it is not true that there exists a coordinate system y^γ in $\mathcal{E}^3$ such that the set $\{\mathbf{e}_1, \mathbf{e}_2, \mathbf{e}_3\}$ is a set of covariant measuring vectors for this coordinate system. In fact, this (*holonomic*) coordinate system will exist if and only if $\overline{\mathscr{F}}^{\delta\gamma} = 0$ for an arbitrary function f. Measuring vectors that do not satisfy this condition are called anholonomic measuring vectors. Arc lengths along the vector lines of these vectors are called anholonomic coordinates. In general, there does not exist a one-to-one mapping between a (holonomic) coordinate system in $\mathcal{E}^3$ and an anholonomic coordinate system in $\mathcal{E}^3$. Thus, given a function f with values $f(x^j)$ for all x^j in $\mathcal{E}^3$, there does not exist a function g such that $f(x^j) = g(y^\gamma(x^j))$ in $\mathcal{E}^3$. Subsequently, $\overline{\mathscr{F}}^{\delta\gamma}$ often appears as part of a "correction term" when identities which are familiar in standard (holonomic) coordinates are derived for the case of anholonomic coordinates.

D.1 Space curve

We recall the position vector in the Euclidean space $\mathcal{E}^3$

$$\mathbf{r} = x^i \mathbf{e}_i = x_j \mathbf{e}^j. \tag{D.9}$$

A curve $\mathscr{C}$ embedded in $\mathcal{E}^3$ can be represented by the pair of scalar equations

$$f_1(x^k, t) = 0 \qquad \text{and} \qquad f_2(x^k, t) = 0. \tag{D.10}$$

Then two unit vectors normal to the curve and to each other are given by

$$\mathbf{a}_1 = \frac{\operatorname{grad} f_1}{|\operatorname{grad} f_1|}, \qquad \mathbf{a}_1 \cdot \mathbf{a}_1 = 1, \tag{D.11}$$

$$\mathbf{a}_2 = \frac{\operatorname{grad} f_2}{|\operatorname{grad} f_2|}, \qquad \mathbf{a}_2 \cdot \mathbf{a}_2 = 1, \tag{D.12}$$

$$\operatorname{grad} f_1 \cdot \operatorname{grad} f_2 = 0, \qquad \mathbf{a}_1 \cdot \mathbf{a}_2 = 0. \tag{D.13}$$

Note that the last requirement restricts $f_2(x^k, t)$ for a given $f_1(x^k, t)$.

Alternatively, $\mathbf{r} = x^i(s, t)\,\mathbf{e}_i$ represents the three-dimensional space curve $\mathscr{C}$ as a function of the arc length parameter s and time t. From now until near the end of this section, we suppress the dependence of the curve on t, since when considering the geometry of a curve, the value of t is fixed. Subsequently, all the following derivatives with respect to s are to be interpreted later as partial derivatives with respect to s while keeping t fixed.

The tangent vector to the curve $\mathscr{C}$ at point s is given by

$$\mathbf{a}_3 = \frac{d\mathbf{r}}{ds} = \frac{dx^i}{ds}\mathbf{e}_i = t^i \mathbf{e}_i, \tag{D.14}$$

where

$$t^i \equiv \frac{dx^i}{ds} = \mathbf{e}^i \cdot \mathbf{a}_3 \tag{D.15}$$

is the contravariant component. It should be noted that the magnitude of t^i is unity since

$$(ds)^2 = d\mathbf{x} \cdot d\mathbf{x} = dx^i dx^j \mathbf{e}_i \cdot \mathbf{e}_j = g_{ij} \frac{dx^i}{ds}\frac{dx^j}{ds}(ds)^2, \tag{D.16}$$

so that

$$g_{ij} t^i t^j = 1, \tag{D.17}$$

and subsequently,

$$\mathbf{a}_3 \cdot \mathbf{a}_3 = 1. \tag{D.18}$$

If we now take the absolute derivative of (D.17) with respect to the arc length s and use Ricci's theorem (2.248), we have

$$g_{ij} \frac{\delta t^i}{\delta s} t^j + g_{ij} t^i \frac{\delta t^j}{\delta s} = 0, \tag{D.19}$$

which implies that

$$g_{ij} t^i \frac{\delta t^j}{\delta s} = 0. \tag{D.20}$$

Thus, the vector with components $\delta t^j/\delta s$ is orthogonal to the tangent vector $\mathbf{a}_3$ with components t^i. Define the unit normal vector

$$\mathbf{a}_1 = n^j\,\mathbf{e}_j = \frac{\operatorname{grad} f_1}{|\operatorname{grad} f_1|} \tag{D.21}$$

to the space curve to be in the same direction as the vector $\delta t^j/\delta s$ and write

$$n^j = \frac{1}{\kappa}\frac{\delta t^j}{\delta s}, \tag{D.22}$$

where κ is a scale factor, called the *curvature*, and is selected such that

$$\mathbf{a}_1 \cdot \mathbf{a}_1 = g_{ij} n^i n^j = 1, \tag{D.23}$$

which implies that

$$g_{ij}\frac{\delta t^i}{\delta s}\frac{\delta t^j}{\delta s} = \kappa^2. \tag{D.24}$$

The reciprocal of the curvature, κ^{-1}, is called the *radius of curvature.* The curvature measures the rate of change of the tangent vector to the curve as the arc length varies. By taking the absolute derivative of (D.20), or equivalently of

$$\mathbf{a}_3 \cdot \mathbf{a}_1 = g_{ij} t^i n^j = 0, \tag{D.25}$$

with respect to the arc length (and use Ricci's theorem (2.248)), we have that

$$g_{ij}\frac{\delta t^i}{\delta s} n^j + g_{ij} t^i \frac{\delta n^j}{\delta s} = 0. \tag{D.26}$$

Consequently, the curvature can be determined from

$$-g_{ij} t^i \frac{\delta n^j}{\delta s} = g_{ij}\frac{\delta t^i}{\delta s} n^j = \kappa g_{ij} n^i n^j = \kappa, \tag{D.27}$$

which also defines the sign of the curvature.

In a similar fashion, if we take the absolute derivative of (D.23) with respect to the arc length parameter s (and use Ricci's theorem (2.248)), we find that

$$g_{ij} n^i \frac{\delta n^j}{\delta s} = 0. \tag{D.28}$$

This indicates that the vector $(\delta n^j/\delta s)\,\mathbf{e}_j$ is orthogonal to the unit normal vector $\mathbf{a}_1$. Equation (D.25) indicates that $\mathbf{a}_3$ is also orthogonal to $\mathbf{a}_1$, and hence, any linear combination of these vectors will also be orthogonal to $\mathbf{a}_1$. The unit binormal vector

$$\mathbf{a}_2 = b^j\,\mathbf{e}_j = \frac{\operatorname{grad} f_2}{|\operatorname{grad} f_2|} \tag{D.29}$$

is defined to be the unit vector chosen to be in the direction of the linear combination of $(\delta n^j/\delta s + \kappa t^j)\,\mathbf{e}_j$, i.e.,

$$b^j = \frac{1}{\tau}\left(\frac{\delta n^j}{\delta s} + \kappa t^j\right), \tag{D.30}$$

where τ is a scalar called the *torsion.* The reciprocal of the torsion is called the *radius of torsion.* The sign of the torsion is selected such that the vectors $\{\mathbf{a}_1, \mathbf{a}_2, \mathbf{a}_3\}$ form a right-handed coordinate system:

$$\mathbf{a}_1 \cdot (\mathbf{a}_2 \times \mathbf{a}_3) = 1 \qquad \text{or} \qquad \varepsilon_{ijk} n^i b^j t^k = 1, \tag{D.31}$$

and the magnitude is selected such that b^i is a unit vector satisfying

$$\mathbf{a}_2 \cdot \mathbf{a}_2 = g_{ij} b^i b^j = 1. \tag{D.32}$$

By using (D.30) it is easily shown that $\mathbf{a}_2$ is orthogonal to both $\mathbf{a}_1$ and $\mathbf{a}_3$ since

$$\mathbf{a}_2 \cdot \mathbf{a}_1 = g_{ij} b^i n^j = 0 \qquad \text{and} \qquad \mathbf{a}_2 \cdot \mathbf{a}_3 = g_{ij} b^i t^j = 0. \tag{D.33}$$

The vectors $\{\mathbf{a}_1, \mathbf{a}_2, \mathbf{a}_3\}$ form a right-handed orthogonal system at a point on the space curve and satisfy the relation

$$\mathbf{a}_2 = \mathbf{a}_3 \times \mathbf{a}_1, \qquad \text{or} \qquad b^i = \varepsilon^{ijk} t_j n_k. \tag{D.34}$$

Additionally, it is easy to show that the binormal vector satisfies the relation $\delta \mathbf{a}_2 / \delta s = -\tau \mathbf{a}_1$.

The triad of vectors $\{\mathbf{a}_1, \mathbf{a}_2, \mathbf{a}_3\}$ form three planes at a point on the curve $\mathscr{C}$. The plane containing $\mathbf{a}_2$ and $\mathbf{a}_3$ is called the *rectifying plane.* The plane containing $\mathbf{a}_1$ and $\mathbf{a}_2$ is called the *normal plane.* The plane containing $\mathbf{a}_1$ and $\mathbf{a}_3$ is called the *osculating plane.* The torsion measures the rate of change of the osculating plane. The three relations

$$\frac{\delta \mathbf{a}_1}{\delta s} = \tau \mathbf{a}_2 - \kappa \mathbf{a}_3, \qquad \frac{\delta \mathbf{a}_2}{\delta s} = -\tau \mathbf{a}_1, \qquad \frac{\delta \mathbf{a}_3}{\delta s} = \kappa \mathbf{a}_1, \tag{D.35}$$

or

$$\frac{\delta n^i}{\delta s} = \tau b^i - \kappa t^i, \qquad \frac{\delta b^i}{\delta s} = -\tau n^i, \qquad \frac{\delta t^i}{\delta s} = \kappa n^i, \tag{D.36}$$

are known as the *Frenet–Serret formulas.*

In general, instead of considering the arc distance as a coordinate, we can take $\mathbf{r} = x^i(\xi)\,\mathbf{e}_i$ to represent the three-dimensional space curve as a function of the convected coordinate ξ along the curve $\mathscr{C}$. In such case, we have the corresponding vectors $\{\mathbf{d}_1, \mathbf{d}_2, \mathbf{d}_3\}$ where

$$\mathbf{d}_1 = \mathbf{a}_1, \qquad \mathbf{d}_2 = \mathbf{a}_2, \quad \text{and} \quad \frac{\mathbf{d}_3}{(d_{33})^{1/2}} = \mathbf{a}_3, \tag{D.37}$$

where

$$d_{33} = \mathbf{d}_3 \cdot \mathbf{d}_3 \tag{D.38}$$

is not unity in this case, and

$$\frac{\delta}{\delta s} = \frac{\delta \xi}{\delta s} \frac{\delta}{\delta \xi} = \frac{1}{(d_{33})^{1/2}} \frac{\delta}{\delta \xi}. \tag{D.39}$$

Subsequently, the Frenet–Serret formulas become

$$\frac{\delta \mathbf{d}_1}{\delta \xi} = \tau (d_{33})^{1/2} \mathbf{d}_2 - \kappa\, \mathbf{d}_3, \qquad \frac{\delta \mathbf{d}_2}{\delta \xi} = -\tau\, (d_{33})^{1/2} \mathbf{d}_1,$$
$$\frac{\delta \mathbf{d}_3}{\delta \xi} = \kappa\, d_{33}\, \mathbf{d}_1 + \frac{1}{2\, d_{33}} \frac{d(d_{33})}{d\xi} \mathbf{d}_3, \tag{D.40}$$

or

$$\frac{\delta n^i}{\delta \xi} = \tau (d_{33})^{1/2} b^i - \kappa t^i, \qquad \frac{\delta b^i}{\delta \xi} = -\tau\, (d_{33})^{1/2} n^i,$$
$$\frac{\delta t^i}{\delta \xi} = \kappa\, (d_{33})^{1/2} n^i + \frac{1}{2\, d_{33}} \frac{d(d_{33})}{d\xi} t^i. \tag{D.41}$$

Now take

$$\mathbf{r} = r(\xi, t)\, \mathbf{d}_3 + \theta^\alpha(\xi, t)\, \mathbf{d}_\alpha, \tag{D.42}$$

where r denotes the component of $\mathbf{r}$ in the $\mathbf{d}_3$ direction, and $\theta^\alpha(\xi, t)$ denotes the component of $\mathbf{r}$ in the $\mathbf{d}_\alpha$ direction. We use the convention that Greek subscripts or superscripts only run from 1 to 2, and take $\mathbf{d}_\alpha$ to have units of length so that θ^α is dimensionless. Now

$$\mathbf{v} = \dot{\mathbf{r}} = \mathbf{u} + \mathbf{w} = u(\xi, t)\, \mathbf{d}_3 + w^\alpha(\xi, t)\, \mathbf{d}_\alpha, \tag{D.43}$$

where

$$u(\xi, t) = \dot{r}(\xi, t) \qquad \text{and} \qquad w^\alpha(\xi, t) = \dot{\theta}^\alpha(\xi, t), \tag{D.44}$$

and the superposed dot denotes the material derivative.

D.2 Balance law for a space curve

Consider an arbitrary material curve $\mathscr{C}(t) : \xi_1(t) \leq \xi \leq \xi_2(t)$, which is separated into two parts $\mathscr{C}^+(t)$ and $\mathscr{C}^-(t)$, or $\mathscr{C}(t) - \gamma(t)$, by a discontinuity located at $\xi_1(t) < \gamma(t) < \xi_2(t)$. The singularity at $\gamma(t)$ moves with velocity $\dot{\boldsymbol{\gamma}} = \dot{\gamma}\, \mathbf{d}_3$. The general balance law for the material curve is given by

$$\frac{d}{dt} \int_{\mathscr{C}(t)} \psi(\xi, t)\, d\xi = [\phi(\xi, t)]_{\xi_1(t)}^{\xi_2(t)} + \int_{\mathscr{C}(t)} g(\xi, t)\, d\xi, \tag{D.45}$$

or more specifically,

$$\frac{d}{dt} \int_{\mathscr{C}(t)-\gamma(t)} \psi_l(\xi, t)\, d\xi + \dot{\psi}_p(\gamma, t) = [\phi(\xi, t)]_{\xi_1(t)}^{\xi_2(t)} + \int_{\mathscr{C}(t)-\gamma(t)} g_l(\xi, t)\, d\xi + g_p(\gamma, t), \tag{D.46}$$

where $\psi(\xi, t)$ is an additive tensor quantity per unit length, $\phi(\xi, t)$ is the flux of ψ through the curve's endpoints, and $g(\xi, t)$ denotes the combined external supply and internal production of ψ. Their specific contributions should be distinguished when the general balance law is applied to different physical balances since the external supply and internal production represent different physical contributions. The tensors are decomposed appropriately between the regions $\mathscr{C}(t) - \gamma(t)$, denoted by the subscript l, and $\gamma(t)$, denoted by the subscript p.

Now using the generalized Leibnitz rule (3.450) and the generalized divergence theorem on a curve (2.301), we can rewrite the above in the following form:

$$\int_{\mathscr{C}(t)-\gamma(t)} \left[\frac{\partial \psi_l}{\partial t} + \frac{\partial (\psi_l v)}{\partial \xi} - \frac{\partial \phi_l}{\partial \xi} - g_l \right] d\xi +$$
$$\left\{ \dot{\psi}_p(\gamma, t) + [\![\psi_l(\gamma, t) \left[v(\gamma, t) - \dot{\gamma}(t) \right] - \phi_l(\gamma, t)]\!] - g_p(\gamma, t) \right\} = 0. \tag{D.47}$$

Evaluating the balance law over an arbitrary point in the continuous region $\mathscr{C}(t) - \gamma(t)$, we deduce the local form of the balance law for the tensor quantity ψ_l over the region:

$$\frac{\partial \psi_l}{\partial t} + \frac{\partial(\psi_l v)}{\partial \xi} - \frac{\partial \phi_l}{\partial \xi} - g_l = 0. \tag{D.48}$$

Alternatively, evaluating the balance law over the singular region $\gamma(t)$, we have the evolution equation for the quantity ψ_p defined at the singular point:

$$\dot{\psi}_p(\gamma,t) + [\![\psi_l(\gamma,t)\,[v(\gamma,t) - \dot{\gamma}(t)] - \phi_l(\gamma,t)]\!] - g_p(\gamma,t) = 0. \tag{D.49}$$

D.3 Space surface

A surface embedded in $\mathcal{E}^3$ can be represented by the scalar equation

$$f(x^k,t) = 0. \tag{D.50}$$

From now until near the end of this section, we suppress the dependence of the surface on the parameter t (which indicates time), since when considering the geometry of a surface, the value of t is fixed. Then the unit vector normal to the surface is given by

$$\mathbf{n} = \frac{\operatorname{grad} f}{|\operatorname{grad} f|}, \qquad \mathbf{n}\cdot\mathbf{n} = 1, \tag{D.51}$$

with components

$$n^i = \mathbf{n}\cdot\mathbf{e}^i = \frac{g^{ij} f_{,j}}{|\operatorname{grad} f|}, \qquad n_i = \mathbf{n}\cdot\mathbf{e}_i = \frac{f_{,i}}{|\operatorname{grad} f|}. \tag{D.52}$$

Alternatively, if we define a *surface coordinate system*, with coordinates (y^1,y^2), say, then a point on the surface can be represented as a function of these surface coordinates:

$$\mathbf{r} = x^i(y^1,y^2)\,\mathbf{e}_i = x_i(y^1,y^2)\,\mathbf{e}^i. \tag{D.53}$$

Note that for (y^1,y^2) to uniquely determine a point on the surface, y^1 should be a coordinate curve on the surface along which y^1 varies while y^2 is fixed, and y^2 a coordinate curve on the surface along which y^2 varies while y^1 is fixed. For any surface, there are an infinite number of coordinate systems that might be used. Any two families of lines may be chosen as coordinate curves, as long as each member of one family intersects each member of the other at one and only one point.

We now introduce a modification to our index notation. An italic index will continue to indicate a value from 1 to 3, while a Greek index will range from 1 to 2 only. Then, the equations of a space surface can be written in the parametric form $x^i = x^i(y^\alpha)$, were y^α is called a curvilinear coordinate of the surface. Now the basis $\{\mathbf{a}_1,\mathbf{a}_2\}$, known as the *surface natural basis*, corresponds to vectors that are tangent to the coordinate system (y^1,y^2) lying on the surface, and is related to the *spatial natural basis* $\{\mathbf{e}_1,\mathbf{e}_2,\mathbf{e}_2\}$ in $\mathcal{E}^3$ with spatial coordinates (x^1,x^2,x^3) by the transformation

$$\mathbf{a}_\alpha = \frac{\partial \mathbf{r}}{\partial y^\alpha} = \frac{\partial x^i}{\partial y^\alpha}\mathbf{e}_i = a^i_\alpha \mathbf{e}_i, \tag{D.54}$$

where it is noted that $[a^i_\alpha]$ is a 3×2 transformation matrix having a rank of 2. The components of the tangent vector to the coordinate curves defining the surface are given by

$$a^i_\alpha = \mathbf{e}^i \cdot \mathbf{a}_\alpha, \tag{D.55}$$

and can be viewed as either the components of a covariant surface vector or the components of a contravariant spatial vector. Alternately, we could have chosen vectors on the surface that are normal to the coordinate lines as a basis: $\{\mathbf{a}^1, \mathbf{a}^2\}$. This surface reciprocal, or dual, basis, when normalized to unit length has the property that

$$\mathbf{a}^\alpha \cdot \mathbf{a}_\beta = \delta^\alpha_\beta, \tag{D.56}$$

where δ^α_β is the Kronecker delta symbol, corresponding to the components of the two-dimensional rank-2 isotropic unit tensor $\widetilde{\mathbf{1}}$. In this section, we will represent tangential tensor fields by a superposed tilde symbol when written in bold notation. The obvious exceptions are surface coordinate bases and the associated surface metric tensor. All other tensor fields are understood to represent spatial fields.

Now since $\mathbf{n}$ is a unit vector normal to the surface, we have $\mathbf{a}^\alpha \cdot \mathbf{n} = \mathbf{a}_\beta \cdot \mathbf{n} = 0$. Because the vector fields $\{\mathbf{a}_1, \mathbf{a}_2, \mathbf{n}\}$ or $\{\mathbf{a}^1, \mathbf{a}^2, \mathbf{n}\}$ are linearly independent, they form a basis for spatial vector fields on the surface. Note that the normal vector and its negative are both orthogonal to $\mathbf{a}_1$ and $\mathbf{a}_2$. We choose $\mathbf{n}$ so that $\{\mathbf{a}_1, \mathbf{a}_2, \mathbf{n}\}$ forms a right-handed system. Subsequently, we can decompose $\mathbf{e}^i$ with respect to the bases $\{\mathbf{a}^1, \mathbf{a}^2, \mathbf{n}\}$:

$$\mathbf{e}^i = a^i_\alpha \mathbf{a}^\alpha + n^i \mathbf{n}. \tag{D.57}$$

We now define the *surface metric tensor* as

$$\mathbf{h} = h_{\alpha\beta}\mathbf{a}^\alpha \mathbf{a}^\beta = h^{\alpha\beta}\mathbf{a}_\alpha \mathbf{a}_\beta, \tag{D.58}$$

where the components are given by

$$h_{\alpha\beta} = \mathbf{a}_\alpha \cdot \mathbf{a}_\beta \qquad \text{and} \qquad h^{\alpha\beta} = \mathbf{a}^\alpha \cdot \mathbf{a}^\beta. \tag{D.59}$$

Furthermore, we note that the matrix of the components is symmetric: $[h_{\alpha\beta}] = [h_{\beta\alpha}]$. We also define the determinant of the surface metric components matrix:

$$h = \det[h_{\alpha\beta}]. \tag{D.60}$$

It is easy to show that the determinant can be obtained, respectively, by expanding by columns or rows:

$$\epsilon^{\alpha\beta} h_{\alpha\gamma} h_{\beta\delta} = h\,\epsilon_{\gamma\delta} \qquad \text{or} \qquad \epsilon_{\alpha\beta} h^{\gamma\alpha} h^{\delta\beta} = h\,\epsilon^{\gamma\delta}, \tag{D.61}$$

where the two-dimensional Levi–Civita, or permutation, symbol is defined as

$$\epsilon_{\alpha\beta} = \epsilon^{\alpha\beta} = \begin{cases} 1 & \text{if } \alpha\beta = 12, \\ -1 & \text{if } \alpha\beta = 21, \\ 0 & \text{if } \alpha = \beta, \end{cases} \tag{D.62}$$

or in matrix form

$$[\epsilon_{\alpha\beta}] = [\epsilon^{\alpha\beta}] = \begin{bmatrix} 0 & 1 \\ -1 & 0 \end{bmatrix}. \tag{D.63}$$

The absolute tangential Levi–Civita tensor is of rank 2 and is given by

$$\widetilde{\boldsymbol{\varepsilon}} = \varepsilon^{\alpha\beta}\mathbf{a}_\alpha\mathbf{a}_\beta = \varepsilon_{\alpha\beta}\mathbf{a}^\alpha\mathbf{a}^\beta, \tag{D.64}$$

where the contravariant and covariant components are related to the permutation symbols as follows:

$$\varepsilon^{\alpha\beta} \equiv \frac{1}{\sqrt{h}}\epsilon^{\alpha\beta} \qquad \text{and} \qquad \varepsilon_{\alpha\beta} \equiv \sqrt{h}\,\epsilon_{\alpha\beta}. \tag{D.65}$$

Now, since

$$dx^i = a^i_\alpha dy^\alpha, \tag{D.66}$$

then a small change in dy^α on the surface coordinates results in change dx^i in the space coordinates. Hence, an element of arc length on the surface can be represented in terms of the curvilinear coordinates of the space or curvilinear coordinates of the surface:

$$ds^2 = g_{ij}\,dx^i dx^j = g_{ij}\,a^i_\alpha a^j_\beta dy^\alpha dy^\beta = h_{\alpha\beta}\,dy^\alpha dy^\beta, \tag{D.67}$$

from which we can relate the spatial and surface metrics:

$$g_{ij}\,a^i_\alpha a^j_\beta = h_{\alpha\beta}. \tag{D.68}$$

The quadratic scalar $\widetilde{A} = h_{\alpha\beta}\,dy^\alpha dy^\beta$ written in surface coordinates is also called the *first fundamental form of the surface*, while the metric tensor $\mathbf{h}$ is also called the *first fundamental tensor of the surface.* The first fundamental form is connected with distance on the surface.

We are particularly interested in describing vector fields defined on a surface. A tangential vector field is a two-dimensional subspace of the spatial vector field on the surface in the sense that every tangential vector field $\widetilde{\mathbf{v}}$ can be expressed as a linear combination of $\mathbf{a}_1$ and $\mathbf{a}_2$ or their duals:

$$\widetilde{\mathbf{v}} = v^\alpha\mathbf{a}_\alpha = v_\beta\mathbf{a}^\beta. \tag{D.69}$$

Now dotting the above equation from the right with $\mathbf{a}^\gamma$, we see that

$$v^\gamma = h^{\alpha\gamma}v_\alpha. \tag{D.70}$$

On the other hand, dotting both sides from the right by $\mathbf{a}_\gamma$, we have that

$$v_\gamma = h_{\alpha\gamma}v^\alpha. \tag{D.71}$$

The above results can be easily shown to remain valid when operating on the surface bases vectors:

$$\mathbf{a}_\gamma = h_{\alpha\gamma}\mathbf{a}^\alpha \qquad \text{and} \qquad \mathbf{a}^\gamma = h^{\alpha\gamma}\mathbf{a}_\alpha. \tag{D.72}$$

Subsequently, it is easy to verify that

$$h_{\alpha\beta}h^{\beta\gamma} = \delta^\gamma_\alpha, \tag{D.73}$$

$$[h_{\gamma\beta}]^{-1} = [h^{\beta\gamma}], \tag{D.74}$$

and

$$\mathbf{h} \cdot \mathbf{h}^{-1} = \mathbf{h}^{-1} \cdot \mathbf{h} = \widetilde{\mathbf{1}}. \tag{D.75}$$

If we define the *surface gradient* by

$$\widetilde{\nabla} \equiv \mathbf{a}^\alpha \frac{\partial}{\partial y^\alpha} = h^{\alpha\beta} \mathbf{a}_\beta \frac{\partial}{\partial y^\alpha}, \tag{D.76}$$

then it is clear that the reciprocal basis is given by the surface gradient of the surface coordinates:

$$\mathbf{a}^\alpha = \widetilde{\nabla} y^\alpha = h^{\beta\gamma} \mathbf{a}_\gamma \frac{\partial y^\alpha}{\partial y^\beta} = h^{\alpha\gamma} \mathbf{a}_\gamma. \tag{D.77}$$

We will now require that physical quantities described by scalars, vectors, and higher order tensors remain invariant when a different surface curvilinear coordinate system is used, i.e., when a new basis $\{\overline{\mathbf{a}}_\alpha\}$ is used instead of $\{\mathbf{a}_\alpha\}$. Note that corresponding to the new basis, we also have its reciprocal basis $\{\overline{\mathbf{a}}^\beta\}$, where as before we require that $\overline{\mathbf{a}}_\alpha \cdot \overline{\mathbf{a}}^\beta = \delta^\beta_\alpha$. Now since a surface vector can be written in the alternate forms

$$\widetilde{\mathbf{v}} = v^\alpha \mathbf{a}_\alpha = v_\beta \mathbf{a}^\beta = \overline{v}^\gamma \overline{\mathbf{a}}_\gamma = \overline{v}_\delta \overline{\mathbf{a}}^\delta, \tag{D.78}$$

it is easy to show that the covariant and contravariant surface bases transform as

$$\overline{\mathbf{a}}_\gamma = \frac{\partial y^\alpha}{\partial \overline{y}^\gamma} \mathbf{a}_\alpha \qquad \text{and} \qquad \overline{\mathbf{a}}^\delta \frac{\partial y^\beta}{\partial \overline{y}^\delta} = \mathbf{a}^\beta \tag{D.79}$$

as long as the transformation is non-singular, i.e.,

$$\det\left[\frac{\partial y^\alpha}{\partial \overline{y}^\gamma}\right] \neq \{0, \pm\infty\}. \tag{D.80}$$

It should be recognized that the natural basis is orthogonal if $\mathbf{a}_\alpha \cdot \mathbf{a}_\beta = 0$ when $\alpha \neq \beta$. In such case, it is usually more convenient to work in terms of an orthonormal basis:

$$\mathbf{a}_{<\alpha>} = \frac{\mathbf{a}_\alpha}{\sqrt{h_{\underline{\alpha\alpha}}}} = \frac{\mathbf{a}^\alpha}{\sqrt{h^{\underline{\alpha\alpha}}}} = \sqrt{h_{\underline{\alpha\alpha}}}\,\mathbf{a}^\alpha, \tag{D.81}$$

where we recall that underlined subscripts are not summed. This basis is referred to as the physical basis for the surface coordinate system. Subsequently, any vector field can be expressed in the orthogonal coordinate system in terms of the physical surface components

$$\widetilde{\mathbf{v}} = v_{<\alpha>} \mathbf{a}_{<\alpha>}, \tag{D.82}$$

where

$$v_{<\alpha>} = \sqrt{h_{\underline{\alpha\alpha}}}\, v^\alpha = \frac{v_\alpha}{\sqrt{h_{\underline{\alpha\alpha}}}}. \tag{D.83}$$

A *surface tensor* $\mathbf{S}$ is a particular type of second-order spatial tensor field that is defined only on the surface, and assigns to each given tangential vector field $\widetilde{\mathbf{v}}$ on a surface another spatial vector field $\mathbf{S}\cdot\widetilde{\mathbf{v}}$ defined on the surface, and transforms every spatial vector field normal to the surface into the zero vector. A surface tensor can be defined by the way it transforms the surface natural basis field $\{\mathbf{a}_1, \mathbf{a}_2\}$:

$$\mathbf{S}\cdot\mathbf{a}_\alpha = S^i_{\ \alpha}\mathbf{e}_i. \tag{D.84}$$

It is easy to show that

$$\mathbf{S} = S^i_{\ \alpha}\mathbf{e}_i\mathbf{a}^\alpha = S^{i\alpha}\mathbf{e}_i\mathbf{a}_\alpha = S_{i\alpha}\mathbf{e}^i\mathbf{a}^\alpha, \tag{D.85}$$

where

$$S^{i\alpha} = h^{\alpha\beta}S^i_{\ \beta} \qquad \text{and} \qquad S_{i\alpha} = g_{ij}S^j_{\ \alpha}. \tag{D.86}$$

Now if we take the surface tensor $\mathbf{S} = a^i_{\ \alpha}\mathbf{e}_i\mathbf{a}^\alpha$, the surface vector $\widetilde{\mathbf{v}}$ can also be viewed as a spatial vector $\mathbf{v}$:

$$\mathbf{S}\cdot\widetilde{\mathbf{v}} = \left(a^i_{\ \alpha}\mathbf{e}_i\mathbf{a}^\alpha\right)\cdot\left(v^\beta\mathbf{a}_\beta\right) = a^i_{\ \alpha}v^\alpha\mathbf{e}_i = v^i\mathbf{e}_i = \mathbf{v}.$$

Thus, the relation between the spatial and surface representations is

$$v^i = a^i_{\ \alpha}v^\alpha. \tag{D.87}$$

The surface and spatial representations define the same magnitude and direction since

$$\mathbf{v}\cdot\mathbf{v} = g_{ij}\,v^iv^j = g_{ij}\,a^i_{\ \alpha}v^\alpha a^j_{\ \beta}v^\beta = g_{ij}\,a^i_{\ \alpha}a^j_{\ \beta}v^\alpha v^\beta = h_{\alpha\beta}\,v^\alpha v^\beta = \widetilde{\mathbf{v}}\cdot\widetilde{\mathbf{v}}. \tag{D.88}$$

A *tangential tensor* $\widetilde{\mathbf{T}}$ is a second-order tangential transformation that projects every tangential vector field $\widetilde{\mathbf{v}}$ on a surface to another tangential vector field $\widetilde{\mathbf{T}}\cdot\widetilde{\mathbf{v}}$ on the surface and every spatial vector field normal to the surface into the zero vector. The tangential tensor has the standard form

$$\widetilde{\mathbf{T}} = T^{\alpha\beta}\mathbf{a}_\alpha\mathbf{a}_\beta = T_{\alpha\beta}\mathbf{a}^\alpha\mathbf{a}^\beta. \tag{D.89}$$

As in the spatial case, it is easy to show that the covariant and contravariant components of this field transform as

$$\overline{T}_{\mu\nu} = \frac{\partial y^\alpha}{\partial \overline{y}^\mu}\frac{\partial y^\beta}{\partial \overline{y}^\nu}T_{\alpha\beta} \qquad \text{and} \qquad \frac{\partial y^\alpha}{\partial \overline{y}^\mu}\frac{\partial y^\beta}{\partial \overline{y}^\nu}\overline{T}^{\mu\nu} = T^{\alpha\beta}. \tag{D.90}$$

It is usually convenient to introduce an orthogonal surface coordinate system. If $\{\mathbf{a}_{<\alpha>}\}$ is the associated physical basis field, then in terms of physical surface components, we can write

$$\widetilde{\mathbf{T}} = T_{<\gamma\mu>}\mathbf{a}_{<\gamma>}\mathbf{a}_{<\mu>}, \tag{D.91}$$

where

$$T_{<\gamma\mu>} = \sqrt{h_{\underline{\gamma\gamma}}}\sqrt{h_{\underline{\mu\mu}}}T^{\gamma\mu} = \frac{T_{\gamma\mu}}{\sqrt{h_{\underline{\gamma\gamma}}}\sqrt{h_{\underline{\mu\mu}}}}. \tag{D.92}$$

Now if $\widetilde{\mathbf{v}}$ is any tangential vector field

$$\widetilde{\mathbf{v}} = v^{\alpha}\mathbf{a}_{\alpha}, \tag{D.93}$$

then the tangential vector field that has the same length as $\widetilde{\mathbf{v}}$ and is orthogonal to it is given by

$$\widetilde{\mathbf{u}} = -\widetilde{\varepsilon}\cdot\widetilde{\mathbf{v}} \qquad \text{or} \qquad u^{\alpha} = -\varepsilon^{\alpha\beta}v_{\beta}. \tag{D.94}$$

This is easily seen, since

$$\begin{aligned}\widetilde{\mathbf{u}}\cdot\widetilde{\mathbf{u}} &= \left(-\varepsilon^{\alpha\beta}v_{\beta}\mathbf{a}_{\alpha}\right)\cdot\left(-\varepsilon_{\gamma\varsigma}v^{\varsigma}\mathbf{a}^{\gamma}\right)\\ &= \varepsilon^{\alpha\beta}\varepsilon_{\alpha\varsigma}v_{\beta}v^{\varsigma}\\ &= \delta^{\beta}_{\varsigma}v_{\beta}v^{\varsigma}\\ &= \widetilde{\mathbf{v}}\cdot\widetilde{\mathbf{v}},\end{aligned}$$

and

$$\begin{aligned}\widetilde{\mathbf{u}}\cdot\widetilde{\mathbf{v}} &= \left(-\varepsilon^{\alpha\beta}v_{\beta}\mathbf{a}_{\alpha}\right)\cdot\left(v_{\gamma}\mathbf{a}^{\gamma}\right)\\ &= -\varepsilon^{\alpha\beta}v_{\alpha}v_{\beta}\\ &= -\frac{1}{2}\left(\varepsilon^{\alpha\beta}+\varepsilon^{\beta\alpha}\right)v_{\alpha}v_{\beta}\\ &= 0.\end{aligned}$$

Because of this last property, $\widetilde{\varepsilon}$ is referred to as the tangential cross tensor. Notice that

$$(-\widetilde{\varepsilon}\cdot\mathbf{a}_1)\cdot\mathbf{a}_2 = \varepsilon^{\alpha\beta}h_{\alpha1}h_{\beta2} = \sqrt{h}. \tag{D.95}$$

It is also easy to show that

$$-\widetilde{\varepsilon}\cdot\mathbf{a}_1 = \sqrt{h}\,\mathbf{a}_2 \qquad \text{and} \qquad \widetilde{\varepsilon}\cdot\mathbf{a}_2 = \sqrt{h}\,\mathbf{a}_1, \tag{D.96}$$

which imply that

$$\left|\mathbf{a}^2\right| = \frac{1}{\sqrt{h}}\left|\mathbf{a}_1\right|, \tag{D.97}$$

that $\mathbf{a}^2$ is orthogonal to $\mathbf{a}_1$, and that the rotation from $\mathbf{a}_1$ to $\mathbf{a}^2$ is positive. Similarly, we have the implications that

$$\left|\mathbf{a}^1\right| = \frac{1}{\sqrt{h}}\left|\mathbf{a}_2\right|, \tag{D.98}$$

that $\mathbf{a}^1$ is orthogonal to $\mathbf{a}_2$, and that the rotation from $\mathbf{a}_2$ to $\mathbf{a}^1$ is negative.

Consider any two surface vectors with components u^{α} and v^{β} and their spatial representations u^i and v^j, where

$$u^i = a^i_{\alpha}u^{\alpha} \qquad \text{and} \qquad v^j = a^j_{\beta}v^{\beta}. \tag{D.99}$$

These vectors are tangent to the surface and so a unit normal vector to the surface can be defined. Actually, the are two normals: they are the negative of each other.

The normal $\mathbf{n}$ is chosen to correspond to the surface normal, which, together with $\mathbf{a}_1$ and $\mathbf{a}_2$, forms a right-handed coordinate system: $(\mathbf{a_1} \times \mathbf{a_2}) \cdot \mathbf{n} > 0$. More explicitly, from the cross product relations

$$\varepsilon_{\alpha\beta} u^\alpha v^\beta n_i = \varepsilon_{ijk} u^j v^k, \tag{D.100}$$

which, for arbitrary surface vectors, implies

$$n_i = \frac{1}{2} \varepsilon^{\alpha\beta} \varepsilon_{ijk} a^j_\alpha a^k_\beta, \tag{D.101}$$

we have the definition of a *surface unit normal vector* in terms of the tangent vectors to the coordinate curves. It is readily seen that

$$g_{ij} n^i n^j = 1. \tag{D.102}$$

The *surface projection tensor* $\widetilde{\mathbf{P}}$ is a second-order tangential tensor field that transforms every tangential vector field into itself:

$$\widetilde{\mathbf{P}} \cdot \mathbf{a}_\beta = \mathbf{a}_\beta = \delta^\alpha_\beta \mathbf{a}_\alpha. \tag{D.103}$$

It is easy to see that

$$\widetilde{\mathbf{P}} = \mathbf{a}_\alpha \mathbf{a}^\alpha = \mathbf{a}^\alpha \mathbf{a}_\alpha = h_{\alpha\beta} \mathbf{a}^\alpha \mathbf{a}^\beta = h^{\alpha\beta} \mathbf{a}_\alpha \mathbf{a}_\beta. \tag{D.104}$$

Note that the covariant and contravariant components of the surface metric tensor $\mathbf{h}$ can also be viewed as covariant and contravariant components of the projection tensor $\widetilde{\mathbf{P}}$.

Spatial vector fields defined on a surface play an important role in continuum mechanics. Thus it is often convenient to think in terms of their tangential and normal components. Subsequently, since the vectors $\mathbf{a}_1$, $\mathbf{a}_2$ and $\mathbf{n}$ are linearly independent, they form a basis for a spatial vector field on the surface:

$$\mathbf{v} = \widetilde{\mathbf{v}} + \mathbf{v}_{(n)} = v^\alpha \mathbf{a}_\alpha + v_{(n)} \mathbf{n} = v_\alpha \mathbf{a}^\alpha + v_{(n)} \mathbf{n}, \tag{D.105}$$

where $v_{(n)}$ is the normal component of the spatial vector field $\mathbf{v}$. Now, this fact can also be represented in the form

$$\mathbf{v} = (\mathbf{v} \cdot \mathbf{a}^\alpha)\, \mathbf{a}_\alpha + (\mathbf{v} \cdot \mathbf{n})\, \mathbf{n}, \tag{D.106}$$

or

$$\mathbf{1} \cdot \mathbf{v} = \left(\widetilde{\mathbf{P}} + \mathbf{P}_{(n)}\right) \cdot \mathbf{v}, \tag{D.107}$$

where $\mathbf{1}$ is the spatial unit second-order tensor and

$$\mathbf{P}_{(n)} = \mathbf{n}\mathbf{n} \tag{D.108}$$

is the normal projection tensor, i.e., it projects a spatial vector field on a surface to one in the direction normal to the surface. Since $\mathbf{v}$ is an arbitrary spatial vector field defined on the surface, this implies that

$$\mathbf{1} = \widetilde{\mathbf{P}} + \mathbf{P}_{(n)}, \tag{D.109}$$

and thus

$$\mathbf{v} = \widetilde{\mathbf{v}} + \mathbf{v}_{(n)}, \qquad \text{where} \qquad \widetilde{\mathbf{v}} = \widetilde{\mathbf{P}} \cdot \mathbf{v} \qquad \text{and} \qquad \mathbf{v}_{(n)} = \mathbf{P}_{(n)} \cdot \mathbf{v}. \tag{D.110}$$

Subsequently, we can also write the surface projection tensor in the alternate form

$$\widetilde{\mathbf{P}} = \mathbf{1} - \mathbf{P}_{(n)}. \tag{D.111}$$

The contravariant component of this relationship is given by

$$h^{\alpha\beta} a^i_\alpha a^j_\beta = g^{ij} - n^i n^j. \tag{D.112}$$

It is important to note that the projection tensor plays the role of the identity or unit tensor for the set of all tangential vector fields. A tensor $\mathbf{P}$ is a projection tensor if it is symmetric and $\mathbf{P}^m = \mathbf{P}$ for m a positive integer. From above, it can be readily verified that our projection tensors have the properties

$$\widetilde{\mathbf{P}} + \mathbf{P}_{(n)} = \mathbf{1}, \tag{D.113}$$

$$\widetilde{\mathbf{P}} \cdot \widetilde{\mathbf{P}} = \widetilde{\mathbf{P}}, \tag{D.114}$$

$$\mathbf{P}_{(n)} \cdot \mathbf{P}_{(n)} = \mathbf{P}_{(n)}, \tag{D.115}$$

$$\widetilde{\mathbf{P}} \cdot \mathbf{P}_{(n)} = \mathbf{0}. \tag{D.116}$$

The normal unit vector is related to the covariant derivative of the surface tangents as will be shown next. If we take the covariant derivative of (D.68) with respect to the surface coordinate y^γ and use Ricci's theorem (2.248), we have

$$g_{ij}\, a^i_{\alpha,\gamma} a^j_\beta + g_{ij}\, a^i_\alpha a^j_{\beta,\gamma} = h_{\alpha\beta,\gamma} = 0. \tag{D.117}$$

Interchanging the indices α and β, it is easy to see that

$$g_{ij} a^i_{\alpha,\beta} a^j_\gamma = 0. \tag{D.118}$$

From the definition of the covariant derivative, we note that

$$a^i_{\alpha,\beta} = \frac{\partial a^i_\alpha}{\partial y^\beta} + \Gamma^i_{jk} a^j_\alpha a^k_\beta - \Gamma^\gamma_{\alpha\beta} a^i_\gamma, \tag{D.119}$$

and that we have two Christoffel symbols, one related to the spatial metric and the other related to the surface metric. Analogous to (2.218), the surface Christoffel symbol can be written as

$$\Gamma^\delta_{\alpha\beta} = \mathbf{a}^\delta \cdot \frac{\partial \mathbf{a}_\alpha}{\partial y^\beta} = \frac{1}{2} h^{\delta\gamma} \left(\frac{\partial h_{\alpha\gamma}}{\partial y^\beta} + \frac{\partial h_{\beta\gamma}}{\partial x^\alpha} - \frac{\partial h_{\alpha\beta}}{\partial x^\gamma} \right). \tag{D.120}$$

The result (D.118) indicates that in terms of space coordinates, the vector $a^i_{\alpha,\beta}$ is orthogonal to the surface tangent vector a^j_γ and so must have the same direction as the unit surface normal n^i. Therefore, there must exist a tensor

$$\widetilde{\mathbf{B}} = B_{\alpha\beta} \mathbf{a}^\alpha \mathbf{a}^\beta = B^{\alpha\beta} \mathbf{a}_\alpha \mathbf{a}_\beta = B^\alpha_\beta \mathbf{a}_\alpha \mathbf{a}^\beta \tag{D.121}$$

whose components $B_{\alpha\beta}$ are such that

$$\mathbf{a}_{\alpha,\beta} = B_{\alpha\beta}\, \mathbf{n} \qquad \text{or} \qquad a^i_{\alpha,\beta} = B_{\alpha\beta}\, n^i. \tag{D.122}$$

This second-order symmetric tensor is called the *curvature tensor*, or the *second fundamental form of the surface.* It is connected with the rate of change of the tangent vectors. By using (D.101) and (D.102), we can rewrite the above equation in the form

$$B_{\alpha\beta} = g_{ij} a^i_{\alpha,\beta} n^j = \frac{1}{2} \varepsilon^{\gamma\delta} \varepsilon_{ijk} a^i_{\alpha,\beta} a^j_\gamma a^k_\delta. \tag{D.123}$$

Now take the covariant derivative of (D.102) with respect to the surface coordinates and use Ricci's theorem (2.248). From it, it readily follows that

$$g_{ij} n^i n^j_{,\alpha} = 0, \tag{D.124}$$

where

$$n^i_{,\alpha} = \frac{\partial n^i}{\partial y^\alpha} + \Gamma^i_{jk} n^j a^k_\alpha. \tag{D.125}$$

The above result shows that the vector $n^i_{,\alpha}$ is orthogonal to n^i and must lie in the tangent plane to the surface. It can therefore be expressed as a linear combination of the surface tangent vector components a^i_α and written in the form

$$n^i_{,\alpha} = \eta^\beta_\alpha a^i_\beta, \tag{D.126}$$

where the coefficients η^β_α can themselves be written in terms of the surface metric components $h_{\alpha\beta}$ and the curvature components $B_{\alpha\beta}$. To see this, first recall that the unit vector n^i is normal to the surface so that

$$g_{ij} n^i a^j_\alpha = 0. \tag{D.127}$$

The covariant derivative of this equation with respect to the surface coordinates, upon using Ricci's theorem (2.248), gives

$$g_{ij} n^i_{,\beta} a^j_\alpha + g_{ij} n^i a^j_{\alpha,\beta} = 0. \tag{D.128}$$

Substituting (D.68), (D.123), and (D.126) in the above equation, we have

$$B_{\alpha\beta} = -h_{\alpha\gamma} \eta^\gamma_\beta, \tag{D.129}$$

or upon solving for the coefficients η^γ_β, we find

$$\eta^\gamma_\beta = -h^{\gamma\alpha} B_{\alpha\beta}. \tag{D.130}$$

Substituting this result into (D.126) produces what is known as the *Weingarten formula:*

$$n^i_{,\alpha} = -h^{\beta\gamma} B_{\gamma\alpha} a^i_\beta. \tag{D.131}$$

This is a relation for the covariant derivative along the surface of the unit normal to the surface in terms of the surface metric, the curvature tensor, and surface tangents.

A *third fundamental form of the surface*, connected with the rate of change of the normal vector, is given by the symmetric surface tensor

$$\widetilde{\mathbf{C}} = C_{\alpha\beta} \mathbf{a}^\alpha \mathbf{a}^\beta, \tag{D.132}$$

where

$$C_{\alpha\beta} = g_{ij} n^i_{,\alpha} n^j_{,\beta}. \tag{D.133}$$

By using the Weingarten formula and (D.68), the components can be rewritten as

$$C_{\alpha\beta} = h^{\gamma\delta} b_{\alpha\gamma} b_{\beta\delta}. \tag{D.134}$$

We would like to investigate the properties of the Riemann–Christoffel tensor given in (2.237) or (2.238) in a two-dimensional space with metric $h_{\alpha\beta}$ and coordinates y^α. In this case, $R_{\alpha\beta\gamma\delta}$ has only four nonzero components. Furthermore, these four components are either $+R_{1212}$ or $-R_{1212}$, since using the symmetry conditions (2.241) in the two-dimensional space, it is easy to see that they are all related:

$$R_{1212} = -R_{2112} = R_{2121} = -R_{1221}. \tag{D.135}$$

From this, it follows that we can write it in terms of the components of the two-dimensional absolute Levi–Civita tensor:

$$R_{\alpha\beta\gamma\delta} = K_G \varepsilon_{\alpha\beta} \varepsilon_{\gamma\delta}. \tag{D.136}$$

The surface scalar invariant K_G is called the *Gaussian curvature* or *total curvature.*

Now consider the two-dimensional form of the curvature equation (2.236) applied to $x^i = x^i(y^\alpha)$:

$$\mathbf{a}_{\alpha,\beta\gamma} - \mathbf{a}_{\alpha,\gamma\beta} = \mathbf{a}_\delta R^\delta_{\alpha\beta\gamma} \quad \text{or} \quad a^i_{\alpha,\beta\gamma} - a^i_{\alpha,\gamma\beta} = a^i_\delta R^\delta_{\alpha\beta\gamma}. \tag{D.137}$$

Using this relation, we now derive interesting relations connected with surface properties. The covariant derivative of (D.122) with respect to surface coordinates is given by

$$a^i_{\alpha,\beta\gamma} = B_{\alpha\beta,\gamma} n^i + B_{\alpha\beta} n^i_{,\gamma}, \tag{D.138}$$

where

$$B_{\alpha\beta,\gamma} = \frac{\partial B_{\alpha\beta}}{\partial y^\gamma} - \Gamma^\delta_{\alpha\gamma} B_{\delta\beta} - \Gamma^\delta_{\beta\gamma} B_{\alpha\delta}. \tag{D.139}$$

By using the Weingarten formula (D.131), we can rewrite the above equation in the form

$$a^i_{\alpha,\beta\gamma} = B_{\alpha\beta,\gamma} n^i - B_{\alpha\beta} h^{\sigma\delta} B_{\sigma\gamma} a^i_\delta, \tag{D.140}$$

or, upon using (D.137) and (D.139), we see that

$$a^i_{\alpha,\beta\gamma} - a^i_{\alpha,\gamma\beta} = (B_{\alpha\beta,\gamma} - B_{\alpha\gamma,\beta}) n^i - h^{\sigma\delta} (B_{\alpha\beta} B_{\sigma\gamma} - B_{\alpha\gamma} B_{\sigma\beta}) a^i_\delta = a^i_\delta R^\delta_{\alpha\beta\gamma}, \tag{D.141}$$

or

$$\mathbf{a}_{\alpha,\beta\gamma} - \mathbf{a}_{\alpha,\gamma\beta} = (B_{\alpha\beta,\gamma} - B_{\alpha\gamma,\beta}) \mathbf{n} - h^{\sigma\delta} (B_{\alpha\beta} B_{\sigma\gamma} - B_{\alpha\gamma} B_{\sigma\beta}) \mathbf{a}_\delta = \mathbf{a}_\delta R^\delta_{\alpha\beta\gamma}. \tag{D.142}$$

Multiplying by $g_{ij} n^j$ and using (D.102) and (D.127), we obtain what is known as the *Mainardi–Codazzi equation*:

$$B_{\alpha\beta,\gamma} - B_{\alpha\gamma,\beta} = 0. \tag{D.143}$$

On the other hand, multiplying (D.141) by $g_{ij} a^j_\sigma$, using (D.127), and simplifying, we obtain the *Gauss equations* of the surface:

$$R_{\sigma\alpha\beta\gamma} = B_{\alpha\gamma} B_{\sigma\beta} - B_{\alpha\beta} B_{\sigma\gamma}. \tag{D.144}$$

By using the Gauss equations, the equation for the Gaussian curvature (D.136) can be rewritten as

$$K_G \varepsilon_{\sigma\alpha} \varepsilon_{\beta\gamma} = B_{\alpha\gamma} B_{\sigma\beta} - B_{\alpha\beta} B_{\sigma\gamma}, \tag{D.145}$$

or

$$K_G = \det [B^\alpha_\beta]. \tag{D.146}$$

Still another form for the Gaussian curvature can be obtained by using (D.134) together with the relation $h_{\alpha\beta} = -\varepsilon_{\sigma\alpha}\varepsilon_{\beta\gamma} h^{\sigma\gamma}$, which can be easily verified,

$$-K_G h_{\alpha\beta} = C_{\alpha\beta} - h^{\sigma\gamma} B_{\sigma\gamma} B_{\alpha\beta}. \tag{D.147}$$

If we define the scalar invariant called the *mean surface curvature* by

$$K_M = \frac{1}{2} h^{\sigma\gamma} B_{\sigma\gamma} = \frac{1}{2} B^\sigma_\sigma, \tag{D.148}$$

then the above equation becomes a relationship between the three surface fundamental tensor forms:

$$\widetilde{\mathbf{C}} - 2K_M \widetilde{\mathbf{B}} + K_G \mathbf{h} = \mathbf{0} \qquad \text{or} \qquad C_{\alpha\beta} - 2K_M B_{\alpha\beta} + K_G h_{\alpha\beta} = 0. \tag{D.149}$$

The surface gradient of a spatial vector field $\mathbf{v}$ is given by

$$\widetilde{\nabla} \mathbf{v} = \frac{\partial \mathbf{v}}{\partial y^\alpha} \mathbf{a}^\alpha. \tag{D.150}$$

Note that if $\widetilde{\mathbf{r}} = \widetilde{\mathbf{r}}(y^\beta) = y^\beta \mathbf{a}_\beta$ is the surface position vector field, then

$$\widetilde{\nabla} \widetilde{\mathbf{r}} = \frac{\partial \widetilde{\mathbf{r}}}{\partial y^\alpha} \mathbf{a}^\alpha = \mathbf{a}_\alpha \mathbf{a}^\alpha = \widetilde{\mathbf{P}}. \tag{D.151}$$

This provides an additional useful expression for the surface projection tensor. The *surface divergence* of the spatial vector field is naturally given by the contraction

$$\widetilde{\nabla} \cdot \mathbf{v} = \operatorname{tr}\left(\widetilde{\nabla} \mathbf{v}\right) = \frac{\partial \mathbf{v}}{\partial y^\alpha} \cdot \mathbf{a}^\alpha. \tag{D.152}$$

In the above expressions, the spatial vector field $\mathbf{v}$ may be an explicit function of position in space or it may be an explicit function of position on the surface. As will be seen, these lead to different expressions.

If $\mathbf{v} = v^i(x^j)\mathbf{e}_i = v_i(x^j)\mathbf{e}^i$, then

$$\widetilde{\nabla} \mathbf{v} = \frac{\partial \mathbf{v}}{\partial x^i} a^i_{\ \alpha} \mathbf{a}^\alpha = \frac{\partial \mathbf{v}}{\partial x^i} \left(\mathbf{e}^i \cdot \mathbf{a}_\alpha\right) \mathbf{a}^\alpha = \widetilde{\mathbf{P}} \cdot \nabla \mathbf{v}. \tag{D.153}$$

For $\mathbf{v}$ written using contravariant and covariant components, the corresponding expressions are

$$\widetilde{\nabla} \mathbf{v} = v^j_{\ ,i} a^i_{\ \alpha} \mathbf{e}_j \mathbf{a}^\alpha = v_{j,i} a^i_{\ \alpha} \mathbf{e}^j \mathbf{a}^\alpha. \tag{D.154}$$

Subsequently, we can also write

$$\widetilde{\nabla} \cdot \mathbf{v} = \operatorname{tr}\left(\widetilde{\nabla} \mathbf{v}\right) = \operatorname{tr}\left(\widetilde{\mathbf{P}} \cdot \nabla \mathbf{v}\right), \tag{D.155}$$

with corresponding expressions when using contravariant and covariant components for $\mathbf{v}$:

$$\widetilde{\nabla}\cdot\mathbf{v} = v^j_{,i}a^i_\alpha a^k_\beta g_{jk}h^{\alpha\beta} = v_{j,i}a^i_\alpha a^j_\beta h^{\alpha\beta}. \tag{D.156}$$

On the other hand, if $\mathbf{w} = w^i(x^j(y^\alpha))\mathbf{e}_i = w_i(x^j(y^\alpha))\mathbf{e}^i$, then from (D.2) we have

$$\widetilde{\nabla}\mathbf{w} = w^i_{,\alpha}\mathbf{e}_i\mathbf{a}^\alpha, \tag{D.157}$$

where

$$w^i_{,\alpha} = \frac{\partial w^i}{\partial y^\alpha} + \Gamma^i_{jk}a^j_\alpha w^k, \tag{D.158}$$

which is the surface covariant derivative of w^i, or from (D.1) we have

$$\widetilde{\nabla}\mathbf{w} = w_{i,\alpha}\mathbf{e}^i\mathbf{a}^\alpha, \tag{D.159}$$

where

$$w_{i,\alpha} = \frac{\partial w_i}{\partial y^\alpha} - \Gamma^k_{ji}a^j_\alpha w_k, \tag{D.160}$$

which is the surface covariant derivative of w_i.

The corresponding expressions for the surface divergence of $\mathbf{w}$ are

$$\widetilde{\nabla}\cdot\mathbf{w} = w^i_{,\alpha}a^j_\beta g_{ij}h^{\alpha\beta} = w_{i,\alpha}a^i_\beta h^{\alpha\beta}. \tag{D.161}$$

It is also useful to write $\mathbf{w}$ in terms of tangential and normal components:

$$\mathbf{w} = \mathbf{w}\cdot\mathbf{1} = \mathbf{w}\cdot\widetilde{\mathbf{P}} + \mathbf{w}\cdot\mathbf{P}_{(n)} = \mathbf{w}\cdot\mathbf{a}^\alpha\mathbf{a}_\alpha + \mathbf{w}\cdot\mathbf{n}\mathbf{n} = w^\alpha\mathbf{a}_\alpha + w_{(n)}\mathbf{n}, \tag{D.162}$$

or equivalently

$$\mathbf{w} = w_\alpha\mathbf{a}^\alpha + w_{(n)}\mathbf{n}. \tag{D.163}$$

In this case, the surface gradient becomes either

$$\widetilde{\nabla}\mathbf{w} = \frac{\partial w^\alpha}{\partial y^\beta}\mathbf{a}_\alpha\mathbf{a}^\beta + w^\alpha\frac{\partial\mathbf{a}_\alpha}{\partial y^\beta}\mathbf{a}^\beta + \frac{\partial w_{(n)}}{\partial y^\beta}\mathbf{n}\mathbf{a}^\beta + w_{(n)}\frac{\partial\mathbf{n}}{\partial y^\beta}\mathbf{a}^\beta \tag{D.164}$$

or

$$\widetilde{\nabla}\mathbf{w} = \frac{\partial w_\alpha}{\partial y^\beta}\mathbf{a}^\alpha\mathbf{a}^\beta + w_\alpha\frac{\partial\mathbf{a}^\alpha}{\partial y^\beta}\mathbf{a}^\beta + \frac{\partial w_{(n)}}{\partial y^\beta}\mathbf{n}\mathbf{a}^\beta + w_{(n)}\frac{\partial\mathbf{n}}{\partial y^\beta}\mathbf{a}^\beta. \tag{D.165}$$

Now

$$\begin{aligned}\frac{\partial\mathbf{a}_\alpha}{\partial y^\beta} &= \frac{\partial}{\partial y^\beta}\left(a^i_\alpha\mathbf{e}_i\right)\\ &= \frac{\partial a^i_\alpha}{\partial y^\beta}\mathbf{e}_i + a^i_\alpha a^j_\beta\frac{\partial\mathbf{e}_i}{\partial x^j}\\ &= \frac{\partial a^i_\alpha}{\partial y^\beta}\mathbf{e}_i + a^i_\alpha a^j_\beta\Gamma^k_{ji}\mathbf{e}_k\\ &= \left(\frac{\partial a^i_\alpha}{\partial y^\beta} + a^j_\beta a^k_\alpha\Gamma^i_{jk}\right)\mathbf{e}_i,\end{aligned} \tag{D.166}$$

and in addition

$$\begin{aligned} \mathbf{e}_i &= \mathbf{1} \cdot \mathbf{e}_i \\ &= \left(\widetilde{\mathbf{P}} + \mathbf{P}_{(n)}\right) \cdot \mathbf{e}_i \\ &= \left(\mathbf{a}_\gamma \mathbf{a}^\gamma + \mathbf{n}\mathbf{n}\right) \cdot \mathbf{e}_i \\ &= \left(\mathbf{a}_\gamma h^{\gamma\delta} a^l_\delta \mathbf{e}_l + \mathbf{n}\mathbf{n}\right) \cdot \mathbf{e}_i \\ &= h^{\gamma\delta} g_{il} a^l_\delta \mathbf{a}_\gamma + n_i \mathbf{n}. \end{aligned} \tag{D.167}$$

Combining the last two expressions, we have

$$\frac{\partial \mathbf{a}_\alpha}{\partial y^\beta} = \Gamma^\gamma_{\beta\alpha} \mathbf{a}_\gamma + B_{\beta\alpha} \mathbf{n}, \tag{D.168}$$

where

$$\Gamma^\gamma_{\beta\alpha} = \left(\frac{\partial a^i_\alpha}{\partial y^\beta} + a^j_\beta a^k_\alpha \Gamma^i_{jk} \right) h^{\gamma\delta} g_{il} a^l_\delta \tag{D.169}$$

is the *surface Christoffel symbol of the second kind*, and

$$B_{\beta\alpha} = \left(\frac{\partial a^i_\alpha}{\partial y^\beta} + a^j_\beta a^k_\alpha \Gamma^i_{jk} \right) n_i = B_{\alpha\beta} \tag{D.170}$$

are the components of the symmetric second fundamental form tangential tensor field.

Now note that since $\mathbf{a}_\gamma \cdot \mathbf{n} = 0$ and $\mathbf{n} \cdot \mathbf{n} = 1$, upon differentiating we have

$$\mathbf{a}_\gamma \cdot \frac{\partial \mathbf{n}}{\partial y^\beta} = -B_{\gamma\beta} \qquad \text{and} \qquad \mathbf{n} \cdot \frac{\partial \mathbf{n}}{\partial y^\beta} = 0. \tag{D.171}$$

Thus we see that $\partial \mathbf{n} / \partial y^\beta$ is the tangential vector field

$$\frac{\partial \mathbf{n}}{\partial y^\beta} = -B_{\gamma\beta} \mathbf{a}^\gamma. \tag{D.172}$$

Similarly, since $\mathbf{a}_\gamma \cdot \mathbf{a}^\alpha = \delta^\alpha_\gamma$ and $\mathbf{n} \cdot \mathbf{a}^\alpha = 0$, upon differentiating we have

$$\mathbf{a}_\gamma \cdot \frac{\partial \mathbf{a}^\alpha}{\partial y^\beta} = -\frac{\partial \mathbf{a}_\gamma}{\partial y^\beta} \cdot \mathbf{a}^\alpha = -\Gamma^\delta_{\beta\gamma} \mathbf{a}_\delta \cdot \mathbf{a}^\alpha = -\Gamma^\alpha_{\beta\gamma} \tag{D.173}$$

and

$$\mathbf{n} \cdot \frac{\partial \mathbf{a}^\alpha}{\partial y^\beta} = -\frac{\partial \mathbf{n}}{\partial y^\beta} \cdot \mathbf{a}^\alpha = B_{\gamma\beta} h^{\gamma\alpha}, \tag{D.174}$$

thus concluding that

$$\frac{\partial \mathbf{a}^\alpha}{\partial y^\beta} = -\Gamma^\alpha_{\beta\gamma} \mathbf{a}^\gamma + h^{\alpha\gamma} B_{\beta\gamma} \mathbf{n}. \tag{D.175}$$

With (D.168) and (D.172), we can rewrite (D.164) in a simpler form:

$$\begin{aligned} \widetilde{\nabla} \mathbf{w} &= \frac{\partial w^\alpha}{\partial y^\beta} \mathbf{a}_\alpha \mathbf{a}^\beta + \Gamma^\gamma_{\beta\alpha} w^\alpha \mathbf{a}_\gamma \mathbf{a}^\beta + B_{\beta\alpha} w^\alpha \mathbf{n} \mathbf{a}^\beta + \frac{\partial w_{(n)}}{\partial y^\beta} \mathbf{n} \mathbf{a}^\beta - B_{\gamma\beta} w_{(n)} \mathbf{a}^\gamma \mathbf{a}^\beta \\ &= \left[\left(w^\alpha_{,\beta} - h^{\gamma\alpha} B_{\gamma\beta} w_{(n)} \right) \mathbf{a}_\alpha + \left(\frac{\partial w_{(n)}}{\partial y^\beta} + B_{\beta\alpha} w^\alpha \right) \mathbf{n} \right] \mathbf{a}^\beta \\ &= w^\alpha_{,\beta} \mathbf{a}_\alpha \mathbf{a}^\beta - w_{(n)} \widetilde{\mathbf{B}} + \mathbf{n} \left(\widetilde{\nabla} w_{(n)} + \widetilde{\mathbf{B}} \cdot \mathbf{w} \right) \\ &= \widetilde{\mathbf{P}} \cdot \widetilde{\nabla} \left(\widetilde{\mathbf{P}} \cdot \mathbf{w} \right) - w_{(n)} \widetilde{\mathbf{B}} + \mathbf{n} \left(\widetilde{\nabla} w_{(n)} + \widetilde{\mathbf{B}} \cdot \mathbf{w} \right), \end{aligned} \tag{D.176}$$

where we have introduced the surface covariant derivative of w^α

$$w^\alpha_{,\beta} = \frac{\partial w^\alpha}{\partial y^\beta} + \Gamma^\alpha_{\beta\gamma} w^\gamma \tag{D.177}$$

and have noted the fact that

$$\widetilde{\mathbf{P}} \cdot \widetilde{\nabla}(\widetilde{\mathbf{P}} \cdot \mathbf{w}) = w^\alpha_{,\beta} \mathbf{a}_\alpha \mathbf{a}^\beta. \tag{D.178}$$

Analogously, using (D.172) and (D.175), we can rewrite (D.165) in the form:

$$\begin{aligned} \widetilde{\nabla}\mathbf{w} &= \left[\left(w_{\alpha,\beta} - B_{\alpha\beta} w_{(n)} \right) \mathbf{a}^\alpha + \left(\frac{\partial w_{(n)}}{\partial y^\beta} + B_{\beta\alpha} w^\alpha \right) \mathbf{n} \right] \mathbf{a}^\beta \\ &= w_{\alpha,\beta} \mathbf{a}^\alpha \mathbf{a}^\beta - w_{(n)} \widetilde{\mathbf{B}} + \mathbf{n} \left(\widetilde{\nabla} w_{(n)} + \widetilde{\mathbf{B}} \cdot \mathbf{w} \right), \end{aligned} \tag{D.179}$$

where we have introduced the surface covariant derivative of w_α,

$$w_{\alpha,\beta} = \frac{\partial w_\alpha}{\partial y^\beta} - \Gamma^\gamma_{\beta\alpha} w_\gamma. \tag{D.180}$$

Now the corresponding expressions for the surface divergence of $\mathbf{w}$ are

$$\widetilde{\nabla} \cdot \mathbf{w} = w^\alpha_{,\alpha} - w_{(n)} h^{\alpha\beta} B_{\alpha\beta} = w^\alpha_{,\alpha} - 2K_M w_{(n)}, \tag{D.181}$$

where K_M is the mean surface curvature. From the above result, it is now easily seen that

$$\widetilde{\mathbf{B}} = -\widetilde{\nabla}\mathbf{n} \tag{D.182}$$

and

$$K_M = \frac{1}{2} \widetilde{\nabla} \cdot \mathbf{n}. \tag{D.183}$$

There is another measure of surface curvature other than the total and mean curvatures. It is the *normal curvature* in the $\boldsymbol{\lambda}$ direction,

$$\kappa_n = \boldsymbol{\lambda} \cdot \widetilde{\mathbf{B}} \cdot \boldsymbol{\lambda}, \tag{D.184}$$

where $\boldsymbol{\lambda}$ is a unit vector tangent to the surface at the point where the normal curvature is measured. *Principal curvatures* κ_1 and κ_2 are the maximum and minimum values of the normal curvature κ_n. It can be shown that the direction $\boldsymbol{\lambda}$, which corresponds to one of the principal curvatures, satisfies

$$\left(\widetilde{\mathbf{B}} - \kappa_n \widetilde{\mathbf{P}} \right) \cdot \boldsymbol{\lambda} = \mathbf{0}, \tag{D.185}$$

or if $\boldsymbol{\lambda}$ is nonzero, then the curvatures must satisfy the characteristic equation

$$\kappa_n^2 - \mathrm{tr}[\widetilde{\mathbf{B}}]\kappa_n + \det[\widetilde{\mathbf{B}}] = 0. \tag{D.186}$$

From this, it can be shown that

$$\kappa_1 + \kappa_2 = 2K_M, \qquad \kappa_1 \kappa_2 = K_G, \quad \text{and} \quad |\widetilde{\mathbf{B}}|^2 = \mathrm{tr}[\widetilde{\mathbf{B}} \cdot \widetilde{\mathbf{B}}] = 4K_M^2 - 2K_G. \tag{D.187}$$

As with vector fields, the surface gradient of a rank-2 spatial tensor field $\mathbf{T}$ is given analogously by

$$\widetilde{\nabla}\mathbf{T} = \frac{\partial \mathbf{T}}{\partial y^\alpha} \mathbf{a}^\alpha. \tag{D.188}$$

Actually, this definition is the same if $\mathbf{T}$ is a spatial tensor field of any rank. The surface divergence is obtained by contraction:

$$\widetilde{\nabla} \cdot \mathbf{T} = \frac{\partial \mathbf{T}}{\partial y^\alpha} \cdot \mathbf{a}^\alpha. \tag{D.189}$$

Also as before, $\mathbf{T}$ can be expressed as an explicit function of space coordinates or an explicit function of surface coordinates. If $\mathbf{T} = T^{ij}(x^i)\,\mathbf{e}_i\mathbf{e}_j = T_{ij}(x^i)\,\mathbf{e}^i\mathbf{e}^j$, then the surface gradient is given by

$$\widetilde{\nabla}\mathbf{T} = \widetilde{\mathbf{P}} \cdot \nabla\mathbf{T} = T^{jk}_{,i} a^i_\alpha \mathbf{e}_j\mathbf{e}_k\mathbf{a}^\alpha = T_{jk,i} a^i_\alpha \mathbf{e}^j\mathbf{e}^k\mathbf{a}^\alpha, \tag{D.190}$$

and the corresponding surface divergence by

$$\widetilde{\nabla} \cdot \mathbf{T} = T^{jk}_{,i} a^m_\alpha a^i_\beta g_{km} h^{\alpha\beta}\mathbf{e}_j = T_{jk,i} a^k_\alpha a^i_\beta h^{\alpha\beta}\mathbf{e}^j. \tag{D.191}$$

On the other hand, if $\widetilde{\mathbf{T}} = T^{\alpha\beta}(y^\gamma)\,\mathbf{a}_\alpha\mathbf{a}_\beta = T_{\alpha\beta}(y^\gamma)\,\mathbf{a}^\alpha\mathbf{a}^\beta$, then the surface gradient is given by

$$\begin{aligned}
\widetilde{\nabla}\widetilde{\mathbf{T}} &= T^{\alpha\beta}_{,\gamma}\mathbf{a}_\alpha\mathbf{a}_\beta\mathbf{a}^\gamma + T^{\alpha\beta}B_{\alpha\gamma}\mathbf{n}\mathbf{a}_\beta\mathbf{a}^\gamma + T^{\alpha\beta}B_{\beta\gamma}\mathbf{a}_\alpha\mathbf{n}\mathbf{a}^\gamma && \text{(D.192)}\\
&= T_{\alpha\beta,\gamma}\mathbf{a}^\alpha\mathbf{a}^\beta\mathbf{a}^\gamma + T_{\delta\beta}B_{\alpha\gamma}h^{\delta\alpha}\mathbf{n}\mathbf{a}^\beta\mathbf{a}^\gamma + T_{\alpha\delta}B_{\beta\gamma}h^{\delta\beta}\mathbf{a}^\alpha\mathbf{n}\mathbf{a}^\gamma, && \text{(D.193)}
\end{aligned}$$

where the surface covariant derivatives of $T^{\alpha\beta}$ and $T_{\alpha\beta}$ are given by

$$\begin{aligned}
T^{\alpha\beta}_{,\gamma} &= \frac{\partial T^{\alpha\beta}}{\partial y^\gamma} + \Gamma^\alpha_{\gamma\delta}T^{\delta\beta} + \Gamma^\beta_{\gamma\delta}T^{\alpha\delta}, && \text{(D.194)}\\
T_{\alpha\beta,\gamma} &= \frac{\partial T_{\alpha\beta}}{\partial y^\gamma} - \Gamma^\delta_{\gamma\alpha}T_{\delta\beta} - \Gamma^\delta_{\gamma\beta}T_{\alpha\delta}. && \text{(D.195)}
\end{aligned}$$

The corresponding surface divergence of $\widetilde{\mathbf{T}}$ takes the forms

$$\begin{aligned}
\widetilde{\nabla} \cdot \widetilde{\mathbf{T}} &= T^{\alpha\beta}_{,\beta}\mathbf{a}_\alpha + T^{\alpha\beta}B_{\alpha\beta}\mathbf{n} && \text{(D.196)}\\
&= T_{\alpha\beta,\gamma}h^{\beta\gamma}\mathbf{a}^\alpha + T_{\delta\beta}B_{\alpha\gamma}h^{\delta\alpha}h^{\beta\gamma}\mathbf{n}. && \text{(D.197)}
\end{aligned}$$

We will have occasion to perform integration over the surface coordinates y^1 and y^2. At any point on the surface $\mathscr{S}$, the unit tangent vectors to the two surface coordinate curves are $\mathbf{a}_1 = \partial\widetilde{\mathbf{r}}/\partial y^1$ and $\mathbf{a}_2 = \partial\widetilde{\mathbf{r}}/\partial y^2$. We take da to be the differential area of the parallelogram formed from $\mathbf{a}_1$ and $\mathbf{a}_2$ with sides of length dy^1 and dy^2, so that using (D.96) we have

$$da = \mathbf{a}_1 \cdot (\widetilde{\boldsymbol{\varepsilon}} \cdot \mathbf{a}_2)\,dy^1dy^2 = \sqrt{h}\;dy^1dy^2, \tag{D.198}$$

and subsequently,

$$\int_{\mathscr{S}} F(y^1,y^2)\,da = \iint_{\mathscr{S}} F(y^1,y^2)\sqrt{h}\,dy^1dy^2. \tag{D.199}$$

We also need to know the rate of change with respect to time of some geometrical objects and physical quantities defined on a surface. If we differentiate the equation $f(\mathbf{x},t) = 0$ with respect to time, then

$$\frac{d}{dt}f(\mathbf{x},t) = \dot{f}(\mathbf{x},t) = \frac{\partial f}{\partial t} + \frac{\partial \mathbf{x}}{\partial t} \cdot \nabla f = \frac{\partial f}{\partial t} + \mathbf{v} \cdot \nabla f = 0, \tag{D.200}$$

where $\mathbf{v}$ is the velocity of a particle located at $\mathbf{x}$ (or equivalently $\mathbf{y}$) on the surface:

$$\mathbf{v} \equiv \dot{\mathbf{x}} = \left. \frac{\partial \mathbf{x}(y^\alpha, t)}{\partial t} \right|_{y^\alpha}, \tag{D.201}$$

so that $\dot{\mathbf{x}}(y^\alpha, t)$ is the material derivative on the surface, or the derivative of $\mathbf{x}(y^\alpha, t)$ with respect to t holding y^α fixed. It is useful to decompose the velocity field into tangential and normal components (see Problem 12):

$$\mathbf{v} = v^\alpha \mathbf{a}_\alpha + v_{(n)} \mathbf{n} \qquad \text{or} \qquad v^i = v^\alpha a^i_\alpha + v_{(n)} n^i, \tag{D.202}$$

where

$$v_{(n)} = \mathbf{v} \cdot \mathbf{n} = -\frac{\partial f / \partial t}{|\mathrm{grad}\, f|}. \tag{D.203}$$

Now the velocity of the surface determines how the surface metric tensor $\mathbf{h}$ changes with time, because

$$\dot{\mathbf{a}}_\beta = \frac{\partial}{\partial t}\left(\frac{\partial \mathbf{x}}{\partial y^\beta} \right) = \frac{\partial \mathbf{v}}{\partial y^\beta}. \tag{D.204}$$

Subsequently, using (D.176), we have

$$\begin{aligned} \dot{\mathbf{a}}_\beta &= \frac{\partial v^\alpha}{\partial y^\beta} \mathbf{a}_\alpha + v^\alpha \frac{\partial \mathbf{a}_\alpha}{\partial y^\beta} + \frac{\partial v_{(n)}}{\partial y^\beta} \mathbf{n} + v_{(n)} \frac{\partial \mathbf{n}}{\partial y^\beta} \\ &= \left(v^\alpha_{,\beta} - h^{\gamma\alpha} B_{\gamma\beta} v_{(n)} \right) \mathbf{a}_\alpha + \left(\frac{\partial v_{(n)}}{\partial y^\beta} + B_{\beta\alpha} v^\alpha \right) \mathbf{n}, \end{aligned} \tag{D.205}$$

or equivalently, using (D.179),

$$\dot{\mathbf{a}}_\beta = \left(v_{\alpha,\beta} - B_{\alpha\beta} v_{(n)} \right) \mathbf{a}^\alpha + \left(\frac{\partial v_{(n)}}{\partial y^\beta} + B_{\beta\alpha} v^\alpha \right) \mathbf{n}. \tag{D.206}$$

Now, since the surface metric $h_{\beta\delta} = \mathbf{a}_\beta \cdot \mathbf{a}_\delta$ is symmetric, the rate of change of the surface metric tensor is given by

$$\dot{h}_{\beta\delta} = 2\dot{\mathbf{a}}_\beta \cdot \mathbf{a}_\delta = 2\left(v^\alpha_{,\beta} h_{\alpha\delta} - B_{\delta\beta} v_{(n)} \right) = 2\left(v_{\delta,\beta} - B_{\delta\beta} v_{(n)} \right). \tag{D.207}$$

In particular, if the surface particles all move in the normal direction, the rate of change of the surface metric tensor is determined by the curvature tensor and the normal speed. In addition, we have that

$$\dot{h} = \frac{\partial h}{\partial h_{\beta\delta}} \dot{h}_{\beta\delta} = h h^{\beta\delta} \dot{h}_{\beta\delta} = 2h \left(v^\alpha_{,\alpha} - 2K_M v_{(n)} \right). \tag{D.208}$$

Subsequently,

$$\dot{\overline{da}} = \frac{\dot{h}}{2\sqrt{h}} dy^1 dy^2 = \left(v^\alpha_{,\alpha} - 2K_M v_{(n)} \right) da. \tag{D.209}$$

This result shows that if the particles move normal to the surface, then da is unchanged if the surface is flat or at rest.

Next, we would like to obtain the equivalent relation as the spatial divergence theorem, but now applied on the surface $\mathscr{S}$. First, it is easy to show that if we have the tangential vector field $\widetilde{\mathbf{w}} = w^\alpha \mathbf{a}_\alpha$, then its surface divergence is given by

$$\widetilde{\nabla} \cdot \widetilde{\mathbf{w}} = w^\alpha_{,\alpha} = \frac{1}{\sqrt{h}} \frac{\partial}{\partial y^\alpha} \left(\sqrt{h} w^\alpha \right). \tag{D.210}$$

Subsequently, using (D.199), we have

$$\begin{aligned} \int_{\mathscr{S}} \widetilde{\nabla} \cdot \widetilde{\mathbf{w}} \, da &= \iint_{\mathscr{S}} w^\alpha_{,\alpha} \sqrt{h} \, dy^1 dy^2 \\ &= \iint_{\mathscr{S}} \frac{\partial(\sqrt{h}\, w^\alpha)}{\partial y^\alpha} \, dy^1 dy^2 \\ &= \iint_{\mathscr{S}} \left[\frac{\partial(\sqrt{h}\, w^1)}{\partial y^1} + \frac{\partial(\sqrt{h}\, w^2)}{\partial y^2} \right] dy^1 dy^2. \end{aligned} \tag{D.211}$$

Now the Gauss–Green theorem for a surface tells us that if $P(y^\alpha)$ and $Q(y^\alpha)$ are continuous functions having continuous first partial derivatives on the surface, then

$$\iint_{\mathscr{S}} \left[\frac{\partial P}{\partial y^1} + \frac{\partial Q}{\partial y^2} \right] dy^1 dy^2 = \int_{\mathscr{C}} \left(P \frac{dy^2}{ds} - Q \frac{dy^1}{ds} \right) ds, \tag{D.212}$$

where $\mathscr{C}$ is a piecewise-smooth simple closed curve bounding $\mathscr{S}$, and s indicates the arc length measured along this curve in the positive direction. Subsequently, using the Gauss–Green theorem, we have

$$\int_{\mathscr{S}} \widetilde{\nabla} \cdot \widetilde{\mathbf{w}} \, da = \int_{\mathscr{C}} \left(w^1 \frac{dy^2}{ds} - w^2 \frac{dy^1}{ds} \right) \sqrt{h} \, ds = \int_{\mathscr{C}} \varepsilon_{\alpha\beta} w^\alpha \frac{dy^\beta}{ds} \sqrt{h} \, ds. \tag{D.213}$$

Now the unit tangent to the curve $\mathscr{C}$ is given by

$$\widetilde{\boldsymbol{\lambda}} = \frac{d\widetilde{\mathbf{r}}}{ds} = \frac{\partial \widetilde{\mathbf{r}}}{\partial y^\beta} \frac{dy^\beta}{ds} = \frac{dy^\beta}{ds} \mathbf{a}_\beta. \tag{D.214}$$

It follows that the unit tangent vector field that is normal to the curve $\mathscr{C}$ is given by

$$\widetilde{\boldsymbol{\mu}} = \widetilde{\boldsymbol{\varepsilon}} \cdot \widetilde{\boldsymbol{\lambda}} = \varepsilon_{\alpha\beta} \frac{dy^\beta}{ds} \mathbf{a}^\alpha. \tag{D.215}$$

We note that the rotation from $\widetilde{\boldsymbol{\lambda}}$ to $\widetilde{\boldsymbol{\mu}}$ is negative. Because of the requirement that s is measured along $\mathscr{C}$ in the positive sense, it is clear that $\widetilde{\boldsymbol{\mu}}$ is directed outward with respect to the closed curve. Thus we can rewrite

$$\int_{\mathscr{S}} \widetilde{\nabla} \cdot \widetilde{\mathbf{w}} \, da = \int_{\mathscr{C}} \widetilde{\mathbf{w}} \cdot \widetilde{\boldsymbol{\mu}} \, ds = \int_{\mathscr{C}} \widetilde{\mathbf{w}} \cdot d\widetilde{\mathbf{s}}. \tag{D.216}$$

This result is known as the *surface divergence theorem.*

If, on the other hand, $\mathbf{v}$ is a spatial vector field defined on the surface, from (D.181) we have

$$\widetilde{\nabla} \cdot \mathbf{v} = \widetilde{\nabla} \cdot \widetilde{\mathbf{v}} - 2K_M \mathbf{v} \cdot \mathbf{n}. \tag{D.217}$$

Subsequently, we arrive at the *generalized surface divergence theorem:*

$$\int_{\mathscr{S}} \widetilde{\nabla} \cdot \mathbf{v} \, da = \int_{\mathscr{C}} \widetilde{\mathbf{v}} \cdot \widetilde{\boldsymbol{\mu}} \, ds - \int_{\mathscr{S}} 2K_M \mathbf{v} \cdot \mathbf{n} \, da = \int_{\mathscr{C}} \mathbf{v} \cdot d\widetilde{\mathbf{s}} - \int_{\mathscr{S}} 2K_M \mathbf{v} \cdot d\mathbf{a}. \tag{D.218}$$

In addition, using (D.194) and (D.195), it can be shown that in general

$$\varepsilon^{\alpha\beta}_{,\gamma} = \varepsilon_{\alpha\beta,\gamma} = 0, \tag{D.219}$$

so that with (D.101) it is not difficult to show that

$$n_i \varepsilon^{ijk} v_{k,j} = \left(\varepsilon^{\alpha\beta} v_\beta\right)_{,\alpha}. \tag{D.220}$$

Subsequently, if $(\mathbf{a}_1 \times \mathbf{a}_2) \cdot \mathbf{n} > 0$, $\widetilde{\boldsymbol{\lambda}}$ and $\widetilde{\boldsymbol{\mu}}$ are unit tangent vectors such that on the surface the rotation from $\widetilde{\boldsymbol{\lambda}}$ to $\widetilde{\boldsymbol{\mu}}$ is positive, and $\widetilde{\boldsymbol{\lambda}}$ is the unit tangent vector to the curve $\mathscr{C}$ bounding $\mathscr{S}$, then

$$\int_{\mathscr{S}} (\nabla \times \mathbf{v}) \cdot \mathbf{n}\, da = \int_{\mathscr{S}} \widetilde{\nabla} \cdot (\widetilde{\boldsymbol{\varepsilon}} \cdot \widetilde{\mathbf{v}})\, da = \int_{\mathscr{C}} (\widetilde{\boldsymbol{\varepsilon}} \cdot \widetilde{\mathbf{v}}) \cdot \widetilde{\boldsymbol{\mu}}\, ds = \int_{\mathscr{C}} \widetilde{\mathbf{v}} \cdot (-\widetilde{\boldsymbol{\varepsilon}} \cdot \widetilde{\boldsymbol{\mu}})\, ds = \int_{\mathscr{C}} \widetilde{\mathbf{v}} \cdot \widetilde{\boldsymbol{\lambda}}\, ds,$$

and thus we obtain *Stokes' theorem:*

$$\int_{\mathscr{S}} (\nabla \times \mathbf{v}) \cdot \mathbf{n}\, da = \int_{\mathscr{C}} \mathbf{v} \cdot \widetilde{\boldsymbol{\lambda}}\, ds, \qquad \text{or} \qquad \int_{\mathscr{S}} (\nabla \times \mathbf{v}) \cdot d\mathbf{a} = \int_{\mathscr{C}} \mathbf{v} \cdot d\widetilde{\mathbf{s}}. \tag{D.221}$$

We will also have occasion to integrate quantities along a curve $\mathscr{C}(t)$ on a surface so that $\widetilde{\mathbf{r}} = \widetilde{\mathbf{r}}(y^\alpha)$. In such case, we consider the integrand to be given as an explicit function of y^2. Subsequently, we have that the arc length s along the curve lying on the surface is given by

$$ds = \left(\frac{\partial \widetilde{\mathbf{r}}}{\partial y^2} \cdot \frac{\partial \widetilde{\mathbf{r}}}{\partial y^2}\right)^{1/2} dy^2 = \sqrt{\mathbf{a}_2 \cdot \mathbf{a}_2}\, dy^2 = \sqrt{h_{22}}\, dy^2. \tag{D.222}$$

Then the line integral takes the form

$$\int_{\mathscr{C}(t)} F(y^2)\, ds = \int_{\mathscr{C}(t)} F(y^2) \sqrt{h_{22}}\, dy^2. \tag{D.223}$$

Next, we would like to obtain the transport theorem of a tensor quantity on the material surface $\mathscr{S}(t)$, which in the reference configuration is given by S. We take the surface density of the quantity to be ψ and note from (D.198) that $da = \sqrt{h/H}\, dA$, where dA and H respectively denote the differential material surface and determinant of the metric tensor in the reference configuration. Then, using either (D.208) or (D.209), we have

$$\begin{aligned}
\frac{d}{dt} \int_{\mathscr{S}(t)} \psi\, da &= \frac{d}{dt} \int_S \psi \sqrt{\frac{h}{H}}\, dA \\
&= \int_S \left[\dot{\psi} + \psi \frac{\dot{h}}{2\sqrt{Hh}}\right] dA \\
&= \int_S \left[\frac{\partial \psi}{\partial t} + v^\alpha \psi_{,\alpha} + \psi \left(v^\alpha_{,\alpha} - 2K_M v_{(n)}\right)\right] \sqrt{\frac{h}{H}}\, dA,
\end{aligned}$$

or

$$\frac{d}{dt} \int_{\mathscr{S}(t)} \psi\, da = \int_{\mathscr{S}(t)} \left[\frac{\partial \psi}{\partial t} + (\psi v^\alpha)_{,\alpha} - 2K_M \psi\, v_{(n)}\right] da. \tag{D.224}$$

This result represents the transport theorem for material surfaces.

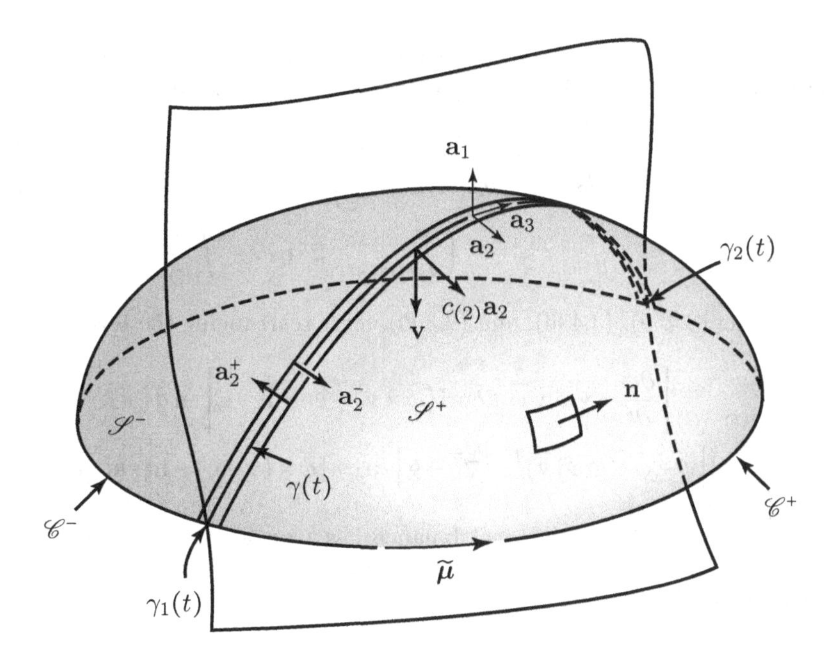

Figure D.1: Arbitrary material surface $\mathscr{S}$ intersected by a discontinuous surface.

D.4 Balance law for a flux through a space surface

Consider an infinitesimally thin material membrane having surface $\mathscr{S}(t)$, which is separated into two parts $\mathscr{S}^+(t)$ and $\mathscr{S}^-(t)$, or $\mathscr{S}(t) - \gamma(t)$, by a singular surface whose intersection with $\mathscr{S}(t)$ is denoted by the line $\gamma(t)$ as shown in Fig. D.1. The unit normal vector to surfaces $\mathscr{S}^+(t)$ and $\mathscr{S}^-(t)$ is denoted by $\mathbf{n}$. The parts of the boundary of $\mathscr{S}(t)$ that are not on the singular line $\gamma(t)$ are denoted as $\mathscr{C}^+(t)$ and $\mathscr{C}^-(t)$, and their unit tangent vector by $\widetilde{\boldsymbol{\mu}}$. The points on the boundary of $\mathscr{S}(t)$ that are on the singular line $\gamma(t)$ will be denoted by $\{\mathscr{P}(t) : (\gamma_1(t), \gamma_2(t))\}$. We denote by $\mathbf{a}_1$ the unit vector normal to the line $\gamma(t)$, which is tangent to the singular surface. The singular line moves on the surface $\mathscr{S}(t)$ with normal velocity $c_{(2)}\mathbf{a}_2$. The unit vector tangent to the singular line $\gamma(t)$ is denoted by $\mathbf{a}_3 = \mathbf{a}_1 \times \mathbf{a}_2$. The material through $\mathscr{S}(t)$ moves with particle velocity $\mathbf{v}$.

The general balance statement of the flux of a tensor quantity Φ flowing through a surface $\mathscr{S}(t)$ is given in the form

$$\frac{d\Phi}{dt} = \mathcal{H}(\Phi) + \mathcal{G}(\Phi), \tag{D.225}$$

where $\mathcal{H}(\Phi)$ denotes the external supply flux of Φ through the surface enclosed by the curve $\mathscr{C}(t)$, and $\mathcal{G}(\Phi)$ denotes the combined external supply of Φ to the surface and internal production of Φ within the surface. One should note that while we represent symbolically the sum of supply and production by a single term in the present derivation, physically supply is different from production because it may be controlled from the exterior of the surface. Subsequently, in applications, one should recognize this difference by writing their contributions separately. We assume that additive surface densities of these tensor quantities exist, and denote the corresponding quantities that are defined on the surface $\mathscr{S}(t) - \gamma(t)$ by a tilde, i.e., $\widetilde{\phi}$, $\widetilde{\mathbf{h}}$, and $\widetilde{\mathbf{g}}$, while those that are defined only on the singular line $\gamma(t)$ by an overbar, i.e., $\overline{\phi}$, $\overline{\mathbf{h}}$, and $\overline{\mathbf{g}}$. Subsequently, (D.225) can be rewritten more explicitly as

$$\frac{d}{dt}\left(\int_{\mathscr{S}(t)-\gamma(t)} \widetilde{\phi}\cdot\mathbf{n}\,ds + \int_{\gamma(t)} \overline{\phi}\cdot\mathbf{a}_3\,dl\right) = \left(\int_{\mathscr{C}(t)-\mathscr{P}(t)} \widetilde{\mathbf{h}}\cdot\widetilde{\mu}\,dl + \left[\overline{h}(\gamma(t))\right]_{\gamma_1(t)}^{\gamma_2(t)}\right) + \left(\int_{\mathscr{S}(t)-\gamma(t)} \widetilde{\mathbf{g}}\cdot\mathbf{n}\,ds + \int_{\gamma(t)} \overline{\mathbf{g}}\cdot\mathbf{a}_3\,dl\right). \tag{D.226}$$

Now using (3.459), (3.436), and (2.300), after rearranging the terms, we obtain

$$\int_{\mathscr{S}(t)-\gamma(t)} \left[\frac{\partial\widetilde{\phi}}{\partial t} + \mathbf{v}\ \mathrm{div}\,\widetilde{\phi} + \mathrm{curl}\left(\widetilde{\phi}\times\mathbf{v}\right) - \mathrm{curl}\,\widetilde{\mathbf{h}} - \widetilde{\mathbf{g}}\right]\cdot\mathbf{n}\,ds + \int_{\gamma(t)} \left\{\left[\dot{\overline{\phi}} + \overline{\phi}\cdot(\mathrm{grad}\,\mathbf{v})^T - \nabla\overline{h} - \overline{\mathbf{g}}\right]\cdot\mathbf{a}_3 + [\![\widetilde{\phi}\times(\mathbf{v}-\mathbf{c}) - \widetilde{\mathbf{h}}]\!]\cdot\mathbf{a}_2\right\} dl = 0. \tag{D.227}$$

If we now apply the integral balance law over an arbitrary regular material surface, then the integrand of the first integral must be zero and thus we obtain the local balance law:

$$\frac{\partial\widetilde{\phi}}{\partial t} + \mathbf{v}\ \mathrm{div}\,\widetilde{\phi} + \mathrm{curl}\left(\widetilde{\phi}\times\mathbf{v}\right) - \mathrm{curl}\,\widetilde{\mathbf{h}} - \widetilde{\mathbf{g}} = \mathbf{0}. \tag{D.228}$$

Subsequently, applying the integral balance law over the singular region, and since $\mathbf{a}_2$ and $\mathbf{a}_3$ are orthogonal to each other, we obtain the jump condition across the singular curve

$$[\![\widetilde{\phi}\times(\mathbf{v}-\mathbf{c}) - \widetilde{\mathbf{h}}]\!]\cdot\mathbf{a}_2 = 0, \tag{D.229}$$

and the local balance law along the singular curve

$$\dot{\overline{\phi}} + \overline{\phi}\cdot(\mathrm{grad}\,\mathbf{v})^T - \nabla\overline{h} - \overline{\mathbf{g}} = 0. \tag{D.230}$$

Note that this last balance law along the singular curve is identical to the balance law (D.48) obtained for a curve in space if $\overline{\phi}$ depends only on one spatial dimension, i.e., $\overline{\phi} = \overline{\phi}(\xi, t)$.

Problems

1. Given a Cartesian coordinate system and a plane surface ξ_3 = const., choose

$$y^1 = \xi_1 \qquad \text{and} \qquad y^2 = \xi_2.$$

   a) Show that

$$h_{11} = \frac{1}{h^{11}} = h_{22} = \frac{1}{h^{22}} = 1, \quad h_{12} = h_{21} = 0,$$

   and

$$h = 1.$$

   b) Prove that all of the surface Christoffel symbols of the second kind are zero.

c) Show that $\widetilde{\mathbf{B}} = \mathbf{0}$.

d) Conclude that $K_M = K_G = 0$.

2. Given the cylindrical coordinate system

$$\begin{aligned}\xi_1 &= x^1 \cos x^2 = r\cos\theta,\\ \xi_2 &= x^1 \sin x^2 = r\sin\theta,\\ \xi_3 &= x^3 = z,\end{aligned}$$

and a plane surface $\xi_3 =$ const., choose

$$y^1 = x^1 = r \qquad \text{and} \qquad y^2 = x^2 = \theta.$$

a) Show that

$$h_{11} = \frac{1}{h^{11}} = 1, \quad h_{22} = \frac{1}{h^{22}} = r^2, \quad h_{12} = h^{21} = 0,$$

and

$$h = r^2.$$

b) Prove that all of the surface Christoffel symbols of the second kind are zero.

c) Show that $\widetilde{\mathbf{B}} = \mathbf{0}$.

d) Conclude that $K_M = K_G = 0$.

3. With the cylindrical coordinates given in Problem 2 and a plane surface $\theta =$ const., choose

$$y^1 = x^1 = r \qquad \text{and} \qquad y^2 = x^3 = z.$$

a) Show that

$$h_{11} = \frac{1}{h^{11}} = h_{22} = \frac{1}{h^{22}} = 1, \quad h_{12} = h^{21} = 0,$$

and

$$h = 1.$$

b) Show that all surface Christoffel symbols of the second kind are zero.

c) Show that $\widetilde{\mathbf{B}} = \mathbf{0}$.

d) Conclude that $K_M = K_G = 0$.

4. Given the cylindrical coordinate system defined in Problem 2 and a cylindrical surface of radius R, choose

$$y^1 = x^2 = \theta \qquad \text{and} \qquad y^2 = x^3 = z.$$

a) Show that

$$h_{11} = \frac{1}{h^{11}} = R^2, \quad h_{22} = \frac{1}{h^{22}} = 1, \quad h_{12} = h^{21} = 0,$$

and

$$h = R^2.$$

c) Show that all surface Christoffel symbols of the second kind are zero.

b) Show that $B_{11} = -R$ and $B_{22} = B_{12} = B_{21} = 0$.

d) Conclude that

$$K_M = -\frac{1}{2R} \qquad \text{and} \qquad K_G = 0.$$

5. Given the spherical coordinate system

$$\begin{aligned}\xi_1 &= x^1 \sin x^2 \cos x^3 = r \sin\theta \cos\phi,\\ \xi_2 &= x^1 \sin x^2 \sin x^3 = r \sin\theta \sin\phi,\\ \xi_3 &= x^1 \cos x^2 = r\cos\theta,\end{aligned}$$

and a spherical surface of radius R, choose

$$y^1 = x^2 = \theta \qquad \text{and} \qquad y^2 = x^3 = \phi.$$

a) Show that

$$h_{11} = \frac{1}{h^{11}} = R^2, \quad h_{22} = \frac{1}{h^{22}} = R^2 \sin^2\theta, \quad h_{12} = h^{21} = 0,$$

and

$$h = R^4 \sin^2\theta.$$

b) Show that the only nonzero surface Christoffel symbols of the second kind are

$$\Gamma^1_{22} = -\sin\theta\cos\theta, \qquad \text{and} \qquad \Gamma^2_{12} = \Gamma^2_{21} = \cot\theta.$$

c) Show that $B_{11} = -R$, $B_{22} = -R\sin^2\theta$, and $B_{12} = B_{21} = 0$.

d) Conclude that

$$K_M = -\frac{1}{R} \qquad \text{and} \qquad K_G = \frac{1}{R^2}.$$

6. Given a Cartesian coordinate system and the surface

$$\xi_3 = g(\xi_1, t),$$

choose

$$y^1 = \xi_1 \qquad \text{and} \qquad y^2 = \xi_2.$$

a) Show that

$$h_{11} = 1 + \left(\frac{\partial g}{\partial \xi_1}\right)^2, \quad h_{22} = 1, \quad h_{12} = h^{21} = 0,$$

and

$$h = 1 + \left(\frac{\partial g}{\partial \xi_1}\right)^2.$$

b) Show that the only nonzero surface Christoffel symbol of the second kind is

$$\Gamma^1_{11} = \frac{\partial g}{\partial \xi_1}\frac{\partial^2 g}{\partial \xi_1^2}\left[1 + \left(\frac{\partial g}{\partial \xi_1}\right)^2\right]^{-1}.$$

c) Show that

$$B_{11} = \frac{\partial^2 g}{\partial \xi_1}\left[1 + \left(\frac{\partial g}{\partial \xi_1}\right)^2\right]^{-1/2},$$

and $B_{22} = B_{12} = B_{21} = 0$.

d) Conclude that

$$K_M = \frac{1}{2}\frac{\partial^2 g}{\partial \xi_1^2}\left[1 + \left(\frac{\partial g}{\partial \xi_1}\right)^2\right]^{-3/2} \qquad \text{and} \qquad K_G = 0.$$

7. Given the cylindrical coordinates defined in Problem 2 and the axially symmetric surface $\xi = g(r)$, choose

$$y^1 = r \qquad \text{and} \qquad y^2 = \theta.$$

a) Show that

$$h_{11} = 1 + \left(\frac{dg}{dr}\right)^2, \quad h_{22} = r^2, \quad h_{12} = h^{21} = 0,$$

and

$$h = r^2\left[1 + \left(\frac{dg}{dr}\right)^2\right].$$

b) Show that the only nonzero surface Christoffel symbols of the second kind are

$$\Gamma^1_{11} = g'g''[1 + (g')^2]^{-1}, \quad \Gamma^1_{22} = -r[1 + (g')^2]^{-1}, \quad \text{and} \quad \Gamma^2_{12} = \Gamma^2_{21} = \frac{1}{r}.$$

c) Show that

$$B_{11} = g''[1+(g')^2]^{-1/2}, \quad B_{22} = rg'[1+(g')^2]^{-1/2}, \quad \text{and} \quad B_{12} = B_{21} = 0.$$

d) Conclude that

$$K_M = \frac{1}{2r}[rg'' + g' + (g')^3][1+(g')^2]^{-3/2} \quad \text{and} \quad K_G = \frac{1}{r}g'g''[1+(g')^2]^{-2}.$$

8. Given the cylindrical coordinates defined in Problem 2 and the axially symmetric surface $r = g(z)$, choose

$$y^1 = x^3 = z \qquad \text{and} \qquad y^2 = x^2 = \theta.$$

   a) Show that

$$h_{11} = 1 + \left(\frac{dg}{dz}\right)^2, \quad h_{22} = g^2, \quad h_{12} = h^{21} = 0,$$

   and

$$h = g^2\left[1 + \left(\frac{dg}{dz}\right)^2\right].$$

   b) Show that the only nonzero surface Christoffel symbols of the second kind are

$$\Gamma^1_{11} = g'g''[1+(g')^2]^{-1}, \quad \Gamma^1_{22} = -g'g''[1+(g')^2]^{-1}, \quad \text{and} \quad \Gamma^2_{12} = \Gamma^2_{21} = \frac{g'}{g}.$$

   c) Show that

$$B_{11} = g''[1+(g')^2]^{-1/2}, \quad B_{22} = -g'[1+(g')^2]^{-1/2}, \quad \text{and} \quad B_{12} = B_{21} = 0.$$

   d) Conclude that

$$K_M = \frac{1}{2g}[gg''-(g')^2-1][1+(g')^2]^{-3/2} \quad \text{and} \quad K_G = -\frac{g''}{g}[1+(g')^2]^{-2}.$$

9. Given a Cartesian coordinate system and the surface $\xi_3 = g(\xi_1, t)$ show that the Cartesian components of $\mathbf{n}$ are

$$n_1 = -\frac{\partial g}{\partial \xi_1}\left[1 + \left(\frac{\partial g}{\partial \xi_1}\right)^2\right]^{-1/2},$$
$$n_2 = 0,$$
$$n_3 = \left[1 + \left(\frac{\partial g}{\partial \xi_1}\right)^2\right]^{-1/2}.$$

10. Given the cylindrical coordinate system described in Problem 2 and the axially symmetric surface $z = g(r)$, show that the cylindrical components of $\mathbf{n}$ are

$$n_r = -\frac{dg}{dr}\left[1 + \left(\frac{dg}{dr}\right)^2\right]^{-1/2},$$
$$n_\theta = 0,$$
$$n_z = \left[1 + \left(\frac{dg}{dr}\right)^2\right]^{-1/2}.$$

11. Given the cylindrical coordinate system described in Problem 2 and the axially symmetric surface $r = g(z)$, show that the cylindrical components of $\mathbf{n}$ are

$$n_r = \left[1 + \left(\frac{dg}{dr}\right)^2\right]^{-1/2},$$
$$n_\theta = 0,$$
$$n_z = -\frac{dg}{dr}\left[1 + \left(\frac{dg}{dr}\right)^2\right]^{-1/2}.$$

12. Prove that

   a)
$$v^\alpha = h^{\alpha\beta} a^i_\beta v_i;$$

   b)
$$v^i = v^\alpha a^i_\alpha + v_{(n)} n^i.$$

13. Let ϕ be an explicit function of position in space. Show that the surface gradient of ϕ is the projection of the spatial gradient of ϕ:

$$\widetilde{\nabla}\phi = \widetilde{\boldsymbol{P}} \cdot \nabla\phi.$$

14. At any point (y^1, y^2) on a surface, two tangential vector fields $\widetilde{\boldsymbol{\lambda}}$ and $\widetilde{\boldsymbol{\mu}}$ may be viewed as forming two edges of a parallelogram. If the rotation from $\widetilde{\boldsymbol{\lambda}}$ to $\widetilde{\boldsymbol{\mu}}$ is positive, show that $\widetilde{\boldsymbol{\lambda}} \cdot (\widetilde{\boldsymbol{\varepsilon}} \cdot \widetilde{\boldsymbol{\mu}})$ determines the area of the corresponding parallelogram.

15. Let $\widetilde{\boldsymbol{\lambda}}$ and $\widetilde{\boldsymbol{\mu}}$ be unit tangent vector fields.

   a) If at any point (y^1, y^2) on a surface the rotation from $\widetilde{\boldsymbol{\lambda}}$ to $\widetilde{\boldsymbol{\mu}}$ is positive, show that

$$\widetilde{\boldsymbol{\lambda}} \cdot (\widetilde{\boldsymbol{\varepsilon}} \cdot \widetilde{\boldsymbol{\mu}}) = \sin\theta,$$

   where θ is the angle measured between the two directions.

   b) Conclude that at this point

$$\mathbf{n}\left[\widetilde{\boldsymbol{\lambda}} \cdot (\widetilde{\boldsymbol{\varepsilon}} \cdot \widetilde{\boldsymbol{\mu}})\right] = \widetilde{\boldsymbol{\lambda}} \times \widetilde{\boldsymbol{\mu}}$$

   or

$$n_i \varepsilon_{\alpha\beta} = \varepsilon_{ijk} a^j_\alpha a^k_\beta,$$

   where $\mathbf{n}$ is the unit vector normal to the surface.

   c) Show that

$$\varepsilon^{\alpha\beta} a^j_\alpha a^k_\beta = \varepsilon^{ijk} n_i.$$

16. Show that, upon a change of surface coordinate systems, $\varepsilon^{\alpha\beta}$ and $\varepsilon_{\alpha\beta}$ transform according to the rules appropriate to the contravariant and covariant components of a tangential second-order tensor field. Hint: Start with

$$\epsilon^{\alpha\beta}\det\left[\frac{\partial\overline{y}^{\phi}}{\partial y^{\gamma}}\right] = \epsilon^{\mu\nu}\frac{\partial\overline{y}^{\alpha}}{\partial y^{\mu}}\frac{\partial\overline{y}^{\beta}}{\partial y^{\nu}} \quad\text{and}\quad \epsilon_{\alpha\beta}\det\left[\frac{\partial y^{\gamma}}{\partial\overline{y}^{\phi}}\right] = \epsilon_{\mu\nu}\frac{\partial y^{\mu}}{\partial\overline{y}^{\alpha}}\frac{\partial y^{\nu}}{\partial\overline{y}^{\beta}}.$$

17. Show that

$$-\widetilde{\varepsilon}\cdot\widetilde{\varepsilon} = \widetilde{\mathbf{P}}.$$

18. Show that

$$-\widetilde{\varepsilon}\cdot\mathbf{a}_1 = \sqrt{h}\,\mathbf{a}^2$$

and

$$-\widetilde{\varepsilon}\cdot\mathbf{a}_2 = -\sqrt{h}\,\mathbf{a}^1.$$

This implies that

$$|\mathbf{a}^2| = \frac{1}{\sqrt{h}}|\mathbf{a}_1|,$$

that $\mathbf{a}^2$ is orthogonal to $\mathbf{a}_1$, and that the rotation from $\mathbf{a}_1$ to $\mathbf{a}^2$ is positive. In a similar manner, we conclude that

$$|\mathbf{a}^1| = \frac{1}{\sqrt{h}}|\mathbf{a}_2|,$$

that $\mathbf{a}^1$ is orthogonal to $\mathbf{a}_2$, and that the rotation from $\mathbf{a}_2$ to $\mathbf{a}^1$ is negative. It is also interesting to observe that the above two relations imply

$$h^{22} = \frac{h_{11}}{h} \quad\text{and}\quad h^{11} = \frac{h_{22}}{h}.$$

19. Prove that $\det[\widetilde{\mathbf{P}}] = 1$.

20. Starting with (D.170), prove that

$$B_{\alpha\beta} = a^{i}_{\alpha,\beta}n_i.$$

21. Show that

$$\left(h_{\alpha\beta}w^{\beta}\right)_{,\gamma} = h_{\alpha\beta}w^{\beta}_{,\gamma},$$

implying that $h_{\alpha\beta}$ can be treated as a constant with respect to surface covariant differentiation (Ricci's theorem).

22. Starting with

$$\left(h_{\alpha\beta}w^{\beta}\right)_{,\gamma} = \frac{\partial(h_{\alpha\beta}w^{\beta})}{\partial y^{\gamma}} - \Gamma^{\delta}_{\gamma\alpha}h_{\delta\beta}w^{\beta},$$

rework Problem 21.

23. a) Prove that

$$\frac{\partial h_{\alpha\beta}}{\partial y^\gamma} = \Gamma^\delta_{\gamma\beta} h_{\alpha\delta} + \Gamma^\delta_{\gamma\alpha} h_{\beta\delta}.$$

b) Deduce that $h_{\alpha\beta,\gamma} = 0$ (Ricci's theorem),

$$\widetilde{\nabla}\widetilde{\mathbf{P}} = B_{\alpha\beta}\left[\mathbf{n}\mathbf{a}^\alpha\mathbf{a}^\beta + \mathbf{a}^\alpha\mathbf{n}\mathbf{a}^\beta\right]$$

and

$$\widetilde{\nabla}\cdot\widetilde{\mathbf{P}} = \mathrm{tr}\,[\widetilde{\mathbf{B}}]\mathbf{n} = 2K_M\mathbf{n},$$

where K_M is the mean curvature.

c) Prove that

$$\widetilde{\nabla}\widetilde{\mathbf{P}} = a^i_{\alpha,\beta}\mathbf{g}_i\mathbf{a}^\alpha\mathbf{a}^\beta + B_{\alpha\beta}\mathbf{a}^\alpha\mathbf{n}\mathbf{a}^\beta.$$

d) Conclude that

$$a^i_{\alpha,\beta} = B_{\alpha\beta}n^i$$

or

$$B_{\alpha\beta} = a^i_{\alpha,\beta}n_i.$$

24. Show that

$$\widetilde{\nabla}\widetilde{\varepsilon} - \varepsilon^{\delta\alpha}B_{\beta\delta}\left[\mathbf{n}\mathbf{a}_\alpha\mathbf{a}^\beta - \mathbf{a}_\alpha\mathbf{n}\mathbf{a}^\beta\right] = \varepsilon_{\alpha\beta,\gamma}\mathbf{a}^\alpha\mathbf{a}^\beta\mathbf{a}^\gamma.$$

Since the right-hand side indicates that the quantity on the left-hand side is a tangential third-order tensor field, the properties of which are independent of the surface coordinate system chosen, subsequently show that

$$\varepsilon^{\alpha\beta}_{,\gamma} = \varepsilon_{\alpha\beta,\gamma} = 0$$

and

$$\widetilde{\nabla}\widetilde{\varepsilon} = \varepsilon^{\delta\alpha}B_{\beta\delta}\left[\mathbf{n}\mathbf{a}_\alpha\mathbf{a}^\beta - \mathbf{a}_\alpha\mathbf{n}\mathbf{a}^\beta\right].$$

25. Show that

$$\frac{\partial \ln\sqrt{h}}{\partial y^\gamma} = \Gamma^\beta_{\gamma\beta}.$$

[Hint: Write out in full $\varepsilon_{\alpha\beta,\gamma} = 0$ and set $\alpha, \beta = 1, 2$.]

26. Let $\widetilde{\mathbf{w}}$ be a tangential vector field. Use the result of Problem 25 to show that

$$\widetilde{\nabla}\cdot\widetilde{\mathbf{w}} = w^\alpha_{,\alpha} = \frac{1}{\sqrt{h}}\frac{\partial}{\partial y^\alpha}\left(\sqrt{h}w^\alpha\right).$$

27. Let $\widetilde{\mathbf{T}}$ be a tangential second-order tensor field. Show that

$$\widetilde{\nabla}\cdot\widetilde{\mathbf{T}} = \left(a^i_\alpha T^{\alpha\beta}\right)_{,\beta}\mathbf{e}_i.$$

28. a) Noting that

$$v^i_{,\alpha\beta} = \left(v^i_{,\alpha}\right)_{,\beta},$$

show that

$$v^i_{,\alpha\beta} = v^i_{,\beta\alpha}.$$

b) Show that for a tangential vector field $\tilde{\mathbf{v}}$

$$\begin{aligned} 0 &= v^i_{,\alpha\beta} - v^i_{,\beta\alpha}, \\ &= n^i_{,\beta} B_{\gamma\alpha} v^\gamma + n^i B_{\gamma\alpha,\beta} v^\gamma + a^i_\gamma v^\gamma_{,\alpha\beta} - n^i_{,\alpha} B_{\gamma\beta} v^\gamma - n^i B_{\gamma\beta,\alpha} v^\gamma + a^i_\gamma v^\gamma_{,\beta\alpha}. \end{aligned}$$

c) Noting that

$$n^i_{,\beta} = -B^\delta_\beta a^i_\delta,$$

and using the above result and

$$g_{ij} a^j_\nu \left(v^i_{,\alpha\beta} - v^i_{,\beta\alpha} \right) = 0,$$

conclude that

$$v_{\alpha,\beta\gamma} - v_{\alpha,\gamma\beta} = R_{\delta\alpha\beta\gamma} v^\delta,$$

which means that $v_{\alpha,\beta\gamma}$ is not symmetric in β and γ. Here

$$R_{\delta\alpha\beta\gamma} = K_G \varepsilon_{\delta\alpha} \varepsilon_{\beta\gamma}$$

are known as the components of the surface Riemann–Christoffel tensor, and K_G is the total curvature.

29. Starting with the result of Problem 28, show that

$$B_{\alpha\beta,\gamma} = B_{\alpha\gamma,\beta}.$$

This is known as the Mainardi–Codazzi equation for the surface.

Bibliography

R. Aris. *Vectors, Tensors and the Basic Equations of Fluid Mechanics.* Dover Publications, Inc., Mineola, NY, 1962.

R.M. Bowen and C.-C. Wang. *Introduction to Vectors and Tensors – Linear and Multilinear Algebra*, volume 1. Plenum Press, New York, NY, 1976.

R.M. Bowen and C.-C. Wang. *Introduction to Vectors and Tensors – Vector and Tensor Analysis*, volume 2. Plenum Press, New York, NY, 1976.

L. Brand. *Vector and Tensor Analysis.* John Wiley & Sons, Inc., New York, NY, 1955.

A.J. McConnell. *Applications of Tensor Analysis.* Dover Publications, Inc., New York, NY, 1957.

I. Müller. *Thermodynamics.* Pitman Publishing, Inc., Boston, MA, 1985.

J.C. Slattery. *Interfacial Transport Phenomena.* Springer-Verlag, New York, NY, 1990.

E Representation of isotropic tensor fields

In this appendix, we wish to illustrate the procedure for obtaining the representation of isotropic tensor fields. We shall assume that the tensor fields are functions of a vector $\mathbf{v}$ and a symmetric tensor $\mathbf{A}$. If the tensor field is also a function of a scalar, the representations are the same. If they are functions of more than one vector or symmetric tensor, or functions of skew-symmetric tensors, then the derivations will be modified accordingly, but the procedure is essentially the same. Comprehensive results are given in Tables 5.1–5.8.

E.1 Scalar function

Let ψ be an isotropic scalar function of the vector $\mathbf{v}$ and symmetric tensor $\mathbf{A}$, i.e., $\psi = \psi(\mathbf{v}, \mathbf{A})$. Then, from (5.116), (5.120), and (5.121), it must satisfy

$$\psi(\mathbf{Q}\cdot\mathbf{v}, \mathbf{Q}\cdot\mathbf{A}\cdot\mathbf{Q}^T) = \psi(\mathbf{v}, \mathbf{A}), \tag{E.1}$$

for all orthogonal tensors $\mathbf{Q}$. If ψ is an isotropic function, then it can only depend on scalar combinations of the components of $\mathbf{v}$ and $\mathbf{A}$. The only invariant scalar combination that can be formed from the components of $\mathbf{v}$ alone is $\mathbf{v}\cdot\mathbf{v}$. The only invariant scalar combinations that can be formed from the components of $\mathbf{A}$ alone are the three invariants $A_{(1)}$, $A_{(2)}$, and $A_{(3)}$, or what are the same $\operatorname{tr} A$, $\operatorname{tr} A^2$, and $\operatorname{tr} A^3$ (see (3.98)–(3.100)). Note that all other possible scalars are related to these through the Cayley–Hamilton theorem. Lastly, we have to consider invariant scalars that can be formed from products of $\mathbf{v}$ and $\mathbf{A}$. There are only two independent such scalars, and they are $\mathbf{v}\cdot\mathbf{A}\cdot\mathbf{v}$ and $\mathbf{v}\cdot\mathbf{A}^2\cdot\mathbf{v}$; all other scalars are related to these by using Cayley–Hamilton's theorem. In summary, the isotropic scalar function ψ can only depend on the following six scalar combinations, which are also listed in Table 5.1:

$$\mathbf{v}\cdot\mathbf{v}, \quad \operatorname{tr} A, \quad \operatorname{tr} A^2, \quad \operatorname{tr} A^3, \quad \mathbf{v}\cdot\mathbf{A}\cdot\mathbf{v}, \quad \mathbf{v}\cdot\mathbf{A}^2\cdot\mathbf{v}. \tag{E.2}$$

E.2 Vector function

Let $\mathbf{h}$ be an isotropic vector function of the vector $\mathbf{v}$ and symmetric tensor $\mathbf{A}$, i.e., $\mathbf{h} = \mathbf{h}(\mathbf{v}, \mathbf{A})$. Then, from (5.117), (5.120), and (5.121), it must satisfy

$$\mathbf{h}(\mathbf{Q}\cdot\mathbf{v}, \mathbf{Q}\cdot\mathbf{A}\cdot\mathbf{Q}^T) = \mathbf{Q}\cdot\mathbf{h}(\mathbf{v}, \mathbf{A}), \tag{E.3}$$

for all orthogonal tensors $\mathbf{Q}$. The problem of finding a representation can be reduced to the problem of finding a scalar function χ of $\mathbf{v}$ and $\mathbf{A}$, and an arbitrary vector $\mathbf{u}$ by writing

$$\chi = \mathbf{u}\cdot\mathbf{h}(\mathbf{v}, \mathbf{A}). \tag{E.4}$$

Now χ is an isotropic scalar function of $\mathbf{v}$, $\mathbf{A}$, and $\mathbf{u}$, so it can depend only on independent scalar combinations of their components. Using the procedure of the previous subsection, we have

$$\mathbf{v}\cdot\mathbf{v}, \quad \mathbf{u}\cdot\mathbf{v}, \quad \mathbf{u}\cdot\mathbf{u}, \tag{E.5}$$

$$\operatorname{tr} A, \quad \operatorname{tr} A^2, \quad \operatorname{tr} A^3, \tag{E.6}$$

$$\mathbf{v}\cdot\mathbf{A}\cdot\mathbf{v}, \quad \mathbf{v}\cdot\mathbf{A}^2\cdot\mathbf{v}, \quad \mathbf{u}\cdot\mathbf{A}\cdot\mathbf{v}, \quad \mathbf{u}\cdot\mathbf{A}^2\cdot\mathbf{v}, \quad \mathbf{u}\cdot\mathbf{A}\cdot\mathbf{u}, \quad \mathbf{u}\cdot\mathbf{A}^2\cdot\mathbf{u}. \tag{E.7}$$

Now if χ was an arbitrary function of $\mathbf{u}$, then it would depend on all the above combinations. However, from the definition (E.4), χ can only be a linear homogeneous function of $\mathbf{u}$, and thus must be such that

$$\chi = \mathbf{u} \cdot \left(\alpha \mathbf{v} + \beta \mathbf{A} \cdot \mathbf{v} + \gamma \mathbf{A}^2 \cdot \mathbf{v} \right), \tag{E.8}$$

where α, β, and γ depend on the scalar combinations given above that do not depend on $\mathbf{u}$, or what is the same, on (E.2). Now, since $\mathbf{u}$ is an arbitrary vector, we must have that the invariant representation of $\mathbf{h}$ is given by

$$\mathbf{h} = \alpha \mathbf{v} + \beta \mathbf{A} \cdot \mathbf{v} + \gamma \mathbf{A}^2 \cdot \mathbf{v}. \tag{E.9}$$

This result is also noted in Table 5.2.

E.3 Symmetric tensor function

Let $\mathbf{T}$ be an isotropic symmetric tensor function of the vector $\mathbf{v}$ and symmetric tensor $\mathbf{A}$, i.e., $\mathbf{T} = \mathbf{T}(\mathbf{v}, \mathbf{A})$. Then, from (5.118), (5.120), and (5.121), it must satisfy

$$\mathbf{T}(\mathbf{Q} \cdot \mathbf{v}, \mathbf{Q} \cdot \mathbf{A} \cdot \mathbf{Q}^T) = \mathbf{Q} \cdot \mathbf{T}(\mathbf{v}, \mathbf{A}) \cdot \mathbf{Q}^T, \tag{E.10}$$

for all orthogonal tensors $\mathbf{Q}$. Following a procedure analogous to that in the previous subsection, we introduce an arbitrary symmetric tensor $\mathbf{S}$ and define

$$\zeta = \mathbf{S} : \mathbf{T}(\mathbf{v}, \mathbf{A}) \tag{E.11}$$

so that ζ is an isotropic scalar function of $\mathbf{v}$, $\mathbf{A}$, and $\mathbf{S}$ which is homogeneous and linear in $\mathbf{S}$. We find that the independent and complete set of such combinations is given by

$$\zeta = \mathbf{S} : \left[\alpha \mathbf{1} + \beta \mathbf{v}\mathbf{v} + \gamma \mathbf{A} + \delta \mathbf{A}^2 + \eta \left(\mathbf{v}\mathbf{A} \cdot \mathbf{v} + \mathbf{A} \cdot \mathbf{v}\mathbf{v} \right) + \lambda \left(\mathbf{v}\mathbf{A}^2 \cdot \mathbf{v} + \mathbf{A}^2 \cdot \mathbf{v}\mathbf{v} \right) \right], \tag{E.12}$$

where α, β, γ, δ, η, and λ depend on the scalar invariants (E.2). Now, since $\mathbf{S}$ is an arbitrary symmetric tensor, we must have that the invariant representation of $\mathbf{T}$ is given by

$$\mathbf{T} = \alpha \mathbf{1} + \beta \mathbf{v}\mathbf{v} + \gamma \mathbf{A} + \delta \mathbf{A}^2 + \eta \left(\mathbf{v}\mathbf{A} \cdot \mathbf{v} + \mathbf{A} \cdot \mathbf{v}\mathbf{v} \right) + \lambda \left(\mathbf{v}\mathbf{A}^2 \cdot \mathbf{v} + \mathbf{A}^2 \cdot \mathbf{v}\mathbf{v} \right). \tag{E.13}$$

This result is also noted in Table 5.3.

In arriving at the linear independent list of scalar combinations that are homogeneous and linear in $\mathbf{S}$, besides making use of the Cayley–Hamilton theorem, use is also made of the following identity, which is called the *generalized Cayley–Hamilton theorem*:

$$\begin{aligned}
&(\mathbf{X} \cdot \mathbf{Y} \cdot \mathbf{Z} + \mathbf{X} \cdot \mathbf{Z} \cdot \mathbf{Y} + \mathbf{Y} \cdot \mathbf{Z} \cdot \mathbf{X} + \mathbf{Y} \cdot \mathbf{X} \cdot \mathbf{Z} + \mathbf{Z} \cdot \mathbf{X} \cdot \mathbf{Y} + \mathbf{Z} \cdot \mathbf{Y} \cdot \mathbf{X}) - \\
&\quad \mathrm{tr}\,\mathbf{X} (\mathbf{Y} \cdot \mathbf{Z} + \mathbf{Z} \cdot \mathbf{Y}) - \mathrm{tr}\,\mathbf{Y} (\mathbf{Z} \cdot \mathbf{X} + \mathbf{X} \cdot \mathbf{Z}) - \mathrm{tr}\,\mathbf{Z} (\mathbf{X} \cdot \mathbf{Y} + \mathbf{Y} \cdot \mathbf{X}) - \\
&\quad [\mathrm{tr}\,(\mathbf{Y} \cdot \mathbf{Z}) - \mathrm{tr}\,\mathbf{Y}\ \mathrm{tr}\,\mathbf{Z}]\mathbf{X} - [\mathrm{tr}\,(\mathbf{Z} \cdot \mathbf{X}) - \mathrm{tr}\,\mathbf{Z}\ \mathrm{tr}\,\mathbf{X}]\mathbf{Y} - [\mathrm{tr}\,(\mathbf{X} \cdot \mathbf{Y}) - \\
&\quad \mathrm{tr}\,\mathbf{X}\ \mathrm{tr}\,\mathbf{Y}]\mathbf{Z} - [\mathrm{tr}\,\mathbf{X}\ \mathrm{tr}\,\mathbf{Y}\ \mathrm{tr}\,\mathbf{Z} - \mathrm{tr}\,\mathbf{X}\ \mathrm{tr}\,(\mathbf{Y} \cdot \mathbf{Z}) - \mathrm{tr}\,\mathbf{Y}\ \mathrm{tr}\,(\mathbf{Z} \cdot \mathbf{X}) - \\
&\quad \mathrm{tr}\,\mathbf{Z}\ \mathrm{tr}\,(\mathbf{X} \cdot \mathbf{Y}) + \mathrm{tr}\,(\mathbf{X} \cdot \mathbf{Y} \cdot \mathbf{Z}) + \mathrm{tr}\,(\mathbf{Z} \cdot \mathbf{Y} \cdot \mathbf{X})]\,\mathbf{1} = \mathbf{0},
\end{aligned} \tag{E.14}$$

where $\mathbf{X}$, $\mathbf{Y}$, and $\mathbf{Z}$ are arbitrary second-order tensors. Note that the standard Cayley–Hamilton theorem is recovered if one sets $\mathbf{X} = \mathbf{Y} = \mathbf{Z}$. If in (E.14) we set $\mathbf{Z} = \mathbf{X}$, we obtain

$$\begin{aligned}\mathbf{X}\cdot\mathbf{Y}\cdot\mathbf{X} + \mathbf{X}^2\cdot\mathbf{Y} + \mathbf{Y}\cdot\mathbf{X}^2 - \operatorname{tr}\mathbf{X}\,(\mathbf{X}\cdot\mathbf{Y} + \mathbf{Y}\cdot\mathbf{X}) - \operatorname{tr}\mathbf{Y}\,\mathbf{X}^2 - \\ \left[\operatorname{tr}(\mathbf{X}\cdot\mathbf{Y}) - \operatorname{tr}\mathbf{X}\,\operatorname{tr}\mathbf{Y}\right]\mathbf{X} - \frac{1}{2}\left[\operatorname{tr}\left(\mathbf{X}^2\right) - (\operatorname{tr}\mathbf{X})^2\right]\mathbf{Y} - \\ \left\{\operatorname{tr}\left(\mathbf{X}^2\cdot\mathbf{Y}\right) - \operatorname{tr}\mathbf{X}\,\operatorname{tr}(\mathbf{X}\cdot\mathbf{Y}) + \frac{1}{2}\operatorname{tr}\mathbf{Y}\left[\operatorname{tr}\mathbf{X}^2 - (\operatorname{tr}\mathbf{X})^2\right]\right\}\mathbf{1} = \mathbf{0}.\end{aligned} \tag{E.15}$$

In our application to obtain (E.13), we subsequently take $\mathbf{X} = \mathbf{A}$ and $\mathbf{Y} = \mathbf{vv}$ in (E.15).

In addition, the following independent general identity can be proved:

$$\begin{aligned}\mathbf{V}\cdot\mathbf{Y}\cdot\mathbf{W} + \mathbf{Y}^T\cdot\mathbf{V}\cdot\mathbf{W} + \mathbf{V}\cdot\mathbf{W}\cdot\mathbf{Y}^T - \operatorname{tr}\mathbf{Y}\,(\mathbf{V}\cdot\mathbf{W}) - \frac{1}{2}\operatorname{tr}(\mathbf{W}\cdot\mathbf{V})\,\mathbf{Y}^T - \\ \left[\operatorname{tr}(\mathbf{W}\cdot\mathbf{V}\cdot\mathbf{Y}) - \frac{1}{2}\operatorname{tr}(\mathbf{W}\cdot\mathbf{V})\operatorname{tr}\mathbf{Y}\right]\mathbf{1} = \mathbf{0},\end{aligned} \tag{E.16}$$

where $\mathbf{Y}$ is an arbitrary second-order tensor, and $\mathbf{V}$ and $\mathbf{W}$ are second-order skew-symmetric tensors. This identity is useful in obtaining invariant representations of quantities that depend on skew-symmetric tensors.

Problems

1. Derive (E.13).

Bibliography

J.P. Boehler. On irreducible representations for isotropic scalar functions. *Zeitschrift fur Angewandte Mathematik und Mechanik*, 57(6):323–327, 1977.

P. Pennisi and M. Trovato. On the irreducibility of Professor G.F. Smith's representation for isotropic functions. *International Journal of Engineering Science*, 25(8):1059–1065, 1987.

R.S. Rivlin. Further remarks on the stress-deformation relations for isotropic materials. *Journal of Rational Mechanics and Analysis*, 4(5):681–702, 1955.

R.S. Rivlin and G.F. Smith. On identities for 3 × 3 matrices. *Rendiconti di Matematica, Universitá degli studi di Roma*, 8:348–353, 1975.

G.F. Smith. On isotropic functions of symmetric tensors, skew-symmetric tensors and vectors. *International Journal of Engineering Science*, 9(10):899–916, 1971.

G.F. Smith. Constitutive equations for anisotropic and isotropic materials. In G.C. Sih, editor, *Mechanics and Physics of Discrete Systems*, volume 3. Elsevier Science B.V., Amsterdam, The Netherlands, 1994.

A.J.M. Spencer. Theory of invariants. In A.C. Eringen, editor, *Continuum Physics*, volume I. Academic Press, New York, 1971.

C.-C. Wang. A new representation theorem for isotropic functions: An answer to professor G.F. Smith's criticism of my papers on representations for isotropic functions. Part 1. Scalar-valued isotropic functions. *Archive for Rational Mechanics and Analysis*, 36(3):166–197, 1970.

C.-C. Wang. A new representation theorem for isotropic functions: An answer to professor G.F. Smith's criticism of my papers on representations for isotropic functions. Part 2. Vector-valued isotropic functions, symmetric tensor-valued isotropic functions, and skew-symmetric tensor-valued isotropic functions. *Archive for Rational Mechanics and Analysis*, 36(3):198–223, 1970.

C.-C. Wang. Corrigendum to my recent papers on "Representations for isotropic functions". *Archive for Rational Mechanics and Analysis*, 43(3):392–395, 1971.

Q.-S. Zheng. On the representations for isotropic vector-valued, symmetric tensor-valued and skew-symmetric tensor-valued functions. *International Journal of Engineering Science*, 31(7):1013–1024, 1993.

Q.-S. Zheng. Theory of representations for tensor functions – A unified invariant approach to constitutive equations. *Applied Mechanics Reviews*, 47(11):545–587, 1994.

F Legendre transformations

The fundamental aspect of Legendre transformations is that we are given an equation (e.g., the fundamental relation) of the form

$$y = y(x_0, x_1, \ldots, x_n) \tag{F.1}$$

and we would like to find a way whereby the derivatives

$$p_i \equiv \frac{\partial y}{\partial x_i} \tag{F.2}$$

could be considered as independent variables without loss of any information (as contained in the fundamental relation). The technique for doing this is provided by *Legendre transformations.*

First, consider the case where we have an equation of only a single independent variable:

$$y = y(x). \tag{F.3}$$

This equation just represents a curve with Cartesian coordinates (x, y) and

$$p = p(x) \equiv \frac{dy}{dx} \tag{F.4}$$

is the slope of the curve, which is a function of x. An intuitive way to consider p as an independent variable would be to eliminate x between the equations for y and p and then we would end up with the equation where p is now the independent variable:

$$y = y(p). \tag{F.5}$$

However, just the knowledge of p and y is not sufficient to reconstruct the curve (F.3). It should be noted that the curve (F.5) is a first-order ordinary differential equation and the solution can only be determined up to an arbitrary constant. Clearly, one such constant, the y intercept, would provide us the unique solution. If we take ψ to be the y intercept, then a relation of the form

$$\psi = \psi(p) \tag{F.6}$$

would enable us to construct a curve such that at x, where the ordinate is y and the slope is p, the intercept of the line with this slope is given by ψ. In such case, (F.6) would provide the same complete information as (F.3) as the two relations would be equivalent. To see that these two representations are equivalent, we note that if ψ is the y intercept and p the slope at x, then we have

$$p = \frac{y - \psi}{x - 0} \tag{F.7}$$

or

$$\psi = y - px. \tag{F.8}$$

Now suppose that we are given (F.3), and by differentiation, we find (F.4). By elimination of x and y among equations (F.3), (F.4), and (F.8), we obtain the desired new relation between ψ and p. In order for us to be able to solve for

Table F.1: Legendre transformation of one variable.

$y = y(x)$	$\psi = \psi(p)$
$p = \dfrac{dy}{dx}$	$-x = \dfrac{d\psi}{dp}$
$\psi = y - p\,x$	$y = \psi + x\,p$
Elimination of y and x from the above equations yields	Elimination of ψ and p from from the above equations yields
$\psi = \psi(p)$	$y = y(x)$

x, we require that $d^2y/dx^2 \neq 0$. Thus, we now have that (F.3) is a fundamental relation in the y representation, whereas (F.6) is a fundamental relation in the ψ representation. Equation (F.8) is taken as the definition of the function ψ, referred to as the Legendre transform of y.

The inverse problem of recovering (F.3) from (F.6) proceeds as follows. Taking the differential of (F.8) and recalling from (F.4) that $dy = p\,dx$, we have

$$d\psi = dy - p\,dx - x\,dp = -x\,dp, \tag{F.9}$$

or

$$-x = \frac{d\psi}{dp}. \tag{F.10}$$

If the two variables ψ and p are eliminated from (F.6), (F.8), and (F.10), we recover equation (F.3). In order for us to be able to solve for p, we require that $d^2\psi/dp^2 \neq 0$. The symmetry between the Legendre transform and its inverse is evident from Table F.1.

The generalization of the Legendre transformation to functions of more than a single independent variable is straightforward. In general, the fundamental relation of the form (F.1) represents a hyper-surface in a $(n+2)$-dimensional space with Cartesian coordinates $y, x_0, x_1, \ldots, x_n$. The derivatives (F.2) are the partial slopes of the hyper-surface. The family of tangent hyper-planes may be characterized by giving the intercept of a hyper-plane, ψ, as a function of the slopes:

$$\psi = y - \sum_{i=0}^{n} p_i x_i. \tag{F.11}$$

Taking the differential of this equation, we find

$$d\psi = -\sum_{i=0}^{n} x_i dp_i; \tag{F.12}$$

thus

$$-x_i = \frac{\partial\psi}{\partial p_i}. \tag{F.13}$$

A Legendre transformation

$$\psi = \psi(p_0, p_1, \ldots, p_n) \tag{F.14}$$

Table F.2: General Legendre transformation.

$y = y(x_0, x_1, \ldots, x_m, x_{m+1}, \ldots, x_n)$	$\psi = \psi(p_0, p_1, \ldots, p_m, x_{m+1}, \ldots, x_n)$
$p_i = \dfrac{dy}{dx_i}, \quad i \le m$	$-x_i = \dfrac{\partial \psi}{\partial p_i}, \quad i \le m$
$\psi = y - \sum_{i=0}^{m} p_i x_i$	$y = \psi + \sum_{i=0}^{m} x_i p_i$
Elimination of y and x_i, $i \le m$, yields	Elimination of ψ and p_i, $i \le m$, yields
$\psi = \psi(p_0, p_1, \ldots, p_m, x_{m+1}, \ldots, x_n)$	$y = y(x_0, x_1, \ldots, x_m, x_{m+1}, \ldots, x_n)$

is obtained by eliminating y and the x_i from (F.1), the set (F.2), and (F.11). In order for us to be able to solve for x_i, we require that $d^2y/dx_idx_j \neq 0$, where $i, j \le n$. The inverse transformation is obtained by eliminating ψ and p_i from (F.14), the set (F.13), and (F.11). In order for us to be able to solve for p_i, we require that $d^2\psi/dp_idp_j \neq 0$, where $i, j \le n$.

Finally, a Legendre transformation may be made in only an $(m+2)$-dimensional subspace, $m \le n$, of the full $(n+2)$-dimensional space of the relation (F.1). Of course, the subspace must contain the y-coordinate but may involve any choice of $m+1$ coordinates from the set $x_0, x_1, \ldots, x_n$. For convenience of notation, the coordinates are ordered so that the Legendre transformation is made in the subspace of the first $m+1$ coordinates (and of y); the coordinates $x_{m+1}, \ldots, x_n$ are just treated as constants and thus left untransformed. Subsequently, we can represent such transformation and inverse as indicated in Table F.2. We note that in the Legendre transformation, the variables x_i and p_i, for $i \le m$, are considered to be conjugate variables.

Bibliography

H.B. Callen. *Thermodynamics.* John Wiley & Sons, Inc., New York, NY, 1962.

M.R. El-Saden. A thermodynamic formalism based on the fundamental relation and the Legendre transformation. *International Journal of Mechanical Sciences*, 8(1):13–24, 1966.

Index